LECTURES ON LOGARITHMIC ALGEBRAIC GEOMETRY

This graduate textbook offers a self-contained introduction to the concepts and techniques of logarithmic geometry, a key tool for analyzing compactification and degeneration in algebraic geometry and number theory. It features a systematic exposition of the foundations of the field, from the basic results on convex geometry and commutative monoids to the theory of logarithmic schemes and their de Rham and Betti cohomology. The book will be of use to graduate students and researchers working in algebraic, analytic, and arithmetic geometry, as well as related fields.

Arthur Ogus is Professor Emeritus of Mathematics at the University of California, Berkeley. His work focuses on arithmetic, algebraic, and logarithmic geometry. He is the author of 35 research publications, and has lectured extensively on logarithmic geometry in Berkeley, France, Italy, and Japan.

CAMBRIDGE STUDIES IN ADVANCED MATHEMATICS

All the titles listed below can be obtained from good booksellers or from Cambridge University Press.
For a complete series listing visit: www.cambridge.org/mathematics.

Already Published

138. C. Muscalu & W. Schlag *Classical and multilinear harmonic analysis, II*
139. B. Helffer *Spectral theory and its applications*
140. R. Pemantle & M. C. Wilson *Analytic combinatorics in several variables*
141. B. Branner & N. Fagella *Quasiconformal surgery in holomorphic dynamics*
142. R. M. Dudley *Uniform central limit theorems (2nd Edition)*
143. T. Leinster *Basic category theory*
144. I. Arzhantsev, U. Derenthal, J. Hausen & A. Laface *Cox rings*
145. M. Viana *Lectures on Lyapunov exponents*
146. J.-H. Evertse & K. Győry *Unit equations in Diophantine number theory*
147. A. Prasad *Representation theory*
148. S. R. Garcia, J. Mashreghi & W. T. Ross *Introduction to model spaces and their operators*
149. C. Godsil & K. Meagher *Erdős-Ko-Rado theorems: Algebraic approaches*
150. P. Mattila *Fourier analysis and Hausdorff dimension*
151. M. Viana & K. Oliveira *Foundations of ergodic theory*
152. V. I. Paulsen & M. Raghupathi *An introduction to the theory of reproducing kernel Hilbert spaces*
153. R. Beals & R. Wong *Special functions and orthogonal polynomials*
154. V. Jurdjevic *Optimal control and geometry: Integrable systems*
155. G. Pisier *Martingales in Banach spaces*
156. C. T. C. Wall *Differential topology*
157. J. C. Robinson, J. L. Rodrigo & W. Sadowski *The three-dimensional Navier-Stokes equations*
158. D. Huybrechts *Lectures on K3 surfaces*
159. H. Matsumoto & S. Taniguchi *Stochastic analysis*
160. A. Borodin & G. Olshanski *Representations of the infinite symmetric group*
161. P. Webb *Finite group representations for the pure mathematician*
162. C. J. Bishop & Y. Peres *Fractals in probability and analysis*
163. A. Bovier *Gaussian processes on trees*
164. P. Schneider *Galois representations and (φ, Γ)-modules*
165. P. Gille & T. Szamuely *Central simple algebras and Galois cohomology (2nd Edition)*
166. D. Li & H. Queffelec *Introduction to Banach spaces, I*
167. D. Li & H. Queffelec *Introduction to Banach spaces, II*
168. J. Carlson, S. Müller-Stach & C. Peters *Period mappings and period domains (2nd Edition)*
169. J. M. Landsberg *Geometry and complexity theory*
170. J. S. Milne *Algebraic groups*
171. J. Gough & J. Kupsch *Quantum fields and processes*
172. T. Ceccherini-Silberstein, F. Scarabotti & F. Tolli *Discrete harmonic analysis*
173. P. Garrett *Modern analysis of automorphic forms by example, I*
174. P. Garrett *Modern analysis of automorphic forms by example, II*
175. G. Navarro Character theory and the McKay conjecture
176. P. Fleig, H. P. A. Gustafsson, A. Kleinschmidt & D. Persson *Eisenstein series and automorphic representations*
177. E. Peterson *Formal geometry and bordism operators*
178. A. Ogus *Lectures on logarithmic algebraic geometry*

Lectures on Logarithmic Algebraic Geometry

ARTHUR OGUS
University of California, Berkeley

CAMBRIDGE
UNIVERSITY PRESS

CAMBRIDGE
UNIVERSITY PRESS

University Printing House, Cambridge CB2 8BS, United Kingdom

One Liberty Plaza, 20th Floor, New York, NY 10006, USA

477 Williamstown Road, Port Melbourne, VIC 3207, Australia

314–321, 3rd Floor, Plot 3, Splendor Forum, Jasola District Centre, New Delhi – 110025, India

79 Anson Road, #06–04/06, Singapore 079906

Cambridge University Press is part of the University of Cambridge.

It furthers the University's mission by disseminating knowledge in the pursuit of education, learning, and research at the highest international levels of excellence.

www.cambridge.org
Information on this title: www.cambridge.org/9781107187733
DOI: 10.1017/9781316941614

First published 2018

A catalogue record for this publication is available from the British Library.

Library of Congress Cataloging-in-Publication Data
Names: Ogus, Arthur, author.
Title: Lectures on logarithmic algebraic geometry /
Arthur Ogus (University of California, Berkeley, USA).
Description: Cambridge : Cambridge University Press, 2018. | Series: Cambridge studies in advanced mathematics ; 178 | Includes bibliographical references and indexes.
Identifiers: LCCN 2018009845 | ISBN 9781107187733 (hardback)
Subjects: LCSH: Geometry, Algebraic–Textbooks. | Logarithmic functions–Textbooks. | Number theory–Textbooks. | Compactifications–Textbooks.
Classification: LCC QA565 .O38 2018 | DDC 516.3/5–dc23
LC record available at https://lccn.loc.gov/2018009845

ISBN 978-1-107-18773-3 Hardback

Contents

1 Introduction

1.1 Motivation

Logarithmic geometry was developed to deal with two fundamental and related problems in algebraic geometry: compactification and degeneration. A key aspect of algebraic geometry is that it is essentially global in nature. Algebraic varieties can be compactified: any separated scheme S of finite type over a field k admits an open embedding $j\colon S \hookrightarrow T$, with T/k proper and with S Zariski dense in T [55, 9]. Since proper schemes are much easier to study than general schemes, it is often convenient to work with T even if it is the original scheme S that is of primary interest. It then becomes necessary to keep track of the complement $Z := T \setminus S$ and to study how functions, differential forms, sheaves, and other geometric objects on T behave near Z, and to have a mechanism to extract S from T. In differential topology, these problems are often addressed by working with manifolds with boundary, and logarithmic geometry can be thought of as a substitute for, or version of, the notion of "algebraic variety with boundary." Indeed, log schemes over the field of complex numbers have "Betti realizations,"[1] and the Betti realizations of logarithmically smooth log schemes are topological manifolds with boundary.

The compactification problem is related to the phenomenon of degeneration. A scheme S often arises as a moduli space, for example, a space parameterizing smooth proper schemes of a certain type. If S is a fine moduli space, there is a smooth proper morphism $f\colon U \to S$ whose fibers are the objects one wants to classify. One can then hope to find a compactification T of S such that the boundary points parameterize "decorated degenerations" of the original objects. In this case there should be a proper and flat (but not smooth) $g\colon X \to T$ extending $f\colon U \to S$. Then one is left with the problem of comparing f to g and in particular of analyzing the behavior of g near $Y := X \setminus U$. In many cases one obtain important information about the original family f by studying the degenerate family over Z. A typical example is the compactification of the moduli stack of smooth curves by the moduli stack of stable curves.

The problems of compactification and degeneration are thus manifest in a

[1] Betti realizations of log schemes were introduced by Kato and Nakayama and are often called "Kato–Nakayama spaces."

diagram of the form:

$$\begin{array}{ccccc} U & \hookrightarrow & X & \hookleftarrow & Y \\ \downarrow{\scriptstyle f} & & \downarrow{\scriptstyle g} & & \downarrow{\scriptstyle g_{|Y}} \\ S & \hookrightarrow & T & \hookleftarrow & Z. \end{array}$$

It turns out that in many such cases there is a natural way to equip X and T with *log structures,* which somehow "remember" U and S and are compatible with g. Then $g\colon X \to T$ becomes a *morphism of log schemes* and inherits many of the nice features of f. The log structures on X and T restrict in a natural way to Y and Z, and the resulting morphism of log schemes $g_{|Y}\colon Y \to Z$ still remembers useful information about f, thanks to the "decoration" provided by the log structures on Y and Z.

In good cases, the log structures on f, X, and T render the morphism f *logarithmically smooth*, which makes it much easier to study than the underlying morphism of schemes. The concept of smoothness for log schemes fits very naturally into Grothendieck's geometric deformation theory. Furthermore, Betti realizations of proper log smooth morphisms behave in some respects like topological fibrations (see [44] and [57]). The fact that this picture works so well both in topological and in arithmetical settings is one of the main justifications for the theory of log geometry.

Let us illustrate how log geometry works in the most basic case, that of a (possibly partial) compactification. Let $j\colon U \to X$ be an open immersion, with complementary closed immersion $i\colon Y \to X$. Then Y (and hence U) is determined by the sheaf $I_Y \subseteq \mathcal{O}_X$ consisting of those local sections of $\mathcal{O}_X$ whose restriction to Y vanishes, a sheaf of ideals of $\mathcal{O}_X$. However, it is not Y but rather U that is our primary interest, so instead we consider the subsheaf $\mathcal{M}_{U/X}$ of $\mathcal{O}_X$ consisting of the local sections of $\mathcal{O}_X$ whose restriction to U is invertible. If f and g are sections of $\mathcal{M}_{U/X}$, then so is fg, but $f+g$ need not be. Thus $\mathcal{M}_{U/X}$ is not a sheaf of rings, but it is a sheaf of *submonoids* of the multiplicative sheaf of monoids underlying $\mathcal{O}_X$. Note that $\mathcal{M}_{U/X}$ contains the sheaf of units $\mathcal{O}_X^*$, and if X is integral, the quotient $\mathcal{M}_{U/X}/\mathcal{O}_X^*$ can be naturally identified[2] with the sheaf of effective Cartier divisors on X with support in the complement Y of U in X. The morphism of sheaves of monoids $\alpha_{U/X}\colon \mathcal{M}_{U/X} \to \mathcal{O}_X$ (inclusion) is a *logarithmic structure*, called the *compactifying log structure* associated to the embedding j. In good cases this log structure "remembers" the inclusion $U \to X$ and furthermore satisfies a technical *coherence* condition that makes

[2] This identification takes the class of a local section m of $\mathcal{M}_{U/X}$ to the inverse of the (invertible) ideal sheaf generated by $\alpha_{U/X}(m)$.

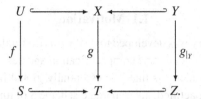

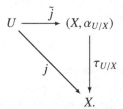

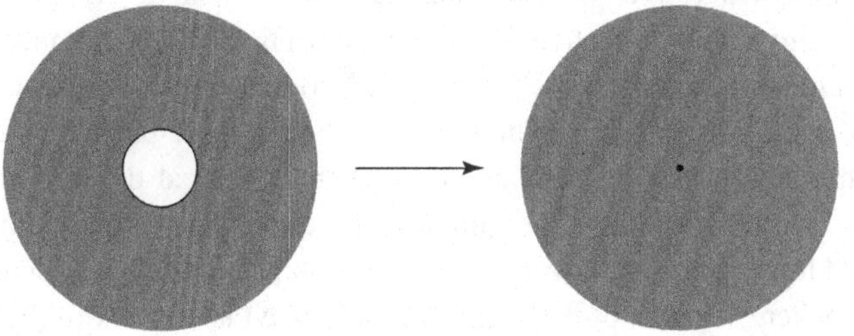

Figure 1.1.1 Compactifying an open immersion

it manageable. In the category of log schemes, the open immersion j fits into a commutative diagram

$$\begin{array}{ccc} U & \xrightarrow{\tilde{j}} & (X, \alpha_{U/X}) \\ & \searrow^{j} & \downarrow{\scriptstyle \tau_{U/X}} \\ & & X. \end{array}$$

This diagram provides a relative compactification of the open immersion j. The map $\tau_{U/X}$ is proper but the map $\tilde{j}$ somehow preserves much of the essential nature of the original open immersion j: in good cases, it behaves like a local homotopy equivalence. We can imagine that the log structure $\alpha_{U/X}$ cuts away or blows up enough of X to make it look like U, but leaves enough of a boundary for it to remain compact. It is in this sense that the log scheme $(X, \alpha_{U/X})$ plays the role of an "algebraic variety with boundary." For example, in the case of the standard embedding of $\mathbf{G_m} \to \mathbf{A}^1$, the corresponding log scheme $(\mathbf{A}^1, \alpha)$ behaves very much like the complex plane in which the origin is blown up to become a circle, as shown in Figure 1.1.1. The morphism in this picture can be identified with the multiplication map $\mathbf{R}_{\geq} \times \mathbf{S}^1 \to \mathbf{C}$, where $\mathbf{R}_{\geq}$ is the set of nonnegative real numbers and $\mathbf{S}^1$ is the set of complex numbers of absolute value one. This "real blowup" resolves the ambiguity of polar coordinates. It serves as a proper model of the inclusion $\mathbf{G_m} \to \mathbf{A}^1$, whose homotopy theory it closely resembles. These ideas will be made more precise in Section 1 of Chapter V, where we discuss Betti realizations of log schemes. In particular, Theorem V.1.3.1 shows that the Betti realization of a (logarithmically) smooth log scheme over $\mathbf{C}$ really is a topological manifold with boundary.

In general, a *log structure* on a scheme X is a morphism of sheaves of commutative monoids $\alpha\colon \mathcal{M} \to \mathcal{O}_X$ inducing an isomorphism $\alpha^{-1}(\mathcal{O}_X^*) \to \mathcal{O}_X^*$. We do not require α to be injective. In particular, sections of $\mathcal{M}$ can map to zero in $\mathcal{O}_X$, although in good cases $\mathcal{M}$ is *integral,* so that on $\mathcal{M}$, multiplication by any local section is injective. The tension between these behaviors accounts for much of the power, as well as many of the technical difficulties, of log geometry, particularly those involving fiber products. The flexibility and functoriality of log structures allow us to restrict a compactifying log structure $\alpha_{U/X}$ to $X \setminus U$, where sections of the sheaf of monoids $\mathcal{M}$ keep track of the "ghosts of vanishing coordinates."

The naturality of these constructions allows them to work in appropriate relative settings, for example, in the context of semistable reduction. Let X be a regular scheme, let T be the spectrum of a discrete valuation ring, and let $f\colon X \to T$ be a flat and proper morphism whose generic fiber X_τ/τ is smooth and whose special fiber is a reduced divisor with normal crossings. Then the compactifying log structures α_X and α_T associated as above to the open embeddings $X_\tau \to X$ and $\tau \to T$ fit into a morphism of log schemes

$$f\colon (X, \alpha_X) \to (T, \alpha_T),$$

which is in fact logarithmically smooth.

The value of the machinery of log geometry must be judged by its applications to problems outside the theory itself. A detailed discussion of any of these would be beyond the scope of this book, and we can only point readers to the literature. Historically, the first (and perhaps still most striking) such application is in the proof, due to Hyodo and Kato [37], Kato [48], Tsuji [75], Faltings [19] [18], and others, of the "C_{st} conjecture" in p-adic Hodge theory. Indeed, log geometry began as an attempt to discern what additional structure on the special fiber of a semistable reduction was needed to define a "limiting crystalline cohomology," in analogy to Steenbrink's construction of limiting mixed Hodge structures in the complex analytic context [73], [74]. In ℓ-adic cohomology, the main applications have been to the Bloch conductor formula [51] and higher dimensional Ogg–Shafarevich formulas [1] and to results on resolution, purity, and duality [42]. Log geometry has also been notably used in the theory of mirror symmetry [24] and the study of compactifications of moduli spaces of curves [47], [68], abelian varieties [64], K3 surfaces [63], and toric Hilbert schemes [65].

1.2 Roots

The development of logarithmic geometry, like that of any organism, began well before its official birth, and was preceded by many classical methods dealing with the problems of compactification and degeneration. These include most notably the theories of toroidal embeddings, of differential forms and equations with log poles and/or regular singularities, and of vanishing cycles and monodromy. Logarithmic geometry was influenced by all these ideas and provides a language that incorporates and extends them in functorial and systematic ways.

Logarithmic structures fit so naturally with the usual building blocks of schemes that it is possible, and in most (but not all) cases, straightforward and natural, to adapt many of the standard techniques and intuitions of algebraic geometry to the logarithmic context. Log geometry seems to be especially compatible with infinitesimal techniques, including Grothendieck's notions of smoothness, differentials, and differential operators. The sheaf of Kähler differentials of a logarithmic scheme (X, α_X), constructed from Grothendieck's deformation-theoretic viewpoint, coincides with the classical sheaf of differential forms of X with log poles along $X \setminus U$; this fact is one justification for the terminology. Furthermore, any toric variety (with the log structure corresponding to the dense open torus it contains) is log smooth, and the theory of toroidal embeddings is essentially equivalent to the study of (logarithmically) smooth log schemes over a field.

1.3 Goals

Our aim in this book is to provide an introduction to the basic notions and techniques of log geometry that is accessible to graduate students with a basic knowledge of algebraic geometry. We hope the material will also be useful to researchers in other areas of geometry, to which we believe the theory can be profitably adopted, as has already been done in the case of complex analytic geometry. For the sake of concreteness, we work systematically with schemes as locally ringed spaces, although it certainly would have been possible and profitable to develop the theory for complex analytic varieties, or for algebraic spaces or stacks. Even in the case of schemes, it is quite valuable to work locally in the étale topology, and we shall allow ourselves to do so, although we do not use the language of topos theory. (That more powerful approach is taken in the very thorough treatment in [22].)

Just as scheme theory starts with the study of commutative rings, log geometry starts with the study of commutative monoids. Much of this foundational material is already available in the literature, but we have decided to offer a self-contained presentation more directly suited to our purposes. In log geometry, in an apparent contrast with toric geometry, the study of the *category* of monoids, and in particular of *homomorphisms of monoids*, plays a fundamental role. This difference was part of our motivation for including this material, and we hope our treatment may be of interest apart from its applications to log geometry per se. Thus Chapter I begins with the study of limits and colimits in the category of monoids, and in particular with the construction of pushouts, which are analogous to tensor products in the category of rings. We then discuss sets endowed with a monoid action (the analogs of modules in ring theory), ideals, localization, and the spectrum of a monoid, with its Zariski topology. After these preliminaries we turn to more familiar constructions in convex geometry, including basic results about finiteness, duality, and cones. Then we discuss monoid algebras and some facts about affine toric varieties. The final sections of Chapter I are devoted to a deeper study of properties of homomorphisms and actions of monoids, and in particular to certain analogs of flatness. Of particular importance is Kato's key concept of *exactness*, which we encounter in Section 1.1. An example of its importance is manifest in the "four point lemma" 4.2.16, where exactness is needed to make fiber products of logarithmically integral log schemes behave well. *Integrality* and *saturation* of morphisms, which we discuss next, are refinements of the notion of exactness. Theorem 4.7.2 reveals the structure of "critically" and "locally" exact homomorphisms and plays an important role throughout log geometry and this text. We finish by showing how locally exact homomorphisms can be made integral and saturated by a suitable base change, which can be viewed as a logarithmic version of semistable reduction. This material is more technical than the rest of our exposition and can be skipped over in a first reading.

Chapter II discusses sheaves of monoids on topological spaces. After disposing of some generalities, we define *monoschemes*, which are constructed by gluing together spectra of commutative monoids, just as schemes are constructed by gluing together spectra of commutative rings. Our monoschemes are sometimes called "schemes over $\mathbf{F}_1$" in the literature [12] and are generalizations of the *fans* used to construct toric varieties. We use this concept to construct *monoidal transformations* (blowups) for monoids (and monoschemes). The main application is Theorem 1.8.1, which explains how a homomorphism of monoids can be made locally exact by a monoidal transformation. Section 1.10 explains the *moment map* for a monoid scheme $\underline{\mathsf{A}}_{\mathsf{Q}}$, which gives a linearized model of the set of its $\mathbf{R}_{\geq}$-valued points. As an application, we show that the "positive part" of each fiber of a monoidal transformation is contractible. The remainder of Chapter II is devoted to Kato's important notions of *charts* and *coherence* for sheaves of monoids, which form the main technical link between logarithmic and toric geometry.

With the preliminaries well in hand, we are ready in Chapter III to turn to logarithmic geometry per se, including two variants of the standard theory:

idealized log schemes and *relatively coherent log structures*. We work with log structures in both the Zariski and étale topologies, since each has its own advantages and disadvantages, and explain the relation between the two. After giving the main definitions and basic constructions, we discuss some examples: log points and dashes, and the compactifying log structures coming from open immersions $U \to X$. We then describe in some detail a precursor of the notion of log structures, due (independently) to Deligne and Faltings. This notion, although less flexible and functorial than the point of view taken here, is convenient for describing the log structures that arise in the context of divisors with normal crossings and semistable reduction. It was in some sense already envisioned in the work of Friedman [20] and Steenbrink [73]. We then discuss *hollow* and especially *solid* log structures. In the first case, the log structure reflects the geometry of the part of a scheme that has been cut away, and in the second the log structure is tightly tied to the part of the geometry of the scheme which can be modelled by a toric structure. The notion of solidity of a log structure is closely related to, and helpful in, the study of Kato's notion of *log regularity*, which we discuss next. Finally, we briefly discuss *frames* for log structures, a weak version of charts that can be quite useful.

The remainder of Chapter III is devoted to the study of morphisms of log schemes, including the rather delicate construction of fibered products. *Exact* morphisms of log schemes play an especially important role, as well as the related notions of *integral* and *saturated* morphisms. We also study the logarithmic versions of immersions, inseparable morphisms, and *Kummer* and *small* morphisms, as well as *logarithmic blowups*.

Chapter IV is devoted to *logarithmic differentials* and *logarithmic smoothness*. We begin with a purely algebraic construction of Kähler differentials for (pre) log schemes, then explain its geometric meaning in terms of deformation theory. Next we discuss smoothness for logarithmic schemes, defined in terms of a logarithmic version of Grothendieck's infinitesimal lifting criterion. Although smooth morphisms in logarithmic geometry are much more complicated than in classical geometry, locally they admit nice toric models. As in the classical case, smoothness and regularity are related notions, the former being a "relative" version of the latter. We next discuss the more general notion of *logarithmic flatness*, which is quite useful but, as of this writing, technically challenging. We explore the relationships among the notions of flatness, smoothness, exactness, and integrality, extending in some cases the fundamental results of Kato.

In Chapter V we discuss topology and cohomology. To provide a geometric intuition, we begin with the construction of the *Betti realization* X_{log} of a log scheme X over **C**. This is a topological space that comes with a natural proper map $\tau_X\colon X_{log} \to X_{an}$ which embodies the picture exemplified in Figure 1.1.1. We explain the definition and basic topological properties of Betti realizations, make them explicit for toric models, and show that the Betti realization of a smooth analytic space is a topological manifold with boundary. (We do not include the proof, but in fact the Betti realization of a smooth proper and exact morphism of log analytic spaces is a topological fibration of manifolds with boundary [57].) We then define the sheaf of rings $\mathcal{O}_X^{log}$ on X_{log}, which is obtained by adjoining logarithms of sections of $\mathcal{M}_X$ in a canonical way and which allows for a generalization of the familiar exponential sequence in classical complex analytic geometry. Our next main topic is logarithmic de Rham cohomology. We begin with an algebraic description of the *logarithmic de Rham complex* of a monoid algebra and some of the natural filtrations (defined by faces and ideals) it carries. Then we explain the sheafification and globalization of these constructions for log schemes. We give several versions of the logarithmic Poincaré lemma in the analytic setting, proving that analytic de Rham cohomology calculates the Betti cohomology of X_{log}. In the algebraic setting, we construct the Cartier isomorphism and the Cartier operator in positive characteristics, and explain how the Cartier operator relates to the restricted Lie-algebra structure on the logarithmic tangent sheaf. Finally we study algebraic de Rham cohomology in characteristic zero, concluding with some finiteness theorems and comparisons with analytic, and hence log Betti, cohomology.

Time and space constraints have prevented us from discussing many important topics which we had earlier hoped to include and for which we can only indicate some references in the literature. Some fundamental results not covered include the resolution of toric singularities [49],[58],[42], the cohomology of log blowups [39], and the fact that normal toric varieties are Cohen–Macaulay [36]. We have also had to omit examples of applications of log geometry and can only suggest that the reader look at work on the moduli of stable curves [47], on the logarithmic Riemann–Hilbert correspondence [40, 61], and on crystalline cohomology [37, 60], as a scattered set of examples.

1.4 Organization

The goals of this text are to introduce the reader to the basic ideas of log geometry and to provide a technical foundation for further work on the theory and its applications. These goals are somewhat contradictory, in that a good deal of the foundational material depends on the algebra of monoids and the geometry of convex bodies, the study of which can impede the momentum toward the ultimate goals coming from algebraic geometry. Although a fair amount of this material can be found in the literature, we have decided to treat it carefully here, partly because the author himself wanted to become comfortable with it and partly because the perspective from log geometry, in which homomorphisms play a central role, is not to be found in the standard texts. We have grouped nearly all this material in the first two chapters and consequently don't arrive at log geometry itself until Chapter III, potentially discouraging a reader eager to try out log geometry in some specific context. Such a reader may find it preferable to skip some of the earlier sections, returning to them as necessary. We hope our exposition will make this possible. In particular, the material on idealized monoids, idealized log schemes, and relative coherence, concepts whose ultimate utility has not yet been convincingly demonstrated, can be skipped on a first reading. Probably the same is true of monoschemes, which are really just an alternative to the classical theory of fans from toric geometry. Readers focused on the essence of log geometry could try reading only Sections 1.1, 4.1, and 4.2 of Chapter I, and then Sections 1.1 and 2.1 of Chapter II, before proceeding to Chapter III. Readers whose primary interest is convex rather than log geometry may find it interesting to concentrate on the material in Chapters I and II, since some of it may be new to them, especially Section 4 of Chapter I. Unfortunately, the key concept of logarithmic smoothness does not appear until well into Chapter IV; fortunately, this concept was already well explained in Kato's original paper [48]. In any case, we hope that impatient readers will find our treatment palatable even if they have not digested all the preceding material.

To facilitate flexibility in reading the text, we have tried to be careful with references. We use the same numbering scheme for definitions, theorems, remarks, etc. within each chapter. When referring to a result from a different chapter, we include the (roman numeral) chapter number in the reference; otherwise we omit it.

It is probably appropriate to remark on the writing style. We have attempted to include a considerable degree of detail, both in motivating and in defining concepts and in writing the proofs. Some readers, especially those familiar with the techniques of toric geometry, may consequently find the presentation ponderous. However, we found no alternative compatible with the goals of solidifying our understanding and of avoiding a plethora of errors, which would otherwise crop up not just in the proofs themselves, but also in statements of theorems and, worse, definitions. It seems easier for the reader to skip some arguments as s/he sees fit rather than to worry about errors hidden in unwritten proofs. Readers who feel the (understandable) desire for exercises can refrain from reading the proofs supplied and provide their own and/or content themselves with the search for errors, of which we fear many may remain. We would be grateful for notifications of any errors, which we hope eventually to correct on a web page available at <https://math.berkeley.edu/~ogus/logpage.html>.

1.5 Acknowledgements

Most of the material presented here is already in the literature in one form or another, often in several places. I have not made a systematic attempt to keep track of the proper original attributions. The main conceptual ideas of the form of logarithmic geometry treated here are due to L. Illusie, J.-M. Fontaine, and K. Kato; a precursor was developed independently by P. Deligne and by G. Faltings. In many places when I got stuck on basics I had invaluable help and guidance from G. Bergman, H. Lenstra, B. Sturmfels, C. Nakayama, and A. Abbes. I would also like to thank IRMAR and the IHES for their very generous support for this project. Many graduate students at U.C. Berkeley have attended courses on this material and helped to clarify the exposition; B. Conrad also gave a course at Stanford using a preliminary version of the manuscript and provided me with much meticulous and invaluable feedback. Many other mathematicians have also pointed out errors and misprints. O. Gabber's corrections and suggestions were particularly subtle and valuable. Special thanks go to Luc Illusie, who provided detailed advice, guidance, and important mathematical content throughout the planning and writing of this book; any defects in presentation or technical accuracy occur only when I failed to follow his advice. Most of the technical drawings were provided by J. Ogus, and the commutative diagrams were produced with Paul Taylor's Commutative Diagrams package. I am very appreciative of the patience, support, and encouragement provided by Kaitlin Leach, acquisition editor of Cambridge University Press, during the very long preparation of this manuscript.

I

The Geometry of Monoids

1 Basics on monoids

1.1 Limits in the category of monoids

A *monoid* is a triple $(M, \star, e_M)$ consisting of a set M, an associative binary operation $\star$, and a two-sided identity element e_M of M. A *homomorphism of monoids* is a function $\theta : M \to N$ such that $\theta(e_M) = e_N$ and $\theta(m \star m') = \theta(m) \star \theta(m')$ for any pair of elements m and m' of M. Note that, although the element e_M is the unique two-sided identity of M, compatibility of θ with e_M is not automatic from compatibility with $\star$. All monoids we consider here will be commutative unless explicitly noted otherwise, and we write **Mon** for the category of commutative monoids and homomorphisms of monoids.

We will often follow the common practice of writing M or $(M, \star)$ in place of $(M, \star, e_M)$ when there seems to be no danger of confusion. Similarly, if a and b are elements of a monoid $(M, \star, e_M)$, we will often write ab (or $a + b$) for $a \star b$, and 1 (or 0) for e_M.

The most basic example of a monoid is the set $\mathbf{N}$ of natural numbers, with addition as the monoid law. If M is any monoid and $m \in M$, there is a unique monoid homomorphism $\mathbf{N} \to M$ sending 1 to m; thus $\mathbf{N}$ is the free monoid with generator 1. More generally, if S is any set, the set $\mathbf{N}^{(S)}$ of functions $I\colon S \to \mathbf{N}$ such that $I_s = 0$ for almost all s, endowed with the pointwise addition of functions as a binary operation, is the free (commutative) monoid with basis $S \subseteq N^{(S)}$. Thus the functor $S \mapsto \mathbf{N}^{(S)}$ is left adjoint to the forgetful functor from the category of monoids to the category of sets.

Arbitrary (projective) limits exist in the category of monoids, and their formation commutes with the forgetful functor to the category of sets. In particular, the intersection of a set of submonoids of M is again a submonoid; hence if S is a subset of M, the intersection of all the submonoids of M that contain

S is the smallest submonoid of M containing S, the *submonoid of M generated by S*. If there exists a finite subset S of M that generates M, one says that M is *finitely generated* as a monoid.

Arbitrary colimits (inductive limits) of monoids also exist. Direct sums are easy to construct: the direct sum $\bigoplus M_i$ of a family $\{M_i : i \in I\}$ of monoids is the submonoid of the product $\prod_i M_i$ consisting of those elements $m.$ such that $m_i = 0$ for almost all i. The general construction is more difficult, and we will first investigate quotients and equivalence relations in the category of monoids.

Let $\theta \colon P \to Q$ be a homomorphism of monoids. Note that the kernel $\theta^{-1}(0)$ of θ is not very useful: for example, the kernel of the homomorphism $\theta \colon \mathbf{N} \oplus \mathbf{N} \to \mathbf{N}$ sending (a, b) to $a + b$ is just $\{(0, 0)\}$, but θ is not injective. Instead we consider the set $E(\theta)$ of pairs $(p_1, p_2) \in P \times P$ such that $\theta(p_1) = \theta(p_2)$, an equivalence relation on P. The fact that θ is also a homomorphism of monoids implies that $E(\theta)$ is a submonoid of $P \times P$. An equivalence relation on P that is also a submonoid of $P \times P$ is called a *congruence* (or *congruence relation*) on P. One checks easily that if E is a congruence relation on P, then the set P/E of equivalence classes has a unique monoid structure making the projection $P \to P/E$ a monoid homomorphism. Thus there is a dictionary between congruence relations on P and isomorphism classes of surjective monoid homomorphisms $P \to P'$. The following proposition, whose proof is immediate, summarizes these considerations.

Proposition 1.1.1. *Let P be a monoid.*

1. *Let $\pi \colon P \to Q$ be a surjective homomorphism of monoids, and let E be the equalizer of the two maps $P \times P \underset{p_2}{\overset{p_1}{\rightrightarrows}} P \xrightarrow{\pi} Q$, i.e., $E := P \times_Q P$.*
   (a) *E is a congruence relation on P.*
   (b) *Q is the coequalizer of the two maps $E \to P \times P \underset{p_2}{\overset{p_1}{\rightrightarrows}} P$. Thus, the diagram*

$$\begin{array}{ccc} E & \longrightarrow & P \\ \big\downarrow & & \big\downarrow \\ P & \longrightarrow & Q \end{array}$$

   is cocartesian, as well as cartesian.
2. *Let $E \subseteq P \times P$ be a congruence relation on P, let $Q := P/E$ be the set of equivalence classes, and let $\pi \colon P \to Q$ be the function taking an element of P to the equivalence class containing it.*

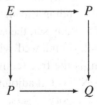

(a) *There is a unique monoid structure on Q such that $\pi\colon P \to Q$ is a monoid homomorphism.*
(b) *The inclusion $e : E \to P{\times}P$ is the equalizer of the two homomorphisms*

$$P \times P \underset{p_2}{\overset{p_1}{\rightrightarrows}} P \xrightarrow{\pi} Q,$$

and Q is the coequalizer of the two homomorphisms

$$E \xrightarrow{e} P \times P \underset{p_2}{\overset{p_1}{\rightrightarrows}} P.$$

Thus, the diagram of (1b) is cartesian and cocartesian.

The passage from E to Q induces a bijection from the set of congruence relations on P to the set of isomorphism classes of surjective homomorphisms whose domain is P. □

In the terminology of [2, 10.3 and 10.8, Exp. I], Proposition 1.1.1 says that every surjective homomorphism of monoids is an "effective epimorphism," and every congruence relation E is an "effective equivalence relation."

Remark 1.1.2. If $P \to Q$ is surjective and $Q' \to Q$ is any homomorphism, then the pullback map $P \times_Q Q' \to Q'$ is again surjective. This implies that $P \to Q$ and E are in fact "universally effective." On the other hand, not every epimorphism in the category of monoids is surjective. In fact, a homomorphism of monoids is universally an epimorphism if and only if it is surjective.

The intersection of a family of congruence relations is a congruence relation, and hence it makes sense to speak of the *congruence relation generated by a subset of $P \times P$*. One says that a congruence relation E is *finitely generated* if there is a finite subset S of $P{\times}P$ that generates E as a congruence relation; this does not imply that S generates E as a monoid.

Here is a useful description of the congruence relation generated by a subset of $P \times P$.

Proposition 1.1.3. *Let P be a (commutative) monoid.*

1. *An equivalence relation $E \subseteq P \times P$ is a congruence relation if and only if $(a + p, b + p) \in E$ whenever $(a, b) \in E$ and $p \in P$.*
2. *If S is a subset of $P \times P$, let*

$$S_P := \{(a + p, b + p) : (a, b) \in S, p \in P\}.$$

Then the congruence relation E generated by S is the equivalence relation generated by S_P. Explicitly, E is the union of the diagonal with the set of pairs (x, y) for which there exists a finite sequence $(p_0, \ldots, p_n)$ with $p_0 = x$

and $p_n = y$ such that, for every $i > 0$, either (p_{i-1}, p_i) or (p_i, p_{i-1}) belongs to S_P.

Proof Let E be an equivalence relation on P that is stable under addition by elements of the diagonal. Suppose that (a, b) and $(c, d) \in E$. Then $(a+c, b+c) \in E$ and $(c + b, d + b) \in E$, and since P is commutative and E is transitive, $(a + c, b + d) \in E$. Thus E is closed under addition. Since E contains the diagonal, the identity element $(0, 0)$ of $P \times P$ belongs to E, so E is a submonoid of $P \times P$, hence a congruence relation. Conversely, if E is a congruence relation, then $(p, p) \in E$ for every $p \in E$; hence $(a + p, b + p) \in E$ whenever $(a, b) \in E$, This proves (1). For (2), let E denote the congruence relation generated by S and let E' denote the equivalence relation generated by S_P. Since $S_P \subseteq E$ and E is an equivalence relation, it follows that $E' \subseteq E$. The associative law implies that S_P is stable under addition by elements of the diagonal of $P \times P$. Hence if $(p_0, \ldots, p_n)$ is a sequence such that (p_{i-1}, p_i) or $(p_i, p_{i-1}) \in S_P$ for all $i > 0$, then $(p_0 + p, \ldots, p_n + p)$ shares the same property. Thus if $(x, y) \in E'$ and $p \in P$, then $(x + p, y + p) \in E'$. Then it follows from (1) that E' is a congruence relation, and so $E' = E$. □

Remark 1.1.4. If P is an abelian group and $E \subseteq P \times P$ is a congruence relation on P, then the image of E under the homomorphism $h \colon P \oplus P \to P$ sending (p_1, p_2) to $p_2 - p_1$ is a subgroup K of P, and $E = h^{-1}(K)$. Conversely the inverse image under h of any subgroup of P is a congruence relation on P. This defines a bijective correspondence between the subgroups of P and the congruence relations on P.

If θ_1 and θ_2 are monoid homomorphisms $P \to Q$, one can construct the *coequalizer* of θ_1 and θ_2 as the quotient of Q by the congruence relation on Q generated by the set of pairs $(\theta_1(p), \theta_2(p))$ for $p \in P$.

The existence of arbitrary colimits follows from the existence of direct sums and coequalizers of pairs of morphisms by the following standard construction (see also [2, 2.3, Exp. I]). Let $\{P_i, \theta_a\}$ be a functor from a small category I to the category of monoids, where i ranges over the objects of I and a over the arrows $a \colon i(a) \to j(a)$ of I. Since the category of monoids has direct sums, we can form $Q := \oplus\{P_i : i \in Ob(I)\}$ and $R := \oplus\{P_{i(a)}, a \in Arr(I)\}$, with canonical homomorphisms

$$\{u_i \colon P_i \to Q : i \in Ob(I)\} \quad \text{and} \quad \{u_a \colon P_{i(a)} \to R : a \in Arr(I)\}.$$

Then there are unique homomorphisms $\theta_1, \theta_2 \colon R \to Q$ such that, for all a,

$\theta_1 \circ u_a = u_{i(a)}$ and $\theta_2 \circ u_a = u_{j(a)} \circ \theta_a$:

$$\begin{array}{ccccccc} P_{i(a)} & \xrightarrow{u_a} & R & \quad & R & \xleftarrow{u_a} & P_{i(a)} \\ & \searrow^{u_{i(a)}} & \downarrow{\scriptstyle \theta_1} & & \downarrow{\scriptstyle \theta_2} & & \downarrow{\scriptstyle \theta_a} \\ & & Q & & Q & \xleftarrow{u_{j(a)}} & P_{j(a)}. \end{array}$$

The colimit of the functor $\{P_i, \theta_a\}$ is the coequalizer of θ_1 and θ_2.

A *presentation* of a monoid M is a coequalizer diagram

$$L_1 \rightrightarrows L_0 \longrightarrow M$$

with L_0 and L_1 free. It is equivalent to the data of a map from a set I to M whose image generates M and a map from a set J to $\mathbf{N}^{(I)} \times \mathbf{N}^{(I)}$ whose image generates the congruence relation on $\mathbf{N}^{(I)}$ defined by the surjective monoid map $\mathbf{N}^{(I)} \to M$ corresponding to the set map $I \to M$. The monoid M is said to be *of finite presentation* if it admits a presentation with L_0 and L_1 free and finitely generated. We shall see in Theorem 2.1.7 that in fact every finitely generated (commutative) monoid is of finite presentation.

The *amalgamated sum* $Q_1 \xrightarrow{v_1} Q \xleftarrow{v_2} Q_2$ of a pair of monoid morphisms $u_i \colon P \to Q_i$, often denoted simply by $Q_1 \oplus_P Q_2$, is the colimit of the diagram $Q_1 \xleftarrow{u_1} P \xrightarrow{u_2} Q_2$. That is, the pair (v_1, v_2) universally makes the diagram

$$\begin{array}{ccc} P & \xrightarrow{u_1} & Q_1 \\ {\scriptstyle u_2}\downarrow & & \downarrow{\scriptstyle v_1} \\ Q_2 & \xrightarrow{v_2} & Q \end{array} \tag{1.1.1}$$

commute. This amalgamated sum can be viewed as the pushout of u_1 along u_2 or the pushout of u_2 along u_1. It can also be viewed as the coequalizer of the two maps $(u_1, 0)$ and $(0, u_2)$ from P to $Q_1 \oplus Q_2$.

The following proposition describes the pushout $Q_1 \oplus_P Q_2$ explicitly. Its calculation is considerably simplified if one of the monoids in question is a group. (See Proposition 4.6.1 for a generalization.)

Proposition 1.1.5. *Let $u_i \colon P \to Q_i$ be a pair of monoid morphisms, let Q be their amalgamated sum, as in Diagram 1.1.1, and let E be the congruence relation on $Q_1 \oplus Q_2$ given by the natural surjection $Q_1 \oplus Q_2 \to Q$.*

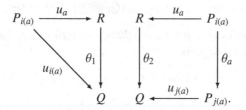

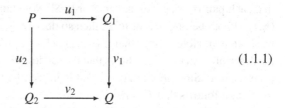

1. *Let S be the set of pairs*

$$((q_1, q_2), (q'_1, q'_2)) \in (Q_1 \oplus Q_2) \times (Q_1 \oplus Q_2)$$

such that there exists a $p \in P$ such that $q'_1 = u_1(p) + q_1$ and $q_2 = u_2(p) + q'_2$. Then E is the set of pairs

$$(a, b) \in (Q_1 \oplus Q_2) \times (Q_1 \oplus Q_2)$$

such that there exists a sequence $(r_0, \ldots, r_n)$ in $Q_1 \oplus Q_2$ such that $(a, b) = (r_0, r_n)$ and such that (r_i, r_{i+1}) belongs to S if i is even and to $S^t := \{(a, b) : (b, a) \in S\}$ if i is odd.

2. *Let E' be the set of pairs $((q_1, q_2), (q'_1, q'_2))$ of elements of $Q_1 \oplus Q_2$ such that there exist p and p' in P with $q_1 + u_1(p') = q'_1 + u_1(p)$ and $q_2 + u_2(p) = q'_2 + u_2(p')$. Then E' is a congruence relation on $Q_1 \oplus Q_2$ containing E, and if any of P, Q_1, or Q_2 is a group, then $E = E'$.*
3. *If P is a group, then two elements of $Q_1 \oplus Q_2$ are congruent modulo E if and only if they lie in the same orbit of the action of P on $Q_1 \oplus Q_2$ defined by $p(q_1, q_2) = (q_1 + u_1(p), q_2 + u_2(-p))$.*
4. *If P and Q_i are groups, then so is $Q_1 \oplus_P Q_2$, which is in fact just the amalgamated sum in the category of abelian groups.*

Proof To prove (1), observe first that S is stable under the action of the diagonal of $(Q_1 \oplus Q_2) \times (Q_1 \oplus Q_2)$ and contains this diagonal. Then, by Proposition 1.1.3, the congruence relation R generated by S is the set of pairs (a, b) such that there exists a sequence $(r_0, \ldots, r_n)$ with $r_0 = a$ and $r_n = b$ and such that each pair (r_i, r_{i+1}) belongs to S or to S^t. Note however that if (r_{i-1}, r_i) and (r_i, r_{i+1}) both belong to S or to S^t, then so does (r_{i-1}, r_{i+1}), so the sequence can be shortened. Note further that if (r_0, r_1) belongs to S^t then (r_0, r_0, r_1) satisfies the description in (1). This shows that the set described in (1) really is a congruence relation. Since E contains S, and is in fact the smallest such congruence relation, it follows that $E = R$.

To prove (2), note first that the set E' is evidently symmetric and reflexive. To prove its transitivity, let us say that a pair (a, b) in $P \times P$ "links" a pair of elements (q_1, q_2) and (q'_1, q'_2) of $Q_1 \oplus Q_2$ if $q_1 + u_1(b) = q'_1 + u_1(a)$ and $q_2 + u_2(a) = q'_2 + u_2(b)$. One checks immediately that if (a, b) links (q_1, q_2) and (q'_1, q'_2) and (a', b') links (q'_1, q'_2) and (q''_1, q''_2), then $(a + a', b + b')$ links (q_1, q_2) and (q''_1, q''_2). Moreover, if (a, b) links (q_1, q_2) and (q'_1, q'_2) then, for any $(\tilde{q}_1, \tilde{q}_2) \in Q_1 \oplus Q_2$, (a, b) links $(q_1 + \tilde{q}_1, q_2 + \tilde{q}_2)$ and $(q'_1 + \tilde{q}_1, q'_2 + \tilde{q}_2)$. Then by Proposition 1.1.3, E' is a congruence relation on $Q_1 \oplus Q_2$. Furthermore, if $p \in P$, $(p, 0)$ links $(u_1(p), 0)$ and $(0, u_2(p))$, and since E is the congruence relation generated by such pairs, $E \subseteq E'$. If P or either Q_i is a group, then

$v := v_i \circ u_i$ factors through the group Q^* of invertible elements of Q. If (a, b) links (q_1, q_2) and (q'_1, q'_2), we find that

$$\begin{aligned} v_1(q_1) + v_2(q_2) + v(a+b) &= v_1(q_1 + u_1(b)) + v_2(q_2 + u_2(a)) \\ &= v_1(q'_1 + u_1(a)) + v_2(q'_2 + u_2(b)) \\ &= v_1(q'_1) + v_2(q'_2) + v(a+b). \end{aligned}$$

Since $v(a + b) \in Q^*$, it follows that

$$v_1(q_1) + v_2(q_2) = v_1(q'_1) + v_2(q'_2).$$

Thus $E' \subseteq E$. This proves (2), and (3) and (4) are immediate consequences. □

Example 1.1.6. Taking $Q_2 = 0$ in Proposition 1.1.5 one obtains the *cokernel* of the morphism $u_1 : P \to Q_1$, or, equivalently, the coequalizer of u_1 and the zero mapping $P \to Q_1$. If P is a submonoid of Q, one writes $Q \to Q/P$ for the cokernel of the inclusion $P \to Q$, and it follows from (2) of the proposition that two elements q and q' of Q have the same image in Q/P if and only if there exist p and p' in P such that $q + p = q' + p'$. For example, the cokernel of the diagonal embedding $\mathbf{N} \to \mathbf{N} \oplus \mathbf{N}$ is the homomorphism

$$\mathbf{N} \oplus \mathbf{N} \to \mathbf{Z} : (a, b) \mapsto a - b.$$

Note that Q/P can be zero even if P is a proper submonoid of Q; this holds, for example, if P is the submonoid of $Q := \mathbf{N} \oplus \mathbf{N}$ generated by $(1, 0)$ and $(1, 1)$. If P' is a submonoid of Q containing P, then P'/P is a submonoid of Q/P and the natural map $(Q/P)/(P'/P) \to Q/P'$ is an isomorphism.

1.2 Monoid actions

If S is a set, then the set of functions from S to itself forms a (not necessarily commutative) monoid $\mathrm{End}(S)$ under composition. If Q is a monoid, an *action of Q on S* is a monoid homomorphism θ_S from Q to $\mathrm{End}(S)$. In this context we often write the monoid law of Q multiplicatively, and, if $q \in Q$ and $s \in S$, we write qs for $\theta_S(q)(s)$. A *Q-set* is a set endowed with an action of Q, and $\mathbf{Ens}_Q$ will denote the category of Q-sets, with the evident notion of morphism. If S is a Q-set and $s \in S$, the image of the map $Q \to S$ sending q to qs is the minimal Q-stable subset of S containing s, called the *trajectory* of s in S.

A *basis* for a Q-set S is a map of sets $i : T \to S$ such that the induced map $Q \times T \to S : (q, t) \mapsto qi(t)$ is bijective. If such a basis exists we say that S is a *free Q-set*. A free Q-set with basis $T \to S$ satisfies the usual universal property of a free object: every map from T to the set underlying a Q-set S'

extends uniquely to a morphism of Q-sets $S \to S'$. If T is any set and if $Q \times T$ is endowed with the action ρ defined by $\rho(q')(q,t) = (q'q,t)$, then the map $T \to Q \times T$ sending t to $(1,t)$ is a basis. Thus the functor taking a set T to the free Q-set $Q \times T$ is left adjoint to the forgetful functor from the category of Q-sets to the category of sets. Note that if G is a group and S is a G-set, then S has a basis as a G-set if and only if the action is free in the sense that $gs = s$ implies $g = 1$. This equivalence is not true for general monoids.

The category $\mathbf{Ens}_Q$ of Q-sets admits arbitrary projective limits, and their formation commutes with the forgetful functor to the category of sets. This is a formal consequence of the fact that the forgetful functor $\mathbf{Ens}_Q \to \mathbf{Ens}$ has a left adjoint. In particular, if S and T are Q-sets, then Q acts on $S \times T$ by $q(s,t) := (qs, qt)$, and this action makes $S \times T$ the product of S and T in $\mathbf{Ens}_Q$.

Colimits in $\mathbf{Ens}_Q$ also exist. The direct sum of a family $S_i : i \in I$ is just the disjoint union with the evident Q-action. To understand the construction of quotients in the category $\mathbf{Ens}_Q$, note that if $\pi\colon S \to T$ is a surjective map of Q-sets, then the corresponding equivalence relation $E \subseteq S \times S$ is a Q-subset of $S \times S$; such an equivalence relation is called a *congruence relation* on S. Conversely, if E is any congruence relation on S, then there is a unique Q-set structure on S/E such that the projection $S \to S/E$ is a morphism of Q-sets. When $S = Q$ acting regularly on itself, the notion of a congruence relation on Q as a monoid coincides with the notion of a congruence relation as a Q-set, thanks to Proposition 1.1.3. Furthermore, the analog of statement (2) of Proposition 1.1.3 holds for Q-sets, and in particular the equivalence relation generated by a subset of $S \times S$ that is stable under the diagonal action of Q is already a congruence relation. If u and v are two morphisms $S' \to S$, the coequalizer of u and v is the quotient of S by the congruence relation generated by $\{(u(s'), v(s')) : s' \in S'\}$. The existence of general colimits follows.

If S and T are Q-sets, the set $\mathrm{Hom}_Q(S,T)$ has a natural action of Q, given by $(qh)(s) := qh(s) = h(qs)$ for $h\colon S \to T$, $q \in Q$, and $s \in S$ There is also a tensor product construction for Q-sets. If S, T, and W are Q-sets, then a *Q-bimorphism* $S \times T \to W$ is by definition a function $\beta\colon S \times T \to W$ such that $\beta(qs,t) = \beta(s,qt) = q\beta(s,t)$ for any $(s,t) \in S \times T$ and $q \in Q$. The *tensor product of S and T* is the universal Q-bimorphism $S \times T \to S \otimes_Q T$. To construct it, begin by regarding $S \times T$ as a Q-set via its action on S, so that $q(s,t) := (qs,t)$, and consider the equivalence relation R on $S \times T$ generated by the set of pairs

$$((qs,t),(s,qt)) \in (S \times T) \times (S \times T) \text{ for } q \in Q, s \in S, t \in T.$$

Note that this set of pairs is stable under the action of Q, since if $q' \in Q$, and $s' := q's$ then $((q'qs,t),(q's,qt)) = ((qs',t),(s',qt))$. It follows that the equivalence relation R is a congruence relation. Then the projection $\pi\colon S \times T \to$

$(S{\times}T)/R$ is a Q-bimorphism and satisfies the universal mapping property of the tensor product. If Q is a (commutative) group, then $S \otimes_Q T$ can be constructed in the usual way as the orbit space of the action of Q on $S{\times}T$ given by $q(s,t) := (qs, q^{-1}t)$.

In general, one has a natural isomorphism of Q-sets

$$\mathrm{Hom}_Q(S \otimes_Q T, W) \cong \mathrm{Hom}_Q(S, \mathrm{Hom}_Q(T, W)),$$

taking a Q bimorphism β to the Q-morphism γ given by $\gamma(s)(t) := \beta(s,t)$. It follows formally that, for fixed T, the functor $S \mapsto S \otimes_Q T$ commutes with colimits.

Let $\theta\colon P \to Q$ be a homomorphism of monoids. Then θ defines an action of P on Q given by $pq := \theta(p)q$. If S is a P-set, the tensor product $Q \otimes_P S$ has the natural action of Q, with $q(q' \otimes s) = (qq' \otimes s)$, and the map $S \to Q \otimes_P S$ sending s to $1 \otimes s$ is a morphism of P-sets over the homomorphism θ. If $\theta_i\colon P \to Q_i$ for $i = 1, 2$ is a pair of monoid homomorphisms, then there is a unique monoid structure on $Q_1 \otimes_P Q_2$ such that

$$(q_1 \otimes q_2)(q_1' \otimes q_2') = (q_1 q_1' \otimes q_2 q_2'),$$

and this is also the unique monoid structure for which the natural maps $Q_i \to Q_1 \otimes_P Q_2$ are homomorphisms. It can be checked that this monoid structure makes $Q_1 \otimes_P Q_2$ into the amalgamated sum of Q_1 and Q_2 along P.

Remark 1.2.1. We have seen that, for a fixed Q-set T, the functor $S \mapsto S \otimes_Q T$ commutes with colimits. It is perhaps no surprise that it does not commute with limits in general. We want to emphasize that this functor need not even commute with finite products, even if T is free. Indeed, if T has basis Λ, then $S \otimes_Q T \cong S \times \Lambda$, and if the cardinality of Λ is greater than one, the functor $S \mapsto S \times \Lambda$ does not commute with products. This fact complicates the calculation of tensor products from generators and relations. Indeed, suppose that $F \to S$ is a surjective morphism and $E := F \times_S F$ is the corresponding equivalence relation on F, where F is free. Then $F \to S$ is the coequalizer of the two maps $E \rightrightarrows F$, and, since $\otimes_Q T$ commutes with colimits, it follows that $F \otimes_Q T \to S \otimes_Q T$ is the coequalizer of $E \otimes_Q T \rightrightarrows F \otimes_Q T$. However, the natural map

$$(F \times F) \otimes_Q T \to (F \otimes_Q T) \times (F \otimes_Q T)$$

is not an isomorphism, and the image of $E \otimes_Q T$ in $(F \otimes_Q T) \times (F \otimes_Q T)$ might not be an equivalence relation. Thus one is left with the often challenging problem of computing the congruence relation it generates.

Definition 1.2.2. *Let Q be a monoid and let S be a Q-set. The* transporter *of S is the category $\mathcal{T}_Q S$ whose objects are the elements of S and for which the* *morphisms from an object s to an object t are the elements q of Q such that $qs = t$, with composition given by the monoid law of Q. The transporter of a monoid Q is the transporter of Q regarded as a Q-set, and is denoted simply by $\mathcal{T}Q$.*

Associated with the category $\mathcal{T}_Q S$ is a partially ordered set that is worth making explicit.

Definition 1.2.3. *Let Q be a monoid and S a Q-set. If s and t are elements of S, we write $s \le t$ if there exists a $q \in Q$ such that $qs = t$, and $s \sim t$ if $s \le t$ and $t \le s$.*

It is clear that $s \le w$ if $s \le t$ and $t \le w$ and that $s \le s$ for every $s \in S$. Thus the relation $\le$ defines a preorder on S. The relation $\sim$ is a congruence relation on S, and the relation $\le$ on the quotient $S/\!\sim$ is a partial order. We shall use this notion especially when $S = Q$ with the regular representation. Since $\sim$ is a congruence relation, it follows from Proposition 1.1.3 that $Q/\!\sim$ inherits a monoid structure.

1.3 Integral, fine, and saturated monoids

If M is any commutative monoid, there is a universal homomorphism λ_M from M to a group M^{gp}. That is, M^{gp} is a group, $\lambda_M \colon M \to M^{\mathrm{gp}}$ is a homomorphism of monoids, and any homomorphism from M to a group factors uniquely through λ_M. Thus, the functor $M \mapsto M^{\mathrm{gp}}$ is the left adjoint of the inclusion functor from the category of groups to the category of monoids; since it has a right adjoint, it automatically commutes with the formation of direct limits. In fact, M^{gp} can be identified with the cokernel (Example 1.1.6) of $M \oplus M = M \times M$ by the diagonal, and λ_M with the composite of $(\mathrm{id}_M, 0)$ and the projection $M \times M \to (M \times M)/\Delta_M$. One can also construct M^{gp} as the set of equivalence classes of pairs (x, y) of elements of M, where (x, y) is equivalent to (x', y') if and only if there exists $z \in M$ such that $x + y' + z = x' + y + z$. The explicit description of the equivalence relation in Example 1.1.6 shows that the two constructions are in fact the same. One writes $x - y$ for the equivalence class containing (x, y), and then $(x - y) + (x' - y') = (x + x') - (y + y')$.

If M is a monoid, let M^* denote the set of all $m \in M$ such that there exists an $n \in M$ such that $m + n = 0$. Then M^* forms a submonoid of M. It is in fact a subgroup—the largest subgroup of M. Any homomorphism from a group to M factors uniquely through M^*, so that $M \mapsto M^*$ is right adjoint to the inclusion functor from groups to monoids. We call M^* the *group of units* of M; it acts naturally on M by translation. If G is any subgroup of M, the orbit space M/G

can be identified with the quotient in the category of monoids discussed in Example 1.1.6. In particular, we write $\overline{M}$ for M/M^*.

Definition 1.3.1. *A (commutative) monoid M is said to be:*

1. *sharp if $M^* = \{0\}$;*
2. *dull if $M^* = M$, i.e., if M is a group;*
3. *u-integral if $m \in M, u' \in M^*$ and $m + u' = m$ implies that $u' = 0$;*
4. *quasi-integral if $m, m' \in M$ and $m + m' = m$ implies that $m' = 0$;*
5. *integral if $m, m', m'' \in M$ and $m + m' = m + m''$ implies that $m' = m''$.*

Evidently every integral monoid is quasi-integral and every quasi-integral monoid is u-integral. If M is u-integral, then M^* acts freely on M and the map $M \to \overline{M}$ makes M an M^*-torsor over $\overline{M}$. The universal map $\lambda_M : M \to M^{gp}$ is injective if and only if M is integral, and the induced map $M^* \to M^{gp}$ is injective if and only if M is u-integral. For any monoid M, the quotient $\overline{M}$ is sharp, and the map $M \to \overline{M}$ is the universal homomorphism from M to a sharp monoid. For any monoid M, the monoid $M/\sim$ (see Definition 1.2.3) is sharp, and if M is quasi-integral, the natural map $M/M^* \to M/\sim$ is an isomorphism. The inverse limit of a family of integral monoids is again integral.

Remark 1.3.2. The formation of M^{gp} commutes with direct products but not with fibered products in general. For example, let $s : \mathbf{N}^2 \to \mathbf{N}$ be the map taking (a, b) to $a + b$ and let t be the map taking (a, b) to 0. Then the equalizer of s and t is zero. However, the equalizer of the associated maps on groups $\mathbf{Z}^2 \to \mathbf{Z}$ is the antidiagonal $\mathbf{Z} \to \mathbf{Z}^2$, sending c to $(c, -c)$. On the other hand, it is true that if $\theta : P \to Q$ is injective and Q is integral, then $\theta^{gp} : P^{gp} \to Q^{gp}$ is also injective.

Proposition 1.3.3. *If Q is an integral monoid and P is a submonoid, the natural map $Q/P \to Q^{gp}/P^{gp}$ is injective. Thus Q/P is integral and can be identified with the image of Q in Q^{gp}/P^{gp}. A monoid Q is integral if and only if it is u-integral and $\overline{Q}$ is integral.*

Proof If q and q' are two elements of Q with the same image in Q^{gp}/P^{gp}, then there exist p and p' such that $q - q' = p - p'$ in Q^{gp}. Since Q is integral, $q + p' = q' + p$ in Q. Then it follows from the discussion in Example 1.1.6 that q and q' have the same image in Q/P. Thus $Q/P \to Q^{gp}/P^{gp}$ is injective, and Q/P is integral. In particular, if Q is integral, so is $\overline{Q}$. Conversely, suppose that Q is u-integral and $\overline{Q}$ is integral, and that q, q' and p are elements of Q with $q + p = q' + p$. Since $\overline{Q}$ is integral, there exists a unit u such that $q' = q + u$. Then $q + p = q + p + u$. Since Q is u-integral, $u = 0$ and hence $q = q'$. This shows that Q is integral. □

Let $\mathbf{Mon}^{\text{int}}$ denote the full subcategory of **Mon** whose objects are the integral monoids. For any monoid M, let M^{int} denote the image of

$$\lambda_M \colon M \to M^{\text{gp}}.$$

Then $M \mapsto M^{\text{int}}$ is left adjoint to the inclusion functor $\mathbf{Mon}^{\text{int}} \to \mathbf{Mon}$.

Proposition 1.3.4. *Let Q be the amalgamated sum of two homomorphisms $u_i \colon P \to Q_i$ in the category* **Mon**. *Then Q^{int} is the amalgamated sum of $u_i^{\text{int}} \colon P^{\text{int}} \to Q_i^{\text{int}}$ in the category $\mathbf{Mon}^{\text{int}}$, and can be naturally identified with the image of Q in $Q_1^{\text{gp}} \oplus_{P^{\text{gp}}} Q_2^{\text{gp}}$. If P, Q_1, and Q_2 are integral and any of these monoids is a group, then Q is integral.*

Proof The fact that Q^{int} is the amalgamated sum of u_i^{int} in $\mathbf{Mon}^{\text{int}}$ is a formal consequence of the fact that $M \mapsto M^{\text{int}}$ preserves colimits. Moreover, since $M \mapsto M^{\text{gp}}$ also preserves colimits, $Q^{\text{gp}} \cong Q_1^{\text{gp}} \oplus_{P^{\text{gp}}} Q_2^{\text{gp}}$. It follows that Q^{int} is the image of Q in $Q^{\text{gp}} \cong Q_1^{\text{gp}} \oplus_{P^{\text{gp}}} Q_2^{\text{gp}}$. Now suppose that any of P and Q_i is a group and that q and q' are two elements of Q with the same image in Q^{gp}. Choose (q_1, q_2) and (q_1', q_2') in $Q_1 \oplus Q_2$ mapping to q and q' respectively. Then $v_1(q_1)+v_2(q_2) = v_1(q_1')+v_2(q_2')$ in Q^{gp}, and so there exist elements a and b in P such that $(q_1'-q_1, q_2'-q_2) = (u_1(a-b), u_2(b-a))$. Then $q_1' + u_1(b) = q_1 + u_1(a)$ and $q_2' + u_2(a) = q_2 + u_2(b)$. It then follows from (2) of Proposition 1.1.5 that $v_1(q_1) + v_2(q_2) = v_1(q_1') + v_2(q_2')$ in Q, i.e., that $q = q'$. Thus the map $Q \to Q_1^{\text{gp}} \oplus_{P^{\text{gp}}} Q_2^{\text{gp}}$ is injective and Q is integral. □

A monoid Q is said to be *fine* if it is finitely generated and integral. A monoid Q is called *saturated* if it is integral and if whenever $q \in Q^{\text{gp}}$ is such that $mq \in Q$ for some $m \in \mathbf{Z}^+$, then $q \in Q$. For example, the monoid of all integers greater than or equal to some natural number d, together with zero, is not saturated if $d > 1$. For another example, let Q be the submonoid of $\mathbf{Z} \oplus \mathbf{Z}/2\mathbf{Z}$ generated by $x := (1, 0)$ and $y := (1, e)$, where e is the nonzero element of $\mathbf{Z}/2\mathbf{Z}$. Then $2x = 2y$, so $z := (0, x - y) \in Q^{\text{sat}} \setminus Q$. In fact z is a nonzero unit of Q^{sat}, but Q is sharp.

Proposition 1.3.5. *Let Q be an integral monoid.*

1. *The natural homomorphism $Q^{\text{gp}}/Q^* \to \overline{Q}^{\text{gp}}$ is an isomorphism.*
2. *If Q is saturated, then $\overline{Q}^{\text{gp}}$ is torsion free.*
3. *The set Q^{sat} of all elements x of Q^{gp} such that there exists $n \in \mathbf{Z}^+$ with $nx \in Q$ is a saturated submonoid of Q^{gp}, and the functor $Q \mapsto Q^{\text{sat}}$ is left adjoint to the inclusion functor from the category $\mathbf{Mon}^{\text{sat}}$ of saturated monoids to $\mathbf{Mon}^{\text{int}}$.*
4. *Q is saturated if and only if $\overline{Q}$ is saturated. An element of Q is a unit if and only if its image in Q^{sat} is a unit.*

5. *The natural map* $Q^{\text{sat}}/Q^* \to \overline{Q}^{\text{sat}}$ *is an isomorphism. Furthermore, every unit of* $\overline{Q}^{\text{sat}}$ *is torsion, and the natural map*

$$\overline{Q^{\text{sat}}} \to \overline{\overline{Q}}^{\text{sat}}$$

is an isomorphism.

Proof Suppose that $q_1, q_2 \in Q$ and $q_2 - q_1$ maps to zero in $\overline{Q}^{\text{gp}}$. Since $\overline{Q} \subseteq \overline{Q}^{\text{gp}}$, $\overline{q}_1 = \overline{q}_2$ in $\overline{Q}$, and hence there exists a $u \in Q^*$ with $q_2 = u + q_1$. Then $q_2 - q_1 = u \in Q^*$. This proves (1). Suppose Q is saturated and $q \in Q^{\text{gp}}$ maps to a torsion element $\overline{q}$ of $\overline{Q}^{\text{gp}}$. Then $nq \in Q^*$ for some $n \in \mathbf{Z}^+$, and since Q is saturated, $q \in Q$. The fact that nq belongs to Q^* now implies that q belongs to Q^*, so $\overline{q} = 0 \in \overline{Q}$. Thus $\overline{Q}^{\text{gp}}$ is torsion free. If q and p are elements of Q^{gp} with $mq \in Q$ and $np \in Q$, then $mn(q+p) \in Q$, and it follows that Q^{sat} is a submonoid of Q^{gp}. Hence $(Q^{\text{sat}})^{\text{gp}} = Q^{\text{gp}}$ and, if $q \in Q^{\text{gp}}$ and $nq \in Q^{\text{sat}}$, then there exists an $m \in \mathbf{Z}^+$ with $mnq \in Q$. It follows that $q \in Q^{\text{sat}}$, so Q^{sat} is saturated. The verification of the adjointness of the functor $Q \mapsto Q^{\text{sat}}$ is immediate, as is the verification of (4).

If $q \in Q^{\text{sat}}$ and $\overline{q}$ is a unit of $\overline{Q}^{\text{sat}}$, then there also exists an element p of Q^{sat} with $q + p \in Q^*$. Then there exist m and n in $\mathbf{Z}^+$ such that mq and np belong to Q. But then $mnq + mnp \in Q^*$, and hence mnq is a unit of Q. This shows that $\overline{q}$ is a torsion element of $\overline{Q}^{\text{sat}}$. It is clear that the map in (5) is surjective. Suppose that q and p are two elements of Q^{sat} with the same image in $\overline{\overline{Q}}^{\text{sat}}$. Then $q - p \in Q^{\text{gp}}$ maps to a unit of $\overline{Q}^{\text{sat}}$, and hence to a torsion element of $\overline{Q}^{\text{sat}} \subseteq \overline{Q}^{\text{gp}}$. Hence $mq - mp \in (Q^{\text{sat}})^*$ for some m. Then $mp - mq \in Q^*$ also, so $q - p$ is a unit of Q^{sat} and q and p have the same image in $\overline{Q^{\text{sat}}}$. This proves the injectivity. □

Monoids that are both fine and saturated are of central importance in logarithmic geometry, and are often called *fs-monoids*. A monoid P is said to be *toric* if it is fine and saturated and in addition P^{gp} is torsion free; in this case P^{gp} can be viewed as the character group of an algebraic torus. The schemes arising from toric monoids form the building blocks of toric geometry.

Proposition 1.3.6. *Let* $\{M_i : i \in I\}$ *be a direct system of monoids each of which satisfies one of the following properties* **P**: *integral, saturated, dull. Then the direct limit* M *also satisfies* **P**.

Proof Suppose that each M_i is integral and let m be an element of M. Then there exist $i \in I$ and $m_i \in M_i$ such that m_i maps to m in M. For each $i \to j$, let m_{ji} denote the image of m_i in M_j. Then multiplication by $m_{ji}\colon M_j \to M_j$ is injective. It follows that the limit of these maps, i.e., multiplication by m, is also injective. Thus M is integral. Suppose further that each M_i is saturated and that $x \in M^{gp}$, with $nx \in M$ for some $n > 0$. Since the formation of M^{gp} commutes with direct limits, there exist $i \in I$ and $x_i \in M_i^{gp}$ mapping to x. Replacing i by some element to which it maps, we may further assume that there is some $m_i \in M_i$ mapping to nx. Again replacing i, we may assume that $nx_i = m_i$ in M_i. Since M_i is saturated, it follows then that $x_i \in M_i$ and hence that $x \in M$. Thus M is saturated. Since the formation of direct limits is the same in the categories of commutative monoids and groups, the direct limit of dull monoids is dull. □

A monoid M is said to be *valuative* if it is integral and if for every $x \in M^{gp}$, either x or $-x$ lies in M. This is equivalent to saying that the preorder relation (Definition 1.2.3) on M^{gp} defined by the action of M is a total preorder. The monoid $\mathbf{N}$ is valuative and, if V is a valuation ring, the submonoid V' of nonzero elements of V is valuative. Every valuative monoid is saturated.

If R is any commutative ring, its underlying multiplicative monoid $(R, \cdot, 1)$ is not quasi-integral unless $R = \{0\}$, since $0 \cdot 0 = 1 \cdot 0$. On the other hand, the set R' of nonzero divisors of R forms an integral submonoid of the multiplicative monoid of R. For example, $\mathbf{Z}'$ is integral, and $\overline{\mathbf{Z}}' = \mathbf{Z}'/(\pm)$ is a free (commutative) monoid, generated by the prime numbers. If R is a discrete valuation ring, $\overline{R}' = R'/R^*$ is freely generated by the image of a uniformizer of R'. Although there is a unique isomorphism of monoids $R'/R^* \cong \mathbf{N}$, this isomorphism is not functorial: if $R \to S$ is a finite extension of valuation rings with ramification index e, the induced map $\overline{R}' \to \overline{S}'$ sends the unique generator of $\overline{R}'$ to e times that of $\overline{S}'$.

If Q is a sharp commutative monoid, free Q-sets are very rigid, as the following simple observation shows.

Proposition 1.3.7. *Let Q be a sharp commutative monoid, and let S be a free Q-set. Then any basis for S is unique up to unique isomorphism. Explicitly, every basis $i\colon T \to S$ induces a bijection between T and $S \setminus Q^+S$.*

Proof Let $i\colon T \to S$ be a basis for S. Since the induced map $Q \times T \to S$ is bijective, i must be injective. Let us verify that if $t \in T$, then $i(t) \in S \setminus Q^+S$. Suppose that $i(t) = qs$ with $q \in Q$ and $s \in S$. Then there is a unique $(q', t') \in Q \times T$ such that $s = q'i(t')$, and so $i(t) = qq'i(t')$. Then $(1, t)$ and (qq', t') are two elements of $Q \times T$ with the same image in S, so $qq' = 1$. Since Q is commutative, $q \in Q^*$ and, since Q is sharp, $q = 1$. Since q was arbitrary, $i(t) \notin Q^+S$. On the other hand, suppose that $s \in S \setminus Q^+S$. Since i is a basis for S, there is some $(q, t) \in Q \times T$ such that $qi(t) = s$ and, since $s \notin Q^+S$, $q = 1$ and $s = i(t)$. Thus the induced map from T to $S \setminus Q^+S$ is also surjective. □

1.4 Ideals, faces, and localization

Definition 1.4.1. *An ideal of a monoid M is a subset I such that $k \in I$ and $q \in M$ implies $q + k \in I$. An ideal I is called prime if $I \neq M$ and $p + q \in I$ implies $p \in I$* or *$q \in I$. A face of a monoid M is a submonoid F such that $p + q \in F$ implies that both p and q belong to F.*

For example, the empty set $\emptyset$ is a prime ideal of M, as is the set M^+ of non-units of M. Note that any ideal I containing a unit must be all of M, so that every proper ideal of M is contained in M^+. Thus M^+ is the unique (proper) maximal ideal of M; moreover, $\emptyset$ is the unique minimal ideal of M. In many respects, a monoid is analogous to a local ring. In particular, a monoid homomorphism $\theta\colon P \to Q$ is said to be *local* if $\theta^{-1}(Q^+) = P^+$ or, equivalently, $\theta^{-1}(Q^*) = P^*$. Observe that a face is just a submonoid whose complement is an ideal (necessarily prime) and a prime ideal is an ideal whose complement is a submonoid (necessarily a face). Thus $\mathfrak{p} \mapsto F_\mathfrak{p} := M \setminus \mathfrak{p}$ gives an order reversing bijection between the set of prime ideals of M and the set of faces of M. The set of units M^* is the smallest face of M (contained in every face), and M is the largest face of M. The notion of a face of a monoid corresponds to the notion of a saturated multiplicative subset of a ring; we do not use this terminology here because of its conflict with the notion of a saturated monoid.

As in the case of rings, the intersection of a family of ideals is an ideal, but for monoids the union of a family of ideals is also an ideal. Furthermore the union of a family of prime ideals is a prime ideal and the intersection of a family of faces is a face. If T is a subset of M, the intersection $\langle T \rangle$ of all the faces containing T is a face of M, called the *face generated by T*. It is analogous to the multiplicatively saturated set generated by a subset of a ring. The *interior ideal* I_M of a monoid M is the set of all elements that do not lie in a proper face of M, i.e., the intersection of all the nonempty prime ideals of M.

We denote by Spec(M) the set of prime ideals of a monoid M, and for each ideal I of M we denote by $Z(I)$ the set of primes ideals of M containing I. Then if (I_λ) is any family of ideals, $\cup I_\lambda$ is an ideal and $Z(\cup I_\lambda) = \cap Z(I_\lambda)$. Also, if I and J are ideals, so is the set IJ of all elements of the form $p+q$ with p in I and q in J, and $Z(I) \cup Z(J) = Z(I \cap J) = Z(IJ)$. Thus the set of all subsets of Spec M of the form $Z(I)$ for variable I is closed under intersections and finite unions, and hence defines a topology (the *Zariski topology*) on $S :=$ Spec(M). Since M has a unique minimal prime ideal, Spec(M) has a unique generic point, and in particular is irreducible. Since M has a unique maximal ideal it also has a unique closed point. If $f \in M$ and F is the face it generates, then

$$D(f) := S_f := \{\mathfrak{p} : f \notin \mathfrak{p}\} = \{\mathfrak{p} : \mathfrak{p} \cap F = \emptyset\}$$

is open in S, and the set of all such sets forms a basis for the topology on S.

Note that Spec(M) is never empty: it contains a unique closed point (the maximal ideal M^+ of M, consisting of all the non-units) and a unique generic point (the empty ideal of M). These two points coincide if and only if every element of M is a unit, that is, if and only if M is a group.

If $\theta \colon P \to Q$ is a homomorphism of monoids, then the inverse image of an ideal is an ideal, the inverse image of a prime ideal is a prime ideal, and the inverse image of a face is a face. Thus θ induces a continuous map

$$\mathrm{Spec}(Q) \to \mathrm{Spec}(P) : \quad \mathfrak{p} \mapsto \theta^{-1}(\mathfrak{p}).$$

The preorder relation (Definition 1.2.3) is useful when describing the ideals and faces of a monoid, as the following proposition shows.

Proposition 1.4.2. *Let S be a subset of a monoid Q and let P be the submonoid of Q generated by S.*

1. *The ideal (S) of Q generated by S is the set of all $q \in Q$ such that $q \geq s$ for some $s \in S$.*
2. *The face $\langle S \rangle$ of Q generated by S is the set P' of elements q of Q for which there exists a $p \in P$ such that $q \leq p$. In particular, the face generated by an element p of Q is the set of all elements $q \in Q$ such that $q \leq np$ for some $n \in \mathbf{N}$.*
3. *If Q is integral, then Q/P is sharp if and only if $P^{\mathrm{gp}} \cap Q$ is a face of Q. In particular, if F is a face of Q, then Q/F is sharp.*

Proof The first statement follows immediately from the definitions. For the second, note that a submonoid F of Q is a face if and only if F contains q whenever $q \leq f$ for some $f \in F$. Hence $\langle S \rangle$ contains P'. Since in fact P' is necessarily a face of Q containing S, it follows that $P' = \langle S \rangle$. If Q is integral, Q/P can be identified with the image of Q in $Q^{\mathrm{gp}}/P^{\mathrm{gp}}$, by Proposition 1.3.3. Thus an element $q \in Q$ maps to 0 in Q/P if and only if $q \in Q \cap P^{\mathrm{gp}}$, and q maps to a unit in Q/P if and only if there exists an element $q' \in Q$ such that $q + q' \in P^{\mathrm{gp}}$, i.e., if and only if $q \in \langle Q \cap P^{\mathrm{gp}} \rangle$. This shows that Q/P is sharp if and only if $Q \cap P^{\mathrm{gp}}$ is a face of Q. Finally, note that if F is a face of Q, and $q \in Q \cap F^{\mathrm{gp}}$, then $q + f \in F$ for some $f \in F$; hence $q \in F$. □

Corollary 1.4.3. *Let K be an ideal of a monoid Q and let*

$$\sqrt{K} := \{q : nq \in K \text{ for some } n \in \mathbf{Z}^+\}$$

be its radical.

1. *$\sqrt{K}$ is a radical ideal, that is, $\sqrt{K} = \sqrt{\sqrt{K}}$.*

2. *$\sqrt{K}$ is the intersection of all the prime ideals of Q containing K.*
3. *The mapping $I \to Z(I)$ induces an order-reversing bijection between the radical ideals of Q and the closed subsets of* Spec(Q)*, with inverse $S \mapsto \cap\{\mathfrak{p} : \mathfrak{p} \in S\}$.*
4. *A closed subset S of Q is irreducible if and only if the corresponding radical ideal is prime.*

Proof If $q \in \sqrt{\sqrt{K}}$, there exists $n \in \mathbf{Z}^+$ such that $nq \in \sqrt{K}$ and then there exists $m \in \mathbf{Z}^+$ such that $mnq \in K$. Thus $\sqrt{\sqrt{K}} \subseteq \sqrt{K}$, and the reverse inclusion is obvious. This proves (1). For (2), it is clear that $\sqrt{K}$ is contained in every prime ideal $\mathfrak{p}$ containing K. Conversely, suppose that $q \in Q \setminus \sqrt{K}$ and let f be an element of the face F of Q generated by q. By Proposition 1.4.2, there exist n and q' such that $nq = f + q'$. Since $nq \notin K$, the same is true of f. This shows that $F \cap K = \emptyset$, and hence that $\mathfrak{p} := Q \setminus F$ is a prime ideal of Q containing K but not q. It is clear that $Z(J) \subseteq Z(I)$ if $I \subseteq J$. If S is any subset of Q, then $\cap S := \cap\{\mathfrak{p} : \mathfrak{p} \in S\}$ is clearly a radical ideal of Q, and $S \subseteq Z(\cap S)$, since for every $\mathfrak{p} \in S$, $\cap S \subseteq \mathfrak{p}$. Moreover, if I is any ideal of Q and $S \subseteq Z(I)$, then $I \subseteq \mathfrak{p}$ for every $\mathfrak{p} \in S$, and hence $I \subseteq \cap S$ and $Z(\cap(S)) \subseteq Z(I)$. Thus $Z(\cap S)$ is the closure of S. In particular, if S is closed, $S = Z(\cap S)$. On the other hand, if K is a radical ideal, statement (2) shows that $K = \cap Z(K)$. This completes the proof of statement (3). If S is closed and $Z(S)$ is a prime ideal $\mathfrak{p}$, then $\mathfrak{p} \in S$ and S is the closure of $\{\mathfrak{p}\}$ and hence is irreducible. Conversely, if S is irreducible and $a + b \in \cap S$, then $a + b$ belongs to every $\mathfrak{p} \in S$; hence every such $\mathfrak{p}$ contains either a or b, and so $S \subseteq Z(a) \cup Z(b)$. Since S is irreducible, it is contained in either $Z(a)$ or $Z(b)$. If, for example, $S \subseteq Z(a)$, it follows that $\sqrt{(a)} \subseteq \cap S$ and, since $\cap S$ is a radical ideal, that $a \in \cap S$. Thus $\cap S$ is prime. □

Proposition 1.4.4. *Let M be a monoid, S a subset of M, and E an M-set. Then there exist an M-set, denoted by $S^{-1}E$ or E_S, on which the elements of S act bijectively and a map of M-sets $\lambda_S : E \to S^{-1}E$ with the following universal property: for any morphism of M-sets $E \to E'$ such that each $s \in S$ acts bijectively on E', there is a unique M-map $S^{-1}E \to E'$ such that*

$$\begin{array}{ccc} E & \xrightarrow{\lambda_S} & S^{-1}E = E_S \\ & \searrow & \downarrow \\ & & E' \end{array}$$

commutes. The morphism λ_S is called the localization of E by S. A morphism *of M-sets $\phi: E \to E'$ induces a morphism $\phi_S: E_S \to E'_S$; if ϕ is injective (resp. surjective), then so is ϕ_S.*

Proof Let T be the submonoid of M generated by S. The set $S^{-1}E$ can be constructed in the familiar way as the set of equivalence classes of pairs $(e,t) \in E \times T$, where $(e,t) \equiv (e',t')$ if and only if $t't''e = tt''e'$ for some t'' in T. Then $\lambda_S(e)$ is the class of $(e,1)$, and the action of an element m of M sends the class of (e,t) to the class of (me,t). The proof of the remainder of the proposition is straightforward. □

Notice that in fact every element of the face F generated by S acts bijectively on $S^{-1}E$, so that in fact $S^{-1}E \cong F^{-1}E$. Indeed, let E' be any M-set such that $\theta_{E'}(s)$ is bijective for every $s \in S$. If $f \in F$, then $f \leq p$ for some p in the submonoid P of M generated by S. Thus $p = fm$ for some $m \in M$. Then $\theta_{E'}(p) = \theta_{E'}(f)\theta_{E'}(m) = \theta_{E'}(m)\theta_{E'}(f)$ and, since $\theta_{E'}(p)$ is bijective, the same is true of $\theta_{E'}(f)$. If $\mathfrak{p} := M \setminus F$ is the prime ideal of M corresponding to F, one often writes $E_{\mathfrak{p}}$ instead of $S^{-1}E$. An M-set E is called *M-regular* if the elements of M act as injections on E. If this is the case, the localization map $\lambda_S: E \to S^{-1}E$ is injective, for every subset S of M.

The most important case of Proposition 1.4.4 is the regular representation, where M acts on $E = M$ by translation. Then $M_S := S^{-1}M$ has a unique monoid structure for which λ_S is a homomorphism compatible with the M-actions. The morphism $\lambda_S: M \to M_S$ is also characterized by a universal property: any homomorphism $\lambda: M \to N$ with the property that $\lambda(s) \in N^*$ for each $s \in S$ factors uniquely through M_S. In fact, every element of the face $\langle S \rangle$ generated by S maps to a unit in $S^{-1}M$, and $\lambda_S^{-1}(M_S^*) = \langle S \rangle$. Indeed, if $m \in \langle S \rangle$, then by (2) of Proposition 1.4.2 there exists $m' \in M$ such that mm' belongs to the submonoid T of M generated by S. Then $\lambda_S(mm')$ is a unit in M_S, and hence so is m'. Conversely, if $\lambda_S(m)$ is a unit of M_S, then there exist $m' \in M$ and $t' \in T$ such that $(m,1)(m',t') \equiv (1,1)$. This means that there is some t such that $mm't = t't$. Since $tt' \in \langle S \rangle$, it follows that $m \in \langle S \rangle$. If M is integral, then the natural map $S^{-1}M \to M^{\mathrm{gp}}$ is injective, and $S^{-1}M$ can be identified with the set of elements of M^{gp} of the form $m - t$ with $m \in M$ and t belonging to the submonoid (or face) of M generated by S. If $\theta: M \to N$ is a morphism of monoids and S is a subset of M we write $S^{-1}N$ to mean the localization of N by the image of S, when no confusion can arise. We should note that if E is an M-set, then the localized monoid M_S acts naturally on E_S, and in fact the natural map $M_S \otimes_M E \to E_S$ is an isomorphism.

Remark 1.4.5. The localization of an integral (resp. saturated) monoid is integral (resp. saturated), but the analog for quasi-integral and u-integral monoids fails. For example, let Q and P be monoids and let K be an ideal of Q. Let E be the subset of $(P \oplus Q)^2$ consisting of those pairs $(p \oplus q, p' \oplus q)$ such that either $p = p'$ or $q \in K$. In fact E is a congruence relation on $P \oplus Q$, and we denote the quotient $(P \oplus Q)/E$ by $P \star_K Q$ (the *join of P and Q along K*). If K is a prime ideal with complement F, then $P \star_K Q$ can be identified with the disjoint union of $P \times F$ with K, and $(p, f) + k = f + k$. Then $\mathbf{N} \star_{\mathbf{N}^+} \mathbf{N}$ is u-integral, but its localization by the element 1 of the "first" $\mathbf{N}$ in the previous expression is $\mathbf{Z} \star_{\mathbf{N}^+} \mathbf{N}$, which is not u-integral.

Definition 1.4.6. *Let Q be a monoid.*

1. *The dimension of Q is the maximum length d of a chain of prime ideals*

$$\emptyset = \mathfrak{p}_0 \subset \mathfrak{p}_1 \subset \cdots \subset \mathfrak{p}_d = Q^+,$$

   i.e., the Krull dimension of the topological space Spec(Q).
2. *If $\mathfrak{p} \in$ Spec(Q), the height of P, denoted by* ht$(\mathfrak{p})$, *is the maximum length of a chain of prime ideals*

$$\mathfrak{p} = \mathfrak{p}_0 \supset \mathfrak{p}_1 \supset \cdots \supset \mathfrak{p}_h.$$

If $\mathfrak{p}$ is a prime ideal of Q, the map $\mathrm{Spec}(Q_\mathfrak{p}) \to \mathrm{Spec}(Q)$ induced by the localization map $\lambda\colon Q \to Q_\mathfrak{p}$ is injective and identifies $\mathrm{Spec}(Q_\mathfrak{p})$ with the subset of $\mathrm{Spec}(Q)$ consisting of those primes contained in $\mathfrak{p}$. Equivalently, $F \mapsto \lambda^{-1}(F)$ is a bijection from the set of faces of $Q_\mathfrak{p}$ to the set of faces of Q containing $Q \setminus \mathfrak{p}$. These bijections preserve the topologies and order. In particular, every ideal of $Q_\mathfrak{p}$ is induced from an ideal of Q, and so $\mathrm{Spec}(Q_\mathfrak{p})$ has the topology induced from its embedding in $\mathrm{Spec}(Q)$. Moreover, we have $\mathrm{ht}(\mathfrak{p}) = \dim(Q_\mathfrak{p})$. If Q is fine, $\mathrm{Spec}(Q)$ is a finite topological space, and is catenary [26, 14.3.2, 14.3.3], as the following proposition implies. We defer its proof until Section 2.3, after Corollary 2.3.8.

Proposition 1.4.7. *Let Q be an integral monoid.*

1. Spec(Q) *is a finite set if Q is finitely generated.*
2. $\dim(Q) \leq \mathrm{rank}(\overline{Q}^{\mathrm{gp}})$, *with equality if Q is fine.*
3. *If Q is fine, every maximal chain $\mathfrak{p}_0 \subset \mathfrak{p}_1 \subset \cdots \subset \mathfrak{p}_d$ of prime ideals has length* $\dim(Q)$. *If $\mathfrak{p} \in$ Spec(Q) and $F := Q \setminus \mathfrak{p}$,*

$$\mathrm{ht}(\mathfrak{p}) = \mathrm{rank}(\overline{Q}_\mathfrak{p}^{\mathrm{gp}}) = \mathrm{rank}(Q^{\mathrm{gp}}/F^{\mathrm{gp}}) = \dim(Q) - \dim(F).$$

Examples 1.4.8.

1. The monoid $\mathbf{N}$ has just two faces, $\{0\}$ and $\mathbf{N}$, with complementary prime ideals $\mathbf{N}^+$ and $\emptyset$, respectively.

2. More generally, let S be a set and let $M = \mathbf{N}^{(S)}$, the free monoid generated by S. If T is any subset of S, $\mathbf{N}^{(T)}$ can be identified with the set of all $I \in \mathbf{N}^{(S)}$ such that $I_s = 0$ for $s \notin T$. This is a face of M, and every face of M is of this form.
3. Let $Q_{2,2}$ be the monoid given by generators x, y, z, w subject to the relation $x+y = z+w$. This is isomorphic to the amalgamated sum $\mathbf{N}^2 \oplus_{\mathbf{N}} \mathbf{N}^2$, where both maps $\mathbf{N} \to \mathbf{N}^2$ send 1 to $(1, 1)$, to the submonoid of $\mathbf{N}^4$ generated by $\{(1, 1, 0, 0), (0, 0, 1, 1), (1, 0, 1, 0), (0, 1, 0, 1)\}$, and to the submonoid of $\mathbf{Z}^3$ generated by
$$\{(1, 1, 1), (-1, -1, 1), (1, -1, 1), (-1, 1, 1)\}.$$
In addition to the faces $\{0\}$ and $Q_{2,2}$, it has four faces of dimension one, corresponding to each of the generators, and four faces of dimension two: $\langle x, z\rangle, \langle x, w\rangle, \langle y, z\rangle$, and $\langle y, w\rangle$.
4. Let $Q_{3,2}$ be the monoid given by generators x, y, z, u, v subject to the relations $x + y + z = u + v$. This four-dimensional monoid has five faces of dimension one, nine of dimension two, and six of dimension three.

When attempting to visualize the relations among the faces of a monoid, it is often helpful to draw a picture not of the monoid itself but rather of its slice with a suitable hyperplane not containing the vertex. Doing so reduces the dimension of the monoid and of each of its faces by one. For example, slices of the monoids $Q_{2,2}$ and $Q_{3,3}$ in the above examples are shown in Figure 1.4.1

1.5 Idealized monoids

A surjective homomorphism of commutative rings $A \to B$ induces a closed immersion $\operatorname{Spec}(B) \to \operatorname{Spec}(A)$, but the analog for monoids is not true. In fact, if $P \to Q$ is any homomorphism of monoids, the generic point of $\operatorname{Spec} P$ lies in the image of $\operatorname{Spec} Q$, so the map $\operatorname{Spec} Q \to \operatorname{Spec} P$ cannot be a closed immersion unless it is bijective. To remedy this we introduce the notion of an *idealized monoid*, which will be useful in studying the stratifications that arise naturally in the context of toric varieties and log schemes.

Definition 1.5.1. *An idealized monoid is a pair (M, K), where M is a monoid and K is an ideal of M. A homomorphism of idealized monoids*
$$\theta\colon (P, I) \to (Q, J)$$
is a monoid homomorphism $P \to Q$ sending I to J. A face of an idealized monoid (M, K) is a face F of M such that $F \cap K = \emptyset$ or, equivalently, such that

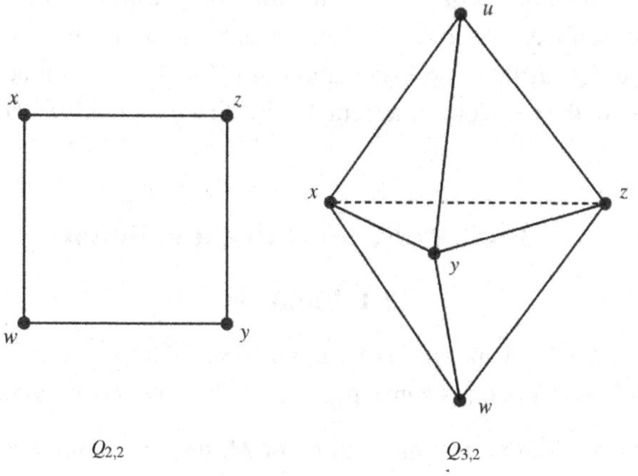

Figure 1.4.1 Slices of monoids and their faces

the corresponding prime ideal $M \setminus F$ *contains* K, *and* Spec(M, K) *is the set of all such prime ideals, with the Zariski topology.*

Note that Spec(M, K) is empty if and only if $K = M$. Let us say therefore say that an idealized monoid (M, K) is *acceptable* if either K is a proper ideal of M or M is the zero monoid. We write **Moni** for the category of acceptable idealized monoids. The functor **Moni** $\to$ **Mon** taking (Q, J) to Q has a fully faithful left adjoint, taking a monoid P to $(P, \emptyset)$. Thus we may view **Mon** as a full subcategory of **Moni**.

Remark 1.5.2. The Krull dimension of (M, K) is as usual the supremum of the set of lengths of the chains of prime ideals of (M, K), or, equivalently, of a chain of faces of (M, K). If C is a maximal chain of such faces, then $F := \cup C$ is another face of (M, K) and hence belongs to C; furthermore, each member of C is a face of F. Thus the set of faces of (M, K) admits maximal elements, and the dimension of (M, K) is the same as the maximum of the dimensions of its faces.

Limits and colimits exist in the category of idealized monoids, and are compatible with the forgetful functor to the category of monoids. For example, the pushout of a pair of morphisms $u_i \colon (P, I) \to (Q_i, J_i)$ in the category of idealized monoids is given by the obvious maps $v_i \colon (Q_i, J_i) \to (Q, J)$, where $Q_i \to Q$ is the pushout of the underlying monoid homomorphisms and J is the ideal of Q generated by the images of J_i. Note that this compatibility is not quite true for acceptably idealized monoids. For example, if (Q, J) is an integral idealized monoid and $f \in J$, then the localization of (Q, J) by f is not acceptable: the localization in the category of acceptable idealized monoids is $(0, 0)$.

2 Finiteness, convexity, and duality

2.1 Finiteness

Proposition 2.1.1. *A monoid is finitely generated as a monoid if and only if M^* is finitely generated (as a group) and $\overline{M}$ is finitely generated (as a monoid).*

Proof If S is a finite set of generators for M, then any nonzero element m of M can be written as a sum $\sum n_i s_i$, with each $n_i > 0$. If m is a unit, so is each s_i, and it follows that M^* is generated by the finite set $S \cap M^*$. Since $M \to \overline{M}$ is surjective, $\overline{M}$ is finitely generated as a monoid. For the converse, suppose $\{s_i\}$ is a finite set of generators for the group M^* and $\{t_j\}$ is a finite subset of M whose images in $\overline{M}$ generate $\overline{M}$ as a monoid. Then the set $\{s_i, -s_i, t_j\}$ generates M as a monoid. □

Recall that if M is a monoid and x and y are elements of an M-set X, we write $x \leq y$ if there exists some $m \in M$ such that $y = m + x$. If S is a subset of X we say that $s \in S$ is an *M-minimal element of S* if, whenever $s' \in S$ and $s' \leq s$, then also $s \leq s'$ (so that $s \sim s'$ in the equivalence relation corresponding to $\leq$). If there seems to be no danger of confusion, we just say "minimal" instead of "M-minimal." For example, with the regular action of M on itself, the units are the minimal elements of M. If M is quasi-integral, the minimal elements of the maximal ideal M^+ are called *irreducible*. Thus an element c of M^+ is irreducible if and only if it is not a unit and, whenever $c = a + b$ in M, a or b is a unit. The set M^{+2} of elements of the form $a + b$ with $a, b \in M^+$ is an ideal of M, and we see that the irreducible elements of M are precisely the elements of $M^+ \setminus M^{+2}$.

Proposition 2.1.2. *Let M be a sharp quasi-integral monoid. Then every set of generators of M contains every irreducible element of M. If in addition M is finitely generated, then the set of irreducible elements of M is finite and generates M.*

Proof The first statement is obvious. Suppose now that M is finitely generated. It is clear that every finite set of generators contains a minimal set of generators. Let S be such a minimal set; we claim that every element x of S is irreducible. If $x = y + z$ with y and z in M, we can write y and z as sums of elements of S, say $y = \sum_s a_s s$ and $z = \sum_s b_s s$, where a_s and $b_s \in \mathbf{N}$ for all $s \in S$. Then $x = \sum_s c_s s$, where $c_s = a_s + b_s$. Let $S' := S \setminus \{x\}$, so that $x = c_x x + m'$ where m' is in the submonoid M' of M generated by S'. If $c_x = 0$, then $x \in M'$, and S' generates M, a contradiction of the minimality of S. Hence $c_x \geq 1$, and we can write $x = x + (c_x - 1)x + m'$. But then the quasi-integrality of M implies $(c_x - 1)x + m' = 0$, and the sharpness implies that $c_x = 1$ and $m' = 0$. Then $y = a_x x$ and $z = b_x x$, where $a_x + b_x = 1$. Thus exactly one of y and z is zero, so x is irreducible, as claimed. Since S generates M by hypothesis and since the elements of S are irreducible, M is generated by its set of irreducible elements. Since S contains all the irreducible elements of M and is finite, there can be only finitely many irreducible elements. □

Corollary 2.1.3. *The automorphism group of a fine sharp monoid is finite and contained in the permutation group of the set of its irreducible elements.* □

Remark 2.1.4. Proposition 2.1.2 shows that every element in a fine sharp monoid can be written as a sum of irreducible elements. In fact a standard argument applies somewhat more generally. Let M be a sharp quasi-integral monoid in which every nonempty subset has a minimal element. Then every element of M can be written as a sum of irreducible elements. (Note that 0 is by definition the sum over the empty set.) Let us recall the argument. We claim that the set S of elements of M^+ that cannot be written as a sum of irreducible elements is empty. If not, by assumption it contains a minimal element s. Since s is not irreducible, $s = a + b$ where a and b are not zero and hence belong to M^+. If both a and b can be written as sums of irreducible elements, then the same is true of s, a contradiction. But if for example a cannot be written as a sum of irreducible elements, then $a \in S$ and $a \leq s$, and, since b is not a unit, $s \not\leq a$, contradicting the minimality of s.

Let M be a monoid and let S be an M-set. We say that S is *noetherian* if every sub-M-set is finitely generated, and we say that M is noetherian if it is so when regarded as an M-set via the regular representation. It is clear that a finite union of noetherian M-sets is again noetherian, that a sub-M-set of a noetherian M-set is noetherian, and that the image of a noetherian M-set is noetherian. It follows that if M is a noetherian monoid, an M-set is noetherian if and only if it is finitely generated as an M-set.

Proposition 2.1.5. *Let M be a monoid and let S be an M-set. Then the following conditions are equivalent:*

1. *Every sub-M-set of S is finitely generated, i.e., S is noetherian.*

2. *Every ascending chain $S_1 \subseteq S_2 \subseteq S_3 \subseteq \cdots$ of sub-M-sets of S is eventually constant.*
3. *Every sequence $(s_1, s_2, s_3 \ldots)$ in S contains an increasing subsequence; that is, there is a strictly increasing sequence $(i_1, i_2, i_3, \ldots)$ of natural numbers such that $(s_{i_1}, s_{i_2}, s_{i_3} \ldots)$ is an increasing sequence in S.*
4. *Every nonempty subset of S contains a minimal element, and there are only finitely many equivalence classes of such elements for the equivalence relation $\sim$.*
5. *Every nonempty set of sub-M-sets of S has a maximal element.*
6. *The quotient of S by the congruence relation $\sim$ is noetherian.*

Proof The equivalence of (1), (2), and (5) is proved in exactly the same way as in the case of modules over a ring. Suppose that (2) holds and that $(s_1, s_2, \ldots)$ is a sequence in S. For each i, let (s_i) be the sub-M-set of S generated by s_i and let S_i denote the union $(s_1) \cup (s_2) \cup \cdots \cup (s_i)$. Then $S_1 \subseteq S_2 \subseteq \cdots$, so by (2) there exists some N such that $S_j = S_N$ for all $j \geq N$. Thus for every $j \geq N$, there exists some $i \leq N$ such that $s_j \geq s_i$. Since there are infinitely many such j and finitely many such i, there must exist an $i \leq N$ and a sequence $j_1 < j_2 < \cdots$ such that $s_i \leq s_{j_k}$ for all k. Thus, replacing $(s_1, s_2, \ldots)$ by the subsequence $(s_i, s_{j_1}, s_{j_2}, \ldots)$, we may assume that $s_1 \leq s_j$ for all $j > 1$. Repeating this process, we may also arrange that $s_2 \leq s_j$ for all $j > 2$, and then that $s_3 \leq s_j$ for all $j > 3$, and so on.

To prove that (3) implies (4), let us first observe that any decreasing sequence $(s_1, s_2, \ldots)$ in S is eventually in a single equivalence class for the relation $\sim$. Indeed, (3) implies that there is an increasing sequence $(i_j : j \in \mathbf{Z}^+)$ such that $(s_{i_1}, s_{i_2}, \ldots)$ is increasing (as well as decreasing). Then all s_i with $i \geq i_1$ are equivalent. Indeed, if $i \geq i_1$, choose j such that $i_j \geq i$, and then

$$s_{i_1} \geq s_i \geq s_{i_j} \geq s_{i_1}.$$

Now if T is a nonempty subset of S, choose any element t_1 of T. If t_1 is M-minimal, we are done; if not there exists an element t_2 of T such that $t_2 \leq t_1$ and $t_2 \not\geq t_1$. If t_2 is M-minimal, we are done, and if not there exists t_3 with $t_3 \leq t_2$ and $t_3 \not\geq t_2$. Continuing in this way, we find a decreasing sequence $(t_1, t_2, \ldots)$ of elements of T with $t_i \not\geq t_{i-1}$ for all i. As we have just seen, such a sequence must terminate, and so we find an M-minimal element of T. If there were an infinite number of equivalence classes of such minimal elements, we could find an infinite sequence $(s_1, s_2, \ldots)$ of elements all belonging to distinct equivalence classes, and by (3) such a sequence would contain an increasing subsequence s. But then $s_1 \leq s_2$ and $s_1 \nsim s_2$, contradicting the minimality of s_2. This proves (4).

Suppose that (4) holds and T is a sub-M-set of S. By (4), there is a finite set T' of minimal elements of T such that every minimal element is equivalent to some element of T'. Now let t be an arbitrary element of T and let $T_t := \{s \in T : s \leq t\}$. Then T_t is not empty and hence by (4) contains a minimal element t''. Note that if $t''' \in T$ and $t''' \leq t''$, then $t''' \in T_t$, and hence $t''' \sim t''$, by the minimality of t''. Thus t'' is in fact a minimal element of T, and hence is equivalent to some element t' of T'. Since $t'' \in T_t$, $t = m + t'' = m' + t'$ for some $m, m' \in M$. Thus the finite set T' generates T. This proves (1).

Let $\pi: S \to S/\sim$ be the natural projection and let S' be a sub-M-set of S. Note that if $s' \in S'$ and $s \in S$ and $s \sim s'$, then $s \in S'$. Thus $S' = \pi^{-1}(\pi(S))$, so that π induces a bijection from the family of sub-M-sets of S to that of $S/\sim$. Hence property (5) holds for S if and only if it does for $S/\sim$. □

Corollary 2.1.6. *Every dull monoid is noetherian, and a monoid is noetherian if and only if its sharpening is noetherian.* □

Theorem 2.1.7. *A finitely generated monoid is noetherian and finitely presented. Conversely, a sharp and quasi-integral noetherian monoid is finitely generated.*

Proof We shall use the following analog of the Hilbert basis theorem.

Lemma 2.1.8. *If P and Q are noetherian, then $P \oplus Q$ is noetherian. In particular if Q is noetherian, then $Q \oplus \mathbf{N}$ is also noetherian.*

Proof We use condition (3) of Proposition 2.1.5. Let $(p_\cdot, q_\cdot)$ be a sequence in $P \oplus Q$. Since P is noetherian, there is a strictly increasing sequence $(n_\cdot)$ in $\mathbf{N}$ such that the subsequence $(p_{n_\cdot})$ of $(p_\cdot)$ is increasing. Replacing the original sequence by the sequence $(p_{n_\cdot}, q_{n_\cdot})$, we may assume that the sequence $(p_\cdot)$ was already increasing. Now since Q is noetherian, we may choose a strictly increasing sequence $(n_\cdot)$ such that $(q_{n_\cdot})$ is increasing. But then $(p_{n_\cdot}, q_{n_\cdot})$ is an increasing subsequence of the original sequence. □

It is immediate to verify that if $M \to M'$ is surjective and M is noetherian, M' is also noetherian. Then it follows from the lemma and induction that every finitely generated monoid is noetherian. The fact that a finitely generated monoid is finitely presented follows from Lemma 2.1.9 below.

For the converse, suppose that M is sharp, quasi-integral, and noetherian. We may assume that M is not the zero monoid. Applying (4) of Proposition 2.1.5 to the nonempty subset M^+ of M, we see that the set S of minimal elements of M^+ is finite and not empty. Furthermore, the argument of Remark 2.1.4 shows that every element of M can be written as a sum of elements of S. Thus M is finitely generated. □

Lemma 2.1.9. *Every congruence relation on a finitely generated monoid is finitely generated (as a congruence relation).*

Proof The following proof is due to Pierre Grillet [23]. Let E be a congruence relation on a monoid P. If $P' \to P$ is a surjective homomorphism of monoids, the inverse image E' of E in $P' \times P'$ is a congruence relation on P', and if $S' \subseteq E'$ generates E' as a congruence relation, then the image of S' in E generates E as a congruence relation. If P is finitely generated, we can find a surjective homomorphism $\mathbf{N}^r \to P$, and thus we are reduced to proving the lemma when $P = \mathbf{N}^r$.

If p and q are elements of P, write $p \preceq q$ if p precedes q in the lexicographical order of $\mathbf{N}^r$, and write $p \prec q$ if in addition $p \neq q$. If $p \preceq q$ and $p' \preceq q'$, then $p + p' \preceq q + q'$, and if $p \leq q$ in the partial order defined by the monoid structure, then $p \preceq q$. Then $\preceq$ well-orders P: every nonempty subset has a unique $\preceq$-minimal element. If $p \in P$, let $E(p)$ denote the E-congruence class of p, and let $\mu(p)$ denote the $\preceq$-minimal element in $E(p)$. Then, if $(x, y) \in P{\times}P$, $(x, y) \in E$ if and only if $\mu(x) = \mu(y)$, and E is the congruence relation generated by the set of pairs $(x, \mu(x))$ for $x \in P$. Then $\mu \circ \mu = \mu$, and the complement K of the image of $\mu \colon P \to P$ is the set of all elements k of P such that $\mu(k) \prec k$. Note that if $p \in P$ and $k \in K$, then $\mu(k) \prec k$, so $\mu(k) + p \prec k + p$. Since $(\mu(k) + p) \equiv_E (k + p)$, $k + p$ is not $\preceq$-minimal in $E(k + p)$. Thus $\mu(k + p) \prec k + p$ and so K is an ideal of P, finitely generated since P is noetherian. Let S be a finite set of generators for K and let E' be the congruence relation on P generated by the set of pairs $(s, \mu(s))$ with $s \in S$. Then E' is finitely generated as a congruence relation and contained in E, so it will suffice to prove that $E \subseteq E'$, i.e., that E' contains $(x, \mu(x))$ for every $x \in P$. If this fails, there exists an x such that $\mu(x)$ does not belong to $E'(x)$ and that is $\preceq$-minimal among all such elements. Note that x does not belong to the image of μ, since otherwise $x = \mu(x)$, which would contradict $\mu(x) \notin E'(x)$. Thus $x \in K$, and hence $x = p + s$ for some $s \in S$ and $p \in P^+$. Since $s \in K$, $\mu(s) \prec s$, so also $x' := p + \mu(s) \prec p + s = x$. Then by the minimality of x, $\mu(x') \in E'(x')$. But $\mu(s) \equiv_{E'} s$, so $x' = p + \mu(s) \equiv_{E'} p + s = x$, and consequently $x' \equiv_E x$. But then $\mu(x') = \mu(x)$ and $\mu(x) \in E'(x)$, a contradiction. This completes the proof. □

Corollary 2.1.10. *A quasi-integral monoid M is noetherian if and only if $\overline{M}$ is finitely generated.* □

Corollary 2.1.11. *If M is a finitely generated monoid, any sub-M-set of a finitely generated M-set is finitely generated, and in fact is generated by a finite set of minimal elements.* □

The following simple result can be viewed as an analog of Nakayama's lemma in commutative algebra.

Proposition 2.1.12. *Let Q be a monoid and let S be a finitely generated Q-set. Assume that $qs \neq s$ for every $s \in S$ and every $q \in Q^+$. Then S is generated by $S \setminus Q^+S$. In particular, if $Q^+S = S$, then in fact $S = \emptyset$.*

Proof Let $\{s_1, \dots, s_n\}$ be a finite set of generators for S. After omitting some of its elements, we may assume that no proper subset also generates S. Then it suffices to prove that each $s_i \notin Q^+S$. For example, if $s_n \in Q^+S$, then there exist $q \in Q^+$ and $s \in S$ with $s_n = qs$, and, since $\{s_1, \dots, s_n\}$ generates S, there exist some $i \leq n$ and some $q' \in Q$ with $s = q's_i$. Then $s_n = qq's_i$, and if $i < n$ the set $\{s_1, \dots, s_{n-1}\}$ generates S, contradicting the minimality assumption. It follows that $i = n$, so that $s_n = qq's_n$. The hypotheses then imply that qq' is unit, contradicting the assumption that $q \in Q^+$. □

It follows from (2) of Proposition 2.1.5 that in a noetherian monoid M, every nonempty set of ideals has a maximal element. We can deduce a primary decomposition theorem for ideals; again the proof is a simple adaption of the standard one in commutative algebra. Here we will use multiplicative notation for the monoid law to exhibit the similarity with the argument from commutative algebra.

Definition 2.1.13. *A proper ideal $\mathfrak{q} \subset M$ in a monoid M is primary if $ax \notin \mathfrak{q}$ whenever $a \in M \setminus \sqrt{\mathfrak{q}}$ and $x \notin \mathfrak{q}$.*

Proposition 2.1.14. *Let M be a noetherian monoid.*

1. *If $\mathfrak{q}$ is a primary ideal of M, its radical $\sqrt{\mathfrak{q}}$ is prime, and there exists some element x of M such that $\sqrt{\mathfrak{q}} = \{a : ax \in \mathfrak{q}\}$.*
2. *Every ideal of M can be written as the intersection of a finite number of primary ideals.*
3. *If $K = \mathfrak{q}_1 \cap \cdots \cap \mathfrak{q}_n$, where each $\mathfrak{q}_i$ is a primary ideal, let $F_i := M \setminus \sqrt{\mathfrak{q}_i}$ and let $F = F_1 \cap \cdots \cap F_n$. Then $M \setminus K$ is stable under multiplication by F.*

Proof Suppose $\mathfrak{q}$ is primary, and $a \notin \sqrt{\mathfrak{q}}$, but $ax \in \sqrt{\mathfrak{q}}$. Then there exists an n such that $a^nx^n \in \mathfrak{q}$, and, since $\mathfrak{q}$ is primary, it follows that $x^n \in \mathfrak{q}$ and so $x \in \sqrt{\mathfrak{q}}$. Thus $\sqrt{\mathfrak{q}}$ is prime. For each $x \in M \setminus \mathfrak{q}$, let $K_x := \{a : ax \in \mathfrak{q}\}$. Then $K_x \subseteq \sqrt{\mathfrak{q}}$. Furthermore, since M is noetherian, there exists an $x \in M \setminus \mathfrak{q}$ such that K_x is not properly contained in any $K_{x'}$. We claim that in fact $K_x = \sqrt{\mathfrak{q}}$. Indeed, if $b \in \sqrt{\mathfrak{q}}$, then some power of b belongs to $\mathfrak{q}$, and hence there is a natural number n such that $b^nx \notin \mathfrak{q}$ but $b^{n+1}x \in \mathfrak{q}$. Let $x' := b^nx$. Then $K_x \subseteq K_{x'}$, and by maximality $K_x = K_{x'}$. Since $b \in K_{x'}$, it follows that $b \in K_x$, as required, completing the proof of (1).

If statement (2) were false, we could find an ideal of M that is maximal among all ideals not admitting a primary decomposition. Such an ideal K would necessarily be proper. For $a \in M \setminus \sqrt{K}$, let $K_n := \{x : a^n x \in K\}$. Then $K \subseteq K_1 \subseteq K_2 \subseteq \cdots$, so there exists an n such that $K_n = K_{n+1}$. Note that if $x \in K_1 \cap (a^n)$, we can write $x = a^n y$ for some $y \in M$, and $ax = a^{n+1} y \in K$. Then $y \in K_{n+1} = K_n$, so in fact $x = a^n y \in K$. This implies that $K_1 \cap (a^n) \subseteq K$, and hence that $K = K_1 \cap K'$, where $K' := K \cup (a^n)$. Since $a \notin \sqrt{K}$, K' strictly contains K and hence admits a primary decomposition. Then K_1 cannot admit such a decomposition, and hence $K_1 = K$. This means that $x \in K$ whenever $ax \in K$, assuming as before that $a \notin \sqrt{K}$. We have in fact proved that K is primary, another contradiction.

To prove (3), suppose that $fm \in K$, where $f \in F$ and $m \in K$. Then $fm \in \mathfrak{q}_i$ for every i and, since $f \notin \sqrt{\mathfrak{q}_i}$, it follows that $m \in \mathfrak{q}_i$ for every i, so $m \in K$. □

Let S be a nonempty subset of a monoid P, and suppose that P is a local submonoid of a fine monoid Q. Since Q is fine, it is noetherian, and Proposition 2.1.5 shows that S contains a Q-minimal element s. Such an element is necessarily also P-minimal: if $s = p + s'$ with $p \in P$ and $s' \in S$, then there exist $q \in Q$ such that $s' = q + s$; hence $p + q = 0$. Then $q \in Q^*$ and $p \in P^*$, so $s' \geq_P s$, and s is P-minimal. In particular, Remark 2.1.4 implies that P is generated by its irreducible elements. On the other hand, P-minimal elements of S need not be Q-minimal, and it could happen that S has an infinite number of minimal elements and that P has an infinite number of irreducible elements. For example, in $Q := \mathbf{N} \times \mathbf{N}$, consider the submonoid E of $\mathbf{N} \times \mathbf{N}$ consisting of $(0, 0)$ together with all pairs (m, n) such that m and n are both positive. (This submonoid is actually a congruence relation on $\mathbf{N}$; the quotient $\mathbf{N}/E$ is the unique (up to isomorphism) monoid with two elements which is not a group.) Then for every $m > 0$, the element $(1, m)$ is irreducible in E, and in particular E is not finitely generated as a monoid. This situation is ameliorated by the notion of *exactness*, which will turn out to be of fundamental importance.

Definition 2.1.15. *A monoid homomorphism $\theta \colon P \to Q$ is exact if the diagram*

$$\begin{array}{ccc} P & \xrightarrow{\theta} & Q \\ \downarrow & & \downarrow \\ P^{\mathrm{gp}} & \xrightarrow{\theta^{\mathrm{gp}}} & Q^{\mathrm{gp}} \end{array}$$

is cartesian.

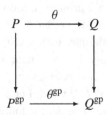

For example, an inclusion of integral monoids $P \to Q$ is exact if and only if $P = P^{gp} \cap Q$ in Q^{gp}. In this case one says that P is an *exact submonoid* of Q.

Proposition 2.1.16. *The following statements hold.*

1. *If M is any monoid, the diagonal map $\Delta_M \colon M \to M \times M$ is exact if and only if M is integral.*
2. *An exact homomorphism $\theta \colon P \to Q$ is local, and is injective if P is sharp.*
3. *Let $\theta \colon P \to Q$ be a homomorphism of integral monoids. Then θ is exact if and only if, for any $p_1, p_2 \in P$, $\theta(p_2) \geq \theta(p_1)$ implies that $p_2 \geq p_1$.*
4. *In the category of integral monoids, the pullback of an exact homomorphism is exact.*
5. *Let P be a submonoid of an integral monoid Q. Then the inclusion $P \to Q$ is exact if and only if $Q \setminus P$ is stable under the action of P on Q.*

Proof A monoid M is integral if and only if the localization map $\lambda_M \colon M \to M^{gp}$ is injective. This is true if and only if the diagram

$$\begin{array}{ccc} M & \xrightarrow{\Delta_M} & M \times M \\ \downarrow{\scriptstyle \lambda_M} & & \downarrow{\scriptstyle \lambda_M \times \lambda_M} \\ M^{gp} & \xrightarrow{\Delta_{M^{gp}}} & M^{gp} \times M^{gp} \end{array}$$

is cartesian. Since $M \times M \cong M \oplus M$, $M^{gp} \times M^{gp} \cong (M \times M)^{gp}$, and thus (1) is clear. Assume that $\theta \colon P \to Q$ is exact and that $p \in P$ with $\theta(p) \in Q^*$. Then there exists a $q \in Q$ such that $q + \theta(p) = 0$, and hence $(-\lambda(p), q) \in P^{gp} \times_{Q^{gp}} Q$. By the exactness of θ, there is a unique $p' \in P$ such that $\lambda(p') = -\lambda(p)$ and $\theta(p') = q$. Then $p' + p$ and 0 are two elements of P with the same image in $P^{gp} \times_{Q^{gp}} Q$, so $p' + p = 0$ and $p \in P^*$. This proves that θ is local. Now suppose that P is sharp and that $\theta(p_1) = \theta(p_2)$. Then $(\lambda(p_1) - \lambda(p_2), 0) \in P^{gp} \times_{Q^{gp}} Q$, so there exists a unique $p \in P$ such that $\lambda(p) = \lambda(p_1) - \lambda(p_2)$ and $\theta(p) = 0$. As we have just observed, θ is local, so $p \in P^*$, and, since P is sharp, $p = 0$. Then $\lambda(p_1) = \lambda(p_2)$, and p_1 and p_2 have the same image in $P^{gp} \times_{Q^{gp}} Q$, hence are equal.

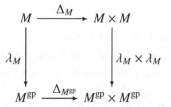

Let $\theta \colon P \to Q$ be a homomorphism of integral monoids. We notate the maps $P \to P^{gp}$ and $Q \to Q^{gp}$ as inclusions. Assume that θ is exact and that $p_1, p_2 \in P$ with $\theta(p_2) \geq \theta(p_1)$. This means that for some $q \in Q$, $\theta(p_2) = \theta(p_1) + q$. Then $\theta^{gp}(p_2 - p_1) = q$ in Q^{gp}, so there is a unique $p \in P$ mapping to $(p_2 - p_1, q)$ in $P^{gp} \times Q$. Since P is integral, this implies that $p_2 = p + p_1$. For the converse, suppose that $(x, q) \in P^{gp} \times Q$ and $\theta^{gp}(x) = q$ in Q^{gp}. Write $x = p_2 - p_1$, with $p_1, p_2 \in P$. Since Q is integral, it follows that $\theta(p_2) + q = \theta(p_1)$, and hence by hypothesis that there exists a $p \in P$ such that $p_2 = p_1 + p$. Then $\theta(p) = q$ by the integrality of Q, and p is unique by the integrality of P. This shows that θ is exact, completing the proof of (3).

Let $\theta\colon P \to Q$ be an exact homomorphism of integral monoids, let $\phi\colon Q' \to Q$ be a homomorphism with Q' integral, and let $\theta'\colon P' := P \times_Q Q' \to Q'$ be the induced homomorphism. Then we have the following commutative diagram:

$$\begin{array}{ccccccc}
 & P' & \xrightarrow{\ \gamma\ } & P'^{\mathrm{gp}} \times_{Q'^{\mathrm{gp}}} Q' & & & \\
 \cong \swarrow & & & \downarrow \alpha & \searrow \beta & & \\
P \times_Q Q' & \cong P^{\mathrm{gp}} \times_{Q^{\mathrm{gp}}} Q \times_Q Q' & \cong & P^{\mathrm{gp}} \times_{Q^{\mathrm{gp}}} Q' & \longrightarrow & P^{\mathrm{gp}} \times_{Q^{\mathrm{gp}}} Q'^{\mathrm{gp}}. &
\end{array}$$

The diagram shows that $\alpha \circ \gamma$ is an isomorphism, and in particular that γ is injective and α is surjective. Since P and Q' are integral, the map

$$P' := P \times_Q Q' \to P^{\mathrm{gp}} \times_{Q^{\mathrm{gp}}} Q'^{\mathrm{gp}}$$

is injective, and hence so is the map $P'^{\mathrm{gp}} \to P^{\mathrm{gp}} \times_{Q^{\mathrm{gp}}} Q'^{\mathrm{gp}}$. Thus the map β is injective, hence so is α, and it follows that α and γ are isomorphisms.

For the last statement, suppose that P is an exact submonoid of Q. If $p \in P$ and $q \in Q$ and $p' := p + q \in P$, then $q = p' - p \in P^{\mathrm{gp}} \cap Q = P$. Thus $Q \setminus P$ is P-invariant. Conversely, if $Q \setminus P$ is P invariant and $p - p' = q \in Q$, then $p' + q \in P$, hence $q \in P$. □

The following theorem contains most of the important finiteness results for fine monoids. The notion of exactness plays a key role.

Theorem 2.1.17.

1. *Let $\theta\colon P \to Q$ be an exact homomorphism of integral monoids. Suppose that S is a sub-P-set of P^{gp} whose image in Q^{gp} is contained in a noetherian Q-set. Then S is a noetherian P-set.*
2. *Every exact submonoid of a fine (resp. saturated, resp. toric) monoid is fine (resp. saturated, resp. toric).*
3. *A face of an integral monoid is an exact submonoid. Every face of a fine monoid is finitely generated (as a monoid), and monogenic (as a face).*
4. *Every localization (see Proposition 1.4.4) of a fine (resp. saturated) monoid is fine (resp. saturated).*
5. *The equalizer E of two homomorphisms of integral monoids $\theta_i\colon P \to Q$ is an exact submonoid of P. If P is fine (resp. saturated), then the same is true of E.*

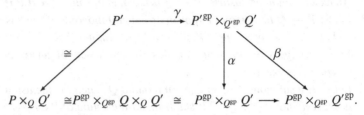

6. *The fiber product of a pair of finitely generated (resp. fine, resp. saturated) monoids over an integral monoid is finitely generated (resp. fine, resp. saturated).*
7. *If $P \to Q$ is a homomorphism of fine monoids, then the congruence relation $E := P \times_Q P$ is finitely generated as a monoid and in particular as a congruence on P.*
8. *Let P and Q be monoids. If Q is fine and P is finitely generated, then* Hom(P, Q) *is also fine. If Q is saturated,* Hom(P, Q) *is also saturated.*

Proof The proof of statement (1) will rely on criterion (4) of Proposition 2.1.5 and the following simple observation.

Lemma 2.1.18. *Let P be an exact submonoid of a fine sharp monoid Q and let Σ be a nonempty subset of P^{gp}. Then an element s of Σ is P-minimal in Σ if and only if $\theta^{gp}(s)$ is Q-minimal in $\theta^{gp}(\Sigma)$.*

Proof We assume for convenience and without loss of generality that P and Q are sharp. Then θ^{gp} is injective, since θ is exact. Let s be an element of Σ. It is clear that s is P-minimal in Σ if $\theta(s)$ is Q-minimal in $\theta^{gp}(\Sigma)$. On the other hand, if $s' \in \Sigma$ and $\theta^{gp}(s') < \theta^{gp}(s)$, then $q := \theta^{gp}(s) - \theta^{gp}(s') \in Q^+$, and it follows from the exactness of θ that $p := s - s' \in P^+$. Then s cannot be P-minimal in S. □

To prove statement (1), we may again assume that P and Q are sharp. Let Σ be a nonempty subset of S. If $\theta^{gp}(S)$ is contained in a noetherian Q-set T, then (4) of Proposition 2.1.5 implies that the set Σ' of Q-minimal elements of $\theta^{gp}(\Sigma)$ is finite and nonempty. The lemma implies that $\theta^{gp^{-1}}(\Sigma')$ is precisely the set of P-minimal elements of Σ. Since θ^{gp} is injective, this set is also finite and nonempty. Then Proposition 2.1.5 implies that S is noetherian.

To prove (2), suppose that P is an exact submonoid of Q. If Q is fine, then by Theorem 2.1.7, Q is noetherian as a Q-set. Then statement (1) implies that P is noetherian as a P-set, and Theorem 2.1.7 implies that $\overline{P}$ is finitely generated as a monoid. Furthermore, P^* is contained in Q^*, a finitely generated group, so it too is finitely generated. Then P is finitely generated by Proposition 2.1.1. If Q is saturated, we can conclude from the following (slightly stronger) lemma that P is also saturated.

Lemma 2.1.19. *If $\theta\colon P \to Q$ is an exact homomorphism of integral monoids and Q is saturated, then P is also saturated,*

Proof Suppose that $x \in P^{gp}$ and $nx \in P$. Then $n\theta(x) = \theta(nx) \in Q$, and, since Q is saturated, it follows that $\theta(x) \in Q$. Since θ is exact we conclude that $x \in P$, so P is saturated. □

In particular, an exact submonoid of a saturated is saturated. Of couse, if Q^{gp} is torsion free, then $P^{gp} \subseteq Q^{gp}$ is also torsion free, and consequently if Q is toric, so is P.

Let F be a face of an integral monoid M, let x and y be elements of F, and suppose $z := x - y \in M$. Then $x = y + z \in F$ and, since F is a face, it follows that $z \in F$. Thus F is an exact submonoid of M, and hence is finitely generated as a monoid if M is fine. If $f_1, \ldots, f_n$ are generators, then $f := f_1 + \cdots + f_n$ generates F as a face of M. This proves (3). For (4), note that if $S \subseteq M$ is a finite set of generators of M, then M_F is generated by the set of elements $\lambda(s)$ for $s \in S$ together with $-\lambda(f)$, where f is any generator of F as a face. Thus M_F is finitely generated as a monoid. If M is saturated and x is an element of M_F^{gp} and n is a positive integer such that $nx \in M_F$, then there exist $y \in M$ and $f \in F$ such that $nx = y - f$. But then $n(x + f) = y + (n - 1)f \in M$, and, since M is saturated, $x + f \in M$ and $x \in M_F$. This proves (4).

Let $E \to P$ be the equalizer of two homomorphisms θ_1 and θ_2 from P to Q, with P and Q integral. Then $E \to P$ is just the pullback of the diagonal Δ_Q via the map $(\theta_1, \theta_2)\colon P \to Q \times Q$. Proposition 4.2.1 implies that Δ_Q is exact, and thus that $E \to P \times P$ is also exact. This proves (5), since an exact submonoid of a fine (resp. saturated) monoid is fine (resp. saturated).

Let $\theta\colon P \to M$ and $\phi\colon Q \to M$ be monoid homomorphisms, where M is integral. If P and Q are fine (resp. saturated), then so is $P \times Q$. In this case, $P \times_M Q$ is an exact submonoid of $P \times Q$, since $\theta(p) = \phi(q)$ if and only if $\theta^{gp}(p) = \phi^{gp}(q)$. Then $P \times_M Q$ is fine (resp. saturated), by statement (2). If P and Q are finitely generated but not necessarily integral, we choose surjections $\mathbf{N}^r \to P$ and $\mathbf{N}^s \to Q$, and observe that the homomorphism $\mathbf{N}^r \times_M \mathbf{N}^s \to P \times_M Q$ is surjective. As we have just seen, $\mathbf{N}^r \times_M \mathbf{N}^s$ is finitely generated, and hence the same is true of $P \times_M Q$. This proves statement (6), and (7) is a special case.

In statement (8), it is clear that $\mathrm{Hom}(P, Q)$ is integral (resp. saturated) if Q is integral (resp. saturated). If P is finitely generated, choose a surjective homomorphism $\mathbf{N}^r \to P$ for some $r \in \mathbf{Z}^+$. Then $\mathrm{Hom}(P, Q)$ can be identified with the equalizer of the two obvious maps

$$\mathrm{Hom}(\mathbf{N}^r, Q) \to \mathrm{Hom}(\mathbf{N}^r \times_P \mathbf{N}^r, Q).$$

Since $\mathrm{Hom}(\mathbf{N}^r, Q) \cong Q^r$ is finitely generated if Q is finitely generated, (5) implies that $\mathrm{Hom}(P, Q)$ is also finitely generated. □

Corollary 2.1.20. *If P is an integral monoid and E is a congruence relation on P, then P/E is integral if and only if $E \to P \times P$ is exact. In particular, if P and P/E are fine, then E is also fine.*

Proof Indeed the congruence relation E determined by a surjective homomorphism $\pi\colon P \to Q$ of integral monoids is just the equalizer of the two maps $P \times P \to Q$, and we saw in (5) of Theorem 2.1.17 that it is then an exact submonoid of $P \times P$. For the converse, let E be a congruence relation on P which is an exact submonoid of $P \times P$ and let $\pi\colon P \to Q$ be the quotient mapping. Then π is the coequalizer of the two maps $E \to P$. To prove that Q is integral, let q_1, q_2, and q be elements of Q such that $q_1 + q = q_2 + q$, and choose p_i and p in P with $\pi(p_i) = q_i$ and $\pi(p) = q$. Then $e := (p_1, p_2) + (p, p) \in E$, and, since $(p, p) \in E$, it follows that $(p_1, p_2) \in E^{gp} \cap (P \times P)$. Since E is an exact submonoid, it follows that $(p_1, p_2) \in E$ and hence that $\pi(p_1) = \pi(p_2)$. If P and P/E are fine, E is also fine by (2) of Theorem 2.1.17. □

In particular, congruence relations on P yielding integral quotients Q correspond to congruence relations on P^{gp}, and hence by (1.1.4) to subgroups of P^{gp}. Of course, the subgroup of P^{gp} corresponding to a surjective homomorphism of integral monoids $P \to Q$ is just the kernel of $P^{gp} \to Q^{gp}$.

Corollary 2.1.21. *Let Q be a finitely generated integral monoid, let L be a finitely generated abelian group, and let $L \to Q^{gp}$ be a homomorphism. Then $L \times_{Q^{gp}} Q$ is a finitely generated monoid.*

Proof This corollary is a special case of statement (6) of Theorem 2.1.17. □

Corollary 2.1.22. *Let L be a finitely generated abelian group and let P be a finitely generated submonoid of L. Then*

$$P^{\vee} := \{w \in \mathrm{Hom}(L, \mathbf{Z}) : \langle w, p\rangle \geq 0 \text{ for all } p \in P\}$$

is a fine monoid.

Proof For each $w \in P^{\vee}$, the restriction $\rho(w)$ of w to P belongs to $H(P) := \mathrm{Hom}(P, \mathbf{N})$. Thus there is a a homomorphism of integral monoids

$$\rho\colon P^{\vee} \to H(P).$$

We claim that ρ is exact. Indeed, suppose that $w \in (P^{\vee})^{gp}$ and that there exists $h \in H(P)$ such that $\rho^{gp}(w) = h$ in $H(P)^{gp}$. Then the same relation holds in $\mathrm{Hom}(P, \mathbf{Z})$, so that in fact $w(p) = h(p) \geq 0$ for all $p \in P$, and indeed $w \in P^{\vee}$, as desired. It follows easily that the induced homomorphism $\overline{\rho}\colon \overline{P^{\vee}} \to H(P)$ is also exact, and hence injective, by Proposition 2.1.16. According to (8) of Theorem 2.1.17, $H(P)$ is fine, and therefore so is its exact submonoid $\overline{P^{\vee}}$. Since $P^{\vee *}$ is a subgroup of a finitely generated group, it is also finitely generated, and hence so is $P^{\vee}$, by Proposition 2.1.1. □

Remark 2.1.23. Let Q be an integral monoid. A subset K of Q^{gp} that is invariant under the action of Q is called a *fractional ideal*. (Sometimes this terminology is reserved for the case in which moreover there exists an element q of Q such that $q + K \subseteq Q$. This is automatically the case if K is finitely generated as a Q-set. Moreover, if Q is fine and $q + K \subseteq Q$, Corollary 2.1.11 implies that $q + K$ is finitely generated as a Q-set and hence so is K.) The natural map $\pi\colon Q \to \overline{Q}$ induces a bijection between the set of fractional ideals of Q and of $\overline{Q}$, and this bijection takes finitely generated fractional ideals to finitely generated fractional ideals.

Proposition 2.1.24. *Suppose that $\theta\colon P \to Q$ is an exact homomorphism of fine monoids and that K is a finitely generated fractional ideal of Q. Then $J := \theta^{\mathrm{gp}^{-1}}(K)$ is a finitely generated fractional ideal of P.*

Proof This proposition is an immediate consequence of statement (1) of Theorem 2.1.17. □

To see that the exactness hypothesis is not superfluous, consider the summation homomorphism $\theta\colon \mathbf{N} \oplus \mathbf{N} \to \mathbf{N}$, and let $K = \mathbf{N} \subseteq \mathbf{Z}$. Then $\theta^{\mathrm{gp}^{-1}}(K) = \{(m, n) : m + n \geq 0\} \subseteq \mathbf{Z} \oplus \mathbf{Z}$, which is not finitely generated as an $\mathbf{N} \oplus \mathbf{N}$-set.

Lemma 1.9.2 of Chapter III gives an application of Theorem 2.1.17 to algebraic geometry.

2.2 Duality

If Q is a monoid, let $H(Q)$ denote the monoid of homomorphisms $Q \to \mathbf{N}$. The following proposition, which guarantees the existence of enough elements of $H(Q)$, plays a fundamental role in the theory of monoids. It also connects our definition (given in Definition 1.4.1) of a face of a monoid with the standard definition in the theory of convex bodies (see for example [21, 1.2].)

Proposition 2.2.1. *Let Q be a fine monoid and let F be a face of Q. Then there exists a homomorphism $h\colon Q \to \mathbf{N}$ such that $h^{-1}(0) = F$.*

Proof Let $\lambda_F\colon Q \to Q_F$ be the localization homomorphism, and recall that $F = \lambda_F^{-1}(Q_F^*)$. Thus it suffices to construct a homomorphism $h\colon Q_F \to \mathbf{N}$ with $h^{-1}(0) = Q_F^*$. Since Q_F is still fine, we may as well replace Q by Q_F, and thus we have reduced to the case in which $F = Q^*$. Replacing Q by Q/Q^*, we may also assume that Q is sharp and so $F = 0$. We must show that there is a local homomorphism $h\colon Q \to \mathbf{N}$.

Let T be a finite set of nonzero generators of Q. We argue by induction on the cardinality of T, using the technique of *Fourier–Motzkin elimination*. If T is empty, $Q = 0$ and there is nothing to prove. If $|T| = 1$, then, since

Q is sharp, the homomorphism $\mathbf{N} \to Q$ sending 1 to the unique element of T is an isomorphism, and its inverse is the desired homomorphism. In the general case, if T is a finite set of nonzero generators for Q, we must construct a homomorphism $h\colon Q^{\mathrm{gp}} \to \mathbf{Z}$ such that $h(t) > 0$ for all $t \in T$. Choose some $t \in T$, let $S := T \setminus \{t\}$, and let P be the submonoid of Q generated by S. The induction hypothesis implies that there exists a local homomorphism $h\colon P \to \mathbf{N}$. Then h induces a homomorphism $P^{\mathrm{gp}} \to \mathbf{Z}$, which we denote again by h. Since $\mathrm{Ext}^1(Q^{\mathrm{gp}}/P^{\mathrm{gp}}, \mathbf{Z})$ is a finite group, there is a positive integer n such that nh extends to all of Q^{gp}. Let us replace h by nh and choose an extension, still denoted by h, of h to Q^{gp}. If $h(t) > 0$, there is nothing more to prove. If $h(t) = 0$, then the submonoid Q' of Q generated by t is isomorphic to $\mathbf{N}$, and some positive multiple of the isomorphism extends to a homomorphism $g\colon Q^{\mathrm{gp}} \to \mathbf{Z}$. Then, if n is a sufficiently large natural number, $nh(s) + g(s) > 0$ for all $s \in S$. Since also $nh(t) + g(t) = g(t) > 0$, $nh + g$ is positive on all the elements of T, as required.

Suppose on the other hand that $h(t) < 0$. Let $\phi\colon Q^{\mathrm{gp}} \to Q^{\mathrm{gp}}$ be the homomorphism defined by

$$\phi(q) := h(q)t - h(t)q.$$

Then $\phi(t) = 0$ and, if $s \in S$, $s' := \phi(s) = h(s)t + |h(t)|s \in Q$. Since Q is sharp, $s' \neq 0$. Furthermore, the image Q' of ϕ is a submonoid of Q, hence is again sharp, and is generated by the image S' of S under ϕ. By the induction hypothesis, there exists a local homomorphism $Q' \to \mathbf{N}$. Let $h'\colon Q \to \mathbf{N}$ be the composition of such a homomorphism with the homomorphism $Q \to Q'$ induced by ϕ. Then $h'(s) > 0$ for all $s \in S$ and $h'(t) = 0$. This reduces us to the previous case. □

Corollary 2.2.2. *Let Q be a fine monoid and let x be an element of Q^{gp}. Then $x \in Q^{\mathrm{sat}}$ if and only if $h^{\mathrm{gp}}(x) \geq 0$ for every $h \in H(Q)$.*

Proof If $x \in Q^{\mathrm{sat}}$ then $nx \in Q$ for some $n \in \mathbf{Z}^+$ and hence $h^{\mathrm{gp}}(x) \geq 0$ for every $h \in H(Q)$. Suppose conversely that $h^{\mathrm{gp}}(x) \geq 0$ for every $h \in H(Q)$. Let Q' be the submonoid of Q^{gp} generated by Q and $-x$, and choose a local homomorphism $h\colon Q' \to \mathbf{N}$. Then $h^{\mathrm{gp}}(x) \geq 0$ and $h(-x) \geq 0$, so that in fact $h(-x) = 0$. Since h is local, $-x \in Q'^{*}$. Thus $x \in Q'$, and writing $x = -mx + q$ with $m \in \mathbf{N}$ and $q \in Q$, we see that $(m+1)x = q$, so $x \in Q^{\mathrm{sat}}$. □

The following result is the main duality theorem for fine monoids.

Theorem 2.2.3. *Let Q be a fine monoid, and let $H(Q) := \mathrm{Hom}(Q, \mathbf{N})$.*

1. *The monoid $H(Q)$ is fine, saturated, and sharp.*

2. *The natural map* $H(Q)^{\mathrm{gp}} \to \mathrm{Hom}(Q^{\mathrm{gp}}, \mathbf{Z})$ *factors through an isomorphism*

$$\epsilon\colon H(Q)^{\mathrm{gp}} \to \mathrm{Hom}(\overline{Q}^{\mathrm{gp}}, \mathbf{Z}).$$

3. *The evaluation mapping* $ev\colon Q \to H(H(Q))$ *factors through an isomorphism*

$$\overline{ev}\colon \overline{Q^{\mathrm{sat}}} \to H(H(Q)).$$

Thus, the functor H induces a contravariant involution of the category of sharp toric monoids. In particular, this category is self dual.

Proof The first statement follows immediately from (8) of Theorem 2.1.17. Since $H(Q) \to \mathrm{Hom}(Q, \mathbf{Z})$ is injective, so is the map $H(Q)^{\mathrm{gp}} \to \mathrm{Hom}(Q, \mathbf{Z})$. Any element h of $H(Q)$ necessarily annihilates Q^*, so this map factors through a map ϵ as claimed, and ϵ is still injective. To prove its surjectivity, let h be a local homomorphism $\overline{Q} \to \mathbf{N}$ and let S be a finite set of generators for $\overline{Q}$. For $g \in \mathrm{Hom}(\overline{Q}, \mathbf{Z})$, there exists $n \in \mathbf{Z}^+$ such that $nh(s) \geq g(s)$ for each $s \in S$. Then $nh(\overline{q}) \geq g(\overline{q})$ for every $\overline{q} \in \overline{Q}$, so $h' := nh - g \in H(\overline{Q})$. Thus $g = nh - h' \in H(\overline{Q})^{\mathrm{gp}} \cong H(Q)^{\mathrm{gp}}$, as required.

Since $H(H(Q))$ is fine saturated and sharp, ev factors through a map $\overline{ev}$ as claimed in the statement (3). Let x_1 and x_2 be two elements of Q^{sat} with $ev(x_1) = ev(x_2)$, and let $x := x_1 - x_2 \in Q^{\mathrm{gp}}$. Then $h(x) = 0$ for every $h \in H(Q)$. It follows from Corollary 2.2.2 that x and $-x$ belong to Q^{sat}, so $x \in (Q^{\mathrm{sat}})^*$. Thus $\overline{x}_1 = \overline{x}_2$ in $\overline{Q^{\mathrm{sat}}}$, and this proves the injectivity of $\overline{ev}$. For the surjectivity, suppose that $g \in H(H(Q))$. Since Q^{gp} is a finitely generated group, the map from Q^{gp} to its double dual is surjective. Thus there exists an element q of Q^{gp} such that $ev(q) = g$, i.e., such that $h(q) = g(h)$ for all $h \in H(Q)$. Then $h(q) \geq 0$ for all h, so $q \in Q^{\mathrm{sat}}$, as required. □

Corollary 2.2.4. *Let Q be a fine monoid. A subset S of Q is a face if and only if there exists an element h of* $H(Q)$ *such that* $S = h^{-1}(0)$. *For each* $S \subseteq Q$, *let* $S^\perp$ *be the set of* $h \in H(Q)$ *such that* $h(s) = 0$ *for all* $s \in S$, *and, for* $T \subseteq H(Q)$, *let* $T^\perp$ *be the set of* $q \in Q$ *such that* $t(q) = 0$ *for all* $t \in T$. *Then* $F \mapsto F^\perp$ *induces an order-reversing bijection between the set of faces of Q and the set of faces of* $H(Q)$, *and* $F = (F^\perp)^\perp$ *for any face F of either.*

Proof The first statement follows from Proposition 2.2.1. It is clear that if S is any subset of Q, then $S^\perp$ is a face of $H(Q)$ and that $T^\perp$ is a face of Q if T is any subset of $H(Q)$. Furthermore, $S_2^\perp \subseteq S_1^\perp$ if $S_1 \subseteq S_2$, and $S \subseteq (S^\perp)^\perp$. Let F be a face of Q. By Proposition 2.2.1, there exists an $h \in H(Q)$ such that $h^{-1}(0) = F$. Then $h \in F^\perp$ and, if $q \in (F^\perp)^\perp$, $h(q) = 0$ so $q \in F$. Thus $F = (F^\perp)^\perp$, and hence the map $\perp$ from the set of faces of Q to the set of faces of $H(Q)$ is injective and the map $\perp$ from the set of faces of $H(Q)$ to the set of faces of Q is surjective. Hence the map from the set of faces of $H(Q)$ to the set of faces of $H(H(Q))$ is also injective. By Corollary 2.3.8, the map $Q \to \overline{Q^{\mathrm{sat}}}$ induces a bijection on the corresponding sets of faces, and hence by (3) of Theorem 2.2.3, $\overline{ev}$ identifies the faces of Q with the faces of $H(H(Q))$. Now it follows that $\perp$ is bijective. □

The next corollary is an analog of the finiteness of integral closure in commutative algebra.

Corollary 2.2.5. *If Q is a fine monoid, then Q^{sat} is again fine. In fact, the action of Q on Q^{sat} defined by the homomorphism $Q \to Q^{\mathrm{sat}}$ makes Q^{sat} a finitely generated Q-set.*

Proof Since $(Q^{\mathrm{sat}})^*$ is contained in Q^{gp}, it is a finitely generated abelian group. Theorem 2.2.3 implies that $\overline{Q^{\mathrm{sat}}}$ is fine, and, since Q^{sat} is integral, it follows from Proposition 2.1.1 that Q^{sat} is finitely generated, hence fine. Choose a finite set of generators T for Q^{sat} as a monoid, and, for each $t \in T$, choose $n_t \in \mathbf{N}^+$ such that $n_t t \in Q$. Then $\{\sum j_t t$ such that $j_t \le n_t$ for all $t \in T\}$ generates Q^{sat} as a Q-set. □

Corollary 2.2.6. *If $\pi\colon Q' \to Q$ is a surjective homomorphism of fine monoids, then $H(\pi)\colon H(Q) \to H(Q')$ is injective and exact.*

Proof It is clear that $H(\pi)$ is injective if π is surjective. Moreover, by (2) of Theorem 2.2.3, we can view an element h of $H(Q)^{\mathrm{gp}}$ as a homomorphism $Q \to \mathbf{Z}$, and we see that $h \in H(Q)$ if and only if $h \circ \pi \in H(Q')$. □

Corollary 2.2.7. *Let Q be a fine sharp monoid. Then Q is isomorphic to a submonoid of $\mathbf{N}^r \oplus T$ for some $r \in \mathbf{N}$ and some finite group T. If Q^{gp} is torsion free, we can take $T = 0$. If Q is also saturated, then it is isomorphic to an exact submonoid of some $\mathbf{N}^r$.*

Proof Suppose that Q is fine and sharp. By (8) of Theorem 2.1.17, the monoid $P := H(Q)$ is fine and sharp, and, by (2) of Theorem 2.2.3, $P^{\mathrm{gp}} := H(Q)^{\mathrm{gp}} \cong \mathrm{Hom}(Q^{\mathrm{gp}}, \mathbf{Z})$. Then $\mathrm{Hom}(P^{\mathrm{gp}}, \mathbf{Z}) \cong H(P)^{\mathrm{gp}}$ is a finitely generated free group. It follows that the kernel of the natural map

$$Q^{\mathrm{gp}} \to \mathrm{Hom}(P^{\mathrm{gp}}, \mathbf{Z}) \cong H(P)^{\mathrm{gp}}$$

is just the torsion subgroup T of Q^{gp}. Choose a splitting

$$Q^{\mathrm{gp}} \cong H(P)^{\mathrm{gp}} \oplus T$$

and a surjection $\mathbf{N}^r \to P$. By Corollary 2.2.6, $H(P)$ is then an exact submonoid of $H(\mathbf{N}^r) \cong \mathbf{N}^r$. Then the natural injection

$$Q \longrightarrow Q^{\mathrm{gp}} \xrightarrow{\cong} H(P)^{\mathrm{gp}} \oplus T$$

factors through the inclusion $H(P) \oplus T \subseteq H(P)^{\mathrm{gp}} \oplus T$, and Q is a submonoid of $\mathbf{N}^r \oplus T$ since $H(P) \subseteq \mathbf{N}^r$. Furthermore, by (3) of Theorem 2.2.3, the natural map $Q \to H(P)$ factors through an isomorphism $\overline{Q^{\mathrm{sat}}} \to H(P)$. Thus if Q also is saturated, $Q \cong H(P)$, an exact submonoid of $\mathbf{N}^r$. □

Remark 2.2.8. Recall that the interior of a monoid is the complement of all its proper subfaces. Let us observe that if Q is fine, then an element h of $H(Q)$ lies in the interior of $H(Q)$ if and only if $h\colon Q \to \mathbf{N}$ is a local homomorphism. Indeed, by Corollary 2.2.4, h lies in the interior of $H(Q)$ if and only if $h^\perp$ does not contain any nontrivial face of Q, i.e., if and only if $h^\perp = Q^*$. This is exactly the condition that $h\colon Q \to \mathbf{N}$ be a local homomorphism.

We shall find the following crude finiteness result useful. More precise variants are available, most of which rely on the theory of Hilbert polynomials in algebraic geometry.

Proposition 2.2.9. *Let Q be a fine sharp monoid, let d be the rank of Q^{gp}, and let $h\colon Q \to \mathbf{N}$ be a local homomorphism. For each real number r, let*

$$B_h(r) := \{q \in Q : h(q) < r\}.$$

Then there are positive real constants c and C such that, for all $r \gg 0$,

$$cr^d < \#B_h(r) < Cr^d.$$

Proof The torsion subgroup T of Q^{gp} is finite, and the quotient Q^{gp}/T is a free abelian group of rank d. By Theorem 2.2.3, $H(Q)$ is finitely generated and sharp, and hence by Proposition 2.1.2 it has a unique set of minimal generators $\{h_1, \ldots, h_m\}$. Since h is local, it belongs to the interior of $H(Q)$, by Remark 2.2.8. Thus the face of $H(Q)$ generated by h is all of $H(Q)$, and in particular contains each h_i. By Proposition 1.4.2, this means that for each i there exists a positive integer n_i such that $n_i h \geq h_i$ in $H(Q)$. Choose $n \geq n_i$ for all i. Then $B_h(r) \subseteq \cap_i B_{h_i}(nr)$ for every $r \in \mathbf{R}^+$. Since Q is sharp, statement (2) of Theorem 2.2.3 implies that $H(Q)^{\mathrm{gp}} \cong \mathrm{Hom}(Q^{\mathrm{gp}}, \mathbf{Z})$, and consequently $\{h_1, \ldots, h_m\}$ spans the $\mathbf{Q}$-vector space $\mathrm{Hom}(Q^{\mathrm{gp}}, \mathbf{Q})$. Since this space has dimension d, the set $\{h_1, \ldots, h_m\}$ contains a basis, which we may assume is $\{h_1, \ldots, h_d\}$. Then the map $Q^{\mathrm{gp}} \otimes \mathbf{Q} \to \mathbf{Q}^d$ sending $x \otimes 1$ to $(h_1(x), h_2(x), \ldots, h_d(x))$ is an isomorphism and induces an injection from the image Q' of Q in Q^{gp}/T to $\mathbf{N}^d$. If $q \in B_h(r)$, its image in $\mathbf{N}^d$ lies in $\{(I_1, \ldots, I_d) : 0 \leq I_i \leq nr\}$, a set of cardinality $(1 + [nr])^d$. Since the cardinality of the fibers of the map $Q \to Q'$ is bounded by the order t of T, it follows that $\#B_h(r) \leq t(1+nr)^d$. Thus if $C := t(1+n)^d$ and $r \geq 1$, the cardinality of $B_h(r)$ is bounded by Cr^d.

On the other hand, any set of generators for the monoid Q also generates the d-dimensional $\mathbf{Q}$-vector space $\mathbf{Q} \otimes Q^{\mathrm{gp}}$, and therefore contains a subset $\{q_1, \ldots, q_d\}$ whose image forms a basis. Thus the homomorphism $\theta \colon \mathbf{N}^d \to Q$ sending $I := (I_1, \ldots, I_d)$ to $\sum I_i q_i$ is injective. Let $n := \max\{h_1(q_1), \ldots, h_d(q_d)\}$ and let $c := (2nd)^{-d}$. Then we claim that $\#B_h(r) \geq cr^d$ for $r \geq nd$. Indeed, if $r \geq nd$, let $m := [r/nd]$, and note that

$$r/nd \geq m \geq r/nd - 1 \geq r/2nd.$$

The set $\{I : I_i \leq m, i = 1, \ldots, d\}$ has cardinality $(1+m)^d$ and its image in Q also has cardinality $(1+m)^d \geq m^d \geq cr^d$ and is contained in $B_h(r)$. □

2.3 Monoids and cones

Just as vector spaces are simpler than abelian groups, cones are simpler than monoids, and it is frequently very helpful to replace a monoid by the cone it spans. Let K be an Archimedean totally ordered field and let $K^{\geq 0}$ denote the set of nonnegative elements of K, regarded as a multiplicative monoid. Since $0 \in K^{\geq 0}$, this monoid is not u-integral, but $K^{>} := K^{\geq 0} \setminus \{0\}$ is a group. In practice, here K will be either $\mathbf{R}$ or $\mathbf{Q}$. The constructions when $K = \mathbf{Q}$ are considerably simpler and suffice for most purposes, but in some cases it is useful to work with the locally compact topology of $\mathbf{R}$. Proposition 2.3.11 will help control the behavior of the constructions when one changes the field K.

Definition 2.3.1. *A K-cone is an integral monoid $(C, +, 0)$ endowed with an action of the monoid $(K^{\geq 0}, \cdot, 1)$, such that*

$$(a+b)x = ax + bx \quad \text{for } a, b \in K^{\geq 0} \text{ and } x \in C, \text{ and}$$
$$a(x+y) = ax + ay \quad \text{for } a \in K^{\geq 0} \text{ and } x, y \in C.$$

A morphism of K-cones is a morphism of monoids compatible with the actions of $K^{\geq 0}$.

In the sequel we shall say "cone" instead of "K-cone," and write $C(S)$ instead of $C_K(S)$, when there seems to be no danger of confusion.

Remark 2.3.2. We could define the notion of a K-cone without the integrality assumption, but there seems to be no need for this extra generality. Note that it follows from the definition (assuming only that C is quasi-integral) that $a0 = 0$ for every $a \in K^{\geq}$ and that $0x = 0$ for every $x \in C$. Thus it really suffices to give the action of the group $K^{>}$ on C, and in fact any such action satisfying the definition extends uniquely to an action of $K^{\geq}$.

Any K-vector space V is a K-cone, and any submonoid of V stable under the action of $K^{\geq}$ forms a subcone. On the other hand, if C is any K-cone, then one verifies immediately that the action of $K^{\geq}$ on C extends to a unique action of K on C^{gp} and that this action defines a K-vector space structure on C^{gp}. Thus every K-cone can be viewed as a subcone of a K-vector space.

If C is a K-cone, then its group of units C^* is automatically stable under the action of $K^{\geq}$ and hence forms a K-vector space. A cone is sharp if and only if $C^* = 0$; some authors call such a C a *strongly convex cone.* If C is a K-cone, then $\overline{C} := C/C^*$ is a sharp K-cone. By the *dimension* of C we mean the dimension of C^{gp} (as a K-vector space), and we call the dimension of $\overline{C}$ (which inherits a cone structure) the *sharp dimension* of C.

If S is any subset of a K-vector space V we can define its *conical hull* $C_K(S)$ to be the set of all elements v of V that can be written $v = \sum a_s s$ with $a_s \in K^{\geq}$ and $s \in S$. Then $C_K(S)$ is the smallest K-cone in V containing S. A K-cone C is called *finitely generated* if it admits a finite subset S such that $C = C_K(S)$.

Let C be a K-cone and let F be a face of C. Then F is automatically a subcone of C. Indeed, if $x \in F$ and $a \in K^{\geq 0}$, then there exists $n \in \mathbf{N}$ with $a \leq n$, since K is Archimedean. Then $ax \leq nx$ and $nx \in F$, and, since F is a face, $ax \in F$ also. If F is a face of a cone C, then C/F is a sharp cone, and we call its dimension the *codimension of* F. If this codimension is one, we say that F is a *facet* of C. A one-dimensional face of C is sometimes called an *extremal ray* of C.

Remark 2.3.3. An ideal J of C need not be invariant under $K^{>}$; if it is we say that is a *conical ideal.* Since the complement of every prime ideal is a face and hence is invariant under $K^{>}$, it follows that every prime ideal of C is necessarily conical. In fact, an ideal is conical if and only if it is a radical ideal. It is clear that a conical ideal is radical. Conversely, suppose that J is a radical ideal of C and that $q \in J$ and $\lambda \in K^{>}$. We claim that $p := \lambda q \in J$. Since K is Archimedean, there exists $\lambda' \in K^{>}$ such that $n := \lambda' + \lambda^{-1}$ is a positive integer. Then $np = \lambda' p + \lambda^{-1} p = \lambda' p + q \in J$. Since J is an ideal and $q \in J$, it follows that $np \in J$, and thus $p \in J$ since J is a radical ideal. It follows that the conical ideal generated by a subset S of C is the intersection of the set of prime ideals of C containing S.

Let us say that an element x of a sharp cone C is *K-indecomposable in C* if it is not a unit and, whenever $x = y + z$ with y and z in C, y and z are K-multiples of x. Thus x is K-indecomposable if and only $\langle x \rangle^{\mathrm{gp}}$ is a one-dimensional K-vector space. (Recall that $\langle x \rangle$ is the face of C generated by x.) For example, in the monoid P given by generators $\{x, y, z\}$ and relation $x + y = 2z$, the elements x, y, and z are irreducible, and in the corresponding cone x and y are indecomposable, but z is not indecomposable.

Proposition 2.3.4. *Let C be a finitely generated sharp cone. Then every element of every minimal set S of generators of C is indecomposable, and every indecomposable element of C is a multiple of some element of S. In particular, C is spanned by a finite number of indecomposable elements.*

Proof The proof is essentially the same as the proof of the analogous result (Proposition 2.1.2) for monoids, but we write it in detail anyway. Let x be an element of a minimal set S of generators for C and let $S' := S \setminus \{x\}$. Write $x = y + z$, with $y = \sum a_s s$, $z = \sum b_s s$, and $a_s, b_s \in K^{\geq 0}$. Then $x = \sum c_s s$, with $c_s = a_s + b_s$, so $(1 - c_x)x = \sum_{s \in S'} c_s s$. If $c_x < 1$ we see that S' generates C, a contradiction, and if $c_x > 1$, x is a unit, contradicting the sharpness of C. Then necessarily $c_x = 1$, so $0 = \sum_{s \in S'} a_s s + b_s s$. Since S is sharp, this implies that $a_s s = b_s s = 0$ for all $s \in S'$. Then $y = a_x x$ and $z = b_x x$, proving that x is indecomposable. On the other hand, if x' is any element of C, we can write $x' = \sum a_s s$, and if x' is indecomposable, $a_s \neq 0$ implies that s is a multiple of x', hence x' is a multiple of s. □

The next result relates the Krull dimension of the spectrum of a cone to the dimension of the corresponding vector space.

Proposition 2.3.5. *Let C be a K-cone and S a set of generators for C.*

1. *Every face F of C is generated as a cone by $F \cap S$, and in particular is finitely generated if C is finitely generated.*
2. *If C is finitely generated, C contains only a finite number of faces.*
3. *The length d of every maximal increasing chain of faces*

$$C^* = F_0 \subset F_1 \subset F_2 \subset \cdots \subset F_d = C$$

is less than or equal to the K-dimension of the vector space $\overline{C}^{\text{gp}}$, with equality if C is finitely generated as a K-cone. In this case, every chain of faces is contained in a chain of length d.

4. *If C is finitely generated, every proper face of C is contained in a facet.*

Proof Let F be a face of C and $x \in F$, $x \neq 0$. Then we can write $x = \sum a_s s$ with $a_s \in K^{\geq 0}$ and $s \in S$. Since F is a face, each s for which $a_s \neq 0$ must belong to F. This shows that in fact F is generated as a cone by $F \cap S$. If S is finite, it has only finitely many subsets, so C can have only finitely many faces. This proves (1) and (2). Since there is a natural bijection between the faces of C and the faces of $\overline{C}$ we may as well assume in the proof of (3) that $C^* = 0$. Let $C := (F_0 \subset \cdots \subset F_d)$ be a maximal chain of faces of C; necessarily $F_0 = \{0\}$

and $F_d = C$. Since each F_i is an exact submonoid of C, the inclusions $F_0^{\mathrm{gp}} \subsetneq F_1^{\mathrm{gp}} \subset \cdots \subset F_d^{\mathrm{gp}}$ of linear subspaces of C^{gp} are all strict. It follows that d is less than or equal to the dimension $\overline{d}$ of C^{gp}. Suppose that C is finitely generated and that $\mathcal{G} := (G_0 \subsetneq \cdots \subsetneq G_e)$ is a chain of faces. We will prove that $\mathcal{G}$ is contained in a chain of length d by induction on the dimension $\overline{d}$ of C^{gp}. Without loss of generality we may assume that $G_0 = 0$ and that $G_e = C$. If $\overline{d} = 0$, $C = \{0\}$ and the result is trivial. Suppose that $\overline{d} > 0$; we may assume by Proposition 2.3.4 that S is a set of indecomposable elements of C. Necessarily $G_1 \neq \{0\}$, and it follows from statement (1) that G_1 must contain a K-indecomposable element c. Then $F := \langle c \rangle$ has dimension one and the quotient C/F has sharp dimension $d - 1$. Thus the induction assumption implies that the chain $\mathcal{G}/F$ in C/F is contained in a chain of length $d - 1$. The inverse image of this chain in C, together with G_0, is then a chain of length d containing $\mathcal{G}$. This proves statement (3), and (4) is an immediate consequence. □

Proposition 2.3.6. *Let Q be an integral monoid and let K be an Archimedean ordered field. Denote by $C_K(Q)$ the K-subcone of $K \otimes Q^{\mathrm{gp}}$ spanned by the image of the map $Q \to K \otimes Q^{\mathrm{gp}}$ sending q to $1 \otimes q$.*

1. *Two elements q_1, q_2 of Q have the same image in $C_K(Q)$ if and only if there exists $n \in \mathbf{Z}^+$ such that $nq_1 = nq_2$ in Q.*
2. *If $K = \mathbf{Q}$, then an element x of $\mathbf{Q} \otimes Q^{\mathrm{gp}}$ lies in $C_{\mathbf{Q}}(Q)$ if and only if there exist $n \in \mathbf{Z}^+$ and $q \in Q$ such that $nx = 1 \otimes q$. For general K, $C_K(Q)$ is the set of elements of $K \otimes Q^{\mathrm{gp}}$ that can be written as a sum of the form $\sum a_i \otimes q_i$, with $a_i \in K^{\geq}$ and $q_i \in Q$.*
3. *If C is any K cone and $Q \to C$ is a homomorphism of monoids, then there is a unique homomorphism of K-cones $C_K(Q) \to C$ such that the diagram*

$$\begin{array}{ccc} Q & \longrightarrow & C_K(Q) \\ & \searrow & \downarrow \\ & & C \end{array}$$

commutes.

4. *Let $H(Q) := \mathrm{Hom}(Q, \mathbf{N}) \subseteq \mathrm{Hom}(Q, K^{\geq})$. If Q is fine, the natural maps*

$$C_K(H(Q)) \to \mathrm{Hom}(Q, K^{\geq}) \quad \text{and} \quad Q^{\mathrm{sat}} \to Q^{\mathrm{gp}} \cap C_K(Q)$$

are isomorphisms.

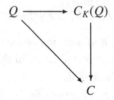

Proof Since Q is integral, the map $Q \to Q^{\mathrm{gp}}$ is injective, and the kernel of the map $Q^{\mathrm{gp}} \to \mathbf{Q} \otimes Q^{\mathrm{gp}}$ consists of the torsion subgroup of Q^{gp}. This proves (1)

when $K = \mathbf{Q}$. Since K is flat over $\mathbf{Q}$, the map $\mathbf{Q} \otimes Q^{gp} \to K \otimes Q^{gp}$ is injective, and the general case follows.

For (2), denote by C the set of all x in $\mathbf{Q} \otimes Q^{g}$ for which $nx = 1 \otimes q$ for some $n \in \mathbf{Z}^+$ and some $q \in Q$. It is clear that $C \subseteq C_{\mathbf{Q}}(Q)$. On the other hand, if $nx = 1 \otimes q$ and $n'x' = 1 \otimes q'$, then $nn'(x + x') = 1 \otimes (n'q + nq)$, so $x + x' \in C$. Since C is stable under the action of $Q^{\geq}$, it follows that C is a Q-cone, and hence $C_{\mathbf{Q}}(Q) \subseteq C$. The statement for general K is an immediate consequence of the definition.

For (3), observe first that a homomorphism $\theta \colon Q \to C$ induces a group homomorphism $\theta^{gp} \colon Q^{gp} \to C^{gp}$, where C^{gp} is now a K-vector space. This homomorphism extends uniquely to a K-linear map $K \otimes Q^{gp} \to C^{gp}$. It then follows from (2) that this map sends $C_K(Q)$ to C. It is clear that the induced map is a map of K-cones and is unique.

For now we just prove (4) when K is the field of rational numbers; see Remark 2.3.13 for the general case. It is clear that the map $C_K(H(Q)) \to \mathrm{Hom}(Q, K^{\geq})$ is injective. For the surjectivity, let h be a homomorphism $Q \to \mathbf{Q}^{\geq}$ and let S be a finite set of generators for Q. For each $s \in S$, choose $n_s \in \mathbf{Z}^+$ such that $n_s h(s) \in \mathbf{N}$ and let $n := \prod n_s$. Then $nh \in H(Q)$. The second part of (4) follows immediately from (2). □

Let us now explain the relationship between the faces of a monoid and the faces of the cone it spans. The following proof of the following result is almost trivial if $K = \mathbf{Q}$, but is slightly more delicate in the general case.

Proposition 2.3.7. *Let Q be an integral monoid and let $c \colon Q \to C_K(Q)$ be the map sending q to $1 \otimes q$.*

1. *If F is a face of Q, then $C_K(F)$ is a face of $C_K(Q)$, and the map $F \mapsto C_K(F)$ is a bijection from the set of faces of F to the set of faces of $C_K(Q)$, with inverse $G \mapsto c^{-1}(G)$. For each face F of Q, $c^{-1}(C_K(F)^{gp}) = F^{gp}$.*
2. *If I is an ideal of Q, let $C^+_K(I)$ denote the set of elements that can be written as a sum $\sum\{a_s s : s \in S\}$, where each $a_s \in K^>$ and S is a nonempty finite subset of I. Then $C^+_K(I)$ is the smallest $K^>$-invariant ideal of $C_K(Q)$ containing the image of I, and $c^{-1}(C^+_K(I))$ is the radical of I. Furthermore, $I \mapsto C^+_K(I)$ induces a bijection from the set of radical ideals of Q to the set of $K^>$-invariant ideals of $C_K(Q)$.*

Proof Since the map from Q to its image Q' in Q^{gp}/Q^{gp}_{tor} induces a bijection between the sets of faces (resp. radical ideals) of Q and of Q' and an isomorphism $C_K(Q) \to C_K(Q')$, we may assume without loss of generality that Q^{gp} is torsion free. In this case, the map c is injective, and we write it as an inclusion and omit the subscript K. Let F be a face of Q. To prove that $C(F)$

is a face of $C(Q)$, suppose that $c := a + b \in C(F)$ with $a, b \in C(Q)$. Write $c = \sum c_f f$, $a = \sum a_q q$, and $b = \sum b_q q$ with $a_q, b_q, c_f \in K^{\geq 0}$ and $q \in Q, f \in F$. There is a fine submonoid Q' of Q containing all q and f such that a_q, b_q, or c_f is nonzero. Then $a, b, c \in C(Q')$, and $F \cap Q'$ is a face of Q'. By Proposition 2.2.1, there is a homomorphism $h \colon Q' \to \mathbf{N}$ such that $h^{-1}(0) = F \cap Q'$. This homomorphism extends to a homomorphism $C(h') \colon C(Q') \to K^{\geq 0}$. Then $C(h')(a) + C(h')(b) = C(h')(c) = 0$, so $C(h')(a) = C(h')(b) = 0$. Since $0 = \sum a_q h(q)$ and each $a_q \geq 0$, it follows that $a_q = 0$ if $h(q) > 0$, i.e., if $q \notin F$. Thus $a \in C(F)$, and similarly for b. This shows that $C(F)$ is indeed a face of $C(Q)$. If $q \in Q \cap C(F)$, then there exists a finite subset S of F such that q can be written $q = \sum\{a_s s : s \in S\}$, with each $a_s \in K^{\geq 0}$. Let Q'' be the submonoid of Q generated by S, let F'' be the face of Q'' generated by S, and choose a homomorphism $h'' \colon Q'' \to \mathbf{N}$ with $h''^{-1}(0) = F''$. Then $h''(q) = C(h'')(q) = 0$, so $q \in F'' \subseteq F$. This shows that $Q \cap C(F) = F$, and a similar argument shows that $Q^{\mathrm{gp}} \cap C(F)^{\mathrm{gp}} = F^{\mathrm{gp}}$. On the other hand, if G is a face of $C(Q)$, then $F := G \cap Q$ is a face of Q. Every element of G can be written as a sum $g = \sum a_s s$ with $s \in Q$, and since G is a face of $C(Q)$, $s \in G$ whenever $a_s > 0$. But then each such s lies in F, and hence $g \in C(F)$, so in fact $G = C(F)$. This proves (1).

For (2), let I be any ideal of Q. It is clear that $C^+_K(I)$ is contained in every $K^>$-invariant ideal containing $c(I)$ and that it is invariant under $K^>$, hence a subcone, but perhaps it is not clear that it is an ideal of $C_K(Q)$. Let us first check that if $a, b \in K^>$, $q \in Q$, and $s \in I$, then $aq + bs \in C^+_K(I)$. If $a \leq b$, this is clear, since then $aq+bs = a(q+s)+(b-a)s \in C^+_K(I)$. On the other hand, if $b < a$, let n be the largest natural number less than a/b, so that $a = bn+a'$ and $a' < b$. (Such a number exists because K is Archimedean.) Then $aq + bs = a'q + b(nq + s)$ and we are reduced to the previous case. Now in general, if $c = \sum a_q q$ and $v = \sum b_s s$, with $a_q, b_s \in K^>$ then it follows, by induction on the numbers of q's appearing, that $c + v \in C^+_K(I)$. It is clear that $c^{-1}(C^+_K(I))$ is a radical ideal of Q containing I, and hence containing its radical $\sqrt{I}$. Suppose that $q \notin \sqrt{I}$. Then there is a prime ideal $\mathfrak{p}$ of Q containing I but not q, i.e., there is a face F of Q containing q and which does not meet I. Then if $c(q) \in C^+_K(I)$, $c(q) = \sum a_s s$ with $a_s \in K^>$ and $s \in I$. Since $c(q) \in C_K(F)$, the same is true of each s, and so each $s \in C_K(F) \cap Q = F$, a contradiction. This shows that $\sqrt{I} = c^{-1}(C^+_K(I))$.

On the other hand, let J be any $K^>$-invariant ideal of $C_K(Q)$ and let $I := c^{-1}(J)$. Then $I = \sqrt{I}$ and furthermore it is clear that $C^+_K(I) \subseteq J$. To prove the reverse inclusion, write an element v of J as a finite sum $v = \sum a_q q$ with $a_q \in K^>$. Choose a natural number n with $n \geq a_q$ for all q and let $w := \sum q$. Then $nw = \sum(n - a_q)q + v \in J$, and, since J is $K^>$ invariant, it follows that $w \in c^{-1}(J) = I$. Moreover, there is some $a \in \mathbf{Q}^> \subseteq K^>$ such that $a < a_q$ for all

q. Then aq belongs to the ideal $C_K^+(I)$ of $C_K(Q)$, and, since $v = \sum(a_q - a)q + aw$, it follows that $v \in C_K^+(I)$. □

Corollary 2.3.8. *Let Q be an integral monoid. Then the natural maps in the commutative diagram:*

$$\begin{array}{ccc} \mathrm{Spec}(C_K(Q^{sat})) & \longrightarrow & \mathrm{Spec}(C_K(Q)) \\ \big\downarrow & & \big\downarrow \\ \mathrm{Spec}(Q^{sat}) & \longrightarrow & \mathrm{Spec}(Q) \end{array}$$

are homeomorphisms (for the Zariski topologies).

Proof It is clear from that the map $C_K(Q) \to C_K(Q^{sat})$ is an isomorphism, and hence so is the top arrow in the diagram. Proposition 2.3.7 implies that the vertical arrows are bijections, and even homeomorphisms. It follows that the bottom arrow is also a homeomorphism, but this could also be easily checked directly. □

Note that Proposition 1.4.7, whose proof we deferred, follows immediately from Corollary 2.3.8 and Proposition 2.3.5. Moreover, if Q is fine, then $S :=$ Spec S is a finite topological space. In fact the topology of S can be recovered from the partial order on S defined by inclusion on the set Spec S of prime ideals of Q, or, equivalently, the set of faces of Q. To make this explicit, let $\mathfrak{p}$ be a point of S. The complement F of $\mathfrak{p}$ is a face of Q, and, since Q is finitely generated, by statement (3) of Theorem 2.1.17 there exists an $f \in Q$ such that $\langle f \rangle = F$. Then

$$\{\mathfrak{p}' : \mathfrak{p} \in \{\mathfrak{p}'\}^-\} = S_F := \{\mathfrak{p}' : F \cap \mathfrak{p}' = \emptyset\} = S_f := \{\mathfrak{p}' : f \notin \mathfrak{p}'\}$$

is open in S. Thus the set of generizations of each point is open, and hence a subset of S is open if and only if it is stable under generization. This shows that the topology of S is entirely determined by the order relation among the primes of Q.

Let s be a point of S corresponding to a prime ideal $\mathfrak{p}$ of Q and let F be the complementary face. The *height* of s is the maximum length of a chain of prime ideals containing $\mathfrak{p}$, or, equivalently, the dimension of Q/F. The space S has a natural stratification defined by the heights of its points. Let d be the Krull dimension of S and for $0 \le i \le d$ let $K_i := \cap\{\mathfrak{p} : \mathrm{ht}\,\mathfrak{p} = i\}$, an ideal of Q. We saw in Definition 1.4.6 that every prime of height $i + 1$ contains a prime of height i, hence $K_i \subseteq K_{i+1}$. We have

$$\emptyset = K_0 \subset I_Q = K_1 \subset \cdots \subset K_d = Q^+,$$

where I_Q is the interior ideal of Q.

Since $\{\mathfrak{p} : \operatorname{ht} \mathfrak{p} = i\}$ is finite,

$$Z_i := Z(K_i) = \cup\{Z(\mathfrak{p}) : \operatorname{ht} \mathfrak{p} = i\} = \{\mathfrak{p} : \operatorname{ht} \mathfrak{p} \geq i\}.$$

Thus we have a chain of closed sets

$$\{Q^+\} = Z_d \subset Z_{d-1} \subset \cdots \subset Z_1 \subset Z_0 = \operatorname{Spec} Q.$$

If $\mathfrak{p} \in \operatorname{Spec} Q$ and $F := Q \setminus \mathfrak{p}$, then $\mathfrak{p}$ belongs to the open subset S_F of S defined by Q, and F is the largest face with this property. Let $\mathcal{F}_i$ denote the set of faces F of Q such that $Q \setminus F$ has height i, i.e., such that the rank of Q/F is i. A prime $\mathfrak{p}$ belongs to some S_F with $F \in \mathcal{F}_i$ if and only if $\operatorname{ht} \mathfrak{p} \leq i$. This shows that

$$\cup\{S_F : F \in \mathcal{F}_i\} = \{\mathfrak{p} : \operatorname{ht} \mathfrak{p} \leq i\} = S \setminus Z_{i+1}$$

If G is a face of Q, then $\operatorname{Spec}(Q/G) \cong \operatorname{Spec}(Q_G)$ is an open subset of $\operatorname{Spec} Q$, and the height of a prime ideal computed in $\operatorname{Spec}(Q/G)$ is the same as its height computed in $\operatorname{Spec}(Q)$. It follows that, for every i, the ideal of Q/G generated by the ideal K_i of Q is the ideal of K_i of Q/G.

The following corollary summarizes this discussion.

Corollary 2.3.9. *Let S be the spectrum of a fine monoid Q. For each natural number i, let*

$$K_i := \cap\{\mathfrak{p} : \operatorname{ht} \mathfrak{p} = i\} \quad \text{and}$$
$$Z_i := Z(K_i) = \{\mathfrak{p} \in S : \operatorname{ht} \mathfrak{p} \geq i\}.$$

1. *The Krull dimension d of Q is the rank of $\overline{Q}$, and we have a chain of closed subsets*
$$\emptyset = Z_{d+1} \subset Z_d = \{Q^+\} \subset Z_{d-1} \subset \cdots \subset Z_0 = S.$$
2. *The complement of Z_i in S is the union of the set of all the open sets S_F as F ranges over the faces of Q such that the rank of Q/F is $i-1$. In particular, $S \setminus Z_2$ is the union of the sets $S_F := \operatorname{Spec}(Q_F)$ as F ranges over the facets of Q, and $S \setminus Z_1 = \operatorname{Spec}(Q^{\mathrm{gp}})$.*
3. *The topological space S is catenary, and Z_i has codimension i in S.*
4. *If G is face of Q, then the ideal of Q/G generated by the ideal K_i of Q is the ideal K_i of Q/G.*

We now discuss a few finiteness and duality results for K-cones. When $K = \mathbf{Q}$, these can be deduced from analogous results for monoids, but in the general case it is necessary to use "Fourier–Motzkin elimination theory." Its key lemma will allow us to compute a set of generators for the intersection of a cone C

with a hyperplane from a set of generators of C. It will also show that formation of this intersection is compatible with field extensions.

Lemma 2.3.10. *Let V be a K-vector space, let w be a linear map $W \to K$, and let $i_w \colon \Lambda^2 V \to \mathrm{Ker}(w)$ signify interior multiplication by w:*

$$i_w(v_1 \wedge v_2) = \langle w, v_1 \rangle v_2 - \langle w, v_2 \rangle v_1.$$

If S is a subset of V, let

$$S_w := \{ i_w(s_1 \wedge s_2) : \langle w, s_1 \rangle > 0 \text{ and } \langle w, s_2 \rangle < 0 \} \cup (S \cap \mathrm{Ker}(w)).$$

Then $C_K(S_w) = C_K(S) \cap \mathrm{Ker}(w)$. If $K \to K'$ is an extension of ordered fields and $V' := K' \otimes_K V$ and $w' := \mathrm{id}_{K'} \otimes w$, then $C_{K'}(S_w) = C_{K'}(S) \cap \mathrm{Ker}(w')$.

Proof It is clear from the definition that $S_w \subseteq C(S) \cap \mathrm{Ker}(w)$ and hence that $C(S_w) \subseteq C(S) \cap \mathrm{Ker}(w)$. To prove the reverse inclusion, suppose that v belongs to $C(S) \cap \mathrm{Ker}(w)$. Since $v \in C(S)$, it can be written in the form

$$v = \sum_{s \in S'} a_s s, \quad \text{where each } a_s > 0,$$

for some finite subset S' of S. We will prove that $v \in C(S_w)$ by induction on the cardinality of S'. Since $S \cap \mathrm{Ker}(w) \subseteq S_w$, we may assume without loss of generality that $S' \cap \mathrm{Ker}(w) = \emptyset$, i.e., that $\langle w, s \rangle \neq 0$ for all $s \in S'$.

If S' is empty there is nothing to prove. Otherwise,

$$0 = \langle w, v \rangle = \sum_{s \in S'} a_s \langle w, s \rangle,$$

and since each $a_s > 0$, there must exist $s, t \in S'$ with $\langle w, s \rangle > 0$ and $\langle w, t \rangle < 0$. Then

$$e := i_w(s \wedge t) = \langle w, s \rangle t - \langle w, t \rangle s \in S_w.$$

Suppose that $a_s \langle w, s \rangle + a_t \langle w, t \rangle \geq 0$. Then

$$a'_s = \langle w, s \rangle^{-1} (a_s \langle w, s \rangle + a_t \langle w, t \rangle) \geq 0, \quad \text{and}$$

$$v' := v - a_t \langle w, s \rangle^{-1} e = a'_s s + \sum_{s'' \in S''} a_{s''} s'',$$

where $S'' := S \setminus \{s, t\}$. Thus $v' \in C(S) \cap \mathrm{Ker}(w)$, and by the induction hypothesis, $v' \in C(S_w)$. Since $v = v' + a_t \langle w, s \rangle^{-1} e$, we conclude that v also belongs to $C(S_w)$, as desired.

If on the other hand, $a_s \langle w, s \rangle + a_t \langle w, t \rangle \leq 0$, let

$$a'_t = \langle w, t \rangle^{-1} (a_s \langle w, s \rangle + a_t \langle w, t \rangle) \geq 0 \quad \text{and}$$

$$v' := v + a_s\langle w,t\rangle^{-1}e = a'_t t + \sum_{s''\in S''} a_{s''}s''.$$

Then the same argument shows that v' belongs to $C(S_w)$ and hence so does v.

If $K \to K'$ is an extension of ordered fields, then, with the notations of the lemma, the same argument with V' in place of V shows that $C_{K'}(S_{w'}) = C_{K'}(S) \cap \mathrm{Ker}(w')$. Since it follows from the construction that $S_{w'}$ is just (the image in V' of) S_w, in fact $C_{K'}(S_w) = C_{K'}(S) \cap \mathrm{Ker}(w')$, as claimed. □

Proposition 2.3.11. *Let $g\colon V' \to V$ be a homomorphism of K-vector spaces such that* $\mathrm{Ker}(g)$ *is finite dimensional.*

1. *If C is a finitely generated K-cone in V, then $g^{-1}(C)$ is a finitely generated K-cone in V'.*
2. *If C is a K-cone in V and $K \to K'$ is an extension of ordered fields, then $g^{-1}(C)$ generates the K'-cone $(\mathrm{id}_{K'} \otimes g)^{-1}(C_{K'}(C))$.*
3. *An exact subcone of a finitely generated K-cone is finitely generated.*

Proof It is clear that $g^{-1}(C)$ is a K-cone in V'. To prove statements (1) and (2), we may factor g as the composition of a surjection (with finite dimensional kernel) and an injection, and it suffices to treat each of these separately.

Assume first that g is injective. Replacing V by the span of C and $g(V')$, we may assume that the cokernel of g is finite dimensional. Then we can factor g into a finite composition of injections each of whose cokernels is one-dimensional, and it suffices to treat each of these. Thus we are reduced to the case in which V' is a subspace of V of codimension one. In this case there exists a linear map $w\colon V \to K$ whose kernel is V'. Then if S is a finite set of generators for C, Lemma 2.3.10 tells us that S_w is a finite set of generators for $g^{-1}(C)$. If $K \to K'$ is a field extension, the second part of the same lemma shows, without assuming that S is finite, that the K'-cone spanned by $g^{-1}(C)$ is $(\mathrm{id}_{K'} \otimes g)^{-1}(C_K(S))$.

Now suppose that g is surjective with finite dimensional kernel. Arguing as before, we may assume that the kernel has dimension one. For each s in a set S of generators for C, choose $s' \in V'$ with $g(s') = s$ and let $S' := \{s' : s \in S\}$. Let t be a nonzero element of the kernel of g, and let $T := S' \cup \{t, -t\}$. Note that T is finite if S is finite. Since $T \subseteq g^{-1}(C)$, it follows that $C(T) \subseteq g^{-1}(C)$. Thus, to complete the proof of statements (1) and (2), it will suffice to show that $g'^{-1}(C_{K'}(C)) = C_{K'}(T)$, where $g' := \mathrm{id} \otimes_{K'} g$. If $v' \in g'^{-1}(C_K(C))$, then $g'(v') = \sum a'_s s$ with $a'_s \in K'^{\geq 0}$. Let $v'' := \sum a'_s s' \in C_{K'}(S') \subseteq C_{K'}(T)$. Since $g'(v' - v'') = 0$, we can find some $a' \in K'$ with $v' - v'' = a't$. If $a' \geq 0$, we have $v' = v'' + a't \in C_{K'}(T)$, and if $a' \leq 0$ we have $v' = v'' + (-a')(-t) \in C(T)$. Thus in either case $v' \in C_{K'}(T)$, as required.

If C is a finitely generated K-cone, then C^{gp} is a finite dimensional K-vector space, and if C' is an exact subcone of C, then $C' = V' \cap C$, where $V' = C'^{\mathrm{gp}} \subseteq V$. It follows that C' is also finitely generated. □

The following statement summarizes the major duality results for cones that we will need.

Theorem 2.3.12. *Let V be a finite dimensional K-vector space and let $W := \mathrm{Hom}_K(V, K)$ be its dual. If C is a finitely generated K-cone contained in V, let*

$$C^\vee := \{w \colon V \to K : w(c) \geq 0 \quad \text{for all } c \in C\}$$
$$C^\perp := \{w \colon V \to K : w(c) = 0 \quad \text{for all } c \in C\}.$$

1. *The K-cone $C^\vee$ is finitely generated. If $K \to K'$ is an extension of ordered fields, then*

$$C_{K'}(C^\vee) := (C_{K'})^\vee = \{w' \colon V \to K' : w'(c) \geq 0 \quad \text{for all } c \in C\}.$$

2. *For every face F of C, there is an element w of $C^\vee$ such that*

$$w^{-1}(0) \cap C = F.$$

The evaluation mapping from V to its double dual induces an isomorphism $C \to (C^\vee)^\vee$.

3. *An element v of V belongs to C (resp. C^*) if and only if $w(v) \geq 0$ (resp. $= 0$) for all $w \in C^\vee$.*
4. *The map $F \mapsto C^\vee \cap F^\perp = (C_F)^\vee$ induces an order-reversing bijection from the set of faces of C to the set of faces of $C^\vee$.*
5. *If F is a face of C and $G := C^\vee \cap F^\perp$, then $F^\vee \subseteq W$ is the localization $C^\vee_G$ of $C^\vee$ by G, and $G^\vee$ is the localization of C by F.*
6. *C is sharp if and only if $C^\vee$ spans* $\mathrm{Hom}(V, K)$.

Proof Let S be a finite set of generators for C and consider the homomorphism $g \colon W \to K^S$ sending w to the function $s \mapsto w(s)$. Then $C^\vee$ is just $g^{-1}((K^{\geq})^S)$, and, since $(K^{\geq})^S$ is a finitely generated K-cone in K^S, statement (1) of Proposition 2.3.11 implies that $C^\vee$ is finitely generated. Moreover $C^\vee_{K'} = (\mathrm{id} \otimes g^{-1})((K'^{\geq})^S)$, and statement (2) of Proposition 2.3.11 implies that this K'-cone is spanned by $C^\vee$.

Statements (2) through (4) are proved by the arguments used in the proofs of the analogous statements for monoids, and we do not repeat them here; see Corollary 2.2.4. For (5), observe first that $G^{\mathrm{gp}} \subseteq F^\vee$, so the natural injection $C^\vee \to F^\vee$ factors naturally through the localization of $C^\vee$ by G. To prove that the resulting (injective) map is surjective, let w be an element of $F^\vee$. By (2), there is some $\phi \in C^\vee$ such that $\phi^{-1}(0) \cap C = F$; in particular $\phi \in G$. Let S

be a finite set of generators for C. If $s \in S \setminus F$, $\phi(s) > 0$, and if $s \in S \cap F$, $w(s) \geq 0$. Thus there exists an $n > 0$ such that $w(s) + n\phi(s) \geq 0$ for all $s \in S$. Then $w + n\phi \in C^\vee$ and $\phi \in G$, so $w \in (C^\vee)_G$. The second part of (5) then follows from (4). For (6), note that evidently $C^\vee \subseteq (C^*)^\perp$, so if $C^\vee$ spans W then $(C^*)^\perp = W$ and hence $C^* = 0$. On the other hand, if $C^\vee$ is contained in a proper subspace W' of W, then $W'^\perp$ is a nonzero subgroup of $C = (C^\vee)^\vee$, so C is not sharp. □

Remark 2.3.13. We can now explain the proof of the general case of (4) of Proposition 2.3.6. Let $V := \mathbf{Q} \otimes Q^{\mathrm{gp}}$ and let C be the cone of V generated by Q. Then (1) of Theorem 2.3.12, applied to the extension $\mathbf{Q} \to K$, implies that the K-cone $\mathrm{Hom}(C, K^{\geq})$ is spanned by $\mathrm{Hom}(C, \mathbf{Q}^{\geq})$ and hence also by $\mathrm{Hom}(Q, \mathbf{N})$.

Corollary 2.3.14. *Let F be a face of a finitely generated cone C (resp. a fine monoid Q) and let $\mathcal{F}_F$ denote the set of facets of C (resp. of Q) containing F. Then*

1. $F = \cap\{G : G \in \mathcal{F}_F\}$,
2. $F^{\mathrm{gp}} = \cap\{G^{\mathrm{gp}} : G \in \mathcal{F}_F\}$, *and*
3. $C_F = \cap\{C_G : G \in \mathcal{F}_F\}$, *(resp.* $Q_F = \cap\{Q_G : G \in \mathcal{F}_F\}$*).*

(The improper face $F = C$ (resp. Q) is viewed as the intersection of the empty set of facets.)

Proof Proposition 2.3.7 shows that the statements for monoids follow from the statements for cones. Moreover, the second statement implies the first, since F is an exact submonoid of C. The homomorphism $C \to C/F$ induces a bijection between the facets of C containing F and the facets of C/F. Thus, replacing C by C/F, we reduce to the case in which C is sharp and $F = 0$. Without loss of generality we may assume that $C \neq \{0\}$. Let $V := C^{\mathrm{gp}}$ and $W := \mathrm{Hom}_K(V, K)$. Then, by (6) of the previous theorem, the cone $C^\vee \subseteq W$ is sharp and nonzero, and hence by Proposition 2.3.4 is generated by a nonempty finite set of indecomposable elements $w_1, \ldots, w_n$. Each $w_i^\perp \cap C$ is a facet G_i of C, and hence if x lies in G^{gp} for every facet G, then $w_i(x) = 0$ for every i. Since C is sharp, $C^\vee$ spans W, hence $w(x) = 0$ for every w, hence $x = 0$. For (3), we can replace C by C_F to reduce to the case in which $F = C^*$. Arguing as before, we see that $C^\vee$ is sharp, hence is generated by the set T of its indecomposable elements. By (3) of Theorem 2.3.12, $C = \cap\{C_t : t \in T\}$, where $C_t := \{v \in V : \langle t, v\rangle \geq 0\}$. By (5) of the theorem, each C_t is the localization of C by $\langle t\rangle^\perp \cap C$, a facet of C. □

The following application is a monoidal analog of the Hauptidealsatz.

Corollary 2.3.15. *If Q is a fine monoid and q is an element of Q, then every prime ideal of Q that is minimal among the primes containing q has height one. In particular, $\sqrt{(q)}$ is a finite intersection of height one primes.*

Proof Applying Corollary 2.3.14 to the complementary faces of prime ideals, we see that every prime ideal $\mathfrak{p}$ of Q is the union of the set of height one primes contained in $\mathfrak{p}$. Suppose $\mathfrak{p}$ is minimal among the set of primes containing q and write $\mathfrak{p}$ as the union $\mathfrak{p}_1 \cup \cdots \cup \mathfrak{p}_k$, where each $\mathfrak{p}_i$ has height one. Then $q \in \mathfrak{p}_i$ for some i, and, since $\mathfrak{p}$ is minimal with this property and $\mathfrak{p}_i \subseteq \mathfrak{p}$, in fact $\mathfrak{p} = \mathfrak{p}_i$. By Corollary 1.4.3, $\sqrt{(q)}$ is the intersection of the prime ideals containing q, and, since this set is finite, it is the intersection of a finite set of minimal such elements. As we have just seen, each of these has height one. Perhaps we should remark that the set of such primes is not empty if and only if $q \in Q^+$; if on the other hand $q \in Q^*$, the ideal it generates is Q, which is the intersection of the empty set of prime ideals of height one. □

Corollary 2.3.16. *Let C_1 and C_2 be cones in a finite dimensional K-vector space V.*

1. *$(C_1 + C_2)^\vee = C_1^\vee \cap C_2^\vee$.*
2. *If C_1 and C_2 are finitely generated, then $C_1 \cap C_2$ and $C_1 + C_2$ are finitely generated, and $C_1^\vee + C_2^\vee = (C_1 \cap C_2)^\vee$.*

Proof The first statement is obvious. If C_1 and C_2 are finitely generated, it is also obvious that $C_1 + C_2$ is finitely generated. Furthermore, it follows from Theorem 2.3.12 that $C_1^\vee$ and $C_2^\vee$ are finitely generated, and hence that $C_1^\vee + C_2^\vee$ is finitely generated. Now Theorem 2.3.12 tells us that $(C_1^\vee + C_2^\vee)^\vee$ is finitely generated, and by (1) this is $(C_1^\vee)^\vee \cap (C_2^\vee)^\vee$, which is $C_1 \cap C_2$, by another application of Theorem 2.3.12. Thus $C_1 \cap C_2$ is finitely generated. Now let us apply (1) to the duals of C_1 and C_2 to obtain the equality $(C_1^\vee + C_2^\vee)^\vee = C_1 \cap C_2$. The dual of this equality then gives the equation in (2). □

Corollary 2.3.17. *Let C be a finitely generated cone in a finite dimensional K-vector space V. Endow V with the topology induced from the ordered field K.*

1. *The cone C is closed in V.*
2. *Every face of C is closed in C.*
3. *The interior I_C of C (the complement in C of the union of its proper faces) is an open dense subset of C.*

Proof The inclusion $C \to V$ factors through an inclusion $C^{\text{gp}} \to V$, and, since C^{gp} and V are finite dimensional, C^{gp} is closed in V. Thus we may as well assume that $V = C^{\mathrm{gp}}$. Each $\phi \in C^\vee$ is a linear and hence continuous map $V \to K$, and hence each $\phi^{-1}(K^{\geq 0})$ is closed in V. By (2) of Theorem 2.3.12, C is the intersection of these closed sets, and hence is closed. Furthermore each face F is the intersection of C with a linear subspace of V, hence is also closed, and since C has only a finite number of proper faces, their union is closed. Thus the complement I_C of this union is open. To prove that I_C is dense, let S be a finite set of generators for C. An element c of C can be written as a sum $\sum a_s$ with $a_s \in K^{\geq}$, and c lies in the interior of C if no a_s vanishes. In any case $c_i := \sum(a_s + i^{-1})s$ lies in the interior, and the sequence $(c_1, c_2, \ldots,)$ converges to c. □

Corollary 2.3.18. *Let C be a finitely generated sharp K-cone and let $h\colon C \to K^{\geq}$ be a local homomorphism. Then, for each $a \in K^{\geq}$,*

$$B_a := \{c \in C : h(c) \leq a\}$$

is a closed and bounded subset of C^{gp}. If $K = \mathbf{R}$, then h is a proper map.

Proof Let $V := C^{\mathrm{gp}}$, a finite dimensional K-vector space, let W be its dual, and let $(w_1, \ldots, w_d)$ be a basis for W. Then a subset B of V is bounded if and only if there exists a constant M such that $|w_i(b)| \leq M$ for all i and all $b \in B$. This condition is independent of the basis. In fact, if $(h_1, \ldots h_m)$ is any finite sequence in W spanning W, then B is bounded if and only if there exists a constant M such that $|h_i(b)| \leq M$ for all i and all $b \in B$. In the situation of corollary, C is finitely generated and sharp, so its dual $C^\vee$ is finitely generated and spans W. Furthermore, since h is local, it belongs to the interior of $C^\vee$. Thus if $(h_1, \ldots, h_m)$ is a finite set of generators for $C^\vee$, each h_i belongs to $\langle h \rangle$, and so there exists an $n_i > 0$ such that $h_i \leq n_i h$ for all i. Replacing h_i by h_i/n_i, we see that $0 \leq h_i(c) \leq h(c) \leq a$ if $c \in B_a$, so B_a is bounded. Since C is finitely generated it is closed in V and, since h is continuous, B_a is closed in C and hence B_a is also closed. If $K = \mathbf{R}$, we see that B_a is a closed and bounded subset of a Euclidean space, hence compact. It follows that the inverse image by h of every compact subset of $\mathbf{R}$ is compact, so h is proper. □

The following is another useful finiteness result, whose proof makes use of the topological properties of the real numbers.

Theorem 2.3.19 (Gordon). *Let L be a finitely generated abelian group, let K be a subfield of $\mathbf{R}$, and let C be the K-cone in $V := K \otimes L$ generated by a finite subset S of L. Then*

$$L(C) := L \times_V C = L \times_{V_{\mathbf{R}}} C_{\mathbf{R}}$$

is a finitely generated monoid.

Proof Let us first treat the case in which L is free, so that it may be identified with its image in V. Since C is a finitely generated K-cone in V, Theorem 2.3.12 implies that its dual is also finitely generated and that there is a finite subset T of $\mathrm{Hom}(V, K)$ such that

$$C = \{v \in V : \langle t, v\rangle \geq 0 \text{ for all } t \in T\}.$$

Then $C_{\mathbf{R}} = \{v \in V_{\mathbf{R}} : \langle t, v\rangle \geq 0 \text{ for all } t \in T\}$ is a finitely generated $\mathbf{R}$-cone and $C_{\mathbf{R}} \cap V = C$. It follows that $L \times_V C = L \times_{V_{\mathbf{R}}} C_{\mathbf{R}}$ and so the proof is reduced to the case when $K = \mathbf{R}$.

Let $S' \subseteq V$ be the set of all linear combinations of elements of S with coefficients in the interval $[0, 1]$. The map $[0, 1]^S \to V$ sending $\{a_s : s \in S\}$ to $\sum a_s s$ is continuous and maps surjectively to S'; hence S' is compact. Then $S'' := L \cap S'$ is compact and discrete, hence finite; furthermore it contains S. Any element x of C can be written as a sum $\sum a_s s$ with $s \in S$ and $a_s \in \mathbf{R}^{\geq 0}$, and a_s can be written $a_s = m_s + a'_s$ with $m_s \in \mathbf{N}$ and $a'_s \in [0, 1)$. Then $x = \sum m_s s + s'$ with $s' \in S'$; if also $x \in L$, in fact $s' \in S''$ and so x is a sum of elements of S''. Thus the monoid $L(C) = L \cap C$ is generated by the finite set S''. For the general case, let L_t be the torsion subgroup of L and let $L_f := L/L_t$. Notice that $L_t \subseteq L(C)^*$, and the natural map $L(C) \to L$ identifies $L(C)/L_t$ with $L_f(C) = L_f \cap C$ and $L(C)^*/L_t$ with L^*_f. Since $L_f(C)$ is a fine monoid, it follows from Proposition 2.1.1 that $L_f(C)^*$ is a finitely generated group, and, since L_t is finitely generated, so is $L(C)^*$. Now Proposition 2.1.1 implies that $L(C)$ is a finitely generated monoid. □

Recall that a subset S of a vector space V is said to be *convex* if $\sum \lambda_i s_i \in S$ whenever $(s_1, \ldots, s_n) \in S^n$ and $(\lambda_1, \ldots, \lambda_n) \in (K^{\geq})^n$ with $\sum \lambda_i = 1$. For example, if C is a cone in V and $h\colon V \to K$ is a linear map, then $C \cap h^{-1}(a)$ (the h-slice of C at a) is convex for every $a \in K$. Conversely, if S is any convex subset of V, consider the cone C in $V \times K$ spanned by $S \times \{1\}$, and let $h\colon C \to K$ be the inclusion followed by the natural projection. Then S identifies with $S \times \{1\} = C \cap h^{-1}(1)$, the h-slice of C at 1. In this way one can often reduce questions about convex sets to questions about cones. Some care is required; for example, if C_1 and C_2 are cones in V, then the convex hull of $(C_1 \cap h^{-1}(1)) \cup (C_2 \cap h^{-1}(1))$ is contained in $(C_1 + C_2) \cap h^{-1}(1)$, but the containment may be strict.

The following result plays a crucial role in the analysis of morphisms of monoids and cones. In particular, it is used in the proof of the main structure theorem (Theorem 4.7.2) for exact morphisms and also to show that log blowups satisfy a strong surjectivity property (Proposition II.1.7.7). It asserts

Figure 2.3.1 Vertices of a convex set

the existence of "vertices" of certain convex sets that are cut out by suitable systems of inequalities.

Let V be a finite dimensional K-vector space and let W be its dual. If $T := \{(w_t, a_t)\}$ is a subset of $W \times K$, let

$$\begin{aligned} S(T) &:= \{v \in V : \langle v, w_t \rangle + a_t \geq 0 \text{ for all } t \in T\} \\ Q(T) &:= \{v \in V : \langle v, w_t \rangle \geq 0 \text{ for all } t \in T\} \qquad (2.3.1) \\ S(T)^\vee &:= \{(w, a) \in W \times K : \langle v, w \rangle + a \geq 0 \text{ for all } v \in S(T)\}. \end{aligned}$$

Then $S(T)$ is a convex subset of V and $Q(T)$ is a cone in V. Note that $S(T)$ depends only on the cone generated by T and that if $q \in Q(T)$ and $v \in S(T)$, then $q + v \in S(T)$. Thus $S(T)$ is a $Q(T)$-set. An element v of $S(T)$ is called a *vertex* if there exists a $(w, a) \in S(T)^\vee$ such that v is the unique element of $S(T)$ such that $\langle v, w \rangle + a = 0$. An element of $S(T)^\vee$ with this property is sometimes called a *supporting equation* of v. See Figure 2.3.1.

Theorem 2.3.20. *Let $T := \{(w_t, a_t) : t \in T\}$ be a finite subset of $W \times K$, or a finitely generated cone in $W \times K$. Assume that $\{w_t : t \in T\}$ spans W and that $S(T)$ is not empty. Then $S(T)$ has at least one vertex. Any such vertex is minimal with respect to the action of $Q(T)$ on $S(T)$.*

Proof Our approach will be to consider the cone spanned by $S(T) \times \{1\}$ in $V \times K$; the edges of this cone will help us find the vertices of $S(T)$. Note that the pairing

$$(V \times K) \times (W \times K) \to K : \langle (v, a), (w, b) \rangle := \langle v, w \rangle + ab$$

identifies $W \times K$ with the dual of $V \times K$.

Lemma 2.3.21. *With the notations (2.3.1) above, let*

$$Q^*(T) := \{(v,r) \in V \times K^{>0} : \langle v, w_t \rangle + ra_t \geq 0 \text{ for all } t \in T\} \cup \{(0,0)\}.$$

Then the following statements hold.

1. $Q^*(T)$ *is the cone in* $V \times K$ *generated by* $S(T) \times \{1\}$, *and*
$$S(T) \times \{1\} = Q^*(T) \cap (V \times \{1\}).$$
2. $Q'(T) := Q^*(T) \cup (Q(T) \times \{0\})$ *is the closure of* $Q^*(T)$.
3. $Q'(T)$ *is the cone in* $V \times K$ *dual to* $T' := T \cup \{(0,1)\} \subseteq W \times K$.
4. $S(T)^\vee$ *is the dual of the cone* $Q'(T)$.

Proof It is clear that $Q^*(T)$ is closed under addition and under multiplication by elements of $K^{\geq 0}$, i.e., is a cone in $V \times K$, and that it contains $S(T) \times \{1\}$. Suppose $(v,r) \in Q^*(T)$ and is not equal to $(0,0)$. Then by the definition of $Q^*(T)$, $r \neq 0$ and $(v/r, 1) \in S(T)\times\{1\}$. Thus (v,r) belongs to the cone generated by $S(T)\times\{1\}$. The remainder of statement (1) is immediate from the definition.

To prove (2), we must show that that $Q(T) \times \{0\}$ is contained in the closure of $Q^*(T)$. Suppose $v \in Q(T)$. Choose $v' \in S(T)$ and note that for every $t \in T$ and every $r > 0$,

$$\langle v + rv', w_t \rangle + ra_t = \langle v, w_t \rangle + r(\langle v', w_t \rangle + a_t) \geq 0.$$

Thus $(v + rv', r) \in Q^*(T)$ for all $r > 0$, and it follows that $(v, 0)$ belongs to the closure of $Q^*(T)$.

If $(v,r) \in V \times K$ and $t \in T$, then

$$\langle (v,r), t \rangle = \langle v, w_t \rangle + ra_t, \text{ and} \tag{2.3.1a}$$
$$\langle (v,r), (0,1) \rangle = r \tag{2.3.1b}$$

If $(v,r) \in Q^*(T)$, both these expressions are nonnegative, and it follows that $Q^*(T)$ is contained in the dual $T'^\vee$ of T'. Since $T'^\vee$ is closed, it follows from (2) that $Q(T)\times\{0\}$ is also contained in $T'^\vee$ and hence that $Q'(T) \subseteq T'^\vee$. Conversely, suppose that $(v,r) \in T'^\vee$, so that the expressions (2.3.1a) and (2.3.1b) above are nonnegative. Then (2.3.1b) implies that $r \geq 0$. If $r > 0$, (2.3.1a) implies that $(v,r) \in Q^*(T)$, and if $r = 0$, it implies that $v \in Q(T)$. Thus $T^\vee \subseteq Q'(T)$, completing the proof of (3).

Since $Q'(T)$ is the closure of the cone generated by $S(T) \times \{1\}$, its dual is the set of (w,r) such that $\langle (v,1), (w,r) \rangle \geq 0$ for all $v \in S(T)$, which is just the definition of $S(T)^\vee$. □

Let us return to the proof of the proposition. If T is a finitely generated cone, we may replace it by a finite set of generators for the sake of the argument.

Since $\{w_t : t \in T\}$ spans W, the finite set T' spans $W \times K$, and it follows that the dual cone $Q'(T)$ is sharp and finitely generated, hence generated by its indecomposable elements. Since $S(T)$ is not empty, $Q'(T)$ is not contained in $W \times \{0\}$. Thus it contains at least one edge E not contained in $W \times \{0\}$, and E contains a unique point of the form $(v, 1)$. Then $v \in S(T)$, and we claim that it is a vertex. Since E is a face of the finitely generated cone $Q'(T)$, there is an element $h = (w, a)$ in its dual $S(T)^\vee$ such that $h^\perp = E$. Then $\langle w, v\rangle + a = 0$, and if v' is another such point, then $(v', 1) \in E$ and hence $v' = v$. This shows that v is a vertex of $S(T)$.

It remains only to prove that every vertex v of $S(T)$ is $Q(T)$-minimal. Let $(w, a) \in S(T)^\vee$ be a supporting equation for v, so that v is the unique element of $S(T)$ with $\langle v, w\rangle + a = 0$. Now $q + v \in S(T)$ if $q \in Q(T)$, hence

$$\langle q, w\rangle = \langle q, w\rangle + \langle v, w\rangle + a = \langle q + v, w\rangle + a \geq 0.$$

This shows that w is in the dual of the cone $Q(T)$. But now if $v = q + v'$ with $q \in Q(T)$ and $v' \in S(T)$,

$$0 = \langle q + v', w\rangle + a = \langle q, w\rangle + \langle v', w\rangle + a \geq \langle v', w\rangle + a \geq 0.$$

Then $\langle v', w\rangle + a = 0$ and, since v is a vertex, it follows that $v' = v$, as required. □

We should point out that in Theorem 2.3.20, the set of vertices of $S(T)$ is nonempty and finite, but it does not necessarily generate $S(T)$ as a $Q(T)$-set. Indeed, the cone $Q(T)$ can be the zero cone, even when $S(T)$ is infinite.

Here is an application of Theorem 2.3.20, which will be used in the proof of the surjectivity result for log blowups alluded to earlier. Its goal is to establish the existence and location of vertices of the convex hull of an ideal in a fine monoid. Note that although such an ideal is necessarily finitely generated as an ideal, its convex hull will not be finitely generated as a convex set.

Proposition 2.3.22. *Let Q be a fine sharp monoid, let I be an ideal of Q, with generators $(s_1, \ldots, s_n)$, and let S be the convex hull of I in $V := K \otimes Q^{gp}$. Then S is an ideal of $C(Q)$, and the set of vertices of S is a nonempty subset of $\{s_1, \ldots, s_n\}$.*

Proof To prove that S is an ideal of $C(Q)$, we first check that $q + s \in S$ if $q \in Q$ and $s \in S$. Write $s = \sum \lambda_i s_i$, where $s_i \in I$ and $0 \leq \lambda_i$ and $\sum \lambda_i = 1$. Then $q + s = \sum \lambda_i(q + s_i) \in S$. It follows that $nq + s \in S$ for every natural number n. Now, if $a \in K^>$, there is a natural number n such that $n \leq a < n + 1$, so $aq + s = bq + t$, where $0 \leq b < 1$ and $t := nq + s \in S$. Then $bq + t = b(q + t) + (1 - b)t \in S$, and thus S is invariant under translation by elements of the form aq. Since every element of $C(Q)$ can be written as a sum $\sum a_i q_i$ with $a_i \in K$ and q_i in Q, it follows that S is invariant under translation by all such elements.

To find the vertices of S, our strategy will be to show that there exists a finitely generated cone $T = \{(w_t, a_t)\}$ in $W \times K$ such that, with the notation of (2.3.1), $S(T) = S$ and then to apply Theorem 2.3.20. Let $(w_1, \dots, w_m)$ be a finite set of generators for the dual cone $C(Q)^\vee \subseteq W$, which is finitely generated by Theorem 2.3.12. For $i = 1, \dots, m$, let $t_i := (w_i, 0) \in W \times K$ and, for each such i and for $j = 1, \dots, n$, let $t_{i,j} := (w_i, -\langle s_j, w_i\rangle)$. Let T_j be the cone in $W \times K$ spanned by $\{t_i, t_{i,j}\}$ for $i = 1, \dots, m$.

Claim 2.3.23. An element (v, λ) of $V \times K$ belongs to the dual of T_j if and only if v belongs to $C(Q) \cap \left(\lambda s_j + C(Q)\right)$.

To check this, suppose that $v = q + \lambda s_j$ for some $q \in C(Q)$ and that v also belongs to $C(Q)$. Then $\langle (v, \lambda), t_i\rangle = \langle v, w_i\rangle \geq 0$ because $v \in C(Q)$ and $w_i \in C(Q)^\vee$. Furthermore, $\langle (v, \lambda), t_{ij}\rangle = \langle v, w_i\rangle - \lambda\langle s_j, w_i\rangle = \langle q, w_i\rangle \geq 0$. Suppose conversely that $(v, \lambda) \in T_j^\vee$. Then the positivity of the pairings with the elements t_i implies that $v \in C(Q)$, and the positivity of the pairings with $t_{i,j}$ implies that $v - \lambda s_j \in C(Q)$, i.e., that $v \in C(Q) + \lambda s_j$.

Lemma 2.3.24. *Let $T = T_1 \cap \dots \cap T_n \subseteq W \times K$. Then T is finitely generated as a cone, and*

$$S(T) := \{v : \langle v, w_t\rangle + a_t \geq 0 \text{ for all } t \in T\}$$

is the convex hull S of I in V. Furthermore, $\{w_t : t \in T\}$ spans W.

Proof The finite generation of T follows from Corollary 2.3.16, which also implies that $T^\vee = T_1^\vee + \dots + T_n^\vee$. Note that $S(T) = \{v : (v, 1) \in T^\vee\}$. In particular, it follows from Claim 2.3.23 that $(s_j, 1) \in T_j^\vee \subseteq T^\vee$ for all j, hence that each $s_j \in S(T)$ and that $S \subseteq S(T)$. On the other hand, if $v \in S(T)$, then $(v, 1) \in T^\vee$ can be written

$$(v, 1) = (v_1, \lambda_1) + \dots + (v_n, \lambda_n), \qquad \text{where } (v_j, \lambda_j) \in T_j^\vee.$$

Let us arrange the indices so that $\lambda_j > 0$ if $j \leq k$ and $\lambda_j \leq 0$ if $j > k$. Since $\lambda_1 + \dots + \lambda_n = 1$, necessarily $k \geq 1$ and $m := \lambda_1 + \dots + \lambda_k \geq 1$. Let $\lambda'_j := \lambda_j/m$ and $\mu_j := \lambda_j - \lambda'_j$. By Claim 2.3.23, we can write $v_j = q_j + \lambda_j s_j$, with $q_j \in C(Q)$ and $s_j \in S$. Now let

$$q'_1 := q_1 + \mu_1 s_1 + v_{k+1} + \dots + v_n \in C(Q)$$
$$s'_1 := \lambda_1'^{-1} q'_1 + s_1 \in S,$$

and, for $j = 2, \cdots, k$, let

$$q'_1 := q_j + \mu_j s_j \in C(Q)$$
$$s'_j := \lambda'^{-1}_j q'_j + s_j \in S.$$

Then

$$\begin{aligned} \lambda'_1 s'_1 + \cdots + \lambda'_k s'_k &= q'_1 + \lambda'_1 s_1 + \cdots + q'_k + \lambda'_k s_k \\ &= q_1 + \lambda_1 s_1 + \cdots + q_k + \lambda_k s_k + v_{k+1} + \cdots + v_n \\ &= v. \end{aligned}$$

Thus v belongs to S, showing that $S(T) \subseteq S$. By construction $\{w_t : t \in T\}$ contains $Q^\vee$ and, since Q is sharp, $C(Q)^\vee$ spans W. □

Thanks to Lemma 2.3.24, we can apply Theorem 2.3.20 to see the existence of a vertex of $S = S(T)$. It remains to prove that the vertices of $S(T)$ lie among the generators $\{s_1, \ldots, s_n\}$. First suppose that $n = 1$. Then S is the ideal of $C(Q)$ generated by $s := s_1$. Let w be a local homomorphism $Q \to K^{\geq}$ and let $a := -\langle s, w\rangle$. Then, for every $q \in C(Q)$, $\langle q + s, w\rangle + a = \langle q, w\rangle \geq 0$, with equality if and only if $q = 0$. Thus s is a vertex of S. If s' is another vertex, with supporting equation (w', a'), write $s' = q + s$ with $q \in C(Q)$. Then

$$0 = \langle s', w'\rangle + a' = \langle q, w'\rangle + \langle s, w'\rangle + a',$$

where $\langle s, w'\rangle + a \geq 0$. If $q \neq 0$, the uniqueness of s' implies that $\langle s, w'\rangle + a' > 0$, and it follows that $\langle q, w'\rangle < 0$. But then $s'' := q + s' \in S$ and $\langle s'', w'\rangle + a' < 0$, a contradiction.

Since the ideal generated by $\{s_1, \ldots, s_n\}$ is the union of the ideals generated by each generator, the general case will follow from the following lemma.

Lemma 2.3.25. *If S_1 and S_2 are convex sets, then every vertex of the convex hull S of $S_1 \cup S_2$ is either a vertex of S_1 or a vertex of S_2.*

Proof Suppose that s is a vertex of S, and write $s = \lambda_1 s_1 + \lambda_2 s_2$ with $s_i \in S_i$ and $\lambda_1 + \lambda_2 = 1$, $\lambda_i \geq 0$. Let (w, a) be a supporting equation for s. Since $S_i \subseteq S$, $\langle s_i, w\rangle + a \geq 0$ for each i, and since

$$0 = \langle s, w\rangle + a = \lambda_1(\langle s_1, w\rangle + a) + \lambda_2(\langle s_2, w\rangle + a),$$

it follows that each $\lambda_i(\langle s_i, w\rangle + a) = 0$. Thus either λ_1 or λ_2 vanishes, and so s belongs either to S_2 or to S_1. In either case it is a vertex of the set to which it belongs. □

This lemma concludes the proof of Proposition 2.3.22. □

2.4 Valuative monoids and valuations

Recall that an integral monoid Q is said to be *valuative* if, for every $x \in Q^{gp}$, either x or $-x$ lies in Q. For example, $\mathbf{N}$, $\mathbf{Z}$, and $(\mathbf{N} \times \mathbf{N}) \cup (\mathbf{Z} \times \mathbf{N}^+)$ are valuative; the last of these is not finitely generated. In general, the monoid $\Gamma_\geq$ of nonnegative elements in any totally ordered group Γ is a valuative monoid, and, conversely if Q is a valuative monoid, the induced order relation (see Definition 1.2.3) on Q^{gp} is a total preordering. If L is an abelian group and $\phi\colon L \to \Gamma$ is a homomorphism, then $\phi^{-1}(\Gamma_\geq)$ is a valuative monoid.

If L is an abelian group and Q_1, Q_2 are submonoids of L, one says that Q_2 *dominates* Q_1 if $Q_1 \subseteq Q_2$ and the inclusion map is a local homomorphism. This defines an ordering on the set of submonoids of L.

The following existence theorem for valuative monoids is a direct analog of the well-known result for commutative rings; however, the existence of fine valuative monoids here is considerably easier than the analog for rings.

Proposition 2.4.1. *Let L be an abelian group. Then a submonoid Q of L is a maximal element for the order relation of domination if and only if it is valuative and $Q^{gp} = L$. Every submonoid of L is dominated by such a monoid. If L is finitely generated, then every finitely generated submonoid of L is dominated by a finitely generated valuative monoid Q with $Q^{gp} = L$.*

Proof Suppose first that Q is a submonoid of L and is maximal under domination. Then $Q \subseteq Q^{sat} \subset L$ and, by (4) of Proposition 1.3.5, the homomorphism $Q \to Q^{sat}$ is local. Thus Q^{sat} dominates Q, and hence $Q = Q^{sat}$. Now to prove that Q is valuative, suppose $x \in L$, and consider the submonoid P of L generated by Q and x. If P dominates Q, then $P = Q$ and of course $x \in Q$. If not, there exist $q \in Q$ and $p \in P$ with $q + p = 0$. Write $p = q' + nx$ with $n > 0$. Then $q + q' + nx = 0$. This shows that $-nx \in Q$, and hence that $-x \in Q^{sat} = Q$, as required. Since x was an arbitrary element of L, it follows also that $Q^{gp} = L$.

Conversely, suppose that Q is valuative and that $Q^{gp} = L$. If P is a submonoid of L dominating Q and $p \in P$, then either p or $-p$ belongs to Q. But if $-p$ belongs to Q, then since $-p$ becomes a unit in P it is already a unit in Q, so p belongs to Q in any case. This shows that $Q = P$, as required.

Suppose P is a submonoid of L. Then the set S_P of submonoids of L that dominate P is not empty, and the union of any chain in S_P again belongs to S_P. By the Hausdorff maximality principle, S_P contains a maximal element. If L and P are finitely generated, then, by Corollary 2.1.22,

$$P^\vee := \{w \in \mathrm{Hom}(L, \mathbf{Z}) : \langle w, p \rangle \geq 0 \text{ for all } p \in P\}$$

is a finitely generated submonoid of $\mathrm{Hom}(L, \mathbf{Z})$. Let $w\colon L \to \mathbf{Z}$ be an element in the interior $I_{P^\vee}$ of $P^\vee$. Then, by Remark 2.2.8, w defines a local homomorphism $P \to \mathbf{N}$. Since $\mathbf{N}$ is valuative, $Q := w^{-1}(\mathbf{N})$ is valuative and clearly $Q^{gp} = L$. Furthermore, $Q = \{w\}^\vee$ and hence is finitely generated, by Corollary 2.1.22. Since w factors through a homomorphism $P \to Q \to \mathbf{N}$ and is local, it follows that $P \to Q$ is also local, so Q dominates P. □

Proposition 2.4.2. *Let Q be a sharp, integral, and nonzero monoid. Then the following conditions are equivalent:*

1. *Q is valuative and finitely generated.*
2. *Q is isomorphic to* $\mathbf{N}$.
3. $\dim(Q) = 1$ *and Q is saturated and finitely generated.*
4. *Q is saturated and* $Q^{gp} \cong \mathbf{Z}$.

Proof Since Q is nonzero and sharp, its maximal ideal Q^+ is not empty. If Q is finitely generated, then by Proposition 2.1.2, Q is generated by the minimal elements of Q^+. If Q is valuative, the order relation on Q is a total order and Q^+ has just one minimal element q. Thus there is a surjective homomorphism $\mathbf{N} \to Q$ and, since Q is integral and sharp, this homomorphism is an isomorphism. Thus (1) implies (2). It is obvious that (2) implies (1) and (3). If (3) holds then, by Proposition 1.4.7, Q^{gp} has rank one. Since Q is sharp and saturated, Q^{gp} is torsion free, and hence (4) holds. If (4) holds, choose an isomorphism $\phi\colon Q^{gp} \cong \mathbf{Z}$; such an isomorphism is unique up to a sign. Since ϕ is injective, $\phi(q) \neq 0$ for $q \in Q^+$; choose $q \in Q^+$ which minimizes $|\phi(q)|$ and adjust the sign of ϕ so that $n := \phi(q) > 0$. Then it follows from the sharpness of Q that ϕ maps Q to $\mathbf{N}$. If $x \in Q^{gp}$ is the element such that $\phi(x) = 1$, then $\phi(nx) = \phi(q)$, hence $nx = q \in Q$. Since Q is saturated, in fact $x \in Q$, and, by the minimality of $|\phi(x)|$, necessarily $n = 1$. Then $\phi_{|Q}\colon Q \to \mathbf{N}$ is injective and surjective, hence an isomorphism. □

Corollary 2.4.3. *Every fine saturated monoid of dimension one is isomorphic to $\Gamma \oplus \mathbf{N}$ for some finitely generated abelian group Γ. If Q is toric, $Q \cong \mathbf{Z}^r \oplus \mathbf{N}$ for some natural number r.*

Proof If Q is such a monoid, the previous result implies that there is an isomorphism $\phi\colon \overline{Q} \to \mathbf{N}$. Let q be an element of Q with $\phi(q) = 1$. Then the obvious homomorphism $Q^* \oplus \mathbf{N} \to Q$ sending 1 to q is an isomorphism. Since Q is fine, Q^* is a finitely generated abelian group and, if Q is toric, Q^* is torsion free, hence free. □

Corollary 2.4.4. *Let $\mathfrak{p}$ be a height one prime ideal in a fine monoid Q. Then*

$Q^{\mathrm{sat}}_{\mathfrak{p}}$ is valuative, and there is a unique epimorphism

$$\nu_{\mathfrak{p}} \colon Q^{\mathrm{gp}} \to \mathbf{Z}$$

such that $\nu_{\mathfrak{p}}^{-1}(\mathbf{N}^+) \cap Q = \mathfrak{p}$. Furthermore, $Q^{\mathrm{sat}}_{\mathfrak{p}} = \{x \in Q^{\mathrm{gp}} : \nu_{\mathfrak{p}}(x) \geq 0\}$.

Proof We know that that $Q^{\mathrm{gp}} \cong (Q^{\mathrm{sat}})^{\mathrm{gp}}$, that Q^{sat} is fine (Corollary 2.2.5) and that $\mathrm{Spec}(Q^{\mathrm{sat}}) \to \mathrm{Spec}(Q)$ is a homeomorphism (Corollary 2.3.8). Thus we may as well replace Q by Q^{sat}, and so we assume that Q is saturated. Then $Q_{\mathfrak{p}}$ is saturated, so $\overline{Q_{\mathfrak{p}}}^{\mathrm{gp}}$ is torsion free, and, since $\mathfrak{p}$ has height one, $\dim(\overline{Q}_{\mathfrak{p}}) = 1$. By the previous proposition, $\overline{Q}_{\mathfrak{p}}$ is valuative and hence so is $Q_{\mathfrak{p}}$. Moreover, there is a unique isomorphism $\overline{Q}_{\mathfrak{p}} \cong \mathbf{N}$, and we let $\nu_{\mathfrak{p}}$ be the composite of $Q^{\mathrm{gp}} \cong Q_{\mathfrak{p}}^{\mathrm{gp}} \to \overline{Q}_{\mathfrak{p}}^{g}$ with the induced isomorphism $\overline{Q}_{\mathfrak{p}}^{\mathrm{gp}} \to \mathbf{Z}$. It is clear that $\nu_{\mathfrak{p}}^{-1}(\mathbf{N}) = Q_{\mathfrak{p}}$ and that $\nu_{\mathfrak{p}}^{-1}(\mathbf{N}^+) = \mathfrak{p}$. The uniqueness of $\nu_{\mathfrak{p}}$ is easily verified. □

The following result is an analog of the fact that a normal domain is the intersection of valuation rings in its fraction field.

Corollary 2.4.5. *Let Q be a fine saturated monoid. Then*

$$Q = \{x \in Q^{\mathrm{gp}} : \nu_{\mathfrak{p}}(x) \geq 0 : \mathrm{ht}(\mathfrak{p}) = 1\}, \text{ and}$$
$$Q^* = \{x \in Q^{\mathrm{gp}} : \nu_{\mathfrak{p}}(x) = 0 : \mathrm{ht}(\mathfrak{p}) = 1\}.$$

In particular, Q is the intersection in Q^{gp} of the set of all its localizations at height one primes.

Proof Let $\mathfrak{p}$ be a height one prime and let G be its complementary face, a facet of Q. Then $\{x \in Q^{\mathrm{gp}} : \nu_{\mathfrak{p}}(x) \geq 0\} = Q_G$ and $\{x \in Q^{\mathrm{gp}} : \nu_{\mathfrak{p}}(x) = 0\} = G^{\mathrm{gp}}$. Thus statements (3) and (2) of Corollary 2.3.14 (applied to the face $F = Q^*$ of Q), respectively imply the two statements of the corollary. □

Proposition 2.4.6. *Let Q be a fine monoid and let $\mathcal{W}_Q^+$ denote the free monoid on the set of height one primes of Q. (Recall from Proposition 1.4.7 that this set is finite.) For $q \in Q$, let*

$$\nu(q) := \sum \{\nu_{\mathfrak{p}}(q)\mathfrak{p} : \mathrm{ht}(\mathfrak{p}) = 1\} \in \mathcal{W}_Q^+.$$

Then $\nu \colon Q \to \mathcal{W}_Q^+$ is a local homomorphism. Furthermore, $\nu(q_1) = \nu(q_2)$ if and only if there is some $n \in \mathbf{Z}^+$ such that $n\overline{q}_1 = n\overline{q}_2$ in $\overline{Q}$, and ν is exact if and only if Q is saturated.

Proof It is apparent that $\nu \colon Q \to \mathcal{W}_Q^+$ is a homomorphism of monoids. To see that ν is local, note that its target is sharp and, by the previous corollary, $\nu(q) = 0$ implies that q is a unit in Q^{sat} and hence also in Q. The same corollary also implies that ν induces an exact homomorphism $Q^{\mathrm{sat}} \to Q$. Thus if $q_1, q_2 \in$

Q with $v(q_1) = v(q_2)$, then $q_1 - q_2$ is a unit in Q^{sat} and there exists some $n \in \mathbf{Z}^+$ such that $nq_1 - nq_2 \in Q^*$. This implies that $n\overline{q}_1 = n\overline{q}_2$. We have already seen that v is exact if Q is saturated, and the converse follows from the fact that an exact submonoid of a saturated monoid is saturated (see Theorem 2.1.17.) □

2.5 Simplicial monoids

A cone C is called *simplicial* if it is finitely generated and free, that is, if there exists a finite set S such that each element of C can be written uniquely as a linear combination of elements $\sum_{s \in S} a_s s$ with $a_s \in K^{\geq 0}$; such a set S necessarily forms a basis for C^{gp}. A monoid is called simplicial if its corresponding cone is simplicial. It is not hard to see that any sharp finitely generated cone in K^1 or K^2 is simplicial. This is false for K^3; for example, the monoid $Q_{2,2}$ in part (3) of Examples 1.4.8 is not simplicial. For a useful criterion, see Proposition 2.5.3.

In fact, every finitely generated cone is a finite union of simplicial cones, as the following result of Carathéodory shows.

Theorem 2.5.1 (Carathéodory). *Let C be a K-cone and let S be a set of generators for C. Then every element of C lies in a cone generated by a linearly independent subset of S.*

Proof If $0 \neq x \in C$, we can write $x = \sum a_i s_i$ with $s_i \in S$ and $a_i > 0$. We may suppose that this has been done in such a way that the number e of terms in the sum is minimal; since x is not zero, $e \geq 1$. Then we claim that the sequence $(s_1, s_2, \ldots, s_e)$ is independent in the vector space C^{gp}. Suppose that $\sum c_i s_i = 0$. We may choose the indexing so that c_i is positive if $1 \leq i \leq m$, negative if $m + 1 \leq i \leq n$, and 0 if $i > n$. Furthermore, we may suppose that $a_i/c_i \leq a_{i+1}/c_{i+1}$ for $1 \leq i < m$ and for $m + 1 \leq i < n$. If $m > 0$ then, for all i, set $a'_i := a_i - (a_1/c_1)c_i \geq 0$. Since $\sum c_i s_i = 0$,

$$x = x - (a_1/c_1) \sum c_i s_i = \sum_{i \neq 1} a'_i s_i,$$

contradicting the minimality of e. Thus $m = 0$. If $n > 0$, then for all i we let $a'_i = a_i - (a_n/c_n)c_i \geq 0$ and

$$x = x - (a_n/c_n) \sum c_i s_i = \sum_{i \neq n} a'_i s_i,$$

again a contradiction. Thus $m = n = 0$, all $c_i = 0$, and $(s_1, \ldots, s_3)$ is independent. □

Corollary 2.5.2. *Let C be a finitely generated sharp K-cone of dimension d. Then C is a finite union of simplicial cones of dimension d.*

Proof Let S be a finite set of generators of C. Since the K-span of S is the d-dimensional vector space C^{gp}, every linearly independent subset T of S is contained in a linearly independent subset T' of S of cardinality d. Theorem 2.5.1 implies that every element of C belongs to some $C(T)$ and hence to some $C(T')$, and each $C(T')$ is a simplicial cone of dimension d. □

The next two results, whose proofs are due to Bernd Sturmfels, give criteria for a fine monoid to be simplicial or free, respectively.

Proposition 2.5.3. *Let C be a finitely generated sharp K-cone and S a finite subset. Suppose that every proper subset of S is contained in a proper face of C and that S spans C^{gp} as a vector space. Then S is linearly independent and spans C as a K-cone. In particular, C is simplicial.*

Proof Suppose that $\sum a_s s = 0$ with $a_s \in K$ and $s \in S$. Let $S' := \{s \in S : a_s > 0\}$, let $S'' := \{s \in S : a_s < 0\}$, and let $T := S \setminus S' \cup S''$. Then let t be the sum of all the elements of T, and let

$$f := \sum_{s \in S'} a_s s + t = \sum_{s \in S''} -a_s s + t;$$

note that $f \in C$. If S'' is not empty, then $S' \cup T$ is a proper subset of S and hence by assumption is contained in a proper face F of C. The first formula for f above implies that f belongs to F, and the second formula then implies that all the elements of S'' also belong to F. Then all of S is contained in F, and so $F^{\mathrm{gp}} = C^{\mathrm{gp}}$ since S spans V. But F is a face of C, hence an exact submonoid, and it follows that $F = C$, a contradiction of the fact that F is a proper face. Thus we must have $S'' = \emptyset$. Similarly $S' = \emptyset$, and it follows that S is linearly independent.

Let c be an element of the interior of C. Then there exist disjoint subsets S' and S'' of S and elements $a_s \in K^{\geq 0}$ such that

$$c = \sum_{s \in S'} a_s s - \sum_{s \in S''} a_s s.$$

It follows that $c + \sum\{a_s s : s \in S''\}$ also belongs to the interior of C. If S' were a proper subset of S, it would be contained in a proper face of C, which would contradict the fact that $\sum\{a_s s : s \in S'\} = c + \sum\{a_s s : s \in S''\}$ is in the interior of C. Hence $S' = S$ and $S'' = \emptyset$. We have thus shown that every element of the interior of C lies in the the $K^{\geq 0}$-span of S. Since this span is closed, and since the interior of C is dense in C by Corollary 2.3.17, S spans C, as claimed. □

The formulation of the next result requires a preliminary definition and lemma.

Lemma 2.5.4. *Let Q be a fine and sharp monoid of dimension d. For each positive integer m, let I_m denote the set of all elements of Q that can be written as a sum $q = q_1 + \cdots + q_m$, where $(q_1, \ldots, q_m)$ is a linearly independent sequence in $Q \subseteq \mathbf{Q} \otimes Q^{gp}$. In fact I_m is an ideal of Q and is contained in the ideal K_{d-m+1} (the intersection of all the primes of height $d - m + 1$).*

Proof If $q = q_1 + \cdots + q_m$ and $(q_1, \ldots, q_m)$ is independent, then $\langle q \rangle$ has rank at least m. If $\mathfrak{p}$ is a prime ideal of height $d - m + 1$, then $F_\mathfrak{p} := Q \setminus \mathfrak{p}$ is a face of codimension $d - m + 1$ and hence of dimension $m - 1$, so $q \notin F_\mathfrak{p}$, i.e., $q \in \mathfrak{p}$.

Suppose that $q \in I_m$ and $p \in Q$. To prove that $p + q \in I_m$, it is enough to prove that the sequence $s_i := (q_1, \ldots, q_{i-1}, p + q_i, q_{i+1}, \ldots, q_m)$ is independent for some i. Suppose on the contrary that s_i is dependent. Then since the sequence s'_i obtained from s_i by omitting $p + q_i$ is independent, it must be the case that $p + q_i$ is in the span of s'_i. Let q'_i be the sum of the sequence s'_i. Then $p + q_i + q'_i = p + q$ is also in the span of s'_i. If this is the case for every i, then $p + q$ is in the intersection of all these spans, which is $\{0\}$ since the original sequence is independent. It follows that $p + q = 0$, which is impossible since Q is sharp and $q \neq 0$. □

Proposition 2.5.5. *A fine sharp and saturated monoid Q of dimension d is free if and only if $I_d = K_1$.*

Proof Suppose that $I_d = K_1$. Choose a local homomorphism $h\colon Q \to \mathbf{N}$ and choose $k \in K_1$ minimizing h. We will show that if $k \in I_d$, then in fact Q is free and k is the sum of its set of minimal generators.

Write $k = q_1 + \cdots + q_d$, where $(q_1, \ldots, q_d)$ is independent, and then write each q_i as a sum of irreducible elements of Q. The collection of irreducible elements thus appearing spans $\mathbf{Q} \otimes Q^{gp}$ and hence contains a linearly independent sequence $(q'_1, \ldots, q'_d)$. Let $k' := q'_1 + \cdots + q'_d$, so that $k = p + k'$ for some $p \in Q$. Since $k' \in I_d \subseteq K_1$, the minimality of $h(k)$ and the sharpness of Q imply that $p = 0$. Thus each q_i is already irreducible. Furthermore, if q is the sum of a proper subset of the q_i, then $h(q) < h(k)$, so $q \notin K_1$ and hence $\langle q \rangle$ is a proper subset of Q. Thus $S := \{q_1, \cdots, q_d\}$ is a set of elements of Q^{gp} spanning $\mathbf{Q} \otimes Q^{gp}$ and every proper subset of S is contained in a proper face of Q. It follows from Proposition 2.5.3 that Q is simplicial and that S is a basis for the cone it spans.

To show that Q is free, it will now suffice to show that every irreducible element q of Q already belongs to S. Since S is a basis for the cone spanned by Q, there exist $n > 0$ and $n_i \in \mathbf{N}$ such that $nq = \sum n_i q_i$. Reorder the sequence in such a way that $n_i > 0$ if $i \leq e$ and $n_i = 0$ if $i > e$. Then $\langle q, q_{e+1}, \ldots, q_d \rangle = Q$, so $q + q_{e+1} + \cdots + q_m \in K_1$. By the minimality of $h(k) = h(q_1) + \cdots + h(q_d)$,

$$\begin{aligned} h(q) + h(q_{e+1}) + \cdots + h(q_d) &\geq h(q_1) + \cdots + h(q_e) + \cdots + h(q_d), \quad \text{so} \\ h(q) &\geq h(q_1) + \cdots + h(q_e) \\ h(nq) &\geq nh(q_1) + \cdots + nh(q_e) \\ n_1 h(q_1) + \cdots + n_e h(q_e) &\geq nh(q_1) + \cdots + nh(q_e) \end{aligned}$$

It follows that $n_i \geq n$ for some i. But then $nq - nq_i \in Q$ and, since Q is saturated, in fact $q - q_i \in Q$. Since q is irreducible, this impies that $q = q_i$, completing the proof that Q is free. The converse is immediate. □

3 Affine toric varieties

3.1 Monoid algebras and monoid schemes

Let R be a fixed commutative ring, usually the ring of integers $\mathbf{Z}$ or a field, and let $\mathbf{Alg}_R$ denote the category of R-algebras. If Q is a monoid, the *R-monoid algebra of Q* is the R-algebra $R[Q]$ whose underlying R-module is free with basis Q, endowed with the unique ring structure making the inclusion map $e: Q \to R[Q]$ a homomorphism from the monoid Q into the multiplicative monoid of $R[Q]$. Thus, if p and q are elements of Q and if we use additive notation for Q, we have $e(p + q) = e(p)e(q)$; for this reason we sometimes write e^p for $e(p)$. For example, $R[\mathbf{N}]$ is the polynomial algebra $R[T]$, where $T = e^1$. More generally, if $\mathbf{N}^{(X)}$ is the free monoid with basis X, then $R[\mathbf{N}^{(X)}]$ is the polynomial algebra $R[X]$: if $I \in \mathbf{N}^{(X)}$ is a multi-index, e^I corresponds to the monomial $X^I := \prod\{x^{I_x} : x \in X\}$.

The functor $Q \mapsto R[Q]$ is left adjoint to the functor $\underline{\mathsf{A}}_{\mathsf{m}}$ taking an R-algebra to its underlying multiplicative monoid. Thus if A is an R-algebra, giving a homomorphism of R-algebras $R[Q] \to A$ is equivalent to giving a homomorphism of monoids $Q \to \underline{\mathsf{A}}_{\mathsf{m}}(A)$. Such a homomorphism is often called an *A-valued character* of Q, and an R-algebra equipped with a character of Q is often called a *Q-algebra* if R is understood. Since the functor $Q \to R[Q]$ is a left adjoint, it automatically commutes with colimits. For example, if Q is the amalgamated sum of a pair of monoid homomorphisms $\theta_i : P \to Q_i$, then $R[Q] \cong R[Q_1] \otimes_{R[P]} R[Q_2]$.

The set $\underline{\mathsf{A}}_{\mathsf{Q}}(A)$ of A-valued characters of Q has a natural monoid structure, with the multiplication law defined by the pointwise product and the identity element given by the constant function whose value is 1. Thus we can view $\underline{\mathsf{A}}_{\mathsf{Q}}$ as a functor

$$\underline{\mathsf{A}}_{\mathsf{Q}} : \mathbf{Alg}_R \to \mathbf{Mon}$$

from the category of R-algebras to the category of monoids. The functor $\underline{\mathsf{A}}_{\mathsf{Q}}$ is represented by the pair $(R[Q], e)$, where $R[Q]$ is the monoid R-algebra of Q and $e\colon Q \to R[Q]$ is the map taking an element of Q to the corresponding basis element of $R[Q]$. We use the same notation for the scheme $\underline{\mathsf{A}}_{\mathsf{Q}}$ and the functor $\underline{\mathsf{A}}_{\mathsf{Q}}$ it represents. The monoid structure on the functor $\underline{\mathsf{A}}_{\mathsf{Q}}$ is given by the natural transformations

$$m_Q\colon \underline{\mathsf{A}}_{\mathsf{Q}} \times \underline{\mathsf{A}}_{\mathsf{Q}} \to \underline{\mathsf{A}}_{\mathsf{Q}}, \quad 1_Q\colon \operatorname{Spec} R \to \underline{\mathsf{A}}_{\mathsf{Q}},$$

and these give the scheme $\underline{\mathsf{A}}_{\mathsf{Q}}$ the structure of a monoid scheme. Since $\underline{\mathsf{A}}_{\mathsf{Q}} = \operatorname{Spec} R[Q]$, the corresponding morphisms of these schemes are given by R-algebra homomorphisms:

$$m_Q^\sharp\colon R[Q] \to R[Q] \otimes R[Q] : \quad e^q \mapsto e^q \otimes e^q$$

$$1_Q^\sharp\colon R[Q] \to R : \quad \sum_q a_q e^q \mapsto \sum_q a_q.$$

Thus the identity section 1_Q of the monoid scheme $\underline{\mathsf{A}}_{\mathsf{Q}}$ is given by the homomorphism $Q \to R$ sending every element to 1. There is another natural section v_Q of $\underline{\mathsf{A}}_{\mathsf{Q}}$, called the *vertex*, given by the homomorphism sending $q \in Q$ to 1 if $q \in Q^*$ and to 0 if $q \in Q^+$. These two sections coincide if and only if Q is dull.

In particular, the functor $\underline{\mathsf{A}}_{\mathsf{m}}$ is isomorphic to the functor $\underline{\mathsf{A}}_{\mathsf{N}}$; for each R-algebra A we have an isomorphism of monoids

$$\underline{\mathsf{A}}_{\mathsf{m}}(A) := (A, \cdot, 1_A) \cong \operatorname{Hom}(\mathbf{N}, A),$$

where an element a of A corresponds to the monoid homomorphism $n \mapsto a^n$. Note in particular that $a^0 = 1$ for any a, including $a = 0$, by the definition of a monoid homomorphism. The vertex of $\underline{\mathsf{A}}_{\mathsf{m}}$ corresponds to the point 0 of $\underline{\mathsf{A}}_{\mathsf{N}}$ and the identity section to the point 1.

The following proposition shows that Q can often be recovered from the functor $\underline{\mathsf{A}}_{\mathsf{Q}}$ (with its monoid scheme structure).

Proposition 3.1.1. *Suppose that* Spec R *is connected. Then the functor*

$$Q \mapsto \underline{\mathsf{A}}_{\mathsf{Q}}$$

from the category of monoids to the category of monoid schemes over R is fully faithful:

$$\operatorname{Hom}(P, Q) \cong \operatorname{Hom}((\underline{\mathsf{A}}_{\mathsf{Q}}, 1, \cdot), (\underline{\mathsf{A}}_{\mathsf{P}}, 1, \cdot)).$$

Proof A morphism of schemes $\underline{\mathsf{A}}_{\mathsf{Q}} \to \underline{\mathsf{A}}_{\mathsf{P}}$ corresponds to a homomorphism of rings $\theta\colon R[P] \to R[Q]$. Since $Q \to R[Q]$ is injective, a homomorphism $P \to Q$ is determined by the corresponding homomorphism $R[P] \to R[Q]$. This proves that our functor is faithful. For the fullness, let θ be a homomorphism and, for each $p \in P$ write,

$$\theta(e^p) = \sum_{q \in Q} a_q(p) e^q,$$

with $a_q(p) \in R$. The statement that θ corresponds to a monoid morphism is the statement that the following diagrams commute:

$$\begin{array}{ccc} R[P] & \xrightarrow{\ \theta\ } & R[Q] \\ \Big\downarrow 1_P^\# & & \Big\downarrow 1_Q^\# \\ R & \xrightarrow{\ \mathrm{id}\ } & R \end{array} \qquad \begin{array}{ccc} R[P] & \xrightarrow{\ \theta\ } & R[Q] \\ \Big\downarrow m_P^\# & & \Big\downarrow m_Q^\# \\ R[P] \otimes R[P] & \xrightarrow{\ \theta \otimes \theta\ } & R[Q] \otimes R[Q]. \end{array}$$

The second diagram says that, for each $p \in P$,

$$\sum_{q,q'} a_q(p) a_{q'}(p) e^q \otimes e^{q'} = \sum_q a_q(p) e^q \otimes e^q,$$

i.e., that $a_q(p)a_{q'}(p)$ equals zero if $q \neq q'$ and equals $a_q(p)$ if $q = q'$. In other words, the $a_q(p)$'s are orthogonal idempotents of the ring R. The first diagram says that $\sum_{q \in Q} a_q(p) = 1$ for each $p \in P$. Since $\operatorname{Spec} R$ is connected, every idempotent is either 0 or 1 and, since $a_q(p)$ and $a_{q'}(p)$ are orthogonal if $q \neq q'$, there is a unique element $\beta(p) \in Q$ such that $a_q(p) = 0$ if $q \neq \beta(p)$ and $a_q(p) = 1$ if $q = \beta(p)$. Thus β is a function $P \to Q$ such that $\theta \circ e = e \circ \beta$. Since θ is a ring homomorphism, β is a monoid homomorphism, as required. □

Corollary 3.1.2. *Suppose that* $\operatorname{Spec} R$ *is connected and* Q *is a monoid.*

1. *The monoid of characters of* $\underline{\mathsf{A}}_{\mathsf{Q}}$*, i.e., of morphisms* $\underline{\mathsf{A}}_{\mathsf{Q}} \to \underline{\mathsf{A}}_{\mathsf{m}}$*, is canonically isomorphic to* Q*.*
2. *The monoid of cocharacters of* $\underline{\mathsf{A}}_{\mathsf{Q}}$*, i.e. of morphisms* $\underline{\mathsf{A}}_{\mathsf{m}} \to \underline{\mathsf{A}}_{\mathsf{Q}}$*, is canonically isomorphic to* $H(Q) := \operatorname{Hom}(Q, \mathbf{N})$*.* □

If Q is a monoid and A is an R-algebra, $\underline{\mathsf{A}}_{\mathsf{Q}^{\mathrm{gp}}}(A)$ is precisely the set of invertible elements of $\underline{\mathsf{A}}_{\mathsf{Q}}(A)$, i.e., $\underline{\mathsf{A}}_{\mathsf{Q}^{\mathrm{gp}}} = \underline{\mathsf{A}}_{\mathsf{Q}}^*$. If Q is fine, the localization $R[Q] \to R[Q^{\mathrm{gp}}]$ is injective and of finite type, and hence $\underline{\mathsf{A}}_{\mathsf{Q}}^* \to \underline{\mathsf{A}}_{\mathsf{Q}}$ is a dominant and affine open immersion.

3.2 Monoid sets and monoid modules

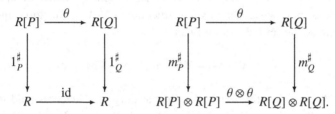

It will also be important to discuss linear representations of monoid schemes. Let us begin with a geometric description. If E is an R-module, let $\mathbf{V}E$ be the functor taking an R-algebra A to the set of R-linear maps $E \to A$. As explained in [31], this functor is represented by the spectrum of the symmetric algebra $S^{\cdot}E$, and the universal element of $\mathbf{V}E(S^{\cdot}E)$ is the inclusion $E \to S^{\cdot}E$. Thus

$$\mathbf{V}E(A) = \mathrm{Hom}_R(E, A) = \mathrm{Hom}_{\mathbf{Alg}_R}(S^{\cdot}E, A).$$

The set $\mathbf{V}(A)$ has a natural structure of an A-module, where the operations are defined pointwise. The module structure on $\mathbf{V}(A)$ is functorial in A, and therefore defines morphisms of functors

$$\mathbf{V}E \times \mathbf{V}E \longrightarrow \mathbf{V}E \quad \text{and} \quad \mathbf{A} \times \mathbf{V}E \to \mathbf{V}E,$$

where $\mathbf{A}$ is the identity functor on the category of R-algebras (and is represented by the affine line over $\operatorname{Spec} R$). Thus the R-scheme $\mathbf{V}E$ has the structure of a vector scheme over $\operatorname{Spec} R$, and one can recover the R-module E from the vector scheme $\mathbf{V}E$ as the R-module of $\mathbf{A}$-linear morphisms $\mathbf{V}E \to \mathbf{A}$. However, it will be enough for our purposes to use only the multiplicative monoid structures of R and $\underline{\mathsf{A}}_{\mathsf{m}}$ acting on E and $\mathbf{V}E$ respectively. Note that an element e of E defines a natural transformation $\hat{e}\colon \mathbf{V}E \to \underline{\mathsf{A}}_{\mathsf{m}}$, where $\hat{e}_A(\alpha) := \alpha(e)$ for each R-algebra A and each $\alpha\colon E \to A \in \mathbf{V}E(A)$. Note also that $\hat{e}$ is compatible with the actions of $\underline{\mathsf{A}}_{\mathsf{m}}$: $a\hat{e}_A(\alpha) = \hat{e}_A(a\alpha)$ for each A and each $a \in A$. The following proposition shows that the R-set underlying the R-module E is determined by the $\underline{\mathsf{A}}_{\mathsf{m}}$-set $\mathbf{V}E$.

Proposition 3.2.1. *For any R-module E, the natural map*

$$E \to \mathrm{Mor}_{\underline{\mathsf{A}}_{\mathsf{m}}}(\mathbf{V}E, \underline{\mathsf{A}}_{\mathsf{m}})$$

is an isomorphism of R-sets. Consequently, the functor $\mathbf{V}$ from the category of R-modules to the category of $\underline{\mathsf{A}}_{\mathsf{m}}$-sets is fully faithful.

Proof The statement that $e \mapsto \hat{e}$ is a morphism of R-sets says that $\widehat{re} = r\hat{e}$ for every e in E and r in R, an immediate verification. To prove that $e \mapsto \hat{e}$ is bijective, let us begin by noting that the action of $\underline{\mathsf{A}}_{\mathsf{m}}$ on $\mathbf{V}E$ is given by the unique homomorphism of R-algebras $\lambda_E\colon S^{\cdot}E \to R[T] \otimes S^{\cdot}E$ sending each e in E to $T \otimes e$ in $R[T] \otimes S^{\cdot}E$. It follows that $\lambda_E(f) = \sum_d T^d \otimes f_d$ for every $f = \sum_d f_d \in S^{\cdot}E$, where each $f_d \in S^d E$.

By Yoneda, any morphism of functors $\eta\colon \mathbf{V}E \to \mathbf{A_m}$ is given by a homomorphism of rings $\theta\colon R[T] \to S^{\cdot}E$, which in turn is determined by the element $f := \theta(T)$ in $S^{\cdot}E$. An element α of $\mathbf{V}E(A)$ induces a homomorphism of R-algebras $\tilde{\alpha}\colon S^{\cdot}E \to A$, and $\eta(\alpha) = \tilde{\alpha}(f)$. Then η is a morphism of $\underline{\mathsf{A}}_{\mathsf{m}}$-sets if and only if the diagram

$$
\begin{array}{ccc}
R[T] & \xrightarrow{\ \theta\ } & S^{\cdot}E \\
\Big\downarrow{\scriptstyle m^{\sharp}} & & \Big\downarrow{\scriptstyle \lambda_E} \\
R[T]\otimes R[T] & \xrightarrow{\ \mathrm{id}\otimes\theta\ } & R[T]\otimes S^{\cdot}E
\end{array}
$$

commutes. But $\lambda_E \circ \theta(T) = \sum_d T^d \otimes f_d$ and $\mathrm{id}\otimes\theta\circ m^{\sharp}(T) = \mathrm{id}\otimes\theta(T\otimes T) = T\otimes f$. Thus the diagram commutes if and only if $f = f_1$, that is, if and only if $f \in E \subseteq S^{\cdot}E$. □

Now if Q is a monoid, an *action of* $\underline{\mathsf{A}}_{\mathsf{Q}}$ *on* $\mathbf{V}E$ is a natural transformation $m_E\colon \underline{\mathsf{A}}_{\mathsf{Q}} \times \mathbf{V}E \to \mathbf{V}E$ satisfying the usual rules for a monoid action. We require also that it be compatible with the $\underline{\mathsf{A}}_{\mathsf{m}}$-set structure on $\mathbf{V}E$, i.e., that, for every A and for every $\alpha \in \underline{\mathsf{A}}_{\mathsf{Q}}(A)$, $a \in \underline{\mathsf{A}}_{\mathsf{m}}(A)$, and $v \in \mathbf{V}E(A)$, we have $\alpha(av) = a\alpha(v)$. In other words, $\mathbf{V}E$ should be an $(\underline{\mathsf{A}}_{\mathsf{Q}}, \underline{\mathsf{A}}_{\mathsf{m}})$-biset. The map m_E is given by a homomorphism $S^{\cdot}E \to R[Q]\otimes S^{\cdot}E$, and the compatibility with the $\underline{\mathsf{A}}_{\mathsf{m}}$-action implies that m_E is induced by a homomorphism $\theta_E\colon E \to R[Q]\otimes_R E$, thanks to Proposition 3.2.1. The morphism θ_E will be an action of $\underline{\mathsf{A}}_{\mathsf{Q}}$ on $\mathbf{V}E$ if and only if θ_E is a *coaction of* $R[Q]$ *on* E, i.e., if the following diagrams commute:

$$
\begin{array}{ccc}
E & \xrightarrow{\ \theta_E\ } & R[Q]\otimes E \\
 & {\scriptstyle \mathrm{id}_E}\searrow & \Big\downarrow{\scriptstyle 1_Q\otimes\mathrm{id}_E} \\
 & & E
\end{array}
$$

$$
\begin{array}{ccc}
E & \xrightarrow{\ \theta_E\ } & R[Q]\otimes E \\
\Big\downarrow{\scriptstyle \theta_E} & & \Big\downarrow{\scriptstyle m_Q^{\sharp}\otimes\mathrm{id}_E} \\
R[Q]\otimes E & \xrightarrow{\ \mathrm{id}_{R[Q]}\otimes\theta_E\ } & R[Q]\otimes R[Q]\otimes E.
\end{array}
$$

For each $e \in E$, write

$$\theta_E(e) = \sum q\otimes\pi_q(e) \in R[Q]\otimes_R E.$$

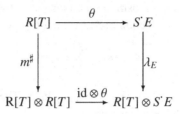

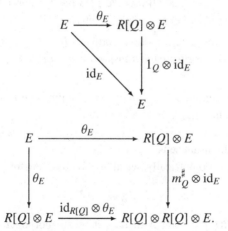

Then each π_q is an R-linear endomorphism of E. The two diagrams assert that $\pi_p\pi_q = 0$ if $p \neq q$ and that $\sum \pi_q = \mathrm{id}_E$. Thus the family $\{\pi_q : q \in Q\}$ defines a direct sum decomposition $E = \oplus\{E_q : q \in Q\}$, where E_q is the image of the idempotent π_q. This decomposition is nothing but a Q-grading of the R-module E, and the corresponding action of $\underline{\mathsf{A}}_{\mathsf{Q}}(A)$ on $\mathbf{V}E(A)$ is given by

$$(\alpha, v)(e) = \sum_q \alpha(q)v(e_q)$$

for $\alpha \in \underline{\mathsf{A}}_{\mathsf{Q}}(A), v \in \mathbf{V}E(A)$, and $e = \sum e_q$ in E. The following proposition summarizes the results of this argument.

Proposition 3.2.2. *Let Q be a monoid. Then the construction $E \mapsto \mathbf{V}E$ induces a fully faithful functor from the category of Q-graded R-modules to the category of $(\underline{\mathsf{A}}_{\mathsf{Q}}, \underline{\mathsf{A}}_{\mathsf{m}})$-bisets.*

We shall find it useful to generalize Proposition 3.2.2 by considering R-modules graded by a Q-set S in place of Q.

Definition 3.2.3. *If S is a Q-set, $R[S]$ is the free R-module with basis S and endowed with the unique $R[Q]$-module structure compatible with the action of Q on S.*

The functor $S \mapsto R[S]$ is left adjoint to the functor taking an $R[Q]$-module to its underlying Q-set. It follows that if $\{s_i : i \in I\}$ is a basis for S as a Q-set, then $\{e^{s_i} : i \in I\}$ is a basis for $R[S]$ as a Q-module, and that if S and S' are Q-sets, there is a natural isomorphism $R[S \otimes_Q S'] \cong R[S] \otimes_{R[Q]} R[S']$.

If S is a Q-set, then the $R[Q]$-module $R[S]$ has a natural S-grading. Let us generalize this notion and explain its geometric significance.

Definition 3.2.4. *Let Q be a monoid and let S be a Q-set. Then an S-graded $R[Q]$-module is a functor from the transporter category $\mathcal{T}S$ of Q to the category of R-modules.*

The data of an S-graded $R[Q]$-module E is equivalent to the data of a collection of R-modules $\{E_s : s \in S\}$ and, for every $q \in Q$, an R-linear map $h_q : E_s \to E_{q+s}$ such that $h_{q'} \circ h_q = h_{q+q'}$ and $h_0 = \mathrm{id}$. Thus $\oplus_s E_s$ becomes an $R[Q]$-module in the usual sense.

To interpret this geometrically, we attach to S the functor

$$\mathbf{V}S : \mathbf{Alg}_{R[Q]} \to \mathbf{Ens}$$

which takes an $R[Q]$-algebra A to the set of all morphisms of Q-sets $S \to A$, where A is viewed as a Q-set via the character $Q \to A$ coming from its $R[Q]$-algebra structure. Note that $\mathbf{V}S(A)$ has a natural structure of an A-module, where the module operations are defined pointwise. In particular, we can and shall view $\mathbf{V}S$ as being endowed with an action of the functor $\underline{\mathsf{A}}_{\mathsf{m}}$.

The functor $\mathbf{V}S$ is representable by an affine scheme over $\underline{\mathsf{A}}_{\mathsf{Q}}$, which we also denote by $\mathbf{V}S$. Indeed, if A is any $R[Q]$-algebra, then to give a homomorphism of $R[Q]$-modules $R[S] \to A$ is equivalent to giving a morphism of Q-sets $S \to A$. Thus we have an isomorphism of functors $\mathbf{V}S \cong \mathbf{V}(R[S])$ on the category $\mathbf{Alg}_{R[Q]}$. Explicitly, $\mathbf{V}S$ is the spectrum of the symmetric algebra $S^{\cdot}(R[S])$ (computed with respect to $R[Q]$), or, equivalently, the quotient of the polynomial algebra $R[Q][X_s : s \in S]$ by the ideal generated by the elements $qX_s - X_{qs}$ for $q \in Q$ and $s \in S$. It follows from Proposition 3.2.1 (applied with $R[Q]$ in place of R) that $R[S]$ can be identified with the set of $\underline{\mathsf{A}}_{\mathsf{m}}$-morphisms $\mathbf{V}S \to \underline{\mathsf{A}}_{\mathsf{m}}$.

We have a natural morphism $\mathbf{V}S \to \underline{\mathsf{A}}_{\mathsf{Q}} \to \operatorname{Spec} R$. Let $\mathbf{A}_S$ denote the scheme $\mathbf{V}S$ viewed over $\operatorname{Spec} R$. Explicitly, if A is an R-algebra,

$$\mathbf{A}_S(A) := \{(\alpha, \sigma) : \alpha \in \underline{\mathsf{A}}_{\mathsf{Q}}(A), \sigma \in \mathbf{V}S(A, \alpha)\}.$$

Pointwise multiplication defines a map

$$m_S \colon \mathbf{A}_S \times \mathbf{A}_S \to \mathbf{A}_S,$$

which is bilinear with respect to the action of $\underline{\mathsf{A}}_{\mathsf{m}}$. The corresponding homomorphism $R[S] \to R[S] \otimes R[S]$ sends $s \in S$ to $s \otimes s$. The constant function $S \to R$ whose value is always 1 defines a section $1_S : \operatorname{Spec} R \to \mathbf{A}_S$, and then $(m_S, 1_S)$ gives $\mathbf{A}_S$ the structure of a monoid scheme. (In fact it also has the structure of an algebra scheme, but we shall not need this extra structure.) The natural morphism $\mathbf{A}_S \to \underline{\mathsf{A}}_{\mathsf{Q}}$ is a morphism of monoid schemes. For example, if $S = Q^{\mathrm{gp}}$ with the natural action of Q, the morphism $\mathbf{A}_S \to \underline{\mathsf{A}}_{\mathsf{Q}}$ identifies $\mathbf{A}_S$ with $\underline{\mathsf{A}}_{\mathsf{Q^{gp}}} = \underline{\mathsf{A}}^*_{\mathsf{Q}} \subseteq \underline{\mathsf{A}}_{\mathsf{Q}}$.

We shall see that an S-grading structure on an $R[Q]$-module E amounts to an equivariant action of the $\underline{\mathsf{A}}_{\mathsf{m}}$-monoid scheme $\mathbf{A}_S$ on $\mathbf{V}E$.

Definition 3.2.5. *Let E be an $R[Q]$-module and let S be a Q-set. Then an equivariant $(\mathbf{A}_S, \underline{\mathsf{A}}_{\mathsf{m}})$-biset structure on $\mathbf{V}E$ is a monoid action*

$$m_E \colon \mathbf{A}_S \times \mathbf{V}E \to \mathbf{V}E$$

that is bilinear with respect to the $\underline{\mathsf{A}}_{\mathsf{m}}$ structures on $\mathbf{A}_S$ and $\mathbf{V}E$ and such that the diagram

$$\begin{array}{ccc} \mathbf{A}_S \times \mathbf{V}E & \xrightarrow{m_E} & \mathbf{V}E \\ \downarrow & & \downarrow \\ \underline{\mathsf{A}}_{\mathsf{Q}} \times \underline{\mathsf{A}}_{\mathsf{Q}} & \xrightarrow{m_Q} & \underline{\mathsf{A}}_{\mathsf{Q}} \end{array}$$

commutes.

Let us explain how an S-grading on E gives rise to such an action. If E is S-graded and A is a Q-algebra, then an element of $\mathbf{V}E(A)$ can be viewed as a collection of R-linear maps $\eta_s\colon E_s \to A$ such that, for each $s \in S$ and $q \in Q$, the diagram

$$\begin{array}{ccc} E_s & \xrightarrow{\eta_s} & A \\ {\scriptstyle h_q}\downarrow & & \downarrow{\scriptstyle \alpha(q)} \\ E_{q+s} & \xrightarrow{\eta_{q+s}} & A \end{array}$$

commutes, where $\alpha\colon Q \to A$ is the Q-algebra structure of A. Let (β, σ) be an element of $\mathbf{A}_S(A)$. Then $\alpha' := \beta\alpha$ is an element of $\underline{\mathsf{A}}_Q(A)$, and we can define

$$\eta'_s\colon E_s \to A := \sigma(s)\eta_s.$$

One can easily check that η' satisfies the diagram above with α' in place of α. Then $(\beta, \sigma)\eta := \eta'$ defines the desired action; the bilinearity and compatibility with m_Q are immediately verified.

Proposition 3.2.6. *Let S be a Q-set. The construction above defines a fully faithful functor from the category of S-graded $R[Q]$-modules to the category of equivariant $(\mathbf{A}_S, \underline{\mathsf{A}}_{\mathsf{m}})$-bisets.*

Proof We give only a sketch. Let E be an $R[Q]$-module and let $\mu\colon \mathbf{A}_S \times \mathbf{V}E \to \mathbf{V}E$ be an equivariant $(\mathbf{A}_S, \underline{\mathsf{A}}_{\mathsf{m}})$-biset structure on $\mathbf{V}E$. Then $\mathbf{A}_S \times \mathbf{V}E$ is represented by $S^{\cdot}(R[S]) \otimes_{R[Q]} S^{\cdot}E$. Since μ is compatible with the action of $\underline{\mathsf{A}}_{\mathsf{m}}$, it follows from Proposition 3.2.1 that μ is induced by a homomorphism of $R[Q]$-modules:

$$\mu^\sharp\colon E \to R[S] \otimes_R E.$$

The compatibility of μ with m_Q implies that this homomorphism is linear over the homomorphism $\mu^\sharp_Q$. The fact that μ is a monoid action implies that the following diagrams commute:

$$\begin{array}{ccc} E & \xrightarrow{\mu^\sharp} & R[S] \otimes_R E \\ & \searrow{\scriptstyle \mathrm{id}_E} & \downarrow{\scriptstyle 1^\sharp_S \otimes \mathrm{id}_E} \\ & & E \end{array}$$

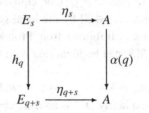

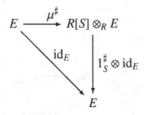

$$\begin{array}{ccc} E & \xrightarrow{\mu^\sharp} & R[S] \otimes_R E \\ {\scriptstyle \mu^\sharp}\downarrow & & \downarrow{\scriptstyle m_S^\sharp \otimes \mathrm{id}_E} \\ R[S] \otimes_R E & \xrightarrow{\mathrm{id}_{R[S]} \otimes \mu^\sharp} & R[S] \otimes_R R[S] \otimes_R E. \end{array}$$

If $e \in E$, write $\mu^\sharp(e) = \sum s \otimes \pi_s(e)$. Then each $\pi_s \colon E \to E$ is an R-linear map, and these diagrams say that $\sum_s \pi_s = \mathrm{id}_E$ and $\pi_s \circ \pi_t = \delta_{s,t}\pi_t$. (Note: for each $e \in E$, $\pi_s(e) = 0$ for almost all $s \in S$.) In others words, $\{\pi_s : s \in S\}$ is the family of projections corresponding to a direct sum decomposition $E = \oplus E_s$. If $e \in E_s$, then $\mu^\sharp(e) = s \otimes e$ and, since $\mu^\sharp$ is linear over m_Q,

$$\mu^\sharp(qe) = (q \otimes q)\mu^\sharp(e) = (q \otimes q)(s \otimes e) = (q + s) \otimes (qe),$$

so that $qe \in E_{q+s}$. Then the map $h_q : E_s \to E_{q+s}$ given by $e \mapsto qe$ is R- linear, and $\{E_s, h_q\}$ defines an S-grading on the $R[Q]$-module E. This shows how the S grading is determined by the biset structure, and it follows that the functor is fully faithful. □

For example, suppose that Q is an integral monoid, so that whenever $q' \geq q$, there is a unique $p \in Q$ such that $q' = p + q$. Then a *Q-filtration* on an R-module V is a family of submodules $F_q \subseteq V$ such that $F_q \subseteq F_{q'}$ whenever $q \leq q'$. Thus F can be viewed as functor from $\mathcal{T}Q$ to the category of R-modules, and $\bigoplus F_q \subseteq V \otimes R[Q]$ is a graded $R[Q]$-submodule. A Q-filtration on R itself defines an ideal of $R[Q]$, and the corresponding closed subscheme of $\underline{\mathsf{A}}_Q$ is stable under the action of Q on itself. If K is an ideal in Q, the free R-module $R[K]$ with basis K can be viewed as an ideal of $R[Q]$ defined by the Q-filtration that equals 0 for $q \notin K$ and equals R if $q \in K$. When R is a field, every Q-filtration of R has this form.

Example 3.2.7. Let Q be an integral monoid. For each $q \in Q$, let $\langle q \rangle$ be the face of Q generated by q (see Proposition 1.4.2). If $q' \in Q$, $\langle q \rangle \subseteq \langle q + q' \rangle$. Hence $q \mapsto \langle q \rangle^{\mathrm{gp}} \subseteq Q^{\mathrm{gp}}$ defines a Q-filtration of Q^{gp} and hence an $\underline{\mathsf{A}}_Q$-invariant submodule of $\mathbf{Z}[Q] \otimes Q^{\mathrm{gp}}$. More generally, for any integer i,

$$q \mapsto \Lambda^i \langle q \rangle^{\mathrm{gp}} \subseteq \Lambda^i Q^{\mathrm{gp}}$$

defines a Q-filtration of $\Lambda^i Q^{\mathrm{gp}}$. When i is the rank of Q^{gp}, this filtration is related to the filtration $R[I_Q] \otimes \Lambda^i Q^{\mathrm{gp}}$, where I_Q is the interior ideal of Q (see Section 1.4, as well as Lemma 2.5.4 and Proposition V.2.3.10).

Remark 3.2.8. Let F be a face of an integral monoid Q, let S be a Q-set, and let E an S-graded $R[Q]$-module. Then the localization E_F of E by F is naturally an

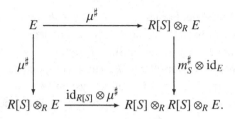

S_F-graded $R[Q_F]$-module. For any $t \in S_F$ and $f \in F$, multiplication by e^f defines an isomorphism from the component of E_F in degree t to the component in degree $f + t$. Since these isomorphisms all commute, we can safely identify all the components $E_{F,f+t}$ for all $f \in F$. These components can be computed as follows. For each $s \in S$, there is a direct system $\{e^f \cdot E_s \to E_{f+s} : f \in F\}$, where F is endowed with the standard monoid order. Then one sees easily that, for any s in S mapping to any $f + t \in S_F$,

$$\varinjlim \{e^f \cdot E_s \to E_{f+s} : f \in F\} \cong E_{F,t}.$$

3.3 Faces, orbits, and trajectories

In this section we discuss how the geometry of the spectrum of a fine monoid Q, or, equivalently, the partially ordered set of its faces, relates to the action of the monoid scheme $\underline{\mathsf{A}}_Q$ on itself.

The functor $\mathbf{Alg}_R \to \mathbf{Mon}$ taking an R-algebra to its underlying multiplicative monoid can be lifted naturally to a functor $\mathbf{Alg}_R \to \mathbf{Moni}$ taking A to the acceptably idealized monoid $(A, \cdot, 1, \{0\})$. If K is an ideal of a monoid Q, then the $R[Q]$-module $R[K]$ is an ideal of $R[Q]$, and the quotient

$$R[Q, K] := R[Q]/R[K]$$

is a free R-module with basis $Q \setminus K$. For any R-algebra A,

$$\mathrm{Hom}_{\mathbf{Moni}}((Q, K), (A, 0)) = \mathrm{Hom}_R(R[Q, K], A),$$

so that the functor $(Q, K) \mapsto R[Q, K]$ is left adjoint to the functor $A \mapsto (A, \{0\})$. We call $R[Q, K]$ the *monoid algebra of the idealized monoid* (Q, K).

Let $\underline{\mathsf{A}}_{Q,K}$ denote the functor taking an R-algebra A to the set of maps $(Q, K) \to (A, 0)$; as we have seen, this functor is representable by $R[Q, K]$. Thus $\underline{\mathsf{A}}_{Q,K}$ is a closed subscheme of the monoid scheme $\underline{\mathsf{A}}_Q$, and Proposition 3.2.2 shows that it is invariant under the action of $\underline{\mathsf{A}}_Q$ on itself. In other words, $\underline{\mathsf{A}}_{Q,K}$ is an ideal scheme of the monoid scheme of $\underline{\mathsf{A}}_Q$: for every A, the image of the injective map

$$i_K(A) \colon \underline{\mathsf{A}}_{Q,K}(A) \to \underline{\mathsf{A}}_Q(A)$$

is an ideal in the monoid $\underline{\mathsf{A}}_Q(A)$.

In particular, let $\mathfrak{p}$ be a prime ideal of Q and let $F := Q \setminus \mathfrak{p}$ be the corresponding face. The inclusion $F \to Q$ defines a morphism of monoid algebras $R[F] \to R[Q]$ and hence a morphism of monoid schemes

$$r_F \colon \underline{\mathsf{A}}_Q \to \underline{\mathsf{A}}_F. \tag{3.3.1}$$

The composition of the map $R[F] \to R[Q]$ with the homomorphism

$$i^\sharp_{\mathfrak{p}} \colon R[Q] \to R[Q, \mathfrak{p}]$$

yields an isomorphism of R-algebras $R[F] \to R[Q, \mathfrak{p}]$, since it induces a bijection on the basis elements. This isomorphism of rings defines an isomorphism of schemes $\underline{\mathsf{A}}_{Q,\mathfrak{p}} \to \underline{\mathsf{A}}_F$, and we let

$$i_F \colon \underline{\mathsf{A}}_F \to \underline{\mathsf{A}}_Q \tag{3.3.2}$$

be the composition of the inverse of this isomorphism with the closed immersion $i_{\mathfrak{p}}$. Thus,

$$i^\sharp_F(e^q) = \begin{cases} e^q & \text{if } q \in F \\ 0 & \text{otherwise.} \end{cases}$$

For example, if Q is sharp, then $\underline{\mathsf{A}}_{Q,Q^+} \cong \operatorname{Spec} R$. The corresponding R-valued point of Q is the homomorphism $v \colon Q \to R$ such that $v_Q(0) = 1$ and $v_Q(q) = 0$ if $q \in Q^+$, i.e., the vertex of $\underline{\mathsf{A}}_Q$.

Proposition 3.3.1. *Let F be a face of an integral monoid Q, let i_F and r_F be the morphisms (3.3.2) and (3.3.1) defined above, and let $i_{Q/F}$ be the closed immersion induced by the surjection $Q \to Q/F$. Let S denote the spectrum of the base ring R.*

1. *These morphisms fit into a commutative diagram with cartesian squares:*

$$\begin{array}{ccccc} S & \xrightarrow{v_{Q/F}} & \underline{\mathsf{A}}_{Q/F} & \xrightarrow{\pi} & S \\ \Big\downarrow{\scriptstyle 1_F} & & \Big\downarrow{\scriptstyle i_{Q/F}} & & \Big\downarrow{\scriptstyle 1_F} \\ \underline{\mathsf{A}}_F & \xrightarrow{i_F} & \underline{\mathsf{A}}_Q & \xrightarrow{r_F} & \underline{\mathsf{A}}_F . \end{array}$$

   In this diagram, 1_F is the S-valued point corresponding to the identity section of the monoid scheme $\underline{\mathsf{A}}_F$ and $v_{Q/F}$ is the vertex of the monoid scheme $\underline{\mathsf{A}}_{Q/F}$. The composition $r_F \circ i_F$ is $\mathrm{id}_{\underline{\mathsf{A}}_F}$.
2. *The map r_F is a morphism of monoid schemes, and the morphism i_F is compatible with the actions of the monoid scheme $\underline{\mathsf{A}}_Q$ on itself and on the ideal scheme $\underline{\mathsf{A}}_F \subseteq \underline{\mathsf{A}}_Q$.*
3. *If Q is fine, then i_F is a strong deformation retract. That is, there exists a morphism*

$$f \colon \underline{\mathsf{A}}_Q \times \underline{\mathsf{A}}_{\mathsf{m}} \to \underline{\mathsf{A}}_Q$$

   such that $f \circ j_0 = i_F \circ r_F$, $f \circ j_1 = \mathrm{id}$, and $f \circ (i_F \times \mathrm{id}) = i_F \circ pr_1$, where

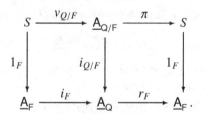

$j_0, j_1 \colon \underline{\mathsf{A}}_{\mathsf{Q}} \to \underline{\mathsf{A}}_{\mathsf{Q}} \times \underline{\mathsf{A}}_{\mathsf{m}}$ are given respectively by the sections 0 and 1 of $\underline{\mathsf{A}}_{\mathsf{m}}$ and $pr_1 \colon \underline{\mathsf{A}}_{\mathsf{F}} \times \underline{\mathsf{A}}_{\mathsf{m}} \to \underline{\mathsf{A}}_{\mathsf{F}}$ is the projection.

Proof The closed immersion i_F preserves the composition law for the monoid schemes $\underline{\mathsf{A}}_{\mathsf{F}}$ and $\underline{\mathsf{A}}_{\mathsf{Q}}$ but not the identity section of the monoid scheme structures, so that $\underline{\mathsf{A}}_{\mathsf{F}}$ cannot be regarded as a submonoid of $\underline{\mathsf{A}}_{\mathsf{Q}}$—in fact it is rather an ideal scheme in the monoid scheme $\underline{\mathsf{A}}_{\mathsf{Q}}$. On the other hand, the inclusion $F \to Q$ defines a homomorphism $R[F] \to R[Q]$ and hence a map $r_F \colon \underline{\mathsf{A}}_{\mathsf{Q}} \to \underline{\mathsf{A}}_{\mathsf{F}}$. Since r_F is induced by a monoid homomorphism, it is a morphism of monoid schemes. It follows from the definitions that $r_F \circ i_F = \mathrm{id}_{\underline{\mathsf{A}}_{\mathsf{F}}}$. Thus r_F and i_F are morphisms of $\underline{\mathsf{A}}_{\mathsf{Q}}$-sets, and $r_F(a) = r_F(a \cdot 1) = ar_F(1_A)$ for every $a \in \underline{\mathsf{A}}_{\mathsf{Q}}(A)$.

One can check immediately that the two squares in the above diagram commute. All the morphisms in the outer rectangle are identity maps, so it is cartesian, and hence the square on the left will automatically be cartesian if the square on the right is cartesian. The latter statement asserts that the ideal of the closed immersion $i_{Q/F}$ is the ideal I generated by the set of all $e^f - 1$ such that $f \in F$. Indeed, it is evident that $i^\sharp_{Q/F}$ annihilates all these elements and hence factors through a map $R[Q]/I \to R[Q/F]$. On the other hand, the map $Q \to R[Q]/I$ sends F to 1, and hence factors through Q/F because of its universal mapping property. This gives the inverse map $R[Q/F] \to R[Q]/I$.

If Q is fine, then by Proposition 2.2.1 there exists a morphism $h \colon Q \to \mathbf{N}$ such that $h^{-1}(0) = F$. This h defines a morphism $t \colon \underline{\mathsf{A}}_{\mathsf{m}} \to \underline{\mathsf{A}}_{\mathsf{Q}}$; on A-valued points $t(a) = a^h$, where a^h is the homomorphism $Q \to A$ sending q to $a^{h(q)}$. Let

$$f \colon \underline{\mathsf{A}}_{\mathsf{Q}} \times \underline{\mathsf{A}}_{\mathsf{m}} \to \underline{\mathsf{A}}_{\mathsf{Q}}$$

be the composition of $\mathrm{id}_{\underline{\mathsf{A}}_{\mathsf{Q}}} \times t$ with the multiplication map m_Q of the monoid structure on $\underline{\mathsf{A}}_{\mathsf{Q}}$. On A-valued points, f sends (α, a) to αa^h. Let i_0 and i_1 be the sections of $\underline{\mathsf{A}}_{\mathsf{m}}$ corresponding to 0 and 1 and let j_0 and j_1 be the corresponding maps $\underline{\mathsf{A}}_{\mathsf{Q}} \to \underline{\mathsf{A}}_{\mathsf{Q}} \times \underline{\mathsf{A}}_{\mathsf{m}}$. We check that $f \circ j_0 = i_F \circ r_F$ and that $f \circ j_1 = \mathrm{id}$ on A-valued points. The second of these equations is obvious and, for the first, we just have to observe that $f(\alpha, 0) = \alpha 0^h$ and remember that 0^n is 0 if $n > 0$ and is 1 if $n = 0$. Finally, if α belongs to the image of i_F then, for every $a \in A$, $f(\alpha, a)(q) = \alpha(q) a^{h(q)} = \alpha(q)$, since $\alpha(q) = 0$ whenever $h(q) \neq 0$. This proves that i_F is a strong deformation retract. □

Corollary 3.3.2. *If Q is a fine monoid, then $\underline{\mathsf{A}}_{\mathsf{Q}}(\mathbf{C})$, endowed with the complex analytic topology, is connected if and only if Q^* is torsion free. If Q is sharp, then $\underline{\mathsf{A}}_{\mathsf{Q}}(\mathbf{C})$ is contractible.*

Proof Proposition 3.3.1 implies that $\underline{\mathsf{A}}_{\mathsf{Q}}(\mathbf{C})$ and $\underline{\mathsf{A}}_{\mathsf{Q}^*}(\mathbf{C})$ have the same homotopy type. Since the torsion subgroup of Q^* identifies with the group of connected components of the algebraic group $\underline{A}_{Q^*}$, it follows that $\underline{A}_Q(\mathbf{C})$ and $\underline{A}_{Q^*}(\mathbf{C})$ are connected if and only if Q^* is torsion free. If Q is sharp, the scheme $\underline{A}_{Q^*}$ reduces to a single point, so $\underline{A}_Q(\mathbf{C})$ is contractible. □

Variant 3.3.3. Proposition 3.3.1 also holds for idealized monoids: if F is a face of (Q, K), then the prime ideal $Q \setminus F$ contains K, so i_F factors through $\underline{A}_{Q,K}$ and the vertex of Q/F lies in $\underline{A}_{Q/F,K/F}$. Furthermore, since $\underline{A}_{Q,K}$ is an ideal of the monoid scheme $\underline{A}_Q$, it is preserved by the homotopy f. As a consequence, $\underline{A}_{Q,K}$ is contractible if Q is sharp and is connected if Q^* is torsion free.

When k is a field and Q is integral, the monoid $\underline{A}_Q(k)$ admits an explicit stratification indexed by the faces of Q. If $x \in \underline{A}_Q(k) = \mathrm{Hom}(Q, k)$, let $F(x) := x^{-1}(k^*)$, a face of Q. If x and z are points of $\underline{A}_Q(k)$, then

$$F(xz) = F(x) \cap F(z).$$

Note that x is zero outside $F(x)$ and induces a map $F^{\mathrm{gp}} \to k^*$ which in fact determines x. Thus we can view a point of $\underline{A}_Q(k)$ as a pair (F, x'), where F is a face of Q and x' is a homomorphism $F^{\mathrm{gp}} \to k^*$.

Proposition 3.3.4. *Let Q be a fine monoid, let k be a field, and let F be a face of Q. Then the set of all $y \in \underline{A}_Q(k)$ such that $F(y) = F$ (or, equivalently, $y^{-1}(0) = Q \setminus F$) is*

$$\underline{A}^*_F(k) := \underline{A}_{F^{\mathrm{gp}}}(k) \subseteq \underline{A}_F(k) \subseteq \underline{A}_Q(k),$$

a Zariski open subset of $\underline{A}_F(k)$. If x and y are two points of $\underline{A}_Q(k)$, then the following are equivalent:

1. *$F(y) \subseteq F(x)$;*
2. *$y \in \underline{A}_{F(x)}(k)$;*
3. *There exists a $z \in \underline{A}_Q(k)$ such that $y = zx$.*

Furthermore, if either k is algebraically closed or $Q^{\mathrm{gp}}/F(x)^{\mathrm{gp}}$ is torsion free, then $F(y) = F(x)$ if and only if there exists a $z \in \underline{A}^*_Q(k)$ with $y = zx$. In particular, if k is algebraically closed or if $Q^{\mathrm{gp}}/F^{\mathrm{gp}}$ is torsion free for every face F of Q, then the partition of $\underline{A}_Q(k) = \sqcup \underline{A}^*_F(k) : F \in \mathcal{F}(Q)$ defined by the faces of Q corresponds to its orbit decomposition under the action of $\underline{A}^*_Q(k)$, and the closed sets $\underline{A}_F(k)$ correspond to the trajectories of $\underline{A}_Q(k)$ under the action of $\underline{A}_Q(k)$ on itself.

Proof We identify a point y of $\underline{A}_{Q(k)}$ with the corresponding character $Q \to k$. Then $F(y) \subseteq F$ if and only if $y(Q \setminus F) = 0$, i.e., if and only if y factors through i_F; hence the equivalence of (1) and (2). By (3) of Theorem 2.1.17, F is fine, so

$\underline{\mathsf{A}}^*_{\mathsf{F}}$ is Zariski dense and open in $\underline{\mathsf{A}}_{\mathsf{F}}$, and the inclusions $\underline{\mathsf{A}}^*_{\mathsf{F}}(k) \subseteq \underline{\mathsf{A}}_{\mathsf{F}}(k) \subseteq \underline{\mathsf{A}}_{\mathsf{Q}}(k)$ identify $\underline{\mathsf{A}}^*_{\mathsf{F}}(k)$ with the set of all y such that $F(y) = F$. If $F(y) \subseteq F(x)$, define $z\colon Q \to k$ by $z(q) := 0$ if $q \in Q \setminus F(x)$ and $z(q) := y(q)/x(q)$ if $q \in F(x)$. Then, using the fact that F is a face of Q, one can check that in fact $z \in \underline{\mathsf{A}}_{\mathsf{Q}}(k)$ and $y = zx$. Thus (2) implies (3), and the converse is obvious. If $F(x) = F(y) = F$, then y/x defines a homomorphism $F^{\mathrm{gp}} \to k^*$. If k is algebraically closed, k^* is divisible, and if $Q^{\mathrm{gp}}/F^{\mathrm{gp}}$ is torsion free, the sequence $F^{\mathrm{gp}} \to Q^{\mathrm{gp}} \to Q^{\mathrm{gp}}/F^{\mathrm{gp}}$ splits. In either case, there exists an extension z of y/x to Q^{gp}, which defines a point of $\underline{\mathsf{A}}^*_{\mathsf{Q}}$ such that $zx = y$. □

Proposition 3.3.4 describes the action of $\underline{\mathsf{A}}^*_{\mathsf{Q}}(k)$ on $\underline{\mathsf{A}}_{\mathsf{Q}}(k)$ when k is a field. We shall need to generalize this discussion somewhat, allowing k to be any ring and furthermore working in a relative situation.

Proposition 3.3.5. *Let $\theta\colon P \to Q$ be a homomorphism of integral monoids, write $Q^{\mathrm{gp}}/P^{\mathrm{gp}}$ for $\mathrm{Cok}(\theta^{\mathrm{gp}})$, and let $\underline{\mathsf{A}}^*_{\mathsf{Q/P}} := \underline{\mathsf{A}}_{\mathsf{Q^{gp}/P^{gp}}} \subseteq \underline{\mathsf{A}}_{\mathsf{Q}}$. Then $\underline{\mathsf{A}}^*_{\mathsf{Q/P}}$ acts naturally on $\underline{\mathsf{A}}_{\mathsf{Q}}$, viewed as an object over $\underline{\mathsf{A}}_{\mathsf{P}}$. For each face F of Q, the subfunctors $\underline{\mathsf{A}}_{\mathsf{Q_F}} \subseteq \underline{\mathsf{A}}_{\mathsf{Q}}$ and $\underline{\mathsf{A}}_{\mathsf{F}} \cong \underline{\mathsf{A}}_{\mathsf{Q},\mathfrak{p}_{\mathsf{F}}} \subseteq \underline{\mathsf{A}}_{\mathsf{Q}}$ are stable under this action. Furthermore, the subgroup $\underline{\mathsf{A}}^*_{\mathsf{Q/(P+F)}}$ of $\underline{\mathsf{A}}^*_{\mathsf{Q/P}}$ acts trivially on the closed subfunctor $\underline{\mathsf{A}}_{\mathsf{F}} \cong \underline{\mathsf{A}}_{\mathsf{Q},\mathfrak{p}_{\mathsf{F}}}$ of $\underline{\mathsf{A}}_{\mathsf{Q}}$.*

Proof Let A be a commutative ring and suppose that $\gamma \in \underline{\mathsf{A}}^*_{\mathsf{Q/P}}(A)$ and $\alpha \in \underline{\mathsf{A}}_{\mathsf{Q}}(A)$. Then $\gamma(\theta(p)) = 1$ for every $p \in P$, and hence $(\gamma\alpha)(\theta(p)) = \alpha(\theta(p))$. In other words, γ acts on $\underline{\mathsf{A}}_{\mathsf{Q}}$ as an object over $\underline{\mathsf{A}}_{\mathsf{P}}$. Furthermore, for all $q \in Q$, $\gamma(q) \in A^*$. In particular, $\gamma(q)\alpha(q) = 0$ if and only if $\alpha(q) = 0$, and applying this to all q in $\mathfrak{p}_F$, we see that $\gamma\alpha \in \underline{\mathsf{A}}_{\mathsf{Q},\mathfrak{p}_{\mathsf{F}}}(A)$ if and only if $\alpha \in \underline{\mathsf{A}}_{\mathsf{Q},\mathfrak{p}_{\mathsf{F}}}(A)$. Similarly, $\gamma(q)\alpha(q) \in A^*$ if and only if $\alpha(q) \in A^*$, and applying this to all $q \in F$ we see that $\gamma\alpha \in \underline{\mathsf{A}}_{\mathsf{Q_F}}(A)$ if and only if $\alpha \in \underline{\mathsf{A}}_{\mathsf{Q_F}}(A)$. Now suppose that $\gamma \in \underline{\mathsf{A}}^*_{\mathsf{Q/P+F}}(A)$ and $\alpha \in \underline{\mathsf{A}}_{\mathsf{F}}(A) \cong \underline{\mathsf{A}}_{\underline{\mathsf{A}}_{\mathsf{Q}},\mathfrak{p}_{\mathsf{F}}}(A)$. We claim that $\gamma(q)\alpha(q) = \alpha(q)$ for all $q \in Q$. But this is clear: if $q \in \mathfrak{p}_F$ then $\alpha(q) = 0$, and if $q \in F$ then $\gamma(q) = 1$. □

Remark 3.3.6. Observe that the homomorphism $Q^{\mathrm{gp}} \to Q^{\mathrm{gp}}/P^{\mathrm{gp}}$ induces a morphism of group schemes $\underline{\mathsf{A}}^*_{\mathsf{Q/P}} \to \underline{\mathsf{A}}^*_{\mathsf{Q}}$ identifying $\underline{\mathsf{A}}^*_{\mathsf{Q/P}}$ with the kernel of the homomorphism $\underline{\mathsf{A}}^*_{\mathsf{Q}} \to \underline{\mathsf{A}}^*_{\mathsf{P}}$. Let us write the action of $\underline{\mathsf{A}}^*_{\mathsf{Q/P}}$ on $\underline{\mathsf{A}}_{\mathsf{Q}}$ on the right, given by a morphism of monoid schemes

$$m_{Q/P}\colon \underline{\mathsf{A}}_{\mathsf{Q}} \times \underline{\mathsf{A}}^*_{\mathsf{Q/P}} \to \underline{\mathsf{A}}_{\mathsf{Q}}\,.$$

This morphism corresponds to the homomorphism of monoids

$$\theta\colon Q \to Q \oplus Q^{\mathrm{gp}}/P^{g} : q \mapsto (q \oplus [q]).$$

Let us check this by computing the effect on A-valued points. If $(\alpha, \gamma) \in$

$\underline{\mathsf{A}}_{\mathsf{Q}}(A) \times \underline{\mathsf{A}}^*_{\mathsf{Q/P}}(A)$ and $q \in Q$,

$$\begin{aligned}(\alpha,\gamma)(m^{\flat}_{Q/P}(q)) &= m_{Q/P}(\alpha,\gamma)(q)\\ &= (\alpha\gamma)(q)\\ &= \alpha(q)\gamma([q])\\ &= (\alpha,\gamma)(q,[q])\\ &= (\alpha,\gamma)(\theta(q)).\end{aligned}$$

The action of $\underline{\mathsf{A}}^*_{\mathsf{Q/P}}$ on $\underline{\mathsf{A}}_{\mathsf{Q}}$ induces a map

$$h_{Q/P}\colon \underline{\mathsf{A}}_{\mathsf{Q}} \times \underline{\mathsf{A}}^*_{\mathsf{Q/P}} \to \underline{\mathsf{A}}_{\mathsf{Q}} \times_{\underline{\mathsf{A}}_{\mathsf{P}}} \underline{\mathsf{A}}_{\mathsf{Q}} : (\alpha,\gamma) \mapsto (\alpha,\alpha\gamma).$$

This map corresponds to the homomorphism of monoids

$$\phi\colon Q \oplus_P Q \to Q \oplus Q^{\mathrm{gp}}/P^{\mathrm{gp}} : [(q_1,q_2)] \mapsto (q_1+q_2,[q_2]). \tag{3.3.3}$$

Note that the induced morphisms

$$Q^{\mathrm{gp}} \oplus_{P^{\mathrm{gp}}} Q^{\mathrm{gp}} \to Q^{\mathrm{gp}} \oplus Q^{\mathrm{gp}}/P^{\mathrm{gp}} \quad \text{and} \quad \underline{\mathsf{A}}^*_{\mathsf{Q}} \times \underline{\mathsf{A}}^*_{\mathsf{Q/P}} \to \underline{\mathsf{A}}^*_{\mathsf{Q}} \times_{\underline{\mathsf{A}}^*_{\mathsf{P}}} \underline{\mathsf{A}}^*_{\mathsf{Q}} \tag{3.3.4}$$

are isomorphisms. Thus $\underline{\mathsf{A}}^*_{\mathsf{Q}}$, viewed as an object over $\underline{\mathsf{A}}^*_{\mathsf{P}}$, is a torsor under the induced action of $\underline{\mathsf{A}}^*_{\mathsf{Q/P}}$. The inverse of the isomorphism ϕ^{gp} is given by the formula

$$Q^{\mathrm{gp}} \oplus Q^{\mathrm{gp}}/P^{\mathrm{gp}} \to Q^{\mathrm{gp}} \oplus_{P^{\mathrm{gp}}} Q^{\mathrm{gp}} : (x,[y]) \mapsto [(x-y,y)]$$

3.4 Local geometry of affine toric varieties

Here we collect some basic facts relating algebraic properties of a fine monoid to the algebra and geometry of its monoid algebra and monoid scheme.

Proposition 3.4.1. *Let P be an integral monoid and let R be a ring.*

1. *If P^{gp} is torsion free and R is an integral domain, then $R[P]$ is an integral domain.*
2. *If in addition P is finitely generated and R is normal, $R[P^{\mathrm{sat}}]$ is the normalization of $R[P]$. Thus in this case $R[P]$ is normal if and only if P is saturated.*
3. *If P is fine, the morphism $\pi\colon \operatorname{Spec} R[P] \to \operatorname{Spec} R$ is faithfully flat and of finite presentation. Furthermore, the Krull dimension of the fibers of π is the rank of P^{gp}. If in addition the order of the torsion subgroup of P^{gp} is invertible in R, then the fibers are geometrically reduced.*

Proof First suppose that P is finitely generated. Then if P^{gp} is torsion free, it is free of finite rank, so

$$R[P^{\mathrm{gp}}] \cong R[T_1, T_1^{-1}, \ldots, T_n, T_n^{-1}]$$

for some n. (Geometrically, $\underline{A}_P^* = \underline{A}_{P^{gp}}$ is a torus over Spec R.) In particular $R[P^{gp}]$ is an integral domain and, since $R[P]$ is contained in $R[P^{gp}]$, it too is an integral domain. In general, P is the union of its finitely generated submonoids P_λ, and each P_λ^{gp} is torsion free if P^{gp} is. Then $R[P]$ is the direct limit of the set of all $R[P_\lambda]$, each of which is an integral domain, and hence it too is an integral domain. Since $R[P^{sat}]$ is generated as an $R[P]$-algebra by P^{sat} and since e^q is integral over $R[P]$ for every $q \in P^{sat}$, $R[P^{sat}]$ is integral over $R[P]$. Since $R[P^{sat}]$ is contained in $R[P^{gp}]$, which in turn is contained in the fraction field of $R[P]$, $R[P^{sat}]$ is contained in the normalization of $R[P]$. It remains only to prove that $R[P^{sat}]$ is normal, assuming that P is fine. Since P^{sat} is then fine, we may and shall assume without loss of generality that P is saturated. By Corollary 2.4.5, P is the intersection in P^{gp} of all its localizations at height one primes $\mathfrak{p}$, and one sees by looking at basis elements that $R[P]$ is the intersection in $R[P^{gp}]$ of the corresponding monoid algebras $R[P_\mathfrak{p}]$. Since the intersection of a family of normal subrings of a ring is normal, it will suffice to prove that each $R[P_\mathfrak{p}]$ is normal. Replacing P by $P_\mathfrak{p}$, we may assume that P is saturated and of dimension one. Then, by Corollary 2.4.3, P is isomorphic to $\mathbf{Z}^r \oplus \mathbf{N}$ for some natural number r. Thus $R[P] \cong R[T_1, T_1^{-1}, \ldots, T_n, T_n^{-1}, T]$, which is normal since R is.

If P is fine, then it is of finite presentation by Theorem 2.1.7, and it follows that $R[P]$ is finitely presented as an R-algebra. Since by construction it is a free R-module, it is necessarily faithfully flat. To prove the statements about the fibers, we may as well assume that R is an algebraically closed field k. Since $k[P] \to k[P^{sat}]$ is finite and injective, $k[P]$ and $k[P^{sat}]$ have the same dimension, and $k[P]$ is reduced if $k[P^{sat}]$ is. Since $P^{gp} = (P^{sat})^{gp}$, we may assume without loss of generality that P is saturated. Then the torsion subgroup P_t of P^{gp} is contained in P^* by (2) of Proposition 1.3.5. Choose a splitting σ of the projection $\pi\colon P^{gp} \to P^{gp}/P_t$. Then $\sigma(\pi(p)) - p \in P_t \subseteq P$ for every $p \in P$, and it follows that σ maps P/P_t to P. Thus $P \cong P/P_t \oplus P_t$ and $k[P] \cong k[P/P_t] \otimes k[P_t]$. Since P_t is a finite group, it follows that the dimension of $k[P]$ is the same as the dimension of $k[P/P_t]$. Thus we may assume that P^{gp} is torsion free. Then $k[P]$ is an integral domain, and its dimension is the same as the transcendence degree of its fraction field, which is $\mathrm{rk}(P^{gp})$. Finally, if the order of P_t is invertible in R, then $k[P_t]$ is étale over k. Then $k[P] \cong [P/P_t] \otimes k[P_t]$ is reduced, since it is étale over the integral domain $k[P/P_t]$. □

To see that the hypothesis that P^{gp} be torsion free is not superfluous, consider the submonoid P of $\mathbf{Z} \oplus \mathbf{Z}/2\mathbf{Z}$ generated by $p := (1, 0)$ and $q := (1, 1)$. This is the free monoid generated by p and q subject to the relation $2p = 2q$. It is sharp and fine, but $R[P] \cong R[x,y]/(x^2 - y^2)$, which is not an integral domain. Note also that its fibers in characteristic 2 are not reduced.

Proposition 3.4.2. *Let Q be a fine monoid, let k be a field, and let $\underline{\mathsf{A}}_{\mathsf{Q}} :=$ Spec $k[Q]$.*

1. *If $\mathfrak{p}$ is a prime ideal of Q and $F := Q \setminus \mathfrak{p}$, then the height of $\mathfrak{p}$ in Q is the codimension of the closed subscheme $\underline{\mathsf{A}}_{\mathsf{F}} \cong \underline{\mathsf{A}}_{\mathsf{Q},\mathfrak{p}}$ of $\underline{\mathsf{A}}_{\mathsf{Q}}$.*
2. *If x is a scheme-theoretic point of $\underline{\mathsf{A}}_{\mathsf{Q}}$, let $\mathfrak{p}$ be the set of elements of Q whose image in $\mathcal{O}_{X,x}$ lies in the maximal ideal m_x of x, and let $F := Q \setminus \mathfrak{p}$, so that $x \in \underline{\mathsf{A}}^*_{\mathsf{F}} \subseteq \underline{\mathsf{A}}_{\mathsf{F}} \subseteq \underline{\mathsf{A}}_{\mathsf{Q}}$. Then $\mathrm{ht}(\mathfrak{p}) \leq \dim(\mathcal{O}_{X,x})$, and equality holds if and only if x has codimension zero in $\underline{\mathsf{A}}^*_{\mathsf{F}}$, that is, if and only if x is a generic point of $\underline{\mathsf{A}}^*_{\mathsf{F}}$. If equality holds and the order of the torsion part of F^{gp} is invertible in k, then m_x is generated by the image of $\mathfrak{p}$ in $\mathcal{O}_{X,x}$.*

Proof Note that in these statements we are using the closed immersion i_F (3.3.2) to identify $\underline{\mathsf{A}}_{\mathsf{F}}$ and $\underline{\mathsf{A}}_{\mathsf{Q},\mathfrak{p}}$. The dimension of $\underline{\mathsf{A}}_{\mathsf{F}}$ is the rank of F^{gp} and the dimension of $\underline{\mathsf{A}}_{\mathsf{Q}}$ is the rank of Q^{gp}, so the codimension of the locally closed subset $\underline{\mathsf{A}}_{\mathsf{F}}$ of $\underline{\mathsf{A}}_{\mathsf{Q}}$ is the rank of $Q^{\mathrm{gp}}/F^{\mathrm{gp}}$. This is the height of $\mathfrak{p}$ as a prime of Q, by Proposition 1.4.7.

It is clear from the definitions that x belongs to $\underline{\mathsf{A}}^*_{\mathsf{F}} \subseteq \underline{\mathsf{A}}_{\mathsf{F}} \subseteq \underline{\mathsf{A}}_{\mathsf{Q}}$. Let η be the generic point of an irreducible component of $\underline{\mathsf{A}}^*_{\mathsf{F}}$ containing x. As we have just seen, the height of η is the same as the height of $\mathfrak{p}$. Since x is a specialization of η, $\mathrm{ht}(x) \geq \mathrm{ht}(\eta)$, with equality if and only if $\eta = x$, that is, if and only if x is a generic point of $\underline{\mathsf{A}}^*_{\mathsf{F}}$. Suppose this is the case and that the order of the torsion part of F^{gp} is invertible in k. Then $\underline{\mathsf{A}}_{\mathsf{F}}$ is an open subscheme of the reduced scheme $\underline{\mathsf{A}}^*_{\mathsf{F}}$ and so is also reduced. Consequently the local ring $\mathcal{O}_{\underline{\mathsf{A}}^*_{\mathsf{F}},x} \cong \mathcal{O}_{\underline{\mathsf{A}}^*_{\mathsf{Q}},x}/\mathfrak{p}\mathcal{O}_{\underline{\mathsf{A}}^*_{\mathsf{Q}},x}$ of $\underline{\mathsf{A}}^*_{\mathsf{F}}$ is a field. Thus $\mathfrak{p}$ generates the maximal ideal of $\mathcal{O}_{\underline{\mathsf{A}}^*_{\mathsf{Q}},x}$. □

Theorem 3.4.3. *If R is a noetherian Cohen–Macaulay ring, and P is a fine saturated monoid, then the monoid ring $R[P]$ is also Cohen–Macaulay.*

Proof When R is a field, this result is a deep theorem of Hochster whose proof [36] we will not give here. The morphism Spec $R[P] \to$ Spec R is flat, and its fibers are Cohen–Macaulay by Hochster's theorem. It follows that $R[P]$ is also Cohen–Macaulay [27, 6.3.5]. □

The following result is an immediate consequence of the analogous Corollary 2.3.9 for monoids.

Proposition 3.4.4. *Let R be a ring and Q a fine monoid of Krull dimension d. For each $i = 0, \ldots, d$, let $K_i := \cap\{\mathfrak{p} \in \operatorname{Spec} R : \mathrm{ht}\,\mathfrak{p} \leq i\}$, an ideal of Q. Then $\underline{\mathsf{A}}_{\mathsf{Q}} \setminus \underline{\mathsf{A}}_{\mathsf{Q},\mathsf{K}_{i+1}}$ is covered by the special affine open subsets $\underline{\mathsf{A}}_{\mathsf{Q}_\mathsf{F}}$, where F ranges* *over the set of faces F such that* $\operatorname{rk} Q/F = i$. *In particular,* $\underline{\mathsf{A}}_Q \setminus \underline{\mathsf{A}}_{Q,K_2}$ *is covered by the open sets of the form* $\underline{\mathsf{A}}_{Q_F}$ *as F ranges over the facets of Q. If Q is toric and $d > 0$, each such open set is a product of a torus with an affine line over* $\operatorname{Spec} R$. □

3.5 Ideals in monoid algebras

Let P be an integral monoid and R a nonzero ring. We shall find it useful to investigate further the relationship between ideals in P and ideals in $R[P]$. Recall that if K is an ideal in P, then $R[K] \subseteq R[P]$ is the set of elements $\sum\{a_p : p \in P\}$ such that $a_p = 0$ for $p \notin K$, an ideal of $R[P]$. It follows that if K is the intersection of a family of ideals K_λ of P, then $R[K]$ is the intersection of the corresponding family of ideals $R[K_\lambda]$ of $R[P]$.

Proposition 3.5.1. *Let Q be an integral monoid, let R be a reduced ring, and let K be an ideal of Q. Assume that the order of the torsion subgroup of Q^{gp} is invertible in R. Then the ring $R[Q, K]$ is reduced if and only if K is a radical ideal of Q.*

Proof Recall that $Q \setminus K \to R[Q, K] : q \mapsto e^q$ is a basis for $k[Q, K]$. Thus it is clear that if $R[Q, K]$ is reduced, then K must be a radical ideal, since otherwise there would exist $q \in Q \setminus K$ and $n \in \mathbf{N}$ such that $nq \in K$, and then e^q would be a nonzero nilpotent of $R[Q, K]$. For the converse, first note that if K is prime, its complement is a face F of Q, and $R[Q, K] \cong R[F]$. Furthermore, $R[F] \subseteq R[F^{\mathrm{gp}}] \cong R[\mathbf{Z}^r \oplus T]$, where T is a finite group whose order is invertible in R. It follows that $R[F^{\mathrm{gp}}]$ and $R[F]$ are reduced. If K is a radical ideal of P, then as we saw in Corollary 1.4.3, it is the intersection of the prime ideals $\mathfrak{p}_\lambda$ containing it, and it follows that $R[K]$ is the intersection of the ideals $R[\mathfrak{p}_\lambda]$. Since each of these ideals is reduced, so is $R[K]$. □

Proposition 3.5.2. *Let R be an integral domain, let P be a toric monoid, and let p be an element of P. Then the irreducible components of* $\operatorname{Spec} R[P, (p)]$ *are precisely the closed sets defined by the height one prime ideals of P containing p. If P is saturated and R is normal, each of these irreducible components is normal.*

Proof Let $\mathfrak{p}$ be any prime of P and let G be its complementary face. Since $R[P, \mathfrak{p}] \cong R[G]$, it is an integral domain and is normal if R is normal and G is saturated, by Proposition 3.4.1. By Corollary 2.3.15, the ideal $\sqrt{(p)}$ of P can be written as a finite intersection of height one primes $\mathfrak{p}_1 \cap \cdots \cap \mathfrak{p}_n$. We conclude that $\operatorname{Spec}(R[P, (p)] = \operatorname{Spec} R[P, \mathfrak{p}_1] \cup \cdots \cup R[P, \mathfrak{p}_n]$. As we have seen, each of these pieces is irreducible and hence these are indeed the irreducible components of $R[P,(p)]$. If P is saturated, so is each face $G_i := P \setminus \mathfrak{p}_i$, and hence if R is normal, so is each $R[P, \mathfrak{p}_i] \cong R[G_i]$. □

This proposition shows that the Weil divisors with support in a closed subset of Spec $R[P]$ defined by an element of P come from ideals in P. To obtain the analogous result for Cartier divisors will require more work. We begin with the following definition.

Definition 3.5.3. *Let P be an integral monoid and let R be a ring. If $f := \sum_P a_p(f)e^p$ is an element of $R[P]$ and S is a subset of $R[P]$, then:*

1. *$\sigma(f) := \{p \in P : a_p(f) \neq 0\}$, and $\sigma(S) := \cup\{\sigma(f) : f \in S\}$;*
2. *$K(f)$ is the ideal of P generated by $\sigma(f)$, and $K(S) := \cup\{K(f) : f \in S\}$.*

The set $\sigma(f)$ is called the *support of* f, and its convex hull is the *Newton polyhedron* of f. Note that $K(S)$ is an ideal of P, since the union of ideals is an ideal. Furthermore, for any $p \in P$ and $f \in R[P]$,

$$\sigma(e^p f) = p + \sigma(f).$$

It follows that if I is an ideal of $R[P]$, then $\sigma(I) = K(I)$, since if $k \in K(I)$, there exist $f \in I$ and $p \in P$ with $k \in p + \sigma(f) = \sigma(e^p f) \in \sigma(I)$. Note that any f in $R[P]$ is contained in $R[K(f)]$, that any ideal I of $R[P]$ is contained in $R[K(I)]$, and that in fact $K(I)$ is the smallest ideal K of P such that $I \subseteq R[K]$. It is clear from the definitions that if $\theta\colon P \to Q$ is an injective homomorphism of monoids and $f \in R[P]$, then $\sigma(R[\theta](f)) = \theta(\sigma(f))$ and that $K(R[\theta](f))$ is the ideal of Q generated by $\theta(K(f))$.

Proposition 3.5.4. *Suppose that f and g are elements of $R[P]$.*

1. *$\sigma(f + g) \subseteq \sigma(f) \cup \sigma(g)$, hence $K(f + g) \subseteq K(f) \cup K(g)$.*
2. *$\sigma(fg) \subseteq \sigma(f) + \sigma(g)$, hence $K(fg) \subseteq K(f) + K(g) \subseteq K(f) \cap K(g)$.*
3. *$K(f) = K((f))$, where (f) is the ideal of $R[P]$ generated by f.*
4. *If I and J are ideals of $R[P]$, $K(IJ) \subseteq K(I) + K(J)$.*

Proof The first two statements follow from the fact that for every $p \in P$,

$$a_p(f+g) = a_p(f) + a_p(g) \quad \text{and}$$
$$a_p(fg) = \sum_{q+q'=p} a_q(f)a_{q'}(g),$$

and statement (4) is an immediate consequence. It is apparent from the definition that $\sigma(f) \subseteq K((f))$, and hence that $K(f) \subseteq K((f))$. On the other hand, for any $h \in (f)$, it follows from (2) that $\sigma(h) \subseteq K(f)$ and hence that $K(h) \subseteq K(f)$. □

We shall be especially interested in determining when $K(f)$ is principal.

Proposition 3.5.5. *Let P an integral monoid and let R be a ring.*

1. *If $f \in R[P]$, $K(f)$ is the unit ideal of P if and only if $f \notin R[P^+]$.*
2. *If $f \in R[P]$, then $K(f)$ is principally generated by an element p of P if and only if $f = e^p \tilde{f}$, where $\tilde{f}$ is some element of $R[P] \setminus R[P^+]$.*
3. *Suppose R is an integral domain and P^* is torsion free. Then if f and g are elements of $R[P]$ such that $K(f)$ and $K(g)$ are principal, the same is true of fg, and $K(fg) = K(f) + K(g)$.*

Proof Statement (1) is a tautology. If $K := K(f)$ is generated by p, then $k - p \in P$ for every element k of $K(f)$. Hence $f = \sum_{k \in K} a_k e^k = e^p \sum_k a_k e^{k-p}$, so $f = e^p \tilde{f}$ where $\tilde{f} := \sum_k a_k e^{k-p}$. Then

$$(p) = K(f) \subseteq K(e^p) + K(\tilde{f}) = (p) + K(\tilde{f}),$$

and it follows that $K(\tilde{f}) = P$. Conversely, if $f = e^p \tilde{f}$ with $K(\tilde{f}) = P$, then certainly $K(f) \subseteq (p)$. But if $\tilde{f} = \sum \tilde{a}_q e^q$, there exists a $q \in P^*$ such that $\tilde{a}_q \neq 0$, and then $p + q \in K(f)$, so $p \in K(f)$. Thus $K(f) = (p)$, and this proves (2). If $K(f)$ is principally generated by p and $K(g)$ is principally generated by q, then $f = e^p \tilde{f}$ and $g = e^q \tilde{g}$, where $\tilde{f}$ and $\tilde{g}$ belong to $R[P] \setminus R[P^+]$. The quotient of $R[P]$ by $R[P^+]$ is isomorphic to $R[P^*]$. If P^* is torsion free and R is an integral domain, then $R[P^*]$ is also an integral domain by Proposition 3.4.1. Hence $R[P^+]$ is a prime ideal, and so $\tilde{f}\tilde{g} \notin R[P^+]$. Since $fg = e^{p+q} \tilde{f}\tilde{g}$, it follows that $K(fg)$ is principally generated by $p + q$. □

Statement (3) shows that if R is a domain and P^* is torsion free, the set of all f such that $K(f)$ is principal is a submonoid of $R[P]$. We shall see that if P is toric, this set is in fact a face of $R[P]$. This is not true in general. Consider for example the submonoid P of $\mathbf{N}$ generated by the elements 2 and 3, and let $f = e^2 + e^3$ and $g = e^2 - e^3$. Then $K(f) = K(g)$ is the ideal $(2, 3)$ of P, which is not principal, but $fg = e^4 - e^6 = e^4(1 - e^2)$, so $K(fg)$ is principally generated by 4.

Recall from Corollary 2.4.4 that associated to each height one prime $\mathfrak{p}$ of a fine monoid P there is a homomorphism $\nu_{\mathfrak{p}} \colon P \to \mathbf{N}$; this homomorphism is surjective if P is saturated, as we shall assume. The image under $\nu_{\mathfrak{p}}$ of a nonempty ideal K of P is then an ideal $K_{\mathfrak{p}}$ of $\mathbf{N}$, principally generated by

$$\nu_{\mathfrak{p}}(K) := \inf\{\nu_{\mathfrak{p}}(k) : k \in K\}.$$

For each $f \in R[P]$, let

$$\nu_{\mathfrak{p}}(f) := \nu_{\mathfrak{p}}(K(f)).$$

That is, $\nu_{\mathfrak{p}}(f)$ is the minimum of the set of all $\nu_{\mathfrak{p}}(p)$ such that $p \in \sigma(f)$.

Proposition 3.5.6. *Let P be a toric monoid and let R be an integral domain.*

1. *If K is a nonempty ideal of P and $p \in K$ is an element of K such that $\nu_{\mathfrak{p}}(p) = \nu_{\mathfrak{p}}(K)$ for every height one prime $\mathfrak{p}$, then $K = (p)$.*
2. *If f and g are elements of $R[P]$, then $\nu_{\mathfrak{p}}(fg) = \nu_{\mathfrak{p}}(f) + \nu_{\mathfrak{p}}(g)$, for every height one prime $\mathfrak{p}$. Moreover, $K(fg)$ is principal if and only if $K(f)$ and $K(g)$ are.*

Proof Suppose the hypotheses of (1) hold and $k \in K$. Then $\nu_{\mathfrak{p}}(k - p) \geq 0$ for every height one prime $\mathfrak{p}$ of P. By Corollary 2.4.5, $k - p \in P$, and it follows that K is principally generated by p. This proves (1). For (2), let G be the facet of P complimentary to $\mathfrak{p}$. The homomorphism $P \to P_G$ is injective, and hence if $\tilde{f} \in R[P_G]$ is the image of f, $K(\tilde{f})$ is the ideal of P_G generated by the image of $K(f)$, and $\nu_{\mathfrak{p}}(\tilde{f}) = \nu_{\mathfrak{p}}(f)$. Since P_G is valuative, $K(\tilde{f})$ is principal, generated by any $p \in \sigma(f)$ such that $\nu_{\mathfrak{p}}(f) = \nu_{\mathfrak{p}}(p)$. Since $P_G^* = G^{\mathrm{gp}}$ is torsion free, (3) of Proposition 3.5.5 implies that for every $f, g \in R[P]$, $K(\tilde{f}\tilde{g}) = K(\tilde{f}) + K(\tilde{g})$ and hence $\nu_{\mathfrak{p}}(fg) = \nu_{\mathfrak{p}}(f) + \nu_{\mathfrak{p}}(g)$. We already know that $K(fg)$ is principal if $K(f)$ and $K(g)$ are. Conversely, if $K(fg)$ is principally generated by r, statement (2) of Proposition 3.5.4 shows that r can be written as a sum $p + q$, with $p \in K(f)$ and $q \in K(g)$. Then for every $\mathfrak{p}$ of height one, $\nu_{\mathfrak{p}}(p) \geq \nu_{\mathfrak{p}}(f)$ and $\nu_{\mathfrak{p}}(q) \geq \nu_{\mathfrak{p}}(g)$. On the other hand, $\nu_{\mathfrak{p}}(p) + \nu_{\mathfrak{p}}(q) = \nu_{\mathfrak{p}}(r) = \nu_{\mathfrak{p}}(fg) = \nu_{\mathfrak{p}}(f) + \nu_{\mathfrak{p}}(g)$. Hence $\nu_{\mathfrak{p}}(p) = \nu_{\mathfrak{p}}(f)$ and $\nu_{\mathfrak{p}}(q) = \nu_{\mathfrak{p}}(g)$ for every $\mathfrak{p}$. By (1), this implies that $K(f)$ and $K(g)$ are principal. □

Corollary 3.5.7. *Let R be an integral domain, P a toric monoid, and F a face of P. Then the set $\mathcal{F}$ of elements f of $R[P]$ such that $K(f)$ is principally generated by an element of F is a face of the multiplicative monoid $R[P]$.*

Proof If f and g belong to $\mathcal{F}$, then $K(f) = (p)$ and $K(g) = (q)$ with p and q in F, so by (3) of Proposition 3.5.5 $K(fg) = (p + q)$ and $p + q \in F$. Thus $\mathcal{F}$ is a submonoid of $R[P]$. Conversely if $fg \in \mathcal{F}$ then, by (2) of Proposition 3.5.6, $K(f)$ and $K(g)$ are principal, say generated by p and q respectively. Then $p + q$ generates $K(fg)$ and lies in F. Since F is a face, each of p and q belongs to F and hence each of f and g belongs to $\mathcal{F}$. Thus $\mathcal{F}$ is a face of $R[P]$. □

We can now describe the monoid of effective Cartier divisors supported on the closed subset of the toric variety $\underline{\mathsf{A}}_{\mathsf{P}}$ defined by an element of P. In Theorem III.1.9.4, we shall use this description to compute certain "compactification log structures," generalizing a theorem of Kato [49, 11.6]. Note that if P is a toric monoid and $p \in P$, then $e^p \in R[P]$ is a nonzero divisor and hence defines an effective Cartier divisor D_p in the scheme $\operatorname{Spec}(R[P])$.

Theorem 3.5.8. *Let P be a fine sharp and toric monoid and let R be a normal* *integral domain. Let f be an element of P, let $Y := D_f$, and let $\underline{\Gamma}_Y(Div_X^+)$ be the sheaf of effective Cartier divisors of $X := \underline{A}_P$ with support in Y. For each $x \in X$, let G_x denote the set of elements of P whose image in $k(x)$ is nonzero, and let F_x denote the face of P_{G_x} generated by f. Then, for each $p \in F_x$, the germ of the Cartier divisor D_p at x belongs to $\underline{\Gamma}_Y(Div_X^+)_x$, and the map $p \mapsto D_p$ induces an isomorphism*

$$\overline{F}_x \cong \underline{\Gamma}_Y(Div_X^+)_x.$$

Proof This statement is local around x, so we may without loss of generality replace P by its localization by G_x. Thus we may and shall assume that $G_x = P^*$, so that every element of P^+ vanishes at x. If p belongs to the face F of P generated by q, then e^p maps to a unit in the ring $R[P_q]$ and hence D_p has support in Y. Thus we find a homomorphism $F \to \Gamma_Y(Div_X^+)$, and we claim that the induced homomorphism $\overline{\eta}\colon \overline{F} \to \underline{\Gamma}_Y(Div_X^+)_x$ is an isomorphism.

Suppose that p and q are elements of F such that e^p and e^q define the same ideal of $\mathcal{O}_{X,x}$. Then there exist u and $v \in R[P]$ not vanishing at x such that $ue^p = ve^q$. But then u and v belong to $R[P] \setminus R[P^+]$, and hence by Proposition 3.5.5 $K(ue^p) = (p)$ and $K(ve^q) = (q)$. Then $(p) = (q)$ as ideals in P, and so p and q have the same image in $\overline{F}$. This proves the injectivity of η. To prove the surjectivity, let I be a principal ideal of $\mathcal{O}_{X,x}$ that becomes the unit ideal after localization by e^f. Then if a generates I, there exist $b \in \mathcal{O}_{X,x}$ and $r \in F$ such that $ab = e^r$. Furthermore, there exist u and $v \in R[P]$ not vanishing at x such that $\alpha := au$ and $\beta := bv$ belong to $R[P]$. Then $uv \notin R[P^+]$ and $\alpha\beta = aubv = e^r uv$. By Proposition 3.5.5, $K(\alpha\beta)$ is generated by r, an element of F. Then it follows from Corollary 3.5.7 that $K(\alpha)$ and $K(\beta)$ are respectively generated by elements p and q of F. Write $\alpha = e^p\tilde{\alpha}$ and $\beta = e^q\tilde{\beta}$ with $\tilde{\alpha}$ and $\tilde{\beta}$ in $R[P] \setminus R[P^+]$. It follows that $(p+q) = (r)$ and then that $(\tilde{\alpha}\tilde{\beta}) = (uv)$, so $\tilde{\alpha}$ and $\tilde{\beta}$ do not vanish at x. Since $a = \tilde{\alpha}u^{-1}e^p$ in $\mathcal{O}_{X,x}$, e^p generates I, i.e., $I = \eta(p)$. □

3.6 Completions and formal power series

Let Q be a fine sharp monoid and let R be a commutative ring. We shall see that the completion of $R[Q]$ with respect to the ideal $R[Q^+]$ can be conveniently viewed as a ring of *formal power series in* Q. Explicitly, we denote by $R[[Q]]$ the set of functions $Q \to R$, viewed as an R-module using the usual pointwise structure and endowed with the product topology induced by the discrete topology on R.

Proposition 3.6.1. *Let Q be a fine sharp monoid and let R be a commutative ring.*

1. *For each $q \in Q$, $\{(p, p') \in Q \times Q : p + p' = q\}$ is finite.*
2. *The topological R-module $R[[Q]]$ defined above admits a unique continuous multiplication with the property that the natural map $\hat{e}\colon Q \to R[[Q]]$ is a homomorphism of monoids.*
3. *The topological ring $R[[Q]]$ is naturally identified with the formal completion of $R[Q]$ along the ideal $R[Q^+]$.*
4. *If Q^{gp} is torsion free and R is an integral domain, $R[[Q]]$ is also an integral domain.*
5. *If R is a local ring with maximal ideal $\mathfrak{m}$, then $R[[Q]]$ is also a local ring, whose maximal ideal is the set of elements of $R[[Q]]$ whose constant term belongs to $\mathfrak{m}$. Furthermore, $Q \to R[[Q]]$ is a local homomorphism.*

Proof Since Q is fine, there exists a local homomorphism $h\colon Q \to \mathbf{N}$. Then $J_{h,n} := \{q : h(q) > n\}$ is an ideal of Q and, by Proposition 2.2.9, $Q \setminus J_{h,n}$ is a finite set. If $q = p + p'$, then p and p' belong to $Q \setminus J_{h,h(q)}$, and hence there are only finitely many such p and p'. Then if $f = \sum a_p e^p$ and $g = \sum b_{p'} e^{p'}$ are elements of $R[[Q]]$, we can define fg to be $\sum c_q e^q$, where

$$c_q := \sum_{p+p'=q} a_p b_{p'}.$$

For each finite subset σ of Q,

$$U_\sigma := \left\{\sum a_q e^q \in R[[Q]] : a_q = 0 \text{ for } q \in \sigma\right\}$$

is open, and the family of such subsets is a basis of open neighborhoods of zero. It follows that the multiplication operation defined above is continuous, and indeed that it is the unique continuous operation compatible with the monoid law on Q.

If K is an ideal of Q and n is a natural number, let K^n denote the set of all elements of Q which can be written as a sum of n or more elements of K. Then K is an ideal of Q and $R[K^n] = (R[K])^n$. Lemma 3.6.2 below implies that the $R[[Q^+]]$-adic topology on $R[[Q]]$ agrees with the weak topology.

Lemma 3.6.2. *Let Q be a fine sharp monoid and let $h\colon Q \to \mathbf{N}$ be a local homomorphism. The following families of subsets of Q are cofinal; that is, given any member of one of these families, there is another member of each of the other families which it contains:*

1. *$\{J_{h,n} : n \in \mathbf{N}\}$, where $J_{h,n} = \{q : h(q) > n\}$;*
2. *$\{(Q^+)^n : n \in \mathbf{N}\}$;*
3. *the set of all subsets of Q whose complement is finite.*

Proof It is clear that $(Q^+)^n \subseteq J_{h,n-1}$ for every n, since if $q = q_1 + \cdots + q_n \in (Q^+)^n$ with each $q_i \in Q^+$, then $h(q) \geq n$. On the other hand, if $(s_1, \ldots, s_r)$ is a finite set of generators for Q^+ and $M := \max\{h(s_1), \ldots, h(s_r)\}$, then $J_{h,n} \subseteq (Q^+)^{n/M}$. Indeed, any $q \in Q^+$ can be written as $\sum m_i s_i$, with $m_i \in \mathbf{N}$, so if $q \in J_{h,n}$,

$$n < h(q) = \sum m_i h(s_i) \leq M \sum m_i$$

and so $\sum m_i > n/M$ and $q \in (Q^+)^{n/M}$. We have already seen that the complement of $J_{h,n}$ is finite. On the other hand, if σ is a finite set and m is a bound for $\{h(q) : q \in s\}$, then $J_{h,m}$ is contained in the complement of σ. □

It is clear from the construction that an injection $Q \to Q'$ induces an injection $R[[Q]] \to R[[Q']]$. If Q is fine and sharp and Q^{gp} is torsion free then, by Corollary 2.2.7, Q can be embedded in $\mathbf{N}^r$ for some r. Then (4) is reduced to the case of $R[[\mathbf{N}^r]]$, a ring of formal power series, easily seen to be an integral domain by induction on r. Statement (5) can also be reduced to the case of a formal power series ring, by taking a surjection $\mathbf{N}^r \to Q$. □

Corollary 3.6.3. *Let Q be a fine monoid, let $h\colon Q \to \mathbf{N}$ be a homomorphism, and for each $n \in \mathbf{N}$ let $J_{h,n} := \{q \in Q : h(q) > n\}$. Then $J_{h,1}$ is a prime ideal $\mathfrak{p}$, and the families $\{J_{h,n} : n \in \mathbf{N}\}$ and $\{\mathfrak{p}^n : n \in \mathbf{N}\}$ are cofinal.*

Proof If Q is sharp and h is local, this corollary follows immediately from Lemma 3.6.2. In any case, it is clear that $\mathfrak{p}$ is prime. Let $F := Q \setminus \mathfrak{p}$, so that h factors through a local homomorphism $\overline{h}\colon Q/F \to \mathbf{N}$. Then the result for Q and h follows from the result for Q/F and $\overline{h}$. □

More generally, if S is a finitely generated Q-set, we can consider the set $R[[S]]$ of formal power series indexed by S and with coefficients in R, which will form a module over $R[[Q]]$ provided that, for every $(q, s) \in Q^+ \times S$, $qs \neq s$. To explain this properly, it is helpful to use a monoid-theoretic analog of the Artin–Rees lemma and the Rees algebra in commutative algebra. This construction is also useful in connection with monoidal transformations discussed in II.1.7.2.

Definition 3.6.4. *Let Q be an integral monoid and let K be an ideal of Q. The* Rees monoid *of (Q, K) is the monoid $B_K(Q)$ whose elements are pairs (m, p), where $m \in \mathbf{N}$ and $p \in K^m$, and whose monoid law is given by*

$$(m, p) + (n, q) := (m + n, p + q).$$

If S is a Q-set, the Rees set *of (S, K) is the set of pairs (n, s), where $n \in \mathbf{N}$ and $s \in K^n S$, with the action of $B_K(Q)$ given by*

$$(m, p) + (n, s) := (m + n, p + s).$$

Note that we have

$$h: B_K(Q) \to \mathbf{N} : (m, p) \mapsto m \quad \text{and} \quad g: B_K(S) \to \mathbf{N} : (n, s) \mapsto n,$$

where h is a homomorphism of monoids and g is a morphism of $B_K(Q)$-sets over h.

Proposition 3.6.5. *Let K be an ideal of a fine monoid Q and let S be a finitely generated Q-set. Assume that $qs \neq s$ whenever $s \in S$ and $q \in Q^+$. Then the following statements hold:*

1. *If T is a sub-Q-set of S, then there exists an integer m such that $T \cap K^{m+n}S \subseteq K^nT$ for all n.*
2. *If K is a proper ideal of Q, then $\cap\{K^nS : n \geq 0\} = \emptyset$.*
3. *If Q is sharp, $R[[S]]$ can be identified with the $R[[Q^+]]$-adic completion of $R[S]$. In particular, if $(s_1, \ldots, s_n)$ is a sequence of generators for S as a Q-set, then $(e^{s_1}, \ldots, e^{s_n})$ is a sequence of generators for $R[[S]]$ as an $R[[Q]]$-module.*

Proof If Q is finitely generated as a monoid, then it is noetherian, so K is finitely generated as an ideal. If $(p_1, \ldots, p_r)$ is a sequence of generators for Q and $(k_1, \ldots, k_s)$ is a sequence of generators for K, then

$$((0, p_1), \ldots, (0, p_r), (1, k_1), \ldots (1, k_s))$$

is a sequence of generators for the monoid $B_K(Q)$, and it follows from Theorem 2.1.7 that $B_K(Q)$ is noetherian. Furthermore, since S is finitely generated as a Q-set, $B_K(S)$ is finitely generated as a $B_K(Q)$-set, and hence is also noetherian. If $T \subseteq S$ is a sub-Q-set, then

$$B_K(S, T) := \{(n, t) : n \in \mathbf{N}, t \in T \cap K^nS\}$$

is naturally a sub-$B_K(Q)$-set of $B_K(S)$, and consequently is finitely generated, say by $((m_1, t_1), \ldots, (m_p, t_p))$. Then any upper bound m for $(m_1, \ldots, m_p)$ satisfies (1). In particular, $T := \cap\{K^nS : n \geq 0\}$ is a sub-Q-set of S, and (1) implies that $T = KT$. Since $K \subseteq Q^+$, Proposition 2.1.12 implies that $T = \emptyset$, proving (2). It follows that for every $s \in S$, there exists an n such that $s \notin (Q^+)^nS$, and that for every finite subset σ of S, there exists an n such that $\sigma \cap (Q^+)^nS = \emptyset$. On the other hand, it is easy to see that the complement of each $(Q^+)^nS$ is finite, for example by writing S as the quotient of a finitely generated free Q-set. Thus the family of subsets of S whose complement is finite is cofinal with the family $\{(Q^+)^nS : n \in \mathbf{N}\}$. It follows that, for each $s \in S$, $\{(q, t) \in Q \times S : q + t = s\}$ is finite. This allows us to define an obvious action of $R[[Q]]$ on $R[[S]]$ and to see that the product topology is the $R[[Q^+]]$-adic topology, □

3.7 Abelian unipotent representations

In this section we give a geometric construction of the universal unipotent representation of a finitely generated free abelian group. This construction is a discrete version of the construction of the sheaf $\mathcal{O}_X^{log}$ on the Betti realization of a log analytic space, explained in Section V.1.4.

Let $\mathbf{I}$ be a finitely generated free abelian group and let $\mathbf{Z}[\mathbf{I}]$ be its group algebra. Recall that $\mathbf{Z}[\mathbf{I}]$ is the free $\mathbf{Z}$-module with basis $\mathbf{I}$, and that if $\gamma \in \mathbf{I}$, we write e^γ for the corresponding basis vector for $\mathbf{Z}[\mathbf{I}]$. The regular representation of $\mathbf{I}$ is the action of $\mathbf{I}$ on $\mathbf{Z}[\mathbf{I}]$ given by $\delta e^\gamma := e^{\delta+\gamma}$. Moreover, $\operatorname{Spec} \mathbf{Z}[\mathbf{I}] = \underline{\mathsf{A}}_{\mathbf{I}}^*$, which has a group scheme structure corresponding to the comultiplication $e^\gamma \mapsto e^\gamma \otimes e^\gamma$. This group scheme is a torus whose character group is $\mathbf{I}$.

We can also form the "vector group scheme" $\mathbf{VI} := \operatorname{Spec} S^{\cdot}\mathbf{I}$. Its comultiplication is given by $\gamma \mapsto \gamma \otimes 1 + 1 \otimes \gamma$. The $\mathbf{C}$-valued points of $\mathbf{VI}$ are the homomorphisms $v \colon \mathbf{I} \to (\mathbf{C}, +)$, and the $\mathbf{C}$-valued points of $\underline{\mathsf{A}}_{\mathbf{I}}^*$ are the homomorphisms $\rho \colon \mathbf{I} \to (\mathbf{C}, \cdot)$. The exponential map $(\mathbf{C}, +) \to (\mathbf{C}, \cdot)$ induces a morphism $\mathbf{VI}(\mathbf{C}) \to \underline{\mathsf{A}}_{\mathbf{I}}^*(\mathbf{C})$ sending a point v to $\exp \circ v$. This morphism has an algebraic incarnation in a suitable neighborhood of the identity, as we shall now explain.

The identity section of $\underline{\mathsf{A}}_{\mathbf{I}}^*$ is given by the augmentation $f_0 \colon \mathbf{Z}[\mathbf{I}] \to \mathbf{Z}$ sending every e^γ to 1. Let J be its kernel, which is the ideal of $\mathbf{Z}[\mathbf{I}]$ generated by the set of elements of the form $e^\gamma - 1$ for $\gamma \in \mathbf{I}$. We note the formula

$$(e^{\gamma_1+\gamma_2} - 1) = (e^{\gamma_1} - 1) + (e^{\gamma_2} - 1) + (e^{\gamma_1} - 1)(e^{\gamma_2} - 1). \tag{3.7.1}$$

This formula implies that J is generated by the set of elements of the form $(e^\gamma - 1)$ as γ ranges over any set of generators for $\mathbf{I}$. It also implies that the map

$$\lambda \colon \mathbf{I} \to J/J^2 : \gamma \mapsto [e^\gamma - 1]$$

is a group homomorphism. In fact this map is an isomorphism. To see this, let $f_1 \colon \mathbf{Z}[\mathbf{I}] \to \mathbf{I}$ be the unique homomorphism of abelian groups taking each e^γ to γ. Formula (3.7.1) shows that

$$\begin{aligned} f_1\left((e^{\gamma_1} - 1)(e^{\gamma_2} - 1)\right) &= f_1\left((e^{\gamma_1+\gamma_2} - 1) - (e^{\gamma_1} - 1) + (e^{\gamma_2} - 1)\right) \\ &= \gamma_1 + \gamma_2 - \gamma_1 - \gamma_2 \\ &= 0, \end{aligned}$$

so f_1 factors through a map $\overline{f}_1 \colon J/J^2 \to \mathbf{I}$. Evidently $\overline{f}_1 \circ \lambda = \mathrm{id}_{\mathbf{I}}$, so λ is injective, and since it is also surjective, it is an isomorphism. Furthermore, if $\gamma, \delta \in \mathbf{I}$, then

$$\delta(e^\gamma - 1) = e^{\delta+\gamma} - e^\delta = e^\gamma - 1 + (e^\gamma - 1)(e^\delta - 1)$$

so the action of $\mathbf{I}$ on J/J^2 is trivial.

We can assemble f_0 and f_1 to form an isomorphism of abelian groups:

$$\epsilon_1 : \mathbf{Z}[\mathbf{I}]/J^2 \to \mathbf{Z} \oplus \mathbf{I} : e^{\gamma} \mapsto 1 \oplus \gamma.$$

This isomorphism preserves the action of $\mathbf{I}$, where the action on $\mathbf{Z} \oplus \mathbf{I}$ is given by $\delta(n \oplus \gamma) = (n, \gamma + n\delta)$. Note that this action of $\mathbf{I}$ on $\mathbf{Z} \oplus \mathbf{I}$ is not compatible with the splitting, and in fact the right hand side should instead be written as an extension of $\mathbf{I}$-modules, as in the following diagram:

$$\begin{array}{ccccccc} \mathbf{Z}[\mathbf{I}]/J^2 & & J/J^2 & \longrightarrow & \mathbf{Z}[\mathbf{I}]/J^2 & \longrightarrow & \mathbf{Z} \\ \Big\downarrow{\scriptstyle \epsilon_1} & = & \Big\downarrow & & \Big\downarrow & & \Big\downarrow \\ \mathbf{Z} \oplus \mathbf{I} & & \mathbf{I} & \longrightarrow & \mathbf{Z} \oplus \mathbf{I} & \longrightarrow & \mathbf{Z}. \end{array} \tag{3.7.2}$$

The isomorphism ϵ_1 can be extended further, with the help of divided powers [67],[6],[7]. (In fact our main applications will take place after tensoring with $\mathbf{Q}$, where symmetric powers would suffice.) Let $\Gamma_n(\mathbf{I})$ be the nth graded piece of the divided power algebra of $\mathbf{I}$, so that $\gamma \mapsto \gamma^{[n]}$ is the "universal polynomial law of degree n" in the sense of [67], and let $f_n : \mathbf{Z}[\mathbf{I}] \to \Gamma_n(\mathbf{I})$ be the map sending e^{γ} to $\gamma^{[n]}$. Let $\hat{\Gamma}(\mathbf{I}) := \prod_n \Gamma_n(\mathbf{I})$, the completion of the divided power algebra with respect to the family of divided power ideals $\oplus_{m \geq n} \Gamma_m(\mathbf{I})$. Then the map

$$\mathbf{Z}[\mathbf{I}] \to \hat{\Gamma}(\mathbf{I}) : e^{\gamma} \mapsto \sum_{n \geq 0} \gamma^{[n]}$$

is a ring homomorphism and sends J to $\hat{\Gamma}_+(\mathbf{I})$. It therefore induces a divided power homomorphism

$$\hat{\epsilon} : \hat{D}_J(\mathbf{Z}[\mathbf{I}]) \longrightarrow \hat{\Gamma}(\mathbf{I}), \tag{3.7.3}$$

where $D_J(\mathbf{Z}[\mathbf{I}])$ is the completed divided power envelope of the ideal J of $\mathbf{Z}[\mathbf{I}]$. This map is also an isomorphism, with inverse

$$\hat{\lambda} : \hat{\Gamma}(\mathbf{I}) \longrightarrow \hat{D}_J(\mathbf{Z}[\mathbf{I}]) : \gamma \mapsto \sum_{i \geq 1} (-1)^{i+1}(e^{\gamma} - 1)^i / i. \tag{3.7.4}$$

Applying Hom($\mathbf{Z}$) to the exact sequences in Diagram (3.7.2), we find an exact sequence of $\mathbf{I}$-modules

$$0 \longrightarrow \mathbf{Z} \xrightarrow{\ \iota\ } \mathbf{L} \xrightarrow{\ \pi\ } \mathbf{I}^{\vee} \longrightarrow 0 \tag{3.7.5}$$

The (right) action of $\mathbf{I}$ on $\mathbf{L} \cong \mathbf{Z} \oplus \mathbf{I}^{\vee}$ is given by $\ell\gamma = \ell + \iota(\langle \pi(\ell), \gamma \rangle)$. For each natural number n, the sequence (3.7.5) induces an exact sequence

$$0 \longrightarrow S^{n-1}\mathbf{L} \longrightarrow S^n\mathbf{L} \longrightarrow S^n\mathbf{I} \longrightarrow 0.$$

This sequence is split as a sequence of $\mathbf{Z}$-modules, but the splitting is not compatible with the $\mathbf{I}$-actions. We find an isomorphism

$$\mathcal{A}_{\mathbf{I}} := \varinjlim S^n\mathbf{L} \cong \oplus S^{\cdot}(\mathbf{I}^\vee). \tag{3.7.6}$$

Let $N_{\cdot}\mathcal{O}_{\mathbf{L}}$ denote the image of $S^n(\mathbf{L})$ in $\mathcal{A}_{\mathbf{I}}$, an $\mathbf{I}$-invariant submodule. Then $\mathbf{I}$ acts trivially on $\mathrm{Gr}^N_{\cdot}\,\mathcal{A}_{\mathbf{I}}$ and, if $\delta \in \mathbf{I}$, the map $\mathrm{Gr}_{\cdot}\,\mathcal{A}_{\mathbf{I}} \to \mathrm{Gr}_{\cdot-1}\,\mathcal{A}_{\mathbf{I}}$ induced by δ identifies with the action of δ on $S^{\cdot}\mathbf{I}$ by interior multiplication. We refer to [61] for an explanation of the sense in which $\mathbf{Q} \otimes \mathcal{A}_{\mathbf{I}}$ is the universal rational unipotent representation of $\mathbf{I}$.

Recall from [7] and[67] that there is a canonical isomorphism

$$S^n(\mathbf{I}^\vee) \cong \mathrm{Hom}(\Gamma_n(\mathbf{I}), \mathbf{Z}).$$

Thus the algebra $\mathcal{A}_{\mathbf{I}} \cong S^{\cdot}(\mathbf{I}^\vee)$ can be identified with the topological dual of the divided power algebra $\hat{\Gamma}(\mathbf{I})$. It elements can be viewed as functions $\mathbf{I} \to \mathbf{Z}$ (given by "polynomial laws"), and the left action of $\mathbf{I}$ on itself defines a right action on the set of such functions. Explicitly, if $f \in \mathcal{A}_{\mathbf{I}}$ and $\gamma, \delta \in \mathbf{I}$, then $(f\delta)(\gamma) = f(\delta + \gamma)$. Interior multiplication by an element δ of $\mathbf{I}$ is a derivation of the algebra $S^{\cdot}(\mathbf{I}^\vee)$, and if $f \in S^{\cdot}(\mathbf{I}^\vee)$, Taylor's theorem implies that

$$(f\delta)(\gamma) := f(\gamma + \delta) = \sum_n (1/n!)(\delta^n f)(\gamma). \tag{3.7.7}$$

An easy way to check this formula is to use the fact that the formula has already been verified for elements of $\mathbf{L}$, that the action of $\mathbf{I}$ is compatible with multiplication, and that, since δ is a derivation,

$$\delta^n(fg) = \sum_{i+j=n} \binom{n}{i} \delta^i(f)\delta^j(g).$$

In characteristic zero, the isomorphism $\hat{\epsilon}$ (3.7.3) can be used to view locally unipotent representations of $\mathbf{I}$ as $\hat{\Gamma}(\mathbf{I})$-modules and to compute their cohomology. We briefly recall the construction, using the notion of "Higgs fields" and "Higgs cohomology."

Definition 3.7.1. *Let R be a commutative ring, let T be a projective R-module of finite rank, let $\Omega := \mathrm{Hom}_R(T, R)$ and $\Omega^i := \Lambda^i\Omega$. If E is an R-module, a T-Higgs field on E is an R-linear map*

$$\theta\colon E \to \Omega \otimes E$$

whose composition with the map

$$(\pi \otimes \mathrm{id}_E) \circ (\mathrm{id}_\Omega \otimes \theta)\colon \Omega \otimes E \to \Omega \otimes \Omega \otimes E \to \Omega^2 \otimes E$$

vanishes. The Higgs complex of θ is the complex

$$\Omega^\cdot \otimes E := E \longrightarrow \Omega \otimes E \longrightarrow \Omega^2 \otimes E \cdots .$$

obtained by prolonging this construction.

One sees easily that a T-Higgs field on E is equivalent to an extension of the R-module structure on E to an $S^\cdot T$-module structure.

Proposition 3.7.2. *If θ is a Higgs field on E, then the Higgs complex $\Omega^\cdot \otimes E$ represents the complex $\mathrm{R}\operatorname{Hom}_{S^\cdot T}(R, E)$, where E is given the $S^\cdot T$-module structure defined by θ and R is given the $S^\cdot T$-module structure defined by the augmentation $S^\cdot T \to R$.*

Proof Suppose that T has rank n. Recall [38, Chapitre I, 4.2] that the $S^\cdot T$-module R has a canonical projective resolution (the "Koszul complex")

$$K.T := \Lambda^n T \otimes S^\cdot T \to \Lambda^{n-1} T \otimes S^\cdot T \to \cdots \to T \otimes S^\cdot T \to S^\cdot \to R,$$

where the differentials are defined by

$$t_1 \wedge \cdots \wedge t_j \otimes f \mapsto \sum_i (-1)^{i+1} t_1 \wedge \cdots \wedge \hat{t}_i \wedge \cdots \wedge t_j \otimes t_i f.$$

Then $\mathrm{R}\operatorname{Hom}_{S^\cdot T}(R, E) \cong \operatorname{Hom}_{S^\cdot T}(K.(T), E)$, which is easily seen to be the Higgs complex of E. □

Theorem 3.7.3. *Let R be a ring containing $\mathbf{Q}$, let $\mathbf{I}$ be a finitely generated free abelian group of rank n, and let E be an R-module equipped with a locally unipotent action ρ of $\mathbf{I}$. For each $\gamma \in \mathbf{I}$, let*

$$\lambda_\gamma := \sum_{i \geq 1} (-1)^{i+1} (\rho_\gamma - 1)^i / i.$$

Then $\gamma \mapsto \lambda_\gamma$ defines an $R \otimes \mathbf{I}$-Higgs field on E, and there is a natural isomorphism

$$H^i(\mathbf{I}, E) \cong H^i(\Omega^\cdot \otimes E)$$

where $\Omega := \operatorname{Hom}_{\mathbf{Z}}(\mathbf{I}, R)$.

Proof To say that the action of ρ is locally unipotent is to say that each x in E is annihilated by $(\rho_\gamma - \mathrm{id})^i$ for $i \gg 0$. Since R contains $\mathbf{Q}$, this condition guarantees that the formula for λ_γ is well-defined. To prove the theorem on cohomology, we interpret the action of $\mathbf{I}$ on E as providing it with the structure of a $\mathbf{Z}[\mathbf{I}]$-module. Then $H^i(\mathbf{I}, E) = \operatorname{Ext}^i_{\mathbf{Z}[\mathbf{I}]}(\mathbf{Z}, E)$, where $\mathbf{Z}$ is viewed as a $\mathbf{Z}[\mathbf{I}]$-module via the augmentation mapping. The R-module structure on E provides it with the structure of an $R[\mathbf{I}]$-module and, since $\mathbf{Z}[\mathbf{I}] \to R[\mathbf{I}]$ is flat, $\mathrm{Ext}^i_{\mathbf{Z}[\mathbf{I}]}(\mathbf{Z}, E) \cong \mathrm{Ext}^i_{R[\mathbf{I}]}(R, E)$. In fact, the unipotence of ρ means that each element of the augmentation ideal J of $R[\mathbf{I}]$ is locally nilpotent on E, and hence that the $R[\mathbf{I}]$-module structure of E extends to an $\hat{R}[\mathbf{I}]$-module structure, where $\hat{R}[\mathbf{I}]$ is the J-adic completion of $R[\mathbf{I}]$. Since $R[\mathbf{I}] \to \hat{R}[\mathbf{I}]$ is again flat, it is also true that $H^i(\mathbf{I}, E) \cong \mathrm{Ext}^i_{\hat{R}[\mathbf{I}]}(R, E)$.

Since R contains $\mathbf{Q}$, the formula (3.7.4) for $\hat{\lambda}$ defines an isomorphism

$$\hat{\lambda}\colon \hat{S}^{\cdot}_R(R \otimes \mathbf{I}) \to \hat{R}[\mathbf{I}],$$

and the action of any $\gamma \in \mathbf{I}$ on $\hat{\lambda}_* E$ is given by the formula λ_γ in the theorem. Since $S^{\cdot}_R(R \otimes I) \to \hat{S}^{\cdot}_R(R \otimes I)$ is flat and $\hat{\lambda}$ is an isomorphism, there are isomorphisms

$$\mathrm{Ext}^i_{S^{\cdot}_R(R\otimes I)}(R, \lambda_* E) \cong \mathrm{Ext}^i_{\hat{S}^{\cdot}_R(R\otimes I)}(R, \lambda_* E) \cong \mathrm{Ext}^i_{\hat{R}[\mathbf{I}]}(R, E) \cong H^i(\mathbf{I}, E).$$

By Proposition 3.7.2, the first of these Ext groups is calculated by the Higgs complex of E. □

4 Actions and homomorphisms

4.1 Local and logarithmic homomorphisms

The following definition is partly a review.

Definition 4.1.1. *A homomorphism of monoids $\theta\colon P \to Q$ is*

1. *local if $\theta^{-1}(Q^*) = P^*$ or, equivalently, if $\theta^{-1}(Q^+) = P^+$,*
2. *sharp if the induced homomorphism $P^* \to Q^*$ is an isomorphism,*
3. *logarithmic if the induced homomorphism $\theta^{-1}(Q^*) \to Q^*$ is an isomorphism,*
4. *strict if the induced homomorphism $\overline{\theta}\colon \overline{P} \to \overline{Q}$ is an isomorphism,*
5. *s-injective if the induced homomorphism $\overline{\theta}\colon \overline{P} \to \overline{Q}$ is injective.*

Proposition 4.1.2. *Let $\theta\colon P \to Q$ be a sharp and strict monoid homomorphism. Then θ is surjective, and is bijective if Q is u-integral.*

Proof Suppose $q \in Q$. Since θ is strict, there exist $p \in P$ and $v \in Q^*$ with $\theta(p) = q + v$. Since θ is sharp, there exists a $u \in P^*$ with $\theta(u) = -v$, and then $\theta(u + p) = q$. Suppose $p_1, p_2 \in P$ with $\theta(p_1) = \theta(p_2)$. Since θ is strict, there exists a $u \in P^*$ such that $p_2 = p_1 + u$. Then $\theta(p_2) = \theta(p_1) + \theta(u) = \theta(p_2) + \theta(u)$, and $\theta(u) \in Q^*$. If Q is u-integral, it follows that $\theta(u) = 0$. Since θ^* is injective, $u = 0$ and $p_2 = p_1$. □

To see that the u-integrality hypothesis is not superfluous, let $Q = \mathbf{Z} \star \mathbf{N}^+$ be the join (1.4.5) of $\mathbf{Z}$ and $\mathbf{N}$ along $\mathbf{N}^+$. Then the morphism from $\mathbf{Z} \oplus \mathbf{N}$ to Q sending (m, n) to n in $\mathbf{N}^+$ if $n > 0$ and to $m \in \mathbf{Z}$ if $n = 0$ is surjective, sharp, and strict but not bijective.

Proposition 4.1.3. *Let $\theta\colon P \to Q$ be a homomorphism of monoids. Then the following conditions are equivalent:*

1. *θ is sharp and local;*
2. *θ is logarithmic;*
3. *$\theta^* : P^* \to Q^*$ is surjective and $\theta^{-1}(0) = 0$.*
4. *$\theta^{-1}(0) = 0$ and $Q^* \subseteq Im(\theta)$.*

Proof If θ is local, $\theta^{-1}(Q^*) = P^*$, and if θ is also sharp, it induces an isomorphism $P^* \to Q^*$, so (2) holds. If (2) holds, then $\theta^{-1}(Q^*)$ is a subgroup of P containing P^*, hence equal to P^*, and since θ induces an isomorphism $\theta^{-1}(Q^*) \to Q^*$, it follows that (3) holds. It is obvious that (3) implies (4). Finally, suppose (4) is true and let p be an element of P with $\theta(p) \in Q^*$. By (4) there is a $p' \in P$ with $\theta(p') = -\theta(p)$. Then $\theta(p + p') = 0$, hence by (4) $p + p' = 0$. Thus $p \in P^*$ and it follows that θ is local. Since $\mathrm{Ker}(\theta^*)$ is zero, θ^* is injective. Since Q^* is contained in the image of θ and θ is local, in fact θ^* is surjective. Hence θ^* is an isomorphism, i.e., θ is also sharp. Thus (1) is also satisfied. □

Proposition 4.1.4. *Let P be a monoid and let P^{int} be the image of P in P^{gp}. The natural homomorphism $P \to P^{\mathrm{int}}$ is local if P is quasi-integral. Conversely, if $P \to P^{\mathrm{int}}$ is local and P is u-integral, then P is quasi-integral.*

Proof Suppose that P is quasi-integral and that $p \in P$ maps to a unit in P^{int}. Then there exists a $p' \in P$ and a $q \in P$ such that $p + p' + q = q$. Since P is quasi-integral, $p + p' = 0$ and p' is a unit. For the converse, suppose that P is u-integral and that $P \to P^{\mathrm{int}}$ is local. If $p + q = q$, p maps to 0 in P^{int} and hence is a unit in P. It follows that $p = 0$, because P is u-integral. □

In the next proposition we work in the category of commutative, but not necessarily integral, monoids. The more important integral case will be discussed later.

Proposition 4.1.5. *Consider a cocartesian diagram*

$$\begin{array}{ccc} Q & \xrightarrow{\phi'} & Q' \\ \uparrow{\scriptstyle\theta} & & \uparrow{\scriptstyle\theta'} \\ P & \xrightarrow{\phi} & P' \end{array}$$

in the category of (commutative) monoids, so that $Q' \cong P' \oplus_P Q$.

1. *If θ and ϕ are local, so are ϕ' and θ', and $Q'^* \cong P'^* \oplus_{P^*} Q^*$.*
2. *The corresponding diagram of topological spaces*

$$\begin{array}{ccc} \operatorname{Spec} Q' & \longrightarrow & \operatorname{Spec} Q \\ \downarrow & & \downarrow \\ \operatorname{Spec} P' & \longrightarrow & \operatorname{Spec} P \end{array}$$

is cartesian (in the category of all topological spaces).

Proof For the first statement, suppose $p' \in P'$ and $\theta'(p') \in Q'^*$, so that there exists a $q' \in Q'$ such that $q' + \theta'(p') = 0$. Choose $q \in Q$ and $p'' \in P'$ with $q' = \phi'(q) + \theta'(p'')$, so that $\phi'(q) + \theta'(p' + p'') = 0$. This means that $(p' + p'', q) \in (P', Q)$ is equivalent to the element $(0, 0)$ with respect to the congruence relation E defining the quotient $P' \oplus Q \to Q'$. We will use the explicit description of E given in Proposition 1.1.5 to see that there is a sequence $(r_0, \ldots, r_n)$ as described there with $r_0 = (0, 0)$ and $r_n = (p' + p'', q)$. We shall show, by induction on i, that each r_i is a unit. This is clear if $i = 0$. For the induction step, write $r_i = (p'_i, q_i)$. First suppose that i is even. Then there exists a $p \in P$ such that $q_i = \theta(p) + q_{i+1}$ and $p'_{i+1} = \phi(p) + p'_i$. The induction hypothesis tells us that q_i is a unit, hence so are q_{i+1} and $\theta(p)$ and, since θ is local, p is a unit. Then $\phi(p)$ is a unit, and since p'_i is a unit, it then follows that the same is true of p'_{i+1}. Thus p'_{i+1} and q_{i+1} are units, so r_{i+1} is a unit. If i is odd, the same argument, with the roles of θ and ϕ interchanged, implies that r_{i+1} is a unit. We conclude that $p' + p''$ is a unit of P', and hence the same is true of p'. This shows that θ' is local, and by symmetry the same is true of ϕ'.

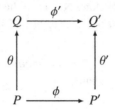

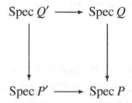

The universal mapping property of a pushout shows the existence of a homomorphism $P'^* \oplus_{P^*} Q^* \to Q'^*$. This is a group homomorphism, so to show it is injective it will suffice to prove that its kernel vanishes. Suppose that $(p', q) \in P'^* \oplus Q^*$ maps to zero in Q'. Then there is a sequence $(r_0, \ldots, r_n)$ as in Proposition 1.1.5 with $r_0 = (0, 0)$ and $r_n = (p', q)$. As we saw in the previous paragraph, in fact each r_i lies in $P'^* \oplus Q^*$, so that the equivalence relation holds within the unit groups, and consequently (p', q) maps to zero in $P'^* \oplus_{P^*} Q'^*$. For the surjectivity, suppose that q' is a unit of Q'. Write $q' = \theta'(p') + \phi'(q)$. Then $\theta(p')$ and $\phi'(q)$ are units, and it follows that p' and q are units, and q' is the image of (p', q).

The second statement is the assertion that the natural map

$$\operatorname{Spec} Q' \to \operatorname{Spec} P' \times_{\operatorname{Spec} P} \operatorname{Spec} Q$$

is a homeomorphism. First we prove the injectivity. We work with faces instead of prime ideals. Let G' be a face of Q', let $G := \phi^{-1}(G')$ and $F' := \theta'^{-1}(G')$. Then $\phi(G) + \theta'(F') \subseteq G'$. On the other hand, any $g' \in G'$ can be written as a sum $g' = \phi(q) + \theta'(p')$, with $q \in Q$ and $p' \in P'$. But then necessarily $\phi(q)$ and $\theta'(p')$ also belong to G' and hence $q \in G$ and $p' \in F'$. This shows that G' is determined by G and F'. Now suppose that F, G, and F' are faces of P, Q and P' respectively and that $F = \theta^{-1}(G) = \phi^{-1}(F')$. We claim that there is a face of Q' which pulls back to F' and to G. Form the pushout diagram

$$\begin{array}{ccc} Q_G & \xrightarrow{\tilde{\phi}'} & Q'' \\ \uparrow{\scriptstyle \tilde{\theta}} & & \uparrow{\scriptstyle \tilde{\theta}'} \\ P_F & \xrightarrow{\tilde{\phi}} & P'_{F'} \end{array}$$

By construction, $\tilde{\theta}$ and $\tilde{\phi}$ are local and, as we saw in part (1), it follows that $\tilde{\theta}'$ and $\tilde{\phi}'$ are local. Then the inverse image of Q''^* in Q' is face of Q' which restricts to G in Q and to F' in P'. Finally we have to check that the Zariski topology of $\operatorname{Spec} Q'$ is induced by the product topology coming from $\operatorname{Spec} P'$ and $\operatorname{Spec} Q'$. But any element q' of Q' can be written as a sum $q' = \phi'(q) + \theta'(p')$, and then $D(q') = D(\phi'(q)) \cap D(\theta'(p'))$, which is the intersection of the inverse images of $D(q)$ and $D(p')$. □

Example 4.1.6. Note that locality of θ' does not follow from the locality of θ alone. For example, let $\theta \colon P \to Q$ be the homomorphism $\mathbf{N} \oplus \mathbf{N} \to \mathbf{N} \oplus \mathbf{N}$ sending (m, n) to $(m + n, n)$. Thus P and Q are freely generated by the standard basis elements, which we denote by $p_1, p_2 \in P$ and $q_1, q_2 \in Q$, and $\theta(p_1) = q_1, \theta(p_2) = q_1 + q_2$. Now let $\phi \colon P \to P'$ be the localization of P by p_2 and form the pushout $Q' := P' \oplus_P Q$. Then, in Q', $\phi'(q_1) + \phi'(q_2) = \theta'(\phi(p_2))$, which is a unit, hence $\phi'(q_1) = \theta'(\phi(p_1))$ is a unit, but $\phi(p_1)$ is not a unit, so θ' is not local.

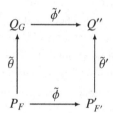

Note also that it is important that the pushout Q' appearing in Proposition 4.1.5 is formed in the category of monoids, not in the category of integral monoids. As the previous remark shows, if Q' fails to be quasi-integral, the map $Q' \to Q'^{\text{int}}$ may not be local, and then θ' could also fail to be local. For example, let $\theta\colon P \to Q$ be as before, but let $\phi\colon P \to P'$ be the homomorphism $\mathbf{N} \oplus \mathbf{N} \to \mathbf{N} \oplus \mathbf{N}$ sending (m, n) to $(m, m + n)$. In terms of generators, $\phi(p_1) = p'_1 + p'_2$ and $\phi(p_2) = p_2$. Then the pushout Q' is generated by $p''_1 := \theta'(p_1), p_2$, and $q'_2 := \phi'(q_2)$, with the relation $p''_1 + p_2 + q'_2 = p_2$. This monoid is sharp, but not quasi-integral and, in Q'^{int}, $p''_1 + q'_2$ maps to zero and hence p''_1 and q'_2 become units. Thus neither $Q \to Q'^{\text{int}}$ nor $P' \to Q'^{\text{int}}$ is local. In fact this issue turns out to be a central technical difficulty in log geometry and will be partially addressed in the next section.

4.2 Exact homomorphisms

Recall from Definition 2.1.15 that a homomorphism of monoids $\theta\colon P \to Q$ is exact if the induced map $P \to P^{\text{gp}} \times_{Q^{\text{gp}}} Q$ is an isomorphism. If P is integral, this map is automatically injective, so it suffices to check the surjectivity. If also Q is integral, then it suffices to check that $(\theta^{\text{gp}})^{-1}(Q) = P$.

The following result collects several useful facts about exactness in the category of integral monoids. In particular, the family of exact homomorphisms is stable under composition, pullbacks, and pushouts.

Proposition 4.2.1. *In the category of integral monoids, the following statements hold.*

1. *The natural homomorphism $\pi\colon Q \to \overline{Q}$ is exact.*
2. *If $\theta\colon P \to Q$ and $\phi\colon Q \to R$ are exact, then so is $\phi \circ \theta$. If $\phi \circ \theta$ is exact, then θ is exact. If in addition $\overline{\theta}^{\text{gp}}$ is surjective, or if $\text{Cok}(\overline{\theta}^{\text{gp}})$ is torsion and Q is saturated, then ϕ is also exact.*
3. *A homomorphism θ is exact if and only if $\overline{\theta}$ is exact.*
4. *A homomorphism $P \to Q$ is local if it is exact, and the converse holds if P is valuative.*
5. *An exact sharp homomorphism is injective. In particular, if θ is exact, then it is s-injective.*
6. *Let $\theta\colon P \to Q$ be an exact homomorphism of integral (resp. saturated) monoids.*
   (a) *If $\alpha\colon P \to P'$ is a homomorphism of integral (resp. saturated) monoids, then the pushout $\theta'\colon P' \to Q'$ of θ along α in the category of integral (resp. saturated) monoids is exact.*
   (b) *If $\beta\colon Q' \to Q$ is a homomorphism of integral monoids, then the pullback $\theta'\colon P' \to Q'$ of θ along β is exact.*

7. *If θ is exact, then $\theta^{\rm sat}$ is also exact.*

Proof To prove (1), let x be an element of $Q^{\rm gp}$ such that $\pi^{\rm gp}(x) \in \overline{Q}$. Since $(Q/Q^*)^{\rm gp} \cong Q^{\rm gp}/Q^*$ there exist $q \in Q$ and $u \in Q^*$ such that $x = q + u$. Then in fact $x \in Q$, so π is exact. If θ and ϕ are exact, the exactness of $\phi\circ\theta$ follows from the fact that the composition of cartesian squares is cartesian. Suppose that $\phi\circ\theta$ is exact, that $x \in P^{\rm gp}$, and that $\theta^{\rm gp}(x) \in Q$. Then $\phi^{\rm gp}\theta^{\rm gp}(x) \in R$, and it follows that $x \in P$, so θ is exact. Suppose that $y \in Q^{\rm gp}$ and $\phi^{\rm gp}(y) \in R$. If $\operatorname{Cok}(\overline{\theta}^{\rm gp})$ is torsion we can write $ny = \theta^{\rm gp}(x) + v$ with $v \in Q^*$ and $x \in P^{\rm gp}$ and $n > 0$; we can do this with $n = 1$ if $\overline{\theta}^{\rm gp}$ is surjective. Then, since $(\phi\circ\theta)^{\rm gp}(x) = \phi^{\rm gp}(y) - \phi(v) \in R$ and $\phi \circ \theta$ is exact, necessarily $x \in P$ and hence $ny \in Q$. If $n = 1$ or if Q is saturated, it follows that $y \in Q$, proving the exactness of ϕ and completing the proof of statement (2).

Note that if $\theta \colon P \to Q$ is any homomorphism of integral monoids there is a commutative diagram

$$\begin{array}{ccc} P & \xrightarrow{\ \theta\ } & Q \\ {\scriptstyle \pi}\big\downarrow & & \big\downarrow{\scriptstyle \pi} \\ \overline{P} & \xrightarrow{\ \overline{\theta}\ } & \overline{Q} \end{array}$$

in which the vertical arrows are exact and surjective. Thus (2) implies that θ is exact if and only if $\overline{\theta}$ is; so statement (3) follows. We have already observed that θ is local if it is exact in Proposition 2.1.16. Suppose that P is valuative, that θ is local, and that $x \in P^{\rm gp}$ with $\theta^{\rm gp}(x) \in Q$. Since P is valuative, either x or $-x$ belongs to P. If $x \in P$ there is nothing to prove. If $-x \in P$, then $\theta(-x) = -\theta(x) \in Q$, hence $\theta(-x) \in Q^*$. Since θ is local, it follows that $-x \in P^*$ and hence that $x \in P$, proving (4). Suppose that θ is exact and sharp, and that $p, p' \in P$ with $\theta(p) = \theta(p')$. Then $\theta^{\rm gp}(p - p') = 0 \in Q$, so $p - p' \in P$. Similarly $p' - p \in P$, so $p - p' \in P^*$. Since $\theta(p - p') = 0$ and θ is sharp, $p = p'$. This proves the injectivity of θ as asserted in (5).

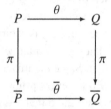

Recall from Proposition 1.3.4 that the integral pushout Q' in (6a) can be identified with the image of $P' \oplus Q$ in $P'^{\rm gp} \oplus_{P^{\rm gp}} Q^{\rm gp}$. Hence if $x' \in P'^{\rm gp}$ and $\theta'^{\rm gp}(x') \in Q'$, there exist $p' \in P'$, $q \in Q$, and $x \in P^{\rm gp}$ such that $x' = p' + \alpha(x)$ and $q = \theta^{\rm gp}(x)$. Since θ is exact, necessarily $x \in P$ and so $x' = p' + \alpha(x) \in P'$. This proves that θ' is exact. Now the saturated pushout $Q'^{\rm sat}$ is the saturation of the integral pushout Q'. Thus if $x' \in P'^{\rm gp}$ and $\theta'^{\rm gp}(x') \in Q'^{\rm sat}$, there exists a positive integer n such that $n\theta'^{\rm gp}(x') \in Q'$. As we just saw, this implies that $nx' \in P'$ and, if P' is saturated, it follows that $x' \in P'$. Thus $P' \to Q'^{\rm sat}$ is also exact. This completes the proof of (6a). Statement (6b) had already been proven as part (4) of Proposition 2.1.16, and the verification of (7) is immediate. □

Proposition 4.2.2. *Let $\theta\colon P \to Q$ be a homomorphism of integral monoids. If θ is exact, then* Spec θ *is surjective. The converse holds if P is fine and saturated. In fact it is enough to check that every prime of height one is contained in the image of* Spec θ.

Proof Suppose that θ is exact and $\mathfrak{p}$ is a prime of P. Let $F := P \setminus \mathfrak{p}$ and let $\theta_F\colon P_F \to Q_F$ be the localization of θ by F. Since Q_F is integral and can be identified with $P_F \oplus_P Q$ it follows from statement (6a) of Proposition 4.2.1 that θ_F is exact and hence local. Thus, $\theta^{-1}(\mathfrak{q}) = \mathfrak{p}$, where $\mathfrak{q} := Q \setminus F$, the prime ideal of Q corresponding to the maximal ideal of Q_F, This proves that Spec θ is surjective.

Conversely, suppose that P is fine and saturated and that Spec θ is surjective. Let x be an element of P^{gp} such that $\theta(x) \in Q$ and let $\mathfrak{p} \in \operatorname{Spec} P$ be a prime of height one. Since Spec θ is surjective, there is a prime $\mathfrak{q}$ of Q lying over $\mathfrak{p}$. Then the map $P_{\mathfrak{p}} \to Q_{\mathfrak{q}}$ is local. Since $P_{\mathfrak{p}}$ is saturated and $\mathfrak{p}$ has height one, it follows from Corollary 2.4.4 that $P_{\mathfrak{p}}$ is valuative. Then by (4) of Proposition 4.2.1, the map $P_{\mathfrak{p}} \to Q_{\mathfrak{q}}$ is exact. Since the image of $\theta(x)$ in Q^{gp} lies in $Q_{\mathfrak{q}}$, it follows that $x \in P_{\mathfrak{p}}$. Thus $x \in P_{\mathfrak{p}}$ for every prime of height one and, since P is saturated, it follows from Corollary 2.4.5 that $x \in P$. □

Exact homomorphisms are convenient for many reasons. For example, we shall see in Proposition 4.2.3 that an exact homomorphism is universally local and in Proposition 4.2.5 that an exact pushout of a local (resp. logarithmic) homomorphism is again local (resp. logarithmic).

Proposition 4.2.3. *Let $\theta\colon P \to Q$ be a homomorphism of integral monoids. If θ is exact, it is universally local, i.e., for every cocartesian square*

$$\begin{array}{ccc} Q & \xrightarrow{\ \psi\ } & Q' \\ {\scriptstyle\theta}\uparrow & & \uparrow{\scriptstyle\theta'} \\ P & \xrightarrow{\ \phi\ } & P' \end{array}$$

in the category of integral monoids, θ' is local. The converse is true if P is saturated. In this case, θ is exact if and only if the pushout θ' is local for all local homomorphisms ϕ. If P is fine it suffices to check local homomorphisms into fine monoids P'.

Proof We know by Proposition 4.2.1 that an exact homomorphism is local and that any pushout (in the category of integral monoids) of an exact homomorphism is exact. Therefore an exact homomorphism is universally local. Conversely, suppose that the pushout θ' of θ along every local ϕ is again local. If x is an element of P^{gp} such that $q := \theta^{\mathrm{gp}}(x)$ belongs to Q, consider the submonoid P' of P^{gp} generated by P and $-x$. If $x \in P$, there is nothing more to prove. Otherwise Lemma 4.2.4 below implies that the inclusion $\phi\colon P \to P'$ is local and that $-x$ is not a unit of P'. However, in Q', $\theta'(-x) + \psi(q) = 0$, so $\theta'(-x)$ is a unit of Q', contradicting the assumption that θ' is local. Note that in Lemma 4.2.4 the monoid P' is fine if P is fine. □

Lemma 4.2.4. *Let P be a saturated monoid, let x be an element of $P^{\mathrm{gp}} \setminus P$, and let P' be the submonoid of P^{gp} generated by P and $-x$. Then the inclusion homomorphism $\phi\colon P \to P'$ is local, and furthermore $-x$ is not a unit of P'.*

Proof Suppose that $p \in P$ and $\phi(p)$ is a unit of P'. The inverse of $\phi(p)$ can be written as $p' - nx$ for some $p' \in P$ and $n \in \mathbf{N}$. Then $nx = p + p'$ in P^{gp}. If $n > 0$, the fact that P is saturated implies that $x \in P$, contradicting the hypothesis of the lemma. It follows that $n = 0$, hence $p \in P^*$, proving that ϕ is indeed local. If $-x$ is a unit of P', there exist $p \in P$ and $n \geq 0$ such that $-x + p - nx = 0$. But this implies that $(n + 1)x = p$, and hence that $x \in P$, another contradiction. □

Proposition 4.2.5. *Let*

$$\begin{array}{ccc} Q & \xrightarrow{\ \psi\ } & Q' \\ {\scriptstyle\theta}\uparrow & & \uparrow{\scriptstyle\theta'} \\ P & \xrightarrow{\ \phi\ } & P' \end{array}$$

be a cocartesian diagram in the category of integral monoids, where θ is exact and ϕ is local.

1. *The homomorphism $\rho\colon Q^* \oplus_{P^*} P'^* \to Q'^*$ is an isomorphism, and the homomorphisms ψ and θ' are local.*
2. *If ϕ is logarithmic then ψ is also logarithmic.*
3. *If ϕ is logarithmic and Q is sharp, then Q' is also sharp.*
4. *If Q and P' are sharp, then Q' is sharp.*
5. *The induced homomorphism $\overline{Q} \oplus_{\overline{P}} \overline{P'} \to \overline{Q'}$ is an isomorphism.*

Proof First let us verify the following assertion.

Claim 4.2.6. With the assumptions of the proposition, suppose that $q \in Q$ and $p' \in P'$ are such that $\psi(q) + \theta'(p') = 0$. Then there exists a $p \in P^*$ such that $q = \theta(p)$ and $p' = \phi(-p)$.

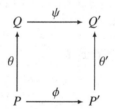

Indeed, the equality $\psi(q)+\theta'(p') = 0$ means that there exists an $x \in P^{\mathrm{gp}}$ such that $q = \theta(x)$ and $p' = -\phi(x)$. The exactness of θ implies that $p := x$ belongs to P. Then $\phi(p) + p' = 0$, so $\phi(p)$ is a unit of P'. Since ϕ is local it follows that $p \in P^*$.

The claim immediately implies that the homomorphism ρ is injective. To see that it is surjective, suppose that q_1' is a unit in Q'. Let $q_2' : -q_1'$ and write $q_i' = \psi(q_i)+\theta'(p_i')$. Then $\psi(q_1+q_2)+\theta'(p_1'+p_2') = 0$. By the claim, there exists a unit p of P such that $q_1 + q_2 = \theta(p)$ and $p_1' + p_2' = \theta(-p)$. It follows that q_1 and p_1' are units. Their sum defines an element of $Q^* \oplus_{P^*} P'^*$ which maps to q_1'. Thus ρ is also surjective.

To prove that ψ is local, suppose that $q \in Q$ and $\psi(q)$ is a unit in Q'. By the surjectivity of ρ, there exist units $q_1 \in Q^*$ and $p' \in P'^*$ such that $\psi(q) = \psi(q_1) + \theta'(p')$. Then $\psi(q - q_1) + \theta'(-p') = 0$ and, by the claim, there exists a unit $p \in P$ such that $q - q_1 = \theta(p)$. It follows that q is a unit. The locality of θ' follows from Proposition 4.2.3, without any assumption on ϕ. This completes the proof of statement (1).

The remaining statements follow easily from the fact that ρ is an isomorphism. If ϕ is logarithmic, it induces an isomorphism $P^* \to P'^*$, and since ρ is an isomorphism, ψ induces an isomorphism $Q^* \to Q'^*$. Since ψ is local, it is in fact logarithmic and, if Q is sharp, necessarily so is Q'. Statement (4) follows immediately from the fact that ρ is an isomorphism. To prove (5), observe that there is a commutative diagram:

$$
\begin{array}{ccc}
Q' := Q \oplus_P P' & \xrightarrow{\ a\ } & \overline{Q} \oplus_{\overline{P}} \overline{P}' \\
 & \searrow^{\pi} & \downarrow b \\
 & & \overline{Q}'.
\end{array}
$$

It is clear that a and b are surjective. If q_1' and q_2' are elements of Q' with the same image in $\overline{Q}'$, there exists a unit u of Q'^* with $q_2' = uq_1'$. By (1), we can write u as a sum of units of Q and P', and it follows that q_1 and q_2 have the same image in $\overline{Q} \oplus_{\overline{P}} \overline{P}'$. Thus b is injective. □

The exactness of a homomorphism $\theta\colon P \to Q$ can be interpreted via the corresponding morphism of affine toric varieties $\underline{\mathsf{A}}_Q \to \underline{\mathsf{A}}_P$. This morphism need not be flat, but does "descend flatness" in the sense of [66, §1, Part II]. The key ideas here are due to Illusie, Nakayama, and Tsuji [41].

Proposition 4.2.7. *Let $\theta\colon P \to Q$ be an injective homomorphism of integral monoids. Then the following are equivalent.*

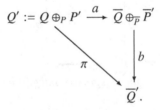

1. *The homomorphism θ is exact.*
2. *The homomorphism $\mathbf{Z}[\theta]\colon \mathbf{Z}[P] \to \mathbf{Z}[Q]$ splits as a sequence of $\mathbf{Z}[P]$-modules.*
3. *The homomorphism $\mathbf{Z}[\theta]$ is universally injective.*

Proof If (1) holds then, by statement (5) of Proposition 2.1.16, the subset $Q \setminus \theta(P)$ of Q is stable under the action of P on Q induced by the homomorphism θ. Thus the P-set Q is a disjoint union of the two P-sets $\theta(P)$ and $Q\setminus\theta(P)$. It follows that the $\mathbf{Z}[P]$-module $\mathbf{Z}[Q]$ is a direct sum $\mathbf{Z}[P]\oplus\mathbf{Z}[Q\setminus\theta(P)]$, proving (2). The implication of (3) by (2) is clear. Suppose (3) holds. To prove that θ is exact, suppose that $p, p' \in P$ and $\theta(p) = q + \theta(p')$ for some $q \in Q$. Let I be the ideal of $\mathbf{Z}[P]$ generated by p'. Then (3) implies that $\mathbf{Z}[P]/I \to \mathbf{Z}[Q]/I\mathbf{Z}[Q]$ is injective. Since $\theta(e^p) \in I\mathbf{Z}[Q]$, it follows that $e^p \in I$, and hence that p belongs to the ideal of P generated by p'. Then $p - p' \in P$, proving the exactness of θ. □

Corollary 4.2.8. *Let $\theta\colon P \to Q$ be an injective and exact homomorphism of integral monoids. For any homomorphism $\alpha\colon \mathbf{Z}[P] \to A$, let $B := A\otimes_{\mathbf{Z}[P]} \mathbf{Z}[Q]$ and let $\theta_A\colon A \to B$ be the induced homomorphism of rings. Then an A-module E is flat over A if and only if $B\otimes_A E$ is flat over B. Thus θ "universally descends flatness."*

Proof We shall need the following general lemma.

Lemma 4.2.9. *Let $\theta\colon A \to B$ be an injective homomorphism of commutative rings which has a left inverse as a homomorphism of A-modules. Then an A-module E is flat over A if and only if $B \otimes_A E$ is flat over B.*

Proof It suffices to prove the "if" part of the lemma. Suppose that $\sigma\colon B \to A$ is an A-linear splitting of θ and that $B \otimes_A E$ is B-flat. Let $i\colon M' \to M$ be an injective homomorphism of A-modules, let K be the kernel of $1_B \otimes i$, and consider the following diagram.

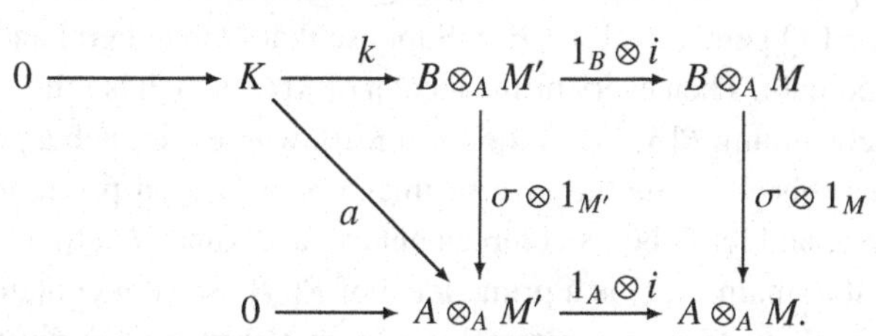

The diagram commutes and i is injective, and it follows that $a = 0$. Now let K'

be the kernel of $i \otimes 1_E$ and consider the diagram

$$\begin{array}{ccccccc}
0 \longrightarrow & K \otimes_A E & \xrightarrow{k \otimes 1_E} & B \otimes_A M' \otimes_A E & \xrightarrow{1_B \otimes i \otimes 1_E} & B \otimes_A M \otimes_A E \\
 & \uparrow{\scriptstyle \theta'} & & \uparrow{\scriptstyle \theta \otimes 1_{M'} \otimes 1_E} & & \uparrow{\scriptstyle \theta \otimes 1_M \otimes 1_E} \\
0 \longrightarrow & K' & \xrightarrow{k'} & A \otimes_A M' \otimes_A E & \xrightarrow{1_A \otimes i \otimes 1_E} & A \otimes_A M \otimes_A E.
\end{array}$$

The top row can be identified with the tensor product of the top row of the previous diagram with the B-module $B \otimes_A E$. Since this B-module is flat, this row is still exact. It follows that $\theta \otimes 1_{M'} \otimes 1_E$ maps K' to $K \otimes_A E$ as shown in the diagram. But $\sigma \circ \theta = \mathrm{id}_A$, so

$$k' = (\sigma \otimes 1_{M'} \otimes 1_E) \circ (\theta \otimes 1_{M'} \otimes 1_E) \circ k' = (\sigma \otimes 1_{M'} \otimes 1_E) \circ (k \otimes 1_E) \circ \theta' = (a \otimes 1_E) \circ \theta'.$$

Since $a = 0$, it follows that $k' = 0$, hence that that $K' = 0$, and hence that $i \otimes 1_E$ is injective. Since i was an arbitrary injection of A-modules, E is flat. □

Corollary 4.2.8 follows easily. Indeed, since θ is exact, Proposition 4.2.7 implies that $\mathbf{Z}[\theta]$ splits as a homomorphism of $\mathbf{Z}[P]$-modules, and it follows that θ_A splits as a homomorphism of A-modules. Then Lemma 4.2.9 implies that θ_A descends flatness. □

A homomorphism of monoids $\theta \colon P \to Q$ induces a continuous map of topological spaces $f \colon \operatorname{Spec} Q \to \operatorname{Spec} P$ and, for each $\mathfrak{q} \in \operatorname{Spec} Q$, a local homomorphism of monoids $\theta_{\mathfrak{q}} \colon P_{f(\mathfrak{q})} \to Q_{\mathfrak{q}}$. To really understand θ we need to understand all these homomorphisms. The most important primes of Q to consider are those lying over the maximal ideal of P. We rephrase this description in terms of faces in the next definition.

Definition 4.2.10. *Let $\theta \colon P \to Q$ be a homomorphism of monoids. Then a face G of Q is θ-critical if $\theta^{-1}(G) = \theta^{-1}(Q^*)$.*

Remark 4.2.11. The maximal θ-critical faces of Q are of special importance; these correspond to the prime ideals $\mathfrak{q}$ of Q minimal among those containing the ideal K_θ of Q generated by $\theta(P^+)$. Suppose that Q is integral and that R is an integral domain. Then every prime ideal $\mathfrak{p}$ of $R[Q]$ which is minimal among the primes containing $R[K_\theta]$ is of the form $R[\mathfrak{q}]$, where $\mathfrak{q}$ is such a prime ideal of Q. To see this, let $\mathfrak{q}$ be the inverse image of $\mathfrak{p}$ in Q, a prime ideal of Q lying over P^+, and let G be its complementary face. Since $R[Q]/R[\mathfrak{q}] \cong R[G]$ is an integral domain, $R[\mathfrak{q}]$ is a prime ideal of $R[Q]$. Since $\mathfrak{q}$ contains P^+ and $R[\mathfrak{q}] \subseteq \mathfrak{p}$, it follows from the minimality of $\mathfrak{p}$ that $R[\mathfrak{q}] = \mathfrak{p}$. Furthermore, the minimality of $\mathfrak{p}$ implies that $\mathfrak{q}$ is minimal among all the primes of Q lying over

$\mathfrak{p}$. Note that, since $\mathfrak{q} \subseteq Q^+$, the prime ideal $\mathfrak{p}$ of $R[Q]$ is necessarily contained in $R[Q^+]$.

In the following definition we introduce a terminology for exact and s-injective homomorphisms which we will also apply to other types of homomorphisms.

Definition 4.2.12. *A homomorphism $\theta: P \to Q$ of integral monoids is*

1. *locally exact (resp. locally s-injective) if for every face G of Q, the localized homomorphism $P_{\theta^{-1}(G)} \to Q_G$ is exact (resp. s-injective);*
2. *critically exact (resp.critically s-injective) if for every θ-critical face G of Q, the homomorphism $P_{\theta^{-1}(Q^*)} \to Q_G$ is exact (resp. s-injective).*

Examples 4.2.13. The inclusion morphism $\mathbf{N} \to \mathbf{Z}$ is not exact, but it is critically and locally exact. The homomorphism $\theta: \mathbf{N} \oplus \mathbf{N} \to \mathbf{N} \oplus \mathbf{N} \oplus \mathbf{N}$ sending (a, b) to $(a, a + b, b)$ is exact but not critically exact. The θ-critical faces are 0, $0 \oplus \mathbf{N} \oplus 0$, $\mathbf{N} \oplus 0 \oplus 0$, $0 \oplus 0 \oplus \mathbf{N}$, and $\mathbf{N} \oplus 0 \oplus \mathbf{N}$, and θ is not exact when localized by any of the last three.

Remark 4.2.14. If X is a topological space and $x \in X$, let $X_{(x)}$ be the set of all generizations of x, i.e., the set of all $x' \in X$ such that $x \in \{x'\}^-$. (For example, if P is a monoid and $X = \operatorname{Spec}(P)$ and if x corresponds to the prime ideal $\mathfrak{p}$ of P, then $X_{(x)}$ identifies with $\operatorname{Spec}(P_\mathfrak{p})$.) Then if $f: X \to Y$ is continuous, f induces a map $f_{(x)}: X_{(x)} \to Y_{(f(x))}$ for each $x \in X$. We say that f is *locally surjective* if, for each x in X, this map $f_{(x)}$ is surjective. If X has a unique closed point x and $y := f(x)$, then we say that f is *critically surjective* if $f_{(x')}: X_{(x')} \to Y_{(y)}$ is surjective for every $x' \in f^{-1}(y)$. It follows from Proposition 4.2.2 that a homomorphism of fine saturated monoids $\theta: P \to Q$ is locally exact (resp. critically exact) if and only if $\operatorname{Spec}(\theta)$ is locally surjective (resp. critically surjective). We shall see in Theorem 4.7.7 that a homomorphism of fine saturated monoids is locally exact if and only if it is critically exact and also if and only if it is critically s-injective.

Corollary 4.2.15. *Let $\theta: P \to Q$ be a locally exact morphism of integral monoids. Let $\mathfrak{q}$ be be a prime ideal of P and let $\mathfrak{p} := \theta^{-1}(\mathfrak{q})$. Then $\operatorname{ht}\mathfrak{p} \leq \operatorname{ht}\mathfrak{q}$.*

Proof Since θ is locally exact, $\operatorname{Spec}(\theta)$ is locally surjective. Let $\mathfrak{p}_0 \supset \mathfrak{p}_1 \supset \cdots \supset \mathfrak{p}_d$ be a maximal chain of prime ideals ideals in $P_\mathfrak{p}$. Since $\operatorname{Spec}(Q_\mathfrak{q}) \to \operatorname{Spec}(P_\mathfrak{p})$ is surjective, there exists a prime ideal $\mathfrak{q}_1$ of $Q_\mathfrak{q}$ lifting $\mathfrak{p}_1$. Since $\operatorname{Spec}(Q_{\mathfrak{q}_1}) \to \operatorname{Spec}(P_{\mathfrak{p}_1})$ is surjective, there exists a prime ideal of $Q_{\mathfrak{q}_1}$ lifting $\mathfrak{p}_2 P_{\mathfrak{p}_1}$. Continuing in this way, we find a lifting $\mathfrak{q} = \mathfrak{q}_0 \supset \mathfrak{q}_1 \supset \cdots \supset \mathfrak{q}_d$ of the chain $\mathfrak{p}_0 \supset \mathfrak{p}_1 \supset \cdots \supset \mathfrak{p}_d$ to $Q_\mathfrak{q}$. It follows that $\operatorname{ht}\mathfrak{p} \leq \operatorname{ht}\mathfrak{q}$. □

Corollary 4.2.16 (The four point lemma). *Let*

$$\begin{array}{ccc} Q & \xrightarrow{\psi} & Q'^{\text{int}} \\ \uparrow{\scriptstyle\theta} & & \uparrow{\scriptstyle\theta'} \\ P & \xrightarrow{\phi} & P' \end{array}$$

be a cocartesian square in the category of integral monoids. Then the corresponding map of fibered products in the category of sets,

$$\operatorname{Spec} Q'^{\text{int}} \to \operatorname{Spec} Q \times_{\operatorname{Spec} P} \operatorname{Spec} P'$$

is injective, and it is bijective if either θ or ϕ is locally exact.

Proof The injectivity follows from the injectivity in Proposition 4.1.5 and the injectivity of the map $\operatorname{Spec} Q'^{\text{int}} \to \operatorname{Spec} Q'$. For the rest of the proof, we just write Q' for Q'^{int}. Suppose that either θ or ϕ is locally exact; by symmetry, we may as well assume that it is θ that is locally exact. A point of the fibered product $\operatorname{Spec} Q \times_{\operatorname{Spec} P} \operatorname{Spec} P'$ corresponds to a pair (G, F'), where G is a face of Q and F' is a face of P' such that $F := \theta^{-1}(G) = \phi^{-1}(F')$. Since θ is locally exact, the map $P_F \to Q_G$ is again exact. Let $Q'' := Q_G \oplus_{P_F} P'_{F'}$. Proposition 4.2.5 asserts that the pushouts $Q_G \to Q''$ and $P'_{F'} \to Q''$ are local. Let G' be the inverse image of Q''^* in Q'. It follows that $\psi^{-1}(G') = G$ and $\theta'^{-1}(G') = F'$. Thus G' corresponds to a point of $\operatorname{Spec}(Q')$ which restricts to the point of $\operatorname{Spec} Q \times_{\operatorname{Spec} P} \operatorname{Spec} P'$ corresponding to (G, F'). □

The following construction is a crude attempt at rendering a homomorphism exact; we will do a better job (see Theorem II.1.8.1) once we have some additional constructions at our disposal.

Proposition 4.2.17. *If $\theta\colon P \to Q$ is a homomorphism of integral monoids, let*

$$P^\theta := P^{\text{gp}} \times_{Q^{\text{gp}}} Q = \{x \in P^{\text{gp}} : \theta(x) \in Q\},$$

so that θ factors:

$$\theta = P \xrightarrow{\tilde\theta} P^\theta \xrightarrow{\theta^e} Q.$$

Then the following properties hold.

1. *The homomorphism $\tilde\theta^{\text{gp}}$ is an isomorphism, and θ^e is exact. The homomorphism $\tilde\theta$ is an isomorphism if and only if θ is exact, and the homomorphism θ^e is an isomorphism if and only if θ^{gp} is an isomorphism.*

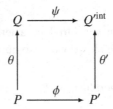

2. *The formation of* P^θ *is functorial in* θ*: given the outer rectangle below, there exists a canonical* ρ *making the following diagram commute.*

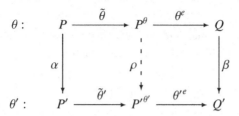

$$\begin{array}{cccccc}
\theta: & P & \xrightarrow{\tilde{\theta}} & P^\theta & \xrightarrow{\theta^e} & Q \\
& \Big\downarrow{\scriptstyle \alpha} & & \Big\downarrow{\scriptstyle \rho} & & \Big\downarrow{\scriptstyle \beta} \\
\theta': & P' & \xrightarrow{\tilde{\theta}'} & P'^{\theta'} & \xrightarrow{\theta'^e} & Q'
\end{array}$$

If θ' *is exact, there is a unique* $\psi\colon P^\theta \to P'$ *such that* $\psi \circ \tilde{\theta} = \alpha$*, and necessarily* $\tilde{\theta}' \circ \psi = \rho$ *and* $\theta' \circ \psi = \beta \circ \theta^e$*.*

3. *In the category of integral monoids, the square A below is cartesian and the square B below is cocartesian.*

$$\begin{array}{ccc}
Q & \longrightarrow & Q^{\mathrm{gp}} \\
\Big\uparrow{\scriptstyle \theta^e} & A & \Big\uparrow{\scriptstyle \theta^{\mathrm{gp}}} \\
P^e & \longrightarrow & P^{\mathrm{gp}}
\end{array}
\qquad\qquad
\begin{array}{ccc}
Q & \xrightarrow{\mathrm{id}} & Q \\
\Big\uparrow{\scriptstyle \theta} & B & \Big\uparrow{\scriptstyle \theta^e} \\
P & \xrightarrow{\tilde{\theta}} & P^\theta
\end{array}$$

4. *If* Q *is fine and* P^{gp} *is finitely generated, then* P^θ *is fine.*

Proof Since $P \subseteq P^\theta \subseteq P^{\mathrm{gp}}$, it follows that $\tilde{\theta}^{\mathrm{gp}}$ is an isomorphism, and the remaining statements in (1) are immediate from the definitions. The existence of the homomorphism ρ in the diagram in (2) is clear from the construction. If θ' is exact, then $\tilde{\theta}'$ is an isomorphism, and we set $\psi := \tilde{\theta}'^{-1} \circ \rho$. Then $\psi \circ \tilde{\theta} = \alpha$ and $\tilde{\theta}' \circ \psi = \rho$. The first of these equalities determines ψ because $\tilde{\theta}$ is an epimorphism, and the second implies that $\theta' \circ \psi = \beta \circ \theta^e$.

The square A in (3) is cartesian by construction. The fact that B is cocartesian follows formally from the fact that $\tilde{\theta}$ is an epimorphism. Namely, suppose that $\phi\colon Q \to R$ and $\psi\colon P^\theta \to R$ are homomorphisms such that $\phi \circ \theta = \psi \circ \tilde{\theta}$. We claim that there is unique homomorphism $\phi'\colon Q \to R$ such that $\phi' \circ \mathrm{id} = \phi$ and $\phi' \circ \theta^e = \psi$. Evidently the only possibility is to take ϕ' to be ϕ, and the equality $\phi \circ \theta^e = \psi$ follows from the fact that $\tilde{\theta}$ is an epimorphism.

If P^{gp} is finitely generated and Q is fine, then Corollary 2.1.21 implies that P^θ is also fine. □

The statements (2) and (3) together say that the homomorphism $\tilde{\theta}\colon P \to P^\theta$ has the following universal property: the pushout of $P \to Q$ along $\tilde{\theta}$ is exact, and any homomorphism $P \to P'$ with this property factors uniquely through $\tilde{\theta}$.

The next result shows the compatibility of exactification and sharpening.

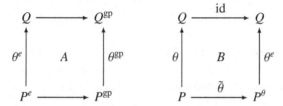

Proposition 4.2.18. *If $\theta\colon P \to Q$ is a homomorphism of integral monoids, the natural map $\overline{P^\theta} \to \overline{P}^{\overline{\theta}}$ is an isomorphism and identifies $\overline{\theta^e}$ with $\overline{\theta}^e$.*

Proof The "natural map" in the statement comes about as follows. By the functoriality of the exactification construction, there is a natural homomorphism $a\colon P^\theta \to \overline{P}^{\overline{\theta}}$. The target monoid is not necessarily sharp, but maps to $\overline{Q}$. Thus there is a commutative diagram

$$\begin{array}{ccccc} P^\theta & \xrightarrow{a} & \overline{P}^{\overline{\theta}} & & \\ \downarrow{\scriptstyle \pi_{P^\theta}} & & \downarrow{\scriptstyle \pi_{\overline{P}^{\overline{\theta}}}} & \searrow{\scriptstyle \overline{\theta}^e} & \\ \overline{P^\theta} & \xrightarrow{\overline{a}} & \overline{\overline{P}^{\overline{\theta}}} & \xrightarrow{\overline{\overline{\theta}^e}} & \overline{Q}. \end{array}$$

The composed arrow along the bottom is $\overline{\theta^e}$, and our claim is that $\overline{a}$ is an isomorphism. Note that it is exact because $\overline{\theta^e}$ is exact and, since $\overline{P^\theta}$ is sharp, it follows that $\overline{a}$ is injective. Since $Q \to \overline{Q}$ is exact, we have

$$P^\theta := P^{\mathrm{gp}} \times_{Q^{\mathrm{gp}}} Q \cong P^{\mathrm{gp}} \times_{Q^{\mathrm{gp}}} Q^{\mathrm{gp}} \times_{\overline{Q}^{\mathrm{gp}}} \overline{Q} \cong P^{\mathrm{gp}} \times_{\overline{Q}^{\mathrm{gp}}} \overline{Q},$$

while $\overline{P}^{\overline{\theta}} := \overline{P}^{\mathrm{gp}} \times_{\overline{Q}^{\mathrm{gp}}} \overline{Q}$. This shows that the map $\overline{a}$ is surjective, hence an isomorphism. □

The construction of exactifications has an interesting form in the case of the diagonal morphism. Again let $\theta\colon P \to Q$ be a homomorphism of integral monoids, and form the integral pushout $Q \oplus_P Q$. The diagonal morphism of monoid schemes $\Delta\colon \underline{\mathsf{A}}_Q \to \underline{\mathsf{A}}_Q \times_{\underline{\mathsf{A}}_P} \underline{\mathsf{A}}_Q$ corresponds to the summation homomorphism

$$\sigma\colon Q \oplus_P Q \to Q : [(q_1, q_2)] \mapsto q_1 + q_2.$$

Recall from Remark 3.3.6 that the action of $\underline{\mathsf{A}}^*_{Q/P}$ on $\underline{\mathsf{A}}_Q$ induces a morphism $\underline{\mathsf{A}}_Q \times \underline{\mathsf{A}}^*_{Q/P} \to \underline{\mathsf{A}}_Q \times_{\underline{\mathsf{A}}_P} \underline{\mathsf{A}}_Q$ corresponding to the homomorphism of monoids

$$\phi\colon Q \oplus_P Q \to Q \oplus Q^{\mathrm{gp}}/P^{\mathrm{gp}} : [(q_1, q_2)] \mapsto (q_1 + q_2, [q_2]), \qquad (4.2.1)$$

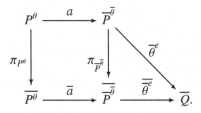

which induces an isomorphism of group envelopes. This homomorphism fits into a commutative diagram:

$$\begin{array}{ccc} Q \oplus_P Q & \xrightarrow{\phi} & Q \oplus Q^{\mathrm{gp}}/P^{\mathrm{gp}} \\ \tilde{\sigma}\downarrow & \overset{\cong}{\nearrow} & \downarrow (\mathrm{id}_Q, 0) \\ (Q \oplus_P Q)^{\sigma} & \xrightarrow{\sigma^e} & Q. \end{array} \tag{4.2.2}$$

Since ϕ^{gp} is an isomorphism and the square commutes, it follows from the constructions that ϕ^{gp} induces an isomorphism $(Q \oplus_P Q)^{\sigma} \to Q \oplus Q^{\mathrm{gp}}/P^{\mathrm{gp}}$. This argument leads to the following statement.

Proposition 4.2.19. *Let $\theta\colon P \to Q$ be a homomorphism of integral monoids, let $\sigma\colon Q \oplus_P Q \to Q$ be the homomorphism $[(q_1, q_2)] \mapsto q_1 + q_2$, and let*

$$\sigma\colon Q \oplus_P Q \xrightarrow{\tilde{\sigma}} (Q \oplus_P Q)^{\sigma} \xrightarrow{\sigma^e} Q$$

be its universal exactification as described in Proposition 4.2.17. Diagram (4.2.2) identifies this construction with the homomorphisms

$$\sigma\colon Q \oplus_P Q \xrightarrow{\phi} Q \oplus Q^{\mathrm{gp}}/P^{\mathrm{gp}} \xrightarrow{(\mathrm{id},0)} Q \quad \square$$

4.3 Small, Kummer, and vertical homomorphisms

The following terminology originated in the literature of log geometry.

Definition 4.3.1. *A homomorphism of integral monoids $\theta\colon P \to Q$ is:*

1. ***Q**-surjective if for every $q \in Q$, there exist $n \in \mathbf{Z}^+$ and $p \in P$ such that $nq = \theta(p)$,*
2. *Kummer if it is injective and **Q**-surjective,*
3. *small if θ^{gp} is **Q**-surjective, i.e., if the cokernel of θ^{gp} is a torsion group,*
4. *vertical if* Cok(θ) *(computed in the category of integral monoids) is dull.*

Remark 4.3.2. If $\theta\colon P \to Q$ is a homomorphism of integral monoids, one often writes Q/P for Cok(θ), which we recall is just the image of Q in Cok(θ^{gp}). Since the faces of Q are exact submonoids of Q, the homomorphism $Q \to Q/P$ induces a bijection between the faces of Q/P and those faces of Q containing the image of θ. Thus θ is vertical if and only if its image is not contained in any proper face of Q. Equivalently, θ is vertical if and only if the generic point of Spec(Q) is the only point of Spec(Q) lying over the generic point of Spec(P). For example, the diagonal homomorphism $\mathbf{N} \to \mathbf{N} \oplus \mathbf{N}$ is vertical, whereas the embedding into either of the coordinate submonoids is not. See Figure 4.3.1.

A small homomorphism need not be **Q**-surjective: for example, the inclusion

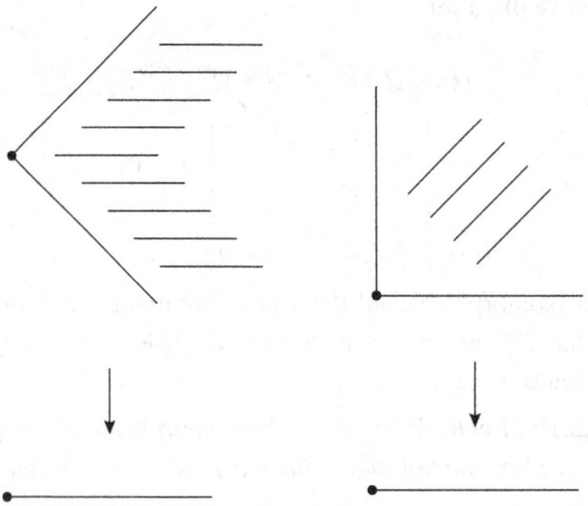

Figure 4.3.1 Vertical and non-vertical homomorphisms

of the submonoid P of $Q := \mathbf{N} \oplus \mathbf{N}$ generated by $(1,0)$ and $(1,1)$ is small but not **Q**-surjective.

We begin with some straightforward "sorites" about these classes of homomorphisms.

Proposition 4.3.3. *Let* **P** *be one of the following classes of homomorphisms of integral monoids:* **Q***-surjective, small, vertical.*

1. *Every* **Q***-surjective homomorphism is small, and every small homomorphism is vertical. If F is a face of P, the localization homomorphism $P \to P_F$ is small and vertical.*
2. *Let $\theta: P \to Q$ and $\phi: Q \to R$ be homomorphisms of integral monoids. If θ and ϕ belong to* **P***, then $\phi \circ \theta$ belongs to* **P***. If $\phi \circ \theta$ belongs to* **P***, then ϕ belongs to* **P***, and the same is true of θ if* **P** *is "***Q***-surjective" or "small" and ϕ is injective.*
3. *The natural homomorphism $P \to P^{\mathrm{sat}}$ belongs to* **P***, and a homomorphism $\theta: P \to Q$ belongs to* **P** *if and only if θ^{sat} belongs to* **P***.*
4. *The natural homomorphism $P \to \overline{P}$ belongs to* **P***. If $\theta: P \to Q$ belongs to* **P***, then $\overline{\theta}$ belongs to* **P***, and the converse holds if* **P** *is "vertical."*
5. *The pushout of a homomorphism in* **P***, formed in the category of integral monoids or in the category of saturated monoids, again belongs to* **P***.*
6. *If $\theta: P \to Q$ belongs to* **P** *and G is a face of Q, then the induced homomorphism $P_{\theta^{-1}(G)} \to Q_G$ belongs to* **P***.*

Proof A **Q**-surjective homomorphism is obviously small. If θ is small, then the cokernel of θ^{gp} is a torsion group, and the submonoid generated by any element is finite and integral, hence a group. Thus Q/P is a group and θ is vertical. If F is a face of P, the homomorphism $P^{gp} \to (P_F)^{gp}$ is an isomorphism, so $P \to P_F$ is small, hence also vertical.

Suppose that θ and ϕ are **Q**-surjective. Then if $r \in R$, there exist $n \in \mathbf{Z}^+$ and $q \in Q$ such that $\phi(q) = nr$, and then there exist $m \in \mathbf{Z}^+$ and $p \in P$ such that $\theta(p) = mq$. Then $\phi(\theta(p)) = mnr$, so $\phi \circ \theta$ is **Q**-surjective. Applying this result to θ^{gp} and ϕ^{gp}, we see that $\phi \circ \theta$ is small if θ and ϕ are small. If $\phi \circ \theta$ is **Q**-surjective and $r \in R$, then there exist $n \in \mathbf{Z}^+$ and $p \in P$ with $\theta(\phi(p)) = nr$, and hence θ is also **Q**-surjective, and we conclude the analogous result for small morphisms. If ϕ is injective and $\phi \circ \theta$ is **Q**-surjective then, for any $q \in Q$, there exist $m \in \mathbf{Z}^+$ and $p \in P$ such that $\phi(\theta(p)) = m\phi(q)$, hence $\theta(p) = mq$, so θ is also **Q**-surjective, and the corresponding result for small morphisms follows. If θ and ϕ are vertical, then the face of Q generated by $\theta(P)$ is all of Q and the face of R generated by Q is all of R. If G is a face of R containing $\phi \circ \theta(P)$, then $\phi^{-1}(G)$ is a face of Q containing $\theta(P)$, hence this face is all of Q and, since $\langle \phi(Q) \rangle = R$, in fact $G = R$, so $\phi \circ \theta$ is vertical. Conversely, if $\langle \phi \circ \theta(P) \rangle = R$, it is clear that $\langle \phi(Q) \rangle = R$, so ϕ is vertical.

It is obvious from the definitions that the natural map $\eta_P \colon P \to P^{sat}$ is **Q**-surjective, hence small and vertical. If θ belongs to **P**, then by (2) so does $\eta_Q \circ \theta = \theta^{sat} \circ \eta_P$, and by (2) it follows that θ^{sat} belongs to **P**. If θ^{sat} is **Q**-surjective (resp. small), we can also conclude from (2) that θ has the same property, because η_Q is injective. It is clear that the same conclusion holds for for the class of vertical homomorphisms, since the faces of a monoid and of its saturation correspond bijectively.

It is also obvious that the natural map $\pi_P \colon P \to \overline{P}$ belongs to **P**, and it follows from (2) that $\overline{\theta}$ belongs to **P** if θ does. If **P** is "vertical," the converse also holds, because the faces of P and of $\overline{P}$ correspond bijectively.

Suppose that $\theta' \colon P' \to Q'$ is the pushout of a homomorphism $\theta \colon P \to Q$ along $\phi \colon P \to P'$ in the category of integral monoids. Any element q' of Q can be written as a sum $q' = \theta'(p') + \phi'(q)$, with $p' \in P'$ and $q \in Q$. If θ is **Q**-surjective, there exist $m > 0$ and $p \in P$ such that $\theta(p) = mq$, and then

$$mq' = \theta'(mp') + \phi'(\theta(p)) = \theta'(mp' + \phi(p)),$$

so θ' is **Q**-surjective. As we saw in (2), it follows that θ'^{sat} is also **Q**-surjective, so the same conclusion holds for pushouts constructed in the category of saturated monoids. If θ is vertical and G' is a face of Q' containing $\theta'(P')$, then $\theta(P) \subseteq \phi'^{-1}(G')$, and since θ is vertical, it follows that $\phi'^{-1}(G') = Q$. Thus

G' contains $\phi'(Q)$ and $\theta'(P)$ and hence is all of Q', so θ' is again vertical. It follows that θ'^{sat} is also vertical.

Suppose that $\theta\colon P \to Q$ belongs to **P** and that G is a face of Q. Let $F := \theta^{-1}(G)$; since $\theta_F\colon P_F \to Q_F$ is the pushout of θ along λ_F, it belongs to **P**. If **P** is the class of small or vertical homomorphisms, the localization homomorphism $Q_F \to Q_G$ also belongs to **P**, and hence so does the homomorphism $P_F \to Q_G$. If **P** is the class of **Q**-surjective homomorphisms, then for every $g \in G$, there exist $n > 0$ and $f \in F$ with $\theta(f) = ng$, and it follows that G is the smallest face of Q containing $\theta(F)$. Then the natural map $Q_F \to Q_G$ is an isomorphism, and it follows that $P_F \to Q_G$ is **Q**-surjective. □

Proposition 4.3.4. *In the category of integral (resp. saturated) monoids, the following statements hold.*

1. *Let $\theta\colon P \to Q$ and $\phi\colon P \to Q$ be homomorphisms. If θ and ϕ are Kummer, then $\phi \circ \theta$ is Kummer. If $\phi \circ \theta$ is Kummer, and ϕ is injective, then θ and ϕ are Kummer.*
2. *The natural homomorphism $P \to P^{\text{sat}}$ is Kummer. A homomorphism θ is Kummer if and only if θ^{sat} is Kummer.*
3. *The pushout of a Kummer homomorphism $\theta\colon P \to Q$ is Kummer. If G is a face of Q, the induced map $P_{\theta^{-1}(G)} \to Q_G$ is Kummer.*

Proof If θ and ϕ are Kummer, then $\phi \circ \theta$ is **Q**-surjective by (2) of Proposition 4.3.3, and is injective because each of θ and ϕ is injective. Thus $\phi \circ \theta$ is Kummer. If $\phi \circ \theta$ is Kummer, then θ is injective. If also ϕ is injective then ϕ and θ are also **Q**-surjective by Proposition 4.3.3; hence both are Kummer.

It is clear from the definitions that $\eta_P\colon P \to P^{\text{sat}}$ is Kummer. If $\theta\colon P \to Q$ is a homomorphism, then $\eta_Q \circ \theta = \theta^{\text{sat}} \circ \eta_P$ and η_Q and η_P are Kummer. If θ is Kummer, it is injective, hence θ^{sat} is also injective and it follows that it is also Kummer. Conversely, if θ^{sat} is Kummer, then the same is true of θ, because $\eta_Q \circ \theta$ is Kummer and η_Q is injective.

Since the pushout of an injective (resp. **Q**-surjective) homomorphism is injective (resp. **Q**-surjective), the pushout of a Kummer homomorphism is Kummer. If θ is Kummer and G is a face of Q, then, as we saw in Proposition 4.3.3, the homomorphism $P_{\theta^{-1}(G)} \to Q_G$ is **Q**-surjective and, since it is injective, it is Kummer. □

The next result shows that, in the category of saturated monoids, a homomorphism is Kummer if and only if it is injective, exact, and small.

Proposition 4.3.5. *Let $\theta\colon P \to Q$ be a homomorphism of integral monoids.*

1. *If θ is exact and small, it is **Q**-surjective.*

2. *If P is saturated, if θ is $\mathbf{Q}$-surjective, and if $\text{Ker}(\theta)$ is torsion, then θ is exact, locally exact, and small.*
3. *If P is saturated and θ is Kummer, then θ is exact, locally exact, and small.*
4. *If θ is exact, sharp, and small, it is Kummer.*

Proof Suppose that θ is exact and small. If $q \in Q$, then since θ is small there exist $x \in P^{\text{gp}}$ and $n \in \mathbf{Z}^+$ such that $\theta^{\text{gp}}(x) = nq$ belongs to Q. Since θ is exact, it follows that $x \in P$, so θ is $\mathbf{Q}$-surjective. This proves statement (1). Suppose that P is saturated and that θ is $\mathbf{Q}$-surjective, and let x be an element of P^{gp} with $\theta(x) \in Q$. Then there exist $p \in P$ and $n \in \mathbf{Z}^+$ with $\theta(p) = n\theta(x)$. If $\text{Ker}(\theta)$ is torsion (and in particular if θ is Kummer), we can conclude that $mp = mnx$ for some $m > 0$. Since P is saturated it follows that $x \in P$ and hence that θ is exact, and it is obviously small. If G is any face of Q, statement (6) of Proposition 4.3.3 shows that the induced homomorphism $P_{\theta^{-1}(G)} \to Q_G$ is again $\mathbf{Q}$-surjective, and it follows that it too is exact. Thus θ is locally exact. Statements (2) and (3) follow. If θ is exact and small it is $\mathbf{Q}$-surjective, and it is injective if it is exact and sharp, by (5) of Proposition 4.2.1. This proves statement (4). □

The next proposition shows that $\mathbf{Q}$-surjective homomorphisms are analogous to integral homomorphisms in the theory of commutative rings.

Proposition 4.3.6. *Let $\theta\colon P \to Q$ be a homomorphism of integral monoids and consider the following conditions.*

1. *The action of P on Q deduced from θ makes Q into a finitely generated P-set.*
2. *θ is $\mathbf{Q}$-surjective.*

Condition (1) implies (2), and the converse holds if Q is fine.

Proof Suppose that (1) holds and that T is a finite set of generators for Q as a P-set. For each $q \in Q$, there exist $t_1 \in T$ and $p_1 \in P$ such that $q = \theta(p_1) + t_1$. Then there exist $t_2 \in T$ and $p_2 \in P$ such that $2t_1 = \theta(p_2) + t_2$. Continuing, we find by induction sequences (t_n) in T and (p_n) in P such that $2t_n = \theta(p_{n+1}) + t_{n+1}$ for $n \geq 1$. Then for every $k > 0$,

$$2^k t_n = \theta(p_{n,k}) + t_{n+k}, \text{ where } p_{n,k} := p_{n+k} + 2p_{n+k-1} + \cdots + 2^{k-1} p_{n+1}.$$

Since T is finite, there exist k and n in $\mathbf{Z}^+$ such that $t_n = t_{n+k}$. Then

$$(2^k - 1)t_n = \theta(p_{n,k}) + t_{n+k} - t_n = \theta(p_{n,k}),$$

and

$$\begin{aligned} t_n + \theta(p_{0,n}) &= t_n + \theta(p_n) + 2\theta(p_{n-1}) + \cdots + 2^{n-1}\theta(p_1) \\ &= 2t_{n-1} + 2\theta(p_{n-1}) + \cdots + 2^{n-1}\theta(p_1) \\ &\dots \\ &= 2^{n-1}t_1 + 2^{n-1}\theta(p_1) \\ &= 2^{n-1}q. \end{aligned}$$

Then

$$(2^k - 1)2^{n-1}q = (2^k - 1)t_n + (2^k - 1)\theta(p_{0,n}) = \theta(p_{n,k} + (2^k - 1)p_{0,n}).$$

This proves that θ is $\mathbf{Q}$-surjective. Conversely suppose that θ is $\mathbf{Q}$-surjective and that S is a finite set of generators for Q as a monoid. For each $s \in S$ there exist $d_s \in \mathbf{Z}^+$ and $p_s \in P$ such that $d_s s = \theta(p_s)$. Then the set T of elements of Q that can be written in the form $\sum\{r_s s : 0 \le r_s < d_s\}$ is finite and generates Q as a P-set. Indeed, any $q \in Q$ can be written as a sum $q = \sum\{n_s s : s \in S\}$, where each $n_s \in \mathbf{N}$, and then each n_s can be written as $n_s = m_s d_s + r_s$, where each $m_s, r_s \in \mathbf{N}$ and $r_s < d_s$. Then

$$q = \sum_s (m_s d_s + r_s)s = \sum_s (\theta(m_s p_s) + r_s s) = \theta\left(\sum_s m_s p_s\right) + \sum_s r_s s \in P + T. \quad \square$$

The following result uses duality to relate the notions of exactness and $\mathbf{Q}$-surjectivity.

Proposition 4.3.7. *Let $\theta\colon P \to Q$ be a homomorphism of integral monoids and let $H(\theta)\colon H(Q) \to H(P)$ be its dual.*

1. *If P and Q are fine and θ is exact, then $H(\theta)$ is $\mathbf{Q}$-surjective. The converse holds if P is saturated.*
2. *If θ is $\mathbf{Q}$-surjective, then $H(\theta)$ is exact, and the converse holds if P and Q are fine and Q is sharp.*

Proof Suppose that P and Q are fine and that θ is exact. Then $\overline{\theta}$ is also exact, and since $H(\theta) = H(\overline{\theta})$ we can and shall assume that P and Q are sharp. Since θ is exact and sharp, it is injective, by (5) of Proposition 4.2.1. We claim that the induced homomorphism of $\mathbf{Q}$-cones

$$C_{\mathbf{Q}}(H(\theta))\colon C_{\mathbf{Q}}(H(Q)) \to C_{\mathbf{Q}}(H(P))$$

is surjective. Note first that $C_{\mathbf{Q}}(\theta)\colon C_{\mathbf{Q}}(P) \to C_{\mathbf{Q}}(Q)$ is exact. Indeed, if $x \in C_{\mathbf{Q}}(Q)$ and $y \in C_{\mathbf{Q}}(P)^{\mathrm{gp}}$, with $C_{\mathbf{Q}}(\theta)(y) = x$, then there exist $m > 0$ and $q \in Q$ with $mx = 1 \otimes q$ and $z \in P^{\mathrm{gp}}$ with $my = 1 \otimes z$. Then $\theta(y)$ and q have the same image in $C_{\mathbf{Q}}(Q)^{\mathrm{gp}}$, so there exists some $n > 0$ such that $n\theta(y) = nq$ in Q^{gp}.

Since θ is exact, it follows that $p := ny$ belongs to P. Then $C_{\mathbf{Q}}(\theta)(n^{-1} \otimes p) = x$, proving the desired exactness of $C_{\mathbf{Q}}(\theta)$. By (4) of Proposition 2.3.6, we may identify $C_{\mathbf{Q}}(H(P))$ with

$$P^\vee := \mathrm{Hom}(P, \mathbf{Q}^{\geq}) = \mathrm{Hom}(C_{\mathbf{Q}}(P), \mathbf{Q}^{\geq}) \subseteq \mathrm{Hom}(P^{\mathrm{gp}}, \mathbf{Q}),$$

and similarly for $C_{\mathbf{Q}}(H(Q))$. Let C denote the image of $\theta^\vee\colon Q^\vee \to P^\vee$. Evidently $C \subseteq P^\vee$; by statement (2) of Theorem 2.3.12, to prove equality it will suffice to prove that $C^\vee \subseteq (P^\vee)^\vee = C_{\mathbf{Q}}(P)$ as subsets of $\mathbf{Q} \otimes P^{\mathrm{gp}}$. Thus suppose that $v \in C^\vee \subseteq \mathbf{Q} \otimes P^{\mathrm{gp}}$. For every $w \in Q^\vee$,

$$\langle C_{\mathbf{Q}}(\theta)(v), w\rangle = \langle v, w \circ C_{\mathbf{Q}}(\theta)\rangle = \langle v, \theta^\vee(w)\rangle \geq 0.$$

Thus $C_{\mathbf{Q}}(\theta)(v) \in (Q^\vee)^\vee = C_{\mathbf{Q}}(Q)$ and, since $C_{\mathbf{Q}}(\theta)$ is exact, it follows that $v \in C_{\mathbf{Q}}(P)$, as claimed.

Suppose conversely that $H(\theta)$ is $\mathbf{Q}$-surjective and that P is saturated. Let x be an element of P^{gp} with $\theta(x) \in Q$. For each $h \in H(P)$, there exist $n \in \mathbf{Z}^+$ and $h' \in H(Q)$ such that $nh = h' \circ \theta$, since $H(\theta)$ is $\mathbf{Q}$-surjective. Then $nh(x) = h'(\theta(x)) \geq 0$, hence $h(x) \geq 0$ and, since this is true for every h, Corollary 2.2.2 implies that $x \in P^{\mathrm{sat}} = P$. Thus θ is exact, completing the proof of statement (1).

Suppose that θ is $\mathbf{Q}$-surjective and that $x \in H(Q)^{\mathrm{gp}} \cong \mathrm{Hom}(Q, \mathbf{Z})$. Then $x \circ \theta \in H(P)$. For every $q \in Q$, there exist $n \in \mathbf{Z}^+$ and $p \in P$ with $nq = \theta(p)$, hence $nx(q) = x(\theta(p)) \geq 0$, and so $x(q) \geq 0$. This shows that $x \in H(Q)$, so $H(\theta)$ is exact. Suppose conversely that P and Q are fine, that Q is sharp, and that $H(\theta)$ is exact. Since θ is $\mathbf{Q}$-surjective if θ^{sat} is and since $H(\theta^{\mathrm{sat}}) = H(\theta)$ is exact, we may assume without loss of generality that P and Q are saturated. We have a commutative diagram:

$$\begin{array}{ccc} P & \xrightarrow{\ \theta\ } & Q \\ \Big\downarrow{\scriptstyle e_P} & & \Big\downarrow{\scriptstyle e_Q} \\ HH(P) & \xrightarrow{\ HH(\theta)\ } & HH(Q) \end{array}$$

Since $H(\theta)$ is exact, it follows from (1) that $HH(\theta)$ is $\mathbf{Q}$-surjective. Theorem 2.2.3 implies that e_Q is an isomorphism and that e_P is surjective. It follows that θ is also $\mathbf{Q}$-surjective □

The following consequence is evident if $K = \mathbf{Q}$, but perhaps less so in the general case.

Corollary 4.3.8. *Let $\theta\colon P \to Q$ be a homomorphism of toric monoids and let*

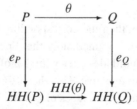

K be an Archimedean field. Then θ is exact if and only if the induced morphism of K-cones $C_K(\theta)\colon C_K(P) \to C_K(Q)$ is exact

Proof Suppose that $C_K(\theta)$ is exact and that $x \in P^{\mathrm{gp}}$ and $\theta^{\mathrm{gp}}(x) \in Q$. Then $C_K(\theta)^{\mathrm{gp}}(x) \in C_K(Q)$ and, since $C_K(\theta)$ is exact, the image of x in $K \otimes P^{\mathrm{gp}}$ lies in $C_K(P)$. By (4) of Proposition 2.3.6, it therefore lies in P^{sat}, hence in P, since we are assuming that P is toric. Suppose conversely that θ is exact. As we have already observed in the proof of Proposition 4.3.5, the exactness of $C_{\mathbf{Q}}(\theta)$ is immediate. For the general case, we use Proposition 4.3.5 to note that the map $H(\theta)\colon H(Q) \to H(P)$ is $\mathbf{Q}$-surjective. Now if $v \in K \otimes P^{\mathrm{gp}}$ and $C_K(\theta)(v) \in C_K(Q)$, it follows that $h(v) \geq 0$ for every $h \in H(P)$. By (4) of Proposition 2.3.6, $H(P)$ generates $P^\vee$, so in fact $v \in (P^\vee)^\vee = C_K(P)$ by Theorem 2.3.12. □

The following proposition relates the $\mathbf{Q}$-surjectivity of a homomorphism θ to the topological properties of $\operatorname{Spec}(\theta)$.

Proposition 4.3.9. *Let $\theta\colon P \to Q$ be a homomorphism of integral monoids. Consider the following conditions.*

1. *The homomorphism θ is $\mathbf{Q}$-surjective.*
2. *The mapping $\operatorname{Spec}(\theta)$ is injective (as a map of sets).*
3. *If G and G' are faces of Q and $\theta^{-1}(G) \subseteq \theta^{-1}(G')$, then $G \subseteq G'$.*
4. *Every face G of Q is generated as a face by $\theta(\theta^{-1}(G))$.*
5. *The topology of $\operatorname{Spec}(Q)$ is the weak topology induced from the topology of $\operatorname{Spec}(P)$ and the map $\operatorname{Spec}(\theta)$.*
6. *The only θ-critical face (4.2.10) of Q is Q^*.*

Then condition (1) implies conditions (2) through (5), which are equivalent, and which imply condition (6). All six conditions are equivalent if Q is fine and sharp.

Proof If θ is $\mathbf{Q}$-surjective, then the corresponding homomorphism of cones $C_{\mathbf{Q}}(\theta)$ is surjective. It follows immediately that $\operatorname{Spec} C_{\mathbf{Q}}(\theta)$ is injective, and, since the spectrum of a monoid identifies with that of the cone it spans, we can conclude that $\operatorname{Spec}(\theta)$ is also injective. This shows that (1) implies (2). Suppose that (2) holds and that G and G' are faces of Q with $\theta^{-1}(G) \subseteq \theta^{-1}(G')$. Then

$$\theta^{-1}(G) = \theta^{-1}(G) \cap \theta^{-1}(G') = \theta^{-1}(G \cap G').$$

Since $\operatorname{Spec}(\theta)$ is injective, it follows that $G = G \cap G'$ and hence that $G \subseteq G'$. Thus (2) implies (3). Suppose that (3) holds, let G be a face of Q, and let G' be the face of Q generated by $\theta(\theta^{-1}(G))$. Then $G' \subseteq G$ and $\theta^{-1}(G') = \theta^{-1}(G)$. Then by (3), $G' = G$, proving (4). Suppose that (4) holds. If $g \in Q$, let $G := \langle g \rangle$ and $F := \theta^{-1}(G)$. Since $\theta(F)$ generates G, there exist f in F and q in Q such that $\theta(f) = g + q$ and, since $\theta(f)$ belongs to $\langle g \rangle$, there exist $n \in \mathbf{N}$ and $q' \in Q$ such that $g^n = \theta(f) + q'$. Then a prime $\mathfrak{q}$ of Q fails to contain g if and only if it fails to contain $\theta(f)$. This implies that $\mathrm{Spec}(\theta)^{-1}(D(f)) = D(g)$, and hence that the topology of $\mathrm{Spec}(Q)$ is induced from the topology of $\mathrm{Spec}(P)$; so (4) implies (5). If (5) holds, let G and G' be faces of Q with $\theta^{-1}(G) \subseteq \theta^{-1}(G')$. Then $\theta^{-1}(\mathfrak{p}') \subseteq \theta^{-1}(\mathfrak{p})$, where $\mathfrak{p}$ and $\mathfrak{p}'$ are the respective complimentary faces. If we write $f\colon X \to Y$ for $\mathrm{Spec}(\theta)\colon \mathrm{Spec}(Q) \to \mathrm{Spec}(P)$, and x and x' for the points of X corresponding to $\mathfrak{p}$ and $\mathfrak{p}'$, this means that $f(x)$ belongs to the closure of $f(x')$, and hence by (5) that x belongs to the closure of x'. But then $\mathfrak{p}' \subseteq \mathfrak{p}$ and $G \subseteq G'$. Thus (5) implies (3), and it is clear that (3) implies (2) and (6).

It remains only to prove that if Q is fine and sharp, then condition (6) implies condition (1). Let G be a one-dimensional face of Q. Then (6) implies that $\theta^{-1}(G) \neq \theta^{-1}(Q^*)$, so there exists some $p \in P$ such that $\theta(p)$ is a nonzero element of G. Then the image of $\theta(p)$ in the cone $C_{\mathbf{Q}}(Q)$ generates the face $C_{\mathbf{Q}}(G)$ of $C_{\mathbf{Q}}(Q)$, which is an arbitrary edge of $C_{\mathbf{Q}}(Q)$. By Proposition 2.3.4, $C_{\mathbf{Q}}(Q)$ is generated by the generators of its edges, so $C_{\mathbf{Q}}(\theta)$ is surjective. □

Corollary 4.3.10. *Let $\theta\colon P \to Q$ be a homomorphism of fine sharp and saturated monoids. Then the following are equivalent.*

1. *θ is exact and small.*
2. *θ is Kummer.*
3. *$\mathrm{Spec}(\theta)$ is a homeomorphism.*
4. *θ is injective and local and $\{0\}$ is its only θ-critical face.*

Proof If θ is exact it is injective, by (5) of Proposition 4.2.1, and if it is also small it is $\mathbf{Q}$-surjective by Proposition 4.3.5, and hence Kummer. Thus (1) implies (2), and the converse was already proved in Proposition 4.3.5. If (1) and (2) hold, θ is exact, so Proposition 4.2.2 implies that $\mathrm{Spec}(\theta)$ is surjective, and then Proposition 4.3.9 implies that $\mathrm{Spec}(\theta)$ is a a homeomorphism. Conversely if (3) holds, it follows from Proposition 4.3.9 again that θ is $\mathbf{Q}$-surjective. Since $\mathrm{Spec}(\theta)$ is surjective, Proposition 4.2.2 implies that θ is also exact. This completes the proof of the equivalence of the first three conditions. If these hold, then it is clear from (3) that $\{0\}$ is the only θ-critical face of Q, from (2) that θ is injective, and from (1) that θ is local, proving (4). Condition (4) here is the same as condition (6) of Proposition 4.3.9, so that proposition implies that θ is $\mathbf{Q}$-surjective. Since θ is injective by assumption, it is Kummer. □

Corollary 4.3.11. *Let $\theta\colon P \to Q$ be a homomorphism of integral monoids. If $\overline{\theta}$ is exact and small, then θ is locally exact and* $\mathrm{Spec}(\theta)$ *is a homeomorphism. In particular these conclusions hold if P is saturated and $\overline{\theta}$ is Kummer,*

Proof If $\overline{\theta}$ is exact, then $\mathrm{Spec}(\theta)$ is surjective by Proposition 4.2.2. If $\overline{\theta}$ is also small, then it is $\mathbf{Q}$-surjective by (1) of Proposition 4.3.5, and it follows from (5) of Proposition 4.3.9 that $\mathrm{Spec}(\overline{\theta})$ is injective and that the topology of $\mathrm{Spec}(\overline{Q})$ is induced from that of $\mathrm{Spec}(\overline{P})$. Thus $\mathrm{Spec}(\overline{\theta})$ is a homeomorphism, and the same is true of $\mathrm{Spec}(\theta)$. Furthermore, if G is any face of Q and $F := \theta^{-1}(G)$, then (4) of Proposition 4.3.9 implies that G is generated by F, and hence the homomorphism $P_F \to Q_G$ can be identified with the localization θ_F of θ by F. Since θ is exact, θ_F is also exact and, since G was arbitrary, θ is locally exact. If P is saturated and $\overline{\theta}$ is Kummer, then it is $\mathbf{Q}$-surjective and exact by Proposition 4.3.5, so the same conclusions hold. □

Corollary 4.3.12. *Let $\theta\colon P \to Q$ be a homomorphism of fine monoids. Suppose that $\overline{\theta}$ is small. Then, with the notation of Proposition 4.2.17, the homomorphisms $\theta^e\colon P^\theta \to Q$, $\overline{\theta}^e\colon \overline{P}^\theta \to \overline{Q}$, and $\overline{\theta^e}\colon \overline{P^\theta} \to \overline{Q}$ are locally exact. Furthermore, $\overline{\theta^e}$ is Kummer.*

Proof Since $\overline{\theta}$ is small, so is $\overline{\theta}^e$ and, the latter is also exact by construction. Then $\overline{\overline{\theta}^e}$ (which by Proposition 4.2.18 can be identified with $\overline{\theta^e}$) is also exact and small, so Corollary 4.3.11 implies that $\overline{\theta}^e$ is locally exact. It follows that θ^e and $\overline{\theta^e}$ are also locally exact. Since $\overline{\theta^e}$ is exact, small, and sharp it is Kummer by (4) of Proposition 4.3.5. □

Proposition 4.3.13. *Let $\theta\colon P \to Q$ be a homomorphism of fine monoids. Assume that θ is exact and $\mathbf{Q}$-surjective and that $K_\theta := P^+ + Q$ is a radical ideal of Q. Then θ is strict.*

Proof Replacing θ by $\overline{\theta}$, we may assume without loss of generality that P and Q are sharp. Then θ is exact, local, and injective. Let us identify P with its image in Q. We claim that $Q \setminus P$ is empty. Since θ is exact, $Q \setminus P$ is a sub-P-set of Q by (5) of Proposition 2.1.16. Since θ is Kummer, it follows from Proposition 4.3.6 that Q is finitely generated as a P-set, and hence by Proposition 2.1.5 that it is noetherian as a P-set. The same proposition tells us that the P-set $Q \setminus P$ is also noetherian and hence, if nonempty, contains a minimal element q_0. Since θ is $\mathbf{Q}$-surjective, there is some positive integer n such that $nq_0 \in P$, and since $q_0 \neq 0$ and Q is torsion free, $nq_0 \in P^+ \subseteq K_\theta$. Since K_θ is a radical ideal, $q_0 \in K_\theta = P^+ + Q$, so we can write $q_0 = p + q$ for some $p \in P^+$ and $q \in Q$. Necessarily $q \in Q \setminus P$ and, by the minimality of q_0, we have $q = p' + q_0$ for some $p' \in P$. Then $q_0 = p + p' + q_0$, hence $p + p' = 0$, contradicting the fact that $p \in P^+$. □

4.4 Toric Frobenius and isogenies

If Q is a (commutative) monoid and n is a natural number, the map

$$n_Q \colon Q \to Q : q \mapsto nq$$

is a monoid homomorphism. The corresponding morphism of monoid schemes $\underline{\mathsf{A}}_{n_Q} \colon \underline{\mathsf{A}}_Q \to \underline{\mathsf{A}}_Q$ is the nth power mapping. If R is a ring of characteristic p, the homomorphism of monoid algebras $R[Q] \to R[Q]$ induced by p_Q is the identity on elements of R and takes a generator e^q to $e^{pq} = (e^q)^p$; this is the homomorphism corresponding to the relative Frobenius morphism of the R-scheme $\underline{\mathsf{A}}_Q$. (Since $\underline{\mathsf{A}}_Q$ is defined over the prime field, we can identify $\underline{\mathsf{A}}_Q$ with its base change with respect to the absolute Frobenius of Spec R.) For this reason, the morphism n_Q is sometimes called the *n-Frobenius endomorphism of Q*.

Fix an integral monoid Q and a natural number n, and let

$$Q^{(n)} := \{x \in Q^{\mathrm{gp}} : nx \in Q\}.$$

The exacification of n_Q, in the sense of Proposition 4.2.17, defines a factorization of n_Q:

$$\begin{array}{ccc} Q & \xrightarrow{\phi_{Q,n}} & Q^{(n)} \\ & \searrow^{n_Q} & \downarrow{\scriptstyle \psi_{Q,n}} \\ & & Q. \end{array}$$

If Q^{gp} is n-torsion free, $\psi_{Q,n}$ is injective and identifies $Q^{(n)}$ with $Q \cap nQ^{\mathrm{gp}}$. If Q is toric, then $\phi_{Q,n}$ is an isomorphism and $\psi_{Q,n}$ can be identified with n_Q.

Of course, the homomorphisms n_Q are compatible with all monoid homomorphisms, fitting into the obvious commutative squares. We relativize these Frobenius homomorphisms in the following way.

Definition 4.4.1. *Let $\theta \colon P \to Q$ be a homomorphism of integral monoids, let*

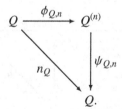

n be a positive integer, and consider the following commutative diagram:

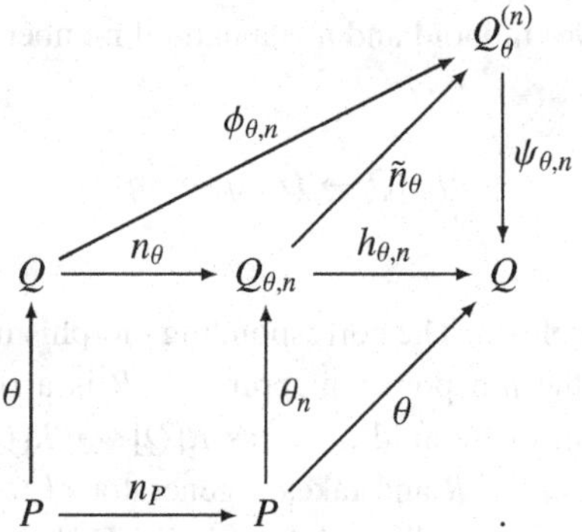

Here the square is cocartesian in the category of integral monoids and $h_{\theta,n}$ is the unique homomorphism such that $h_{\theta,n} \circ n_\theta = n_Q$ and $h_{\theta,n} \circ \theta_n = \theta$. The homomorphism $h_{\theta,n}$ is called the relative n-Frobenius homomorphism *of θ. The homomorphism $\psi_{\theta,n}$ is the exactification of the homomorphism h_{θ_n} in the sense of Proposition 4.2.17 and is called the* exact relative n-Frobenius homomorphism *of θ.*

Recall from Proposition 4.2.17 that by definition

$$Q_\theta^{(n)} := \{x \in Q_{\theta,n}^{\mathrm{gp}} : h_{\theta,n}(x) \in Q\}$$

and that $\tilde{n}_\theta$ induces an isomorphism

$$Q_{\theta,n}^{\mathrm{gp}} \cong (Q_\theta^{(n)})^{\mathrm{gp}}.$$

We shall be especially interested in the subdiagram (the *exact relative Frobenius diagram*)

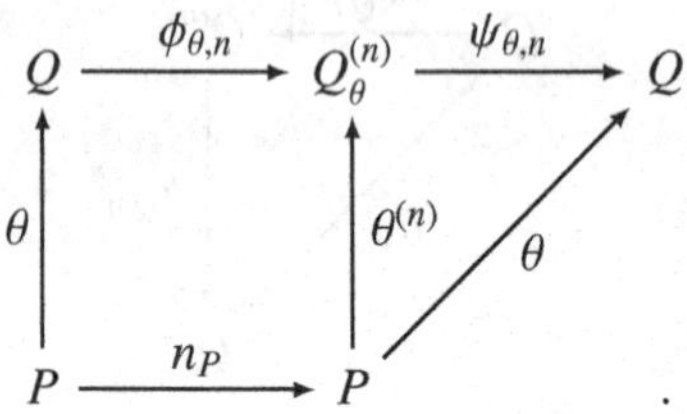

(4.4.1)

Suppose that P and Q are saturated and let $Q_{\theta,n}^{\mathrm{sat}}$ denote the saturation of the integral monoid $Q_{\theta,n}$. Then the homomorphism $h_{\theta,n}$ factors uniquely through a homomorphism $h'_{\theta,n} : Q_{\theta,n}^{\mathrm{sat}} \to Q$. Let us remark that this homomorphism is necessarily exact, so that $Q_\theta^{(n)} \subseteq Q_{\theta,n}^{\mathrm{sat}}$. In fact, the following more precise statement holds.

Proposition 4.4.2. *Let $\theta\colon P \to Q$ be a homomorphism of integral monoids. Then $Q_\theta^{(n)} \subseteq \{z \in Q_{\theta,n}^{\mathrm{gp}} : nz \in Q_{\theta,n}\} \subseteq Q_{\theta,n}^{\mathrm{sat}}$, with equality if Q is saturated.*

Proof To verify this claim, write an element z of $Q_\theta^{(n)}$ as $z = n_\theta(x) + \theta_n(y)$, where $x \in Q^{\text{gp}}$ and $y \in P^{\text{gp}}$. Then $h_{\theta,n}(z) = nx + \theta(y)$ is an element q of Q, and

$$nz = n_\theta(nx) + \theta_n(ny) = n_\theta(q - \theta(y)) + \theta_n(ny) = n_\theta(q).$$

Thus $nz \in Q_{\theta,n}$, as claimed. It is obvious that $\{z \in Q^{\text{gp}}_{\theta,n} : nz \in Q_{\theta,n}\} \subseteq Q^{\text{sat}}_{\theta,n}$. If Q is saturated, Lemma 2.1.19 implies that $Q_\theta^{(n)}$ is already saturated, since $\psi_{\theta,n}$ is exact by construction. Thus equality holds in this case. □

Proposition 4.4.3. *Let $\theta\colon P \to Q$ be an injective homomorphism of integral monoids, and consider the diagram in Definition 4.4.1. If A is an abelian group, we denote by $T_n(A)$ the subgroup of A killed by n.*

1. *The group $\text{Ker}(h^{\text{gp}}_{\theta,n})$ is killed by n, and the natural maps*

$$\text{Ker}(h^{\text{gp}}_{\theta,n}) \to T_n(\text{Cok}(\theta_n^{\text{gp}}))$$
$$\text{Cok}(\theta^{\text{gp}}) \to \text{Cok}(\theta_n^{\text{gp}})$$

   are isomorphisms.
2. *The sequence*

$$0 \to T_n(P^{\text{gp}}) \to T_n(Q^{\text{gp}}_{\theta,n}) \to T_n(\text{Cok}(\theta_n^{\text{gp}})) \to 0$$

   is exact.
3. *The homomorphism $\tilde{n}_\theta^{\text{gp}} : Q^{\text{gp}}_{\theta,n} \to Q_\theta^{(n)\text{gp}}$ is an isomorphism, and the image of $Q_\theta^{(n)}$ in Q is $Q \cap (nQ^{\text{gp}} + P^{\text{gp}})$.*
4. *The homomorphism $\overline{\psi}_{\theta,n}$ is Kummer, the group $\text{Ker}(\psi^{\text{gp}}_{\theta,n})$ is contained in $Q_\theta^{(n)*}$, and $\psi_{\theta,n}$ induces an isomorphism*

$$Q_\theta^{(n)} / \text{Ker}(\psi^{\text{gp}}_{\theta,n}) \to Q \cap (nQ^{\text{gp}} + P^{\text{gp}}).$$

5. *The homomorphism $\psi_{\theta,n}$ induces a bijection between the set of faces of Q and the set of faces of $Q_\theta^{(n)}$. For every face F of Q containing P,*

$$(nF^{\text{gp}} + P^{\text{gp}}) \cap F = (nF^{\text{gp}} + P^{\text{gp}}) \cap Q,$$

   and if Q is saturated,[1]

$$(nF^{\text{gp}} + P^{\text{gp}}) \cap F = (nQ^{\text{gp}} + P^{\text{gp}}) \cap F.$$

Proof The natural map $\text{Cok}(\theta^{\text{gp}}) \to \text{Cok}(\theta_n^{\text{gp}})$ is an isomorphism because the square in the defining diagram is cocartesian. Suppose $\tilde{x} \in \text{Ker}(h^{\text{gp}}_{\theta,n})$, and write $\tilde{x} = n_\theta(x) + \theta_n(y)$, with $x \in Q^{\text{gp}}$ and $y \in P^{\text{gp}}$. Then $nx + \theta(y) = 0$, and so $n\tilde{x} = n_\theta(nx) + \theta_n(ny) = n_\theta(nx + \theta(y)) = 0$. Thus $\tilde{x}$ is killed by n,

[1] In fact it suffices that Q/F^{gp} be n-torsion free for every face F. For example, this is the case if Q is n-saturated (see Definition 4.8.1).

and its image in $\mathrm{Cok}(\theta_n^{\mathrm{gp}})$ identifies with the image of x in $\mathrm{Cok}(\theta^{\mathrm{gp}})$ and is independent of the choices of x and y. If this image is zero, then there exists some $z \in P^{\mathrm{gp}}$ such that $\theta(z) = x$, and we have $\theta(nz + y) = 0$. Since θ is injective, $-nz = y$, and then $\tilde{x} = n_\theta\theta(z) + \theta_n(-nz) = 0$. Finally, if $x \in Q^{\mathrm{gp}}$ is an element whose image in $Q^{\mathrm{gp}}/P^{\mathrm{gp}}$ is killed by n, then there exists a $y \in P^{\mathrm{gp}}$ such that $nx = \theta(y)$, in which case $\tilde{x} := n_\theta(x) - \theta_n(y)$ belongs to $\mathrm{Ker}(h_{\theta,n})$ and maps to the image of x in $\mathrm{Cok}(\theta^{\mathrm{gp}})$. This completes the proof of statement (1). The exactness of the sequence in (2) is clear, except for the surjectivity of the map $T_n(Q_{\theta_n}^{\mathrm{gp}}) \to T_n(\mathrm{Cok}(\theta_n^{\mathrm{gp}}))$, which follows from the first part of statement (1). The homomorphism $\tilde{n}_\theta^{\mathrm{gp}}$ is an isomorphism by construction, and the image of $h_{\theta_n,n}^{\mathrm{gp}}$ is $nQ^{\mathrm{gp}} + \theta(P^{g})$. Since $\psi_{\theta,n}$ is exact by construction, its image is the intersection of Q with the image of $\psi_{\theta,n}^{\mathrm{gp}}$, proving (3). Since $\psi_{\theta,n}$ is exact and small, $\overline{\psi}_{\theta,n}$ is Kummer, by (4) of Proposition 4.3.5. If $\tilde{x} \in \mathrm{Ker}(\psi_{\theta,n})^{\mathrm{gp}}$, write x as $\tilde{q}_2 - \tilde{q}_1$, with $\tilde{q}_i \in Q_\theta^{(n)}$. Then $\tilde{q}_2$ and $\tilde{q}_1$ have the same image in $\overline{Q}$ and, since $\overline{\psi}_{\theta,n}$ is injective, they have the same image in $\overline{Q}_\theta^{(n)}$. It follows that $\tilde{x} \in Q_\theta^{(n)*}$. Since $\overline{\psi}_{\theta,n}$ is Kummer, it and the homomorphism $\psi_{\theta,n}$ induce bijections between the faces of Q and the faces of $Q_\theta^{(n)}$, by Corollary 4.3.10. If F is a face of Q containing P, the equality $(nF^{\mathrm{gp}} + P^{\mathrm{gp}}) \cap F = (nF^{\mathrm{gp}} + P^{\mathrm{gp}}) \cap Q$ follows from the fact that F is an exact submonoid of Q. If $x \in Q^{\mathrm{gp}}, y \in P^{\mathrm{gp}}$, and $f := nx + y \in F$, then $nx \in F^{\mathrm{gp}}$ and, if Q is saturated, $Q^{\mathrm{gp}}/F^{\mathrm{gp}}$ is torsion free, so $x \in F^{\mathrm{gp}}$. It follows that $(nF^{\mathrm{gp}} + P^{\mathrm{gp}}) \cap F = (nQ^{\mathrm{gp}} + P^{\mathrm{gp}}) \cap F$. □

Corollary 4.4.4. *Suppose that the hypotheses of Proposition 4.4.3 hold and that* $\mathrm{Cok}(\theta^{\mathrm{gp}})$ *is n-torsion free. Then* ψ_{θ_n} *factors:*

$$\psi_{\theta,n} \colon Q_\theta^{(n)} \xrightarrow{\cong} (nQ^{\mathrm{gp}} + P^{\mathrm{gp}}) \cap Q \hookrightarrow Q.$$

Assume in addition that for every face F of Q containing P, the group $Q^{\mathrm{gp}}/F^{\mathrm{gp}}$ is n-torsion free. Then for every such face, $F_\theta^{(n)} = \psi_{\theta,n}^{-1}(F)$ and is mapped isomorphically to $(nF^{\mathrm{gp}}+P^{\mathrm{gp}}) \cap Q$ by $\psi_{\theta,n}$. The isomorphism $\mathrm{Cok}(\theta^{\mathrm{gp}}) \to \mathrm{Cok}(\theta_n^{\mathrm{gp}})$ takes $F^{\mathrm{gp}}/P^{\mathrm{gp}}$ isomorphically to $F_\theta^{(n)\mathrm{gp}}/P^{\mathrm{gp}}$, and $F_\theta^{(n)}$ is the face of $Q_\theta^{(n)}$ generated by $\phi_{\theta,n}(F)$.

Proof The first statement follows immediately from (1) and (4) of Proposition 4.4.3. If F is a face of Q containing P, then $F^{\mathrm{gp}}/P^{\mathrm{gp}}$ is also n-torsion free, and hence the map $F_\theta^{(n)} \to (nF^{\mathrm{gp}} + P^{\mathrm{gp}}) \cap F$ is again an isomorphism. By (5) of Proposition 4.4.3, $(nF^{\mathrm{gp}} + P^{\mathrm{gp}}) \cap F = (nQ^{\mathrm{gp}} + P^{\mathrm{gp}}) \cap F$, so $F_\theta^{(n)} = \psi_{\theta,n}^{-1}(F)$. Furthermore, $(nF^{\mathrm{gp}}+P^{\mathrm{gp}}) \cap F = (nF^{\mathrm{gp}}+P^{\mathrm{gp}}) \cap Q$, again by (5) of Proposition 4.4.3. Statement (1) of this same proposition, applied to the homomorphism $P \to F$, implies that the homomorphism $F^{\mathrm{gp}}/P^{\mathrm{gp}} \to F_\theta^{(n)\mathrm{gp}}/P^{\mathrm{gp}}$ is an isomorphism. It is clear that the face F' of $Q_\theta^{(n)}$ generated by $\phi_{\theta,n}(F)$ is contained in $F_\theta^{(n)}$.

Moreover, F' contains $n_P(P)$ and hence also P. Since F^{gp} maps surjectively to $F_\theta^{(n)\mathrm{gp}}/P^{\mathrm{gp}}$, and since F' contains P, it follows that $F'^{\mathrm{gp}} = F_\theta^{(n)\mathrm{gp}}$ and hence that $F' = F_\theta^{(n)}$, since both are exact submonoids of $Q_\theta^{(n)}$. □

Toric Frobenius homomorphisms are examples of the following more general class of homomorphisms. In practice, the number n appearing in the definition below will be prime.

Definition 4.4.5. *If n is a positive natural number, a homomorphism of commutative monoids $\theta\colon P \to Q$ is an n-isogeny if the following conditions are verified:*

1. *For every $q \in Q$, there exist an $r > 0$ and a $p \in P$ such that $\theta(p) = n^r q$.*
2. *For every pair p_1, p_2 of elements in P such that $\theta(p_1) = \theta(p_2)$, there exists an $r > 0$ such that $n^r p_1 = n^r p_2$.*

Proposition 4.4.6. *A homomorphism of integral monoids $\theta\colon P \to Q$ is an n-isogeny if for some $r > 0$ there exists a homomorphism $\theta'\colon Q \to P$ fitting into the following commutative diagram:*

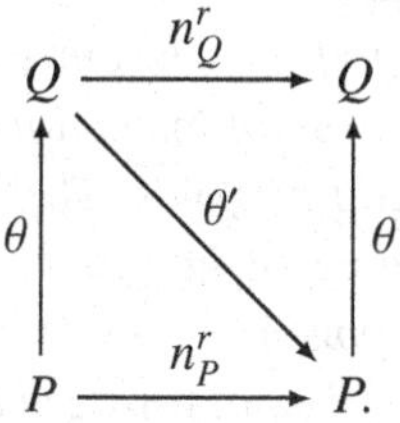

The converse is true if P and Q are fine. If P is saturated (or just n-saturated, see Definition 4.8.1) and θ is an n-isogeny, then θ is exact.

Proof It is clear that the existence of the diagram implies that θ is an n-isogeny. For the converse, suppose that that θ is an n-isogeny and that Q is fine. Then Q^{gp} is a finitely generated group, and the hypothesis implies that the map $\theta_S^{\mathrm{gp}}\colon P_S^{\mathrm{gp}} \to Q_S^{\mathrm{gp}}$ is an isomorphism, where the subscript S means localization by the multiplicative set S of powers of n. Let $\phi \in \mathrm{Hom}(Q_S^{\mathrm{gp}}, P_S^{\mathrm{gp}})$ be the inverse of θ_S^{gp}. Since Q^{gp} is finitely presented, the map $\mathrm{Hom}(Q^{\mathrm{gp}}, P^{\mathrm{gp}})_S \to \mathrm{Hom}(Q_S^{\mathrm{gp}}, P_S^{\mathrm{gp}})$ is an isomorphism, and hence there exist $r > 0$ and $\phi' \in \mathrm{Hom}(Q^{\mathrm{gp}}, P^{\mathrm{gp}})$ such that $\phi'_S = n^r\phi$. Then $\theta_S^{\mathrm{gp}} \circ \phi'_S = n^r_{Q_S^{\mathrm{gp}}}$, and again since Q is finitely generated, there is a $k > 0$ such that $n^k\theta \circ \phi' = n^{k+r}_{Q^{\mathrm{gp}}}$. Replacing ϕ' by $n^k\phi'$ and r by $r+k$, we have $\theta^{\mathrm{gp}} \circ \phi' = n^r_{Q^{\mathrm{gp}}}$. When P is also finitely generated, the same argument applied to $\phi' \circ \theta^{\mathrm{gp}}$ allows us to assume that $\phi' \circ \theta^{\mathrm{gp}} = n^r_P$. Now, for each of a finite set of generators q_i of Q, there exist $p_i \in P$ and $k_i \in \mathbf{N}$ such that $\theta(p_i) = n^{k_i}q_i$. Without loss of generality, we may assume that $k_i = k + r$ for some k. Then $\theta(p_i) = \theta(n^k\phi'(q_i))$ for every i, and hence there exists some j such that $n^j p_i = n^j \phi'(q_i)$ for every i. In particular, $n^j \phi'(q_i) \in P$ for every i, and $\theta' := n^j \phi'$ defines a map $Q \to P$. Replacing r by $r + j$, we find the desired commutative diagram. When P is saturated, n^r_P is exact, and since $\theta' \circ \theta = n^r_P$, it follows that θ is exact. □

Proposition 4.4.7. *Let $\theta \colon P \to Q$ be an n-isogeny of fine monoids and let $P \to P^\theta \to Q$ be its exactification (see Proposition 4.2.17). Then if Q is saturated, $P^\theta = P^{\mathrm{sat}}$, as submonoids of P^{gp}.*

Proof Since $\theta^e \colon P^\theta \to Q$ is exact and Q is saturated, Lemma 2.1.19 implies that P^θ is also saturated, and hence $P^{\mathrm{sat}} \subseteq P^\theta$. Conversely, if $x \in P^\theta$, then $\theta^{\mathrm{gp}}(x) \in Q$, and hence there exist $r \in \mathbf{N}$ and $p \in P$ such that $\theta(p) = n^r \theta^{\mathrm{gp}}(x)$. Then $n^r x - p \in \mathrm{Ker}(\theta)$. Since θ is an n-isogeny, $\mathrm{Ker}(\theta^{\mathrm{gp}})$ is n^∞-torsion, so there exists $s \in \mathbf{N}$ such that $n^{r+s} x = n^s p$. Thus $x \in P^{\mathrm{sat}}$ as claimed. □

Proposition 4.4.8. *The class of n-isogenies enjoys the following properties.*

1. *If $\theta \colon P \to Q$ and $\phi \colon Q \to R$ are homomorphisms of monoids and any two of ϕ, θ, and $\phi \circ \theta$ are n-isogenies, so is the third.*
2. *If θ is an n-isogeny of integral monoids, then θ^{sat} is also an n-isogeny.*
3. *Let $\theta \colon P \to Q$ be an n-isogeny of integral monoids and let $\phi \colon P \to P'$ be a homomorphism of integral monoids. Then the pushout homomorphism $\theta' \colon P' \to Q'$ in the category of integral monoids is an n-isogeny. If P' is saturated, the homomorphisms $P' \to Q'^{\mathrm{sat}}$ and $Q' \to Q'^{\mathrm{sat}}$ are also n-isogenies. In particular, the class of n-isogenies is closed under pushouts in the category of integral monoids and in the category of saturated monoids.*
4. *A homomorphism θ of fine monoids is an n-isogeny for some n if and only if it is $\mathbf{Q}$-surjective and $\theta^{\mathrm{gp}} \otimes \mathrm{id}_{\mathbf{Q}}$ is injective.*
5. *Every homomorphism which is an n-isogeny is local.*
6. *A homomorphism of integral monoids $\theta \colon P \to Q$ is an n-isogeny if and only if θ^* and $\overline{\theta}$ are n-isogenies.*

Proof The proof of the first statement is straightforward and we omit it. Suppose that $\theta \colon P \to Q$ is an n-isogeny of integral monoids and $x \in Q^{\mathrm{sat}}$. Since θ is an n-isogeny, so is θ^{gp}, and hence there exist $r > 0$ and $y \in P^{\mathrm{gp}}$ such that $\theta^{\mathrm{gp}}(y) = n^r x$. Since $x \in Q^{\mathrm{sat}}$, there exists $m > 0$ such that $mx \in Q$, and since θ is an n-isogeny, there exist $s > 0$ and $p \in P$ such that $\theta(p) = n^s mx$. Without loss of generality we may assume that $r = s$. Then $\theta^{\mathrm{gp}}(p) = \theta^{\mathrm{gp}}(my)$, and hence there exists $t > 0$ such that $n^t p = n^t my$. But then $y \in P^{\mathrm{sat}}$, and thus $n^r x$ is in the image of θ^{sat}. Since $(\theta^{\mathrm{sat}})^{\mathrm{gp}} = \theta^{\mathrm{gp}}$, its kernel is n^∞-torsion, and hence θ^{sat} is an n-isogeny.

Suppose that $\theta \colon P \to Q$ is an n-isogeny and $P' \to Q'$ is its pushout along

$P \to P'$ in the category of integral monoids. Any $q' \in Q'$ can be written as a sum $\theta'(p') + \phi'(q)$ with $p' \in P'$ and $q \in Q$. Choose $r > 0$ and $p \in P$ with $\theta(p) = n^r q$. Then $n^r(\theta'(p') + \phi'(q)) = \theta'(n^r p' + \phi(p))$. On the other hand, if $p'_1, p'_2 \in P'$ and $\theta'(p'_1) = \theta'(p'_2)$, then, by the construction of pushouts in the category of integral monoids, there exist $p_1, p_2 \in P$ such that $p'_2 - p'_1 = \phi(p_2) - \phi(p_1)$ and $\theta(p_2) = \theta(p_1)$. Since θ is an n-isogeny, it follows that there exists some $r > 0$ such that $n^r p_2 = n^r p_1$, hence also $n^r p'_2 = n^r p'_1$. If P' is saturated, the homomorphism $P' \to P^{\mathrm{sat}}$ is an isomorphism, hence an n-isogeny, and hence $P' \to Q' \to Q'^{\mathrm{sat}}$ is an n-isogeny. It follows that $Q' \to Q'^{\mathrm{sat}}$ is also an n-isogeny.

Suppose θ is $\mathbf{Q}$-surjective and Q is finitely generated. For each q_i in a finite set of generators $q_1, \ldots, q_m$ for Q, choose $n_i > 0$ and $p_i \in P$ such that $\theta(p_i) = n_i q_i$. Let $n := n_1 n_2 \cdots n_m$. Then, for all $q \in Q$, there exists $p \in P$ such that $\theta(p) = nq$. Suppose that P is finitely generated and that $\theta^{\mathrm{gp}} \otimes \mathrm{id}_{\mathbf{Q}}$ is injective. Then $\mathrm{Ker}(\theta^{\mathrm{gp}})$ is a finite group, hence killed by some positive integer n'. Hence if $\theta(p_1) = \theta(p_2)$, it follows that $n' p_1 = n' p_2$. It follows that θ is an nn'-isogeny.

Suppose that $\theta \colon P \to Q$ is an n-isogeny and $p \in P$ with $\theta(p) \in Q^*$. Then there exists $q' \in Q$ such that $q' + \theta(p) = 0$, and hence there exist $r > 0$ and $p' \in P$ with $n^r q' = \theta(p')$. Then $\theta(p' + n^r p) = 0$, so there exists some $s > 0$ with $n^s p' + n^{r+s} p = 0$. It follows that p is a unit, as required.

Suppose θ is an n-isogeny and $q \in Q^*$. Then there exist $r > 0$ and $p \in P$ such that $\theta(p) = n^r q$. It follows that $\theta(p) \in Q^*$ and hence that $p \in P^*$. This proves that θ^* satisfies condition (1) of Definition 4.4.5. Condition (2) is obvious, so θ^* is an n-isogeny. Suppose that p_1 and p_2 are two elements of P and $\overline{\theta}(p_1) = \overline{\theta}(p_2)$ in $\overline{Q}$. Then there exists $q \in Q^*$ such that $\theta(p_2) = q + \theta(p_1)$ and, since θ is an n-isogeny, there exist $r > 0$ and $p \in P$ such that $n^r q = \theta(p)$. Then $p \in P^*$, and $\theta(n^r p_2) = \theta(p + n^r p_1)$. Hence there exists $s > 0$ such that $n^{s+r} p_2 = n^s p + n^{s+r} p_1$. It follows that $n^{s+r}\overline{p}_2 = n^{s+r}\overline{p}_1$ in $\overline{P}$. Thus $\overline{\theta}$ satisfies condition (2) of Definition 4.4.5. Condition (1) is obvious, so $\overline{\theta}$ is an n-isogeny. Suppose on the other hand that θ^* and $\overline{\theta}$ are n-isogenies, and that $q \in Q$. Then there exist $r > 0$ and $p \in P$ such that $\overline{\theta}(\overline{p}) = n^r \overline{q}$. Hence there exists $v \in Q^*$ such that $n^r q = v + \theta(p)$, and then there exist $s > 0$ and $u \in P^*$ such that $n^s v = \theta(u)$. Then $n^{r+s} q = \theta(u + n^s p)$. If $p_1, p_2 \in P$ with $\theta(p_1) = \theta(p_2)$ then, since $\overline{\theta}$ is an n-isogeny, there exist $r > 0$ and $p \in P^*$ such that $n^r p_2 = u + n^r p_1$. Since Q is integral, it follows that $\theta(u) = 0$ and, since θ^* is an n-isogeny, that $n^s u = 0$ for some $s > 0$. But then $n^{r+s} p_2 = n^{r+s} p_1$, and θ is an n-isogeny. □

Although n-isogenies are not necessarily exact, they are universally local, as the following proposition shows.

Proposition 4.4.9. *Let $\theta \colon P \to Q$ and $\phi \colon P \to P'$ be homomorphisms of* *monoids, and let* $\phi' : Q \to Q'$ *and* $\theta' : P' \to Q'$ *be the pushout maps, in the category of integral monoids. If* θ *is an* n*-isogeny, then* θ' *is also an* n*-isogeny, and in particular it is local. If* ϕ *is local, then* ϕ' *is also local. The kernel* K *of the natural map* $P'^* \oplus_{P^*} Q^* \to P'^{\mathrm{gp}} \oplus_{P^{\mathrm{gp}}} Q^{\mathrm{gp}}$ *is* n^∞*-torsion.*

Proof The fact that θ' is an n-isogeny and hence local follows from Proposition 4.4.8. Assume that ϕ is local and suppose that $q \in Q$ and $\phi'(q)$ is a unit of Q'. Then there exist $q' \in Q$, $p' \in P'$, such that $\phi'(q) + \phi'(q') + \theta'(p') = 0$. This implies that there is an $x \in P^{\mathrm{gp}}$ such that $q + q' = \theta(x)$ and $p' = -\phi(x)$. Since θ is an n-isogeny, there exist $r > 0$ and $p \in P$ such that $n^r(q + q') = \theta(p)$. Then $\theta(p) = \theta(n^r x)$, and hence there exists $s > 0$ such that $n^r p = n^{s+r} x$. Then $\phi(n^s p) = -n^{s+r} p' \in P'$, and hence $\phi(n^s p) \in P'^*$. Since ϕ is local, we can conclude that $p \in P^*$ and hence that $n^{s+r}(q + q') \in Q^*$. It follows that $q + q'$ is a unit of Q, hence the same is true of q.

To prove the second statement, let θ_S^{gp} (resp. θ_S^*) denote the localization of θ^{gp} (resp. of θ^*) by $S := \{n^r : r \geq 0\}$. Since the formation of kernels and pushouts commutes with localization, we can identify K_S with the kernel of the map

$$P_S'^* \oplus_{P_S^*} Q_S^* \to P_S'^{\mathrm{gp}} \oplus_{P_S^{\mathrm{gp}}} Q_S^{\mathrm{gp}}.$$

Since θ is an n-isogeny, both θ_S^* and θ_S^{gp} are isomorphisms, and the above map identifies with the map $P_S'^* \to P_S'^{\mathrm{gp}}$, which is injective. Thus $K_S = 0$, and hence each element of K is killed by some power of n.

□

4.5 Flat and regular monoid actions

Flatness is a key and subtle condition in algebraic geometry, and in this section we begin an investigation of its analog in monoidal geometry. In commutative algebra, an R-module E is flat if $M \mapsto M \otimes_R E$ is an exact functor on the category of R-modules. If P is a monoid, we shall define the flatness of a P-set S in terms of a suitable exactness property of the functor $T \mapsto T \otimes_P S$. To prepare for this definition, let us recall the following terminology from [2, I, 2.7].

Definition 4.5.1. *A category is said to be filtering if it satisfies the following conditions:*

F0. It is nonempty and connected. That is, there is at least one object and, given any two objects a *and* b*, there exists a finite sequence of morphisms con-*

necting a and b:

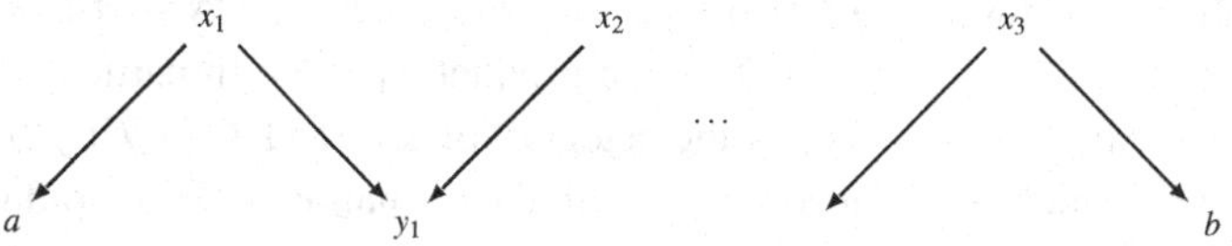

or

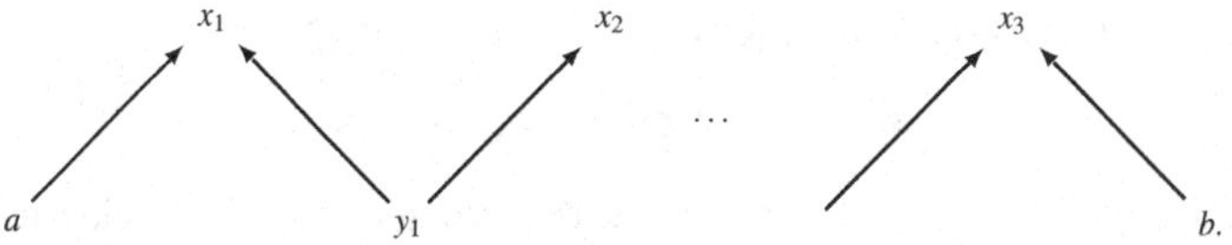

F1. Given arrows u_1 and u_2 as shown in the following diagram

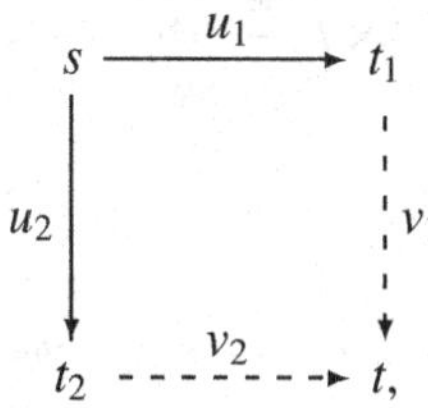

there exist an object t and morphisms $v_1: t_1 \to t$ and $v_2: t_2 \to t$ such that $v_1 u_1 = v_2 u_2$.

F2. Given arrows u_1 and u_2 as in the following diagram,

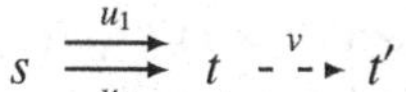

there exists a morphism $v: t \to t'$ such that $v \circ u_1 = v \circ u_2$.

Condition F2 says that, in the situation of F1, if $t_1 = t_2$, then one can choose (v_1, v_2) with $v_1 = v_2$. We should remark that in the presence of condition F2, conditions F0 and F1 can be replaced by

F0$'$. There is at least one object.

F1$'$. Given any two objects a_1 and a_2, there exists an object b and morphisms $u_i: a_i \to b$.

A category is said to be *cofiltering* if its opposite is filtering.

Definition 4.5.2. *Let P be a commutative monoid and let S be a P-set. Then S is said to be flat if for every functor F from a finite connected category Λ to* **Ens**$_P$*, the natural map*

$$(\lim F) \otimes_P S \to \lim(F \otimes_P S)$$

is an isomorphism.

Note that this can fail if Λ is not connected, even if S is free. For example, if Λ is a discrete category with two objects $1, 2$, a functor $F\colon \Lambda \to \mathbf{Ens}_P$ is just a pair of P-sets F_1, F_2, and $\lim F$ is the product $F_1 \times F_2$. In particular, when $P = 0$, the category $\mathbf{Ens}_P$ is just the category of sets, and $T \mapsto T \otimes_P S$ is just $T \mapsto T \times S$, which does not commute with the formation of finite products.

Example 4.5.3. With the definition above, every free P-set S is flat. Indeed if $\{s_i : i \in I\}$ is a basis for S then, for any P-set T, the map

$$T \times I \longrightarrow T \otimes_P S \quad (t, i) \mapsto t \otimes s_i$$

is an isomorphism. (Here $T \times I$ is given the structure of a P-set via the action of P on T.) If $F\colon \Lambda \to \mathbf{Ens}_P$ is a functor, we have a commutative diagram

$$\begin{array}{ccc} (\lim F) \times I & \xrightarrow{\cong} & (\lim F) \otimes_P S \\ \downarrow & & \downarrow \\ \lim(F \times I) & \xrightarrow{\cong} & \lim(F \otimes_P S). \end{array}$$

(Here $F \times I$ is the functor which takes λ to $F_\lambda \times I$ and which takes a morphism $u\colon \lambda \to \lambda'$ to $F_u \times \mathrm{id}_I$.) The left vertical arrow takes $(f., i)$ to the family $(f., i.)$, where $i_\lambda = i$ for all $\lambda \in \Lambda$. It is clear that this map is injective. We claim that if Λ is connected, it is also surjective. Indeed, if $(f., i.) \in \lim(F \times I)$ and if there is a morphism $u\colon \lambda \to \lambda'$, then $(F \times I)_u(f_\lambda, i_\lambda) = (f_{\lambda'}, i_{\lambda'})$, and hence $i_\lambda = i_{\lambda'}$. Thus if Λ is connected, i_λ is independent of λ, and $(f., i.)$ belongs to $(\lim F)\times I$. It follows that the right vertical map is also bijective.

Example 4.5.4. If G is an abelian group, every G-set S satisfies a weak form of flatness. Namely, if $\theta\colon T \to T'$ is an injective map of G-sets then, for every G-set S, the induced map $T \otimes_G S \to T' \otimes_G S$ is also injective. Indeed, suppose that $\theta(t_1) \otimes s_1 = \theta(t_2) \otimes s_2$. Then since G is a group, there exists a $g \in G$ such that $\theta(t_2) = g\theta(t_1)$ and $s_1 = gs_2$. Since θ is injective, it follows that $t_2 = gt_1$ and hence that $t_1 \otimes s_1 = t_2 \otimes s_2$. On the other hand, not every G-set is flat. For example, if S is a singleton, then $T \otimes_G S$ identifies with the orbit space T/G, whose formation does not commute with equalizers if G is not trivial. We shall see as a consequence of Theorem 4.5.7 and Proposition 4.5.10 that a G-set S is flat if and only if the action of G on S is free.

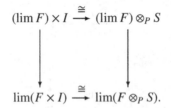

We shall see that the notion of flatness for P-sets can be understood in terms of the transporter category $\mathcal{T}_P S$ of S. Note first that the transporter category of any P-set S satisfies F1, and that the transporter category $\mathcal{T}_P P$ of P is filtering. Next, observe that a P-set S can be written as a disjoint union of its connected components, which are just the connected components of its transporter category $\mathcal{T}_P S$, or, equivalently, of its opposite. The following result describes these more directly.

Proposition 4.5.5. *Let S be a P-set and $\lambda\colon S \to S_P$ be the localization of S by P, so that P^{gp} acts on S_P. Then, if $(s_1, s_2) \in S \times S$, the following are equivalent.*

1. *$\lambda(s_1)$ and $\lambda(s_2)$ lie in the same orbit of the action of P^{gp} on S_P.*
2. *There exist $p_1, p_2 \in P$ with $p_1 s_1 = p_2 s_2$.*
3. *s_1 and s_2 lie in the same connected component of S under the action of P, i.e., in the same connected component of the category $\mathcal{T}_P S$.*

Proof If (1) holds, then there exists $x \in P^{\text{gp}}$ such that $\lambda(s_2) = x\lambda(s_1)$ in S_P. Write $x = p_1 p_2^{-1}$. Then $\lambda(p_2 s_2) = \lambda(p_1 s_1)$ and, by the construction of S_P, there exists a $p \in P$ such that $pp_2 s_2 = pp_1 s_1$. Then (2) holds with p_i replaced by pp_i. It is clear that (2) implies (3). Finally, if s_1 and s_2 lie in the same connected component of S under the action of P, then the same is true of $\lambda(s_1)$ and $\lambda(s_2)$, and it follows that $\lambda(s_1)$ and $\lambda(s_2)$ are in the same P^{gp}-orbit of S_P. □

The next proposition is an immediate consequence of the definitions.

Proposition 4.5.6. *Let P be a monoid and let S be a P-set. Then the transporter category $\mathcal{T}_P S$ of S is cofiltering if and only if S satisfies the following conditions.*

I0. S is connected, that is, the category $\mathcal{T}_P S$ satisfies condition F0 of Definition 4.5.1.
I1. Given $s_1, s_2 \in S$ and $p_1, p_2 \in P$ such that $p_1 s_1 = p_2 s_2$, there exist $s' \in S$ and $p'_1, p'_2 \in P$ such that $s_i = p'_i s'$ and $p_1 p'_1 = p_2 p'_2$.
I2. Given $s \in S$ and $p_1, p_2 \in P$ such that $p_1 s = p_2 s$, there exist $s' \in S$ and $p' \in P$ such that $s = p' s'$ and $p_1 p' = p_2 p'$. □

Condition I1 says that, given the solid arrows in the diagram below, there exist an object s' and dashed arrows making the diagram commute.

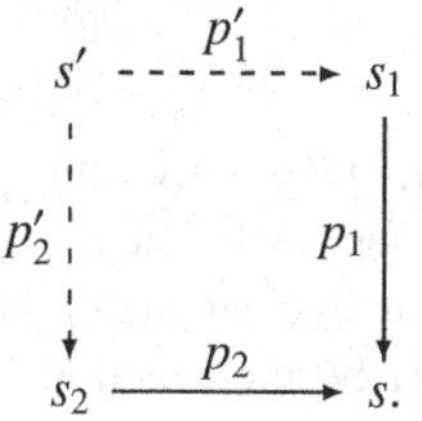

Condition I2 says that when $s_1 = s_2$, one can in fact choose $p'_1 = p'_2$. Note that S satisfies I1 (resp. I2) if and only if each of its connected components does, that is, if and only if each of its connected components is cofiltering. Finally, note that $\mathcal{T}_P P$ is always cofiltering as well as filtering.

The following proposition can be thought of as an analog for monoids of Lazard's theorem in commutative algebra [71].

Theorem 4.5.7. *Let P be a monoid and S a P-set. Then the following conditions are equivalent.*

1. *S satisfies I1 and I2 of Proposition 4.5.6.*
2. *S is a filtered direct limit of free P-sets.*
3. *S is flat.*

Proof The proof that (1) implies (2) depends on the following general way of writing an arbitrary P-set S as a colimit of free P-sets. Let

$$G\colon (\mathcal{T}_P S)^{op} \to \mathbf{Ens}_P$$

be the functor which sends each s in S to the P-set P and for each $p \in P$ sends the morphism $p\colon ps \to s$ to multiplication by $p\colon P \to P$. For each $s \in S$, let

$$f_s\colon G(s) = P \to S$$

denote the unique morphism of P-sets sending 1_P to s. Thus for each $p \in P$ and $s \in S$ we have a commutative diagram

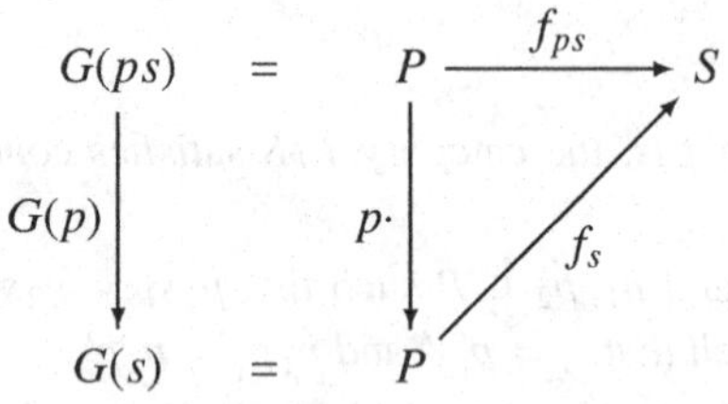

and hence a morphism $f\colon \operatorname{colim} G \to S$.

Lemma 4.5.8. *Let P be a monoid and let S be a P-set, and let $G\colon (\mathcal{T}_P)^{op} \to S$ be the functor described above. Then the corresponding morphism of P-sets*

$$f\colon \operatorname{colim} G \to S$$

is an isomorphism.

Proof For each $s \in S$, let $\eta_s\colon G(s) \to \operatorname{colim} G$ be the natural map. Then f is the unique morphism of P-sets such that $f \circ \eta_s = f_s$ for all s. In particular, $f(\eta_s(1_P)) = f_s(1_P) = s$, so it is clear that f is surjective. To see that it is injective, let $g\colon S \to \operatorname{colim} G$ be the function sending s to $\eta_s(1_P)$. Then, if $p \in P$,

$$g(ps) = \eta_{ps}(1_P) = \eta_s(G(p)(1_P)) = \eta_s(p) = p\eta_s(1_P) = pg(s),$$

so g is a morphism of P-sets. For each $s \in S$, $g \circ f_s(1_P) = g(s) = \eta_s(1_P)$, and hence $g \circ f_s = \eta_s$. Moreover,

$$g \circ f \circ \eta_s = g \circ f_s = \eta_s,$$

hence $g \circ f = \mathrm{id}$. It follows that f is indeed injective. □

Now we can prove that (1) implies (2). First suppose that S satisfies I1 and I2 and additionally that it is connected. Then $(\mathcal{T}_P S)^{op}$ is filtering, so the colimit in Lemma 4.5.8 is in fact a filtered direct limit, hence S is a filtered direct limit of free Q-sets.

In general, S is the disjoint union of its connected components and, if S satisfies I1 and I2, each of these connected components is a filtered direct limit of free P-sets, and it follows that the same is true of S. To see this directly, let $\{S_c : c \in C\}$ be the set of connected components of S, for each $c \in C$, let Λ_c be the opposite of the transporter category of S_c, and let Λ be the product category $\Lambda = \prod\{\Lambda_c : c \in C\}$. An object of Λ is a family $s_\cdot := \{s_c \in S_c : c \in C\}$ and a morphism $s_\cdot \to s'_\cdot$ is a family $p_\cdot := \{p_c : c \in C\}$ such that $s_c = p_c s'_c$ for all $c \in C$. The category Λ is filtering if each Λ_c is filtering. For each object $s_\cdot$ of Λ, let $G(s_\cdot)$ be the free P-set $P \times C$ and, for each morphism $p_\cdot\colon s_\cdot \to s'_\cdot$, let $G(p_\cdot)\colon G(s_\cdot) \to G(s'_\cdot)$ be the map sending (p, c) to $(p_c p, c)$. Then G is a functor from Λ to the category of P-sets. For each object $s_\cdot$ of Λ, let $f_{s_\cdot} : P \times C \to S$ be the morphism sending (p, c) to ps_c. Then $f_{s_\cdot} \circ G(p_\cdot) = f_{p_\cdot s_\cdot}$, and the argument used in the proof of Lemma 4.5.8 shows that the map $\operatorname{colim} G \to S$ induced by f is an isomorphism.

To prove that (2) implies (3), recall from Example 4.5.3 that every free P-set is flat. Thus it will suffice to show that a direct limit of flat P-sets is flat. The argument is standard. Suppose that S is a functor from a filtering category Λ to the category of P-sets and F is a functor from a finite connected category to the category of P-sets. We have the following commutative diagram:

$$\begin{array}{ccc}
(\lim F) \otimes_P \varinjlim S & \xrightarrow{a} & \lim(F \otimes \varinjlim S) \\
\Big\downarrow b & & \Big\uparrow c \\
\varinjlim ((\lim F) \otimes S) & \xrightarrow{d} & \varinjlim (\lim(F \otimes S)).
\end{array}$$

Here b is an isomorphism because the functor $T \to (\lim F) \otimes T$ commutes with all colimits, since it has a right adjoint. The map d is an isomorphism because each object S_λ is flat. Finally, all projective limits and all direct limits (i.e., colimits over filtering categories) in the category $\mathbf{Ens}_P$ are the same as in the category **Ens**. Thus, formation of direct limits in $\mathbf{Ens}_P$ is exact, i.e., commutes with formation of finite limits, and so c is also an isomorphism. It follows that a is an isomorphism, and hence that $\varinjlim S$ is flat.

Next we prove that if S is flat then it satisfies I2. Note first that if p_1 and p_2 are elements of P, then

$$K(p_1, p_2) := \{p \in P : p_1 p = p_2 p\}$$

is an ideal of P, and the inclusion $K(p_1, p_2) \to P$ is the equalizer of multiplication by p_1 and p_2 on P. Since $\otimes_P S$ commutes with formation of equalizers, it follows that $K(p_1, p_2) \otimes_P S \to S$ is the equalizer of multiplication by p_1 and p_2 on S. Thus if $p_1 s = p_2 s$, there exist $s' \in S$ and $p' \in K(p_1, p_s)$ such that $s = p's'$. This is exactly I2.

Finally, we prove that if S is flat then it satisfies I1. Given $p_1, p_2 \in P$, consider the fiber squares

$$\begin{array}{ccc} E_P & \longrightarrow & P \\ \downarrow & & \downarrow{\scriptstyle p_2} \\ P & \xrightarrow{p_1} & P \end{array} \qquad \begin{array}{ccc} E_S & \longrightarrow & S \\ \downarrow & & \downarrow{\scriptstyle p_2} \\ S & \xrightarrow{p_1} & S. \end{array}$$

Thus E_P consists of the set of pairs (p'_1, p'_2) such that $p_1 p'_1 = p_2 p'_2$. The flatness of S implies that the natural map $E_P \otimes_P S \to E_S$ is an isomorphism. If $(s_1, s_2) \in S \times S$ and $p_1 s_1 = p_2 s_2$, then in fact $(s_1, s_2) \in E_S$, and it follows that there exist a pair $(p'_1, p'_2) \in E_P$ and an $s' \in S$ such that $s_i = p'_i s'$. This is exactly I1. □

Let us investigate conditions I1 and I2 in the context of integral monoids. It will also be convenient to introduce a variant of condition I2.

Definition 4.5.9. *Let P be a monoid and let S be a P-set.*

1. *An element p of P is S-regular if the endomorphism of S induced by the action of p is injective.*
2. *The P-set S is P-regular if every p in P is S-regular.*
3. *The P-set S satisfies I2* if whenever $p_1, p_2 \in P$ and $s \in S$ are such that $p_1 s = p_2 s$, there exists a $u \in P^*$ such that $us = s$ and $p_2 = up_1$.*

Proposition 4.5.10. *Let P be an integral monoid and let S be a P-set.*

1. *If P is a group, conditions I1 and I2* are automatically satisfied.*
2. *If I1 holds, then S is P-regular, and the converse holds if P is valuative.*
3. *Condition I2 holds if and only if $p_1 s = p_2 s$ implies that $p_1 = p_2$ or, equivalently, if and only if the action of P^{gp} on the localization S_P of S is free.*

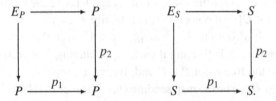

Proof Suppose P is a group. If $p_1 s_1 = p_2 s_2$, then we can take $s' := s_2$, $p'_1 := p_1^{-1} p_2$, and $p'_2 := 1_P$ to satisfy I1. If $s_1 = s_2$, this argument shows that I2* is satisfied as well, with $u := p'_1$.

Suppose only that P is integral and that $ps_1 = ps_2$. If S satisfies I1, then there exist $s' \in S$ and $p'_1, p'_2 \in P$ with $s_i = p'_i s'$ and $pp'_1 = pp'_2$. Since P is integral, it follows that $p'_1 = p'_2$ and hence that $s_1 = s_2$. Thus S is P-regular. Conversely, suppose that S is P-regular and that P is valuative. If $p_1 s_1 = p_2 s_2$, either $p_2 \geq p_1$ or $p_1 \geq p_2$; let us assume the former. Then $p_2 = pp_1$ for some $p \in P$, and $p_1 s_1 = p_1 p s_2$. Since S is P-regular, it follows that $s_1 = ps_2$, and I1 is satisfied, with $s' := s_2$, $p'_1 := p$, and $p'_2 = 1_P$.

Finally suppose that P is integral and that $p_1 s = p_2 s$. Then I2 says that there exist s' and p' with $p' p_1 = p' p_2$ and $s = p' s'$. But then in fact $p_1 = p_2$, as claimed. The converse is immediate, as is the equivalence with the freeness of the action of P^{gp} on S_P. □

Statement (3) of the next proposition is an analog of the fact that a finitely generated flat module over a noetherian local ring is free.

Proposition 4.5.11. *Let P be an integral monoid and let S be a P-set.*

1. *If S satisfies I2* and each of its connected components is monogenic, then S also satisfies I1.*
2. *If S satisfies I1, then an element s of S generates its connected component $S_s(P)$ if and only if it is P-minimal. In particular, if P is fine and S is finitely generated and satisfies I1, then each of its connected components is monogenic.*
3. *If P is fine, a finitely generated P-set S is flat if and only if it is free. In this case each connected component of S is isomorphic to P.*

Proof Suppose that each connected component of S is monogenic and that $p_i \in P$ and $s_i \in S$ with $p_1 s_1 = p_2 s_2$. Then s_1 and s_2 belong to the same connected component of S; suppose that s is a generator of this component. Then there exist $p''_1, p''_2 \in P$ with $s_i = p''_i s$, and it follows that $p_1 p''_1 s = p_2 p''_2 s$. If condition I2* is satisfied, there exists a $u \in P^*$ such that $us = s$ and $p_2 p''_2 = p_1 p''_1 u$. Let $p'_1 := up''_1$ and $p'_2 := p''_2$, so that $s_i = p'_i s$ and $p_1 p'_1 = p_2 p'_2$, as in condition I1.

Suppose that S satisfies I1 and $s \in S$. It is clear that s is minimal in its connected component $S_s(P)$ if it generates $S_s(P)$. Conversely suppose that s_1 is minimal in $S_s(P)$. (Note that such an element will also be minimal in S.) Choose any $s_2 \in S_s(P)$. By Proposition 4.5.5, there exist p_1, p_2 in P such that $p_1 s_1 = p_2 s_2$. By condition I1, there exist $s' \in S$ and $p'_i \in P$ such that $s_i = p'_i s'$

and $p_1p_1' = p_2p_2'$. Necessarily s' also belongs to $S_s(P)$, so that, by the minimality of s_1, there exists some p'' such that $s' = p''s_1$. But then we have $s_2 = p_2'p''s_1$ and hence s_2 is in the trajectory of s_1. Thus s_1 generates $S_s(P)$. If P is fine and S is finitely generated, then it is noetherian and hence by Corollary 2.1.11 each of its connected components contains a minimal element. This completes the proof of (2).

We have already observed that a free P-set is flat. Suppose that S is flat and finitely generated and that P is fine. To prove that S is free it is enough to prove that each of its connected components S' is free. By Theorem 4.5.7, S satisfies I1 and I2, and hence so does S'. As we have just seen, condition I1 already implies that S' is monogenic, say generated by s. Then the map $P \to S'$ sending p to ps is surjective. Condition I2 and the integrality of P imply that this map is also injective, hence an isomorphism. □

The next proposition relates the flatness of a P-set to the flatness of the corresponding $\mathbf{Z}[P]$-module (Definition 3.2.3) and helps to justify our definition of the former notion.

Proposition 4.5.12. *Let P be an integral monoid and let S be a P-set. Then the following conditions are equivalent.*

1. *S is flat as a P-set.*
2. *The $\mathbf{Z}[P]$-module $\mathbf{Z}[S]$ is flat as a $\mathbf{Z}[P]$-module.*
3. *For every prime field k, the $k[P]$-module $k[S]$ is flat as a $k[P]$-module.*

Proof If S is flat, then by Theorem 4.5.7 it is a direct limit of free P-sets, and hence $\mathbf{Z}[S]$ is a direct limit of free $\mathbf{Z}[P]$-modules, and therefore is flat. Thus (1) implies (2). The implication of (3) by (2) is trivial, so it remains only to prove that (3) implies (1). This will follow from Lemma 4.5.14 below. That lemma depends on the following result, which deals with the simpler case of group actions. □

Lemma 4.5.13. *Let G be a group and let S be a G-set. Suppose that for every prime field k, $k[S]$ is flat over $k[G]$. Then S is a free G-set.*

Proof Suppose that $g \in G$, $s \in S$, and $gs = s$. Then $(e^g - 1)e^s = 0$ in the $k[G]$-module $k[S]$ and, since $k[S]$ is flat over $k[G]$, we can write $e^s = \sum_i \alpha_i\sigma_i$ where each $\alpha_i \in k[G]$ is killed by $e^g - 1$ and $\sigma_i \in k[S]$. Write $\alpha_i := \sum c_h e^h$. Then $\sum c_h e^{gh} = \sum c_h e^h$, since α_i is annihilated by $e^g - 1$. This says that $c_{g^{-1}h} = c_h$ for all h, i.e., that α_i is a linear combination of g-orbits of the regular representation of G on itself. Since only finite sums are allowed, either α_i is zero or g has finite order. But if all $\alpha_i = 0$, we could conclude that $e^s = 0$, which is impossible.

Thus g has finite order, say n. Then each α_i is a multiple of $\alpha := \sum_{i=0}^{n-1} e^{g^i}$, and hence we can write $e^s = \alpha\sigma$ for some $\sigma := \sum c_t e^t \in k[S]$. Then

$$e^s = \sum_{i,t} c_t e^{g^i t} = \sum_t c'_t e^t$$

where $c'_t := \sum_i c_{g^i t}$. Comparing the coefficients of e^s, we find that $1 = c'_s := \sum_i c_{g^i s}$. Since $gs = s$, we find that $1 = nc_s$. Thus n is invertible in k and, since this is true for every prime field k, necessarily $n = 1$ and g is the identity, as required. □

Lemma 4.5.14. *Let P be an integral monoid and let S be a P-set.*

1. *If $k[S]$ is flat over $k[P]$ for some field k, then S satisfies condition I1 of Proposition 4.5.6.*
2. *If $k[S]$ is flat over $k[P]$ for every field k, then S also satisfies condition I2 of Proposition 4.5.6 and hence is a flat P-set.*

Proof Suppose that $k[S]$ is flat over $k[P]$ for some field k. Let us first observe that S is P-regular. If p is any element of P, translation by p induces an injective endomorphism of P, since P is integral. Since the elements of P form a basis for $k[P]$, it follows that multiplication by e^p is injective on $k[P]$ and, since $k[S]$ is flat, multiplication by e^p is also injective on $k[S]$. It follows that the action of each $p \in P$ on S is injective, so that S is P-regular. Now, to verify that S satisfies condition I1, suppose that s_1 and s_2 are elements of S and p_1 and p_2 are elements of P with $p_1 s_1 = p_2 s_2$. Let E be the fiber product in the category of $k[P]$-modules in the following diagram:

$$\begin{array}{ccc} E & \longrightarrow & k[P] \\ \big\downarrow & & \big\downarrow{\scriptstyle e^{p_2}} \\ k[P] & \xrightarrow{e^{p_1}} & k[P]. \end{array}$$

Tensoring by $k[S]$ we get by the flatness of $k[S]$ a cartesian diagram:

$$\begin{array}{ccc} E \otimes_{k[P]} k[S] & \longrightarrow & k[S] \\ \big\downarrow & & \big\downarrow{\scriptstyle e^{p_2}} \\ k[S] & \xrightarrow{e^{p_1}} & k[S]. \end{array}$$

Hence (e^{s_1}, e^{s_2}) comes from an element of $E \otimes_{k[P]} k[S]$, and we can find elements

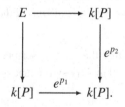

(α_i, β_i) of $E \subseteq k[P] \times k[P]$ and σ_i of $k[S]$ with $e^{s_1} = \sum \alpha_i \sigma_i$ and $e^{s_2} = \sum \beta_i \sigma_i$. Write

$$\alpha_i := \sum_p a_{i,p} e^p, \quad \beta_i := \sum_p b_{i,p} e^p, \quad \sigma_i := \sum_s c_{i,s} e^s.$$

From the equation $e^{s_1} = \sum \alpha_i \sigma_i$, we see that there exist i, p'_1, and s' such that $a_{i,p'_1} \neq 0$, $c_{i,s'} \neq 0$, and $s_1 = p'_1 s'$. Since $e^{p_1}\alpha_i = e^{p_2}\beta_i$, there exists p'_2 such that $b_{i,p'_2} \neq 0$ and $p_1 p'_1 = p_2 p'_2$. But then $p_2 s_2 = p_1 s_1 = p_1 p'_1 s' = p_2 p'_2 s'$ and, since S is P-regular, $s_2 = p'_2 s'$. This gives us I1.

Now suppose that $k[S]$ is flat for every prime field k. Let $S \to S_P$ be the localization of S by P, so that the action of P on S_P extends to an action of P^{gp}. As we remarked in the paragraph following Definition 3.2.3, the functor $S \mapsto k[S]$ is compatible with formation of tensor products, and hence $k[S_P] \cong k[P^{gp}] \otimes_{k[P]} k[S]$. The flatness of $k[S]$ over $k[P]$ implies that $k[S]$ injects into $k[S_P]$ and that $k[S_P]$ is flat over $k[P^{gp}]$. Since this is true for every k, it follows that the action of P^{gp} on S_P is free, as we saw in Lemma 4.5.13. Then it follows from (3) of Proposition 4.5.10 that the action of P on S satisfies I2. □

We shall be especially interested in P-regularity and its stability under the formation of pushouts. We begin with the following generality.

Proposition 4.5.15. *Let P be a monoid.*

1. *The inclusion functor from the category of P-regular P-sets to the category of all P-sets has a left adjoint $S \mapsto S^{reg}$. Explicitly, $S^{reg} = S/E$, where E is the congruence relation on S consisting of the set of pairs (s_1, s_2) of elements of S such that there exists some $p \in P$ with $ps_1 = ps_2$.*
2. *If S is a P-set, then S^{reg} can be identified with the image of the localization mapping $S \to S_P$, and S is regular if and only if the localization map $S \to S_P$ is injective.*

Proof We must first verify that the set E described in (1) really is a congruence relation on S. It is clear that E is symmetric and reflexive. If $ps_1 = ps_2$ and $p's_2 = p's_3$, then

$$pp's_1 = p'ps_1 = p'ps_2 = pp's_2 = pp's_3,$$

so $(s_1, s_3) \in E$ and E is transitive. Furthermore for any $q \in P$,

$$pqs_1 = qps_1 = qps_2 = pqs_2$$

so $(qs_1, qs_2) \in E$, and so by the analog of Proposition 1.1.3 for P-sets, E is a congruence relation. If s_1 and s_2 are elements of S and if there exists a $p' \in P$ such that $p's_1 \equiv p's_2 \pmod{E}$, then by definition there exists $p \in P$ such that $pp's_1 = pp's_2$, and hence $s_1 \equiv s_2 \pmod{E}$. Thus S/E is P-regular. It is evident that any morphism from S to a P-regular S-set factors uniquely through S/E, so that S/E has the desired universal property, i.e., $S/E = S^{\text{reg}}$. This proves statement (1). The identification of S/E with the image of S in S_P follows from the explicit construction of S_P explained in Proposition 1.4.4. The last part of (2) follows, but it can also be easily checked directly from the definitions. □

Corollary 4.5.16. *Let P be an integral monoid, let S be a P-regular P-set, and let J be an ideal of P. Then the natural map $J \otimes_P S \to S$ factors through an injection $(J \otimes_P S)^{\text{reg}} \to S$. In particular, if $\theta \colon P \to Q$ is a homomorphism of integral monoids and J is an ideal of P, then $(Q \otimes_P J)^{\text{reg}}$ can be identified with an ideal of Q.*

Proof We use additive notation in this proof. Thus the "natural map" in the statement sends (j, s) to $j + s$. Since S is P-regular, this map factors through $(J \otimes_P S)^{\text{reg}}$. Suppose that (j_1, s_1) and (j_2, s_2) are elements of $J \times S$ such that $j_1 + s_1 = j_2 + s_2$. Let $p_i := j_i$, regarded now as an element of P. Then $p_1 + s_1 = p_2 + s_2$ in S and $p_2 + j_1 = p_1 + j_2$ in P. In S_P we can write $s_2 = (p_1 - p_2) + s_1$ and in J_P we have $j_2 = (p_2 - p_1) + j_1$. Hence in $J_P \otimes_{P^{\text{gp}}} S_P$ we have

$$j_2 \otimes s_2 = (p_2 - p_1 + j_1) \otimes (p_1 - p_2 + s_1) = j_1 \otimes s_1.$$

By (2) of Proposition 4.5.15, $(J \otimes_P S)^{\text{reg}} \subseteq (J \otimes_P S)_P \cong (J_P \otimes_{P^{\text{gp}}} S_P)$, so (j_1, s_1) and (j_2, s_2) have the same image in $(J \otimes_P S)^{\text{reg}}$. □

The next result shows that a P-set S satisfies condition I1 if and only if it is "universally" P-regular.

Theorem 4.5.17. *Let P be an integral monoid and let S be a P-set. Then the following conditions are equivalent.*

1. *S satisfies I1.*
2. *For every P-regular P-set T, $T \otimes_P S$ is P-regular.*
3. *For every homomorphism of integral monoids $P \to P'$, the P'-set $P' \otimes_P S$ is P'-regular.*

Lemma 4.5.18. *Let $\theta \colon P \to P'$ be a homomorphism of integral monoids and let S be a P-set. Then $S' := P' \otimes_P S$ is P-regular as a P-set if and only if it is P'-regular as a P'-set.*

Proof It is clear that S' is P-regular if it is P'-regular. Suppose conversely that S' is P-regular. Since P' is integral, the localization mapping

$$P^{\text{gp}} \otimes_P P' = P'_P \to P'^{\text{gp}}$$

is an injective map of P^{gp}-sets. As we saw in Example 4.5.4, injectivity is preserved under tensor product with any P^{gp}-set, and it follows that the morphisms a and b in the diagram below are also injective.

$$
\begin{array}{ccc}
(P^{\mathrm{gp}} \otimes_P P') \otimes_{P^{\mathrm{gp}}} (P^{\mathrm{gp}} \otimes_P S) & \xrightarrow{a} & P'^{\mathrm{gp}} \otimes_{P^{\mathrm{gp}}} (P^{\mathrm{gp}} \otimes_P S) \\
\downarrow \cong & & \downarrow \cong \\
P^{\mathrm{gp}} \otimes_P P' \otimes_P S & \xrightarrow{b} & P'^{\mathrm{gp}} \otimes_P S \\
\uparrow c & & \uparrow \cong \\
P' \otimes_P S & \xrightarrow{d} & P'^{\mathrm{gp}} \otimes_{P'} P' \otimes_P S.
\end{array}
$$

The morphism c is injective because $P' \otimes_P S$ is P-regular, and it follows that the morphism d is injective. Then statement (2) of Proposition 4.5.15 shows that $P' \otimes_P S$ is P'-regular. □

Proof of Theorem 4.5.17 Suppose that S satisfies I1 and that T is a P-regular P-set. To prove that $T \otimes_P S$ is P-regular, suppose that we have $p \in P$, $s_1, s_2 \in S$, and $t_1, t_2 \in T$ with $p(t_1 \otimes s_1) = p(t_2 \otimes s_2)$. Then $(t_1 \otimes ps_1) = (t_2 \otimes ps_2)$ in $T \otimes_P S$ and hence also in $(T \otimes_P S)^{\mathrm{reg}} \subseteq T_P \otimes_{P^{\mathrm{gp}}} S_P$. Hence there exist $p_1, p_2 \in P$ such that $p_1 ps_1 = p_2 ps_2$ and $p_1 t_2 = p_2 t_1$. Since S is P-regular, the first of these equations implies that $p_1 s_1 = p_2 s_2$, Since S satisfies I1, there then exist $s' \in S$ and $p'_i \in P$ such that $s_i = p'_i s'$ and $p_1 p'_1 = p_2 p'_2$. Then

$$p_1 p'_1 t_1 = p_2 p'_2 t_1 = p'_2 p_1 t_2,$$

and since T is P-regular, it follows that $p'_1 t_1 = p'_2 t_2$. Then

$$t_1 \otimes s_1 = t_1 \otimes p'_1 s' = p'_1 t_1 \otimes s' = p'_2 t_2 \otimes s' = t_2 \otimes p'_2 s' = t_2 \otimes s_2.$$

Thus $T \otimes_P S$ is P-regular, proving that (1) implies (2).

If $\theta \colon P \to P'$ is a homomorphism of integral monoids, then the action of P on P' defined by θ is regular. Hence if a P-set S satisfies (2), the tensor product $P' \otimes_P S$ is regular as a P-set and, by Lemma 4.5.18, it is also regular as a P'-set. Thus S also satisfies (3).

The proof that (3) implies (1) is more difficult; we follow the method of Kato's proof in [48, 4.1]. Suppose that S is universally P-regular and that x and y are elements of S and a and b are elements of P such that $ax = by$. We construct a homomorphism of integral monoids $P \to P'$ as follows. Let E be the subset of $(P \oplus \mathbf{N}^2) \times (P \oplus \mathbf{N}^2)$ consisting of those pairs $((c, m, n), (c', m', n'))$

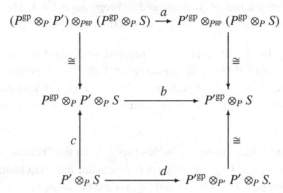

such that $m+n = m'+n'$ and $ca^m b^n = c'a^{m'} b^{n'}$. In fact E is a congruence relation and an exact submonoid of $(P \oplus \mathbf{N}^2) \times (P \oplus \mathbf{N}^2)$, so, by Corollary 2.1.20, the quotient $P' := (P \oplus \mathbf{N}^2)/E$ is an integral monoid. Let $[c, m, n]$ denote the class in P' of an element (c, m, n) of $P \oplus \mathbf{N}^2$ and let $P \to P'$ be the composite $P \to P \oplus \mathbf{N}^2 \to P'$, sending p to $[p, 0, 0]$. Then if $a' := [1_P, 0, 1]$ and $b' := [1_P, 1, 0]$, we have $a'a = [a, 0, 1]$ and $b'b = [b, 1, 0]$, and it follows from the definition of E that $p' := a'a = b'b'$ in P'. Hence in $P' \otimes_P S$ we have

$$p'b' \otimes x = a'ab' \otimes x = a'b' \otimes ax = a'b' \otimes by = a'b'b \otimes y = p'a' \otimes y.$$

By hypothesis $P' \otimes_P S$ is P'-regular, and it follows that $b' \otimes x = a' \otimes y$ in $P' \otimes_P S$.

Now let P act on $S \times \mathbf{N}^2$ via its action on S, and let R be the subset of $(S \times \mathbf{N}^2) \times (S \times \mathbf{N}^2)$ consisting of those pairs $((s, m, n), (s', m', n'))$ such that:

1. $m + n = m' + n'$;
2. there exist c,c' in P and t in S such that $s = ct$, $s' = c't$, and $ca^m b^n = c'a'^{m'} b'^{n'}$.

This subset is symmetric, contains the diagonal, and is invariant under the action of P. It follows from the analog of Proposition 1.1.3 for P-sets that the congruence relation E' that it generates is just the set of pairs (e, f) such that there exists a sequence $(r_0, \ldots, r_k)$ with $(r_{i-1}, r_i) \in R$ for $i > 0$ and $r_0 = e$, $r_k = f$. Write $[s, m, n]$ for the class in $S' := (S \times \mathbf{N}^2)/E'$ of (s, m, n). Then the map $(P \oplus \mathbf{N}^2) \times S \to S'$ sending (c, m, n, s) to $[cs, m, n]$ factors through $P' \times S$, and furthermore the corresponding map $P' \times S \to S'$ is a P-bimorphism. Thus there is a map $g\colon P' \otimes_P S \to S'$ sending each $[c, m, n] \otimes s$ to $[cs, m, n]$. (We shall not need to check that g is an isomorphism.)

We saw above that $b' \otimes x = a' \otimes y$ in $P' \otimes_P S$, and it follows that $[x, 1, 0] = [y, 0, 1]$ in S'. Hence there exists a sequence $r := (r_0, \ldots r_k)$ as above, with $r_i = (s_i, m_i, n_i)$ and $r_0 = (x, 1, 0)$ and $r_k = (y, 0, 1)$. Then, for all i, we have $m_i + n_i = 1$, so that $(m_i, n_i) = (1, 0)$ or $(0, 1)$. Suppose that, for some i, $(m_{i-1}, n_{i-1}) = (m_i, n_i)$. Then there exist c, c', t with $s_{i-1} = ct$, $s_i = c't$ and $ca = c'a$ or $cb = c'b$. But then $c = c'$ and hence $s_{i-1} = s_i$, $r_{i-1} = r_i$, and in fact r_i can be omitted from the sequence r. Consequently we may assume that $m_{i-1} \neq m_i$ for all i. Since $m_0 = 1$, it follows that $m_i = 1$ if i is even and $n_i = 1$ if i is odd. If $k \geq 2$, choose an odd $i \in [0, k]$. Then $r_{i-1} = (ct, 1, 0)$, $r_i = (c't, 0, 1) = (dt', 0, 1)$, and $r_{i+1} = (d't', 1, 0)$, with $ca = c'b$ and $db = d'a$. But then in S we have

$$act = cat = c'bt = bc't = bdt' = dbt' = d'at' = ad't'.$$

Since S is P-regular, it follows that $ct = d't'$ and $r_{i-1} = r_{i+1}$. In this case r_i and r_{i+1} can be omitted from r. Thus we may assume without loss of generality that $k = 1$. Then there exist c, c', t such that $x = ct$, $y = c't$, and $ac = bc'$. This proves that S satisfies I1. □

Here is an another consequence of condition I1.

Proposition 4.5.19. *Let P be a monoid and S a P-set satisfying condition I1. Let $\mathfrak{p}$ be a prime ideal of P and $F := P \setminus \mathfrak{p}$. Then $T := S \setminus \mathfrak{p}S$ is stable under the action of F, and the action of F on T also satisfies I1.*

Proof To prove that T is F-stable, suppose that $t \in T$, $f \in F$, and $ft = ps$ with $p \in \mathfrak{p}$ and $s \in S$. Then, by I1, there exist $s' \in S$ and $p', q \in P$ such that $s = p's'$, $t = qs'$, and $pp' = fq$. Since $p \in \mathfrak{p}$ and $f \in F$, necessarily $q \in \mathfrak{p}$, hence $t \in \mathfrak{p}S$, a contradiction. To prove that the F-set T satisfies I1, suppose that $f_i \in F$ and $t_i \in T$ with $t_1 f_1 = t_2 f_2$. Since S satisfies I1, there exist $t' \in S$ and $p_i \in P$ such that $f_1 p_1 = f_2 p_2$ and $t_i = p_i t'$. Since $t_i \in T$, necessarily $p_i \in F$ and $t' \in S \setminus \mathfrak{p}S = T$, as required. □

4.6 Integral homomorphisms

The fact that the coproduct of integral monoids need not be integral is the source of important technical problems in log geometry. In this section we investigate this issue in some detail. Recall from the discussion in Section 1.2 that if $P \to Q_1$ and $P \to Q_2$ are monoid homomorphisms, then the coproduct $Q_1 \oplus_P Q_2$ can be identified with the tensor product $Q_1 \otimes_P Q_2$, and we use that notation here.

Proposition 4.6.1. *Let $P \to Q_1$ and $P \to Q_2$ be homomorphisms of integral monoids. and let $Q := Q_1 \otimes_P Q_2$ be their coproduct in the category of commutative monoids. Then Q is integral if and only if its underlying P-set is P-regular.*

Proof If Q is integral then for any $p \in P$, multiplication by p on Q is injective and hence Q is P-regular. Suppose conversely that Q is is P-regular, According to Lemma 4.5.18, it is also Q_1-regular and Q_2-regular. Since any $q \in Q$ can be written as a sum of elements coming from Q_1 and Q_2, Q is also Q-regular, that is, it is integral. □

Definition 4.6.2. *A homomorphism $\theta\colon P \to Q$ of integral monoids is integral if it satisfies the following equivalent conditions.*

1. *The action of P on Q defined by θ satisfies condition I1 of Proposition 4.5.6. That is, whenever $q_1, q_2 \in Q$ and $p_1, p_2 \in P$ satisfy*

$$\theta(p_1) + q_1 = \theta(p_2) + q_2,$$

there exist $q' \in Q$ and $p'_i \in P$ such that

$$q_i = \theta(p'_i) + q' \text{ and } p_1 + p'_1 = p_2 + p'_2.$$

2. *The action of P on Q determined by θ makes Q a universally P-regular P-set.*
3. *For every homomorphism $P \to P'$ of integral monoids, the pushout $P' \otimes_P Q$ is an integral monoid.*

The equivalence of conditions (1) and (2) in the definition is just a restatement of Theorem 4.5.17, and the equivalence of (2) and (3) follows from Proposition 4.6.1.

Proposition 4.6.3. *Let $\theta \colon P \to Q$ and $\phi \colon Q \to R$ be homomorphisms of integral monoids.*

1. *If θ and ϕ are integral, then $\phi \circ \theta$ is integral. If $\phi \circ \theta$ is integral and ϕ is exact, then θ is integral. If $\phi \circ \theta$ is integral and θ is surjective, then ϕ is integral. The pushout of any integral homomorphism is integral.*
2. *The homomorphism θ is integral if and only if $\overline{\theta}$ is integral.*
3. *If F is a face of P, the homomorphisms $P \to P_F$ and $P \to P/F$ are integral. If $\theta \colon P \to Q$ is integral and G is a face of Q, containing $\theta(F)$, then the induced homomorphisms $P_F \to Q_G$ and $P/F \to Q/G$ are integral.*
4. *If θ is integral, then it is exact if and only if it is local. In particular, an integral homomorphism is locally exact.*
5. *Every $\theta \colon P \to Q$ is integral if P is valuative, or if Q is dull.*

Proof The fact that the composite and the pushout of integral homomorphisms are integral follows immediately from from characterization (3) in Definition 4.6.2. Suppose that $\phi \circ \theta$ is integral and that ϕ is exact. To verify that θ satisfies I1, suppose that $q_1, q_2 \in Q$ and $p_1, p_2 \in P$ with $q_1 + \theta(p_1) = q_2 + \theta(p_2)$. Then $\phi(q_1) + \phi(\theta(p_1)) = \phi(q_2) + \phi(\theta(p_2))$, and since $\phi \circ \theta$ is integral there exist $r \in R$ and $p'_1, p'_2 \in P$ with $p_1 + p'_1 = p_2 + p'_2$ and $\phi(q_i) = r + \phi(\theta(p'_i))$. Then $\phi(q_i - \theta(p'_i)) = r \in R$, and since ϕ is exact, $q_i - \theta(p'_i) \in Q$. In fact,

$$q := q_1 - \theta(p'_1) = q_1 + \theta(p_1) - \theta(p_2) - \theta(p'_2) = q_2 - \theta(p'_2).$$

It follows that $q_i = q + \theta(p'_i)$ in Q, and since $p_1 + p'_1 = p_2 + p'_2$, that θ is integral. Suppose on the other hand that θ is surjective. Then one checks immediately that if $\phi \circ \theta$ satisfies I1, so does ϕ. This completes the proof of (1).

For (2), we first check that the natural map $\pi \colon P \to \overline{P}$ is integral. Suppose that $p_i \in P$ and $q_i \in \overline{P}$ with $p_1 + q_1 = p_2 + q_2$. Choose $p'_i \in P$ such that $\pi(p'_i) = q_i$. Then $\pi(p_1 + p'_1) = \pi(p_2 + p'_2)$, so there exists a $u \in P^*$ such that

$p_1 + p'_1 = p_2 + p'_2 + u$. Replace p'_2 by $p'_2 + u$, and let $q := 0 \in \overline{P}$. Then $q_i = \pi(p'_i) + q$, as required. Statement (2) follows, since in the diagram

$$\begin{array}{ccc} P & \xrightarrow{\theta} & Q \\ \downarrow & & \downarrow \\ \overline{P} & \xrightarrow{\overline{\theta}} & \overline{Q}, \end{array}$$

the vertical arrows are surjective, integral, and exact.

For (3), first observe that a localization homomorphism $\lambda_F : P \to P_F$ is integral. Indeed, if $P \to P'$ is any homomorphism of integral monoids, then the pushout $P' \otimes_P P_F$ identifies with P'_F, which is necessarily integral. Since $P/F \cong \overline{P_F}$, it follows that $P \to P/F$ is also integral. If $\theta : P \to Q$ is integral, so is its pushout $P_F \to Q_F$, and if $\theta(F) \subseteq G$, the map $Q_F \to Q_G$ is a localization map, hence is integral, and hence so are $P_F \to Q_G$ and $P/F \to Q/G$.

Suppose that $\theta : P \to Q$ is local and integral and $p_1 - p_2 \in P^{gp}$ is such that $\theta(p_1 - p_2)$ is an element q of Q. Then in Q we have $\theta(p_1) + 0 = \theta(p_2) + q$, and since θ satisfies I1, there exist $q' \in Q$ and $p'_i \in P$ with $0 = \theta(p'_1) + q'$, $q = \theta(p'_2) + q'$, and $p'_1 + p_1 = p'_2 + p_2$. But then q' is a unit of Q, and since θ is local p'_1 is a unit of P. Then $p_1 - p_2 = p'_2 - p'_1 \in P$, so θ is exact. Since every exact homomorphism is local (Proposition 4.2.1), we see that an integral homomorphism is exact if and only if it is local.

If Q is dull then $\overline{Q} = 0$ and it is obvious that $\overline{P} \to 0$ is integral. More generally, Q is certainly P-regular since it is integral, and hence if P is valuative it satisfies I1 by Proposition 4.5.10. □

Corollary 4.6.4. *Let $\theta : P \to Q$ be a morphism of integral monoids. Then the following are equivalent.*

1. *$\theta : P \to Q$ is integral.*
2. *$\theta_{Q^*} : P^{loc} := P_{\theta^{-1}(Q^*)} \to Q$ is integral.*
3. *$\theta_G : P_{\theta^{-1}(G)} \to Q_G$ is integral for every face G of Q.*

Proof Statement (3) of Proposition 4.6.3 implies that the localization homomorphism $P \to P^{loc}$ is integral. Since the composition of integral homomorphisms is integral, it follows that if $P^{loc} \to Q$ is integral, then so is $P \to Q$. Conversely, suppose $P \to Q$ is integral and let $P^{loc} \to P'$ be any homomorphism of integral monoids. It follows from the universal mapping properties of pushouts and localizations that the natural homomorphism $P' \otimes_P Q \to P' \otimes_{P^{loc}} Q$ is an isomorphism. Since $P \to Q$ is integral, $P' \otimes_P Q$ is integral, and hence so is $P' \otimes_{P^{loc}} Q$. This proves the equivalence of (1) and (2).

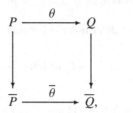

Suppose that θ is integral and that G is a face of Q. Then $Q \to Q_G$ is integral, and hence so is $P \to Q_G$. It follows from the implication of (2) by (1) that $P_F \to Q_G$ is integral. Suppose conversely that each such localization is integral. Taking $G = Q^*$, we see that $P^{loc} \to Q$ is integral, and hence that θ is integral. □

Example 4.6.5. Proposition 4.6.3 shows that the family of integral homomorphisms of monoids is stable under pushouts in the category of integral monoids. The reader is cautioned that this is no longer true in the category of *saturated* monoids: the saturated pushout of an integral homomorphism $\theta: P \to Q$ of fine saturated monoids need not be integral. For an example, let P and Q be freely generated by p and q respectively, and let $\theta: P \to Q$ be the homomorphism sending p to $2q$. Then let P' be the monoid given by generators p'_0, p'_1, p'_2 with the relation $p'_1 + p'_2 = 2p'_0$, and let $P \to P'$ be the homomorphism sending p to p'_1. Then the pushout $Q' := P' \oplus_P Q$ has generators q'_1, p'_2, p'_0, with the relation $2q'_1 + p'_2 = 2p'_0$. Hence $q'_2 := p'_0 - q'_1$ belongs to Q'^{sat}, which is in fact the free monoid generated by q'_1 and q'_2. The homomorphism θ' is given by

$$p'_0 \mapsto q'_1 + q'_2, p'_1 \mapsto 2q'_1, p'_2 \mapsto 2q'_2.$$

Then

$$p'_0 + q'_1 = 2q'_1 + q'_2 = p'_1 + q'_2,$$

but q'_1 and q'_2 are irreducible, and hence θ' is not integral. It is true however that the saturated pushout of an exact homomorphism is exact.

Remark 4.6.6. If $\theta: P \to Q$ is homomorphism of integral monoids, then the corresponding action of P on Q satisfies condition I2 of Proposition 4.5.6 if and only if θ is injective. Thus by Theorem 4.5.7 and Proposition 4.5.12, θ is injective and integral if and only if $\mathbf{Z}[\theta]: \mathbf{Z}[P] \to \mathbf{Z}[Q]$ is flat if and only if the action of P on Q defined by θ is flat. Note that If $\theta: P \to Q$ is local and integral, then by (4) of Proposition 4.6.3 it is exact, and then $\mathrm{Ker}(\theta^{\mathrm{gp}}) \subseteq P^*$, and θ factors through an injective homomorphism $P/\mathrm{Ker}(\theta^{\mathrm{gp}}) \to Q$.

Proposition 4.6.7. *Let $\theta: P \to Q$ be a homomorphism of integral monoids, and consider the following conditions.*

1. *The action of P on Q defined by θ is flat (see Definition 4.5.2).*
2. *The homomorphism $\mathbf{Z}[\theta]: \mathbf{Z}[P] \to \mathbf{Z}[Q]$ is flat.*
3. *For some field k, the homomorphism $k[\theta]: k[P] \to k[Q]$ is flat.*
4. *The homomorphism θ is integral.*

Conditions (1) and (2) are equivalent and imply condition (3). Condition (3) implies condition (4), and if θ is local and P is sharp, all four conditions are equivalent.

Proof We have already seen in Proposition 4.5.12 that (1) implies (2), and it is obvious that (2) implies (3). Lemma 4.5.14 shows that that (3) implies condition I1, so (3) implies (4). If $\theta \colon P \to Q$ is integral and local, then by Proposition 4.6.3 it is exact, and if P is sharp, θ must be injective. The integrality of Q then implies that the action of P on Q satisfies condition I2 of Proposition 4.5.6. Thus the P-set underlying Q satisfies I1 and I2, hence by Theorem 4.5.7 is flat. This proves that (4) implies (1). □

The following result can be viewed as an analog of the local criterion for flatness in commutative algebra.

Proposition 4.6.8. *Let $\theta \colon P \to Q$ be an injective and local homomorphism of fine monoids and let k be a field. Then the following conditions are equivalent.*

1. *The natural map $P^+ \otimes_P Q \to Q$ is injective.*
2. $\mathrm{Tor}_1^{k[P]}(k[P, P^+], k[Q]) = 0$.
3. *For every ideal J of P, the map $J \otimes_P Q \to Q$ is injective.*
4. $\mathrm{Tor}_1^{k[P]}(k[P, J], k[Q]) = 0$ *for every ideal J of P.*
5. *The homomorphism θ is integral.*
6. *The action of P on Q is flat.*
7. *The homomorphism $k[P] \to k[Q]$ is flat.*

If P and Q are sharp, these conditions are also equivalent to

8. *The homomorphism $k[[P]] \to k[[Q]]$ is flat.*
9. $\mathrm{Tor}_1^{k[[P]]}(k[[P, P^+]], k[[Q]]) = 0$.

Proof If J is an ideal of P, the map $Q \otimes_P J \to Q$ is injective if and only if the induced homomorphism $k[Q \otimes_P J] \to k[Q]$ is injective. Since $k[Q \otimes_P J] \cong k[Q] \otimes_{k[P]} k[J]$, this is the case if and only if $\mathrm{Tor}_1^{k[P]}(k[P, J], k[Q]) = 0$. It follows that statements (1) and (2) (resp. (3) and (4)) are equivalent.

Let us prove that (2) implies (4). Let C denote the family of ideals J of P such that $\mathrm{Tor}_1^{k[P]}(k[P, J], k[Q]) \neq 0$. If C is not empty, it has a maximal element, since P is noetherian. We first observe that such a maximal element J is prime. Otherwise there exist elements p_1 and p_2 of $P \setminus J$ such that $p_1 + p_2 \in J$. Let $J_i := J + (p_i)$. Then there is an exact sequence of $k[P]$-modules:

$$0 \to k[J_1, J] \to k[P, J] \to k[P, J_1] \to 0,$$

and hence also an exact sequence:

$$\mathrm{Tor}_1^{k[P]}(k[J_1,J],k[Q]) \to \mathrm{Tor}_1^{k[P]}(k[P,J],k[Q]) \to \mathrm{Tor}_1^{k[P]}(k[P,J_1],k[Q]).$$

Note that $\mathrm{Tor}_1^{k[P]}(k[P,J_1],k[Q])$ vanishes, since J_1 strictly contains J. Let $J' := \{p \in P : p+p_1 \in J\}$, an ideal of P. Since J' contains J_2 which strictly contains J, it follows also that $\mathrm{Tor}_1^{k[P]}(k[P,J'],k[Q])$ vanishes. Multiplication by e^{p_1} induces an isomorphism $k[P,J'] \to k[J_1,J]$, so $\mathrm{Tor}_1^{k[P]}(k[J_1,J],k[Q])$ vanishes as well. Then the exact sequence above shows that $\mathrm{Tor}_1^{k[P]}(k[P,J],k[Q])$ also vanishes, a contradiction. Thus thus our maximal element J of C is a prime ideal. By hypothesis (2), $J \neq P^+$, so there is some $p \in P^+ \setminus J$. Let $J' := J+(p)$. Since J is prime, multiplication by e^p is injective on $k[P,J]$, and there are exact sequences:

$$0 \to k[P,J] \xrightarrow{e^p} k[P,J] \to k[P,J'] \to 0 \quad \text{and}$$

$$\mathrm{Tor}_1^{k[P]}(k[P,J],k[Q]) \xrightarrow{e^p} \mathrm{Tor}_1^{k[P]}(k[P,J],k[Q]) \longrightarrow \mathrm{Tor}_1^{k[P]}(k[P,J'],k[Q]).$$

But $\mathrm{Tor}_1^{k[P]}(k[P,J'],k[Q]) = 0$ since J' strictly contains J, and thus multiplication by e^p induces a surjection on $\mathrm{Tor}_1^{k[P]}(k[P,J],k[Q])$. As we shall see, this will imply that this module in fact vanishes, which will conclude the proof.

Since Q is a fine monoid, there exists a local homomorphism $h\colon Q \to \mathbf{N}$. Since θ is local, so is $g := h\circ\theta$. We define a function $J\times Q \to \mathbf{N}$ by sending (j,q) to $g(j)+h(q)$. This function is a P-bimorphism,and hence factors through a map $s\colon J\otimes_P Q \to \mathbf{N}$. Furthermore, for $(j,q) \in J\times Q$, we have $s(j\otimes q) = h(\theta(j)+q))$. The function h (resp. s) defines an $\mathbf{N}$-grading on $k[Q]$ (resp. on $k[J\otimes Q]$), and the homomorphism $k[J\otimes Q] \to k[Q]$ is compatible with the grading. It follows that its kernel $T := \mathrm{Tor}_1^{k[P]}(k[P,J],k[Q])$ inherits a grading. Since $p \in P^+$, necessarily $h(p) > 0$. Since multiplication by e^p is surjective on T, for every i, multiplication by p induces a surjection $T_{i-h(p)} \to T_i$. Since $T_i = 0$ for $i < 0$, it follows that $T = 0$, as required.

To prove that (4) implies (5), let p_1 and p_2 be two elements of P and let $E := \{(p'_1,p'_2) : p'_1 + p_1 = p'_2 + p_2\}$, a sub-$P$-set of $P\times P$. Then we have a sequence

$$0 \to k[E] \xrightarrow{g} k[P]\oplus k[P] \xrightarrow{f} k[J] \to 0, \tag{4.6.1}$$

where f sends (α,β) to $\alpha e^{p_1} - \beta e^{p_2}$ and g sends $e^{(p'_1,p'_2)}$ to $(e^{p'_1},e^{p'_2})$. It is clear that f is surjective, that g is injective, and that $f\circ g = 0$. Suppose that $f(\alpha,\beta) = 0$. Write $\alpha := \sum_s a_s e^s$ and $\beta := \sum_t a_t e^t$. We prove by induction on the size of the support of α that (α,β) belongs to the image of g. Since $\alpha e^{p_1} = \beta e^{p_2}$, we have $\sum_s a_s e^{p_1+s} = \sum_t b_t e^{p_2+t}$. Suppose that $a_s \neq 0$. Then there is a (necessarily unique) t such that $p_2+t = p_1+s$ and necessarily $b_t = a_s$. Then $(s,t) \in E$ and

$a_s e^{(s,t)} \in k[E]$. It follows that $(\alpha,\beta) - g(a_s e^{(s,t)})$ lies in the kernel of f and hence in the image of g by the induction hypothesis. We conclude that (α,β) also lies in the image of g. It follows that the the sequence (4.6.1) is exact. Tensoring with $k[Q]$, we find an exact sequence

$$k[E] \otimes_{k[P]} k[Q] \longrightarrow k[Q] \oplus k[Q] \longrightarrow k[J] \otimes_{k[P]} k[Q] \to 0.$$

Condition (4) implies that the map $k[J] \otimes_{k[P]} k[Q] \to k[Q]$ is injective, so the sequence

$$k[E] \otimes_{k[P]} k[Q] \xrightarrow{g\otimes \mathrm{id}} k[Q] \oplus k[Q] \xrightarrow{f\otimes \mathrm{id}} k[Q]$$

is also exact. Then if $(q_1, q_2) \in Q \times Q$ and $p_1 + q_1 = p_2 + q_2$, we can argue as in the proof of statement (2) of Lemma 4.5.14 to see that I1 is satisfied, so θ is integral.

As we saw in Remark 4.6.6, an integral and injective homomorphism is flat, so (5) implies (6), and Proposition 4.5.12 shows that (6) implies (7). Since (7) implies (2), we see that (1)–(7) are equivalent.

Now suppose that P and Q are sharp. Then $k[[P]] \to k[[Q]]$ is a local homomorphism of noetherian local rings, so the local criterion for flatness tells us that (8) and (9) are equivalent. If (7) holds, $k[P] \to k[Q]$ is flat, then so is the map $k[P] \to k[Q] \otimes_{k[P]} k[[P]]$, by base change. The ring $k[Q] \otimes_{k[P]} k[[P]]$ is of finite type over $k[[P]]$ and hence noetherian, and hence the map from it to its completion $k[[Q]]$ is also flat, It follows that $k[[P]] \to k[[Q]]$ is flat. Finally, suppose that (8) holds. Then $k[P] \to k[[Q]]$ is flat, and we can again argue as in the proof of (1) of Lemma 4.5.14, computing in $k[[Q]]$ instead of $k[Q]$, to see that condition I1 holds.

To be careful, let us write out the argument. Again we let J be the ideal of P generated by elements p_1 and p_2 of P and let $E := \{(p'_1, p'_2) \in P \times P : p_1 + p'_1 = p_2 + p'_2\}$. Since $k[[Q]]$ is flat over $k[[P]]$, the map $k[[J]] \otimes_{k[[P]]} k[[Q]] \to k[[Q]]$ is injective, and we find an exact sequence:

$$0 \to k[E] \otimes_{k[P]} k[[Q]] \xrightarrow{g\otimes \mathrm{id}} k[[Q]] \oplus k[[Q]] \xrightarrow{h} k[[Q]],$$

where $h(\alpha,\beta) = \alpha e^{p_1} - \beta e^{p_2}$. Then if $(q_1, q_2) \in Q \times Q$ and $p_1 + q_1 = p_2 + q_2$, the element (e^{q_1}, e^{q_2}) of $k[[Q]] \oplus k[[Q]]$ lies in the kernel of h and hence also in the image of $g \otimes \mathrm{id}$. Then there exist $\gamma_i \in k[E]$ and $\delta_i \in k[[Q]]$ such that $(e^{q_1}, e^{q_2}) = \sum_i \delta_i g(\gamma_i)$ Write $\gamma_i = \sum_e a_{i,k} e^{(p'_{1,k}, p'_{2,k})}$, where each $(p'_{1,k}, p'_{2,k}) \in E$ and write $\delta_i = \sum_q b_{i,q} e^q$. Here $a_{i,k} = 0$ for all but finitely many k, but there is no such restriction on the support of δ_i. Then $\delta_i g(\gamma_i) = (\sum_q c_q e^q, \sum_q d_q e^q)$, where

$$c_q = \sum_{\theta(p'_1)+q'=q} a_{i,k} b_{i,q'} \quad \text{and} \quad d_q = \sum_{\theta(p'_2)+q'=q} a_{i,k} b_{i,q'}.$$

Since $\sum_i \delta_i g(\gamma_i) = (e^{q_1}, e^{q_2})$, there must be at least one triple (i, k, q') such that $a_{i,k} b_{i,q'} \neq 0$ and $\theta(p'_1) + q' = q_1$. Then

$$\theta(p_2) + q_2 = \theta(p_1) + q_1 = \theta(p_1 + p'_1) + q' = \theta(p_2) + \theta(p'_2) + q',$$

and hence $q_2 = \theta(p'_2) + q'$. □

Proposition 4.6.9. *Let $\theta \colon P \to Q$ be a homomorphism of integral monoids, and view Q as a P-set via θ. Then, for each $q \in Q$,*

$$S_q(\theta) := (\theta(P^{\text{gp}}) + q) \cap Q$$

is the connected component of the P-set Q containing q. Consider the following conditions:

1. *θ is integral and local;*
2. *θ is exact and, for every $q \in Q$, the P-set $S_q(\theta)$ is monogenic as a P-set.*

Then (2) implies (1), and the converse is true if P and Q are fine. Moreover, if (2) holds, then $\text{Ker}(\theta^{\text{gp}}) \subseteq P^*$, and each $S_q(\theta)$ is a isomorphic as a P-set to $P/\text{Ker}(\theta^{\text{gp}})$.

Proof It is apparent from the definitions that $S_q(\theta)$ is the connected component of Q viewed as a P-set. Suppose that (2) holds. Since θ is exact, it is local by Proposition 4.6.3, and furthermore $\text{Ker}(\theta^{\text{gp}})$ is contained in P^*. Let us prove that θ satisfies I2* of Definition 4.5.9. Suppose that $q \in Q$ and $p_1, p_2 \in P$ with $\theta(p_1) + q = \theta(p_2) + q$. Since Q is integral, it follows that $\theta(p_1) = \theta(p_2)$, hence that $u := p_2 - p_1 \in \text{Ker}(\theta^{\text{gp}}) \subseteq P^*$. Then $\theta(u) + q = q$ and $p_2 = u + p_1$ with $u \in P^*$, as required. If in addition each $S_q(\theta)$ is monogenic, then by (1) of Proposition 4.5.11 it follows that θ satisfies I1, proving that (2) implies (1).

Conversely, suppose that P and Q are fine and that θ is integral and local. By Proposition 4.6.3, θ is exact. Let

$$(Q : q) := \{y \in Q^{\text{gp}} : y + q \in Q\},$$

the principal fractional ideal of Q^{gp} generated by $-q$, and let

$$K_q := \{x \in P^{\text{gp}} : \theta^{\text{gp}}(x) + q \in Q\},$$

the inverse image of $(Q : q)$ in P^{gp}. Then θ^{gp} followed by translation by q induces an isomorphism of P-sets $K_q/Ker(\theta^{\text{gp}}) \to S_q(\theta)$. Since θ is exact and P and Q are fine, Proposition 2.1.24 implies that K_q is finitely generated as a P-set, and hence so is $S_q(\theta)$. Since θ is integral, it satisfies I1, and hence so does each connected component $S_q(\theta)$ of the P-set Q. Then Proposition 4.5.11 implies that $S_q(\theta)$ is monogenic, so (2) is proved. We have already seen that

(2) implies that $\mathrm{Ker}(\theta^{\mathrm{gp}}) \subseteq P^*$ and condition I2*, and it follows that $S_q(\theta)$ is isomorphic to $P/\mathrm{Ker}(\theta^{\mathrm{gp}})$.

□

Remark 4.6.10. Let $\theta\colon P \to Q$ be a local and integral homomorphism of fine monoids and let y be an element of the localization Q_P of Q by P. Choose p such that $p + y \in Q$. Then

$$S_y(\theta) := (P^{\mathrm{gp}} + y) \cap Q = (P^{\mathrm{gp}} + (p + y)) \cap Q,$$

so Proposition 4.6.9 applies equally well to S_y. As we saw in the proof of Proposition 4.6.9, S_y is finitely generated as a P-set and hence contains minimal elements. For each $q \in S_y(\theta)$, the following are equivalent:

1. q is P-minimal as an element of Q.
2. q is P-minimal as an element of $S_y(P)$.
3. q generates $S_y(\theta)$ as a P-set.
4. $P = \{x \in P^{\mathrm{gp}} : \theta^{\mathrm{gp}}(x) + q \in Q\}$.

Indeed, it is obvious that (1) implies (2). Since θ is integral and local, Proposition 4.6.9 implies that $S_y(\theta)$ is monogenic as a P-set and, if q is minimal, it must be a generator. Thus (2) implies (3). Suppose that q generates $S_y(\theta)$ and that $x \in P^{\mathrm{gp}}$ is such that $q' := \theta^{\mathrm{gp}}(x) + q \in Q$. Then $q' \in S_y(Q)$ and hence $q' = \theta(p) + q$ for some $p \in P$. But then $\theta^{\mathrm{gp}}(x) = \theta(p) \in Q$ and, since θ is exact, $x \in P$ and thus (4) holds. Finally, suppose that (4) holds and that $q = p + q'$ for some $q' \in Q$. Then $-p \in \{x \in P^{\mathrm{gp}} : \theta^{\mathrm{gp}}(x) + q \in Q\}$, hence $-p \in P$, and thus q is a minimal element in the P-set Q.

The following corollary shows that a homomorphism which is injective, local, and integral has a very restricted structure.

Corollary 4.6.11. *Let $\theta\colon P \to Q$ be an injective and local homomorphism of fine monoids. Then θ is integral if and only if Q is free as a P-set. If this is the case and P is sharp, the summation map induces a bijection*

$$P \times (Q \setminus K_\theta) \to Q,$$

where K_θ is the ideal of Q generated by P^+; furthermore, for every θ-critical face G of Q, the homomorphism

$$P \oplus G \to P + G$$

is an isomorphism of monoids.

Proof If θ makes Q into a free P-set, it satisfies condition I1 of Proposition 4.5.6 and hence is integral. Conversely, if θ is integral and local, Proposition 4.6.9 implies that each connected component $S_q(\theta)$ of the P-set Q is isomorphic to P, so that Q is free as a P-set. Suppose also that P is sharp, and note that the ideal K_θ is the set $P^+ + Q$ of elements of Q which can be written in the form $p + q$, where $p \in P^+$ and $q \in Q$. Then as we saw in Proposition 1.3.7, $Q \setminus K_\theta$ is a basis for Q. In other words, the map $P \times (Q \setminus K_\theta) \to Q$ is a bijection. If G is a θ-critical face of Q, then $G \subseteq (Q \setminus K_\theta)$, so the last statement follows. □

4.7 The structure of critically exact homomorphisms

Corollary 4.6.11 has a very important analog for cones, Theorem 4.7.2 below. This is the key "structure theorem" (A.3.2.2) of [40]. It plays a crucial role in many places in log geometry, especially with regard to the topological structure of log smooth maps and their Betti realizations [57]. The key is statement (3), illustrated by Figure 4.7.1 for the homomorphism $\mathbf{N} \to \mathbf{N} \oplus \mathbf{N}$ sending n to (n, n).

Remark 4.7.1. If $\theta\colon P \to Q$ is a local homomorphism of fine monoids, then the union $C_K(Q, P)$ of the set of $C_K(\theta)$-critical faces of $C_K(Q)$ is the complement of the conical ideal $C_K(K_\theta)$ of $C_K(Q)$ generated by $\theta(P^+)$. Indeed, an element q of $C_K(Q)$ lies in $C_K(Q, P)$ if and only if the face $\langle q \rangle$ it generates does not meet $\theta(P^+)$, and this is the case if and only if $C_K(Q) \setminus \langle q \rangle$ contains $\theta(P^+)$. Thus the complement $C_K(Q, P)$ is the intersection of all of the primes containing $\theta(P^+)$, which is the conical ideal of $C_K(Q)$ generated by $\theta(P^+)$ (see Remark 2.3.3).

Theorem 4.7.2. *Let $\theta\colon P \to Q$ be a local homomorphism of fine monoids, where P is sharp. Let K be an Archimedean field, let*

$$C_K(\theta)\colon C_K(P) \to C_K(Q)$$

be the induced homomorphism of K-cones, and let $C_K(Q, P)$ be the union of the set of θ-critical faces of $C_K(Q)$. Then the following conditions are equivalent:

1. *$C_K(\theta)\colon C_K(P) \to C_K(Q)$ is locally exact;*
2. *$C_K(\theta)\colon C_K(P) \to C_K(Q)$ is critically exact (Definition 4.2.12);*
3. *the summation map $\sigma\colon C_K(P) \times C_K(Q, P) \to C_K(Q)$ is bijective;*
4. *$C_K(\theta)\colon C_K(P) \to C_K(Q)$ is integral.*

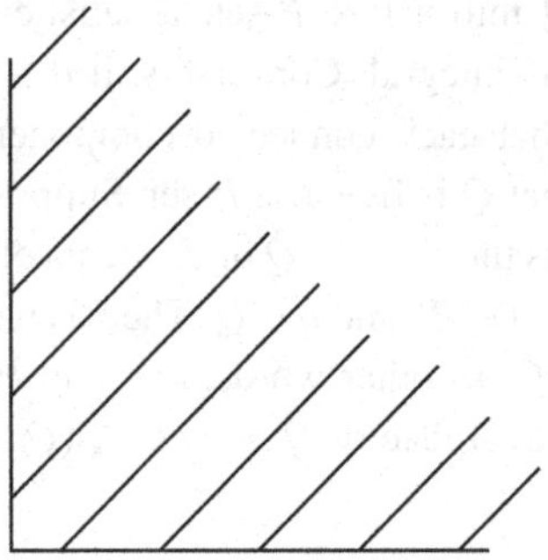

Figure 4.7.1 $C(P) \times C(Q,P) \cong C(Q)$

Proof Our proof will rely on the following result about exact homomorphisms.

Lemma 4.7.3. *Let $\theta \colon P \to Q$ be an injective and exact homomorphism of fine monoids, let K be an Archimedean field, and view $C_K(Q)$ as a $C_K(P)$-set via the homomorphism $C_K(\theta)$.*

1. *An element q of $C_K(Q)$ is $C_K(P)$-minimal if and only if the face it generates is θ-critical.*
2. *Each connected component of the $C_K(P)$-set $C_K(Q)$ has a minimal element.*

Proof To simplify the notation, we view P as a submonoid of Q. For each $q \in C_K(Q)$, recall that

$$\langle q \rangle = \{q' \in C_K(Q) : \lambda q' \leq q \text{ for some } \lambda \in K^{>}\}.$$

Thus if $p \in \langle q \rangle \cap C_K(P)$, then there exist $\lambda \in K^{>}$ and $q' \in C_K(Q)$ such that $\lambda p + q' = q$. If q is $C_K(P)$-minimal it follows that $\lambda p \in C_K(P)^*$ and hence $p \in C_K(P)^*$. This proves that $\langle q \rangle \cap C_K(P) = C_K(P)^*$, i.e., that $\langle q \rangle$ is θ-critical. Suppose for the converse that $\langle q \rangle$ is θ-critical. Then if there exists some $p \in C_K(P)$ with $q = p + q'$, necessarily $p \in \langle q \rangle \cap C_K(P) = C_K(P)^*$ and thus q is minimal. This proves statement (1) of the lemma.

For statement (2), note that for each $q \in C_K(Q)$, the connected component of the $C_K(P)$-set $C_K(Q)$ containing q is $S_q := (C_K(P)^{\text{gp}} + q) \cap C_K(Q)$. Let $J_q = \{x \in C_K(P)^{\text{gp}} : x + q \in S_q\}$. Then the map $J_q \to S_q$ sending x to $x + q$ is an isomorphism of $C_K(P)$-sets, so it suffices to show that J_q has a minimal element.

Since $C_K(Q)$ is a finite generated K-cone, Theorem 2.3.12 says that there exists a finite set of linear maps $h_i \colon C_K(Q)^{\mathrm{gp}} \to K$ such that

$$C_K(Q) = \{y \in C_K(Q)^{\mathrm{gp}} : h_i(y) \geq 0 \text{ for all } i\}.$$

Then

$$J_q = \{x \in C_K(P)^{\mathrm{gp}} : h_i(x) + h_i(q) \geq 0 \text{ for all } i\}.$$

In particular, $J_0 = C_K(P)^{\mathrm{gp}} \cap C_K(Q) = C_K(P)$, since $C_K(P)$ is an exact submonoid of $C_K(Q)$. Let $V := C_K(P)^{\mathrm{gp}}$, let W be its dual, and let $w_i \in W$ be the restriction of h_i to V. For each i, let $a_i := h_i(q)$, and let $T := \{(w_i, a_i)\} \subseteq W \times K$. For $t = (w_i, a_i) \in T$, write w_t for w_i and a_t for a_i. Then

$$J_q = \{v \in V : \langle v, w_t \rangle + a_t \geq 0 \text{ for all } t \in T\}.$$
$$C_K(P) = \{v \in v : \langle v, w_t \rangle \geq 0 \text{ for all } t \in T\}.$$

Furthermore, $\{w_t : t \in T\}$ spans W since $C_K(P)$ is sharp, and $0 \in J_q$, so J_q is not empty. Thus we can apply Theorem 2.3.20, which asserts that the $C_K(P)$-set J_q has a vertex v_0 and that v_0 is $C_K(P)$-minimal. □

We now turn to the proof of the theorem. The implication of (2) by (1) is trivial. Suppose that (2) holds. Since P is sharp, the same is true for $C_K(P)$ by Proposition 2.3.7, and $C_K(\theta)$ is exact and locally exact by Corollary 4.3.8. It follows that $C_K(\theta)$ is injective; to simplify the notation we view $C_K(P)$ as a subcone of $C_K(Q)$.

Now let q be an element of $C_K(Q)$. By statement (2) of Lemma 4.7.3, its connected component S_q has a minimal element q_0; by (1) of the same lemma, $q_0 \in C_K(Q, P)$. We claim that q_0 generates S_q as a $C_K(P)$-set. If $q' \in S_q$ then $S_{q'} = S_q$, so there exist $p, p_0 \in C_K(P)$ such that $p + q' = p_0 + q_0$. Since $\langle q_0 \rangle \cap C_K(P) = 0$ and θ is critically exact, the map $C_K(P) \to C_K(Q)_{\langle q_0 \rangle}$ is again exact. But then

$$p' := p_0 - p = q' - q_0 \in C_K(Q)_{\langle q_0 \rangle} \cap C_K(P)^{\mathrm{gp}} = C_K(P),$$

and $q' = p' + q_0$. Thus q' is in the trajectory of q_0, proving that q_0 generates S_q. as claimed. Since $q_0 \in C_K(Q, P)$, it follows that the summation map σ is surjective. For the injectivity, suppose that $(p_i, q_i) \in C_K(P) \times C_K(Q, P)$ and $p_1 + q_1 = p_2 + q_2$. Then $S_{q_1} = S_{q_2}$ and, since each $\langle q_i \rangle$ is θ-critical, q_1 and q_2 are each minimal generators of S_{q_i}. Since $C_K(P)$ is sharp, these generators must be equal, and it follows that also that $p_1 = p_2$. This completes the proof that (2) implies (3).

Suppose that (3) holds. We shall verify that θ satisfies I1. Note first that (3) certainly implies that θ is injective, and hence again we view $C_K(P)$ as a subset of $C_K(Q)$. Suppose that $q_i \in C_K(Q)$ and $p_i \in C_K(P)$ satisfy $p_1 + q_1 = p_2 + q_2$.

Write $q_i = p'_i + q'_i$ with $p'_i \in C_K(P)$ and $q'_i \in C_K(Q,P)$. Then $p_1 + p'_1 + q'_1 = p_2 + p'_2 + q'_2$, and, by the injectivity of σ, $q'_1 = q'_2$ and $p_1 + p'_1 = p_2 + p'_2$. Let $q' := q'_1 = q'_2$, so that $q_i = p'_i + q'$, showing that I1 holds. Statement (4) of Proposition 4.6.3 says that an integral homomorphism of integral monoids is locally exact, so (4) implies (1). □

Let us reap some consequences for monoids.

Definition 4.7.4. *A homomorphism of integral monoids $\theta\colon P \to Q$ is said to be* **Q**-*integral if the induced homomorphism of* **Q**-*cones*

$$C_{\mathbf{Q}}(\theta)\colon C_{\mathbf{Q}}(P) \to C_{\mathbf{Q}}(Q)$$

is integral. Explicitly, θ is **Q**-*integral if and only if, whenever $p_i \in P$ and $q_i \in Q$ satisfy $\theta(p_1) + q_1 = \theta(p_2) + q_2$, there exist $n \in \mathbf{Z}^+$, $p'_i \in P$ and $q' \in Q$ such that $nq_i = \theta(p'_i) + q'$ and $np_1 + p'_1 = np_2 + p'_2$.*

To check the equivalence, suppose that $C_{\mathbf{Q}}(\theta)$ is integral and that $\theta(p_1)+q_1 = \theta(p_2) + q_2$. Then since $C_{\mathbf{Q}}(\theta)$ satisfies condition I1 in Proposition 4.5.6, there exist $q'' \in C_{\mathbf{Q}}(Q)$ and $p''_i \in C_{\mathbf{Q}}(P)$ such that $q_i = \theta(p''_i) + q'$ and $p_1 + p''_1 = p_2 + p''_2$. Multiplying by a large enough n, we can arrange for $p'_i := np''_i$ to lie in P, and so the equations in the definition hold up to torsion, say killed by $m > 0$. Multiplying by m and replacing p'_i by mp'_i and q' by mq', we may arrange for them to hold exactly. The converse implication is clear.

Proposition 4.7.5. *Let $\theta\colon P \to Q$ be a homomorphism of fine monoids. If P is free and Q is saturated, then θ is integral if and only if it is* **Q**-*integral.* [2]

Proof Suppose that $p_1, p_2 \in P$ and $q_1, q_2 \in Q$ with $\theta(p_1) + q_1 = \theta(p_2) + q_2$. Choose $n \in \mathbf{Z}^+$, $p'_i \in P$, and $q' \in Q$ such that $nq_i = \theta(p'_i) + q'$ and $np_1 + p'_1 = np_2+p'_2$. Then for every $\phi\colon P \to \mathbf{N}$, $\phi(p'_1) \equiv \phi(p'_2) \pmod n$. Let $(e_1, \dots e_r)$ be a basis for P and let $(\phi_1, \dots \phi_r)$ be the dual basis for $\operatorname{Hom}(P, \mathbf{N})$. For each i and j, write $\phi_j(p'_i) = nm_{i,j} + r_{i,j}$ with $m_{i,j}, r_{i,j} \in \mathbf{N}$ and $r_{i,j} < n$. Since $\phi_j(p'_1) \equiv \phi_j(p'_2)$ (mod n), in fact $r_j := r_{1,j} = r_{2,j}$ for all j. Let $p'' := \sum r_j e_j$ and $p''_i = \sum m_{i,j} e_j$ in P. Then $p'_i = np''_i + p''$, and $np_1 + np''_1 + p'' = np_2 + np''_2 + p''$. Since P is integral and torsion free, it follows that $p''_1 + p_1 = p''_2 + p_2$. Now let $x_i := q_i - \theta(p''_i) \in Q^{\mathrm{gp}}$. Note that

$$x_1 + \theta(p_1) + \theta(p''_1) = q_1 + \theta(p_1) = q_2 + \theta(p_2) = x_2 + \theta(p_2) + \theta(p''_2),$$

and hence $x_1 = x_2$. Furthermore,

$$nx_1 = nq_1 - n\theta(p''_1) = nq_1 - \theta(p'_1 - p'') = nq_1 - \theta(p'_1) + \theta(p'') = q' + \theta(p'') \in Q.$$

[2] This proposition is due to Aaron Gray.

Since Q is saturated, $q := x_1 = x_2 \in Q$. Now $q_i = p_i'' + q$ and $p_1 + p_1'' = p_2 + p_2''$, proving that θ satisfies condition I1. □

Example 4.7.6. To see that the hypothesis on P is not superfluous, let $Q := \mathbf{N} \oplus \mathbf{N}$ and let P be the submonoid of Q generated by $(2,0), (1,1), (0,2)$. Then the inclusion homomorphism $P \to Q$ is $\mathbf{Q}$-integral but not integral. This homomorphism is also an example of a locally exact homomorphism which is not integral.

Theorem 4.7.7. *Let $\theta\colon P \to Q$ be a local homomorphism of fine saturated monoids and let K_θ denote the ideal of Q generated by the image of P^+. Then*

$$\dim(Q) \le \dim(P) + \dim(Q, K_\theta).$$

Furthermore, the following conditions are equivalent:

1. $\dim(Q) = \dim(P) + \dim(Q, K_\theta)$.
2. *θ is critically exact.*
3. *θ is critically s-injective.*
4. *θ is $\mathbf{Q}$-integral.*
5. *θ is locally exact.*
6. *θ is locally s-injective.*

Proof The following lemmas will also be useful elsewhere.

Lemma 4.7.8. *Let $\theta\colon P \to Q$ be a local homomorphism of fine monoids, and let G be a θ-critical face of Q. Then G is maximal among all θ-critical faces of Q if and only if the homomorphism $\theta'\colon P \to Q/G$ induced by θ is $\mathbf{Q}$-surjective.*

Proof Suppose that G is a θ-critical face of Q and let $\theta'\colon P \to Q/G$ be the induced homomorphism. The θ'-critical faces of Q/G correspond to the θ-critical faces of Q containing G. Thus G is maximal if and only if Q/G has a unique θ'-critical face. Since Q/G is sharp and fine, Proposition 4.3.9 tells us that this is the case if and only if θ' is $\mathbf{Q}$-surjective. □

Lemma 4.7.9. *Let $\theta\colon P \to Q$ be a local homomorphism of fine monoids, and let G be a θ-critical face of Q which is maximal among the θ-critical faces. Then the following conditions are equivalent:*

1. *The induced homomorphism $\overline{P} \to Q/G$ is an isogeny.*
2. $\dim P + \dim G = \dim Q$.
3. *The induced homomorphism $\mathbf{Q} \otimes \overline{P}^{\mathrm{gp}} \oplus \mathbf{Q} \otimes G^{\mathrm{gp}} \to \mathbf{Q} \otimes Q^{\mathrm{gp}}$ is an isomorphism.*

If in addition P is saturated, these conditions are also equivalent to the following:

4. *The homomorphism $\overline{P} \to Q/G$ is exact.*
5. *The homomorphism $\overline{P} \to Q/G$ is injective.*
6. *The homomorphism $\overline{P} \to Q/G$ is Kummer.*

Proof If (1) holds, then $\dim P = \dim Q/G = \dim Q - \dim G$, so (2) also holds. According to Lemma 4.7.8, the homomorphism $\overline{P} \to Q/G$ is $\mathbf{Q}$-surjective. and hence the homomorphism $\mathbf{Q} \otimes \overline{P}^{\mathrm{gp}} \to \mathbf{Q} \otimes (Q^{\mathrm{gp}}/G^{\mathrm{gp}})$ is surjective. The dimension of the target is $\dim Q - \dim G$, so if (2) holds, this homomorphism is an isomorphism, and it follows that the homomorphism in (3) is also an isomorphism. Condition (3) implies that the map $\mathbf{Q} \otimes \overline{P}^{\mathrm{gp}} \to \mathbf{Q} \otimes (Q^{\mathrm{gp}}/G^{\mathrm{gp}})$ is injective, and hence that the homomorphism of cones $C_{\mathbf{Q}}(\overline{P}) \to C_{\mathbf{Q}}(Q/G)$ is injective. Since it is also surjective, it is in fact an isomorphism. Thus (3) implies (1). Now suppose that P is saturated. If (1) holds, the homomorphism $C_{\mathbf{Q}}(\overline{P}) \to C_{\mathbf{Q}}(Q/G)$ is an isomorphism, hence exact. Since P is saturated, the homomorphism $\overline{P} \to C_{\mathbf{Q}}(P)$ is exact, hence so is the homomorphism $\overline{P} \to C_{\mathbf{Q}}(Q/G)$, and it follows from Proposition 4.2.1 that $\overline{P} \to Q/G$ is also exact. Thus (1) implies (4). Since a sharp exact homomorphism is injective, (4) implies (5). Since $\overline{P} \to Q/G$ is $\mathbf{Q}$-surjective, it is by definition Kummer if it is injective, so (5) implies (6). Since (6) trivially implies (1), all six conditions are equivalent. □

Let us return to the proof of Theorem 4.7.7. The prime ideals of (Q, K_θ) are the primes $\mathfrak{q}$ of Q such that $P^+ \subseteq \theta^{-1}(\mathfrak{q})$, and so the faces of (Q, K_θ) are the faces G of Q such that $\theta^{-1}(G) \subseteq P^*$, i.e., the θ-critical faces of Q. Recall from Remark 1.5.2 that the dimension of (Q, K_θ) is the maximum of the dimensions of its faces. Thus the claimed inequality just means that $\dim(Q) \leq \dim(P) + \dim(G)$ for every maximal θ-critical face G of Q. If G is such a face, Lemma 4.7.8 implies that the induced homomorphism $\overline{P} \to Q/G$ is $\mathbf{Q}$-surjective, hence $\dim P \geq \dim Q - \dim G$, as claimed.

Now suppose that $\dim(Q) = \dim(P) + \dim(Q, K_\theta)$. Then for every θ-critical face G, we have $\dim(G) \leq \dim(Q) - \dim(P)$. Let G' be a maximal θ-critical face containing G. Then as we saw in Lemma 4.7.8, the maximality of G' implies that $\overline{P} \to Q/G'$ is $\mathbf{Q}$-surjective, and hence that $\dim G' \geq \dim Q - \dim P$. Thus equality holds, and it follows from Lemma 4.7.9 that $\overline{P} \to Q/G'$ is exact. Then $P \to Q_{G'}$ is also exact. Since $G \subseteq G'$, we have a factorization $P \to Q_G \to Q_{G'}$, and so $\theta_G \colon P \to Q_G$ is also exact, by (2) of Proposition 4.2.1. This shows that θ is critically exact.

It is clear that any critically exact homomorphism is critically s-injective, since every exact homomorphism is s-injective. On the other hand, suppose θ is critically s-injective, and let G be a maximal θ-critical face of Q. Since θ is critically s-injective, the homomorphism $\overline{P} \to Q/G$ is injective, and Lemma 4.7.9 tells us that $\dim(G) = \dim(Q) - \dim(P)$, proving condition (1).

We have thus proved the equivalence of conditions (1)–(3). If (2) holds, then the induced homomorphism $\overline{\theta}\colon \overline{P} \to \overline{Q}$ is critically exact, and hence so is $C_{\mathbf{Q}}(\overline{\theta})$. Then Theorem 4.7.2 implies that $C_{\mathbf{Q}}(\overline{P}) \to C_{\mathbf{Q}}(\overline{Q})$ is integral. Using (2) of Proposition 4.6.3, we see that the homomorphism $C_{\mathbf{Q}}(P) \to C_{\mathbf{Q}}(Q)$ is integral. This proves (4). If (4) holds then, by (4) of Proposition 4.6.3, $C_{\mathbf{Q}}(P) \to C_{\mathbf{Q}}(Q)$ is locally exact. Since $P \to C_{\mathbf{Q}}(P)$ is locally exact if P is saturated, it follows that $P \to Q \to C_{\mathbf{Q}}(Q)$ is locally exact and hence that $P \to Q$ is locally exact, so (5) holds. Statement (5) of Proposition 4.2.1 shows that condition (5) above implies condition (6) and it is clear that (6) implies (3). The proof of the theorem is complete. □

Proposition 4.7.10. *Let F be a face of a sharp toric monoid Q. Then the following conditions are equivalent:* [3]

1. *F is a direct summand of Q.*
2. *For every face G of Q, $F + G$ is a face of Q.*
3. *For every face G of Q, $F + G^{\mathrm{gp}}$ is face of Q_G.*

Proof Suppose that F is a direct summand of Q, say $Q = F \oplus Q'$. Let G be a face of Q. Observe that if $f \in F$ and $q' \in Q'$ and $f + q' \in G$, then $f \in F \cap G$ and $q' \in Q' \cap G$, since G is a face of Q. To prove that $F + G$ is a face of Q, suppose that $q_i \in Q$ and $q_1 + q_2 \in G$. Write $q_i = f_i + q'_i$, so that $q_1 + q_2 = f + g$, where $f \in F$ and $g \in G \cap Q'$. It follows that $f = f_1 + f_2$ and $g = q'_1 + q'_2$, and hence that each q'_i belongs to G. But then each $q_i \in F + G$, as required. This shows that (1) implies (2).

To prove that (2) implies (3), suppose that F, G, and $F + G$ are faces of Q. We claim that $F + G^{\mathrm{gp}}$ is a face of Q_G. Suppose that $x_1, x_2 \in Q_G$ and $x_1 + x_2 \in F + G^{\mathrm{gp}}$. Then there exist $g_1, g_2, g \in G$ such that $x_i + g_i \in Q$ and $x_1 + x_2 + g \in F + G$. Then $x_1 + g_1 + x_2 + g_2 + g \in F + G$ and, since $F + G$ is a face of Q, each $x_i + g_i \in F + G$. Hence each $x_i \in F + G^g$, as claimed.

Now suppose that (3) holds. Every face G of Q is an exact submonoid of Q, and hence $F \cap G = F \cap G^{\mathrm{gp}}$. Then the natural map

$$F/F \cap G \to (F + G^{\mathrm{gp}})/G^{\mathrm{gp}}$$

is an isomorphism. Since $F + G^{\mathrm{gp}}$ is a face of Q_G, it is an exact submonoid, and it follows that $F_{F \cap G} \to Q_G = Q + G^{\mathrm{gp}}$ is also exact. Since this is true for every G, the inclusion homomorphism $\theta\colon F \to Q$ is locally exact. Now let G be a maximal θ-critical face of Q and let $Q' := F \oplus G$. By (3) of Lemma 4.7.9

[3] This proof that (3) implies (1) is due to Bernd Sturmfels.

the map $\mathbf{Q} \otimes Q''^{\mathrm{gp}} \to \mathbf{Q} \otimes Q^{\mathrm{gp}}$ is an isomorphism. We shall show that in fact the map $Q' \to Q$ is an isomorphism.

First we consider the corresponding rational cones $C_{\mathbf{Q}}(Q') \subseteq C_{\mathbf{Q}}(Q)$ and their duals

$$C_{\mathbf{Q}}(Q)^{\vee} \subseteq C_{\mathbf{Q}}(Q')^{\vee} \subseteq \mathrm{Hom}(Q^{\mathrm{gp}}, \mathbf{Q}).$$

Let ϕ be an indecomposable element of $C_{\mathbf{Q}}(Q')^{\vee} \cong C_{\mathbf{Q}}(F)^{\vee} \oplus C_{\mathbf{Q}}(G)^{\vee}$. Since ϕ is indecomposable, ϕ belongs either to $C_{\mathbf{Q}}(F)^{\vee}$ or to $C_{\mathbf{Q}}(G)^{\vee}$. In the first case, $\phi^{\perp}$ contains all of G^{gp}, and so ϕ factors through $C_{\mathbf{Q}}(Q)/C_{\mathbf{Q}}(G)$. Lemma 4.7.9 implies that the map $C_{\mathbf{Q}}(F) \to C_{\mathbf{Q}}(Q)/C_{\mathbf{Q}}(G)$ is bijective, and ϕ is nonnegative on $C_{\mathbf{Q}}(F)$, hence on all of $C_{\mathbf{Q}}(Q)$. In other words, $\phi \in C_{\mathbf{Q}}(Q)^{\vee}$. In the second case, $G' := \phi^{\perp} \cap G$ is a facet of G, and by assumption, $F + G'^{\mathrm{gp}}$ is a face, hence a facet, of $Q + G'^{\mathrm{gp}}$. Then F is a facet of Q/G', and hence $Q/(F + G')$ is a one-dimensional sharp monoid. Since $Q'/(F + G')$ is also one-dimensional, the map

$$C_{\mathbf{Q}}(Q')/C_{\mathbf{Q}}(F + G') \to C_{\mathbf{Q}}(Q)/C_{\mathbf{Q}}(F + G')$$

is an isomorphism. This again implies that $\phi(q) \geq 0$ for every $q \in C_{\mathbf{Q}}(Q)$. We conclude that $C_{\mathbf{Q}}(Q')^{\vee} = C_{\mathbf{Q}}(Q)^{\vee}$ and hence that $C_{\mathbf{Q}}(Q') = C_{\mathbf{Q}}(Q)$.

We have proved that the homomorphism $Q' \to Q$ is $\mathbf{Q}$-surjective. Thus for every $q \in Q$, there exist $m > 0$, $f \in F$, and $g \in G$ such that $mq = f + g$. Our assumption also implies that F is a face of Q/G, and since $F \to Q/G$ is $\mathbf{Q}$-surjective, it is in fact surjective. Thus we can also write $q = f' + x$ with $f' \in F$ and $x \in G^{\mathrm{gp}}$. But then $mq = mf' + mx$, so $mx = g$ and hence belongs to G. Since G is a face of Q and Q is saturated, G is also saturated, so $x \in G$. This proves that $Q = F + G \cong F \oplus G$. □

Example 4.7.11. The saturation hypothesis is not superfluous. To see this, consider the submonoid of $\mathbf{N} \oplus \mathbf{N}$ generated by $\{(2, 0), (3, 0), (1, 1), (0, 1)\}$, and the face F generated by $(0, 1)$.

Proposition 4.7.12. *Let Q be a fine, sharp, and saturated monoid. Then Q is free if and only if every face of Q is a direct summand.*

Proof Suppose that every face of Q is a direct summand. We prove that Q is free by induction on its dimension. If the dimension of Q is one, the result follows from Proposition 2.4.2. Assume the result is true for all monoids of smaller dimension and choose a face G of Q of dimension one. Then we can write $Q = G \oplus P$, and P is necessarily a face of Q. Every face F of P is also a face of Q and hence is a direct summand of Q: $Q = F \oplus Q'$. In particular, any $p \in P$ can be written as $p = f + q'$ with $f \in F$ and $q' \in Q'$; since P is a face of Q, $q' \in P' := P \cap Q$. Thus in fact we have $P = F \oplus P'$, so F is a direct summand of P. Thus P enjoys the same property as Q, and hence is free by the induction hypothesis. Since G is free and $Q = G \oplus P$, Q is also free. Conversely, suppose Q is the free monoid generated by a finite set S. Then because Q is free, the mapping taking a subset of S to the face it generates establishes a bijection between the set of subsets of S and the set of faces of Q, as we saw in Example 1.4.8. Furthermore, if T is a subset of S, then $Q = \langle T \rangle \oplus \langle S \setminus T \rangle$. □

Corollary 4.7.13. *A direct summand of a finitely generated free monoid is free.*

Proof Suppose that Q is a direct summand of $P := \mathbf{N}^r$, so that there is a retraction $\pi \colon P \to Q$. If $x \in Q^{\mathrm{gp}}$ and $nx \in Q$ then, since P is saturated, $x \in P$, hence $\pi^{\mathrm{gp}}(x) \in Q$. Since π^{gp} is the identity on Q^{gp}, in fact $x \in Q$. Thus Q is saturated. Every face of Q is a direct summand of P and hence also a direct summand of Q, and so the previous proposition implies that Q is free. □

Theorem 4.7.2 allows us to say more about the topology of cones. Recall that if $\theta \colon P \to Q$ is a monoid homomorphism, then $C_K(Q, P)$ is by definition the union of the θ-critical faces of $C_K(Q)$.

Corollary 4.7.14. *Let $\theta \colon P \to Q$ be a critically exact and local homomorphism of fine monoids, with P sharp. Then the following statements are verified.*

1. *The natural map $\pi \colon C_K(Q, P) \to C_K(Q)/C_K(P)$ is a homeomorphism.*
2. *The summation map $\sigma \colon C_K(P) \times C_K(Q, P) \to C_K(Q)$ is a homeomorphism.*

Proof Recall that $C_K(Q)/C_K(P)$ is the image of $C_K(Q)$ in the vector space $C_K(Q)^{\mathrm{gp}}/C_K(P)^{\mathrm{gp}}$. It follows from Theorem 4.7.2 that π is bijective, and its continuity is obvious. To prove that its inverse is continuous, let G be a maximal θ-critical face of Q. Then by (3) of Lemma 4.7.9, the map $\phi \colon C_K(G^{\mathrm{gp}}) \to C_K(Q)^{\mathrm{gp}}/C_K(P)^{\mathrm{gp}}$ induced by π is an isomorphism of finite dimensional K-vector spaces, and hence it is a homeomorphism. By Corollary 2.3.17, the cone $C_K(G)$ is a closed subset of the vector space $C_K(G^{\mathrm{gp}})$ and hence its image $\pi(C_K(G))$ is a closed subset of $C_K(Q)^{\mathrm{gp}}/C_K(P)^{\mathrm{gp}}$. Since ϕ is a homeomorphism, so is the induced map $\phi_{|G} \colon C_K(G) \to \pi(C_K(G))$. Thus the restriction of π^{-1} to $\pi(C_K(G))$ is continuous. Since the family of sets $\pi(C_K(G))$ as G ranges over the maximal critical faces of Q is a finite closed cover of $C_K(Q)^{\mathrm{gp}}/C_K(P)^{\mathrm{gp}}$, π^{-1} is continuous. This proves (1).

We already know that σ is bijective and continuous. To prove that its inverse is continuous, consider the following commutative diagram:

$$\begin{array}{ccc}
C_K(Q) & \xrightarrow{(\rho,\eta)} & C_K(P)\times C_K(Q,P) \\
\Big\downarrow{\scriptstyle \pi_0} & \searrow{\scriptstyle \eta} & \Big\downarrow{\scriptstyle \pi_2} \\
C_K(Q)/C_K(P) & \xleftarrow{\pi} & C_K(Q,P),
\end{array}$$

where the top arrow is the inverse of σ. We have just seen that π is a homeomorphism and, since π_0 is continuous, it follows that η is continuous. But $q = \theta(\rho(q)) + \eta(q)$ for all $q \in C_K(Q)$, so $\theta \circ \rho = \mathrm{id}_Q - \eta$ is continuous. Since θ is a closed embedding, it follows that ρ is continuous, hence that (ρ, η) is continuous. Thus σ is a homeomorphism. □

Corollary 4.7.15. *Let C be a finitely generated K-cone, and let I_C be its interior, that is, the complement of the union of its proper faces. Assume that C is not dull. Then there exist a K-vector space V and a homeomorphism of pairs*

$$(K^{\geq} \times V, K^{>} \times V) \longrightarrow (C, I_C).$$

In particular, if $K = \mathbf{R}$, C is a topological manifold with boundary.

Proof Let Q be the submonoid of C generated by a finite set of generators, so that $C = C_K(Q)$. Let p be an element of I_Q. Since C is not dull, C^* is a proper face and p is not a unit of Q. Let $\theta\colon P := \mathbf{N} \to Q$ be the homomorphism taking 1 to p. Since P is valuative, statement (5) of Proposition 4.6.3 implies that θ is local and integral. Since p belongs to the interior of Q, every proper face of Q is θ-critical, and thus $C_K(Q, P) = C \setminus I_C$. By (1) of Corollary 4.7.14, the map $\pi\colon C_K(Q, P) \to C_K(Q)/C_K(P)$ is a homeomorphism. Since p is not contained in any proper face of Q, the homomorphism $C_K(\theta)$ is vertical, and hence the quotient $C_K(Q)/C_K(P)$ is a group, hence a vector space V. Replacing $C_K(Q, P)$ by V and identifying $C_K(P)$ with $K^{\geq}$ in statement (2) of Corollary 4.7.14, we find a homeomorphism $\tilde{\sigma}\colon K^{\geq} \times V \to C$. By construction, $\tilde{\sigma}^{-1}(C_K(Q, P)) = 0 \times C_K(Q, P)$, and since $C \setminus C_K(Q, P) = I_C$, we see that $\tilde{\sigma}$ induces the desired homeomorphism of pairs. □

4.8 Saturated homomorphisms

Just as the coproduct of integral monoids need not be integral, the coproduct of saturated monoids need not be saturated. This complication is also quite important, and we study how to overcome it in this section and the one that follows. The main ideas in this section are due to K. Kato and T. Tsuji [76].

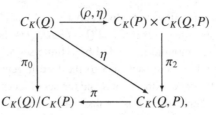

Here are some examples. Let $P = Q = \mathbf{N}$ and let $P \to Q$ be the homomorphism sending 1 to 2. Then $Q \oplus_P Q$ is given by generators a, b satisfying the relation $2a = 2b$ and is not saturated. More generally, let m and n be integers, and consider the coproduct $Q \oplus_P Q$, where the left homomorphism $P \to Q$ is multiplication by m and the right multiplication is multiplication by n. Then $Q \oplus_P Q$ has generators a, b satisfying the relation $na = mb$. Here P and Q are saturated, but $Q \oplus_P Q$ is not, provided both m and n are at least 2. Note that if m and n are relatively prime, $(Q \oplus_P Q)^{\mathrm{gp}}$ is torsion free. For another example, consider the monoid P' given by generators a, b, c with relation $a + b = 2c$, let $\mathbf{N} = P \to P'$ be the homomorphism sending 1 to a, and let $P \to Q = \mathbf{N}$ be multiplication by 2. Then the coproduct $P' \oplus_P Q$ is given by generators a', b, c and relation $2a' + b = 2c$, which is not saturated.

An integral monoid P is saturated if and only the endomorphism n_P of P given by multiplication by n is exact for every $n \in \mathbf{Z}^+$. This motivates the following definition.

Definition 4.8.1. *Let P be an integral monoid and let n be a positive integer. Then P is n-saturated if the endomorphism*

$$n_P \colon P \to P : p \mapsto np$$

is exact.

It is clear from Proposition 4.2.1 that if P is n-saturated then it is also n^k-saturated for every natural number k, and that if P is n-saturated and m divides n, then P is also m-saturated. Thus P is n-saturated if and only if it is p-saturated for every prime p dividing n, and in this case it is n'-saturated, where n' is any number with the same set of prime divisors as n.

Definition 4.8.2. *Let $\theta \colon P \to Q$ be a homomorphism of integral monoids and let n be a positive integer. Consider the following subdiagram of the relative n-Frobenius diagram of Definition 4.4.1:*

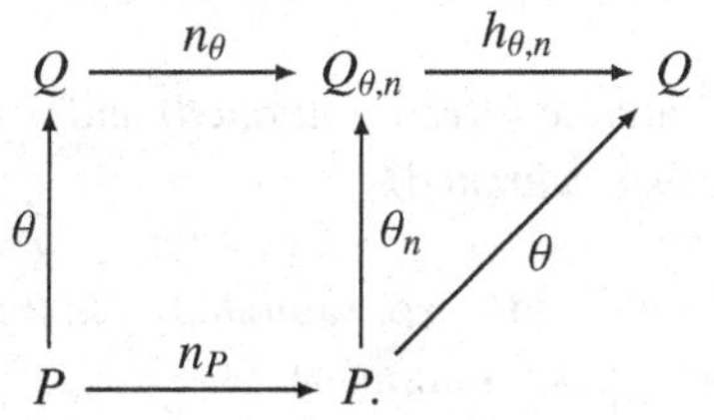

Then θ is said to be n-quasi-saturated if $h_{\theta,n}$ is exact, and θ is said to be n-saturated if it is integral and n-quasi-saturated. Finally, θ is said to be quasi-saturated if it is n-quasi-saturated for all n and to be saturated if it is integral and quasi-saturated.

Example 4.8.3. Let $\theta \colon P \to Q$ be the homomorphism $m_{\mathbf{N}} \colon \mathbf{N} \to \mathbf{N}$. As we saw in the second paragraph of this section, $Q_{\theta,n}$ is generated by a, b, where

$a := n_\theta(1)$ and $b := \theta_n(1)$, satisfying the relation $ma = nb$. Then $h_{\theta,n}$ takes a to n and b to m and is not exact whenever m and n are both greater than 1. Thus if $m > 1$, $m_{\mathbf{N}}$ is not n-quasi-saturated for any $n > 1$.

Proposition 4.8.4. *If P is an integral monoid and n is a natural number, the homomorphism $0 \to P$ is n-saturated if and only if P is n-saturated. If $\theta\colon P \to Q$ is n-quasi-saturated and P is n-saturated, then Q is also n-saturated.*

Proof The zero homomorphism $0 \to P$ is integral, so it is n-saturated if and only if it is n-quasi-saturated. The homomorphism $n_0\colon P \to P_{0,n} := P \oplus_0 0$ is in fact id_P, and the homomorphism $h_{0,n}\colon P_{0,n} \to P$ is n_P. Thus P is n-saturated if and only if $0 \to P$ is n-quasi-saturated. This proves the first statement, and the second statement will follow once we know that the composite of two n-quasi-saturated homomorphisms is n-quasi-saturated, as we will verify in the next proposition. However it is easy to give a direct proof. Indeed, if P is n-saturated, it follows from (6) of Proposition 4.2.1 that n_θ is exact, since it is the pushout of the exact homomorphism n_P. It follows that the the composite $n_Q = h_{\theta,n} \circ n_\theta$ is also exact, i.e., that Q is n-saturated. □

Proposition 4.8.5. *Let n be a natural number.*

1. *The class of n-quasi-saturated (resp. saturated) homomorphisms is closed under composition and pushout in the category of integral monoids.*
2. *If $\psi \circ \theta$ is n-quasi-saturated (resp. saturated) and ψ is exact, then θ is also n-quasi-saturated (resp. saturated).*
3. *If $\psi \circ \theta$ is n-quasi-saturated (resp. saturated) and θ is surjective, then ψ is also n-quasi-saturated (resp. saturated).*
4. *A homomorphism of integral monoids $\theta\colon P \to Q$ is n-quasi-saturated (resp. saturated) if and only if $\overline{\theta}\colon \overline{P} \to \overline{Q}$ is.*

Proof Let $\theta\colon P \to Q$ and $\psi\colon Q \to R$ be n-quasi-saturated homomorphisms of integral monoids, and consider the following diagram.

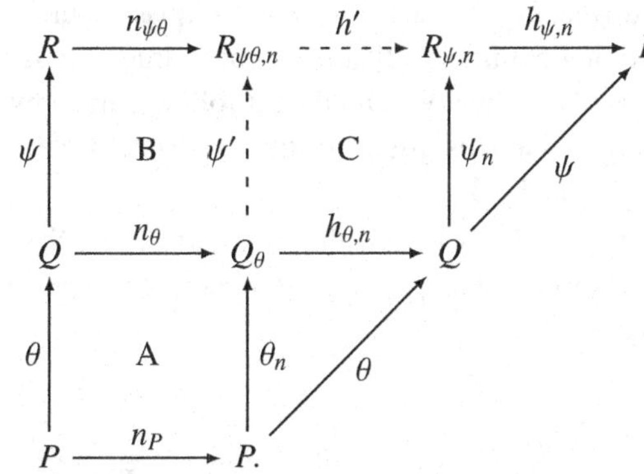

Here the square A and the rectangles BA and BC are, by construction, cocartesian in the category of integral monoids. The homomorphism ψ' is the unique homomorphism such that $\psi' \circ n_\theta = n_{\psi\theta} \circ \psi$ and $\psi' \circ \theta_n = (\psi\theta)_n$, and the homomorphism h' is the unique one such that $h' \circ n_{\psi\theta} = n_\psi$ and $h' \circ \psi' \circ \theta_n = \psi_n \circ \theta$.

Let us observe that

$$h_{\psi,n} \circ h' = h_{\psi\theta,n} \tag{4.8.1}$$

To verify this we compute

$$\begin{aligned} h_{\psi,n} \circ h' \circ n_{\psi\theta} &= h_{\psi,n} \circ n_\psi \\ &= n_R \\ &= h_{\psi\theta,n} \circ n_{\psi\theta}, \end{aligned}$$

and

$$\begin{aligned} h_{\psi,n} \circ h' \circ \psi' \circ \theta_n &= h_{\psi,n} \circ \psi_n \circ \theta \\ &= \psi \circ \theta \\ &= h_{\psi\theta} \circ (\psi\theta)_n \\ &= h_{\psi\theta} \circ \psi' \circ \theta_n. \end{aligned}$$

Since the rectangle BA and the square A are cocartesian, it follows that the square B is also cocartesian and, since the rectangle CB and the square B are cocartesian, the square C is also cocartesian. Now if θ is n-quasi-saturated, $h_{\theta,n}$ is exact and hence so is its pushout h'. If ψ is n-quasi-saturated, then $h_{\psi,n}$ is also exact. Since $h_{\psi,n} \circ h' = h_{\psi\theta,n}$ we can conclude that $h_{\psi\theta,n}$ is the composition of exact homomorphisms, hence is exact, and so $\psi \circ \theta$ is n-quasi-saturated. If θ and ψ are n-saturated, they are also integral, and hence the same is true of $\psi \circ \theta$ by Proposition 4.6.3.

Suppose that $\psi\theta$ is n-quasi-saturated, i.e., that $h_{\psi\theta,n}$ is exact. If also ψ is exact, so are its pushouts ψ' and ψ_n. Then $\psi \circ h_{\theta,n} = h_{\psi,n} \circ h' \circ \psi' = h_{\psi\theta,n} \circ \psi'$ is the composition of exact homomorphisms, hence exact. By (2) of Proposition 4.2.1, we conclude that $h_{\theta,n}$ is exact, so that θ is n-quasi-saturated. If $\psi\theta$ is also integral, Proposition 4.6.3 implies that the same is true of θ. If θ is surjective, then $h_{\theta,n}$ and h' are also surjective, and then it follows, again by (2) of (4.2.1), that $h_{\psi,n}$ is exact, i.e., that ψ is n-quasi-saturated. The saturated case follows, again using Proposition 4.6.3

We have now proved (2), (3), and the first part of (1). Statement (4) will follow easily if we know that map $\pi\colon P \to \overline{P}$ is saturated (and similarly for Q). Consider the diagram

$$\begin{array}{ccccc} \overline{P} & \xrightarrow{n_\pi} & \overline{P}_{\pi,n} & \xrightarrow{h_{\pi,n}} & \overline{P} \\ \Big\uparrow{\scriptstyle \pi} & & \Big\uparrow{\scriptstyle \pi_n} & \nearrow{\scriptstyle \pi} & \\ P & \xrightarrow{n} & P. & & \end{array}$$

Here π_n is surjective because it is the pushout of the surjective homomorphism π. Since $h_{\theta,n} \circ \pi_n = \pi$ is exact, it follows from Proposition 4.2.1 that $h_{\pi,n}$ is also exact. Thus π is n-quasi-saturated and, since it is integral it is also saturated.

To finish the proof, we must show that the pushout of an n-quasi-saturated homomorphism is again n-quasi-saturated. We shall need the following lemma, which asserts that the functors $\theta \mapsto \theta_n$ and $\theta \mapsto h_{\theta,n}$ are compatible with pushouts.

Lemma 4.8.6. *Consider a cocartesian diagram in the category of integral monoids):*

$$\begin{array}{ccc} Q & \xrightarrow{\phi'} & Q' \\ \Big\uparrow{\scriptstyle \theta} & & \Big\uparrow{\scriptstyle \theta'} \\ P & \xrightarrow{\phi} & P'. \end{array}$$

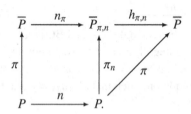

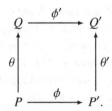

Then for every n, the following squares, in which ϕ_θ is induced by functoriality, are also cocartesian; that is, θ'_n is the pushout of θ_n along ϕ, and $h_{\theta',n}$ is the

pushout of $h_{\theta,n}$ *along* ϕ_θ*:*

$$
\begin{array}{ccc}
Q_{\theta,n} & \xrightarrow{\phi_\theta} & Q'_{\theta',n} \\
{\scriptstyle \theta_n}\uparrow & & \uparrow{\scriptstyle \theta'_n} \\
P & \xrightarrow{\phi} & P'
\end{array}
\qquad\qquad
\begin{array}{ccc}
Q'_{\theta',n} & \xrightarrow{h_{\theta',n}} & Q' \\
{\scriptstyle \phi_\theta}\uparrow & & \uparrow{\scriptstyle \phi'} \\
Q_{\theta,n} & \xrightarrow{h_{\theta,n}} & Q.
\end{array}
$$

Proof In the diagram below, the two squares are cocartesian by construction. It follows that the outer rectangle is also cocartesian:

$$
\begin{array}{ccccc}
Q & \xrightarrow{\phi'} & Q' & \xrightarrow{n_{\theta'}} & Q'_{\theta',n} \\
{\scriptstyle \theta}\uparrow & & \uparrow{\scriptstyle \theta'} & & \uparrow{\scriptstyle \theta'_n} \\
P & \xrightarrow{\phi} & P' & \xrightarrow{n_{P'}} & P'.
\end{array}
$$

In the next diagram, the square on the left is cocartesian by construction, and, ϕ_θ is by definition the unique homomorphism such that $\phi_\theta \circ n_\theta = n_{\theta'} \circ \phi'$ and $\phi_\theta \circ \theta_n = \theta'_n \circ \phi$:

$$
\begin{array}{ccccc}
Q & \xrightarrow{n_\theta} & Q_{\theta,n} & \xrightarrow{\phi_\theta} & Q'_{\theta,n} \\
{\scriptstyle \theta}\uparrow & & \uparrow{\scriptstyle \theta_n} & & \uparrow{\scriptstyle \theta'_n} \\
P & \xrightarrow{n_P} & P & \xrightarrow{\phi} & P'.
\end{array}
$$

Thus the outer rectangles in both diagrams are the same. We conclude that the outer rectangle in the second rectangle is also cocartesian and, since the square on the left is cocartesian, so is the square on the right. Now we claim that there is a commutative diagram:

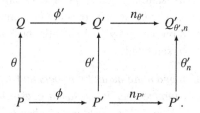

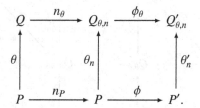

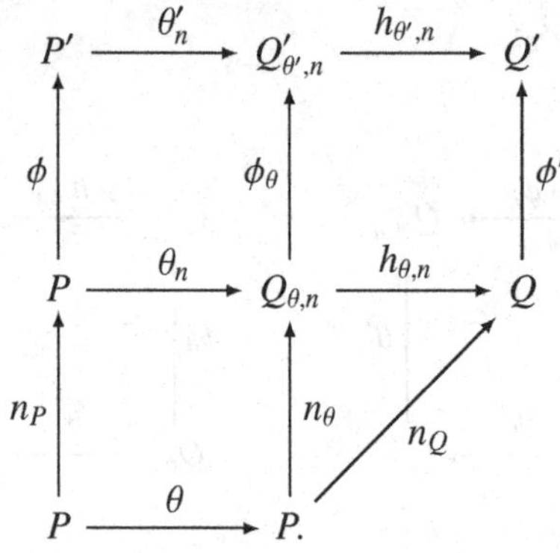

The outer rectangle is just the original cocartesian square, and we have just seen that the square on the top left is also cocartesian. It follows that the square on the top right is also cocartesian. □

The stability of n-quasi-saturation under pushouts follows. Indeed, if θ is n-saturated, $h_{\theta,n}$ is exact, and hence so is its pushout $h_{\theta',n}$. Since the integrality of homomorphisms is also preserved by pushouts, the pushout of an n-saturated homomorphism is also n-saturated. □

Proposition 4.8.7. *If m and n are natural numbers and $\theta\colon P \to Q$ is m-quasi-saturated and n-quasi-saturated, then it is also mn-quasi-saturated. Conversely, if θ is mn-quasi-saturated and P is m-saturated, then θ is n-quasi-saturated. Thus θ is quasi-saturated if and only if it is p-quasi-saturated for every prime number p.*

Proof Consider the following diagram:

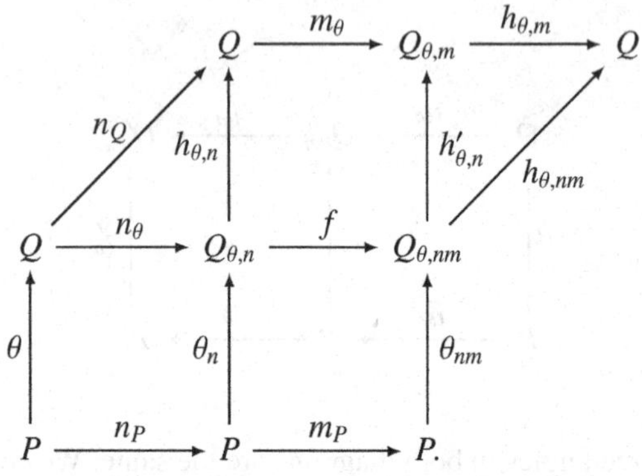

Here f is the unique homomorphism such that $f \circ n_\theta = (nm)_\theta$ and $f \circ \theta_n = \theta_{nm} \circ m_P$, and $h'_{\theta,n}$ is the unique homomorphism such that $h'_{\theta,n} \circ (nm)_\theta = m_\theta \circ n_Q$ and $h'_{\theta,n} \circ \theta_{nm} = \theta_m$. The bottom square on the left and the bottom rectangle are cocartesian by construction, and it follows that the square at the bottom right is also cocartesian. The large rectangle on the right is cocartesian, and so it follows that the top square on the right is cocartesian.

Now suppose that θ is n and m-quasi-saturated, i.e., that $h_{\theta,n}$ and $h_{\theta,m}$ are exact. Since $h_{\theta,n}$ is exact, it follows that its pushout $h'_{\theta,n}$ is exact and, since $h_{\theta,m}$ is exact, it follows that the composition $h_{\theta,nm} = h_{\theta,m} \circ h'_{\theta,n}$ is exact.

Suppose on the other hand that θ is mn-quasi-saturated and that P is m-saturated. Then m_P is exact and hence so is its pushout f. By hypothesis $h_{\theta,nm}$ is exact, and hence so is $h_{\theta,nm} \circ f = h_{\theta,m} \circ m_\theta \circ h_{\theta,n}$. By (2) of Proposition 4.2.1, it follows that $h_{\theta,n}$ is also exact. □

Proposition 4.8.8. *Let $\theta\colon P \to Q$ be a homomorphism of integral monoids sending a face F of P to a face G of Q and let n be a natural number.*

1. *The localization maps $\lambda_F\colon P \to P_F$ and $\lambda_G\colon Q \to Q_G$ are n-saturated, as are the quotient maps $P \to P/F$ and $Q \to Q/G$.*
2. *If θ is n-quasi-saturated (resp. saturated), then so also are the maps $P_F \to Q_G$ and $P/F \to Q/G$.*

Proof We claim that the diagram

$$\begin{array}{ccc} P_F & \xrightarrow{n_{P_F}} & P_F \\ \uparrow{\scriptstyle \lambda_F} & & \uparrow{\scriptstyle \lambda_F} \\ P & \xrightarrow{n_P} & P \end{array}$$

is cocartesian in the category of integral monoids. Indeed, if $\alpha\colon P_F \to R$ and $\beta\colon P \to R$ are homomorphisms with $\alpha \circ \lambda_F = \beta \circ n_{P_F}$ then, for every $f \in F$,

$$n\beta(f) = \beta(n_{P_F}(f)) = \alpha(\lambda_F(f)).$$

Thus $n\beta(f)$ is a unit in R, hence so is $\beta(f)$, and hence β factors uniquely through P_F. This implies that the map h_{λ_F} in the diagram in Definition 4.8.2 is an isomorphism, hence exact. It follows that λ_F is n-quasi-saturated and hence saturated, since it is integral by (3) of Proposition 4.6.3. This proves the first part of (1), and the second part follows from (4) of Proposition 4.8.5.

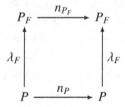

Now suppose $\theta\colon P \to Q$ is n-quasi-saturated. The square in the diagram

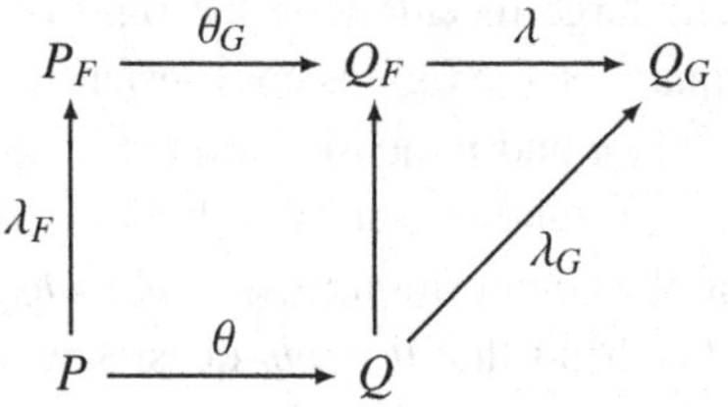

is cocartesian, and thus θ_G is the pushout of an n-quasi-saturated homomorphism and hence is n-quasi-saturated by (1) of Proposition 4.8.5. The previous paragraph implies that the localization map λ is n-quasi-saturated, and hence so is the composite $\lambda \circ \theta_G$. □

We can now explain how a saturated homomorphism of saturated monoids preserves the saturation of its target under pushouts.

Proposition 4.8.9. *Let n be a positive integer and let $\theta\colon P \to Q$ be a homomorphism of n-saturated monoids. The following conditions are equivalent.*

1. *The homomorphism θ is n-quasi-saturated.*
2. *For every homomorphism $\phi\colon P \to P'$ from P to an n-saturated monoid, the pushout $Q' := P' \oplus_P Q$ in the category of integral monoids is n-saturated.*
3. *The monoid $Q_{\theta,n}$ in the diagram of Definition 4.8.2 is n-saturated.*

Proof If θ is n-quasi-saturated and $\phi\colon P \to P'$ is any homomorphism of n-saturated monoids then, by Proposition 4.8.5, the pushout morphism $\theta'\colon P' \to Q'$ is again n-quasi-saturated. Since P' is n-saturated, it then follows from Proposition 4.8.4 that Q' is again n-saturated. This shows that (1) implies (2), and the implication of (3) by (2) is trivial. Condition (3) means that the homomorphism n_{Q_θ} is exact. Since $n_{Q_\theta} = n_\theta \circ h_{\theta,n}$, it follows that $h_{\theta,n}$, is also exact, by (2) of Proposition 4.2.1. □

Corollary 4.8.10. *Let $\theta\colon P \to Q$ be a homomorphism of saturated monoids. The following conditions are equivalent.*

1. *The homomorphism θ is quasi-saturated.*
2. *For every homomorphism $\phi\colon P \to P'$ from P to a saturated monoid P', the integral pushout $P' \oplus_P Q$ is saturated.*
3. *For every natural number n, the monoid $Q_{\theta,n}$ in Definition 4.8.2 is saturated.*
4. *For every prime number p, the monoid $Q_{\theta,p}$ in Definition 4.8.2 is saturated.*

Proof The equivalence of conditions (1) through (3) follows from Proposition 4.8.9, and the equivalence of (1) and (4) follows from this and Proposition 4.8.7 □

The examples at the beginning of this section suggest that the failure of a homomorphism θ to be quasi-saturated is related to the possible existence of torsion in $\mathrm{Cok}(\theta^{\mathrm{gp}})$. This is indeed the case, but one must dig a little deeper, as we shall see in Proposition 4.8.11 and Theorem 4.8.14.

Proposition 4.8.11. *Let $\theta\colon P \to Q$ be a quasi-saturated homomorphism of fine saturated monoids. Then for every face G of Q, the cokernel of the map $P^{\mathrm{gp}} \to (Q/G)^{\mathrm{gp}}$ is torsion free.*

Proof By Proposition 4.8.8, the induced map $P/\theta^{-1}(G) \to Q/G$ is also quasi-saturated, so we may assume without loss of generality that P and Q are sharp and that θ is local. Let π be the natural projection $Q \to \mathrm{Cok}(\theta^{\mathrm{gp}})$, let Γ be the torsion subgroup of $\mathrm{Cok}(\theta^{\mathrm{gp}})$, and let $Q' := \{q \in Q : \pi(q) \in \Gamma\}$. Then θ factors: $\theta = \phi \circ \theta'$, where $\theta' : P \to Q'$ and $\phi\colon Q' \to Q$. Since ϕ is exact, it follows from Proposition 4.8.5 that θ' is also quasi-saturated. Furthermore, $\Gamma = \mathrm{Cok}(\theta'^{\mathrm{gp}})$. Thus we may as well assume that $Q' = Q$. In other words, we may assume that θ is local, that P and Q are sharp, and that $\mathrm{Cok}(\theta^{\mathrm{gp}})$ is a finite group Γ.

Now consider the following diagram:

$$\begin{array}{ccc} Q & \xrightarrow{(\mathrm{id},\pi)} & Q \oplus \Gamma \\ \uparrow{\scriptstyle \theta} & & \uparrow{\scriptstyle (\mathrm{id},0)} \\ P & \xrightarrow{\theta} & Q. \end{array}$$

The associated diagram in the category of abelian groups is cocartesian, and so the integral pushout $Q \oplus_P Q$ can be identified with the submonoid of $Q \oplus \Gamma$ generated by the images of (id, π) and $(\mathrm{id}, 0)$. Since $\Gamma \subseteq Q^{\mathrm{gp}} \oplus_{P^{\mathrm{gp}}} Q^{\mathrm{gp}}$ is torsion, it is contained in the $(Q \oplus_P Q)^{\mathrm{sat}}$. But θ is quasi-saturated, so Corollary 4.8.10 implies that $Q \oplus_P Q$ is saturated. This means that every element (q, g) of $Q \oplus \Gamma$ can be written as $(q_1 + q_2, \pi(q_1))$ for some $q_1, q_2 \in Q$. When $q = 0$ this is only possible when $q_1 = q_2 = 0$, since Q is sharp. Thus $\Gamma = 0$. □

Corollary 4.8.12. *A saturated local homomorphism $\theta\colon P \to Q$ of fine, sharp, and saturated monoids of the same dimension is an isomorphism.*

Proof Since θ is local and integral it is exact, and since it is also sharp, θ^{gp} is injective. Since P and Q are sharp, the ranks of P^{gp} and Q^{gp} are equal to their respective dimensions and hence to each other, and it follows that $\Gamma := \mathrm{Cok}(\theta^{\mathrm{gp}})$ is a finite group. By the previous result, $\Gamma = 0$, so θ^{gp} is an isomorphism. Since θ is exact, it also is an isomorphism. □

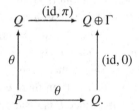

Our next goal is to characterize quasi-saturated homomorphisms more explicitly by giving an analog of the condition I1 for integral homomorphisms in Definition 4.6.2.

Proposition 4.8.13. *Let $\theta\colon P \to Q$ be a homomorphism of integral monoids and let n be a positive integer. If P and Q are n-saturated, then θ is n-quasi-saturated if and only if it satisfies the condition:*

QS_n. *Whenever $p \in P$ and $q \in Q$ satisfy $\theta(p) \leq nq$, there exists a $p' \in P$ such that $p \leq np'$ and $\theta(p') \leq q$, where in each case the inequality sign means is to be taken with respect to the standard ordering within the respective monoid.*

More generally, the following statements hold.

1. *If P is n-saturated and θ is n-quasi-saturated, then θ satisfies condition QS_n.*
2. *If Q is n-saturated and θ satisfies condition QS_n, then θ is n-quasi-saturated.*

Proof We use the notation of the diagram in Definition 4.8.2, recalling that $Q^{\mathrm{gp}}_{\theta,n}$ is the pushout of the corresponding diagram of abelian groups and that $Q_{\theta,n}$ is the image of $Q \oplus P$ in $Q^{\mathrm{gp}}_{\theta,n}$. Suppose that P is n-saturated, that θ is n-quasi-saturated, and that $p \in P$ and $q \in Q$ with $\theta(p) \leq nq$. Then $z := n_\theta(q) - \theta_n(p) \in Q^{\mathrm{gp}}_\theta$ and $h_\theta(z) = nq - \theta(p) \in Q$. Since by hypothesis h_θ is exact, it follows that $z \in Q_{\theta,n}$. Hence there exist $p'' \in P$ and $q'' \in Q$ such that $n_\theta(q'') + \theta_n(p'') = n_\theta(q) - \theta_n(p)$. Then there exists an $x \in P^{\mathrm{gp}}$ such that $q - q'' = \theta^{\mathrm{gp}}(x)$ and $p'' + p = nx$. Since P is n-saturated, it follows that in fact $p' := x \in P$ and $p \leq np'$. Since $q'' = -\theta(p') + q \in Q$, it also follows that $q \geq \theta(p')$. Thus condition QS_n is satisfied.

Now suppose that Q is n-saturated, that θ satisfies condition QS_n, and that $z \in Q^{\mathrm{gp}}_{\theta,n}$ is such that $h^{\mathrm{gp}}_{\theta,n}(z) \in Q$. Choose $x \in P^{\mathrm{gp}}$ and $y \in Q^{\mathrm{gp}}$ such that $z = n^{\mathrm{gp}}_\theta(y) + \theta^{\mathrm{gp}}_n(x)$, and write $x = p_1 - p_2$ with $p_i \in P$. Letting $x' := x - np_1 = -(p_2 + (n-1)p_1)$ and $y' := y + \theta(p_1)$, we see that it is still true that $z = n^{\mathrm{gp}}_\theta(y') + \theta^{\mathrm{gp}}_n(x')$. Changing notation, we may assume without loss of generality that $p := -x \in P$. Then $ny = h_{\theta,n}(z) + \theta(p) \in Q$ and, since Q is n-saturated, it follows that $y \in Q$ and $ny \geq \theta(p)$. Then, by condition QS_n, there exists a $p' \in P$ such that $p \leq np'$ and $\theta(p') \leq y$. Then $p'' := np' - p \in P$, $q := y - \theta(p') \in Q$, and

$$\begin{aligned} z &= n_\theta(y) + \theta_n(-p) \\ &= n_\theta(y - \theta(p')) + \theta_n(np' - p) \\ &= n_\theta(q) + \theta_n(p'') \in Q_{\theta,n}. \end{aligned}$$

This shows that h_θ is exact. □

We can now state and prove the main structure theorem for saturated homomorphisms. Recall from Theorem 4.7.7 that a homomorphism of fine saturated monoids is **Q**-integral if and only if it is locally exact.

Theorem 4.8.14. *Let $\theta\colon P \to Q$ be a locally exact homomorphism of toric monoids. Then the following conditions are equivalent:*

1. *θ is saturated.*
2. *For every $\mathfrak{q} \in \operatorname{Spec} Q$ such that $\mathfrak{q}$ and $\mathfrak{p} := \theta^{-1}(\mathfrak{q})$ have the same height, the homomorphism $\overline{\theta}_{\mathfrak{q}} : \overline{P}_{\mathfrak{p}} \to \overline{Q}_{\mathfrak{q}}$ is an isomorphism.*
3. *θ is integral and, for every $\mathfrak{q} \in \operatorname{Spec} Q$ such that both $\mathfrak{q}$ and $\mathfrak{p} := \theta^{-1}(\mathfrak{q})$ have height one, the homomorphism $\overline{\theta}_{\mathfrak{q}} : \overline{P}_{\mathfrak{p}} \to \overline{Q}_{\mathfrak{q}}$ is an isomorphism.*
4. *For every face G of Q, the cokernel of the homomorphism*
$$P^{\mathrm{gp}} \to Q^{\mathrm{gp}}/G^{\mathrm{gp}}$$
is torsion free.
5. *The ideal $K_\theta := \theta(\theta^{-1}(Q^+)) + Q$ is a radical ideal.*
6. *For every maximal θ-critical face G of Q, the cokernel of the homomorphism $P^{\mathrm{gp}} \to Q^{\mathrm{gp}}/G^{\mathrm{gp}}$ is torsion free.*

If also θ is local and P and Q are sharp, then these conditions are also equivalent to

7. *The summation map $\sigma\colon P \times (Q \setminus \sqrt{K_\theta}) \to Q$ is bijective.*

Proof Recall from Proposition 4.8.7 that θ is saturated if and only if $\overline{\theta}$ is saturated. Furthermore, conditions (2) through (6) also hold for θ if and only if they hold for $\overline{\theta}$. Thus without loss of generality we may assume that P and Q are sharp. Let $F := \theta^{-1}(0)$. Then θ factors:

$$\theta = P \xrightarrow{\ \pi\ } P/F \xrightarrow{\ \theta'\ } Q,$$

and π is saturated by Proposition 4.8.8. Thus it follows from Propositions 4.8.5 and 4.6.3 that θ is saturated if and only if θ' is saturated. Furthermore, properties (2)–(6) are true for θ if and only if they are true for θ'. Thus we may also assume that θ is local, and we are reduced to proving that (1)–(7) are equivalent assuming that P and Q are sharp and that θ is local. In this case we shall prove the implications shown in the following diagram:

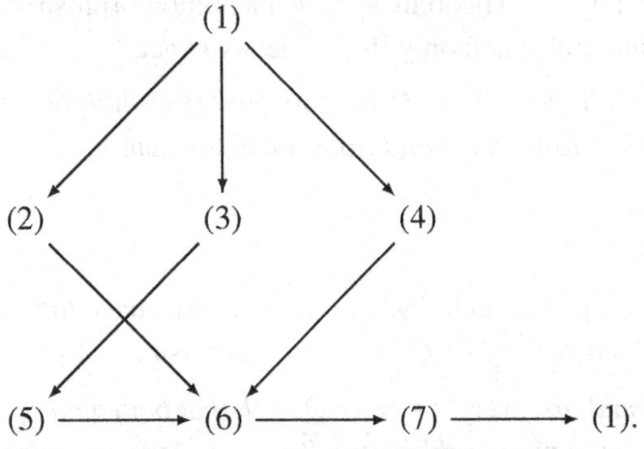

The proofs follow.

(1) implies (2) and (3): If θ is saturated, then by Proposition 4.8.8 so is $\overline{\theta}_{\mathfrak{q}}$. Since $\overline{P}_{\mathfrak{p}}$ and $\overline{Q}_{\mathfrak{q}}$ have the same dimension, it then follows from Corollary 4.8.12 that $\overline{\theta}_{\mathfrak{q}}$ is an isomorphism.

(1) implies (4): This is an immediate consequence of Proposition 4.8.11.

(2) implies (6): If G is a face of Q and $\mathfrak{q} := Q \setminus G$, then the height of $\mathfrak{q}$ is the dimension of Q/G. If G is θ-critical then $\mathfrak{p} := \theta^{-1}(\mathfrak{q}) = P^+$, so the height of $\mathfrak{p}$ is the dimension of P. If G is a maximal θ-critical face, then Lemma 4.7.9 implies that $\dim P = \dim Q/G$, since θ is locally exact. Thus (2) implies that $P \to Q/G$ is in fact an isomorphism.

(3) implies (5): Suppose that (3) holds and that $q \in Q \setminus K_\theta$. We must show that $q \notin \sqrt{K_\theta}$, i.e., that if $n \in \mathbf{Z}^+$, $p_0 \in P$, and $q_0 \in Q$ with $nq = \theta(p_0) + q_0$, then necessarily $p_0 \in P^*$. Since P is fine and saturated, it is enough to prove that, for every facet F of P, p_0 maps to a unit in P_F, by Corollary 2.4.5. Let us first observe that $\lambda_F(q) \notin K_{\theta_F} \subseteq Q_F$. To see this, suppose that $p_1 \in P_F$ and $q_1 \in Q_F$ with $\lambda_F(q) = \theta(p_1) + q_1$. Choose $f \in F$ such that $f + p_1 \in P$ and $\theta(f) + q_1 \in Q$, so that we have, after replacing p_1 by $f + p_1$, that $\theta(f) + q = \theta(p_1) + q_1$, where $f \in F$, $p_1 \in P$, and $q_1 \in Q$. Since θ is integral, there exist $q' \in Q$ and $p', p'_1 \in P$ such that $q = \theta(p') + q'$, $q_1 = \theta(p'_1) + q'$, and $f + p' = p_1 + p'_1$. Since $q \notin K_\theta$, necessarily $p' \in P^*$, and it follows that $p_1 \in F$, so $\lambda(p_1) \in (P_F)^*$, as desired. Let $\nu_{\mathfrak{p}} \colon P \to \mathbf{N}$ be the valuation of P associated with the facet F, and choose $p \in P$ such that $\nu_{\mathfrak{p}}(p) = 1$, i.e., so that p generates the maximal ideal $(P_F)^+$ of P_F. Then in particular $\lambda_F(q)$ does not belong to the ideal of Q_F generated by $\theta_F(p)$, i.e., $\lambda_F(q) - \theta_F(p) \notin Q_F$. Since Q_F is saturated, Corollary 2.4.5 tells us that there is some height one prime $\mathfrak{q}$ of Q such that $\nu_{\mathfrak{q}}(\lambda_F(q)) < \nu_{\mathfrak{q}}(\theta_F(p))$. Then $\theta^{-1}(\mathfrak{q}) = \mathfrak{p}$, and assumption (3) implies that $\nu_{\mathfrak{q}}(p) = \nu_{\mathfrak{p}}(p) = 1$, and hence that $\nu_{\mathfrak{q}}(q) = 0$. But then also $\nu_{\mathfrak{p}}(p_0) = \nu_{\mathfrak{q}}(p_0) = 0$, as required.

(5) implies (6): Let G be a maximal θ-critical face of Q and consider the

homomorphism $\overline{\theta}_G\colon P \to Q/G$. Since P is saturated and θ is locally exact, Lemma 4.7.9 implies that this homomorphism is exact and **Q**-surjective. Assumption (5) implies that $K_{\overline{\theta}_G} + Q/G$ is a radical ideal. Then Proposition 4.3.13 implies that $\overline{\theta}_G$ is strict, hence an isomorphism, since P is sharp.

(6) implies (7). Since θ is locally exact, Theorem 4.7.2 implies that the mapping on cones induced by σ is bijective. Thus for each $q \in Q$ there exist $n > 0$, $p \in P$, and $g \in Q \setminus \sqrt{K_\theta}$ such that $nq = g + p$. Then $\langle g \rangle$ is θ-critical and hence contained in a maximal θ-critical face G. Let $\pi\colon Q \to Q/G$ be the natural projection. Since the cokernel of $P^{\rm gp} \to Q/G^{\rm gp}$ is torsion free, $\pi(q) \in P^{\rm gp}$, and since θ is locally exact and $G \cap P = \{0\}$, in fact $\pi(q)$ lies in the image of P. Thus there exist $p' \in P$ and $g_1, g_2 \in G$ such that $p' + g_2 = q + g_1$. Then $np' + ng_2 = nq + ng_1 = g + p + ng_1$. Hence $np' - p = ng_1 + g - ng_2$, and since $P^{\rm gp} \cap G^{\rm gp} = \{0\}$, $p = np'$ and $g = ng_2 - ng_1$. Since G is saturated, $g' := g_2 - g_1 \in G$, and we can write $nq = ng' + np'$. Since Q is sharp and saturated, it follows that $q = g' + p'$. This proves that the map $\sigma\colon P \times (Q \setminus K_\theta) \to Q$ is surjective. The injectivity of σ follows from the case of cones, since $P^{\rm gp}$ and $Q^{\rm gp}$ are torsion free.

(4) implies (6): This implication is obvious.

(7) implies (1): Let us check first that θ is integral. Suppose that $q_i \in Q$ and $p_i \in P$ with $\theta(p_1) + q_1 = \theta(p_2) + q_2$. Write $q_i = \theta(p'_i) + g_i$, with $p'_i \in P$ and $g_i \in Q \setminus \sqrt{K_\theta}$. Then $\theta(p_1 + p'_1) + g_1 = \theta(p_2 + p'_2) + g_2$, and by the injectivity of σ, $p' := p_1 + p'_1 = p_2 + p'_2$ and $q' := g_1 = g_2$. Then $q_i = \theta(p'_i) + q'$, with $p_1 + p'_1 = p_2 + p'_2$, as required. Next we verify that for every positive integer n, the condition QS_n is satisfied. Suppose $p \in P$ and $q \in Q$ with $\theta(p) \le nq$, say $nq = \theta(p) + q_1$ Write $q = \theta(p') + g$ and $q_1 = \theta(p'_1) + g_1$ with $p', p'_1 \in P$ and $g, g_1 \in Q \setminus \sqrt{K_\theta}$. Then $\theta(np') + ng = nq = \theta(p + p_1) + g_1$. It follows that $np' = p + p_1$, so indeed $\theta(p') \le q$ and $p \le np'$. □

Example 4.8.15. Let $Q := \mathbf{N} \oplus \mathbf{N}$ and let P be the submonoid generated by $(2, 0), (1, 1), (0, 2)$. Then the inclusion $P \to Q$ satisfies (3) with "**Q**-integral" in place of "integral," but it does not satisfy (5).

Corollary 4.8.16. *A locally exact homomorphism of fine saturated monoids which is n-quasi-saturated for some $n > 1$ is in fact saturated.*

Proof Let $\theta\colon P \to Q$ be such a homomorphism, let $I_\theta := \theta^{-1}(Q^+)$ and let $K_\theta := I_\theta + Q$. Since Q^+ is a radical ideal, the same is true of I_θ. Suppose that $q \in Q$ and $nq \in K_\theta$. Then there is some $p \in I_\theta$ such that $\theta(p) \le nq$, and by condition QS_n, there is some $p' \in P$ such that $\theta(p') \le q$ and $p \le np'$. Then $p' \in I_\theta$, and consequently $q \in K_\theta$. It follows by induction on k that, for every positive integer k, $n^k q \in K_\theta$ implies that $q \in K_\theta$. Now if m is any positive integer, we can choose k so that $n^k \geq m$. and then $mq \in K_\theta$ implies that $n^k q \in K_\theta$ and hence $q \in K_\theta$. Thus K_θ is a radical ideal. Since θ is locally exact, Theorem 4.8.14 implies that θ is saturated. □

The following result gives a characterization of saturated homomorphisms in terms of the associated map of toric varieties.

Corollary 4.8.17. *Let $\theta\colon P \to Q$ be an integral and local homomorphism of fine saturated monoids. Assume that P is sharp. Then the following conditions are equivalent.*

1. *θ is saturated.*
2. *For every field k in which the order of the torsion part of Q^{gp} is invertible, $k[Q, K_\theta]$ is reduced.*
3. *There exists a field such that $k[Q, K_\theta]$ satisfies R_0.*

Proof If θ is integral then by Theorem 4.8.14, it is saturated if and only if K_θ is a radical ideal. If k is a field and the order of the torsion part of Q^{gp} is invertible in k, Proposition 3.5.1 tells us that this is the case if and only if the ring $k[Q, K_\theta]$ is reduced. Thus (1) and (2) are equivalent. If (2) holds, there is a field such that $k[Q, K_\theta]$ is reduced. Then the local ring at each of its generic points is a field, hence regular, so $k[Q, K_\theta]$ satisfies condition R_0. Thus (2) implies (3). Suppose that (3) holds. Since $P \to Q$ is integral and local and P is sharp, it is injective, and hence by Remark 4.6.6 the homomorphism $k[P] \to [Q]$ is flat. By Theorem 3.4.3, $k[Q]$ and $k[P]$ are Cohen–Macaulay, and it follows from [27, 6.3.2] that the ring $k[Q, K_\theta]$ of the fiber over the vertex of $\underline{\mathsf{A}}_{\mathsf{P}}$ is also Cohen–Macaulay. In particular it satisfies condition S_1, and so injects into the product of the local rings at its generic points. The condition R_0 says that these rings are regular, hence reduced, and it follows that $k[Q, K_\theta]$ is also reduced. Then by Proposition 3.5.1 it follows that K_θ is a radical ideal and hence θ is saturated. □

4.9 Saturation of monoid homomorphisms

The previous section showed that saturated homomorphisms of monoids are relatively easy to understand. The goal of this section is to show how one can often reduce to this case. We shall see that, given a locally exact homomorphism of fine saturated monoids $\theta\colon P \to Q$, the homomorphism obtained by base change along n_P in the category of saturated monoids is saturated, for suitable n. In geometric terms, this process can be thought of as adjoining nth roots of the parameters in the base and then normalizing. As Theorem 4.8.14 suggests, the key is to control, for each face G of Q, the torsion subgroup

Γ_G of the cokernel of the homomorphism θ_G^{gp} associated to the monoid homomorphism:

$$\theta_G \colon P \xrightarrow{\theta} Q \xrightarrow{\pi_G} Q/G.$$

Theorem 4.9.1. *Let $\theta\colon P \to Q$ be a locally exact homomorphism of fine saturated monoids. Let n be a positive integer which is divisible by the order of Γ_{θ_G} for every θ-critical face G of Q, and form the diagram*

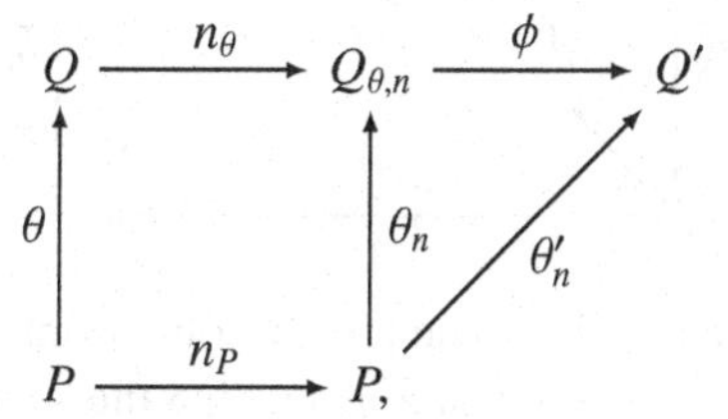

where $Q' := Q^{\mathrm{sat}}_{\theta,n}$. Then θ'_n is saturated and, in particular, integral.

Proof Our first step is to reduce to the case in which θ is local. Let

$$P \xrightarrow{\lambda} P^{loc} \xrightarrow{\theta^{loc}} Q$$

be the factorization of θ through the localization of P by $\theta^{-1}(Q^*)$ and consider the diagram:

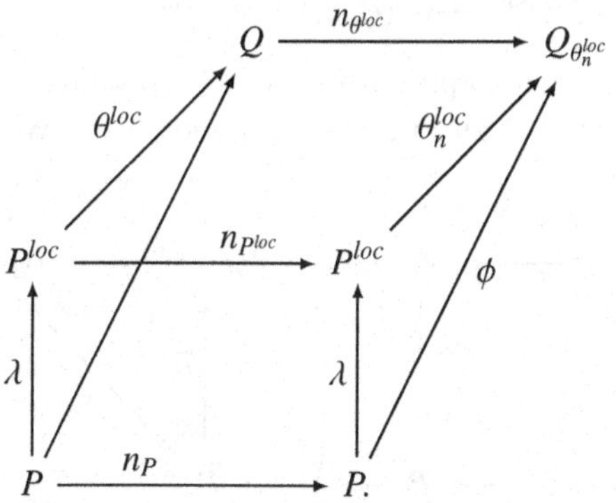

We saw in the proof of Proposition 4.8.8 that the bottom square is cocartesian, and the top rectangle is cocartesian by construction. It follows that the back rectangle is also cocartesian and hence the map ϕ can be identified with θ_n. Proposition 4.8.8 shows that λ is saturated, and hence θ'_n is saturated if and only if $\theta_n^{loc\,\prime}$ is saturated. Thus we may and shall assume that θ is local.

Next we reduce to the case in which P and Q are sharp. Form the diagram

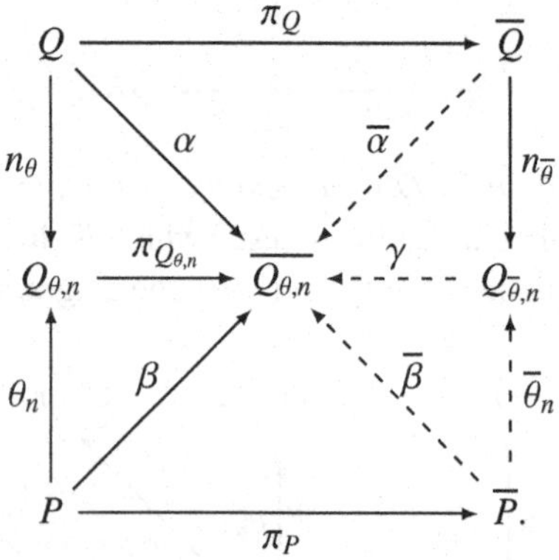

Here the solid horizontal and vertical arrows come from the constructions, and the arrows α and β are the compositions making the solid triangles commute. Then α and β factor as shown, through $\overline{Q}$ and $\overline{P}$ respectively. One checks easily that $\overline{\alpha} \circ \overline{\theta} = \overline{\beta} \circ n_{\overline{P}}$ so there is a unique homomorphism γ making the triangles on the right commute. There is also a unique homomorphism $\delta\colon Q_{\theta,n} \to Q_{\overline{\theta},n}$ such that $\delta \circ n_\theta = n_{\overline{\theta}} \circ \pi_Q$ and $\delta \circ \theta_n = \overline{\theta}_n \circ \pi_P$. The homomorphism $n_{\overline{P}}$ is exact because P is saturated, and $\overline{\theta}$ is local by assumption, so it follows from Proposition 4.2.5 that the coproduct $Q_{\overline{\theta},n}$ is also sharp. Thus δ factors through a homomorphism $\overline{\delta}\colon \overline{Q_{\theta,n}} \to Q_{\overline{\theta},n}$, which is easily seen to be inverse to γ. Then we find homomorphisms

$$\overline{Q^{\mathrm{sat}}_{\theta,n}} \longrightarrow \overline{(Q_{\theta,n})^{\mathrm{sat}}} \xleftarrow{\gamma^{\mathrm{sat}}} \overline{(Q_{\overline{\theta},n})^{\mathrm{sat}}}.$$

The left arrow is an isomorphism by (5) of Proposition 1.3.5, and the right arrow is an isomorphism because γ is an isomorphism. We now have a commutative diagram:

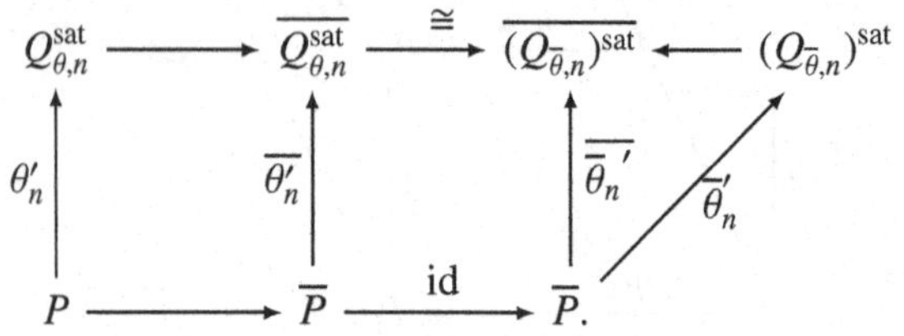

Since a homomorphism ϕ is saturated if and only if $\overline{\phi}$ is saturated, we see that θ'_n will be saturated if $\overline{\theta}_n{}'$ is saturated.

The key idea of the proof is to relate the torsion groups Γ_θ to the splitting of the homomorphism θ^{gp}.

Lemma 4.9.2. *Let $\theta\colon P \to Q$ be an injective homomorphism of fine monoids.*

1. *If the torsion subgroup Γ_θ of Q^{gp}/P^{gp} vanishes, then θ^{gp} is split.*
2. *If θ^{gp} is split, then Γ_θ is contained in the image of $(Q^{sat})^*$.*
3. *If θ^{gp} is split and Q^* is torsion, then the cokernel of the homomorphism $P^{gp} \to \overline{Q^{sat}}^{gp}$ is torsion free.*

Proof Since Q^{gp}/P^{gp} is a finitely generated abelian group, it is free if it is torsion free. In this case the exact sequence

$$0 \to P^{gp} \to Q^{gp} \to Q^{gp}/P^{gp} \to 0$$

splits. This proves (1).

Suppose conversely that θ^{gp} is split. Then every torsion element of Q^{gp}/P^{gp} lifts to a torsion element of Q^{gp}. Such elements are contained in $(Q^{sat})^*$, proving (2).

Now suppose that θ^{gp} is split and that Q^* is a torsion group. If $u \in (Q^{sat})^*$, there is some positive n such that both nu and $n(-u)$ belong to Q. Then in fact $nu \in Q^*$ and hence has finite order. Thus $(Q^{sat})^*$ is also a torsion group. Let P' be the image of P in $\overline{Q^{sat}}$, and consider the diagram

$$\begin{array}{ccccccc}
 & & (Q^{sat})^* & \xrightarrow{a} & K & & \\
 & & \downarrow & & \downarrow & & \\
P^{gp} & \xrightarrow{\theta^{gp}} & Q^{gp} & \longrightarrow & Q^{gp}/P^{gp} & \longrightarrow & 0 \\
\downarrow & & \downarrow & & \downarrow{\scriptstyle c} & & \\
P'^{gp} & \xrightarrow{\theta'^{gp}} & (\overline{Q^{sat}})^{gp} & \longrightarrow & \overline{(Q^{sat})}^{gp}/P'^{gp} & \longrightarrow & 0.
\end{array}$$

Here K is by definition the kernel of c. Thus the columns are exact and the bottom vertical arrows are surjective. It follows from the snake lemma that a is surjective and, since $(Q^{sat})^*$ is a torsion group, so is K. Thus the torsion subgroup Γ_θ of Q^{gp}/P^{gp} maps surjectively to the torsion subgroup Γ of $\overline{(Q^{sat})}^{gp}/P'^{gp}$. If θ^{gp} is split then, by (2), $(Q^{sat})^*$ maps surjectively to Γ_θ. Since $(Q^{sat})^*$ maps to zero in $\overline{(Q^{sat})}^{gp}/P'^{gp}$, it follows that $\Gamma = 0$. □

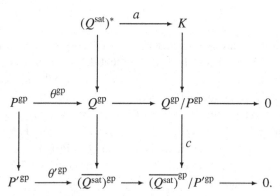

Now we are return to the proof of Theorem 4.9.1, under the assumption that θ is local and that P and Q are sharp. Since θ is locally exact, the same is true for θ_n, by (6) of Proposition 4.2.1. Since $Q_{\theta,n} \to Q'$ induces an isomorphism of cones and P is saturated, the homomorphism θ' is also locally exact. Thus, according to Theorem 4.8.14, it is enough to prove that for every θ'-critical face G' of Q' the torsion subgroup $\Gamma_{\theta',G'}$ of $(Q'/G')^{\text{gp}}/P^{\text{gp}}$ vanishes. Let G' be such a face and let and $G_\theta := G' \cap Q_{\theta,n}$, let $G := n_\theta^{-1}(G_\theta)$, and let $G_n := n_\theta(G)$. Then $G_n \subseteq G_\theta$ and, since θ_n is small, statement (4) of Proposition 4.3.9 implies that G_n generates G_θ as a face of $Q_{\theta,n}$. In fact we shall see that G_n is already a face and hence equal to G_θ. Notice that n_θ is exact because it is the pushout of the exact homomorphism n_P and, since G is an exact submonoid of Q, it follows that its image G_n is an exact submonoid of $Q_{\theta,n}$.

Since G' is θ'-critical, G is θ-critical. Consider the following diagram:

$$
\begin{array}{ccccc}
Q/G & \xrightarrow{\ \psi\ } & Q_{\theta,n}/G_n & \longrightarrow & Q'/G' \\
\uparrow{\scriptstyle \pi} & & \uparrow{\scriptstyle \pi_n} & \nearrow{\scriptstyle \pi'_n} & \\
Q & \xrightarrow{\ n_\theta\ } & Q_{\theta,n} & & \\
\uparrow{\scriptstyle \theta} & & \uparrow{\scriptstyle \theta_n} & & \\
P & \xrightarrow{\ n_P\ } & P. & &
\end{array}
\tag{4.9.1}
$$

Here the bottom square is cocartesian in the category of integral monoids by construction, and one checks immediately that the top square is also cocartesian in this category. It follows that the exterior rectangle is cocartesian and hence $\operatorname{Cok}(\pi^{\text{gp}}\theta^{\text{gp}})$ identifies with $\operatorname{Cok}(\pi_n^{\text{gp}}\theta_n^{\text{gp}})$. The homomorphism n_P is necessarily local, and the homomorphism $\pi\theta$ is exact because θ is locally exact and G is θ-critical. Since P and Q/G are sharp, it follows from Proposition 4.2.5 that $Q_{\theta,n}/G_n$ is also sharp. Since G_n is exact in $Q_{\theta,n}$, Proposition 1.4.2 implies that G_n is a face of $Q_{\theta,n}$. Note that, by (5) of Proposition 1.3.5, we have the following isomorphisms:

$$Q'/G' = \overline{Q'_{G'}} = \overline{(Q_{\theta,n\,G_n})^{\text{sat}}} \cong \overline{(Q_{\theta,n}/G_n)^{\text{sat}}}.$$

The exact sequence

$$0 \to P^{\text{gp}} \to (Q/G)^{\text{gp}} \to Q^{\text{gp}}/(G^{\text{gp}} + P^{\text{gp}}) \to 0$$

defines an element ξ of $\operatorname{Ext}^1(Q^{\text{gp}}/(G^{\text{gp}} + P^{\text{gp}}), P^{\text{gp}})$. Because the rectangle of diagram 4.9.1 is cocartesian, the exact sequence

$$0 \to P^{\text{gp}} \to (Q_{\theta,n}/G_n)^{\text{gp}} \to Q_{\theta,n}^{\text{gp}}/(G_n^{\text{gp}} + P^{\text{gp}}) \to 0$$

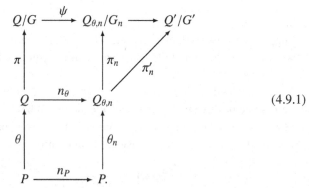

corresponding to the right vertical arrow is obtained by pushout along n_P of the left vertical arrow. Thus the corresponding extension class ξ_n is $n\xi$. Since the order of the torsion subgroup of $(Q/G)^{\text{gp}}/P^{\text{gp}}$ divides n, this extension class $\xi_n = n\xi$ vanishes. Thus the homomorphism : $P^{\text{gp}} \to (Q_{\theta,n}/G_n)^{\text{gp}}$ splits. Since $(Q_{\theta,n}/G_n)^{\text{gp}}$ is sharp, Lemma 4.9.2 implies that the cokernel of the homomorphism

$$P^{\text{gp}} \to \overline{(Q_{\theta,n}/G_n)^{\text{sat}}}^{\text{gp}} = (Q'/G')^{\text{gp}}$$

is torsion free. This completes the proof. □

4.10 Homomorphisms of idealized monoids

In this section we discuss some analogs of exactness, integrality, and saturation in the context of idealized monoids. The material in this section is somewhat provisional.

Recall that if (Q, J) is an idealized monoid, the prime ideals of (Q, J) are by definition the prime ideals of Q that contain J, and the faces of (Q, J) are the faces of Q that do not meet J. Recall also that if R is a ring, the monoid algebra $R[Q, J]$ is the quotient of $R[Q]$ by the ideal $R[J]$ of $R[Q]$, and that the set $Q \setminus J$ is a basis for $R[Q, J]$. Note that if $a + b$ belongs to $Q \setminus J$, then a and b belong to $Q \setminus J$.

Definition 4.10.1. *A homomorphism $\theta\colon (P, I) \to (Q, J)$ of idealized monoids is:*

1. *local if it takes the closed point of* Spec(Q, J) *(if there is one) to the closed point of* Spec(P, I);
2. *exact (resp. integral, saturated) if $\theta^{-1}(J) = I$ and for every face P' of (P, I); the restriction θ' of θ to P' is exact (resp. integral, saturated),*
3. *strict if the induced homomorphisms $\overline{P} \to \overline{Q}$ and $\overline{I} \to \overline{J}$ are isomorphisms.*

Care is required with these notions, which do not always behave consistently with their non-idealized analogs. For example, if F is a face of a monoid P, the localization homomorphism $\lambda_F\colon P \to P_F$ is integral, but the analogous result does not hold for idealized monoids. For example, let $P := \mathbf{N} \oplus \mathbf{N}$ and let $I := \mathbf{N}^+ \oplus \mathbf{N}^+$ be its interior ideal. Then $F := (0, \mathbf{N})$ is a face of (P, I), and $(P_F, I_F) = (\mathbf{N} \oplus \mathbf{Z}, \mathbf{N}^+ \oplus \mathbf{Z})$, so $I \neq \lambda_F^{-1}(I_F) = \mathbf{N}^+ \oplus \mathbf{N}$. On the other hand, with the definitions we have made here, an exact homomorphism is local (as we shall see in the next proposition), and a local integral homomorphism is exact (as follows immediately from the definitions and Proposition 4.6.3).

Proposition 4.10.2. *Let $\theta\colon (P, I) \to (Q, J)$ be a homomorphism of idealized monoids. Consider the following conditions.*

1. *$\theta\colon (P, I) \to (Q, J)$ is exact.*
2. Spec θ: Spec$(Q, J) \to$ Spec(P, I) *is surjective.*

Either condition implies that θ is local. Furthermore, (1) implies (2), and the converse holds if P and Q are fine and saturated and J and K are radical ideals.

Proof To prove that θ is local, we may assume that J is a proper ideal of Q, since otherwise the statement is vacuous. Suppose that (1) is satisfied and that $p \in \theta^{-1}(Q^*)$. Then $\theta(np) \in Q^*$ for every $n \in \mathbf{N}$ and, since θ maps I to the proper ideal J, $np \notin I$ for every n. It then follows from Corollary 1.4.3 that there is a prime ideal $\mathfrak{p}$ of P containing I such that $p \notin \mathfrak{p}$. Then $P' := P \setminus \mathfrak{p}$ is a face P' of (P, I) containing p. By hypothesis, $\theta' \colon P' \to Q$ is exact, hence by Proposition 4.2.1 it is local. It follows that p is a unit of P' and hence also of P. Thus θ is local. Next suppose that θ satisfies condition (2). Then there is some prime $\mathfrak{q}$ of $\operatorname{Spec}(Q, J)$ such that $\theta^{-1}(\mathfrak{q}) = P^+$, and it follows that $\theta^{-1}(Q^+) = P^+$ also, so again θ is local.

Next we prove that (1) implies (2). Suppose that (1) holds and that F is a face of (P, I). Let us check that that $\theta_F \colon (P_F, I_F) \to (Q_F, J_F)$ is still exact. If $p \in P_F$ then there exists some $f \in F$ such that $p + f \in P$ and, if $\theta_F(p) \in J_F$, there exists some $f' \in F$ such that $\theta(p + f) + \theta(f') \in J$. Then $\theta(p + f + f') \in J$ and, since θ is exact, it follows that $p + f + f' \in I$ and hence that $p \in I_F$. Moreover, if P' is a face of (P, I), then $P'_F \to Q_F$ is exact, by Proposition 4.2.1. This proves that θ_F is exact and hence local, as we have already seen. Furthermore, since θ is an exact homomorphism of idealized monoids, $I = \theta^{-1}(J)$ and $J \cap \theta(F) = \emptyset$, since $I \cap F = \emptyset$. It follows that J_F is a proper ideal of Q_F and hence that $(Q_F)^*$ is a face of (Q_F, J_F). Its inverse image G in Q is a face (Q, J), and $\theta^{-1}(G) = F$ because θ_F is local. This proves the surjectivity of $\operatorname{Spec}\theta$. Conversely, suppose that (2) holds and that P and Q are fine and saturated. Let F be a face of (P, I), and choose a face G of (Q, J) lying over F. Every face F' of F is a face of (P, I), so there is a face G' of (Q, J) lying over F'. Then $G \cap G'$ is a face of G lying over F'. This shows that $\operatorname{Spec} G \to \operatorname{Spec} F$ is surjective. Since F and G are fine and saturated, it follows from Proposition 4.2.2 that $F \to G$ is exact. Since G is a face of Q, it is exact in Q, by (3) of Theorem 2.1.17, and it follows that $F \to Q$ is also exact. Now, since I is a radical ideal, it is the intersection of all the primes $\mathfrak{p}$ of $\operatorname{Spec}(P, I)$ and, since each such prime comes from a prime of Q containing J, in fact $I = \theta^{-1}(J)$. □

Proposition 4.10.3. *Let $\theta \colon P \to Q$ be an exact morphism of integral monoids, let I be an ideal of P, and let J be the ideal of Q generated by the image of I. Then $\theta \colon (P, I) \to (Q, J)$ is exact.*

Proof Suppose that $p \in \theta^{-1}(J)$. Then there exists an element q of Q and an element p' of J such that $\theta(p) = q + \theta(p')$. Thus $\theta^{gp}(p - p') \in Q$ and, since θ is exact, $p - p' \in P$. Since $p' \in J$, this implies that $p \in J$. □

Definition 4.10.4. *A homomorphism of idealized monoids* $\theta\colon (P, I) \to (Q, J)$ *is:*

1. *locally exact if for every face* G *of* (Q, J)*, the homomorphism of idealized monoids*
$$(P_{\theta^{-1}(G)}, I_{\theta^{-1}(G)}) \to (Q_G, J_G)$$
is exact,
2. *critically exact if for every face* G *of* (Q, J) *such that* $\theta^{-1}(G) = P^*$*, the homomorphism of idealized monoids*
$$(P, I) \to (Q_G, J_G)$$
is exact.

Lemma 4.10.5. *Let* $\theta\colon (P, I) \to (Q, J)$ *be a local and critically exact homomorphism of fine saturated idealized monoids.*

1. *For every face* G *of* $\operatorname{Spec}(Q, J)$*, the induced homomorphism* $\theta^{-1}(G) \to G$ *is locally exact.*
2. *The homomorphism* $\theta\colon (P, I) \to (Q, J)$ *is locally exact and* **Q***-integral.*

Proof There is nothing to check if $J = Q$, so we assume that J is a proper ideal. To prove (2), let G be a face of (Q, J), let $F := \theta^{-1}(G)$, and let $\phi\colon F \to G$ be the induced homomorphism. Since $F \to P$ and $P \to Q$ are local, so is ϕ. Recall that a face G' of G is called ϕ-critical if $\phi^{-1}(G') = F^*$. Such a G' is also a face of Q; let us note that it is also automatically θ-critical, because $F = \theta^{-1}(G)$. Furthermore, since $G \cap J = \emptyset$ and $G' \subseteq G$, necessarily $G' \cap J = \emptyset$, so G' is a face of (Q, J). The assumption that θ is critically exact says that $(P, I) \to (Q_{G'}, J_{G'})$ is exact, and since F is a face of (P, I), this says in particular that the composite $F \to Q_{G'}$ is exact. Then $F \to G_{G'}$ is also exact. Thus $\phi\colon F \to G$ is is critically exact, and hence by Theorem 4.7.7 it is locally exact. This proves (1).

Statement (2) is more subtle. Since θ is local, Q^* is a θ-critical face of (Q, J) and, since θ is critically exact, it is exact. In particular, it follows that $I = \theta^{-1}(J)$. Now, to prove that θ is locally exact, let G be a face of (Q, J) and let $F := \theta^{-1}(G)$. We claim that the induced homomorphism of idealized monoids
$$\theta_F\colon (P_F, I_F) \to (Q_G, J_G)$$
is exact. First we show that its restriction to any face of (P_F, I_F) is exact. Such a face is necessarily the localization of a face P' of (P, I) containing F. Suppose that $x \in P'^{\mathrm{gp}}$ and $\theta^{\mathrm{gp}}(x) \in Q_G$. Then there exists some $g \in G$ with $g + \theta(x) \in Q$. According to statement (1), the homomorphism $\phi\colon F \to G$ is locally exact, and hence so is the induced homomorphism $\overline{\phi}\colon \overline{F} \to \overline{G}$. Then by Theorem 4.7.2,

there exist $f \in C_{\mathbf{Q}}(F)$ and $g'\ C_{\mathbf{Q}}(G)$ such that $G' := \langle g' \rangle$ is a ϕ-critical face of G and $\overline{g} = \overline{g}' + \theta(\overline{f}) \in C_{\mathbf{Q}}(\overline{G})$. Choose $n \in \mathbf{Z}^+$ such that $n\overline{g}' \in \overline{G}$ and $n\overline{f} \in \overline{F}$. Thus $ng + u = n\theta(f) + ng'$ for some $u \in G^*$. Then $(ng + u) + \theta^{\mathrm{gp}}(x) \in Q$, and so, replacing f by nf and g' by ng', we find that $\theta^{\mathrm{gp}}(f + x) + g' \in Q$. Then $\theta^{\mathrm{gp}}(f + x) \in Q_{G'}$. Since G' is ϕ critical and $F = \theta^{-1}(G)$, G' is also θ-critical, and hence by assumption the homomorphism $P' \to Q_{G'}$ is exact. We conclude that $f + x \in P'$ and hence that $x \in P'_F$, as required. Finally, we claim that the inverse image of J_G in P_F is I_F. Suppose that $p \in P, f \in F, k \in J$, and $g \in G$, with $\theta(p - f) = k - g$. Arguing as before, we can use the local exactness of $\phi \colon F \to G$ to show that there exists such a g that can be written as $g = g' + \theta(f')$, where $\langle g' \rangle$ is θ-critical. Then by assumption $(P, I) \to (Q_{G'}, J_{G'})$ is exact, and so $\theta^{-1}(J_{G'}) = I$. Since $\theta(p + f') = \theta(f) + k - g' \in J_{G'}$, it follows that $p+f' \in I$ and then that $p-f \in I_F$. Thus $\theta \colon (P, I) \to (Q, J)$ is locally exact. In particular, for each face F of (P, I), the induced homomorphism $F \to Q$ is locally exact and hence $\mathbf{Q}$-integral by Theorem 4.7.7. □

We now turn to an idealized version of the key structure theorem 4.7.2. If (Q, J) is an idealized monoid and K is an Archimedean field, let

$$C_K(Q, J) := \cup\{C_K(G) : G \text{ is a face of } (Q, J)\}.$$

If J is a radical ideal of Q, then the ideal $C_K(J)$ of $C_K(Q)$ it generates is also radical, hence is the intersection of set of prime ideals which contain it, and hence in this case $C_K(Q, J) = C_K(Q) \setminus C_K(J)$. If $\theta \colon (P, I) \to (Q, J)$ is a local homomorphism of idealized monoids, let J_θ be the ideal generated by J and K_θ, where $K_\theta := (P^+ + Q)$. Then

$$C_K(Q, J_\theta) = \cup\{C_K(G) : G \text{ is a face of } (Q, J) \text{ with } \theta^{-1}(G) = P^* \}.$$

Theorem 4.10.6. *Let $\theta \colon (P, I) \to (Q, J)$ be a local homomorphism of toric idealized monoids, where P is sharp.*

1. *If θ is locally exact, the addition map induces a homeomorphism*

$$\sigma \colon C_K(P, I) \times C_K(Q, J_\theta) \to C_K(Q, J).$$

   For each pair (F, G), where F is a face of (P, I), and G is a θ-critical face of (Q, J), the map σ induces an isomorphism of K-cones

$$C_K(F) \oplus C_K(G) \to C_K(F) + C_K(G).$$

2. *If θ is integral and I and J are radical ideals, the addition map σ induces a bijection*

$$\sigma \colon (P \setminus I) \times (Q \setminus J_\theta) \to Q \setminus J.$$

For each pair (F, G) as in (2), σ induces an isomorphism of monoids

$$F \oplus G \to F + G.$$

Proof We begin by checking the existence of the map σ, which is not obvious. Supposing that $q \in C_K(Q, J_\theta)$ and $p \in C_K(P, J)$, we must check that $\theta(p) + q \in C_K(Q, J)$. Let $G := \langle q \rangle$ and $F := \langle p \rangle$. Then $G \cap J = F \cap I = \emptyset$ and $\theta^{-1}(G) = P^* = \{0\}$. By hypothesis, $\theta_G : (P, I) \to (Q_G, J_G)$ is exact, so by Proposition 4.10.2, $\mathrm{Spec}(\theta_G)$ is surjective. Hence there exists a face G' of (Q, J) containing G such that $\theta^{-1}(G') = F$. Then G' contains $\theta(p) + q$, and hence $\theta(p) + q \in C_K(Q, J)$, as required.

To see that σ is injective, suppose that $(p_i, q_i) \in C_K(P, I) \times C_K(Q, J_\theta)$ for $i = 1, 2$, and that $q := q_1 + \theta(p_1) = q_2 + \theta(p_2)$. Let G' be the face of C_Q generated by q; note that $\theta(p_i)$ and q_i belong to G'. We have seen that G' is a face of (Q, J), so $F' := \theta^{-1}(G')$ is a face of (P, I) and $\theta' : F' \to G'$ is locally exact, by (1) of Lemma 4.10.5. Since each $\langle q_i \rangle$ is θ-critical and $F' = \theta^{-1}(G')$, each is also θ'-critical. Then it follows from Theorem 4.7.2 that $q_1 = q_2$ and $p_1 = p_2$.

To see that σ is surjective, suppose that $q \in C_K(Q, J)$. Then there is a face G of (Q, J) containing q, and $F := \theta^{-1}(G)$ is a face of (P, I). Again by Lemma 4.10.5, $F \to G$ is locally exact, and it follows from Theorem 4.7.2 that there exist an element q' of G with $\theta^{-1}(\langle q' \rangle) = \{0\}$ and an element p of F such that $\theta(p) + q' = q$. Then $(p, q') \in C_K(P, I) \times C_K(Q, P)$ and $\sigma(p, q') = q$, as required. The proof that σ is a homeomorphism is done in the same way as in the non-idealized case, in Corollary 4.7.14.

Finally, suppose that G is a θ-critical face of (Q, J). It is clear that σ takes $C_K(F) \times C_K(G)$ bijectively to $C_K(F) + C_K(G)$, and that the induced map is a morphism of K-cones.

Now suppose that θ is integral. Then it is locally exact, and the argument above for cones applies to show that the summation map takes $(P \setminus I) \times (Q \setminus J_\theta)$ to $Q \setminus J$. If $q \in Q \setminus J$, let $G' := \langle q \rangle$; since J is by assumption a radical ideal, $G' \cap J = \emptyset$, i.e., G' is a face of (Q, J). Then $F' := \theta^{-1}(G')$ is a face of (P, I), and by assumption the homomorphism $\theta' : F' \to G'$ induced by θ is integral. Let K'_θ be the ideal of G' generated by F'^{+}. Using the fact that G' is a face of Q that does not meet J, one can easily check that $K'_\theta = J_\theta \cap G'$. Since θ' is integral and local and F' is sharp, Corollary 4.6.11 implies that the summation map induces a bijection $\sigma' : F' \times (G' \setminus K'_\theta) \to G'$. It follows that there exists a unique $(p_1, q_1) \in F' \times (G' \setminus K'_\theta) \subseteq (P \setminus I) \times (Q \setminus J_\theta)$ such that $q = \theta(p_1) + q_1$. Moreover, if $(p_2, q_2) \in (P \setminus I) \times (Q \setminus J_\theta)$ and $\theta(p_2) + q_2 = q$, then $\theta(p_2)$ and q_2 necessarily belong to G', hence $p_2 \in F'$ and $q_2 \in G' \setminus K'_\theta$, hence $(p_2, q_2) = (p_1, q_1)$. Furthermore, if F is a face of (P, I) and G is a θ-critical face of (Q, J), then the uniqueness shows that the map $F \oplus G \to F + G$ is an isomorphism. □

Example 4.10.7. The homomorphism

$$\theta\colon (P, I) := (\mathbf{N}, \emptyset) \to (Q, J) := (\mathbf{N} \oplus \mathbf{N}, \mathbf{N}^+ \oplus \mathbf{N}^+) : n \mapsto (n, 0)$$

is exact, local, satisfies condition (1) of Lemma 4.10.5 and $I = \theta^{-1}(J)$. However it is not critical exact: the last condition no longer holds after localization by the θ-critical face $(0, \mathbf{N})$. The ideals I and J are radical ideals, but in this example the addition map does not take $C_K(P, I) \times C_K(Q, J_\theta)$ to $C_K(Q, J)$. In fact, $1 \in P \setminus I$ and $(0, 1) \in Q \setminus J_\theta$, but $\theta(1) + (0, 1) = (1, 1) \notin (Q \setminus J)$.

As another application of the Theorem 4.7.2, we prove that certain slices of idealized cones are contractible.

Proposition 4.10.8. *Let Q be a fine monoid, let $h\colon Q \to \mathbf{N}$ be a homomorphism, and let $C_{\mathbf{R}}(h)\colon C_{\mathbf{R}}(Q) \to C_{\mathbf{R}}(\mathbf{N})$ be the corresponding homomorphism of real cones. If K is a proper ideal of Q, let*

$$C_{\mathbf{R}}(Q, K, h) := C_{\mathbf{R}}(Q, K) \cap \left(C_{\mathbf{R}}(h)^{-1}(1)\right).$$

If K is generated by $K \cap h^{-1}(0)$, then $C_{\mathbf{R}}(Q, K, h)$ is contractible

Proof Choose a sequence of generators $(k_1, \ldots, k_n)$ for K with each $h(k_i) = 0$; the proof will be by induction on n. If $n = 0$, K is empty, and $C_{\mathbf{R}}(Q, K, h) = C_{\mathbf{R}}(Q, h)$ is contractible because it is convex. For the induction step, let K' be the ideal generated by $(k_1, \ldots, k_{n-1})$ and let $k := k_n$. Consider the homomorphism $\theta\colon \mathbf{N} \to Q$ sending 1 to k. Since $k \in Q^+$, this homomorphism is local, and it is integral because $\mathbf{N}$ is valuative. Thus Theorem 4.7.2 implies that the summation map

$$\sigma\colon \mathbf{R}_{\geq} \times C_{\mathbf{R}}(Q, (k)) \to C_{\mathbf{R}}(Q)$$

is a homeomorphism. Let $(\eta, \rho)\colon C_{\mathbf{R}}(Q) \to \mathbf{R}_{\geq} \times C_{\mathbf{R}}(Q, (k))$ be its inverse. Note that if $q = q' + rk$ and $q \in C_{\mathbf{R}}(Q, K')$, then $\langle q' \rangle \subseteq \langle q \rangle$. Thus $q' \in C_{\mathbf{R}}(Q, K')$ if $q \in C_{\mathbf{R}}(Q, K')$, so ρ maps $C_{\mathbf{R}}(Q, K')$ to $C_{\mathbf{R}}(Q, K') \cap C_{\mathbf{R}}(Q, (k)) = C_{\mathbf{R}}(Q, K)$. Moreover, $h(k) = 0$, so $h(q + rk) = h(q)$ and therefore ρ induces a map

$$\rho'\colon C_{\mathbf{R}}(Q, K', h) \to C_{\mathbf{R}}(Q, K, h).$$

Moreover, ρ is the identity on $C_{\mathbf{R}}(Q, K)$, so ρ' is a retraction of $C_{\mathbf{R}}(Q, K', h)$ onto $C_{\mathbf{R}}(Q, K, h)$. For $t \in [0, 1]$ and $q \in C_{\mathbf{R}}(Q, K', h)$, let $\rho'_t(q) := \rho'(q) + t\eta(q)k$. Then $\rho'_t(q) \leq q \in C_{\mathbf{R}}(Q, K', h)$, so $\rho'_t(q) \in C_{\mathbf{R}}(Q, K', h)$. Since $\rho'_0 = \rho$ and ρ'_1 is the identity, we see that $C_{\mathbf{R}}(Q, K, h)$ is a strong deformation retract of

$C_{\mathbf{R}}(Q, K', h)$. The latter is contractible by the induction hypothesis, and hence so is the former. □

For another perspective on this result, see [54, Lemma 4.18 and Exercise 4.7].

II

Sheaves of Monoids

In order to make global geometric use of monoids, it is necessary to bundle them together into sheaves. For example, an analytic space is conveniently described as a pair $(X, \mathcal{O})$, where X is a topological space and $\mathcal{O}$ is a sheaf of rings on X (the sheaf of "holomorphic functions"). The multiplication on $\mathcal{O}$ gives rise in particular to a sheaf of monoids; in logarithmic analytic geometry we consider in addition a logarithmic homomorphism of sheaves of monoids $(\mathcal{M}, +) \to (\mathcal{O}, \cdot)$ (see Definition 1.1.4). Similarly, a logarithmic structure on a scheme X is a logarithmic homomorphism of sheaves of monoids $(\mathcal{M}_X, +) \to (\mathcal{O}_X, \cdot)$. We should note that in practice it is often more convenient to work with the étale topology on a scheme than with the Zariski topology, so strictly speaking we should be working with sheaves of monoids in a topos rather than on a topological space. To improve the accessibility of our presentation we will not insist on this point and will instead simply point out any issues as they arise. If we are dealing with the étale topology on a scheme instead of the Zariski topology, then wherever we say "point" the reader should think "geometric point," and if we say "open subset of X," think "étale map $U \to X$." A more systematic presentation in the context of topos theory can be found in [22].

1 Monoidal spaces

1.1 Generalities

Definition 1.1.1. *A monoidal space is a pair $(X, \mathcal{M}_X)$, where X is a topological space and $\mathcal{M}_X$ is a sheaf of commutative monoids on X. A morphism of monoidal spaces*

$$(f, f^\flat) \colon (X, \mathcal{M}_X) \to (Y, \mathcal{M}_Y)$$

is a pair $(f, f^{\flat})$, *where* $f\colon X \to Y$ *is a continuous function and*

$$f^{\flat}\colon f^{-1}(\mathcal{M}_Y) \to \mathcal{M}_X$$

is a homomorphism of sheaves of monoids such that for each point $x \in X$, *the stalk* $f^{\flat}_x\colon \mathcal{M}_{Y,f(x)} \to \mathcal{M}_{X,x}$ *of* $f^{\flat}$ *at* x *is a local homomorphism of monoids (Definition I.4.1.1).*

Sometimes we will write "locally monoidal space" instead of "monoidal space" for emphasis. There is an obvious functor from the category of locally ringed spaces to the category of locally monoidal spaces taking an object $(X, \mathcal{O}_X)$ to $(X, \mathcal{O}^m_X)$, where $\mathcal{O}^m_X$ is the underlying sheaf of multiplicative monoids on X. (One must note that a local homomorphism of local rings induces a local homomorphism of their underlying multiplicative monoids.)

Variant 1.1.2. An *idealized monoidal space* is a triple $(X, \mathcal{M}_X, \mathcal{K}_X)$, where X is a topological space, $\mathcal{M}_X$ is a sheaf of commutative monoids on X, and $\mathcal{K}_X$ is a sheaf of ideals in $\mathcal{M}_X$ such that $\mathcal{K}_{X,x}$ is a proper ideal of $\mathcal{M}_{X,x}$ for every $x \in X$. A *morphism of idealized monoidal spaces* $(X, \mathcal{M}_X, \mathcal{K}_X) \to (Y, \mathcal{M}_Y, \mathcal{K}_Y)$ is a morphism $(f, f^{\flat})$ of monoidal spaces such that $f^{\flat}$ maps $f^{-1}(\mathcal{K}_Y)$ to $\mathcal{K}_X$.

In general, if **P** is a property of monoids, we say that *a sheaf of monoids* $\mathcal{M}$ *on* X *has* **P** *at a point* x *of* X if $\mathcal{M}_x$ has **P**, and that X or $\mathcal{M}$ *has* **P** if $\mathcal{M}$ does so for each $x \in X$. There are unfortunately some exceptions to this convention, for example the notion of "fine sheaf of monoids" in Definition 2.1.5. Similarly, if **P** is a property of homomorphisms of monoids and $\theta\colon \mathcal{P} \to Q$ is a homomorphism of sheaves of monoids, we say that θ *has* **P** *at* x if $\theta_x\colon \mathcal{P}_x \to Q_x$ has **P**, and simply that θ *has* **P** if it does so at each $x \in X$.

Proposition 1.1.3. *Let* $(X, \mathcal{M}_X)$ *be a monoidal space, and let* **P** *be one of the following properties of monoids: integral, saturated, dull. Then* $\mathcal{M}_X$ *has* **P** *if and only if* $\mathcal{M}_X(U)$ *has* **P** *for every open subset* U *of* X.

Proof Suppose that for every open set U, $\mathcal{M}_X(U)$ has **P**. Then if x is a point of X, the stalk of $\mathcal{M}_X$ at x is the direct limit of the system of $\mathcal{M}_X(U)$ as U ranges over the neighborhoods of x in X. The fact that the stalk also has **P** then follows from Proposition I.1.3.6.

Let us check the converse when **P** is the property "integral." Suppose that each stalk is integral and that $m \in \mathcal{M}_X(U)$. Then multiplication by m defines a morphism of sheaves of monoids $\mathcal{M}_{X|_U} \to \mathcal{M}_{X|_U}$. Since the stalks of $\mathcal{M}_X$ are integral, it follows that this map is injective on the stalks and hence also on sections. This implies that $\mathcal{M}_X(U)$ is integral.

Before proving the converse for saturated monoids, let us observe that if $\mathcal{M}_X$ is a sheaf of integral monoids, then for each open set U, the natural map

$\mathcal{M}_X(U)^{\mathrm{gp}} \to \mathcal{M}_X^{\mathrm{gp}}(U)$ is injective (but not necessarily surjective). Indeed, any element of the kernel can be written as $m - m'$ with $m, m' \in \mathcal{M}_X(U)$, and then the image of $m - m'$ in $\mathcal{M}_{X,x}^{\mathrm{gp}}$ is zero for each $x \in U$. Since $\mathcal{M}_X$ is integral, this implies that $m = m'$ in $\mathcal{M}_X(U)$. Now suppose that each stalk $\mathcal{M}_{X,x}$ is saturated and that $q \in \mathcal{M}_X(U)^{\mathrm{gp}}$ with $nq \in \mathcal{M}_X(U)$ for some $n > 0$. Let q' be the image of q in $\mathcal{M}_X^{\mathrm{gp}}(U)$. For every $x \in U$, the germ of q' at x lies in $\mathcal{M}_{X,x} \subseteq \mathcal{M}_{X,x}^{\mathrm{gp}}$, and it follows that q' in fact lies in $\mathcal{M}_X(U)$. Then q' and q have the same image in $\mathcal{M}_X^{\mathrm{gp}}(U)$ and hence also in $\mathcal{M}_X(U)^{\mathrm{gp}}$. Hence $q \in \mathcal{M}_X(U)$.

We leave the dull case to the reader. □

Because of its importance, we restate below the definition of a logarithmic homomorphism. The equivalence of the two conditions follows immediately from Proposition I.4.1.3.

Definition 1.1.4. *A homomorphism $\theta \colon \mathcal{P} \to \mathcal{Q}$ of sheaves of monoids is logarithmic if and only it is sharp and local, or equivalently if and only if it induces an isomorphism $\theta^{-1}(\mathcal{Q}^*) \to \mathcal{Q}^*$.*

The next result shows how an arbitrary morphism of sheaves of monoids factors canonically through a logarithmic one.

Proposition 1.1.5. *Let $\theta \colon \mathcal{P} \to \mathcal{M}$ be a homomorphism of sheaves of monoids and let ℓ be one of the following properties: "local, sharp, logarithmic." Then there exists a factorization*

$$\theta = \mathcal{P} \longrightarrow \mathcal{P}^\ell \xrightarrow{\theta^\ell} \mathcal{M}$$

where θ^ℓ has the property ℓ and also has the following universal property: if $\mathcal{P} \longrightarrow \mathcal{P}' \xrightarrow{\theta'} \mathcal{M}$ is another factorization of θ and if θ' has ℓ, then there is a unique homomorphism $\mathcal{P}^\ell \to \mathcal{P}'$ such that the left triangle in the diagram

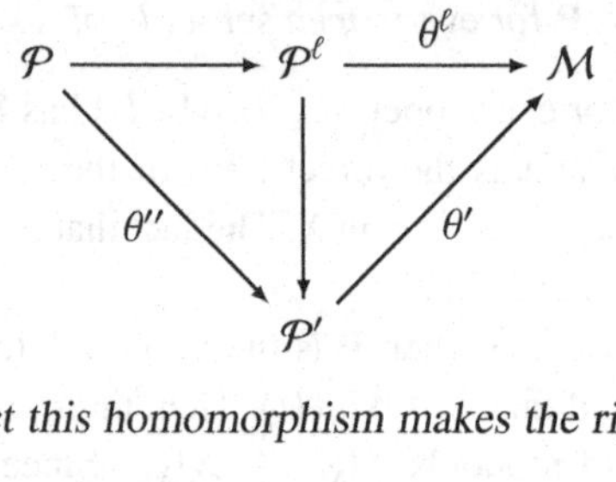

commutes, and in fact this homomorphism makes the right triangle commute as well.

Proof First suppose that ℓ is "local." Let $\mathcal{F} := \theta^{-1}(\mathcal{M}^*)$, a sheaf of faces in $\mathcal{P}$, and let $\lambda \colon \mathcal{P} \to \mathcal{P}_{\mathcal{F}}$ be the localization of $\mathcal{P}$ by $\mathcal{F}$ in the category of sheaves of monoids. This is the sheaf associated to the presheaf $U \mapsto \mathcal{P}(U)_{\mathcal{F}(U)}$ or equivalently, the pushout:

$$\begin{array}{ccc} \mathcal{F} & \longrightarrow & \mathcal{P} \\ \big\downarrow & & \big\downarrow \lambda \\ \mathcal{F}^{\mathrm{gp}} & \longrightarrow & \mathcal{P}_{\mathcal{F}} \end{array}$$

in the category of sheaves of monoids. Then since $\theta(\mathcal{F}) \subseteq \mathcal{M}^*$, the map θ factors uniquely through a map $\theta^\ell \colon \mathcal{P}_{\mathcal{F}} \to \mathcal{M}$, and θ^ℓ is local because $\mathcal{P}^*_{\mathcal{F}} = \mathcal{F}^{\mathrm{gp}}$. Moreover, if θ' in a factorization diagram as shown above is local, then θ'' takes $\mathcal{F}$ to $\mathcal{P}'^*$, and hence θ'' factors uniquely through $\mathcal{P}_{\mathcal{F}}$. The right triangle also commutes because of the universal mapping property of λ (which is an epimorphism).

Next suppose that ℓ is the property "sharp." Take $\mathcal{P}^{sh}$ to be the pushout

$$\begin{array}{ccc} \mathcal{P}^* & \longrightarrow & \mathcal{P} \\ \big\downarrow & & \big\downarrow \\ \mathcal{M}^* & \longrightarrow & \mathcal{P}^{sh} \end{array}$$

in the category of sheaves of monoids. This is just the quotient of $\mathcal{P} \oplus \mathcal{M}^*$ by the antidiagonal action of $\mathcal{P}^*$. It is immediate to check that the natural map $\mathcal{M}^* \to (\mathcal{P}^{sh})^*$ is an isomorphism, so that the map $\mathcal{P}^{sh} \to \mathcal{M}$ is sharp and that $\mathcal{P} \to \mathcal{P}^{sh}$ has the desired universal property.

Finally suppose that ℓ is "logarithmic." Then we can take $\mathcal{P}^{log}$ to be $(\mathcal{P}^{loc})^{sh}$, or, alternatively, the pushout

$$\begin{array}{ccc} \theta^{-1}(\mathcal{M}^*) & \longrightarrow & \mathcal{P} \\ \big\downarrow & & \big\downarrow \\ \mathcal{M}^* & \longrightarrow & \mathcal{P}^{log}. \end{array} \tag{1.1.1}$$

In either case, it is straightforward to check that the construction has the desired properties. □

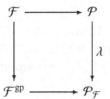

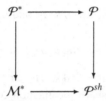

Remark 1.1.6. Since one of the corners of the pushout square in diagram (1.1.1) is a group, the computation of $\mathcal{P}^{log}$ is relatively easy: Proposition I.1.1.5 shows that it is the quotient of $\mathcal{M}^* \oplus \mathcal{P}$ in the category of sheaves of monoids by the equivalence relation that identifies (u, p) with (u', p') if and only if locally

there exist sections v and v' of $\theta^{-1}(\mathcal{M}^*)$ such that $u + \theta(v') = u' + \theta(v)$ and $v + p = v' + p'$.

Remark 1.1.7. If $\theta' : \mathcal{P} \to \mathcal{M}'$ and $\alpha : \mathcal{M}' \to \mathcal{M}$ are homomorphisms and $\theta := \alpha \circ \theta'$, then the inclusion $\theta'^{-1}(\mathcal{M}'^*) \to \theta^{-1}(\mathcal{M})$ induces a homomorphism of sheaves of monoids $\mathcal{P}^{log}_{\theta'} \to \mathcal{P}^{log}_{\theta}$. If α is logarithmic, this map is an isomorphism.

Proposition 1.1.8. *Let $\theta : \mathcal{P} \to Q$ be a homomorphism of sheaves of monoids and $\theta^{log} : \mathcal{P}^{log} \to Q$ the associated logarithmic homomorphism.*

1. *The map $\mathcal{P} \to \overline{\mathcal{P}^{log}}$ factors through an isomorphism*
$$\mathcal{P}/\theta^{-1}(Q^*) \to \overline{\mathcal{P}^{log}}.$$
In particular, the map $\overline{\mathcal{P}} \to \overline{\mathcal{P}^{log}}$ is surjective, and if θ is local it is an isomorphism.
2. *$\mathcal{P}^{log}$ is integral (resp. saturated) if $\mathcal{P}$ is, and the converse holds if θ is local and $\mathcal{P}$ is u-integral.*
3. *$\mathcal{P}^{log}$ is u-integral if $\mathcal{P}$ is u-integral and θ is local.*

Proof It suffices to check the stalks. The first statement follows from the construction of $\mathcal{P}^{log}$ as the sharp localization of $\mathcal{P}$ by θ. If θ is local, then $\mathcal{P}^{log} \cong \mathcal{P}^{sh}$, so $\overline{\mathcal{P}} \to \overline{\mathcal{P}^{log}}$ is an isomorphism. If $\mathcal{P}$ is integral, then by Proposition I.1.3.4, so is $\mathcal{P}^{log}$. If $\mathcal{P}$ is saturated, then so is its localization $\mathcal{P}^{loc}$, and hence so is $\overline{\mathcal{P}^{loc}} \cong \overline{\mathcal{P}^{log}}$, and it follows that $\mathcal{P}^{log}$ is saturated. Conversely, if θ is local, then $\overline{\mathcal{P}} \cong \overline{\mathcal{P}^{log}}$. Moreover, if $\mathcal{P}$ is u-integral and $\mathcal{P}^{log}$ is integral, $\mathcal{P}$ is integral by Proposition I.1.3.3, and is saturated if $\mathcal{P}^{log}$ is saturated. If θ is local, then $\mathcal{P}^{log}$ is just the pushout of $\mathcal{P}$ along the homomorphism $\mathcal{P}^* \to Q^*$, which is easily seen to be u-integral. □

A warning: If $\mathcal{P}$ is u-integral, it does not follow that $\mathcal{P}^{log}$ is also u-integral, since localization can destroy u-integrality, as we saw in Remark I.1.4.5.

Corollary 1.1.9. *Let $\theta : \mathcal{P} \to \mathcal{M}$ be a homomorphism of sheaves of integral monoids such that $\mathcal{P}^{log} \to \mathcal{M}$ is an isomorphism. Then the following are equivalent:*

1. *$\overline{\theta} : \overline{\mathcal{P}} \to \overline{\mathcal{M}}$ is an isomorphism.*
2. *$\theta : \mathcal{P} \to \mathcal{M}$ is exact.*
3. *$\theta : \mathcal{P} \to \mathcal{M}$ is local.*

Proof It again suffices to work at the level of stalks. If $\overline{\theta}$ is an isomorphism, then θ is exact, and if θ is exact it is local, by Proposition I.4.2.1. If θ is local, then by Proposition 1.1.8 the map $\overline{\mathcal{P}} \to \overline{\mathcal{P}^{log}}$ is an isomorphism. Since by hypothesis $\mathcal{P}^{log} \cong \mathcal{M}$, it follows that $\overline{\theta}$ is a isomorphism. □

According to our conventions, a monoidal space $(X, \mathcal{M}_X)$ is dull if $\mathcal{M}_X$ is dull. If $(X, \mathcal{M}_X)$ is any monoidal space, then $\underline{X} := (X, \mathcal{M}_X^*)$ is dull, and there is an obvious morphism of monoidal spaces

$$d_X := (\mathrm{id}_X, inc) \colon X := (X, \mathcal{M}_X) \to \underline{X} := (X, \mathcal{M}_X^*).$$

This morphism is universal: any morphism from X to a dull monoidal space factors uniquely through d_X.

Definition 1.1.10. *Let $f \colon (X, \mathcal{M}_X) \to (Y, \mathcal{M}_Y)$ be a morphism of monoidal spaces.*

1. *$f^a \colon f^*_{log}(\mathcal{M}_Y) \to \mathcal{M}_X$ is the universal logarithmic homomorphism associated to $f^\flat \colon f^{-1}(\mathcal{M}_Y) \to \mathcal{M}_X$. In particular, $f^*_{log}(\mathcal{M}_Y) := \left(f^{-1}(\mathcal{M}_Y)\right)^{log}$ and contains $\mathcal{M}_X^*$.*
2. *$\mathcal{M}_{X/Y}$ is the cokernel of the homomorphism $f^*_{log}(\mathcal{M}_Y) \to \mathcal{M}_X$, computed in the category of sheaves of monoids.*

The inverse image $\mathcal{M}_X^v$ of $\mathcal{M}_{X/Y}^*$ in $\mathcal{M}_X$ is called the *vertical part* of $\mathcal{M}_X$ relative to Y, and $\overline{\mathcal{M}}_{X/Y}$ is called the *horizontal part* (see Definition I.4.3.1). Notice that $\overline{\mathcal{M}}_X \cong \mathcal{M}_{X/\underline{X}}$. More generally, since $f^*_{log}\mathcal{M}_Y$ contains $\mathcal{M}_X^*$, in fact $\mathcal{M}_{X/Y}$ is canonically isomorphic to the cokernel of the natural map $f^*_{log}\overline{\mathcal{M}}_Y \to \overline{\mathcal{M}}_X$. It follows from Propositions I.1.3.3 and 1.1.3 that if $\mathcal{M}_Y$ and $\mathcal{M}_X$ are integral, so is $\mathcal{M}_{X/Y}$. In this case $\mathcal{M}_{X/Y}^{gp}$ is isomorphic to the cokernel of $f^*_{log}\mathcal{M}_Y^{gp} \to \mathcal{M}_X^{gp}$, and $\mathcal{M}_{X/Y}$ can be identified with the image of $\mathcal{M}_X$ in this sheaf of groups. The sheaf $\mathcal{M}_{X/Y}$ is sometimes called the *relative characteristic* of the morphism f.

Definition 1.1.11. *Let* **P** *be a property of homomorphisms of monoids. and let $f \colon X \to Y$ be a morphism of monoidal spaces. Then f has* **P** *at a point x of X if the homomorphism $f_x^\flat \colon (f^*_{log}\mathcal{M}_Y)_x \to \mathcal{M}_{X,x}$ has* **P**. *If $y \in Y$, then we say that f has* **P** *over y if it has* **P** *as every point x mapping to y, and that it has* **P** *if it does so at every point of X.*

This terminological convention for morphisms of monoidal spaces differs somewhat from the convention for homomorphisms of sheaves of monoids, because here we need to suppress the role of the units. However, if **P** is such that a logarithmic homomorphism θ has **P** if and only $\overline{\theta}$ has **P**, then a morphism f has **P** at x if and only if $\overline{f}^\flat_x \colon \overline{\mathcal{M}}_{Y,f(x)} \to \overline{\mathcal{M}}_{X,x}$ has **P**. For example, this is the case for small and **Q**-surjective morphisms. Moreover, if *every* homomorphism θ has **P** if and only if $\overline{\theta}$ has **P**, then f has **P** at x if and only if

$f_x^\flat \colon \mathcal{M}_{Y,f(x)} \to \mathcal{M}_{X,x}$ has **P**. For example, this is the case for exact and integral homomorphisms.

Definition 1.1.12. *A morphism of monoidal spaces $i \colon Y \to Z$ is an immersion if it is set-theoretically injective and the homomorphism $\mathcal{M}_Z \to i_*\mathcal{M}_Y$ is surjective. An open immersion of monoidal spaces is an immersion that induces an isomorphism onto an open subset of its target.*

We should remark that an immersion of monoidal spaces is a monomorphism.

Definition 1.1.13. *If X is a monoidal space and n is a positive integer, the n-Frobenious endomorphism of X is the morphism $X \to X$ that is the identity map on X and is given by multiplication by n on $\mathcal{M}_X$.*

This definition makes sense because if P is a monoid and n is a positive integer, then multiplication by n induces a local homomorphism $P \to P$.

1.2 Monoschemes

Recall from Section I.1.4 that if Q is a monoid, then the topological space $S :=$ Spec(Q) is the set of prime ideals of Q, endowed with the Zariski topology. For $f \in Q$, the topological space Spec(Q_f) identifies with the open set S_f of all $\mathfrak{q} \in$ Spec(Q) not containing f, and the set of all such S_f forms a base $\mathcal{B}$ for the topology of S. Sets of the form S_f are called the *special affine open subsets* of S. If g is another element of Q, then $S_g \subseteq S_f$ if and only if $f \in \langle g \rangle$. If this is the case, then there is a unique homomorphism $Q_f \to Q_g$ making the diagram

$$\begin{array}{ccc} Q & \longrightarrow & Q_f \\ & \searrow & \downarrow \\ & & Q_g \end{array}$$

commute. Thus $S_f \mapsto Q_f$ defines a presheaf of monoids on $\mathcal{B}$, and we let $\mathcal{M}_S$ denote the corresponding sheaf on Spec(Q). For each $f \in Q$, the prime $\mathfrak{p} := Q \setminus \langle f \rangle$ is the unique closed point of S_f, and it follows that

$$\Gamma(S_f, \mathcal{M}_S) = \mathcal{M}_{S,\mathfrak{p}} = Q_{\langle f \rangle} = Q_f.$$

In particular, the face generated by 0 is Q^*, whose complementary prime ideal Q^+ is the unique closed point s of S, and $S_0 = S$. Consequently

$$\Gamma(S, \mathcal{M}_S) = \mathcal{M}_{S,s} = Q. \qquad (1.2.1)$$

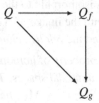

Definition 1.2.1. *If Q is a monoid, then*

$$\mathbf{a}_Q := \operatorname{spec}(Q)$$

is the locally monoidal space consisting of the set Spec(Q) *of prime ideals of Q, endowed with the Zariski topology and the sheaf of monoids defined above.*

A morphism of monoids $\theta\colon P \to Q$ induces a morphism of locally monoidal spaces

$$\operatorname{spec}(Q) \to \operatorname{spec}(P)$$

and thus one obtains a functor from the category of monoids to the category of locally monoidal spaces.

Proposition 1.2.2. *Let Q be a monoid and let X be a locally monoidal space. Then the natural map*

$$\operatorname{Mor}(X, \mathbf{a}_Q) \to \operatorname{Hom}(Q, \Gamma(X, \mathcal{M}_X))$$

is an isomorphism. Consequent to this fact and the isomorphism (1.2.1), the functor spec *from the category of monoids to the category of locally monoidal spaces is fully faithful.*

Proof This a standard argument from scheme theory, which we repeat for fun. Let $S := \operatorname{spec}(Q)$. Given a homomorphism $\theta\colon Q \to \Gamma(X, \mathcal{M}_X)$, we define a map $f\colon X \to S$ as follows. If $x \in X$, we have a homomorphism of monoids

$$\theta_x\colon Q \to \Gamma(X, \mathcal{M}_X) \to \mathcal{M}_{X,x},$$

and we set $f(x) := \theta^{-1}(\mathcal{M}^+_{X,x})$, a prime ideal of Q. The set of special affines S_q such that $q \in Q \setminus f(x)$ is a neighborhood basis of $f(x)$ in S. For any such q, $\theta(q)$ is a unit in $\mathcal{M}^*_{X,x}$ and hence also in a neighborhood of x. Thus $f^{-1}(S_q)$ is open in X and f is continuous. Since $\theta(q)$ is a unit in $\Gamma(f^{-1}(S_q), \mathcal{M}_X)$, θ factors uniquely through a homomorphism

$$\theta_q\colon \Gamma(S_q, \mathcal{M}_S) = Q_q \to \Gamma(f^{-1}(S_q), \mathcal{M}_X),$$

and in this way we find a homomorphism of sheaves of monoids $f^\flat\colon \mathcal{M}_S \to f_*(\mathcal{M}_X)$. Taking $q = 0$, we see that $\Gamma(f^\flat) = \theta$. It follows from the definition of $f(x)$ that for each $x \in X$, the homomorphism

$$f^\flat_x\colon \mathcal{M}_{S,f(x)} = Q_{F_x} \to \mathcal{M}_{X,x}$$

is local.

Conversely, let $f\colon X \to S$ be a morphism of monoschemes. Such an f includes the data of a homomorphism of sheaves of monoids $f^\flat\colon \mathcal{M}_S \to f_*(\mathcal{M}_X)$

and hence a homomorphism of monoids

$$\theta_f \colon Q = \Gamma(S, \mathcal{M}_S) \to \Gamma(X, \mathcal{M}_X).$$

Let us check that f agrees with the morphism f_θ constructed from θ_f. Indeed, for each $x \in X$, we have a commutative diagram

$$\begin{array}{ccc} \Gamma(X, \mathcal{M}_X) & \longrightarrow & \mathcal{M}_{X,x} \\ \uparrow {\scriptstyle \theta_f} & \nearrow {\scriptstyle \theta_x} & \uparrow {\scriptstyle f^\flat_x} \\ Q & \xrightarrow{\lambda_x} & \mathcal{M}_{S,f(x)}. \end{array}$$

Here λ_x identifies $\mathcal{M}_{S,f(x)}$ with the localization of Q at its prime ideal $f(x)$. The homomorphism $f^\flat_x$ is local by assumption, and then it follows from the commutativity of the diagram that $\theta_x^{-1}(\mathcal{M}^+_{X,x}) = f(x)$. This shows that f and f_θ agree set theoretically, and the agreement of the corresponding homomorphisms of sheaves of monoids follows from the universal property of localization.

Now if P and Q are monoids, we use the isomorphism (1.2.1) to see that that the natural map

$$\mathrm{Mor}(\mathbf{a}_P, \mathbf{a}_Q) \to \mathrm{Hom}(Q, \Gamma(\mathbf{a}_P, \mathcal{M}_{\mathbf{a}_P})) \cong \mathrm{Hom}(Q, P)$$

is an isomorphism, and hence the (contravariant) functor spec is fully faithful. □

Definition 1.2.3. *A locally monoidal space is* affine *if it is isomorphic to* spec(Q) *for some monoid Q. A* monoscheme *is a locally monoidal space that admits an open cover by affines, and the category of* monoschemes *is the full subcategory of the category of locally monoidal spaces consisting of such spaces.*[1]

Variant 1.2.4. Recall from Definition I.1.5.1 that if (Q, K) is an idealized monoid, then $S := \mathrm{spec}(Q, K)$ consists of the set of prime ideals of Q that contain K, or equivalently, the set of faces of Q that do not meet K, endowed with a natural Zariski topology. A base for this topology is given by sets of the form S_f, where now $f \in Q \setminus K$. One defines a sheaf of monoids $\mathcal{M}_S$ and a sheaf of ideals $\mathcal{K}_S$ by localizing, as in the non-idealized case. Note that if K is the unit ideal, then S is empty and $\Gamma(S, \mathcal{M}_S) = 0$. Recall that (Q, K) is said to be *acceptable* if either $Q = 0$ or K is a proper ideal of Q. A homomorphism of idealized monoids $\theta \colon (P, J) \to (Q, K)$ induces a morphism of idealized monoidal

[1] Locally monoidal spaces locally of the form spec(Q) are sometimes called "schemes over $\mathbf{F}_1$" (see [12]). The terminology "monoscheme" is intended to suggest both the "field with one element $\mathbf{F}_1$" and the fact that the theory is built out of monoids.

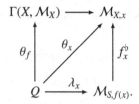

spaces

$$\operatorname{spec}(Q,K) \to \operatorname{spec}(P,J).$$

An *idealized monoscheme* is an idealized monoidal space that admits an open cover $\{U_i : i \in I\}$ such that each U_i is isomorphic to some $\operatorname{spec}(Q_i, K_i)$.

We denote by **Msch** and **Mschi** the categories of monoschemes and idealized monoschemes, respectively.

The analog of Proposition 1.2.2 for idealized monoids requires a slight modification.

Proposition 1.2.5. *Let (Q, K) be an idealized monoid, let $S := \mathbf{a}_{(Q,K)}$ be the corresponding idealized monoidal space, and let $X := (X, \mathcal{M}_X, \mathcal{K}_X)$ be any idealized monoidal space.*

1. *The idealized monoid $\Gamma(X, (\mathcal{M}_X, \mathcal{K}_X))$ is acceptable. The canonical homomorphism*

$$\rho\colon (Q,K) \to \Gamma(S, (\mathcal{M}_S, \mathcal{K}_S))$$

is an isomorphism if and only if (Q, K) is acceptable.

2. *The natural map*

$$\operatorname{Mor}(X, \mathbf{a}_{(Q,K)}) \to \operatorname{Hom}((Q,K), \Gamma(X, (\mathcal{M}_X, \mathcal{K}_X))$$

taking f to

$$\theta_f := (Q,K) \xrightarrow{\ \rho\ } \Gamma(S, (\mathcal{M}_S, \mathcal{K}_S)) \xrightarrow{\Gamma(f^\flat)} \Gamma(X, (\mathcal{M}_X, \mathcal{K}_X)).$$

is bijective.

3. *The functor* spec *from the category of acceptably idealized monoids to the category of idealized locally monoidal spaces is fully faithful.*

Proof If X is an idealized monoidal space and $x \in X$, then by hypothesis $\mathcal{K}_{X,x}$ is contained in the maximal ideal of $\mathcal{M}_{X,x}$ and hence does not contain 0. Thus $\Gamma(X, \mathcal{K}_X)$ must be a proper ideal of $\Gamma(X, \mathcal{M}_X)$ and hence $\Gamma(X, (\mathcal{M}_X, \mathcal{K}_X))$ is acceptable. On the other hand, if X is empty, $\Gamma(X, (\mathcal{M}_X, \mathcal{K}_X)) = (0, 0)$, which is also acceptable. The existence of the homomorphism ρ is clear from the construction of S. If K is a proper ideal of Q, then it is contained in the maximal ideal Q^+ of Q, which then corresponds to the unique closed point s of S, so that $\Gamma(S, (\mathcal{M}_S, \mathcal{K}_S)) = (\mathcal{M}_{S,s}, \mathcal{K}_{S,s}) = (Q, K)$. If (Q, K) is acceptable and K is not a proper ideal, then by definition $Q = 0$, and again ρ is an isomorphism. This concludes the proof of statement (1).

The proof of statement (3) is similar to the proof of Proposition 1.2.2. If $\theta\colon (Q,K) \to \Gamma(X, (\mathcal{M}_X, \mathcal{K}_X))$ is a homomorphism of idealized monoids, then for each x in X we have a homomorphism $\theta_x\colon Q \to \mathcal{M}_{X,x}$, and $f(x) := \theta_x^{-1}(\mathcal{M}^+_{X,x})$ is a prime ideal of Q. By assumption θ maps K to $\mathcal{K}_X$ and $\mathcal{K}_x \subseteq \mathcal{M}^+_{X,x}$, and it follows that $K \subseteq f(x)$, that is, $f(x) \in \mathrm{Spec}(Q, K)$. As before one constructs a homomorphism of sheaves of idealized monoids to obtain a morphism $f_\theta\colon X \to \mathbf{a}_{(Q,K)}$ with $\Gamma(f^\flat_\theta) = \theta$. On the other hand, a morphism $f\colon X \to \mathbf{a}_{(Q,K)}$ induces a homomorphism

$$\theta_f := \Gamma(f^\flat)\colon (Q, K) \to \Gamma(S, (\mathcal{M}_S, \mathcal{K}_S)) \to \Gamma(X, (\mathcal{M}_X, \mathcal{K}_X)),$$

which induces f.

This completes the proof of the first two statements, and (3) follows from (1) and (2). □

Proposition 1.2.6. *Let X be a (possibly idealized) monoscheme and let U be an open subset of X. Then U, with the restriction of $\mathcal{M}_X$ to U as idealized structure sheaf, is a monoscheme.*

Proof Each point x of X admits an affine neighborhood V. Then $V \cap U$ is open in V, and it suffices to prove that there is an open subset of $V \cap U$ that contains x and is affine. Thus we may as well assume that X itself is affine, say $X = \mathrm{spec}(Q)$. Recall that then the set of special affine sets form a base for the topology of X, and hence $U \cap V$ contains a neighborhood of x of this form. □

If X is a monoscheme and x is a point of X, there is a natural morphism of monoschemes

$$\mathrm{spec}(\mathcal{M}_{X,x}) \to X,$$

sending the closed point of $\mathrm{spec}(\mathcal{M}_{X,x})$ to x. If x and ξ are points of X and ξ is a generization of x, there is a cospecialization map

$$\lambda_{x,\xi}\colon \mathcal{M}_{X,x} \to \mathcal{M}_{X,\xi},$$

and the diagram

$$\begin{array}{ccc} \mathrm{spec}(\mathcal{M}_{X,\xi}) & \xrightarrow{\mathrm{spec}(\lambda_{x,\xi})} & \mathrm{spec}(\mathcal{M}_{X,x}) \\ & \searrow & \downarrow \\ & & X \end{array}$$

is commutative.

It is convenient to use the Yoneda formalism to identify a monoscheme X with the contravariant functor F_X taking a monoscheme T to the set of morphisms $T \to X$. It is clear from the definitions that the restriction of F_X

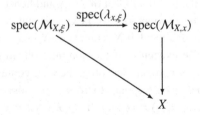

to the category of open subsets of T is a sheaf. Since each such T is covered by affines, in fact F_X is determined by its restriction to the category of affine monoschemes. By Yoneda, it follows that X is determined by the functor $Q \mapsto \mathrm{Mor}(\mathbf{a}_Q, X)$. If Q is a monoid and X is a monoscheme, we write $X(Q)$ for $\mathrm{Mor}(\mathbf{a}_Q, X)$. The next proposition shows that this functor can be computed from the underlying topological space X and its set of cospecialization maps.

Proposition 1.2.7. *If X is a monoscheme and Q is a monoid, we let $h_X(Q)$ denote the set of pairs (x, ϕ_x), where x is a point of X and ϕ_x is a local homomorphism $\mathcal{M}_{X,x} \to Q$.*

1. *If (x, ϕ_x) belongs to $h_X(Q)$ and $\theta\colon Q \to Q'$ is a homomorphism of monoids, there is a unique (ξ, ϕ_ξ) in $h_X(Q')$ such that ξ is a generization of x and $\theta \circ \phi_x = \phi_\xi \circ \lambda_{x,\xi}$. Thus h_X becomes a functor from the category of monoids to the category of sets.*
2. *If (x, ϕ_x) belongs to $h_X(Q)$, then the composite*
$$\mathrm{spec}(Q) \xrightarrow{\mathrm{spec}(\phi_x)} \mathrm{spec}(\mathcal{M}_{X,x}) \to X$$
is a morphism of monoschemes mapping the unique closed point of spec(Q) *to x.*
3. *The correspondence in (2) defines an isomorphism of functors $h_X \to X$.*

Proof Given (x, ϕ_x) and θ, let U be an affine neighborhood of x in X. To prove the existence of (ξ, ϕ_ξ), we may replace X by U, and in fact since every generization of x belongs to U, it also suffices to prove uniqueness when $X = U$. Thus we suppose without loss of generality that $X = \mathrm{spec}(P)$. Then $\mathcal{M}_{X,x}$ is the localization of P at the face F corresponding to x, and ϕ_x is a local homomorphism $P_F \to Q$. Then $G := \phi_x^{-1}(\theta^{-1}(Q'^*))$ is the the unique face of P such that $\theta \circ \phi_x$ factors through a local homomorphism $\phi'\colon P_G \to Q'$. The factorization is unique, and if ξ is the point of P corresponding to G, then (ξ, ϕ') is the unique pair we seek. It is clear that this construction is compatible with composition and hence defines a functor, proving (1). Since by assumption ϕ_x is local, the morphism $\mathrm{spec}(\phi_x)$ sends the closed point of $\mathrm{Spec}(Q)$ to the closed point of $\mathrm{Spec}(\mathcal{M}_{X,x})$, and statement (2) follows. On the other hand, if $f\colon \mathrm{spec}(Q) \to X$ is a morphism of monoschemes, let x be the image in X of the unique closed point of $\mathrm{spec}(Q)$. Then $(x, f_x^\flat) \in h_X(Q)$; this defines the inverse to the correspondence in (2). □

Remark 1.2.8. An affine monoscheme T has a unique closed point t, and T is the unique open neighborhood of t in T. Thus if $\mathcal{F}$ is any presheaf on T, the canonical map $\mathcal{F}(T) \to \mathcal{F}_t$ is an isomorphism. Consequently, if $\{\mathcal{F}_i : i \in I\}$ is a direct system of sheaves on T and $\mathcal{F}$ is their sheaf-theoretic direct limit, then

$\varinjlim \mathcal{F}_i(T) \cong \mathcal{F}(T)$. For example, if X is a monoidal space and $\{X_i : i \in I\}$ is an open cover of X, then $\mathrm{Mor}(T, X) = \cup\{\mathrm{Mor}(T, X_i) : i \in I\}$. Of course, these results do not hold if T fails to be affine.

Examples 1.2.9. Here are some examples of monoschemes. Note that the monoscheme corresponding to any dull monoid contains a unique point, the empty ideal, and the corresponding functor is the one point functor. In fact, the empty monoscheme is not representable in the category of affine monoschemes, although in the category of idealized monoschemes it is represented by the idealized monoid $(0, 0)$. The monoscheme $\mathbf{a} := \mathrm{spec}(\mathbf{N})$ is the analog of the affine line. It contains just two points, a closed point s and an open point σ, and $\mathcal{M}_{\mathbf{a},s} = \mathbf{N}$ and $\mathcal{M}_{\mathbf{a},\sigma} = \mathbf{Z}$. The cospecialization map $\mathcal{M}_{\mathbf{a},s} \to \mathcal{M}_{\mathbf{a},\sigma}$ is the natural inclusion $\mathbf{N} \to \mathbf{Z}$. For any Q,

$$h_{\mathbf{a}}(Q) = \{s\} \times Q^+ \cup \{\sigma\} \times Q^* \cong Q.$$

The origin of $\mathbf{a}$ can be doubled by identifying two copies of $\mathbf{a}$ along σ, with two differing outcomes. The monoscheme $\mathbf{d}$ consists of two closed points s, t and one open point σ, with $\mathcal{M}_{\mathbf{d},s} = \mathcal{M}_{\mathbf{d},t} = \mathbf{N}$ and $\mathcal{M}_{\mathbf{d},\sigma} = \mathbf{Z}$, and the cospecialization maps are again the inclusions. We have

$$h_{\mathbf{d}}(Q) = \{s\} \times Q^+ \cup \{t\} \times Q^+ \cup \{\sigma\} \times Q^*.$$

The analog $\mathbf{p}$ of the projective line has the same topological space as $\mathbf{d}$, with two closed points, now denoted by $0, \infty$, and one open point σ, and again $\mathcal{M}_{\mathbf{p},0} = \mathcal{M}_{\mathbf{p},\infty} = \mathbf{N}$ and $\mathcal{M}_{\mathbf{p},\sigma} = \mathbf{Z}$. However, the cospecialization maps are different: $\mathcal{M}_{\mathbf{p},0} \to \mathcal{M}_{\mathbf{p},\sigma}$ is the inclusion but $\mathcal{M}_{\mathbf{p},\infty} \to \mathcal{M}_{\mathbf{p},\sigma}$ is its negative. We have

$$h_{\mathbf{p}}(Q) = \{0\} \times Q^+ \cup \{\sigma\} \times Q^* \cup \{\infty\} \times Q^+$$

Although this looks the same set-theoretically as $h_{\mathbf{d}}$, the functor is different. For example, for the map $Q \to Q^{\mathrm{gp}}$, one finds the following diagram:

$$\begin{array}{ccc} h_{\mathbf{p}}(Q) & = & \{0\} \times Q^+ \;\cup\; \{\sigma\} \times Q^* \;\cup\; \{\infty\} \times Q^+ \\ \big\downarrow & & \searrow^{j_0} \quad \big\downarrow^{j_\sigma} \quad \swarrow^{j_\infty} \\ h_{\mathbf{p}}(Q^{\mathrm{gp}}) & = & \{\sigma\} \times Q^{\mathrm{gp}}. \end{array}$$

Here j_0 takes $(0, q)$ to (σ, q), j_σ takes (σ, q) to (σ, q), and j_∞ takes (∞, q) to $(\sigma, -q)$. Note that $\Gamma(\mathbf{d}, \mathcal{M}_{\mathbf{d}}) = \mathbf{N}$ but $\Gamma(\mathbf{p}, \mathcal{M}_{\mathbf{p}}) = 0$; on the other hand, $\Gamma(\mathbf{p}, \mathcal{M}^{\mathrm{gp}}_{\mathbf{p}}) \cong \mathbf{Z}$ and $\Gamma(\mathbf{p}, \overline{\mathcal{M}}_{\mathbf{p}}) \cong \mathbf{N} \oplus \mathbf{N}$.

For a slightly more complicated-looking example, consider the monoscheme

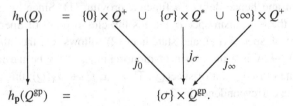

u obtained by obtained by omitting the origin of spec(**N**⊕**N**). This monoscheme has two closed points and one open point, and the corresponding cospecialization morphisms of stalks form the following diagram, in which we write i for the identity map and j for the inclusion.

$$\begin{CD} @. \mathcal{M}_{\mathbf{u},x} @. @. \mathbf{N}\oplus\mathbf{Z} \\ @. @VVV @. @VV(j,i)V \\ \mathcal{M}_{\mathbf{u},y} @>>> \mathcal{M}_{\mathbf{u},\eta} @. \mathbf{Z}\oplus\mathbf{N} @>(i,j)>> \mathbf{Z}\oplus\mathbf{Z}. \end{CD}$$

Then

$$h_{\mathbf{u}}(Q) = \{x\} \times Q^* \times Q^+ \sqcup \{\eta\} \times Q^* \times Q^* \sqcup \{y\} \times Q^+ \times Q^*.$$

In fact there is a morphism of monoschemes $\mathbf{u} \to \mathbf{p}$ making $\mathbf{p}$ the quotient of $\mathbf{u}$ by the diagonal action of $\mathbf{a}^*$. To see this, consider the summation homomorphism $\deg\colon \mathbf{Z} \oplus \mathbf{Z} \to \mathbf{Z}$, and observe that the stalks of $\mathcal{M}_{\mathbf{p}}$ correspond to the elements of the stalks of $\mathcal{M}_U$ of degree zero. Thus $\mathbf{u} \to \mathbf{p}$ is given by the following commutative diagram of cospecializations:

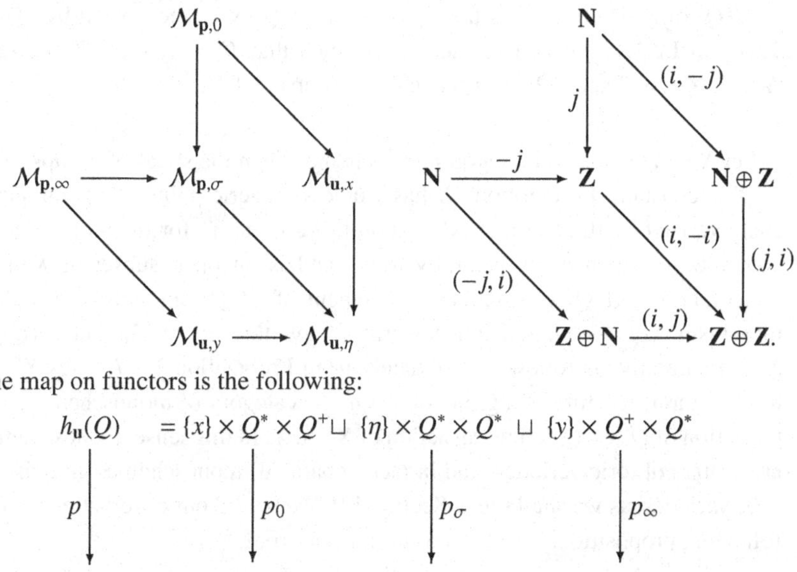

The map on functors is the following:

$$\begin{array}{ccccccc} h_{\mathbf{u}}(Q) & = \{x\} \times Q^* \times Q^+ & \sqcup & \{\eta\} \times Q^* \times Q^* & \sqcup & \{y\} \times Q^+ \times Q^* \\ \downarrow p & \downarrow p_0 & & \downarrow p_\sigma & & \downarrow p_\infty \\ h_{\mathbf{p}}(Q) & = \{0\} \times Q^+ & \sqcup & \{\sigma\} \times Q^* & \sqcup & \{\infty\} \times Q^+, \end{array}$$

where $p_0(u, q) = q - u$, $p_\sigma(u, q) = q - u$, and $p_\infty(q, u) = q - u$. Notice that this diagram identifies $h_{\mathbf{p}}(Q)$ with the quotient of $h_{\mathbf{u}}(Q)$ by the action of $h_{\mathbf{a}}(Q) = Q^*$. This example is a special case of an analog of the Proj construction in algebraic geometry, which we will discuss in Section 1.5.

Recall that a topological space is said to be *quasi-compact* if every open cover has a finite subcover, and to be *quasi-separated* if the intersection of every pair of quasi-compact sets is quasi-compact. Every affine monoscheme X is quasi-compact and quasi-separated. Indeed, if $X = \operatorname{spec} Q$, then X has a unique closed point x, every open cover of X contains an open U containing x, and necessarily $U = X$. Note that the family of special affine open sets is closed under finite intersection and, since each such set is quasi-compact, it follows that X is quasi-separated.

We shall say that a monoscheme X is *integral* (resp. *saturated*) if the sheaf of monoids $\mathcal{M}_X$ is integral (resp. saturated). We shall say that X is *noetherian* if the space X is noetherian and the stalks $\mathcal{M}_{X,x}$ are noetherian, and that X is *fine* (resp. *toric*) if it is noetherian and the stalks $\mathcal{M}_{X,x}$ are fine (resp. toric).

Remark 1.2.10. If X is a fine monoscheme then, for every affine open subset U of X, the monoid $Q := \mathcal{M}_X(U)$ is finitely generated. Indeed, since U is affine, it is isomorphic to $\operatorname{spec}(Q)$, hence has a unique closed point x, and $Q \cong \mathcal{M}_{X,x}$, which is fine by assumption. Moreover, if x is any point of X, then $\operatorname{spec}(\mathcal{M}_{X,x})$ is an open subset of X. To see this, replace X by an affine open neighborhood $\operatorname{Spec}(Q)$ of x. Then $\mathcal{M}_{X,x}$ is the localization of Q by a face F, and by (3) of Theorem I.2.1.17, there is some $f \in F$ such that $F = \langle f \rangle$. It follows that $\mathcal{M}_{X,x} = Q_f$, and $\operatorname{spec}(Q_f)$ is a special affine subset of U.

Let X be a connected integral monoscheme. Then the sheaf of groups $\mathcal{M}_X^{\mathrm{gp}}$ on X is constant. Furthermore, X has a unique generic point ξ; it is uniquely characterized by the fact that $\mathcal{M}_{X,\xi}$ is dull. We write X^* for the monoscheme $\operatorname{spec}(\mathcal{M}_{X,\xi}^{\mathrm{gp}})$, which maps naturally to X (and is an open subset of X if X is noetherian). If Q is a monoid, an element of $X^*(Q)$ amounts to a homomorphism $\mathcal{M}_{X,\xi}^{\mathrm{gp}} \to Q$, which necessarily factors through Q^*. In fact $X^*(Q) \cong X(Q^*)$ naturally, as follows immediately from Proposition 1.2.7. Thus $X^*(Q)$ has a natural structure of a group object in the category of monoschemes, and the action of Q^* on Q defines an action of X^* on X. In this sense, monoschemes are avatars of toric varieties, and in fact (separated) monoschemes give rise to toric varieties, as we shall see in Section 1.9. We record our observations in the following proposition, which needs no further proof.

Proposition 1.2.11. *Let X be a connected integral monoscheme. Then for every Q, there is a natural identification $X^*(Q) \cong X(Q^*)$, and hence a natural action $X^* \times X \to X$ of X^* on X.* □

Proposition 1.2.12. *Let $\theta\colon P \to Q$ be a homomorphism of integral monoids and let $\operatorname{spec}(\theta)\colon T \to S$ be the corresponding morphism of monoschemes. If $\overline{\theta}$*

is Kummer and exact, then the diagram

$$\begin{array}{ccc} T^* & \longrightarrow & T \\ \downarrow & & \downarrow \\ S^* & \longrightarrow & S \end{array}$$

is cocartesian in the category of monoschemes.

Proof Recall from Corollary I.4.3.11 that the morphism $T \rightarrow S$ is a homeomorphism, and we can and shall identify the underlying topological spaces of T and S. Since $\overline{\theta}$ is exact, the same is true of θ, and it follows that $\mathcal{M}_S \cong \mathcal{M}_S^{\mathrm{gp}} \times_{\mathcal{M}_T^{\mathrm{gp}}} \mathcal{M}_T$. Suppose now that we are given morphisms $g\colon T \rightarrow X$ and $f\colon S^* \rightarrow X$ that agree on T^*. Then $g^\flat\colon g^{-1}(\mathcal{M}_X) \rightarrow \mathcal{M}_T$ and $f^\flat\colon f^{-1}(\mathcal{M}_X) \rightarrow \mathcal{M}_S^{\mathrm{gp}}$ agree when mapped to $\mathcal{M}_T^{\mathrm{gp}}$, and hence define a unique map to $\mathcal{M}_S$. □

1.3 Some universal constructions

We follow methods from the theory of schemes. Let $F\colon \mathbf{Mon} \rightarrow \mathbf{Ens}$ be a functor. For each monoid Q, F induces a presheaf F_Q on the category of special affine open subsets of spec Q. We say that F is a *sheaf on* **Mon** if F_Q is a sheaf on this base for the topology of spec(Q) for every Q. Evidently every representable F is a sheaf in this sense.

Definition 1.3.1. *Let $\eta\colon F' \rightarrow F$ be a morphism of sheaves of sets on the category of monoids (or, equivalently, of sheaves on the category of affine monoschemes). Then η is an open immersion if, for every affine monoscheme T and every $\xi\colon T \rightarrow F$, the functor $F' \times_F T \rightarrow T$ is represented (on the category of monoschemes over T) by an open sub-monoscheme of T. An open covering of a sheaf F on* **Mon** *is a family of open immersions $F_i \rightarrow F$ such that the induced map from the sheaf-theoretic union $\cup F_i$ to F is an epimorphism. A sheaf on F is a presheaf P on the category of sheaves $F' \rightarrow F$ over F such that, for every sheaf $F' \rightarrow F$ over F and every open covering $\{F'_i \rightarrow F'\}$ of F', the sequence*

$$P(F') \rightarrow \prod P(F'_i) \rightrightarrows \prod_{i,j} P(F'_i \times F'_j)$$

is exact (that is, $P(F')$ is the equalizer of the two arrows).

For example, if G is a sheaf on **Mon**, then the presheaf

$$h_G : F' \mapsto \mathrm{Mor}(F', G)$$

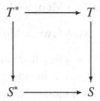

on the category of sheaves over F is in fact a sheaf on F, as is easily checked. In particular, if F is any presheaf on **Mon**, we let

$$\mathcal{M}(F) := \mathrm{Mor}(F, \mathbf{a}) := \mathrm{Mor}(F, h_{\mathbf{a}}),$$

with its natural monoid structure coming from the monoid structure on $\mathbf{a}$. If F is a sheaf, the restriction $\mathcal{M}_F$ of $\mathcal{M}$ to the category of sheaves over F forms a sheaf of monoids on F. If T is a monoscheme, there are canonical isomorphisms

$$\mathcal{M}(T) := \mathrm{Mor}(h_T, h_{\mathbf{a}}) \cong h_{\mathbf{a}}(T) \cong \Gamma(T, \mathcal{M}_T),$$

so our notation is not ambiguous.

Proposition 1.3.2. *A sheaf of sets on* **Mon** *is representable in the category of monoschemes if and only if it admits an open cover by affine monoschemes.*

Proof We give only a sketch, since the argument is standard. Let F be a sheaf of sets on **Mon** admitting an open cover $\{F_i \to F\}$, where $F_i = h_{S_i}$ and each S_i is an affine monoscheme. Let $\tilde{S}$ be the disjoint union of the S_i, and let $\tilde{F}$ denote the disjoint union (in the category of sheaves) of the F_i. Then $\tilde{S}$ is a monoscheme, and $\tilde{F} \cong h_{\tilde{S}}$. Furthermore, the two projections $\tilde{F} \times_F \tilde{F} \to \tilde{F}$ are open immersions. It follows that $\tilde{F} \times_F \tilde{F}$ is representable by an open submonoscheme E of $\tilde{S}$, and that F is the coequalizer of the two maps $\tilde{F} \times_F \tilde{F} \to \tilde{F}$ in the category of sheaves. Let S be the coequalizer of the two maps $E \to \tilde{S}$ in the category of topological spaces. If U is an open subset of S, let $\tilde{U}$ be its inverse image in $\tilde{S}$ and let F_U be the sheaf-theoretic image of $h_{\tilde{U}} \to h_{\tilde{S}} \to F$. Then $F_U \to F$ is an open immersion. Let $\mathcal{M}_S(U) := \Gamma(F_U, \mathcal{M}_F)$. Then $\mathcal{M}_S$ is sheaf of monoids on S. The natural map $S_i \to S$ is an open immersion, and the restriction of $\mathcal{M}_S$ to S_i is the structure sheaf of the affine monoscheme S_i. This shows that $(S, \mathcal{M}_S)$ is a monoscheme. Finally, the map $h_{\tilde{S}} \to F$ descends to a map $h_S \to F$, which is in fact an isomorphism. □

Variants 1.3.3. The evident analogs of the constructions in Proposition 1.3.2 work in exactly the same way if we replace the category of monoids by the category of integral (resp. saturated) monoids, or even the category of acceptable idealized monoids. For this last case, we should describe how to recover the sheaf of ideals. Notice first that if (Q, K) is an idealized monoid, then there are bijections

$$\mathrm{Mor}((\mathbf{N}, \emptyset), (Q, K)) \cong Q$$
$$\mathrm{Mor}((\mathbf{N}, \mathbf{N}^+), (Q, K)) \cong K.$$

Let $\mathbf{a} := \mathrm{spec}(\mathbf{N}, \emptyset)$ and $\mathbf{a}^+ := \mathrm{spec}(\mathbf{N}, \mathbf{N}^+)$, and if F is a sheaf on the category of acceptable idealized monoids, let $\mathcal{M}(F') := \mathrm{Mor}(F', h_{\mathbf{a}})$ and

$\mathcal{K}(F') := \mathrm{Mor}(F', h_{\mathbf{a}^+})$ for every open subsheaf F' of F. Then $(\mathcal{M}, \mathcal{K})$ is a sheaf of idealized monoids on F, and if X is an idealized monoscheme and $F = h_X$, then $(\mathcal{M}, \mathcal{K})$ is the idealized structure sheaf of X. Then the gluing construction of Proposition 1.3.2 can be carried out as before.

We can use Proposition 1.3.2 to construct global examples of some universal constructions we carried out in the category of monoids.

Proposition 1.3.4. *Let* **Msch** *denote the category of monoschemes and let* $\mathbf{Msch}^{\mathrm{int}}$ *be the full subcategory consisting of the integral monoschemes. Then the inclusion functor* $\mathbf{Msch}^{\mathrm{int}} \to \mathbf{Msch}$ *admits a right adjoint* $X \mapsto X^{\mathrm{int}}$*. The analogous statement holds in the category of saturated monoschemes.*

Proof Let X be a monoscheme and let h'_X be the restriction of h_X to the category of integral monoids. If $\{X_i = \operatorname{spec} Q_i\}$ is an affine open cover of X, then $\{h'_{X_i} \to h'_X\}$ is an open cover of h_X. But $h'_{X_i} = h_{X'_i}$, where $X'_i := \operatorname{spec} Q_i^{\mathrm{int}}$, by the universal property of $Q \mapsto Q^{\mathrm{int}}$. Then the representability of h'_X in the category of integral monoschemes follows from the analog of Proposition 1.3.2 for integral monoschemes. The proof in the saturated case is analogous. □

Proposition 1.3.5. *Let* $X \to Y$ *and* $Y \to Z$ *be morphisms of monoschemes. Then the product* $X \times_Z Y$ *exists in the category of monoschemes. Similarly, fiber products exist in the category of integral (resp. saturated) monoschemes, and in the category of idealized monoschemes.*

Proof The proposition asserts that the presheaf $h_X \times_{h_Z} h_Y$ is representable. Since h_X, h_Z, and h_Y are sheaves, so is this fiber product. Let $\{Z_k \to Z\}$ be an affine open cover of Z. Then $X_k := X \times_Z Z_k$ is an open submonoscheme of X and hence admits an affine open cover $\{X_{ik}\}$; similarly, we construct an affine open cover $\{Y_{jk}\}$ of $Y_k := Y \times_Z Z_k$. Then $\{h_{X_{i,k}} \times_{h_{Z_k}} h_{Y_{jk}}\}$ is an affine open cover of $h_X \times_{h_Z} h_Y$. The variants are proved in a similar manner. □

Remark 1.3.6. It follows from Proposition 4.1.5 that, in the category of all monoschemes, the underlying topological space of $X \times_Z Y$ is homeomorphic to the corresponding fiber product in the category of topological spaces. In the category of integral monoschemes this is no longer true: $X \times_Z Y$ is a subspace of the topological fiber product.

1.4 Quasi-coherent sheaves on monoschemes

In this section we adapt some standard constructions in the theory of schemes to the context of monoschemes. The arguments are in general straightforward.

Let $X = \operatorname{spec}(Q)$ and let S be a Q-set. For each $f \in Q$, we have a Q_f-set S_f, and if $X_g \subseteq X_f$, there is a natural map $S_f \to S_g$, just because f maps to a unit in Q_g. Let $\tilde{S}$ denote the sheaf of sets on X associated to the presheaf $X_f \mapsto S_f$. This sheaf has a natural action of the sheaf of monoids $\mathcal{M}_X$.

Definition 1.4.1. *Let X be a monoscheme and let $\mathcal{S}$ be a sheaf of $\mathcal{M}_X$-sets on X. Then $\mathcal{S}$ is quasi-coherent if for every quasi-compact and quasi-separated submonoscheme U of X and every $f \in \Gamma(U, \mathcal{M}_X)$, the natural map*

$$\Gamma(U, \mathcal{S})_f \to \Gamma(U_f, \mathcal{S})$$

is an isomorphism. If X is locally noetherian, $\mathcal{S}$ is said to be coherent if it is coherent and for every $x \in X$, $\mathcal{S}_x$ is finitely generated as an $\mathcal{M}_{X,s}$-set.

Proposition 1.4.2. *Let X be a monoscheme and let $\mathcal{S}$ be a sheaf of $\mathcal{M}_X$-sets.*

1. *If $X = \operatorname{spec}(Q)$ and $\mathcal{S} = \tilde{S}$ for some Q-set S, then $\mathcal{S}$ is quasi-coherent. Moreover, the functor $S \to \tilde{S}$ gives an equivalence from the category of Q-sets to the category of quasi-coherent sheaves of $\mathcal{M}_X$-sets, with quasi-inverse $\mathcal{S} \mapsto \Gamma(X, \mathcal{S})$.*
2. *If there exists a covering $\mathcal{U}$ of X such that $\mathcal{S}_{|_U}$ is quasi-coherent for each $U \in \mathcal{U}$, then $\mathcal{S}$ is quasi-coherent.*

Proof First suppose that X is affine, and let x be its unique closed point. Let $\mathcal{P}$ be a presheaf defined on the family of basic open subsets of X and let $\mathcal{F}$ be its associate sheaf. As we saw in Remark 1.2.8, the maps $\mathcal{P}(X) \to \mathcal{P}_x$ and $\mathcal{F}(X) \to \mathcal{F}_x$ are isomorphisms. It follows that the natural map $\mathcal{P}(X) \to \mathcal{F}(X)$ is an isomorphism, and the same holds for any basic open subset of X. In particular, if $X = \operatorname{Spec}(Q)$ and S is a Q-set, then $S \cong \tilde{S}(X)$ and $S_f \cong \tilde{S}(X)_f \cong \tilde{S}(X_f)$ for every $f \in Q$.

Now suppose that $\mathcal{S}$ is a sheaf of $\mathcal{M}_X$-sets on a monoscheme X, that $\{U_i : i \in I\}$ is a finite open cover of a subset U of X, and that a is an element of $\mathcal{M}_X(U)$. Since $\mathcal{S}$ is a sheaf, the sequence

$$(\Gamma, (U, \mathcal{S})) \longrightarrow \left(\prod_i \Gamma(U_i, \mathcal{S})\right) \rightrightarrows \left(\prod_{i,j} \Gamma(U_i \cap U_j, \mathcal{S})\right)$$

is left exact (that is, the term on the left is the equalizer of the two arrows). Since localization commutes with finite limits, localization by a yields the exactness of the top row of the commutative diagram below; the bottom row is exact just because $\mathcal{S}$ is a sheaf:

$$\begin{array}{ccccc}
(\Gamma,(U,\mathcal{S}))_a & \longrightarrow & \left(\prod_i \Gamma(U_i,\mathcal{S})\right)_a & \rightrightarrows & \left(\prod_{i,j} \Gamma(U_i \cap U_j,\mathcal{S})\right)_a \\
\downarrow & & \downarrow & & \downarrow \\
\Gamma(U_a,\mathcal{S}) & \longrightarrow & \prod_i \Gamma(U_{ia},\mathcal{S}) & \rightrightarrows & \prod_{i,j} \Gamma(U_{ia} \cap U_{ja},\mathcal{S}).
\end{array}$$

Now to prove (1), suppose that $X = \operatorname{spec}(Q)$, that $\mathcal{S} = \tilde{S}$, and that U is a quasi-compact subset of spec Q. Then U admits a finite covering by special affine open subsets U_i. For each of these, the map $\Gamma(U_i,\mathcal{S})_a \to \Gamma(U_{ia},\mathcal{S})$ is an isomorphism, as we have seen. Since the intersection of two special affines is again special affine, the two vertical maps on the right in the diagram above are isomorphisms, and it follows that the vertical map on the left is also an isomorphism. This proves that $\mathcal{S}$ is quasi-coherent. The fact that the functor $S \mapsto \tilde{S}$ is fully faithful follows easily from the isomorphism $S \to \Gamma(X,\tilde{S})$. Suppose that $\mathcal{S}$ is quasi-coherent and let $S := \Gamma(X,\mathcal{S})$. Then X_f is quasi-compact and quasi-separated for every $f \in Q$, so the map $S_f \to \mathcal{S}(X_f)$ is an isomorphism. It follows that the map $\tilde{X} \to \mathcal{S}$ is an isomorphism.

To prove (2), let U be a quasi-compact and quasi-separated open subset of X. Then the hypothesis of (2) implies that U admits a finite and affine open covering $\{U_i : i \in I\}$ such that each $\mathcal{S}_{|U_i}$ is quasi-coherent. Then in the diagram above, the central vertical arrow is an isomorphism, and it follows that the left vertical arrow is injective. Since each $U_i \cap U_j$ is quasi-compact, our argument applies to show that the right vertical arrow is injective, and then it follows that the left vertical arrow is bijective. □

Remark 1.4.3. If $\mathcal{S}$ and $\mathcal{T}$ are sheaves of $\mathcal{M}_X$ sets on a monoidal space X, one can form their tensor product $\mathcal{S} \otimes \mathcal{T}$ by taking the sheaf associated to the presheaf $U \mapsto \mathcal{S}(U) \otimes_{\mathcal{M}_X(U)} \mathcal{S}(T)$. Since localization commutes with taking tensor products, the tensor product of two quasi-coherent sheaves on a monoscheme is quasi-coherent. Furthermore, the fiber product of quasi-coherent sheaves is quasi-coherent, since localization commutes with fiber products.

Remark 1.4.4. A morphism of monoschemes $f\colon X \to Y$ is *quasi-compact* (resp. *quasi-separated*) if for every quasi-compact (resp. quasi-separated) open V in Y, $f^{-1}(V)$ is quasi-compact (resp. quasi-separated). Then one sees easily that if $f\colon X \to Y$ is quasi-compact and quasi-separated, the functor f_* takes quasi-coherent sheaves on X to quasi-coherent sheaves on Y. For example, if f is quasi-compact and quasi-separated and $\mathcal{K}$ is a quasi-coherent sheaf of ideals in $\mathcal{M}_X$, then $f_*(\mathcal{M}_X)$ and $f_*(\mathcal{K}_X)$ are quasi-coherent, and it follows that $f_*(\mathcal{K}_X) \times_{f_*(\mathcal{M}_X)} \mathcal{M}_Y$ is again quasi-coherent. In particular, if f is an open immersion of noetherian monoschemes, any coherent sheaf of ideals on X has a canonical extension to a coherent sheaf of ideals on Y.

Let X be a monoscheme and let Z be a closed subset of X. For each open subset U of X, let $\mathcal{I}_Z(U)$ denote the set of sections of $\mathcal{M}_X(U)$ whose stalk at each point z of Z lies in the maximal ideal of $\mathcal{M}_{X,z}$. It is clear that $\mathcal{I}_Z(U)$ is an ideal of $\mathcal{M}_X(U)$ and that $\mathcal{I}_Z$ forms a sheaf of ideals in $\mathcal{M}_X$. For example, suppose that $X = \operatorname{spec}(Q)$ and that Z is the closed set defined by some ideal I of Q. Then $\mathcal{I}_Z(X) = \cap\{\mathfrak{p} : \mathfrak{p} \in Z\}$, which by Corollary I.1.4.3 is the radical of I. Moreover, $\mathcal{I}_Z(X_f) = \sqrt{I_f} = \sqrt{I}_f$ for every $f \in Q$. Thus $\mathcal{I}_Z = \sqrt{I}^\sim$, and hence is quasi-coherent.

Proposition 1.4.5. *If X a quasi-compact and quasi-separated monoscheme, the following conditions are equivalent.*

1. *X is quasi-affine, i.e., isomorphic to an open subset of an affine monoscheme.*
2. *Every quasi-coherent sheaf on X is generated by its global sections.*
3. *For every closed subset Z of X and every $x \in X \setminus Z$, there exists a global section f of $\mathcal{I}_Z$ such that $f_x \notin \mathcal{M}^+_{X,x}$.*
4. *The natural map $h\colon X \to \operatorname{spec}(\Gamma(X, \mathcal{M}_X))$ is an open immersion.*

Proof Suppose that X is quasi-affine, so that there exists an open immersion $j\colon X \to Y$, where $Y := \operatorname{spec}(P)$. Let Y' be a special affine open subset of Y and let $\{X_i : i \in I\}$ be a finite affine open cover of X. Then each $X_i \cap j^{-1}(Y')$ is special affine in X_i and hence also quasi-compact, and it follows that $j^{-1}(Y')$ is again quasi-compact. Thus the morphism j is quasi-compact, and one can show similarly that it is quasi-separated. Then if $\mathcal{S}$ is quasi-coherent on X, it follows from Remark 1.4.4 that $j_*(\mathcal{S})$ is quasi-coherent on Y. Hence $j_*(\mathcal{S}) \cong \tilde{S}$, where $S := \Gamma(Y, j_*(\mathcal{S})) = \Gamma(X, \mathcal{S})$. Since $\tilde{S}$ is generated by its global sections and $\mathcal{S} = j^*(\tilde{S})$, it too is generated by its global sections, proving that (1) implies (2).

Suppose that (2) holds and that Z is a closed subset of X. Then $\mathcal{I}_Z$ is a quasi-coherent sheaf of ideals on X, hence by hypothesis is generated by its global sections. If $x \in X \setminus Z$, then $\mathcal{I}_x = \mathcal{M}_{X,x}$, and hence there must exist a global section f of $\mathcal{I}$ such that $f_x \notin \mathcal{M}^+_{X,x}$. Thus (2) implies (3).

To prove that (3) implies (4), it will suffice to exhibit a family $\mathcal{V}$ of open subsets of $Y := \operatorname{spec}(\Gamma(X, \mathcal{M}_X))$ such that, for each $V \in \mathcal{V}$, h induces an isomorphism $h^{-1}(V) \to V$ and such that $\{h^{-1}(V) : V \in \mathcal{V}\}$ covers X. Since X is a monoscheme, each $x \in X$ admits an open affine neighborhood U. Then

$Z := X \setminus U$ is a closed subset of X not containing x. By hypothesis, there exists a global section f of $\mathcal{I}_Z$ whose stalk at x is not in $\mathcal{M}^+_{X,x}$. Then $x \in X_f \subseteq U$; furthermore $X_f = U_{f_{|U}}$, hence is affine, and hence is isomorphic to $\operatorname{spec}(\Gamma(X_f, \mathcal{M}_X))$. Since X is quasi-compact and quasi-separated, the map

$$\Gamma(X, \mathcal{M}_X)_f \to \Gamma(X_f, \mathcal{M}_X)$$

is an isomorphism. Then $V := \operatorname{spec}\Gamma(X, \mathcal{M}_X)_f$ is an affine open subset of Y, and in fact $X_f = h^{-1}(V)$. The set $\mathcal{V}$ of such V is the desired family of open subsets of Y. This completes the proof that (3) implies (4), and the implication of (1) by (4) is trivial. □

Definition 1.4.6. *A sheaf of $\mathcal{M}_X$-sets on a locally monoidal space X is invertible if, locally on X, there exists an isomorphism $\mathcal{M}_X \to \mathcal{S}$ of sheaves of $\mathcal{M}_X$-sets.*

It follows from Proposition 1.4.2 that an invertible sheaf of $\mathcal{M}_X$-sets on a monoscheme X is quasi-coherent. The tensor product of two invertible sheaves is again invertible and in particular, for every natural number n, the nth tensor power $\mathcal{S}^n$ of $\mathcal{S}$ is again invertible. The sheaf of generators $\mathcal{S}^*$ of an invertible $\mathcal{S}$ is a torsor under $\mathcal{M}^*_X$, and the natural map $\mathcal{S}^* \otimes_{\mathcal{M}^*_X} \mathcal{M}_X \to \mathcal{S}$ is an isomorphism. Thus there is an equivalence between the category $\mathcal{P}ic(X)$ of invertible $\mathcal{M}_X$-sheaves and the category of $\mathcal{M}^*_X$-torsors, and hence a bijection between the set $\operatorname{Pic}(X)$ of isomorphism classes of invertible sheaves on X and the set of isomorphism classes of $\mathcal{M}^*_X$-torsors. This bijection is compatible with the natural group structures on both sets, and in this way we find an isomorphism of groups

$$\operatorname{Pic}(X) \cong H^1(X, \mathcal{M}^*_X) \tag{1.4.1}$$

and an exact sequence of groups

$$\Gamma(X, \mathcal{M}^*_X) \to \Gamma(X, \mathcal{M}^{\mathrm{gp}}_X) \to \Gamma(X, \overline{\mathcal{M}}^{\mathrm{gp}}_X) \to \operatorname{Pic}(X) \to H^1(X, \mathcal{M}^{\mathrm{gp}}_X).$$

Recall that if X is integral, X is irreducible and $\mathcal{M}^{\mathrm{gp}}_X$ is constant, hence flasque, and $H^1(X, \mathcal{M}^{\mathrm{gp}}_X)$ vanishes. Thus $\operatorname{Pic}(X)$ can be understood as the quotient of the group $\Gamma(X, \overline{\mathcal{M}}^{\mathrm{gp}}_X)$ of divisors on X by the group $\Gamma(X, \mathcal{M}^{\mathrm{gp}}_X)/\Gamma(X, \mathcal{M}^*_X)$ of principal divisors. In particular, $\operatorname{Pic}(\mathbf{p}) \cong \mathbf{Z}$ (see Example 1.2.9).

Now suppose that X is a monoidal space, that $\mathcal{S}$ is an invertible sheaf on X, and that $\mathcal{T}$ is a sheaf of $\mathcal{M}_X$-sets on X. If s is a global section of $\mathcal{S}$, then

$$X_s := \{x \in X : s \notin \mathcal{M}^+_{X,x}\mathcal{S}\}$$

is an open subset of X. Furthermore, multiplication by s defines a map $\mathcal{T} \to \mathcal{S} \otimes \mathcal{T}$ and indeed a direct system $\mathcal{S}^n \otimes \mathcal{T} \to \mathcal{S}^{n+1} \otimes \mathcal{T}$ for $n \in \mathbf{N}$. The restriction of each of these maps to X_s is an isomorphism, and there is a commutative diagram

$$\begin{array}{ccc} \Gamma(X, \mathcal{S}^n \otimes \mathcal{T}) & \xrightarrow{s} & \Gamma(X, \mathcal{S}^{n+1} \otimes \mathcal{T}) \\ & {\scriptstyle s^{-n}} \searrow & \downarrow {\scriptstyle s^{-n-1}} \\ & & \Gamma(X_s, \mathcal{T}). \end{array}$$

Proposition 1.4.7. *With the notation above, suppose that X is a quasi-compact and quasi-separated monoscheme and that $\mathcal{T}$ is quasi-coherent. Then the map*

$$\varinjlim\{\Gamma(X, \mathcal{S}^n \otimes \mathcal{T}), \cdot s : n \in \mathbf{N}\} \to \Gamma(X_s, \mathcal{T})$$

induced by the diagrams above is an isomorphism.

Proof If $\mathcal{S}$ is isomorphic to $\mathcal{M}_X$ the result follows from Proposition 1.4.2. In the general case we can cover X by a finite number of open sets on which this is true, and then use the technique of that proposition to reach the same conclusion. □

1.5 Proj for monoschemes

In this section we show how a version of the Proj construction for schemes can be used to generalize the example of $\mathbf{p}$ discussed in Example 1.2.9. We shall later use this construction to perform monoidal transformations (blowups) for monoschemes and log schemes.

Definition 1.5.1. *A graded monoid is a monoid P together with a homomorphism $h\colon P \to \mathbf{N}$. If (P, h) is a graded monoid, a graded (P, h)-set is a P-set S endowed with a function $g\colon S \to \mathbf{Z}$ such that $g(p + s) = h(p) + g(s)$ for $p \in P$ and $s \in S$. For $p \in P$ (resp. for $s \in S$), one refers to $h(p)$ (resp. $g(s)$) as the degree of p (resp. of s).*

If (P, h) is a graded monoid, the homomorphism h defines a morphism of monoid-valued functors $\mathbf{a} \to \mathbf{a}_P$ and hence an action of $\mathbf{a}$ on $\mathbf{a}_P$. If Q is any monoid, this action is given explicitly by

$$\mathbf{a}(Q) \times \mathbf{a}_P(Q) \to \mathbf{a}_P(Q) : (q, \theta) \mapsto q\theta, \text{ where}$$

$$(q\theta)(p) := h(p)q + \theta(p).$$

Let $P_h^+ := h^{-1}(\mathbf{N}^+)$ and let $U_h := \operatorname{spec} P \setminus Z(P_h^+)$, a quasi-affine monoscheme.

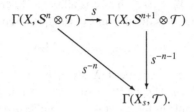

We denote again by h the homomorphism of groups $P^{\text{gp}} \to \mathbf{Z}$ as well as the homomorphism of sheaves of monoids $\mathcal{M}_{U_h} \to \mathbf{Z}$ induced by h. Finally, we let

$$\mathcal{M}_h := h^{-1}(0) \subseteq \mathcal{M}_{U_h},$$

a sheaf of monoids on U_h.

Theorem 1.5.2. *Let (P, h) be a graded monoid.*

1. *With the notations of the previous paragraph, the monoidal space*

$$\text{proj}(P, h) := (U_h, \mathcal{M}_h)$$

is a monoscheme.

2. *For each $p \in P_h^+$, the open set $U_p \subseteq \text{spec}(P)$ is contained in U_h, and*

$$D^+(p) := (U_p, \mathcal{M}_{h|U_p}) \cong \text{spec}(P_{(p)}),$$

where $P_{(p)}$ is the degree zero part of P_p. In particular, the family of all $D^+(p)$ for $p \in P_h^+$ is an affine open cover of $\text{proj}(P, h)$.

3. *If P_h^+ is generated by elements of degree one, then $\text{proj}(P, h)$ can be identified with the quotient of (U, h) by the action of $\mathbf{a}^*$. That is, for every monoid Q, the map $U_h(Q) \to \text{proj}(P, h)(Q)$ is surjective, and two elements of $U_h(Q)$ have the same image in $\text{proj}(P, h)(Q)$ if and only if they are in the same orbit under the action of the group Q^*.*

Proof Let us first establish the following:

Claim 1.5.3. If $p \in P_h^+$, the inclusion map $\iota\colon P_{(p)} \to P_p$ is exact and $\bar{\iota}$ is Kummer. If $h(p) = 1$, the homomorphism:

$$\iota'\colon P_{(p)} \oplus \mathbf{Z} \to P_p : (q', n) \mapsto q' + np$$

is an isomorphism.

The exactness of ι is clear, and it follows from Proposition I.4.2.1 that $\bar{\iota}$ is injective. Now if $q \in P$, let $q' := h(p)q - h(q)p$. Then $q' \in P_{(p)}$ and $h(p)q = q' + h(q)p$. Since $p \in P_p^*$, we have $h(p)\bar{q} = \bar{\iota}(\overline{q'})$ and, since $h(p) > 0$, it follows that $\bar{\iota}$ is Kummer. Note also that if $h(p) = 1$, the same argument shows that ι' is an isomorphism.

Since $\bar{\iota}$ is Kummer, $\text{Spec}(\iota)$ is a homeomorphism, by Corollary I.4.3.11. Note that there is a morphism of locally monoidal spaces

$$(U_h, \mathcal{M}_{U_h}) \to (U_h, \mathcal{M}_h),$$

simply because $\mathcal{M}_h \subseteq \mathcal{M}_{U_h}$. Restricted to U_p, this morphism induces an isomorphism of locally monoidal spaces $(U_p, \mathcal{M}_h) \cong \operatorname{spec}(P_{(p)})$, proving statement (2). Since the family of all U_p for $p \in P_h^+$ covers U_h, it follows that $(U_h, \mathcal{M}_h)$ has an open covering by affines, and hence is a monoscheme.

Suppose that p is an element of degree one of P_h^+. Then $D(p) := \operatorname{spec}(P_p) \subseteq U_h \subseteq \operatorname{Spec}(P)$. The isomorphism ι' of the claim induces an isomorphism of functors

$$D(p) \cong D^+(p) \times \mathbf{a}^*$$

that is compatible with the actions of $\mathbf{a}^*$. It follows that $D^+(p)$ can be identified with the orbit space of the action of $\mathbf{a}^*$ on $D(p)$. Remark 1.2.8 implies that every Q-valued point of $\operatorname{proj}(P, h)$ lies in some $D(p)$, and statement (3) follows. □

Remark 1.5.4. Let (P, h) and (P', h') be a graded monoids and let $\theta\colon P \to P'$ be a homomorphism such that $h = h' \circ \theta$. Then θ maps P_h^+ into $P_{h'}^+$ and hence the map $f := \operatorname{Spec}(\theta)$ sends $Z(P_{h'}^+)$ to $Z(P_h^+)$. If the radical of the ideal of P' generated by $\theta(P_h^+)$ equals the radical of $P_{h'}^+$, then in fact f maps $U'_{h'}$ to U_h and induces a morphism of monoschemes $\operatorname{proj}(P', h') \to \operatorname{proj}(P, h)$. This morphism is an isomorphism if there exists an integer d such that θ_n is an isomorphism for all n divisible by d.

Remark 1.5.5. Let (P, h) be a graded monoid and let $p \in P$ be an element of degree one. The following description of $P_{(p)}$ is also useful. For $n \in \mathbf{N}$, write P_n for the set of elements of P of degree n. Then multiplication by p induces a mapping $P_n \xrightarrow{p} P_{n+1}$. These maps are compatible with the maps $P_n \to P_{(p)}$ sending an element q to $q - np$, and hence the latter assemble into a map

$$\varinjlim\{P_n, \cdot p : n \in \mathbf{N}\} \to P_{(p)}.$$

It is easy to check that this map is a bijection.

Let (S, g) be a graded (P, h)-set. For each $p \in P_h^+$, P_p operates on the localization S_p of S by p, and $S_{(p)} := \{s \in S_p : g(s) = 0\}$ is stable under the action of $P_{(p)}$. One checks that these fit together to form a sheaf of $\mathcal{M}_h$-sets on $\operatorname{proj}(P, h)$. For $d \in \mathbf{Z}$ define $g_d : S \to \mathbf{Z}$ by $g_d(p) = p - d$. Then (S, g_d) is again a graded (P, h)-set, which we also denote by $(S, g)(d)$. Let $\mathcal{S}(d)$ be the corresponding sheaf of $\mathcal{M}_h$-sets, and note that, for $p \in P_h^+$, $\mathcal{S}(d)(U_p)$ is the set of elements s of S_p such that $g(s) = d$.

We are especially interested in the sheaves of sets $\mathcal{M}_h(d)$ on $\operatorname{proj}(P, h)$. Note that if $p \in P_h^+$ and if $h(p)$ divides d, say $d = eh(p)$, then ep is a section of $\mathcal{M}(d)$ on the open set $\operatorname{spec}(P_{(p)})$ of $\operatorname{proj}(P, h)$, and in fact is a basis for the set of such sections. In particular, if P_h^+ is generated by elements p with $h(p)$ dividing d, then $\mathcal{M}_h(d)$ is invertible.

Let us now describe the monoscheme analog of projective space, as well as its universal mapping property in terms of invertible quotients. Let $\mathbf{e}$ be a finite set and let $P_{\mathbf{e}} := \mathbf{N}^{\mathbf{e}}$ be the free commutative monoid with basis $\mathbf{e}$. Let $h\colon P_{\mathbf{e}} \to \mathbf{N}$ be the homomorphism with $h(e) = 1$ for all $e \in \mathbf{e}$. We denote by $\mathbf{pe}$ the monoscheme $\operatorname{proj}(P_{\mathbf{e}}, h)$. Each $e \in \mathbf{e}$ defines an element of degree one of (P, h) and hence a global section of the invertible sheaf $\mathcal{M}_{\mathbf{pe}}(1)$. Since $\mathbf{e}$ generates $P_{\mathbf{e}}$, the induced morphism

$$\mathcal{M}_{\mathbf{pe}} \times \mathbf{e} \to \mathcal{M}_{\mathbf{pe}}(1) \tag{1.5.1}$$

is surjective.

Definition 1.5.6. *If T is a monoscheme and $\mathbf{e}$ is a finite set, an invertible quotient of $\mathbf{e}$ on T is a pair $(\mathcal{S}, \mathbf{s})$, where $\mathcal{S}$ is an invertible sheaf of $\mathcal{M}_T$-sets and $\mathbf{s}$ is an $\mathbf{e}$-indexed sequence of sections of $\mathcal{S}$, such that the corresponding morphism of sheaves*

$$\mathcal{M}_T \times \mathbf{s} \to \mathcal{S}$$

is surjective.

It is clear that if $T' \to T$ is a morphism, an invertible quotient of $\mathbf{e}$ on T pulls back to an invertible quotient on T'.

Proposition 1.5.7. *The invertible quotient (1.5.1) of $\mathbf{e}$ on $\mathbf{pe}$ is universal. That is, if T is any monoscheme, then every invertible quotient of $\mathbf{e}$ on T is isomorphic to the pullback of (1.5.1) via a unique morphism $T \to \mathbf{pe}$.*

Proof Let $(\mathcal{S}, \mathbf{s})$ be an invertible quotient of $\mathbf{e}$ on T. For each $e \in \mathbf{e}$, let T_e be the set of all points t of T such that s_e forms a basis for the stalk $\mathcal{S}_t$ of $\mathcal{S}$ at t. Then T_e is an open subset of T, and $\{T_e : e \in \mathbf{e}\}$ is an open cover of T.

If $t \in T_e$, then for each $e' \in \mathbf{e}$ there is a unique $m_{e',t} \in \mathcal{M}_{T,t}$ such that $s_{e'} = m_{e',t}s_e$ in $\mathcal{S}_t$. The uniqueness implies that these sections patch to a global section $m_{e'}$ of $\mathcal{M}_T$ on T_e. Now, $P_{(e)}$ is isomorphic to the free commutative monoid with basis $\{e'e^{-1} : e' \neq e\}$, and so there is a unique homomorphism of monoids

$$\theta_e\colon P_{(e)} \to \Gamma(T_e, \mathcal{M}_T)$$

sending each $e'e^{-1}$ to $m_{e'}$. Let $f_e\colon T_e \to D^+(e) \subseteq \mathbf{pe}$ be the corresponding morphism of monoschemes. There is a unique morphism of $\mathcal{M}_S$-sets

$$h_e\colon f_e^*(\mathcal{M}_{\mathbf{pe}}(1)) \to \mathcal{S}_{|T_e}$$

sending e to s_e, and then in fact

$$h_e(e') = \theta_e((e'e^{-1})e) = m_{e'}s_e = s_{e'}.$$

It follows that $f_e^*(\mathcal{M}_{\mathbf{pe}}(1), \mathbf{e}) \cong (\mathcal{S}, \mathbf{s})$. Furthermore, it is clear that f_e is the unique morphism $T_e \to \mathbf{pe}$ with this property. Thus the morphisms $\{f_e : e \in \mathbf{e}\}$ patch to a morphism $f\colon T \to \mathbf{pe}$ such that $f^*(\mathcal{M}_{\mathbf{pe}}(1), \mathbf{e}) \cong (\mathcal{S}, \mathbf{s})$, and that f is unique. □

We write $\mathbf{p}^n$ for $\mathbf{pe}$ when $\mathbf{e} := \{0, 1, \ldots, n\}$.

Variant 1.5.8. Let (P, h) be a graded monoid and let K be an ideal in P. Assume that $P_h^+ \not\subseteq \sqrt{K}$. Then $U_{K,h} := \operatorname{spec}(P, K) \cap U_h$ is not empty. We let $\operatorname{proj}(P, K, h)$ denote the idealized monoidal space $U_{K,h}$ endowed with the sheaf of idealized monoids $(\mathcal{M}_h, \mathcal{K}_h)$, obtained by taking the degree zero part of the sheaves $\mathcal{M}$ and $\mathcal{K} := \tilde{K}$, restricted to $U_{K,h}$. Then $\operatorname{proj}(P, K, h)$ is an idealized monoscheme. For example, if P is freely generated by elements x, y of degree one and $K = (xy)$, then $\operatorname{proj}(P, K, h)$ consists of two points, and if P is freely generated by elements x, y, z of degree one and $K = (xyz)$, then $\operatorname{proj}(P, K, h)$ is the union of three copies of $\mathbf{p}^1$, each meeting in two points. More generally, if P is freely generated by $x_0, \ldots, x_n$ of degree one and $K = (x_0, \ldots, x_n)$, then $\operatorname{proj}(P, K, h)$ is the union of the "coordinate hyperplanes" in $\mathbf{p}^n$.

Definition 1.5.9. *Let X be a noetherian monoscheme and let $\mathcal{S}$ be an invertible sheaf on X. Then $\mathcal{S}$ is:*

1. *base-point-free if, for every $x \in X$, there exists a global section s of $\mathcal{S}$ such that $s_x \notin \mathcal{M}_{X,x}^+ \mathcal{S}_x$;*
2. *ample if for every closed subset Z of X and every $x \in X \setminus Z$, there exist a natural number n and a global section s of $\mathcal{S}^n \otimes \mathcal{I}_Z$ such that $s_x \notin \mathcal{M}_{X,x}^+ \mathcal{S}_x^n$;*
3. *extremely ample if (2) is true with $n = 1$ for every x and Z.*

Remark 1.5.10. An invertible sheaf on a noetherian monoscheme is ample if and only if some positive tensor power of it is extremely ample. Indeed, suppose that $\mathcal{S}$ is ample. If $s \in \mathcal{S}^n \otimes \mathcal{I}_Z$ and $s_x \notin \mathcal{M}_{X,x}^+ \mathcal{S}_x^n$ then, for every $d \in \mathbf{N}$, s^d is a global section of $\mathcal{S}^{nd}$ with the same property. Since X is noetherian, it has only finitely many points and finitely many closed sets, and hence finitely many pairs (x, Z) with $x \notin Z$. For each such pair, there exists an n and a section of $\mathcal{S}^n$ as above. Hence if N is the product of all such n, then for every (x, Z) there exists a global section of $\mathcal{S}^n$ such that $s_x \notin \mathcal{M}_{X,x}\mathcal{S}_x^N$. Thus $\mathcal{S}^N$ is extremely ample. The converse is trivial.

Proposition 1.5.11. *Let $\mathcal{S}$ be an invertible sheaf on a quasi-compact and quasi-separated monoscheme X. Then $\mathcal{S}$ is ample if and only if there exists* *a finite set of sections $s_1, \ldots, s_n$ of some power of $\mathcal{S}$ such that the corresponding family of open sets $X_{s_1}, \ldots, X_{s_n}$ is an affine covering of X. In particular, if (P, h) is a graded monoid and P_h^+ is generated by elements $p_1, \ldots, p_n$ with all $h(p_i)$ dividing $d > 0$, then $\mathcal{M}_h(d)$ is an ample invertible sheaf on* $\operatorname{proj}(P, h)$.

Proof Suppose $\mathcal{S}$ is ample. For each $x \in X$, choose an open affine neighborhood U with complement Z. Since $\mathcal{S}$ is ample, there exist $n > 0$ and $s \in \Gamma(X, \mathcal{I}_Z \mathcal{S}^n)$ such that $s_x \notin \mathcal{M}^+_{X,x} \mathcal{S}^n_x$. Then $X_s = U_s$ is an affine neighborhood of x contained in U. Since X is covered by the family of these sets X_s and is quasi-compact, a finite number of them will suffice. Conversely, suppose that there exists an affine open cover $X_{s_1}, \ldots, X_{s_n}$ of this form. Let x be a point of X not contained in some closed subset Z of X and let $s := s_i$ be such that $x \in X_{s_i}$. Since $x \notin Z$, $\mathcal{I}_{Z,x} = \mathcal{M}_{X,x}$, and since X_s is affine, there exists $m \in \Gamma(X_s, \mathcal{I}_Z)$ such that $m_x \notin \mathcal{M}^+_{X,x}$. Applying Proposition 1.4.7 with $\mathcal{T} = \mathcal{I}_Z$, we see that there exist $n > 0$ and $s \in \Gamma(X, \mathcal{S}^n \mathcal{I}_Z)$ such that $s_x \notin \mathcal{M}^+_{X,x} \mathcal{S}_x$.

Now suppose that $X = \operatorname{proj}(P, h)$ and that $p_1, \ldots p_n$ and d are as in the statement of the proposition. Let $d_i := d/h(p_i)$, and $s_i := d_i p_i \in \Gamma(X, \mathcal{M}_h(d))$. Then $X_{s_i} = D^+(p_i)$ is affine, and $X_{s_1}, \ldots, X_{s_n}$ covers X. It follows that $\mathcal{M}_h(d)$ is ample. □

Somewhat more generally, if $f \colon X \to Y$ is a morphism of monoschemes, an invertible sheaf $\mathcal{S}$ on X is said to be *ample relative to* f if Y admits a covering by affine open sets V such that the restriction of $\mathcal{S}$ to each $f^{-1}(V)$ is ample. There is an evident analog of Proposition 1.5.11 in this relative situation, which we leave to the reader.

If $\mathcal{S}$ is an invertible sheaf on a monoscheme X, let Q denote the disjoint union $\{\sqcup \Gamma(X, \mathcal{S}^n) : n \in \mathbf{N}\}$, and let $h \colon Q \to \mathbf{N}$ map $\Gamma(X, \mathcal{S}^n)$ to $\mathbf{N}$. Thus (Q, h) is a graded monoid. Fix a global section s of $\mathcal{S}$. Recalling the maps constructed in Remark 1.5.5 and Proposition 1.4.7, we find a diagram

$$\begin{array}{ccc} \varinjlim\{\Gamma(X, \mathcal{S}^n), \cdot s : n \in \mathbf{N}\} & \xrightarrow{\cong} & Q_{(s)} \\ \big\downarrow & \swarrow_{\theta} & \\ \Gamma(X_s, \mathcal{M}_X) & & \end{array}$$

and hence a morphism

$$t_s \colon X_s \to D^+(s) \to \operatorname{proj}(Q, h). \tag{1.5.2}$$

Explicitly, an element $q \in \Gamma(X, \mathcal{S}^n)$ defines an element $s^{-n}q \in Q_{(s)}$, and $\theta(s^{-n}q)$ is the unique element of $\Gamma(X_s, \mathcal{M}_X)$ such that $\theta(s^{-n}q)s^n = q$.

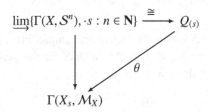

Theorem 1.5.12. *Let X be a noetherian monoscheme and let $\mathcal{S}$ be an invertible sheaf on X. Let $Q := \sqcup\Gamma(X, \mathcal{S}^n)$ with the natural monoid structure and map $h\colon Q \to \mathbf{N}$.*

1. *If $\mathcal{S}$ is base-point-free, then there exists a unique map*
$$f\colon X \to \mathrm{proj}(Q, h)$$
such that for each $s \in \Gamma(X, \mathcal{S})$, the restriction of f to X_s agrees with the morphism $t_s\colon X_s \to D^+(s)$ (1.5.2).
2. *If $\mathcal{S}$ is base-point-free and ample, the morphism f is an open immersion.*

Proof For the first statement, we must show that if s and $\tilde{s}$ are sections of $\mathcal{S}$, the morphisms
$$f\colon X_s \to \mathrm{proj}(Q, h) \quad \text{and} \quad \tilde{f}\colon X_{\tilde{s}} \to \mathrm{proj}(Q, h),$$
defined respectively by θ and $\tilde{\theta}$, agree on $X_s \cap X_{\tilde{s}}$. Observe that $\tilde{\theta}(s\tilde{s}^{-1})\tilde{s} = s$, and so $\tilde{\theta}(s\tilde{s}^{-1})$ restricts to a unit on $X_{\tilde{s}} \cap X_s$. It follows that $\tilde{\theta}$ induces a homomorphism
$$(Q_{(\tilde{s})})_{s/\tilde{s}} \to \Gamma(X_s \cap X_{\tilde{s}}, \mathcal{M}_X).$$
Similarly, θ induces a homomorphism
$$(Q_{(s)})_{\tilde{s}/s} \to \Gamma(X_s \cap X_{\tilde{s}}, \mathcal{M}_X).$$
We must show that, with the canonical identification
$$(Q_{(\tilde{s})})_{s/\tilde{s}} \cong (Q_{(s)})_{\tilde{s}/s} : q\tilde{s}^{-h(q)} \mapsto qs^{-h(q)}(s/\tilde{s})^{h(q)},$$
these two homomorphisms agree. On $X_s \cap X_{\tilde{s}}$, we have $\tilde{s} = \tilde{\theta}(s\tilde{s}^{-1})^{-1}s$. If $n := h(q)$, we can write $\theta(qs^{-n}) = qs^{-n}$ and $\tilde{\theta}(q\tilde{s}^{-n}) = q\tilde{s}^{-n}$. Then the identification above sends $q\tilde{s}^{-n}$ to $qs^{-n}(s\tilde{s}^{-1})^n$), and
$$\tilde{\theta}(q\tilde{s}^{-n}(s\tilde{s}^{-1})^n) = q\tilde{s}^{-n}\tilde{\theta}(s\tilde{s}^{-1})^n = qs^{-n},$$
as required.

Now suppose that $\mathcal{S}$ is base-point-free and ample. Fix a global section s of $\mathcal{S}$, and on the open set X_s, identify $\mathcal{M}_X$ with $\mathcal{S}$ via the basis element s for $\mathcal{S}$. By Proposition 1.4.7, thc map
$$Q_{(s)} = \varinjlim \Gamma(X, \mathcal{S}^n) = \Gamma(X_s, \mathcal{M}_X)$$
is an isomorphism. Because $\mathcal{S}$ is ample, hypothesis (3) of Proposition 1.4.5 is satisfied, and it follows that the map $X_s \to \mathrm{spec}(\Gamma(X_s, \mathcal{M}_X))$ is an open immersion. Thus the map $X_s = f^{-1}(D^+(s)) \to D^+(s)$ is an open immersion, and since the set of all sets of the form $D^+(s)$ covers $\mathrm{Proj}(Q, h)$, f is an open immersion. □

Remark 1.5.13. If T is a finite set of generators for the ideal $P_h^+ := h^{-1}(\mathbf{N}^+)$ of P, then P is generated as a monoid by $F := h^{-1}(0)$ and T. To see this, let Q be the submonoid of P generated by T and, for each $p \in P$, write $p = p' + q$ where $q \in Q$ and $p' \in P$, chosen so that $h(p')$ is minimized. If $h(p') > 0$, then $p' \in P_h^+$ and hence there exist $p'' \in P$ and $t \in T$ such that $p' = p'' + t$. Then $p = p'' + t + q$, with $t + q \in Q$ and $h(p'') = h(p') - h(t) < h(p')$, a contradiction. Thus in fact $p' \in F$ and so $p = p' + q$ belongs to the submonoid of P generated by F and Q. We will use an analog of this remark in our discussion of the moment map for projective monoschemes.

1.6 Separated and proper morphisms

The underlying topology of a monoscheme is not sufficiently rich to characterize either separation or properness. (It is likely that Berkovich's theory of "analytic spaces over $\mathbf{F}_1$" would remedy this defect.) Instead we use the valuative criteria as the definition. For simplicity we restrict our attention to fine monoschemes.

Recall that $\mathbf{a} = \text{spec}(\mathbf{N})$ and $\mathbf{a}^* := \text{spec}(\mathbf{Z})$.

Definition 1.6.1. *Let $f: X \to Y$ be a quasi-compact and quasi-separated morphism of fine monoschemes. Then f is separated (resp. proper) if, for every commutative square*

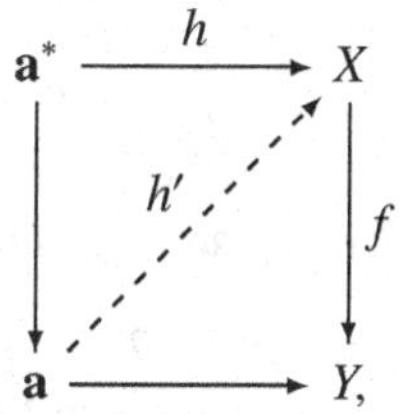

there is a at most one (resp. exactly one) morphism $h': \mathbf{a} \to X$ making the triangles in the diagram commute.

Let us observe that in contrast to the situation in algebraic geometry, a proper morphism need not be closed. For example, the summation map $\mathbf{N} \oplus \mathbf{N} \to \mathbf{N}$ induces a proper morphism of monoschemes, but the image of this map is not closed. It does not seem to be possible to characterize the properness of a morphism of monoschemes by a topological property. For example, $\text{spec}(\mathbf{Z}) \to \text{spec}(0)$ is not proper, but it has the same topology as the identity map $\text{spec}(0) \to \text{spec}(0)$ (even universally).

Remark 1.6.2. It is enough to prove that h' in the diagram in Definition 1.6.1 is unique (resp. exists) after some covering $\mathbf{a} \to \mathbf{a}$ given by multiplication by

n. Indeed, since $\mathbf{N}$ is saturated, multiplication by n is exact and Kummer. Thus, by Proposition 1.2.12, the diagram

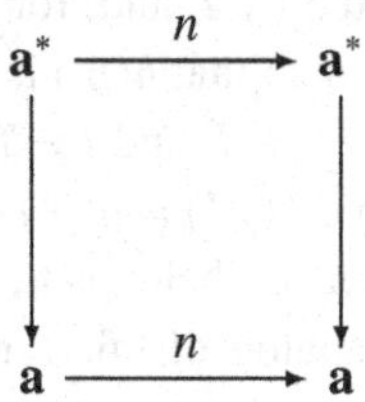

is cocartesian.

Let us also point out that the classes of separated and proper morphisms of monoschemes are closed under composition and base change, just as in the case of schemes. Furthermore, if $g \circ f$ is proper and g is separated, then f is proper. All these verifications are purely formal. The next result is more substantial.

Theorem 1.6.3. *Let $f\colon X \to Y$ be a quasi-compact and quasi-separated morphism of fine monoschemes, let T be an integral monoscheme, and consider a diagram of the form*

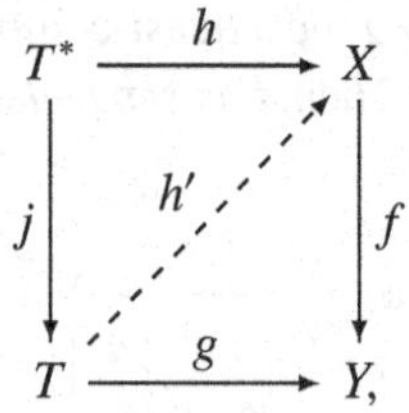

in which the outside commutative square is given.

1. *If h' exists (resp. is unique) whenever $T := \mathbf{a}$, then the same is true whenever T is the spectrum of a valuative monoid V.*
2. *If f is separated, then h' is uniquely determined by h.*

Proof Statement (1) is easy to see if V is finitely generated. In this case Proposition I.2.4.2 implies that $V/V^* \cong \mathbf{N}$, and hence $V \cong V^* \oplus \mathbf{N}$ and $V^{\mathrm{gp}} \cong V^* \oplus \mathbf{Z}$. Then to give a morphism $\operatorname{spec}(V) \to X$ (resp. $\operatorname{spec}(V^{\mathrm{gp}}) \to X$) is the same as to give a morphism $\operatorname{spec}(V^*) \to X$ and a morphism $\mathbf{a} \to X$ (resp. a morphism $\mathbf{a}^* \to X$).

The general case is more difficult. Let us explain how to prove the existence; the uniqueness will follow by applying this surjectivity to the diagonal morphism $X \to X \times_Y X$.

Suppose we are given a diagram as in the theorem and that no lifting h' exists. We will show that there is a similar diagram with $T = \mathbf{a}$. The image under g of the closed point of T is a point of Y, and without loss of generality we may replace Y by an affine neighborhood of this point and assume that $Y = \mathrm{spec}(P)$. Then the morphism g is given by a homomorphism $\theta\colon P \to V$, and h by a homomorphism $\phi\colon \mathcal{M}_{X,\xi} \to V^{\mathrm{gp}}$, where ξ is the generic point of X. Replacing P by the image P' of θ and X by $X \times_Y \mathrm{spec}(P')$, we may assume without loss of generality that θ is injective. Let Γ denote the image of $\mathcal{M}_{X,\xi}$ in V^{gp}. Then $\Gamma \cap V$ is a valuation monoid of Γ through which θ factors. Replacing V by $\Gamma \cap V$, we may as well assume that $\Gamma = V^{\mathrm{gp}}$, a finitely generated group, since X is fine.

Let x be a point of X and let ϕ_x be the composition

$$\phi_x\colon \mathcal{M}_{X,x} \longrightarrow \mathcal{M}_{X,\xi} \xrightarrow{\ \phi\ } \Gamma.$$

If ϕ_x factored through a local homomorphism $\mathcal{M}_{X,x} \to V$, then this homomorphism would induce a morphism h' filling in the square in the diagram, contradicting our hypothesis. Thus there exists $q_x \in \mathcal{M}_{X,x}$ such that either $\phi(q_x) \notin V$, or $q_x \in \mathcal{M}^+_{X,x}$ and $\phi(q_x) \in V^*$. Since V is valuative, in either case $\phi(q_x) \in -V$. Since X is finite, the submonoid Q of Γ generated by $\{\phi(q_x) : x \in X\}$ is finitely generated, and $Q \subseteq -V$. By Proposition I.2.2.1, there exists a local homomorphism $\gamma_0\colon Q \to \mathbf{N}$. After replacing γ_0 by $n\gamma_0$ for some $n > 0$, we may assume that γ_0 can be extended to a homomorphism $\Gamma \to \mathbf{Z}$. On the other hand, if $p \in P^+$, then $\theta(p) \in V^+$ and, since $Q \subseteq -V$, no multiple of $\theta(p)$ lies in Q, i.e., $\theta(p) \notin Q^{\mathrm{sat}}$. Hence by Corollary I.2.2.2 there exists a homomorphism $\gamma_p\colon Q \to \mathbf{N}$ such that $\gamma_p(p) < 0$, which we also may assume extends to a homomorphism $\Gamma \to \mathbf{Z}$. Let $\gamma := \gamma_0 + \sum_p \gamma_p$, where the sum is taken over a finite set of generators for P^+. Then γ induces a local homomorphism $Q \to \mathbf{N}$, and $-\gamma \circ \theta$ induces a local homomorphism $P \to \mathbf{N}$. We have then have the diagram

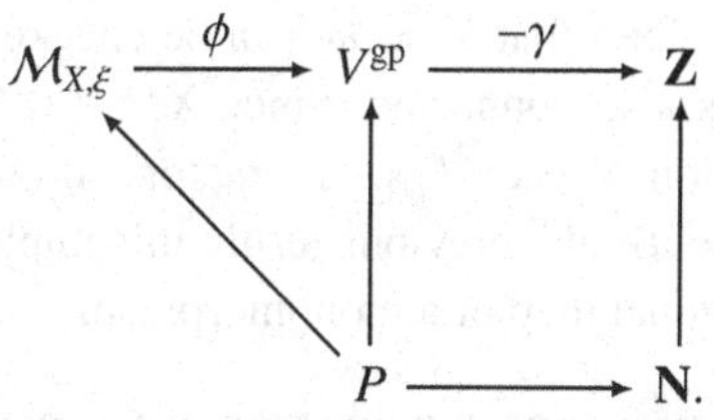

We claim that there is no $x \in X$ such that $-\gamma\circ\phi$ induces a local homomorphism $\mathcal{M}_{X,x} \to \mathbf{N}$. Suppose the contrary, and let $q_x \in \mathcal{M}_{X,x}$ be the element chosen above. If $\phi(q_x) \in Q^+$, then $-\gamma(\phi(q_x)) < 0$, a contradiction. On the other hand, if $\phi(q_x) \in Q^*$, then $-\gamma(\phi(q_x)) = 0$. But $\phi(q_x) \in Q^* \subseteq -V^* = V^*$, and in this case q_x was chosen to lie in $\mathcal{M}^+_{X,x}$. Since $-\gamma(\phi(q_x))$ is a unit, $-\gamma \circ \phi$ is not a local homomorphism $\mathcal{M}_{X,x} \to \mathbf{N}$. This completes the proof of (1).

To prove (2), suppose that $h'_1, h'_2 \colon T \to X$ make the triangles commute, and let $g \colon T \to X \times_Y X$ be the corresponding morphism. Then $g \circ j$ factors through the diagonal Δ_X, and we claim that the same is true of g. Without loss of generality, we may assume that T is affine, say $T := \operatorname{spec}(Q)$. Let t be the closed point of T and let τ be the generic point. By Proposition I.2.4.1, there exists a valuation monoid V of Q^{gp} which dominates Q. Writing S for spec V, we find that the morphism j factors through a morphism $T^* = S^* \to S$, and we find the diagram

$$\begin{array}{ccc} S^* & \longrightarrow & X \\ \Big\downarrow & & \Big\downarrow \Delta_X \\ S & \longrightarrow & X \times_Y X. \end{array}$$

If S were equal to $\mathbf{a}$, then the separation hypothesis on f would assert that the two morphisms $S \to X$ defined by the bottom arrow would coincide, hence that the diagram could be filled in with a map $S \to X$. But we saw in the proof of (1) that the same conclusion holds also for S, since V is valuative. It follows from this that the restrictions of h'_1 and h'_2 to S coincide, and hence that $x := h'_1(t) = h'_2(t)$ in X. Furthermore, the homomorphisms $h'^{\flat}_{1,t}, h'^{\flat}_{2,t} \colon \mathcal{M}_{X,x} \to \mathcal{M}_{T,t} \subseteq \mathcal{M}_{T,\tau}$ also agree and, since t was arbitrary, it follows that $h'_1 = h'_2$. □

Corollary 1.6.4. *Let $f \colon X \to Y$ be a separated morphism of of connected and fine monoschemes. If f induces a monomorphism $X^* \to Y^*$, then f is a monomorphism in the category of integral monoschemes, and the diagonal morphism $X \to X \times_Y X$, computed in that category, is an isomorphism.*

Proof Let $h_1, h_2 \colon T \to X$ be morphisms of integral monoschemes such that $f \circ h_1 = f \circ h_2$. The claim that $h_1 = h_2$ can be checked locally on T, so we may assume that T is also connected. Since $X^* \to Y^*$ is a monomorphism and since the restriction of each h_i to T^* factors through X^* it follows that these restrictions agree. By the previous result, this implies that $h_1 = h_2$. It is a formality that the diagonal map of a monomorphism is an isomorphism. □

Proposition 1.6.5. *A proper open immersion of fine connected and nonempty monoschemes is an isomorphism.*

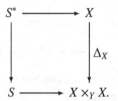

Proof Let $j \colon X \to Y$ be a proper open immersion, with X nonempty. It will suffice to prove that j is surjective. This condition is local on the target, so we may as well assume that $Y = \operatorname{spec}(P)$. Choose a local homomorphism $P \to \mathbf{N}$, which induces a morphism $h \colon \mathbf{a} \to Y$. The morphism $\operatorname{spec}(P^{\mathrm{gp}}) \to Y$ factors through $Y^* = X^*$, and we find the commutative diagram

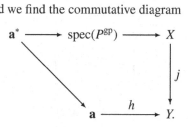

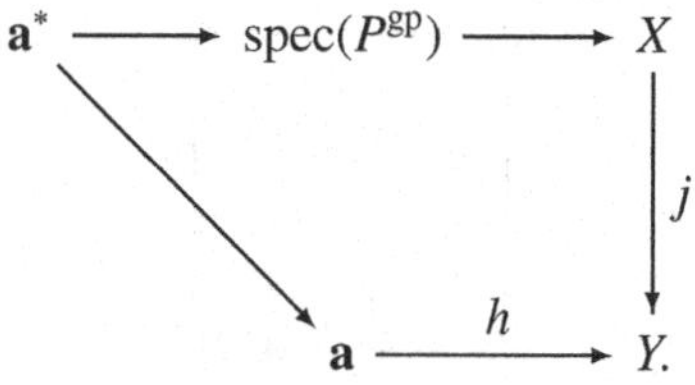

Then the properness of j implies that h lifts to a map $\mathbf{a} \to X$, and hence that the closed point y of Y lies in X. Since X is open in Y, it is stable under generization, so all the generizations of y, i.e., all the points of Y, lie in X. □

Let us agree to say that a morphism of connected and integral monoschemes $f: X \to Y$ is *birational* if it induces an isomorphism $X^* \to Y^*$, or, equivalently, an isomorphism $f^{-1}\mathcal{M}_Y^{\mathrm{gp}} \to \mathcal{M}_X^{\mathrm{gp}}$.

Proposition 1.6.6. *Let $f: X \to Y$ be a birational morphism of connected fine monoschemes.*

1. *If f is separated and exact, it is an open immersion.*
2. *If f is proper and exact, it is an isomorphism.*

Proof Assume that f is exact and birational. Let x be a point of X and recall from Remark 1.2.10 that $U := \operatorname{spec}(\mathcal{M}_{X,x})$ is an affine neighborhood of x and that $V := \operatorname{spec}(\mathcal{M}_{Y,y})$, is an affine neighborhood of $y := f(x)$. By hypothesis, the homomorphism $f_x^\flat: \mathcal{M}_{Y,y} \to \mathcal{M}_{X,x}$ is exact and induces an isomorphism $\mathcal{M}_{Y,y}^{\mathrm{gp}} \to \mathcal{M}_{X,x}^{\mathrm{gp}}$. Then by the definition of exactness, $f_x^\flat$ is an isomorphism. We conclude that, for each $x \in X$, there exists a neighborhood V of $f(x)$ and a section $g: V \to X$ of f. Assume in addition that f is separated. Consider the set of pairs (V, g), where V is open in Y and $g: V \to X$ is a section of f. If (V', g') is another such pair, then necessarily g and g' agree on Y^*. Since f is separated, they also agree on $V \cap V'$, by Theorem 1.6.3, and hence they patch uniquely to a section of f over $V \cup V'$. It follows that there is a maximal pair (W, g). Necessarily such a maximal W contains the image of f, and $g: Y \to X$ is surjective. Since $f \circ g$ is injective, g is bijective. Furthermore, for each $x \in X$, the map $f_x^\flat$ is an isomorphism, and it follows that f is an open immersion. This concludes the proof of statement (1). Statement (2) then follows from Proposition 1.6.5. □

Theorem 1.6.7. *Let $\theta: P \to Q$ be a homomorphism of fine monoids. Then the following conditions are equivalent.*

1. *The morphism* $\operatorname{spec}(\theta): \operatorname{spec}(Q) \to \operatorname{spec}(P)$ *is proper.*

2. *The diagram*

$$\begin{array}{ccc} Q^\vee & \longrightarrow & P^\vee \\ \big\downarrow & & \big\downarrow \\ W_Q & \xrightarrow{W_\theta} & W_P \end{array}$$

is cartesian, where $W_Q := \operatorname{Hom}(Q, \mathbf{Q})$ *and* $Q^\vee$ *is the cone in* W_Q *dual to* Q*, with similar notations for* P*.*

3. *The homomorphism* θ *is* **Q**-*surjective (see Definition I.4.3.1).*
4. *The action of* P *on* Q *defined by* θ *makes* Q *a finitely generated* P*-set.*

Proof Suppose that $\operatorname{spec}(\theta)$ is proper. Let $w \in W_Q$ be such that $W_\theta(w) \in P^\vee$. Choose $n > 0$ such that nw maps Q to $\mathbf{Z}$ and hence P to $\mathbf{N}$. Then we find the diagram

$$\begin{array}{ccc} \mathbf{a}^* & \xrightarrow{\operatorname{spec}(nw)} & \operatorname{spec}(Q) \\ \big\downarrow & & \big\downarrow \operatorname{spec}(\theta) \\ \mathbf{a} & \xrightarrow{f} & \operatorname{spec}(P). \end{array}$$

By the definition of properness, the morphism f factors uniquely through $\operatorname{spec} Q$, and hence in fact nw belongs to $Q^\vee$, and consequently the same is true of w. This implies that the diagram in (2) is cartesian. Now suppose that (2) holds, and let P' be the image of θ. We have a diagram:

$$\begin{array}{ccccc} Q^\vee & \longrightarrow & P'^\vee & \longrightarrow & P^\vee \\ \big\downarrow & & \big\downarrow & & \big\downarrow \\ W_Q & \xrightarrow{W_{\theta'}} & W_{P'} & \longrightarrow & W_P. \end{array}$$

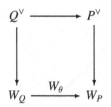

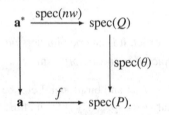

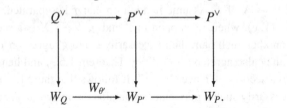

Since the rectangle is cartesian and the map $W_{P'} \to W_P$ is injective, the square on the left is also cartesian. Recall from Theorem I.2.3.12 that $Q^\vee$ is sharp, since Q spans W_Q. Since the square is cartesian, the kernel K of $W_{\theta'}$ is contained in $Q^\vee$, and since the latter is sharp and K is a group, it follows that $K = 0$. Thus $W_{\theta'}$ is injective. Since by construction $\theta' \colon P' \to Q$ is injective, $W_{\theta'}$ is also surjective, hence an isomorphism. Since the square is cartesian, it follows that $Q^\vee \to P'^\vee$ is an isomorphism. Then Theorem I.2.3.12, implies that $C(P') \to C(Q)$ is an isomorphism. It follows that $C(P) \to C(Q)$ is surjective, i.e., θ is $\mathbf{Q}$-surjective. The equivalence of (3) and (4) was proved in Proposition I.4.3.6. Finally, suppose that (3) holds. Then if $h\colon Q \to \mathbf{Z}$ is such that $h \circ \theta$ maps to $\mathbf{N}$, h also sends Q to $\mathbf{N}$. This shows that $\mathrm{spec}(\theta)$ is proper. □

Proposition 1.6.8. *Let $f\colon X \to Y$ be a proper morphism of fine noetherian monoschemes, where $Y := \mathrm{spec}(P)$. Then $f_*(\mathcal{M}_X)$ is a coherent sheaf of $\mathcal{M}_Y$-sets.*

Proof Without loss of generality, we may and shall assume that X is connected and that Y is affine, say $\mathrm{spec}\, P$. Since X is noetherian, f is quasi-compact and quasi-separated, so $f_*(\mathcal{M}_X)$ is quasi-coherent, by Remark 1.4.4. Thus it suffices to prove that $Q := \Gamma(X, \mathcal{M}_X)$ is a finitely generated P-set. Let $S := \mathrm{spec}\, Q$, so that we have a factorization $X \xrightarrow{g} S \xrightarrow{f'} Y$ of f. By Theorem 1.6.7, it will suffice to show that f' is proper. Let $h\colon \mathbf{a} \to Y$ be a morphism whose restriction h^* to $\mathbf{a}^*$ lifts to S. Since $Q^{\mathrm{gp}} \subset \Gamma(X, \mathcal{M}_X^{\mathrm{gp}})$ and the latter is a finitely generated group, nh factors through a map $\mathbf{a}^* \to X^*$ for some $n > 0$. Then the properness of f implies that nh lifts to X and hence to S, as required. □

Theorem 1.6.9. *Let (P, h) be a finitely generated graded monoid and let $F := h^{-1}(0)$. Then the natural map $\mathrm{proj}(P, h) \to \mathrm{spec}(F)$ is proper.*

Proof Suppose we are given a diagram of the form

$$\begin{array}{ccc} \mathbf{a}^* & \xrightarrow{\ g\ } & \mathrm{proj}(P, h) \\ \big\downarrow & & \big\downarrow \\ \mathbf{a} & \longrightarrow & \mathrm{spec}(F). \end{array}$$

Since $\mathbf{a}^*$ has only one point, the morphism g factors through $D^+(s)$ for some $s \in P_h^+$. Let K be the kernel of the homomorphism $h\colon P^{\mathrm{gp}} \to \mathbf{Z}$, and observe that $(P_{(s)})^{\mathrm{gp}} = K$. Indeed, any element of K can be written in the form $p - q$, where $p, q \in P$ and $h(p) = h(q)$. Let $d := h(p)$, let $e := h(s)$, and let $p' := p + (e-1)q$. Then

$$p - q = p' - eq = p' - ds - (eq - ds) \in P_{(s)}^{\mathrm{gp}}.$$

This shows that $K \subseteq (P_{(s)})^{\mathrm{gp}}$, and the reverse inclusion is trivial. Thus g corresponds to a homomorphism $K \to \mathbf{Z}$. Since P^{gp}/K is contained in $\mathbf{Z}$, it is free, and so the homomorphism h extends to a homomorphism $\theta\colon P^{\mathrm{gp}} \to \mathbf{Z}$.

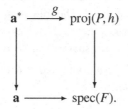

Let S be a finite set of generators for P, and choose an $s \in S \setminus F$ which minimizes $\theta(s)/h(s)$. We shall show that θ maps $P_{(s)}$ into $\mathbf{N}$, so that g extends to a morphism $\mathbf{a} \to D^+(s)$. Every element of $P_{(s)}$ has the form $p - ns$ with $h(p) = nh(s)$, and p can be written in the form $p = \sum n_i s_i + f$ with $s_i \in S \setminus F$ and $f \in F$. Then $h(p) = nh(s) = \sum n_i h(s_i)$. Furthermore, $\theta(s_i) \geq h(s_i)\theta(s)/h(s)$ for all i, so

$$\begin{aligned}\theta(p) &= \sum n_i\theta(s_i) + \theta(f)\\ &\geq \sum n_i h(s_i)\theta(s)/h(s)\\ &\geq h(p)\theta(s)/h(s)\\ &\geq n\theta(s).\end{aligned}$$

Thus $\theta(p - ns) \geq 0$, as required.

Now suppose that g and g' are morphisms $\mathbf{a} \to \operatorname{proj}(P, h)$ with the same restriction to $\mathbf{a}^*$. Let s (resp. s') be an element of P_h^+ containing $g(t)$ (resp $g'(t)$), where t is the closed point of $\mathbf{a}$. Then g factors through $D^+(s)$ and hence is given by a homomorphism $\theta\colon P_{(s)} \to \mathbf{N}$; similarly g' is given by a homomorphism $\theta'\colon P_{(s')} \to \mathbf{N}$. By hypothesis, θ and θ' induce the same homomorphism $P_{(s'+s)} \to \mathbf{Z}$. Now $s'' := h(s)s' - h(s')s \in P_{(s)}$ and hence $\theta(s'') \in \mathbf{N}$. Since $-s'' \in P_{(s')}$, it is also true that $\theta'(-s'') \in \mathbf{N}$. Thus $\theta(s'') = \theta'(s'') = 0$, so that θ extends to a homomorphism $P_{(s+s')} = (P_{(s)})_{s''} \to \mathbf{N}$, and similarly for θ'. That is, g and g' both factor through $D^+(s + s')$, which is affine and hence separated. Since g and g' agree on $\mathbf{a}^*$, they are equal. □

A monoscheme X is said to be *separated* if the unique morphism from X to the final object spec(0) is separated. The following result gives a characterization of separated monoschemes that is useful in comparing monoschemes to the classical notion of fans, as we shall see in Section 1.9

Proposition 1.6.10. *Let X be a connected and fine monoscheme. Then the following conditions are equivalent.*

1. *X is separated.*
2. *The diagonal morphism $\Delta_X\colon X \to X \times X$ is proper.*
3. *For every pair U, V of affine subsets of X, the intersection $U \cap V$ is affine, and the natural map*

$$\Gamma(U, \mathcal{M}_X) + \Gamma(V, \mathcal{M}_X) \to \Gamma(U \cap V, \mathcal{M}_X)$$

   is an isomorphism.

Proof If X is separated, the properness of Δ_X follows immediately from the definitions and the universal property of products. Suppose that Δ_X is proper and that U and V are affine open subsets of X. Then $U \cap V = \Delta_X^{-1}(U \times V)$ is evidently quasi-affine, and so Proposition 1.4.5 shows that the natural map $j\colon U \cap V \to Y := \operatorname{spec}(\Gamma(U \cap V, \mathcal{M}_X))$ is an open immersion. The map

$g: U \cap V \to U \times V$ is proper, since it is obtained by base change from the proper map Δ_X. Since $U \times V$ is affine, the map g factors: $g = g' \circ j$, and the map g' is separated because Y is affine. It follows that j is proper, and hence an isomorphism by Proposition 1.6.5. Thus $U \cap V$ is affine. Let x be the unique closed point of U, let y be the unique closed point of V, and let z be the unique closed point of $U \cap V$. Then

$$\Gamma(U, \mathcal{M}_X) = \mathcal{M}_{X,x}, \quad \Gamma(V, \mathcal{M}_X) = \mathcal{M}_{X,y}, \quad \text{and} \quad \Gamma(U \cap V, \mathcal{M}_X) = \mathcal{M}_{X,z}.$$

Since $U \cap V \to U \times V$ is proper, Theorem 1.6.7 implies that the natural homomorphism $\Gamma(U \times V, \mathcal{M}_{U \times V}) \to \Gamma(U \cap V, \mathcal{M}_X)$ is $\mathbf{Q}$-surjective, and hence the same is true of the homomorphism $\theta: P := \mathcal{M}_{X,x} + \mathcal{M}_{Y,y} \to Q := \mathcal{M}_{X,z}$. Since the latter is also injective, it is Kummer, and it follows from Corollary I.4.3.10 that the induced map $\operatorname{Spec}(\theta)$ is a homeomorphism. Consequently θ is a local homomorphism. Since X is a monoscheme, the cospecialization map $\lambda_{x,z}$ identifies $\mathcal{M}_{X,z}$ with the localization of $\mathcal{M}_{X,x}$ by $F := \lambda_{x,z}^{-1}(\mathcal{M}^*_{X,z})$. Then it follows that $P_F \to Q$ is surjective, and hence an isomorphism. Since $P \to Q$ is local, F maps into P^*. Thus $P_F = P$ and so $P = Q$, as claimed.

Finally suppose that (3) holds. To prove that X is separated, let h_1, h_2 be morphisms $\mathbf{a} \to X$ that agree when restricted to $\mathbf{a}^*$. Let $x_i := h_i(t)$, where t is the closed point of $\mathbf{a}$. It will suffice to prove that $x_1 = x_2$. Since X is noetherian, $U_i := \operatorname{spec}(\mathcal{M}_{X,x_i})$ is an affine open subset of X, and, by hypothesis, so is $U := U_1 \cap U_2$. Let x be the unique closed point of U, so that $\Gamma(U, \mathcal{M}_X) = \mathcal{M}_{X,x}$. By hypothesis, $\mathcal{M}_{X,x} = \mathcal{M}_{X,x_1} + \mathcal{M}_{X,x_2} \subseteq \mathcal{M}^{\mathrm{gp}}_{X,x}$. Since $h_i: \mathcal{M}_{X,x_i} \to \mathbf{N}$, $i = 1, 2$, induce the same homomorphism $h: \mathcal{M}^{\mathrm{gp}}_{X,x_i} \to \mathbf{Z}$, it follows that h maps $\mathcal{M}_{X,x}$ to $\mathbf{N}$. Since h_i is a local homomorphism and factors through the cospecialization map $\lambda_{x_i,x}$, the latter must also be local, But then $x = x_i$, and hence $x_1 = x_2$. □

1.7 Monoidal transformations

Here we discuss the analog of blowing up for monoschemes. For more details and applications, we refer to the discussion in [58]. The figure (1.7) illustrates two examples of blowing up.

Let $(X, \mathcal{M}_X)$ be a monoidal space and let $\mathcal{I}$ be a sheaf of ideals in $\mathcal{M}_X$. We say that $\mathcal{I}$ is *invertible* if it is so as a sheaf of $\mathcal{M}_X$-sets (Definition 1.4.6). When we work in the category of integral monoidal spaces, this condition is especially manageable.

Lemma 1.7.1. *Let X be an integral monoidal space.*

1. *A sheaf of ideals is invertible if and only if it is locally generated by a single element.*

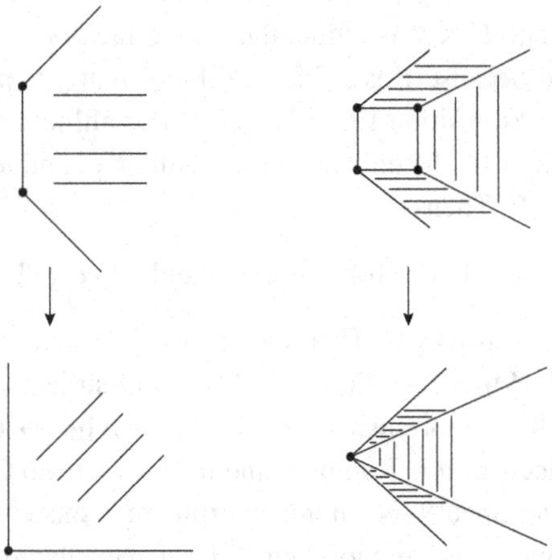

Figure 1.7.1 Blowing up the maximal ideal in $\mathbf{N}^2$ and in $Q_{2,2}$ (Example I.1.4.8).

2. *If $\mathcal{I}$ and $\mathcal{J}$ are sheafs of ideals, then $\mathcal{I}+\mathcal{J}$ is invertible if and only if $\mathcal{I}$ and $\mathcal{J}$ are invertible.*
3. *If $f: X' \to X$ is a morphism of integral monoidal spaces and $\mathcal{I}$ is an invertible sheaf of ideals on X, then the ideal $\mathcal{I}'$ of $\mathcal{M}_{X'}$ generated by the image of $f^{-1}(\mathcal{I})$ is again invertible, and the natural map*

$$f^*(\mathcal{I}) := \mathcal{M}_{X'} \otimes_{f^{-1}(\mathcal{M}_X)} f^{-1}(\mathcal{I}) \to \mathcal{I}'$$

is an isomorphism.

Proof If k is any section of $\mathcal{I}$, the integrality of $\mathcal{M}_X$ implies that multiplication by k defines an injection $\mathcal{M}_X \to \mathcal{I}$. Thus if k generates $\mathcal{I}$, this map is an isomorphism and $\mathcal{I}$ is invertible, proving (1). If $\mathcal{I}$ and $\mathcal{J}$ are invertible, we can choose local generators s of $\mathcal{I}$ and t of $\mathcal{J}$, and then $s+t$ will be a local generator of $\mathcal{I}+\mathcal{J}$. Suppose that $\mathcal{I}+\mathcal{J}$ is invertible; let us check that $\mathcal{I}$ is also invertible. Let k be a local generator of $\mathcal{I}+\mathcal{J}$ and write $k = s+t$ with $s \in \mathcal{I}$ and $t \in \mathcal{J}$. Then, if q is a section of $\mathcal{I}$, $q+t$ is a section of $\mathcal{I}+\mathcal{J}$ and hence $q+t = p+s+t$ for some local section p of $\mathcal{M}_X$. It follows that $q = p+s$ and that s generates $\mathcal{I}$. Thus $\mathcal{I}$ is invertible, and the same argument applies to $\mathcal{J}$. If $f: X' \to X$ is a morphism of integral monoidal spaces and $\mathcal{I}$ is an invertible sheaf of ideals in $\mathcal{M}_X$, it is clear that the ideal $\mathcal{I}'$ is locally monogenic and hence invertible. Furthermore, the map in statement (3) is surjective by definition, and its injectivity can be checked locally using a basis for $\mathcal{I}$. (In the terminology of Corollary I.4.5.16, one notes from the invertibility of $\mathcal{I}$ that $f^*(\mathcal{I})$ is $\mathcal{M}_X$-regular.) □

Theorem 1.7.2. *Let $\mathcal{I}$ be a quasi-coherent sheaf of ideals on an integral monoscheme X. If T is an integral monoidal space, let*

$$F_{\mathcal{I}}(T) := \{f : T \to X : \mathcal{I}_T := f^{-1}(\mathcal{I})\mathcal{M}_T \text{ is invertible}\}.$$

Then $F_{\mathcal{I}}$ is a subfunctor of the functor h_X and is representable in the category of integral monoschemes. Thus there exists a morphism of monoschemes $\beta \colon X_{\mathcal{I}} \to X$ such that:

1. *the ideal sheaf $\mathcal{I}_{\mathcal{I}}$ of $\mathcal{M}_{X_{\mathcal{I}}}$ generated by $\beta^{-1}(\mathcal{I})$ is invertible;*
2. *any morphism of monoidal spaces $\beta' : T \to X$ with this property factors uniquely through β.*

Furthermore, the ideal sheaf $\mathcal{I}_{\mathcal{I}}$ is ample, relative to the morphism $X_{\mathcal{I}} \to X$.

Proof If $g \colon T' \to T$ is a morphism of integral monoschemes over X and $\mathcal{I}_T$ is invertible, then (3) of Lemma 1.7.1 implies that $\mathcal{I}_{T'}$ is also invertible. Thus $F_{\mathcal{I}}$ is indeed a subfunctor of the functor h_X. (Note that this is not the case in the category of schemes, or without the integrality assumption.)

To construct $X_{\mathcal{I}}$, we first suppose that $X = \operatorname{spec}(P)$. Let $I := \Gamma(X, \mathcal{I})$ and let $B_I((P)$ be the Rees monoid of (P, I). Recall from Definition I.3.6.4 that this is the monoid $B_I(P)$ whose elements are pairs (n, p), where $n \in \mathbf{N}$ and $q \in I^n$, and whose monoid law is given by $(m, p) + (n, q) := (m + n, p + q)$. Let $h \colon B_I(P) \to \mathbf{N}$ be the homomorphism sending (n, p) to n, and let $X_I := \operatorname{proj}(B_I(P))$. The degree zero part of $B_I(P)$ is just P, and the ideal $h^{-1}(\mathbf{N}^+)$ is generated by the set of elements of the form $(1, q)$ with $q \in I$. Thus on X_I the sheaf $\mathcal{M}_{X_I}(1)$ is invertible. Consider the ideal IB_I of $B_I(P)$ generated by $I \subseteq P \subseteq B_I(P)$. The degree-$n$ part of IB_I is I^{n+1} if $n \geq 0$ and is empty otherwise. Thus there is an evident inclusion $IB_I(P) \to B_I(P)(1)$ of graded $B_I(P)$-sets, which is an isomorphism in positive degrees and hence induces an isomorphism of sheaves of $\mathcal{M}_{X_I}$-sets. It follows that $I\mathcal{M}_{X_I} \cong \mathcal{M}_{X_I}(1)$ and hence is invertible. It is ample by Proposition 1.5.11.

We claim that $\beta \colon X_I \to X$ represents the functor F_I. Observe first that β is a monomorphism in the category of integral monoschemes. To see this, suppose that h_1 and h_2 are morphisms $T \to X_I$ such that $\beta \circ h_1 = \beta \circ h_2$. To prove that $h_1 = h_2$ we may argue locally on T, hence we may assume that T is affine. If I is empty, X_I is empty and there is nothing to prove. Otherwise β is birational, hence induces an isomorphism $X_I^* \to X^*$, and hence h_1 and h_2 induce the same map $T^* \to X_I$. Since β is projective, it is separated and hence, by Theorem 1.6.3, $h_1 = h_2$.

The following lemma gives a useful description of certain open subsets of X_I.

Lemma 1.7.3. *Let I be an ideal in an integral monoid P, let q be an element of I, and let $b := (1, q)$ be the corresponding element of the Rees monoid $B_I(P)$. Then $B_I(P)_{(b)}$ is naturally isomorphic to the submonoid of P_q generated by elements of the form $p - q$ where $p \in I$. In particular, if I is generated by q, this monoid is naturally isomorphic to P.*

Proof The homomorphism $P \to B_I(P)$ induces an isomorphism from P^{gp} to the degree zero part of $(B_I(P))^{\mathrm{gp}}$. If $p \in I$, then $p - q$ maps to the element $(0, p - q) = (1 - p) - (1 - q) \in B_I(P)_{(b)}$. On the other hand, each element in $B_I(P)_{(b)}$ can be written in the form $(n, p) - (n, nq) = (0, p - nq)$, where $n \in \mathbf{N}$ and $p \in I^n$. Since $p \in I^n$, there exist $p_1, \ldots, p_n \in I$ such that $p = p_1 + \cdots + p_n$, and so $p - nq = \sum(p_i - q)$, which belongs to the submonoid P^{gp} generated by elements of the form $p - q$ with $p \in I$. □

We claim that β is an isomorphism if $\mathcal{I}$ is invertible. Since this assertion can be checked locally on X and since the construction of β is compatible with localization, we may assume that $\mathcal{I}$ is monogenic, say generated by q. Then the ideal $B_I^+(P)$ is generated by $b := (1, q)$, and so $X_{\mathcal{I}} = D^+(1, q)$, which is isomorphic to the degree zero part of the localization of $B_I(P)$ by b. Then Lemma 1.7.3 implies that the natural map $P \to B_I(P)_{(b)}$ is an isomorphism.

Now suppose that $f \colon T \to X$ and $\mathcal{I}_T$ is invertible. If T factors through $X_{\mathcal{I}}$, such a factorization will be unique, so to prove the existence of such a factorization we may argue locally on T. Thus we may assume that T is affine, say $T = \operatorname{spec}(Q)$, and that f is given by a homomorphism $\theta \colon P \to Q$. Let J be the ideal of Q generated by the image of I. Then θ induces a homomorphism of graded monoids $B_P(I) \to B_Q(J)$, and the ideal $B_Q^+(J)$ is generated by the ideal $B_P^+(I)$. By Remark 1.5.4, we conclude that there is a morphism $f_I \colon T_J \to X_I$ over the morphism f. By Lemma 1.7.3 the map T_J to T is an isomorphism, and hence f factors through X_I, as desired.

This concludes the proof of Theorem 1.7.2 when X is affine. The general case follows by a straightforward gluing argument, which we leave to the reader. □

Remark 1.7.4. It may be useful to give a more explicit description of a natural open cover of X_I and the corresponding open cover of the functor F_I. Recall that $X_I = \operatorname{proj}(B_I(P))$ admits an affine cover by sets of the form $D^+(b)$, where b ranges over a set of generators for the ideal $B_I^+(P)$ of $B_I(P)$. In this case, such a set of generators is furnished by a set of elements of the form $(1, q)$, where q ranges over a set of generators for I. Then $D^+((1, q))$ is the spectrum of the monoid obtained by taking the degree zero part of the localization of B_I by $(1, q)$, which, as Lemma 1.7.3 shows, can be identified with the submonoid of P^{gp} generated by elements of the form $p - q$ with $p \in I$. If T is an integral monoidal space, let

$$F_{\mathcal{I},q}(T) := \{f : T \to X : \mathcal{I}_T := f^{-1}(\mathcal{I})\mathcal{M}_T \text{ is generated by } f^\flat(q)\}.$$

Then it is not difficult to see that $F_{\mathcal{I},q}$ is an open subfunctor of $F_\mathcal{I}$ and is representable by the affine monoscheme $D^+((1, q))$.

Definition 1.7.5. *Let $\mathcal{I}$ be a quasi-coherent sheaf of ideals of a monoscheme X. Then the blowup, or monoidal transformation, of X along $\mathcal{I}$ is the map $\beta\colon X_\mathcal{I} \to X$ described in Theorem 1.7.2.*

The following result is an immediate consequence of Lemma 1.7.1 and the universal mapping property of the monoidal transformation.

Proposition 1.7.6. *Let X be an integral monoscheme and let $\mathcal{I}$ be a quasi-coherent sheaf of ideals on X.*

1. *If $g\colon X' \to X$ is a morphism of integral monoschemes and $\mathcal{I}'$ is the ideal of $\mathcal{M}_{X'}$ generated by $g^{-1}(\mathcal{I})$, then the natural map*

$$X'_\mathcal{I} := X'_{\mathcal{I}'} \to X' \times_X X_\mathcal{I}$$

is an isomorphism, where the fiber product is taken in the category of integral monoschemes.

2. *If $\mathcal{J}$ is another quasi-coherent sheaf of ideals on X, there are natural isomorphisms*

$$X_{\mathcal{I}+\mathcal{J}} \cong X_\mathcal{I} \times_X X_\mathcal{J} \cong (X_\mathcal{I})_\mathcal{J} \cong (X_\mathcal{J})_\mathcal{I}. \quad \square$$

The next result shows that monoidal transformations are surjective in a very strong sense.

Proposition 1.7.7. *Let X be a fine (resp. fine and saturated) connected monoscheme, let $\mathcal{I}$ be a nonempty and coherent sheaf of ideals in $\mathcal{M}_X$, and let $\beta\colon X_\mathcal{I} \to X$ be the monoidal transformation defined by $\mathcal{I}$. For each $x \in X$, there is a point x' of $X_\mathcal{I}$ lying over x such that the induced homomorphism*

$$\overline{\beta}^\flat_{x'} : \overline{\mathcal{M}}^{\text{gp}}_{X,x} \to \overline{\mathcal{M}}^{\text{gp}}_{X_\mathcal{I},x'}$$

is surjective with finite kernel (resp. is an isomorphism).

Proof Without loss of generality we may assume that X is affine, say $X = \operatorname{spec}(P)$, and let $I := \Gamma(X, \mathcal{I})$.

Assume for the moment that P is sharp. Let V be the vector space $\mathbf{Q} \otimes P^{\text{gp}}$ and let W be its dual. Since P is fine, I is finitely generated and, according to

Proposition I.2.3.22, there is a generator k of I that is a vertex of the convex hull S of I in V. This means that there exists a (w, a) in $W \oplus \mathbf{Q}$ such that $\langle s, w\rangle + a \geq 0$ for all $s \in S$ and such that k is the unique element of S for which equality holds. In particular, $\langle k', w\rangle + a \geq 0$ if $k' \in I$, and it follows that $\langle k', w\rangle + ma \geq 0$ if $k' \in I^m$. Then $D^+((1, k))$ is an open subset of X_I, which by Lemma 1.7.3 can be identified with the spectrum of the submonoid Q of P^{gp} generated by elements of the form $k' - k$ with $k' \in I$. We claim that Q^* is torsion, and in fact that it vanishes if P is saturated. Any $u \in Q^*$ can be written as a sum of elements of the form $k' - k$ as above and, since u is a unit, so is each such $k' - k$. Thus it will suffice to prove that if $k' - k$ is a unit, then it is torsion (resp. is zero). If $k' - k \in Q^*$, there exists an element q of Q with $k' - k + q = 0$, and we can write $q = k'' - mk$ for some $k'' \in I^m$ and some $m \in \mathbf{N}$. Then $k' + k'' = (m + 1)k$, $\langle k', w\rangle + a \geq 0$, $\langle k'', w\rangle + ma \geq 0$, and

$$\begin{aligned}\langle k', w\rangle + a + \langle k'', w\rangle + ma &= \langle k' + k'', w\rangle + (m + 1)a \\ &= \langle (m + 1)k, w\rangle + (m + 1)a = 0.\end{aligned}$$

It follows that $\langle k', w\rangle + a = 0$, and hence by the uniqueness of k that k' and k have the same image in $C(P)$. Then $nk = nk'$ for some positive integer n, and so $k' - k$ is torsion. If P is saturated and sharp, P^{gp} is torsion free, so $k = k'$, proving that Q is sharp.

It follows that the inclusion $P \to Q$ is local. Indeed, if $p \in P$ maps to a unit in Q, then $np = 0$ for some $n > 0$, hence p is a unit in P. Thus the ideal Q^+ of Q defines a point x' of X_I mapping to x. Then the homomorphism $\overline{\beta}^{\flat}_x$ identifies with the homomorphism $P^{\text{gp}} \to Q^{\text{gp}}/Q^* = P^{\text{gp}}/Q^*$, which is surjective with finite kernel (resp. an isomorphism), as claimed.

To deduce the general case, we argue as follows. Observe first that if $\theta\colon P \to Q$ is a homomorphism of integral monoids, then the pushout $Q \oplus_P \overline{P}$ is obtained by dividing Q by the subgroup $\theta(P^*)$ of Q^*. In particular, the natural map $\overline{Q} \to \overline{Q \oplus_P \overline{P}}$ is an isomorphism. To globalize this observation, write $X' \to X$ for the morphism of monoschemes corresponding to $P \to \overline{P}$, let $\tilde{X} \to X$ be a morphism of integral monoschemes, and let $\tilde{X}' := \tilde{X} \times_X X'$. Then it follows from the affine case just considered that the natural map $\tilde{X}' \to \tilde{X}$ is a homeomorphism and induces an isomorphism $\overline{\mathcal{M}}_{\tilde{X}} \to \overline{\mathcal{M}}_{\tilde{X}'}$. Now let $\mathcal{I}'$ be the ideal of X' generated by $\mathcal{I}$. As we have just seen, the natural map $X'_{\mathcal{I}'} \cong X_{\mathcal{I}} \times_X X' \to X_{\mathcal{I}}$ is a homeomorphism and the homomorphism $\overline{\mathcal{M}}_{X_{\mathcal{I}}} \to \overline{\mathcal{M}}_{X'_{\mathcal{I}'}}$ is an isomorphism. Thus statement (1) for $X'_{\mathcal{I}'} \to X'$ implies the analogous result for $X_{\mathcal{I}} \to X$. □

We should remark that the saturation hypothesis in Proposition 1.7.7 is not superfluous. For example, if P is the monoid given by generators p, q and relation $2p = 2q$, then the monoidal transformation of the ideal (p, q) is given by the saturation homomorphism $P \to P^{\text{sat}}$. The maximal ideal of P^{sat} is the only prime lying over P^+, and the map $\overline{P^{\text{gp}}} \to \overline{(P^{\text{sat}})^{\text{gp}}}$ is not an isomorphism. However, the proof of the proposition shows that it would be enough to assume that $\overline{\mathcal{M}}_X^{\text{gp}}$ is torsion free.

Corollary 1.7.8. *Let X be a connected and fine monoscheme and let β be the monoidal transformation defined by a nonempty sheaf of ideals of X. Then in the category of fine monoschemes, the morphism β is universally surjective.*

Proof Let $\mathcal{I}$ be an nonempty sheaf of ideals on X, let X^{sat} be the saturation of X, and let $\mathcal{J}$ be the ideal of X^{sat} generated by $\mathcal{I}$. Since the pullback of $\mathcal{I}$ to $(X^{\text{sat}})_{\mathcal{J}}$ is invertible, we have a commutative diagram:

$$\begin{array}{ccc} (X^{\text{sat}})_{\mathcal{J}} & \longrightarrow & X_{\mathcal{I}} \\ {\scriptstyle \beta'}\downarrow & & \downarrow{\scriptstyle \beta} \\ X^{\text{sat}} & \longrightarrow & X. \end{array}$$

Then X^{sat} is fine and and $\mathcal{J}$ is nonempty, and Proposition 1.7.7 implies that the monoidal transformation β' defined by $\mathcal{J}$ is surjective. Since $X^{\text{sat}} \to X$ is also surjective, it follows that β is surjective. By Proposition 1.7.6, the base change of a monoidal transformation is again a monoidal transformation, and hence the surjectivity is universal. □

Remark 1.7.9. Although monoidal transformations are universally surjective in the category of fine monoschemes, care must exercised in the context of fiber products. Since monoidal transformations are not exact, the "four point lemma" (see Corollary 4.2.16) does not hold. In fact, if $\mathcal{I}$ is a coherent sheaf of ideals in $\mathcal{M}_X$, the isomorphism (1) of Proposition 1.7.6 implies that the two projections $X_{\mathcal{I}} \times_X X_{\mathcal{I}} \to X_{\mathcal{I}}$ are isomorphisms. Hence if x_1 and x_2 are two distinct points of $X_{\mathcal{I}}$ with the same image in X, there is no point of $X_{\mathcal{I}} \times_X X_{\mathcal{I}}$ mapping to (x_1, x_2). Furthermore, the map $X_{\mathcal{I}} \setminus \{x_1\} \to X$ is surjective but not universally surjective, since

$$X_{\mathcal{I}} \setminus \{x_1\} \times_X X_{\mathcal{I}} \setminus \{x_2\} = X_{\mathcal{I}} \setminus \{x_1, x_2\}.$$

1.8 Monoidal transformations and exactification

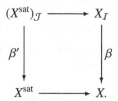

We have seen that monoidal transformations are not exact. In fact, their nonexactness is strong enough to swallow the failure of a general morphism to be exact: any morphism of monoschemes can be "exactified" by a suitable blowup. This result is one of the main applications of blowing up. Its proof is based on arguments in [40] and [46]. First, a warning: recall that a homomorphism θ of integral monoids can be exact but not locally exact, or locally exact but not exact (Examples I.4.2.13), and the corresponding morphism of monoschemes is exact if and only if θ is locally exact.

In the following theorem and its proof, we use the following notation. If $f: X \to Y$ is a morphism of fine monoschemes and $\mathcal{I}$ is a coherent sheaf of ideals in Y, then $X_\mathcal{I} \to X$ is the monoidal transformation defined by the ideal of $\mathcal{I}_X$ of $\mathcal{M}_X$ generated by $\mathcal{I}$. Recall from Proposition 1.7.6 that $X_\mathcal{I} \cong X \times_Y Y_\mathcal{I}$, where the fiber product is taken in the category of integral monoschemes.

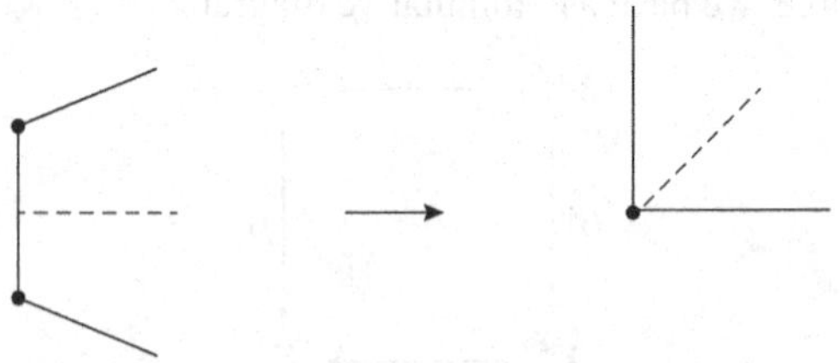

Figure 1.8.1 Exactifying the diagonal map $\underline{A}_\mathbb{N} \to \underline{A}_{\mathbb{N}^2}$.

Theorem 1.8.1 (exactification). *Let $f: X \to Y$ be a quasi-compact morphism of quasi-compact and fine monoschemes. Then there exists a coherent sheaf of ideals $\mathcal{I}$ on Y such that the morphism*

$$f_\mathcal{I}: X_\mathcal{I} \cong X \times_Y Y_\mathcal{I} \to Y_\mathcal{I}$$

is exact. If V is an open subset of Y contained in the image of f with the property that f is exact over V, then $\mathcal{I}$ may be chosen so that its restriction to V is the unit ideal.

Proof We begin with some preliminary lemmas. The first says that any birational map of fine affine monoschemes is an open subset of a blowup.

Lemma 1.8.2. *Let $f: Y' \to Y$ be a morphism of fine affine monoschemes that induces an isomorphism $Y'^* \to Y^*$. Then there exists a coherent sheaf of ideals $\mathcal{I}$ in Y such that the morphism $Y'_\mathcal{I} \to Y_\mathcal{I}$ is an open immersion and the morphism $Y'_\mathcal{I} \to Y'$ is an isomorphism. Furthermore, if V is an open subset of Y over which f is an isomorphism, $\mathcal{I}$ may be chosen to be invertible on V.*

Proof Let $\theta: P \to Q$ be the monoid homomorphism defining f. By hypothesis, θ^{gp} is an isomorphism. Note that f is a monomorphism in the category of integral monoschemes, and in particular that $Y' \cong Y' \times_Y Y'$. Let $(q_1, \ldots, q_n)$ be a finite sequence of elements of Q such that Q is generated as a monoid by P together with $(q_1, \ldots, q_n)$. For each i we can find $p_i, p'_i \in P$ such that $q_i = p_i - p'_i$. In fact we arrange this so that all p'_i are equal, say to $p_0 \in P$. Let I be the ideal of P generated by $(p_0, p_1, \ldots, p_n)$ and let $\mathcal{I}$ be the corresponding sheaf of ideals on Y. Then $Y_{\mathcal{I}}$ (resp. $Y'_{\mathcal{I}}$) is covered by affine open sets $Y_i := D^+((1, p_i)) \subseteq \operatorname{Proj}(B_I(P), h)$ (resp., by open sets $Y'_i := Y_i \times_Y Y'$). In particular, $D^+((1, p_0)) = \operatorname{spec}(P_{((1,p_0))})$ where, by Lemma 1.7.3, $P_{(1,(p_0))}$ is the submonoid of $P^{\mathrm{gp}} = Q^{\mathrm{gp}}$ generated by P and $q_1, \ldots, q_n$. But this is just the monoid Q, and so the maps $Y'_0 \to Y'$ and $Y'_0 \to Y_0$ are isomorphisms. It follows that, for every i, the maps $Y'_0 \cap Y'_i \to Y'_i$ and $Y'_0 \cap Y'_i \to Y_0 \cap Y_i$ are isomorphisms. Since $Y_0 \cap Y_i \to Y_i$ is an open immersion, so is the map $Y'_i \to Y_i$. It follows that $Y'_{\mathcal{I}} \to Y_{\mathcal{I}}$ is an open immersion. Since $Y'_0 \to Y'$ is already an isomorphism, in fact the map $Y'_{\mathcal{I}} \to Y'$ is an isomorphism. Let us check that the restriction of the sheaf of ideals $\mathcal{I}$ to V is invertible. We may cover V by special affine open sets Y_p where $p \in P$ is such that the localization $P_p \to Q_p$ is an isomorphism. Then for each i, there exists an $n \geq 0$ such that $p'_i := np + q_i \in P$. But then $p_i = p'_i + p_0 - np$, so the ideal $I_p \subseteq P_p$ is principally generated by p_0. Since the sets Y_p cover Y, it follows that $\mathcal{I}$ is invertible. □

Lemma 1.8.3. *Let $f\colon X \to Y$ be a morphism of fine affine monoschemes, and let x be the unique closed point of X. Then there exists a coherent sheaf of ideals $\mathcal{I}$ on Y such that the map $X_{\mathcal{I}} \to X$ is an isomorphism and the morphism $X_{\mathcal{I}} \to Y_{\mathcal{I}}$ is exact at the unique point of $X_{\mathcal{I}}$ lying over x. Furthermore, $\mathcal{I}$ can be chosen so that $\mathcal{I}_y$ is invertible at each point y of the image of f over which f is exact.*

Proof The morphism f is is given by a homomorphism of fine monoids $\theta\colon P \to Q$, and x corresponds to the face Q^* of Q. Recall that in Proposition I.4.2.17 we defined a factorization

$$\theta = P \xrightarrow{\tilde{\theta}} P^{\theta} \xrightarrow{\theta^e} Q,$$

where P^{θ} is the inverse image of Q in P^{gp}. Here θ^e is automatically exact, the homomorphism $P \to P^{\theta}$ induces an isomorphism on associated groups, and P^{θ} is again fine. Let

$$f = X \xrightarrow{f^e} Y^e \xrightarrow{\tilde{f}} Y$$

be the factorization of f corresponding to the factorization of θ, and apply Lemma 1.8.2 to the morphism $\tilde{f}$. Then the map $Y^e_{\mathcal{I}} \to Y^e$ is an isomorphism and hence so is the map $X_{\mathcal{I}} \cong X \times_{Y^e} Y^e_{\mathcal{I}} \to X$. Since f^e is exact at x, the map

$X_{\mathcal{I}} \to Y^e_{\mathcal{I}}$ is exact at the unique point x' of $X_{\mathcal{I}}$ lying over x. Since the map $Y^e_{\mathcal{I}} \to Y_{\mathcal{I}}$ is an open immersion, it follows that $X_{\mathcal{I}} \to Y_{\mathcal{I}}$ is also exact at x'.

Suppose $\mathcal{I}$ is chosen so that it is invertible at each point of Y over which $\tilde{f}$ is an isomorphism. Let y be a point in the image of f over which f is exact. Then y corresponds to a face F of P, and by hypothesis there exists a face G of Q such that $\theta^{-1}(G) = F$ and such that the induced homomorphism $P_F \to Q_G$ is exact. Then $P_F \to Q_F$ is also exact, and it follows that $P_F \to P^\theta_F$ is an isomorphism. Hence $\tilde{f}$ is an isomorphism over y and $\mathcal{I}_y$ is invertible. □

Lemma 1.8.4. *With the same hypothesis as in the previous lemma, there exists an ideal $\mathcal{J}$ so that $X_{\mathcal{J}} \to Y_{\mathcal{J}}$ is exact at every point of $X_{\mathcal{J}}$ lying over x and such that $\mathcal{J}_y$ is the unit ideal at every point in the image of f over which f is exact.*

Proof We continue with the notation of the previous lemma. Let I be the ideal constructed there, and let $I^\vee \subseteq P^{\mathrm{gp}}$ denote the set of all $g \in P^{\mathrm{gp}}$ such that $g + I^\vee \subseteq P$. Then $J := I^\vee + I$ is an ideal of P containing I, and $\mathcal{J}_y$ is the unit ideal at every y at which $\mathcal{I}_y$ is invertible. Furthermore, the pullback of $\mathcal{J}$ to $Y_{\mathcal{J}}$ is invertible, and hence by (2) of Lemma 1.7.1 so is the pullback of $\mathcal{I}$. Then the map $Y_{\mathcal{J}} \to Y$ factors uniquely through $Y_{\mathcal{I}}$, and in fact $Y_{\mathcal{J}+\mathcal{I}} \cong Y_{\mathcal{J}}$. It follows that the map $X_{\mathcal{J}} \to Y_{\mathcal{J}}$ is the pullback of the map $X_{\mathcal{I}} \to Y_{\mathcal{I}}$ along $Y_{\mathcal{J}} \to Y_{\mathcal{I}}$ and hence that it is exact at each point of $X_{\mathcal{J}}$ lying over x. □

Now we turn to the proof of the theorem. First assume that Y is affine. Since X is quasi-compact and fine, it contains only finitely many points, say $x_1, \ldots, x_n$. For each i, let $X_i := \operatorname{spec}(\mathcal{M}_{X,x_i})$, an affine open subset of X containing x_i as its unique closed point. Let $\mathcal{I}_i$ be a sheaf of ideals as in the previous lemma, constructed for the restriction f_i of f to X_i. Note that we can arrange for $\mathcal{I}_i$ to be the unit ideal over V. Then $X_{\mathcal{I}_i} \to Y_{\mathcal{I}_i}$ is exact at every point of $X_{\mathcal{I}_i}$ lying over x_i. Then for any $Y' \to Y_{\mathcal{I}_i}$ the base-changed map $X'_{\mathcal{I}_i} \to Y'_{\mathcal{I}_i}$ is exact at every point of $X'_{\mathcal{I}_i}$ lying over x_i. Let $\mathcal{I} := \mathcal{I}_1 + \cdots + \mathcal{I}_n$ and apply (2) of Proposition 1.7.6 to interpret this blowup as an iterated sequence of blowups by each of the ideals $\mathcal{I}_i$. Then the map $X_{\mathcal{I}} \to Y_{\mathcal{I}}$ is exact at every point lying over each of the points $x_1, \ldots, x_n$, and hence everywhere, and it remains the unit ideal over V.

For the general case, choose a finite affine open covering $Y_1, \ldots, Y_m$ of Y and, for each Y_i, a coherent sheaf of ideals $\mathcal{I}_i$ on Y_i which makes the restriction of $f_{\mathcal{I}_i}$ to $f^{-1}_{\mathcal{I}_i}(Y_i)$ exact. We can also arrange for $\mathcal{I}_i$ to be the unit ideal on $Y_i \cap V$, and hence to extend to $V \cup Y_i$. By Remark 1.4.4, we may find a coherent sheaf of ideals $\mathcal{J}_i$ on Y whose restriction to $V \cup Y_i$ is $\mathcal{I}_i$. Let $J := \sum \mathcal{J}_i$, Then the map $X_{\mathcal{J}} \to Y_{\mathcal{J}}$ can be viewed as the successive blowup of Y by the ideals $\mathcal{J}_i$ in any order, and we thus see that its restriction to each $f_{\mathcal{J}}^{-1}(Y_i)$ is exact. Hence it is exact everywhere. Furthermore, the restriction of $\mathcal{J}$ to V is the unit ideal. □

Remark 1.8.5. Blowing up can destroy saturation: $Y_{\mathcal{I}}$ may fail to be saturated even if Y is saturated. In order to apply Theorem 1.8.1 to the category of saturated monoschemes, one needs to observe that if $X \to Y$ is exact, so is the corresponding map of saturations $X^{\mathrm{sat}} \to Y^{\mathrm{sat}}$, by (7) of Proposition I.4.2.1. Thus if $f\colon X \to Y$ is a quasi-compact morphism of fine monoschemes and Y is affine, there is a coherent sheaf of ideals $\mathcal{I}$ of Y such that the morphism $X_{\mathcal{I}}^{\mathrm{sat}} \to Y_{\mathcal{I}}^{\mathrm{sat}}$ is exact.

As an application of Theorem 1.8.1, we have the following global version of Lemma 1.8.2.

Corollary 1.8.6. *Let $f\colon X \to Y$ be a quasi-compact birational morphism of fine monoschemes, with Y quasi-compact. Assume that f is separated (resp. proper). Then there exists a coherent sheaf of ideals $\mathcal{I}$ on Y such that $f_{\mathcal{I}}\colon X_{\mathcal{I}} \to Y_{\mathcal{I}}$ is an open immersion (resp. an isomorphism). Furthermore, if V is an open subset of $Y_{\mathcal{I}}$ over which f is an isomorphism, $\mathcal{I}$ may be chosen so that its restriction to V is the unit ideal.*

Proof Theorem 1.8.1 tells us that there is an $\mathcal{I}$ such that the map $X_{\mathcal{I}} \to Y_{\mathcal{I}}$ is exact. Since it is still birational and separated (resp. proper), it is an open immersion (resp. an isomorphism), by Proposition 1.6.6. Since f is exact over the open set V, and V is contained in the image of f, we may choose $\mathcal{I}$ so that its restriction to V is the unit ideal. □

Theorem 1.8.7. *Let $f\colon X \to Y$ be a a morphism of fine monoschemes. If f is exact then it is* **Q***-integral. If in addition X is quasi-compact, then there exists a natural number $n > 0$ such that the morphism f'' in the following diagram is saturated:*

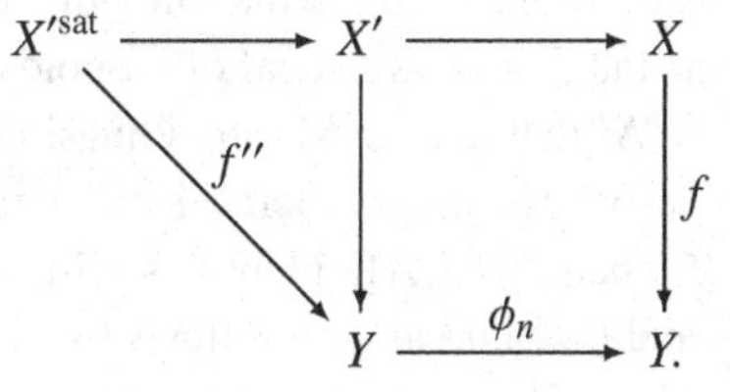

Here $\phi_n\colon Y \to Y$ is the n-Frobenius morphism (Definition 1.1.13), and the square is cartesian.

Proof If $f\colon X \to Y$ is exact then, for every point x of X, the homomorphism $f_x^{\flat}\colon \mathcal{M}_{Y,f(x)} \to \mathcal{M}_{X,x}$ is exact. Since every face of $\mathcal{M}_{X,x}$ corresponds to a point of X, it follows in fact that $f_x^\flat$ is locally exact and hence is **Q**-integral, by Theorem I.4.7.7.

Now suppose that f is exact and that X is quasi-compact. For each $x \in X$, the homomorphism $f_x^\flat : \mathcal{M}_{Y,f(x)} \to \mathcal{M}_{X,x}$ is **Q**-integral. Let

$$f_x : X_x := \operatorname{spec}(\mathcal{M}_{X,x}) \to Y_{f(x)} := \operatorname{spec}(\mathcal{M}_{Y,f(x)})$$

be the corresponding morphism of monoschemes. Then Theorem I.4.9.1 implies that there exists an $n(x) \in \mathbf{N}$ such that the morphism $f'' : X''_x \to Y_{f(x)}$ obtained as in the diagram is saturated for every $n \in \mathbf{N}$ that is divisible by $n(x)$. Since X is quasi-compact, it has only finitely many points and, since the family of saturated morphisms is stable under base change, the theorem holds if n is the product of all the integers $n(x)$. □

The notion of closed immersion is not useful in the context of monoschemes, since the image of any map of nonempty connected monoschemes contains the generic point of the target. It therefore does not make sense to ask for the "monoscheme-theoretic closure" of the image of a morphism. We consider instead "proper immersions" and "proper images." A "proper immersion" of monoschemes is, as the name suggests, an immersion (Definition 1.1.12) which is proper (Definition 1.6.1), and we give the definition of a "proper image" below.

Definition 1.8.8. *The proper image of a morphism of monoschemes $f : X \to Z$ is a factorization $f = i \circ g$, where $i : Y \to Z$ is a proper immersion, with the following universal property: for any factorization $f = i'g'$ with i' a proper immersion, there is a morphism $h : Y \to Y'$ such that $i = i' \circ h$.*

Note that the morphism h is unique, since i' is a monomorphism, and furthermore $h \circ g = g'$, for the same reason. The proper image of a morphism, if it exists, is unique up to unique isomorphism, by the universal mapping property.

In Proposition 1.8.9 below, we shall use the following construction. Suppose that Z is a monoscheme and that $\mathcal{M}$ is a sheaf of monoids on Z endowed with a homomorphism $\mathcal{M}_Z \to \mathcal{M}$ that makes $\mathcal{M}$ into a quasi-coherent sheaf of $\mathcal{M}_Z$-sets. Then there is an associated monoscheme $Y := \operatorname{spec}_Z(\mathcal{M})$, endowed with a morphism $i : Y \to Z$, such that $i_*\mathcal{M}_Y$ identifies with $\mathcal{M}$. If Z is affine, Y is just $\operatorname{spec}(\Gamma(Z, \mathcal{M}(Z))$, and the general case follows by gluing.

Proposition 1.8.9. *Every morphism of connected fine monoschemes admits a proper image, unique up to unique isomorphism. This image has the following properties.*

1. The proper image of a morphism $\operatorname{Spec}(\theta : P \to Q)$ *of affine integral* monoschemes, is again affine, given by the factorization $P \to M \to Q$, where M is the image of P in Q;

2. The proper image of f coincides with the proper image of the restriction of f to X^*.
3. In general, if $f\colon X \to Z$ is a homomorphism of integral monoids, let $\mathcal{M}$ be the sheaf-theoretic image of the homomorphism $\mathcal{M}_Z \to f_*\mathcal{M}_X$. Then the natural map $\mathcal{M}_Z \to \mathcal{M}$ makes $\mathcal{M}$ into a quasi-coherent sheaf of $\mathcal{M}_Z$-sets, and $Y = \operatorname{spec}_Z(f_*\mathcal{M}_X)$. A point z of Z lies in Y if and only if the homomorphism $\mathcal{M}_{Z,z} \to \mathcal{M}_z$ is local, and in this case $\mathcal{M}_{Y,z} = \mathcal{M}_z$.

Proof The proof will rely on the following useful property of proper immersions.

Lemma 1.8.10. *Let $f\colon X \to Z$ and $i\colon Y \to Z$ be morphisms of connected and fine monoschemes. Suppose that i is a proper immersion and that there exists a morphism $g\colon X^* \to Y$ such that $i \circ g = f|_{X^*}$. Then g extends uniquely to a morphism $g'\colon X \to Y$ such that $i \circ g' = f$.*

Proof Since $Y \to Z$ is separated and X is integral, any such g' is unique if it exists, by (2) of Theorem 1.6.3. Thus it suffices to prove that X admits an open cover on which such a g' exists.

Let x be a point of X and choose a local homomorphism $\theta\colon \mathcal{M}_{X,x} \to \mathbf{N}$. Then θ defines a morphism $h\colon \mathbf{a} \to X$ taking the closed point t of $\mathbf{a}$ to x. Since i is proper, there is a unique morphism $h'\colon \mathbf{a} \to Y$ such that $h'_{|\mathbf{a}^*} = g \circ h^*$ and $i \circ h' = f \circ h$. Then $f(x) = i(y)$ where $y := h'(t)$. We now have the solid arrows in the following commutative diagram

$$\begin{array}{ccccc}
\mathcal{M}_{Z,f(x)} & \xrightarrow{i^\flat_y} & \mathcal{M}_{Y,y} & \xrightarrow{\lambda_{y,\eta}} & \mathcal{M}_{Y,\eta} \\
\downarrow{\scriptstyle f^\flat_x} & \swarrow{\scriptstyle g'^\flat_x} & & \swarrow{\scriptstyle g^\flat_\xi} & \\
\mathcal{M}_{X,x} & \xrightarrow{\lambda_{x,\xi}} & \mathcal{M}_{X,\xi}, & &
\end{array}$$

where ξ and η are, respectively, the generic points of X and Y. Since X is integral, the map $\lambda_{x,\xi}$ is injective and, since $i^\flat_y$ is surjective, there is a unique homomorphism $g'^\flat_x\colon \mathcal{M}_{Y,y} \to \mathcal{M}_{X,x}$ making the diagram commute. Since X is fine, $\operatorname{spec}(\mathcal{M}_{X,x})$ is an open subset of X, and we have proved that the desired morphism g' exists in this neighborhood of x. As we have already observed, it follows that g' exists and is unique on all of X, completing the proof of the lemma. □

It is clear that the question of the existence of the proper image of a morphism $f\colon X \to Z$ can be checked locally on Z. Thus we may assume that Z is affine, say $Z = \operatorname{spec}(P)$. We can also reduce to the case in which X is affine. Indeed, suppose that $i\colon Y \to Z$ is the proper image of $f|_{X^*}$. Then it follows from the lemma that the map $X^* \to Y$ extends uniquely to X, and in fact then $Y \to Z$ is also the proper image of f. Thus we may assume that $X = \operatorname{spec}(Q)$, and we may even assume that Q is dull. Then f is given by a homomorphism $\theta\colon P \to Q$. Let M be its image, so that we have $\theta = \gamma \circ \iota$, where ι is surjective and γ is injective. We claim that the corresponding factorization $f = i \circ g$ gives a proper image of f. Since ι is surjective, Theorem 1.6.7 implies that $i := \operatorname{spec}(\iota)$ is proper, and it is also an immersion. Now suppose that $f = i' \circ g'$, where $i'\colon Y' \to Z$ is a proper immersion. We claim that i factors through i'. The morphisms i, g, i', g' induce the solid arrows in the following diagram, where η' is the generic point of Y':

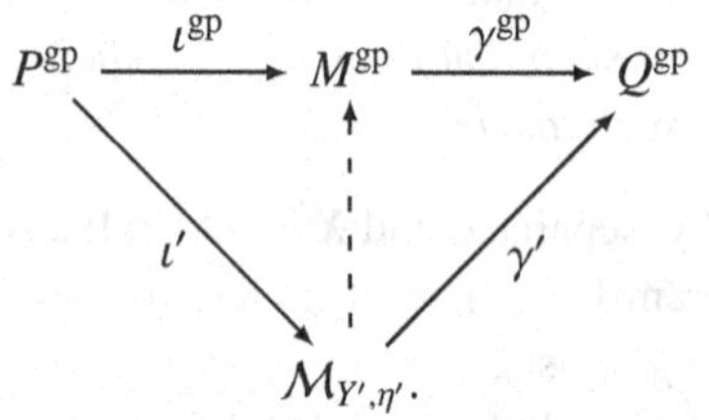

Since γ^{gp} is injective, it follows that the kernel of ι' is contained in the kernel of ι^{gp}, and since ι' is surjective, there is a unique dashed arrow as shown making the diagram commute. It follows that $i_{|Y^*}$ factors through i'. Applying the lemma to the proper immersion i', we see that in fact i factors through i', as claimed.

We have now proved statements (1) and (2). For (3), we drop the assumption that X and Z are affine. Note however that, since X is integral, the sheaf-theoretic image $\mathcal{M}$ of $\mathcal{M}_Z \to f_*\mathcal{M}_X$ is the same as the sheaf-theoretic image of $\mathcal{M}_Z$ in the constant sheaf $\mathcal{M}_X^{gp}$, and its formation is compatible with localization. It therefore follows from our local construction that $Y = \operatorname{spec}_Z(\mathcal{M})$. Note that if $y \in Y$, the map $i_y^\flat\colon \mathcal{M}_{Z,i(y)} \to \mathcal{M}_{Y,y}$ is local and surjective, so $\mathcal{M}_{Y,y} = \mathcal{M}_y$. Conversely, if $\mathcal{M}_{Z,z} \to \mathcal{M}_z$ is local, and if y is the point of $\operatorname{spec}(M_z)$ corresponding to the maximal ideal of $\mathcal{M}_z$, then $i(y) = z$ and $\mathcal{M}_{Y,y} = \mathcal{M}_z$. □

Remark 1.8.11. As another application of exactification, let us explain how the indeterminacy of a morphism can be eliminated by a monoidal transformation. Let X and Y be connected and fine monoschemes, where Y is proper, let U be an open subset of X, and let $f\colon U \to Y$ be a morphism. There exist a sheaf of ideals $\mathcal{I}$ of X whose restriction to U is the unit ideal and a morphism $f'\colon X_{\mathcal{I}} \to Y$ whose restriction to $U_{\mathcal{I}} = U$ agrees with f. To see this, let X' be the proper image of the morphism $U \to X \times Y$ defined by f. Since $X' \to X \times Y$ and $X \times Y \to X$ are proper, so is the map $X' \to X$. Since Y is separated, the map $U \to U \times Y$ is proper. Since $X'_{|_U} \to U \times Y$ is the proper image of $U \to U \times Y$, the map $U \to X'_{|_U}$ is an isomorphism. Thus the map $X' \to X$ is proper and is an isomorphism over U. By Corollary 1.8.6, there is a coherent sheaf of ideals $\mathcal{I}$ on X whose restriction to U is the unit ideal and such that $X'_{\mathcal{I}} \to X_{\mathcal{I}}$ is an open immersion, hence an isomorphism by Proposition 1.6.5. The composition $X_{\mathcal{I}} \cong X'_{\mathcal{I}} \to X' \to Y$ is the desired morphism f'.

1.9 Monoschemes, toric schemes, and fans

In this section we show how (toric) monoschemes give rise to (toric) schemes. The construction is very direct and straightforward. We also review the classical construction of toric varieties from fans [21] and explain the relationships among monoschemes, classical fans, and Kato's fans [49].

If X is a locally monoidal space X, denote by H_X the functor from the category of schemes to the category of sets taking a scheme T to the set of morphisms of locally monoidal spaces $T \to X$. For example, if Q is a monoid, it follows from Proposition 1.2.2 that $H_{\mathbf{a}_Q}(T) = \operatorname{Hom}(Q, \Gamma(T, \mathcal{O}_T))$. Thus $H_{\mathbf{a}_Q}$ is represented by $\underline{\mathsf{A}}_Q$, where $\underline{\mathsf{A}}_Q = \operatorname{Spec}(\mathbf{Z}[Q])$. The next proposition is a straightforward generalization of this to the case of monoschemes.

Proposition 1.9.1. *If X is a monoscheme, the functor H_X is representable by a scheme, which we denote by $\underline{\mathsf{A}}_X$. Thus there is a universal morphism $\mu_X : \underline{\mathsf{A}}_X \to X$, and if T is any scheme and $f : (T, \mathcal{O}_T) \to (X, \mathcal{M}_X)$ is any morphism of locally monoidal spaces, there is a unique morphism of schemes $\tilde{f} : T \to \underline{\mathsf{A}}_X$ such that $f = \mu_X \circ \tilde{f}$.*

Proof It suffices to prove that the functor H_X has an open cover by affine schemes. First note that if $X' \to X$ is an open immersion of locally monoidal spaces, then the corresponding morphism of functors $H_{X'} \to H_X$ is open, i.e., if $f : T \to X$ is any morphism, $h_{X'} \times_{h_X} T$ is represented by the open subscheme $f^{-1}(X')$ of T. Now if X is a monoscheme, it admits an open cover by affine monoschemes X', and if $X' \cong \operatorname{spec}(Q)$, then

$$h_{X'}(T) \cong \operatorname{Hom}(Q, \Gamma(T, \mathcal{O}_T)) \cong \operatorname{Mor}(T, \underline{\mathsf{A}}_Q).$$

Thus, $h_{X'}$ is represented by the affine scheme $\underline{\mathsf{A}}_Q$. □

Now suppose that X is a connected toric monoscheme. Recall that $\mathcal{M}_X^{\mathrm{gp}}$ is a constant sheaf, with value $\Gamma := \mathcal{M}_{X,\xi}$, a finitely generated free abelian group. As we saw in Proposition 1.2.11, $X^* := \operatorname{spec}(\Gamma)$ is a group object in the category of monoschemes and acts on X. It follows that the group scheme $\underline{\mathsf{A}}^*_\Gamma \cong \underline{\mathsf{A}}_{X^*}$

acts naturally on the scheme $\underline{\mathsf{A}}_X$. Furthermore, $\underline{\mathsf{A}}_{X^*}$ is dense and open in $\underline{\mathsf{A}}_X$. Thus $\underline{\mathsf{A}}_X$ can be viewed as a *toric scheme*, a mild generalization of a toric variety. (Note that $\underline{\mathsf{A}}_X$ need not be separated.)

To explain the relationship between toric monoschemes and fans, we begin by reviewing the definition of the latter.

Definition 1.9.2. *Let L be a lattice, i.e., a finitely generated free abelian group, and let $V := \mathbf{Q} \otimes L$. Then a fan in V is a finite set Δ of finitely generated cones in V satisfying the following conditions.*

1. *If σ belongs to Δ, then every face of σ also belongs to Δ.*
2. *If σ and τ belong to Δ, then $\sigma \cap \tau$ is a face of σ and a face of τ, and in particular belongs to Δ.*

A morphism of fans $f\colon (L, \Delta) \to (L', \Delta')$ is a homomorphism $h\colon L \to L'$ such that for each $\sigma \in \Delta$, $h(\sigma)$ is contained in some element of Δ'.

Fans are traditionally used to construct toric varieties, as explained for example in [21]. But in fact there is a more fundamental construction that relates fans to monoschemes. If X is a connected and toric monoscheme with generic point ξ, and if x is a point of X, the cospecialization map allows us to view $\mathcal{M}_{X,x}$ as a submonoid of $\mathcal{M}_{X,\xi}$, and we use the following notation:

$$
\begin{aligned}
\Gamma(X) &:= \Gamma(X, \mathcal{M}_X^{\mathrm{gp}}) = \mathcal{M}_{X,\xi} \text{ (a finitely generated free abelian group)},\\
L(X) &:= \operatorname{Hom}(\Gamma(X), \mathbf{Z}) \cong X^*(\mathbf{a}) \cong X(\mathbf{a}^*),\\
V(X) &:= \mathbf{Q} \otimes L(X),\\
\sigma_x &:= M_{X,x}^{\vee} := \{\phi \in V(X) : \phi(m) \geq 0 \text{ for all } m \in \mathcal{M}_{X,x}\}, \text{ a cone in } V(X),\\
\Delta(X) &:= \{\sigma_x : x \in X\}.
\end{aligned}
$$

Theorem 1.9.3. *Let X be a separated, connected, and toric monoscheme. Then with the notation above, $(L(X), \Delta(X))$ is a fan. A morphism of monoschemes $f\colon X \to X'$ induces a morphism of the corresponding fans*

$$(L(X), \Delta(X) \to (L(X'), \Delta(X')),$$

and this functor gives an equivalence from the category of separated, connected, and toric monoschemes to the category of fans.

Proof Each $x \in X$, $\mathcal{M}_{X,x}$ is a finitely generated toric monoid, and it follows from Theorem I.2.3.12 that its $\mathbf{Q}$-dual σ_x is a finitely generated cone in $V(X)$ and that $\mathcal{M}_{X,x} = \sigma_x^{\vee} \cap \Gamma(X)$. We claim that the map $x \to \sigma_x$ is injective, and hence bijective. Indeed, suppose that x and y are points of X and that $\sigma = \sigma_x = \sigma_y$. Then $M := \mathcal{M}_{X,x} = \mathcal{M}_{X,y}$ as submonoids of $\Gamma(X)$. Furthermore, since σ is a finitely generated $\mathbf{Q}$-cone, its interior contains an element h in

$L(X)$. Then h defines a local homomorphism $M \to \mathbf{N}$ (see Remark I.2.2.8). By Proposition 1.2.7, (x, h) and (y, h) define respective elements f_x and f_y of $X(\mathbf{a})$. Both of these define the same element h^{gp} of $X(\mathbf{a}^*)$ and, since X is separated, it follows that $f_x = f_y$ and hence that $x = y$.

Next suppose that x and y are points of X and that y is a generization of x. Then $\mathcal{M}_{X,y}$ is the localization of $\mathcal{M}_{X,x}$ by a face F, and (4) of Theorem I.2.3.12 tells us that $\sigma_y := \mathcal{M}^\vee_{X,y} = \mathcal{M}^\vee_{X,x} \cap F^\perp$, a face of σ_x. Conversely, if σ_y is a face of σ_x, then by (5) of Theorem I.2.3.12, $\sigma_y^\vee \cap \Gamma(X) = \mathcal{M}_{X,y}$ is the localization of $\sigma_X^\vee \cap \Gamma(X) = \mathcal{M}_{X,x}$ by a face F of $\mathcal{M}_{X,x}$. This F corresponds to a generization y' of x; then $\sigma_{y'} = \sigma_y$ and hence $y = y'$. We conclude that y is a generization of x if and only if σ_y is a face of σ_x.

Let x and y be any two points of X. Then $U_x := \operatorname{spec}(\mathcal{M}_{X,x})$ and $U_y := \operatorname{spec}(\mathcal{M}_{X,y})$ are affine open subsets of X, and since X is separated, Proposition 1.6.10 implies that $U_x \cap U_y$ is affine. Its unique closed point z is a generization both of x and of y, and hence σ_z is a face both of σ_x and of σ_y. Proposition 1.6.10 also implies that $\mathcal{M}_{X,z} = \mathcal{M}_{X,x} + \mathcal{M}_{X,y}$ and hence by Corollary I.2.3.16 that $\sigma_z = \sigma_x \cap \sigma_y$. This shows that the intersection of two elements of $\Delta(X)$ is a face of each, and hence that $(L(X), \Delta(X))$ is indeed a fan.

Now suppose that $f\colon X \to X'$ is a morphism of connected separated toric monoschemes. Then f induces a homomorphism $f^\flat_\xi\colon \Gamma(X', \mathcal{M}^{gp}_{X'}) \to \Gamma(X, \mathcal{M}^{gp}_X)$ and hence a morphism $h\colon L(X) \to L(X')$. Furthermore, if $x' = f(x)$, then $f^\flat_\xi$ maps $\mathcal{M}_{X',x'}$ to $\mathcal{M}_{X,x}$ and hence h maps σ_x into $\sigma_{f(x)}$. Thus h determines a morphisms of fans

$$(L(X), \Delta(X)) \to (L(X'), \Delta(X').$$

Conversely suppose that h is a morphism

$$(L(X), \Delta(X)) \to (L(X'), \Delta(X')).$$

By assumption, for each $x \in \Delta(X)$, $\Sigma_x := \{\sigma' \in \Delta(X') : h(\sigma_x) \subseteq \sigma'\}$ is finite and nonempty and, since $\Delta(X')$ is a fan, $\cap\Sigma_x$ belongs to $\Delta(X')$. Then there is a unique $x' \in X'$ such that $\sigma_{x'} = \cap\Sigma_x$, and we get a function $f\colon X \to X'$ by sending x to this x'. Note that if y is a generization of x, then $\sigma_y \subseteq \sigma_x$, hence $\Sigma_x \subseteq \Sigma_y$ and so $\cap\Sigma_y \subseteq \cap\Sigma_x$ and $f(y)$ is a generization of $f(x)$. This implies that f is a continuous function. Since $\mathcal{M}_{X,x} = \sigma_x^\vee \cap \mathcal{M}^{gp}_X$, $h^\vee$ maps $\mathcal{M}^{gp}_{X'}$ to $\mathcal{M}^{gp}_X$ and induces a homomorphism $f^\flat_x\colon \mathcal{M}_{X',f(x)} \to \mathcal{M}_{X,x}$ for all $x \in X$. Then $f^{\flat^{-1}}(\mathcal{M}^*_{X,x})$ is a face of $\mathcal{M}_{X',f(x)}$, which corresponds to a generization y' of $f(x)$. Then $\sigma_{y'} \subseteq \sigma_{f(x)}$ and $f^\flat_x$ factors through $\mathcal{M}_{X',y'}$. Hence $h(\sigma_x) \subseteq \sigma_{y'}$, so $\sigma_{y'} \in \Sigma_x$. But then $\sigma_{f(x)} := \cap\Sigma_x \subseteq \sigma_{y'}$, and hence $\sigma_{y'} = \sigma_{f(x)}$. Then $f(x) = y'$ and hence in fact $f^\flat_x$ is a local homomorphism. Thus h is subordinate to a (unique) morphism of monoschemes $f\colon X \to X'$.

Finally we sketch the construction of a (separated) monoscheme from a fan (L, Δ). We build a topological space X whose points X are the elements of Δ, and, for each $\sigma \in \Delta$, the smallest open set containing σ is $U_\sigma := \{\tau : \tau \subseteq \sigma\}$. Let $M := \mathrm{Hom}(L, \mathbf{Z})$, and for each $\sigma \in \Delta$, let

$$\mathcal{M}_\sigma := \sigma^\vee \cap M,$$

where $\sigma^\vee$ is the cone in $\mathrm{Hom}(V, \mathbf{Q})$ dual to σ, in the sense of Theorem I.2.3.12. Then $\tau \subseteq \sigma$ if $\tau \in U_\sigma$, and it follows from condition (2) in Definition 1.9.2 that τ is a face of σ, and hence $\tau^\perp \cap \sigma^\vee$ is a face F of $\sigma^\vee$. By Theorem I.2.3.12, $\tau^\vee$ is the localization of $\sigma^\vee$ by this face, and hence $\mathcal{M}_\tau$ is the localization of $\mathcal{M}_\sigma$ by $F \cap \mathcal{M}_\sigma$. Moreover, if G is any face of $\mathcal{M}_\sigma$, $G^\perp \cap \sigma$ is a face of σ, hence corresponds to a point τ of Δ lying in U_σ. Thus the set U_σ becomes homeomorphic to $\mathrm{Spec}(\mathcal{M}_\sigma)$, and the collection of maps $\mathcal{M}_\sigma \to \mathcal{M}_\tau$ corresponding to inclusions in Δ identifies with the cospecialization maps of the structure sheaf of $\mathrm{spec}(\mathcal{M}_\sigma)$. Then, if X is endowed with the sheaf associated to the presheaf taking each U_σ to $\mathcal{M}_\sigma$ (with the corresponding restriction maps), X becomes a toric monoscheme. Furthermore, a morphism $\mathbf{a} \to X$ corresponds to a morphism of fans $(\mathbf{Z}, \mathbf{R}_\geq, \{0\}) \to (L, \Delta)$, which is by its very definition determined by the map $\mathbf{Z} \to L$, that is, its restriction to $\mathbf{a}^*$. This shows that the monoscheme X is separated. □

Remark 1.9.4. In [49], Kato considers monoidal spaces in which the sheaf of monoids is sharp. For example, one can attach to any monoscheme $(X, \mathcal{M}_X)$ the sharpened sheaf of monoids $\overline{\mathcal{M}}_X$. A monoidal space that is locally isomorphic to the *sharpened* monoidal space of a monoid is what Kato terms a fan. Since this notion is not as closely related to the classical one as ours, we suggest calling Kato's notion an *s-fan*. The data of an s-fan is not as precise as those of a monoscheme, and in particular there seems to be no notion of separation for s-fans. However, if X is an integral and connected monoscheme, the natural map $\mathcal{M}_X \to \overline{\mathcal{M}}_X$ is exact, so $\mathcal{M}_X = \Gamma(X) \times_{\overline{\mathcal{M}}_X^{\mathrm{gp}}} \overline{\mathcal{M}}_X$. Thus $(X, \mathcal{M}_X)$ can be recovered from $(X, \overline{\mathcal{M}}_X)$ by specifying the map $\Gamma(X) \to \overline{\mathcal{M}}_X^{\mathrm{gp}}$. For example, the monoscheme versions $\mathbf{d}$ of the doubled affine line and of the projective line $\mathbf{p}$ (Examples 1.2.9) give rise to the same s-fan and the same $\Gamma(X)$, but the corresponding maps $\Gamma(X) \to \overline{\mathcal{M}}_X^{\mathrm{gp}}$ are different.

Proposition 1.9.5. *Let X be a monoscheme, let $\underline{\mathsf{A}}_X$ be the associated scheme, and let $\mu \colon \underline{\mathsf{A}}_X \to X$ be the natural map of monoidal spaces (Proposition 1.9.1).*

1. *If $\mathcal{E}$ is a quasi-coherent sheaf of $\mathcal{O}_{\underline{\mathsf{A}}_X}$-modules on $\underline{\mathsf{A}}_X$, then $\mu_*(\mathcal{E})$ is a quasi-coherent sheaf of $\mathcal{M}_X$-sets on the monoscheme X.*

2. *The functor μ_* of (1) admits a left adjoint μ^*, from the category of quasi-coherent sheaves of $\mathcal{M}_X$-sets to the category of quasi-coherent sheaves of $\mathcal{O}_{\underline{\mathsf{A}}_X}$-sets.*
3. *If $\mathcal{S}$ is an invertible sheaf of $\mathcal{M}_X$-sets on S, then $\mu^*(\mathcal{S})$ is an invertible sheaf of $\mathcal{O}_{\underline{\mathsf{A}}_X}$-modules on $\underline{\mathsf{A}}_X$. If X is quasi-compact and separated and $\mathcal{S}$ is ample on X, then $\mu^*(\mathcal{S})$ is ample on $\underline{\mathsf{A}}_X$.*

Proof Statement (1) can be verified locally on X, so we may assume without loss of generality that $X = \operatorname{spec}(P)$, for some monoid P. Then $\mathsf{A}_X = \operatorname{Spec}(\mathbf{Z}[P])$, and $\mathcal{E}$ is the quasi-coherent sheaf associated to a $\mathbf{Z}[P]$-module E. To show that $\mu_*(\mathcal{E})$ is quasi-coherent, we must show that, for every $p \in P$, the natural map $E := \mu_*(\mathcal{E})(X) \to \mu_*(\mathcal{E})(X_p)$ factors through an isomorphism $E_p \to \mu_*(X_p)$. But we have

$$\mu_*(\mathcal{E})(X_p) = \mathcal{E}(\mu^{-1}(X_p)) = \mathcal{E}(\underline{\mathsf{A}}_{P_p}).$$

Since $\mathbf{Z}[P_p]$ is the localization of $\mathbf{Z}[P]$ by e^p, the desired result follows from the quasi-coherence of $\mathcal{E}$.

Let $\mathcal{S}$ be a sheaf of $\mathcal{M}_X$-sets on a monoscheme X. For every affine open set U of X, the free abelian group $\mathbf{Z}[\mathcal{S}(U)]$ generated by $\mathcal{S}[U]$ has a natural structure of a $\mathbf{Z}[\mathcal{M}_X(U)] = \mu_*(\mathcal{O}_{\underline{\mathsf{A}}_X})(U)$-module. Thus the sheaf $\mathbf{Z}[\mathcal{S}]$ on X associated to this presheaf has a natural structure of a $\mu_*(\mathcal{O}_{\underline{\mathsf{A}}_X})$-module, and $\mu^*(\mathcal{S}) := \mu^{-1}(\mathbf{Z}[\mathcal{S}]) \times_{\mu^{-1}\mathcal{M}_X} \mathcal{O}_{\underline{\mathsf{A}}_X}$ has a natural structure of $\mathcal{O}_{\underline{\mathsf{A}}_X}$-module. It is easily verified that the functor μ^* just defined is left adjoint to the functor μ_*. To see that μ^* takes quasi-coherent sheaves to quasi-coherent sheaves, we may work locally on X, and thus may assume that $X = \operatorname{spec}(P)$ and $\mathcal{S}$ is the sheaf associated to a P-set S. Then $\mu^*(\mathcal{S})$ is the sheaf associated to the $\mathbf{Z}[P]$-module $\mathbf{Z}[S]$, since these two sheaves enjoy the same universal mapping property. If $\mathcal{S}$ is isomorphic to $\mathcal{M}_X$, then $\mu^*(\mathcal{S})$ is isomorphic to $\mathcal{O}_{\underline{\mathsf{A}}_X}$, and hence if $\mathcal{S}$ is invertible, so is $\mu^*(\mathcal{S})$. If $\mathcal{S}$ is ample, then it follows from the definition that X can be covered by affine open sets of the form X_s for sections s of $\mathcal{S}^n$ and some $n > 0$. Then $\underline{\mathsf{A}}_X$ can be covered by affine open sets of the form $(\underline{\mathsf{A}}_X)_s$ as s ranges over sections of $\mu^*(\mathcal{S})$, and [25, 4.5.2] implies that $\mu^*(\mathcal{S})$ is ample. □

1.10 The moment map

The moment map is an important tool that illuminates the topology of the set of complex points of an affine toric variety. If Q is a fine monoid, we use the following notation:

$$
\begin{aligned}
\mathsf{X}_{\mathsf{Q}} &:= \mathrm{Hom}(Q, \mathbf{C}) = \underline{\mathsf{A}}_{\mathsf{Q}}(\mathbf{C}),\\
\mathsf{X}^*_{\mathsf{Q}} &:= \mathrm{Hom}(Q^{\mathrm{gp}}, \mathbf{C}) = \underline{\mathsf{A}}^*_{\mathsf{Q}}(\mathbf{C}),\\
\mathsf{T}_{\mathsf{Q}} &:= \mathrm{Hom}(Q, \mathbf{S}^1), \text{ where } \mathbf{S}^1 := \{z \in \mathbf{C} : |z| = 1\},\\
\mathsf{R}_{\mathsf{Q}} &:= \mathrm{Hom}(Q, \mathbf{R}_{\geq}), \text{ where } \mathbf{R}_{\geq} := \{r \in \mathbf{R} : r \geq 0\},\\
\mathsf{R}^*_{\mathsf{Q}} &:= \mathrm{Hom}(Q, \mathbf{R}_{>}), \text{ where } \mathbf{R}_{>} := \{r \in \mathbf{R} : r > 0\}.
\end{aligned}
$$

There is a natural map $\mathsf{X}_{\mathsf{Q}} \to \mathsf{R}_{\mathsf{Q}}$ sending a homomorphism x to $|x| := \mathrm{abs} \circ x$, and it is easily seen that this map identifies R_{Q} as the orbit space of X_{Q} under the natural action of T_{Q}. Recall from Proposition I.3.3.4 that X_{Q} can be partitioned into the disjoint union of the locally closed subsets $\mathsf{X}^*_{\mathsf{F}}$, where F ranges over the faces of Q. If $f \in F$ and $x \in \mathsf{X}^*_{\mathsf{F}}$ then $x(f) \in \mathbf{C}^* = \mathbf{S}^1 \times \mathbf{R}_{>}$, and we have a corresponding product decomposition: $\mathsf{X}^*_{\mathsf{F}} \cong \mathsf{T}_{\mathsf{F}} \times \mathsf{R}^*_{\mathsf{F}}$. Furthermore, $\rho(f) > 0$ if $\rho \in \mathsf{R}^*_{\mathsf{F}}$ and $f \in F$, and the logarithm $\mathbf{R}_{>} \to \mathbf{R}$ produces a homeomorphism

$$\mathsf{R}^*_{\mathsf{F}} \to V_F := \mathrm{Hom}(F^{\mathrm{gp}}, \mathbf{R}).$$

Thus $\mathsf{X}^*_{\mathsf{F}}$ becomes a product of the compact torus T_{F} and the linear space V_F. The moment map will enable us to glue together the linear spaces V_F to produce a linearization of R_{Q}. Specifically, it produces a homeomorphism from R_{Q} to C_{Q}, the real cone spanned by Q in the vectors space $\mathbf{R} \otimes Q^{\mathrm{gp}}$. Our exposition roughly follows the treatment in [21], with improvements suggested by O. Gabber. We refer also to the more recent discussions in [57] and [10, 12.2.5].

Let $\mathbf{R}[Q]$ be the real monoid algebra of Q. Its underlying vector space is just the set of real linear combinations of elements of Q, which we also refer to as *cycles* in Q. We say a cycle $A := \sum a_q q$ is *effective* if its coefficients are all nonnegative; the set of effective cycles is a semi-ring in $\mathbf{R}[Q]$. The *support* of A is the set of q such that $a_q \neq 0$.

Definition 1.10.1. *Let $A := \sum a_q q$ be an effective cycle in Q. Define*

$$
\begin{aligned}
e_A \colon \mathsf{R}_{\mathsf{Q}} \to \mathbf{R}_{\geq} \quad & \text{by } \rho \mapsto \sum_{q \in Q} a_q \rho(q)\\
\mu_A \colon \mathsf{R}_{\mathsf{Q}} \to \mathsf{C}_{\mathsf{Q}} \quad & \text{by } \rho \mapsto \sum_{q \in Q} a_q \rho(q) q.
\end{aligned}
$$

The map μ_A is called the moment map associated to the cycle A.

We should remark that any $\rho \in \mathsf{R}_{\mathsf{Q}}$ extends uniquely to a continuous homomorphism $\mathsf{C_Q} \to \mathbf{R}_{\geq}$, so e_A and μ_A can also be defined for cycles $a_q q$ with $q \in \mathsf{C_Q}$.

Theorem 1.10.2. *Let Q be a fine monoid and A an effective cycle in $\mathbf{R}[Q]$ whose support S generates Q. Then the moment map μ_A of Definition 1.10.1 is a homeomorphism*

$$\mu_A\colon \mathsf{R_Q} \to \mathsf{C_Q}$$

that is compatible with the stratifications of $\mathsf{R_Q}$ and $\mathsf{C_Q}$ by faces. Thus, for each face F of Q, μ_Q induces homeomorphisms $\mathsf{R_F} \to \mathsf{C_F}$ and $\mathsf{R^*_F} \to \mathsf{C^\circ_F}$, where $\mathsf{R^*_F} := \mathrm{Hom}(F, \mathbf{R}_{>})$ and $\mathsf{C^\circ_F}$ is the interior of $\mathsf{C_F}$.

Proof Let us first explain the "compatibility with the faces." Since the group $\mathbf{R}_{>}$ is divisible, it follows as in Proposition I.3.3.4 that $\mathsf{R_Q}$ is the disjoint union of the subsets $\mathsf{R^*_F}$ for $F \in \mathcal{F}(Q)$, where here we are identifying $\mathsf{R^*_F}$ with the set of $\rho \in \mathsf{R_Q}$ such that $\rho^{-1}(\mathbf{R}_{>}) = F$. Then, if $\rho \in \mathsf{R^*_F}$,

$$\mu_A(\rho) = \sum_{q \in F \cap S} a_q \rho(q) q \in \mathsf{C_F} \subseteq \mathsf{C_Q}\,.$$

For every $q \in F \cap S$, the coefficient of q in the above sum is nonzero, and hence $q \in \langle \mu_A(\rho) \rangle$. Since S generates Q, its intersection with F generates F, and hence $\langle \mu_A(\rho) \rangle = F$ and $\mu_A(\rho) \in \mathsf{C^\circ_F}$. This shows that μ_A maps $\mathsf{R_F}$ to $\mathsf{C_F}$ and $\mathsf{R^*_F}$ to $\mathsf{C^\circ_F}$.

Here is the first of the two main ingredients in the proof that μ_A is a homeomorphism.

Lemma 1.10.3. *If the support S of A generates Q, then $\mu_A\colon \mathsf{R_Q} \to \mathsf{C_Q}$ is proper.*

Proof We assume without loss of generality that $0 \notin S$. If S generates $\mathsf{C_Q}$, then the evaluation map $\mathsf{R_Q} \to \mathbf{R}^S$ is a closed immersion. Since $\mathbf{R}^S$ is a Euclidean space, a subset of $\mathsf{R_Q}$ is compact if and only if its image in $\mathbf{R}^S$ is closed and bounded. Thus to prove that μ_A is proper, it will suffice to show that, for every compact subset K of $\mathsf{C_Q}$, $\mu_A^{-1}(K)$ is bounded in $\mathbf{R}^S$.

First let us suppose that Q is sharp. In this case, we choose a local homomorphism $h\colon \mathsf{C_Q} \to \mathbf{R}_{\geq}$. If K is a compact subset of $\mathsf{C_Q}$, $h(K)$ is a compact subset of $\mathbf{R}_{\geq}$, say bounded by M. Since h is local and S is finite, there exists a positive number ϵ such that $\epsilon < a_s h(s)$ for all $s \in S$. Then, if $\rho \in \mu_A^{-1}(K)$ and $s \in S$,

$$\epsilon\rho(s) < a_s h(s)\rho(s) \leq \sum_{q \in S} a_q \rho(q) h(q) = h(\mu_A(\rho)) \leq M.$$

It follows that each $\rho(s) \leq M/\epsilon$, so $\mu_A^{-1}(K)$ is bounded, hence compact.

For the general case, we argue as follows[2]. Let Σ denote the set of all subsets σ of S such that the subcone C_σ of C_Q generated by σ is sharp. For each such σ we can choose a local homomorphism $h_\sigma\colon \mathsf{C}_\sigma \to \mathbf{R}_\geq$, which we can then extend to a linear map $V_Q \to \mathbf{R}$. Then $h_\sigma(s) > 0$ if $s \in \sigma$. Since S and Σ are finite, there exist positive numbers ϵ and m such that

1. $|a_s h_\sigma(s)| < m$ for all $\sigma \in \Sigma$ and all $s \in S$,
2. $\epsilon < a_s h_\sigma(s)$ for all $\sigma \in \Sigma$ and all $s \in \sigma$.

For $\rho \in \mathsf{R}_\mathsf{Q}$, let $\sigma_\rho := \{s \in S : \rho(s) > 1\}$. Then ρ defines a homomorphism $\mathsf{C}_{\sigma_\rho} \to (\mathbf{R}_>, \cdot)$ and $\log \circ\rho$ is a homomorphism $\mathsf{C}_{\sigma_\rho} \to (\mathbf{R}, +)$. Since in fact this homomorphism is positive on the generators of C_{σ_ρ}, it follows that C_{σ_ρ} is sharp and hence that $\sigma_\rho \in \Sigma$.

Now if K is a compact subset of C_Q, each $h_\sigma(K)$ for $\sigma \in \Sigma$ is bounded. Since Σ is finite, there exists a positive M such that $h_\sigma(k) \leq M$ for all $k \in K$ and all $\sigma \in \Sigma$. If $\rho \in \mu_A^{-1}(K)$, let $\sigma = \sigma_\rho$, and compute as follows:

$$M \geq h_\sigma(\mu_A(x)) = \sum_{s\in\sigma} a_s h_\sigma(s)\rho(s) + \sum_{s\notin\sigma} a_s h_\sigma(s)\rho(s).$$

Recall that $|a_s h_\sigma(s)| \leq m$ and that $0 \leq \rho(s) < 1$ for $s \notin \sigma$, so that

$$\left|\sum_{s\notin\sigma} a_s h_\sigma(s)\rho(s)\right| \leq \sum_{s\notin\sigma} |a_s h_\sigma(s)\rho(s)| \leq m|S|.$$

It follows that

$$\sum_{s\in\sigma} a_s h_\sigma(s)\rho(s) \leq M + m|S|.$$

Each term in the sum on the left is positive, so for each $s \in \sigma$,

$$a_s h_\sigma(s)\rho(s) \leq M + m|S|.$$

Hence

$$\rho(s) \leq \begin{cases} (M + m|S|)/\epsilon & \text{if } \rho \in \sigma \\ 1 & \text{otherwise.} \end{cases}$$

Thus $\mu_A^{-1}(K)$ is bounded, hence compact. □

The second main ingredient in the proof is a study of the maps $\mu_A\colon \mathsf{R}^*_\mathsf{F} \to \mathsf{C}^\circ_\mathsf{F}$ for each face F of Q. The source and the target of these maps are differentiable manifolds, and the proof will exploit this structure. In fact each R^*_F is naturally a Lie group. If $f \in F$ and $\rho \in \mathsf{R}_\mathsf{Q}$, $e_f(\rho) > 0$ if $\rho \in \mathsf{R}^*_\mathsf{F}$. Then $\log(e_f)$ is a

[2] This argument is due to O. Gabber.

well-defined function on R^*_F, and its differential $d\log(e_f)$ is an invariant differential form. The map $f \mapsto d\log(e_f)$ induces a natural isomorphism from $V_F := \mathbf{R} \otimes F^{\mathrm{gp}}$ to the space of invariant differential forms on R^*_F, and hence an isomorphism from the Lie algebra of R^*_F to $V_F^\vee := \mathrm{Hom}(F^{\mathrm{gp}}, \mathbf{R})$. To simplify the notation we write these isomorphisms as identifications. Thus if $f \in F$, we view $1 \otimes f \in V_F$ as an invariant differential form on R^*_F, and if $\phi \in V_F^\vee$ we view ϕ as an invariant vector field on R^*_F. With this notation, we have the formula:

$$de_f = e_f \otimes f. \qquad (1.10.1)$$

Similarly, the interior $\mathsf{C}^\circ_\mathsf{F}$ of the cone spanned by F has a natural structure of a C^∞ manifold, induced from the inclusion $\mathsf{C}^\circ_\mathsf{F} \subseteq V_F$, and the invariant vector fields on the ambient space V_F are naturally identified with elements of V_F.

The following result describes the differential properties of the moment map.

Proposition 1.10.4. *Let A be an effective cycle in C_Q, let $S \subseteq \mathsf{C}_\mathsf{Q}$ be its support, and let F be a face of Q.*

1. *The restriction of the moment map μ_A to R^*_F is the differential of the restriction of the function e_A to R^*_F.*
2. *Let ρ be a point of R^*_F and consider the derivative of μ_A at ρ:*

$$\tau_\rho := T_\rho(\mu_A)\colon T_\rho(\mathsf{R}^*_\mathsf{F}) \to T_{\mu_A(\rho)}(\mathsf{C}^\circ_\mathsf{F}) \quad \text{i.e.,} \quad V_F^\vee \to V_F.$$

Then the associated bilinear form β_ρ on $V_F^\vee$

$$\beta_\rho(\phi, \psi) := \psi(\tau_\rho(\phi))$$

is symmetric and positive semi-definite. If $S \cap \mathsf{C}_\mathsf{F}$ generates C_F, then β_ρ is positive definite, and τ_ρ is an isomorphism.

Proof As we have seen,

$$e_A \circ i_F = \sum_{f \in F \cap S} a_f e_f,$$

so by the formula (1.10.1) for de_f on R^*_F,

$$de_{A_{|F}} = \sum_{f \in F \cap S} a_f de_f = \sum_{f \in F \cap S} a_f e_f \otimes f = \mu_{A_{|F}}.$$

Then

$$T_\rho(\mu_A) = \sum a_f de_f \otimes f = \sum a_f e_f \otimes f \otimes f \in \mathbf{R}[\mathsf{C}_\mathsf{F}] \otimes \mathrm{Hom}(V_F^\vee, V_F).$$

In particular, for $\rho \in \mathsf{R}^*_\mathsf{F}$ and $\phi \in V_F^\vee$,

$$\tau_\rho(\phi) = \sum_f a_f \rho(f)\phi(f) f$$

and

$$\beta_\rho(\phi, \psi) = \sum_f a_f \rho(f)\phi(f)\psi(f).$$

Thus β_ρ is symmetric. Since a_f and $\rho(f)$ are positive for $\rho \in \mathsf{R}^*_\mathsf{F}$ and $f \in S \cap F$, and since

$$\beta_\rho(\phi, \phi) = \sum_f a_f \rho(f)\phi(f)^2,$$

we conclude that β_ρ is positive semi-definite. Suppose $S \cap C_F$ spans C_F. Then if $\beta_\rho(\phi, \phi) = 0$, necessarily $\phi(f) = 0$ for all $f \in S$ and hence $\phi = 0$. Thus in this case β_ρ is positive definite, and consequently τ_ρ is an isomorphism. □

We have already observed that if F is a face of Q, μ_A maps R_F to C_F and R^*_F to C^o_F, so that μ_A is compatible with the stratification by faces. Thus to prove that μ_A is injective it will suffice to prove that, for each face F, μ_A induces an injection $\mathsf{R}^*_\mathsf{F} \to \mathsf{C}^\circ_\mathsf{F}$. To simplify the notation, we may and shall assume that $F = Q$. We have an isomorphism

$$\exp_Q \colon V^\vee_Q \to \mathsf{R}^*_\mathsf{Q} \quad : \quad \phi \mapsto \exp \circ \phi.$$

If ϕ and ϕ' are distinct points of $V^\vee_Q$, let $\psi := \phi' - \phi$ and, for each real number t, let $\phi_t := \phi + t\psi$ and $\rho_t = \exp_Q(\phi_t)$. Then it follows from (2) of Proposition 1.10.4 that the derivative of $\psi(\mu_A(\rho_t))$ with respect to t is $\beta_{\rho_t}(\psi, \psi)$, which is positive. Explicitly,

$$\psi(\mu_A(\rho_t)) = \sum_s a_s \psi(s) \exp\left(\phi(s) + t\psi(s)\right),$$

and the derivative with respect to t is

$$\sum_s a_s (\psi(s))^2 \rho_t(s) > 0.$$

Thus $\psi(\mu_A(\rho_t))$ is an increasing and hence injective function of t. This implies that $\mu_A(\exp \circ \phi) \neq \mu_A(\exp \circ \phi')$.

It is now easy to see that, when S generates Q, μ_A is a homeomorphism. For each face F of Q, $\mathsf{R}^*_\mathsf{F} = \mu_A^{-1}(\mathsf{C}^\circ_\mathsf{F})$ and, in particular, μ_A induces a proper map $\mu_{A,F} \colon \mathsf{R}^*_\mathsf{F} \to \mathsf{C}^\circ_\mathsf{F}$. As we have seen in Proposition 1.10.4, $\mu_{A,F}$ is a differentiable map of differentiable manifolds which induces an isomorphism on tangent spaces at every point, and it follows from the implicit function theorem that its image is open. Since it is also proper, its image is also closed, and since $\mathsf{C}^\circ_\mathsf{F}$ is connected, it follows that $\mu_{A,F}$ is surjective. We conclude that μ_A is bijective, continuous, and proper, and it follows that it is a homeomorphism, since a proper map of Hausdorff spaces is closed. □

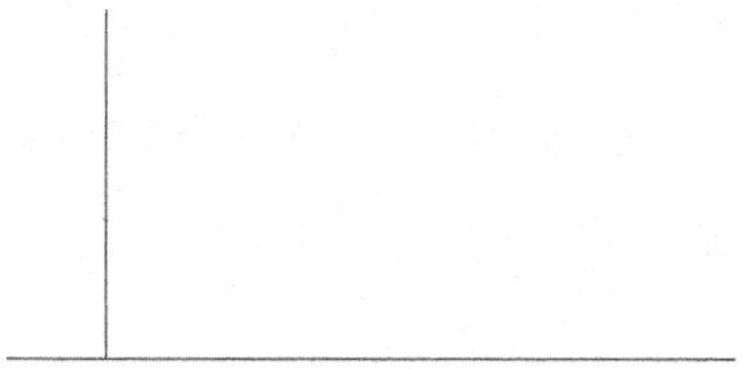

Figure 1.10.1 R_Q for a non-integral Q.

Variant 1.10.5. The fact that the moment map μ is compatible with the stratification by faces implies that, for every ideal K of P, μ induces a homeomorphism $\mathsf{R}_{(\mathsf{P},\mathsf{K})} \to \mathsf{C}_{(\mathsf{P},\mathsf{K})}$. Thus, the analog of Theorem 1.10.2 is also valid for idealized monoids.

Corollary 1.10.6. *If Q is a fine monoid and is not dull, then there is a homeomorphism*

$$(\mathsf{R}_\mathsf{Q}, \mathsf{R}^*_\mathsf{Q}) \to (\mathbf{R}_\geq \times \mathbf{R}^{d-1}, \mathbf{R}_> \times \mathbf{R}^{d-1}),$$

where d is the rank of Q^{gp}. In particular, R_Q is homeomorphic to a manifold with boundary.

Proof This is an is an immediate consequence of Corollary I.4.7.15 and Theorem 1.10.2. □

Remark 1.10.7. Corollary 1.10.6 need not be true if Q is not integral. For example, if Q is given by generators p and q with relations $p + q = q$, then $\mathbf{R}_Q$ is the set of pairs $(x, y) \in \mathbf{R}_\geq \times \mathbf{R}_\geq$ such that $xy = y$, or equivalently, such that $(x - 1)y = 0$. This set, pictured in Figure 1.10.1, is not a manifold with boundary in a neighborhood of the point $(1, 0)$.

The moment map can be used to prove a version of the structure theorem I.4.7.2 for a morphism $\mathsf{R}_\theta \colon \mathsf{R}_\mathsf{Q} \to \mathsf{R}_\mathsf{P}$. This is the "rounding theorem" of [57].

Theorem 1.10.8. *Let $\theta \colon P \to Q$ be a locally exact and local homomorphism of fine monoids, where P is sharp. Then R_θ is a trivial fibration. That is, there*

exists a commutative diagram

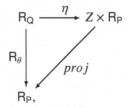

$$\begin{array}{ccc} \mathsf{R}_\mathsf{Q} & \xrightarrow{\eta} & Z \times \mathsf{R}_\mathsf{P} \\ {\scriptstyle \mathsf{R}_\theta}\downarrow & \swarrow {\scriptstyle proj} & \\ \mathsf{R}_\mathsf{P}, & & \end{array}$$

where η *is a homeomorphism.*

Proof We take Z to be $\mathsf{C}_{\mathsf{Q},\mathsf{P}}$, the union of the θ-critical faces of Q. Choose an effective generating cycle A for Q, and let $\eta_Z \colon \mathsf{R}_\mathsf{Q} \to \mathsf{C}_{\mathsf{Q},\mathsf{P}}$ be the composite of the inverse of the moment map $\mu_A \colon \mathsf{R}_\mathsf{Q} \to \mathsf{C}_\mathsf{Q}$ with the inverse of the isomorphism $\mathsf{C}_{\mathsf{Q},\mathsf{P}} \times \mathsf{C}_\mathsf{P} \to \mathsf{C}_\mathsf{Q}$ of Theorem 4.7.2, followed by the projection $\mathsf{C}_{\mathsf{Q},\mathsf{P}} \times \mathsf{C}_\mathsf{P} \to \mathsf{C}_{\mathsf{Q},\mathsf{P}}$. The theorem follows from the fact that $\eta := (\eta_Z, \mathsf{R}_\theta)$ is a homeomorphism. We refer to [57] for the proof, which is rather similar to the proof of Theorem 1.10.2, (but again uses the local exactness of θ). □

It is also possible to define moment maps for projective monoschemes. Note first that the maps $\mathbf{R} \times \mathbf{S}^1 \to \mathbf{C}$ (multiplication) and $\mathbf{C} \to \mathbf{R}_\geq$ (absolute value) define, for any monoscheme X, the commutative diagram

$$\begin{array}{ccc} \mathsf{R}_\mathsf{X} \times \mathsf{T}_\mathsf{X} & \longrightarrow & X(\mathbf{C}) \\ & \searrow & \downarrow \\ & & \mathsf{R}_\mathsf{X}, \end{array}$$

where $\mathsf{R}_\mathsf{X} := X(\mathbf{R}_\geq)$ and $\mathsf{T}_\mathsf{X} := X(\mathbf{S}^1) = X^*(\mathbf{S}^1)$. Furthermore, $X(\mathbf{C})$ (resp. R_X) is the disjoint union of the sets $X^*_x(\mathbf{C})$ (resp. $\mathsf{R}_{\mathsf{X}^*_\mathsf{x}} = X^*_x(\mathbf{R}_\geq)$) as x ranges over the points of X, and X^*_x is a torsor under $\operatorname{spec}(\mathcal{M}^*_{X,x})$. When X is projective, we can fit linearizations of the spaces $\mathsf{R}_{\mathsf{X}^*_\mathsf{x}}$ together as we did in the affine case.

Let (P, h) be a fine graded monoid, let $F := h^{-1}(0)$, $P^+_h := h^{-1}(\mathbf{N}^+)$, and $U_h := \operatorname{spec}(P) \setminus Z(P^+_h)$, so that $\operatorname{Proj}(P, h)$ is the quotient of U_h by $\mathbf{a}^*$. We will modify our construction of the moment map so as to be compatible with this action and at the same time compatible with a moment map defined on R_F.

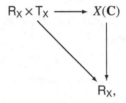

In this construction it will slightly more convenient to work with cones than monoids. We will view h as a homomorphism $\mathsf{C}_\mathsf{P} \to \mathbf{R}_\geq$, and let $\mathsf{C}_{(\mathsf{P},\mathsf{h})} := h^{-1}(1)$, the intersection of C_P with an affine hyperplane. Then $h^{-1}(\mathbf{R}_>) = \mathsf{C}_\mathsf{P} \setminus \mathsf{C}_\mathsf{F}$ is the $\mathbf{R}_>$-invariant ideal $\mathsf{C}_{\mathsf{P}^+_\mathsf{h}}$ of C_P generated by P^+_h and in particular is finitely generated. Let $B := \sum b_t t \in \mathbf{R}[\mathsf{C}_\mathsf{P}]$ be an effective cycle whose support is contained in $\mathsf{C}_{(P,h)}$ and which generates $\mathsf{C}_{P_h^+}$ as an invariant ideal of P. For example, if $(t_1, \ldots, t_m)$ is a finite set of generators for the ideal P_h^+ of P and $b_i \in \mathbf{R}_>$ for each i, then $B := \sum b_i \tilde{t}_i$, will do, where $\tilde{t}_i := t_i/h(t_i) \in \mathsf{C}_P$.

Let T be the support of B and recall that

$$e_B(\rho) := \sum_t b_t \rho(t) \quad \in \mathbf{R}_\geq$$

$$\mu_B(\rho) := \sum_t b_t \rho(t) t \quad \in \mathsf{C}_P .$$

The homomorphism h defines an action of $\mathbf{R}_\geq$ on R_P given by $(r\rho)(p) = r^{h(p)}\rho(p)$. Since the support of B is contained in $h^{-1}(1)$, we have $e_B(r\rho) = re_B(\rho)$ and $\mu_B(r\rho) = r\mu_B(\rho)$. Note further that $h(\mu_B(\rho)) = e_B(\rho)$ and that $e_B(\rho) > 0$ if $\rho \in \mathsf{R}_{U_h}$, since $\rho(t) \neq 0$ for some t in the support of B.

If A is an effective cycle with support S in $h^{-1}(0)$, we define

$$\mu_{A,B}(\rho) := \mu_A(\rho) + e_B(\rho)^{-1}\mu_B(\rho) \;\in\; \mathsf{C}_F + \mathsf{C}_{P_h^+} \subseteq \mathsf{C}_P . \tag{1.10.2}$$

Since $e_B(\rho) = h(\mu_B(\rho))$ and $h(\mu_A(\rho)) = 0$, in fact $\mu_{A,B}(\rho) \in \mathsf{C}_{(P,h)}$. Furthermore, $\mu_{A,B}(r\rho) = \mu_{A,B}(\rho)$ for all r and ρ.

Proposition 1.10.9. *If (P, h) is a fine graded monoid, let $\mathsf{R}_{(P,h)} := \operatorname{proj}(P, h)(\mathbf{R}_\geq)$ and let $\mathsf{C}_{(P,h)} := \mathsf{C}_P \cap h^{-1}(1)$. Then if A and B are as described in the preceding paragraphs, the map $\mu_{A,B}$ (1.10.2) induces a homeomorphism*

$$\mu_{A,B} \colon \mathsf{R}_{(P,h)} \to \mathsf{C}_{(P,h)} .$$

For each ideal K of P, $\mu_{A,B}$ induces a homeomorphism $\mathsf{R}_{(P,K,h)} \to \mathsf{C}_{(P,K,h)}$.

Proof We begin with the following analog of Remark 1.5.13.

Lemma 1.10.10. *If T is a finite set of generators for the $\mathbf{R}_+$-invariant ideal $\mathsf{C}_{P_h^+}$, then the cone C_P is generated by C_F and T.*

Proof Let $C \subseteq \mathsf{C}_P$ be the cone generated by T. Since each $h(t)$ is positive, C is sharp, and the restriction of $h_{|C}$ to C is a local homomorphism. Then, by Corollary I.2.3.18, $h_{|C}$ is a proper morphism. If p is any element of C_P, the restriction of h to $(p - C) \cap \mathsf{C}_P$ is also proper. Let p' be a member of this set that minimizes h, say $p' = p - c$. If $h(p') > 0$, then $p' \in \mathsf{C}_{P_h^+}$ and hence $p' = p'' + \sum r_i t_i$, where $p'' \in \mathsf{C}_P$ and $r_i > 0$ for at least one i. But then $h(p'') < h(p')$ and $p'' = p - c - \sum r_i t_i \in (p - C) \cap \mathsf{C}_P$, a contradiction. It follows that $h(p') = 0$, so $p = p' + c \in \mathsf{C}_F + C$ as claimed. □

It follows from the lemma that the support of $A + B$ generates C_P, and then from Theorem 1.10.2 that the map

$$\mu_{A+B}\colon \mathsf{R}_{\mathsf{P}} \to \mathsf{C}_{\mathsf{P}} : \rho \mapsto \sum_s a_s\rho(s)s + \sum_t b_t\rho(t)t$$

is a homeomorphism.

We have a continuous map $\pi\colon \mathsf{R}_{(\mathsf{U}_\mathsf{h})} \to \mathsf{R}_{(\mathsf{P},\mathsf{h})}$, which makes $\mathsf{R}_{(\mathsf{P},\mathsf{h})}$ the quotient of $\mathsf{R}_{(\mathsf{U},\mathsf{h})}$ by the group $\mathbf{R}_>$, with the action defined by h. Since $\mu_{A,B}$ is invariant under this action, it descends to define the map $\overline{\mu}_{A,B}$.

Let $U_B := \{\rho \in \mathsf{R}_{\mathsf{P}} : e_B(\rho) = 1\}$. Since $h(\mu_{A+B}(\rho)) = e_B(\rho)$, in fact U_B is the inverse image of $\mathsf{C}_{(\mathsf{P},\mathsf{h})}$ by the map μ_{A+B}. Furthermore, μ_{A+B} and $\mu_{A,B}$ agree on U_B. Thus we find the commutative diagram

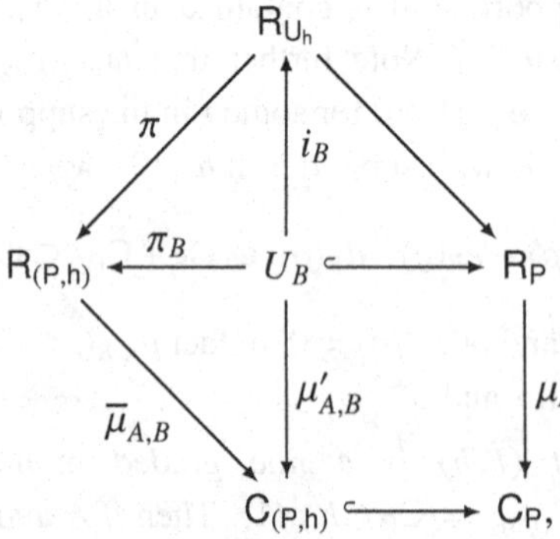

in which the square is Cartesian. It follows that $\mu'_{A,B}$ is also a homeomorphism. Moreover, the map from $\mathsf{R}_{(\mathsf{U}_\mathsf{h})}$ to U_B sending ρ to $(e_B(\rho))^{-1}\rho$ factors through π to define an inverse to the map π_B. Thus π_B is a homeomorphism, and hence so is $\overline{\mu}_{A,B}$. □

Corollary 1.10.11. *Let (P, h) be a fine graded monoid and let K be an ideal of P generated by elements of degree zero. Then $\mathsf{R}_{(\mathsf{P},\mathsf{K},\mathsf{h})}$ is contractible.*

Proof Proposition 1.10.9 produces a homeomorphism $\mathsf{R}_{(\mathsf{P},\mathsf{K},\mathsf{h})} \to \mathsf{C}_{(\mathsf{P},\mathsf{K},\mathsf{h})}$, and the latter is contractible by Proposition I.4.10.8. □

Corollary 1.10.12. *Let X be a fine monoscheme, let $\mathcal{I}$ be a coherent sheaf of ideals in X, and let $\beta\colon X_{\mathcal{I}} \to X$ be the monoidal transformation of X with respect to $\mathcal{I}$. Then the associated map $\mathsf{R}_{\mathsf{X}_{\mathcal{I}}} \to \mathsf{R}_{\mathsf{X}}$ is proper, and its fibers are contractible.*

Proof The properness follows from the facts that β induces a projective morphism of toric varieties and that $\mathsf{R}_{\mathsf{X}_{\mathcal{I}}}$ and R_{X} can be viewed as closed subsets of the corresponding complex varieties. To prove the claim about the fibers, we can reduce to the case in which X is affine, say $X = \operatorname{spec} Q$, and $\mathcal{I}$ is the sheaf of ideals associated to an ideal I of Q. Then $X_I = \mathrm{Proj}(P, h)$, where (P, h) is the Artin-Rees monoid associated to the ideal I. Then $P = h^{-1}(0)$ and the fiber of β over the a point of $\mathsf{A}_\mathfrak{p}$ is subset of $\mathrm{Proj}(P, h)$ defined by the corresponding ideal $\mathfrak{p}$ of P. Thus Corollary 1.10.11 implies the desired result. □

2 Charts and coherence

We shall be primarily interested in sheaves of monoids that are *coherent*. Roughly speaking, this means that, up to units, they are locally controlled by a finitely generated constant monoid. The notion of coherence for log structures is due to K. Kato, but in fact the definition makes sense for any sheaf of monoids.

2.1 Charts

If X is a topological space and Q is a monoid, we denote by Q_X, or just Q or Q, the constant sheaf on X associated to Q, that is, the sheaf associated to the presheaf that takes every open set U of X to Q. If $\mathcal{M}$ is a sheaf of monoids on X, then to give a homomorphism of monoids $Q \to \Gamma(X, \mathcal{M})$ is equivalent to giving a homomorphism of sheaves of monoids $Q_X \to \mathcal{M}$. By Proposition 1.2.2 this is also equivalent to giving a morphism of monoidal spaces $f\colon (X, \mathcal{M}) \to S := \mathrm{spec}(Q)$.

Definition 2.1.1. *Let $\mathcal{M}$ be a sheaf of monoids on a topological space X and let Q be a monoid. A chart for $\mathcal{M}$ subordinate to Q is a monoid homomorphism $\beta\colon Q \to \Gamma(X, \mathcal{M})$ such that the associated logarithmic map $\beta^{log}\colon Q^{log} \to \mathcal{M}$ (see Proposition 1.1.5) is an isomorphism.*

One says that a chart $Q \to \Gamma(X, \mathcal{M})$ is *coherent* (resp. *integral, fine, saturated*) if Q is finitely generated (resp. integral, fine, saturated).

Proposition 2.1.2. *Let Q be a monoid, let $\mathcal{M}$ be a sheaf of monoids on a topological space X, let $\beta\colon Q \to \Gamma(X, \mathcal{M})$ be a homomorphism of monoids, and let $f\colon X \to S :=$ spec(Q) be the corresponding morphism of monoidal spaces. For each point x of X, let $\beta_x^{log}\colon Q_x^{log} \to \mathcal{M}_x$ be the logarithmic homomorphism associated to $\beta_x\colon Q \to \mathcal{M}_x$ (see Definition 1.1.10). Then there is a natural isomorphism:*

$$Q_x^{log} \to f_{log}^*(\mathcal{M}_{S,f(x)}).$$

Furthermore, the following conditions are equivalent.

1. β is a chart for $\mathcal{M}$.

2. *For every $x \in X$, the homomorphism β_x^{log} is an isomorphism.*
3. *The homomorphism $f_{log}^{\flat} : f_{log}^{*}(\mathcal{M}_S) \to \mathcal{M}_X$ is an isomorphism.*

Proof Let x be a point of X. Then $s := f(x)$ is the point in spec(Q) corresponding to $F_x := \beta_x^{-1}(\mathcal{M}_x^*)$, and $\mathcal{M}_{S,s} = Q_{F_x}$. Hence the homomorphism $\beta_x : Q \to \mathcal{M}_x$ factors: $Q \to Q_{F_x} = \mathcal{M}_{S,s} \to \mathcal{M}_x$. We saw in Proposition 1.1.5 that Q_x^{log} can be obtained by first localizing Q at F_x and then sharpening with respect to the morphism $Q_{F_x} \to \mathcal{M}_x$, and one gets the same answer if one does this with Q_{F_x} in place of Q. Thus the map $Q_x^{log} \to f_{log}^{*}(\mathcal{M}_{S,s})$ is an isomorphism,

As we saw in diagram 1.1.1 in the proof of Proposition 1.1.5, there is a cocartesian diagram

$$\begin{array}{ccc} \beta_x^{-1}(\mathcal{M}_x^*) & \longrightarrow & \mathcal{M}_x^* \\ \downarrow & & \downarrow \\ Q & \longrightarrow & Q_x^{log}. \end{array} \tag{2.1.1}$$

Thus there is a unique homomorphism $Q_x^{log} \to \mathcal{M}_x$ compatible with the inclusion $\mathcal{M}_x^* \to \mathcal{M}_x$ and β. Then it follows immediately from the definitions that statements (1) and (2) are equivalent. Since $Q_x^{log} \cong f_{log}^{*}(\mathcal{M}_{S,f(x)})$, statements (2) and (3) are also equivalent. □

For example, if $S = \text{spec}(Q)$, the tautological map $Q \to \Gamma(S, \mathcal{M}_S)$ is a chart for $\mathcal{M}_S$.

Remark 2.1.3. Thanks to Proposition 2.1.2, we can safely generalize the notion of a chart by replacing the monoid Q there by an arbitrary monoscheme S. That is, we can say that a chart for a sheaf $\mathcal{M}$ of monoids on X subordinate to S is a morphism f of monoidal spaces from $(X, \mathcal{M})$ to the monoscheme S such that the associated map $f_{log}^{\flat} : f_{log}^{*}(\mathcal{M}_S) \to \mathcal{M}_X$ is an isomorphism.

Proposition 2.1.4. *Let $f : (X, \mathcal{M}) \to S$ be a morphism of locally monoidal spaces, where S is a monoscheme. Then if f is a chart for $\mathcal{M}$, the map*

$$\overline{f}^{\flat} : \overline{\mathcal{M}}_{S,f(x)} \to \overline{\mathcal{M}}_{X,x}$$

is an isomorphism for every point x of X. The converse is true if $\mathcal{M}$ is u-integral. In particular, if $\mathcal{M}_X$ is u-integral, a homomorphism $\beta : Q \to \mathcal{M}_X$ is a chart if and only if, for every $x \in X$, the natural map $Q/F_x \to \overline{\mathcal{M}}_{X,x}$ is an isomorphism, where $F_x := \beta_x^{-1}(\mathcal{M}_X^*)$.

Proof If f is a chart, then $f_{log}^{\flat} : (f_{log}^{*}\mathcal{M}_S)_x \to \mathcal{M}_{X,x}$ is an isomorphism, hence so is $\overline{f}^{\flat}_{log}$. Since f is a morphism of locally monoidal spaces, the homomorphism $f^{\flat}_x$ is local, so by (1) of Proposition 1.1.8, $\overline{\mathcal{M}}_{S,f(x)} \cong \overline{(f^*_{log}\mathcal{M}_S)_x} \cong \overline{\mathcal{M}}_x$. Conversely, if $\mathcal{M}$ is u-integral and $\overline{f}^{\flat}_{log}$ is an isomorphism, then the same is true of $f^{\flat}_{log}$, by Proposition I.4.1.2. □

Definition 2.1.5. *A sheaf of monoids $\mathcal{M}$ on a topological space X is quasi-coherent (resp. coherent) if X admits an open covering $\mathcal{U}$ such that the restriction of $\mathcal{M}$ to each U in $\mathcal{U}$ admits a chart (resp. a chart subordinate to a finitely generated monoid). A sheaf of monoids $\mathcal{M}$ is fine if it is coherent and integral.*

We remark that this use of the terminology "fine sheaf of monoids" conflicts with the convention used in Proposition 1.1.3: the stalks of a fine sheaf of monoids $\mathcal{M}$ need not be finitely generated (since there is no control over $\mathcal{M}^*$), and a sheaf of monoids all of whose stalks are fine need not be coherent.

Corollary 2.1.6. *Let $\mathcal{M}$ be a coherent sheaf of monoids on a topological space X and let n be an integer. Then*

$$X^{(n)} := \{x \in X : \mathrm{rk}(\overline{\mathcal{M}}^{\mathrm{gp}}_{X,\overline{x}}) \leq n\}$$

is open in X. In particular,

$$X^*_{\mathcal{M}} := \{x \in X : \mathcal{M}^*_{X,x} = \mathcal{M}_{X,x}\}$$

is open in X.

Proof We can check the openness of $X^{(n)}$ locally on X, and thus we may assume without loss of generality that there is a chart $f\colon X \to S$, where S is the spectrum of a finitely generated monoid Q. It follows from Proposition 2.1.4 that $X^{(n)} = f^{-1}(S^{(n)})$, and so we are reduced to the case when $(X, \mathcal{M}) = \mathrm{spec}(S)$. This case follows from Corollary I.2.3.9. □

2.2 Construction and comparison of charts

Let $\mathcal{M}$ be a sheaf of monoids on a topological space X. A *morphism of charts* for $\mathcal{M}$ is a commutative diagram

$$\begin{array}{ccc} Q & \longrightarrow & Q' \\ & \searrow^{\beta} & \downarrow{\scriptstyle \beta'} \\ & & \Gamma(X, \mathcal{M}), \end{array}$$

where β and β' are charts for $\mathcal{M}$. Thus the class of charts for $\mathcal{M}$ forms a category. If x is a point of X, a *germ of a chart at x* is a chart of the restriction of $\mathcal{M}$ to some open neighborhood of x in X, and a morphism of such germs $\beta \to \beta'$ is an element of the direct limit $\varinjlim \operatorname{Hom}(\beta_{|_U}, \beta'_{|_U})$, where U ranges over the neighborhoods of x in X.

Proposition 2.2.1. *Let $\beta\colon Q \to \Gamma(X, \mathcal{M})$ be a chart for a sheaf of monoids $\mathcal{M}$ on X. Suppose that β factors as follow,*

$$\beta = Q \xrightarrow{\theta} Q' \xrightarrow{\beta'} \Gamma(X, \mathcal{M}),$$

where Q' is a finitely generated monoid. Then, locally on X, β' can be factored as

$$\beta' = Q' \xrightarrow{\theta'} Q'' \xrightarrow{\beta''} \Gamma(X, \mathcal{M}),$$

where β'' is a chart for $\mathcal{M}$ and Q'' is finitely generated. In particular, $\mathcal{M}$ is coherent.

Proof Let $\{q'_i : i \in I\}$ be a finite system of generators for Q' and let x be a point of X. Because β is a chart, it follows from Proposition 2.1.4 that $\overline{\beta}_x$ is surjective. Hence for each $i \in I$ there exist an element $q_i \in Q$, a neighborhood U_i of x, and a section u_i of $\mathcal{M}^*(U_i)$ such that $\beta'(q'_i) = \beta(q_i) + u_i$. Replacing X by $\cap\{U_i : i \in I\}$, we may assume that the u_i are global sections of $\mathcal{M}^*$. Let Q'' be the quotient of $Q' \oplus \mathbf{Z}^I$ by the relation identifying $(q'_i, 0)$ with $(\theta(q_i), e_i)$. Then there is a unique homomorphism $\beta'' : Q'' \to \mathcal{M}$ sending the class of any $(q', 0)$ to $\beta'(q')$ and the class of $(0, e_i)$ to u_i, inducing the commutative diagrams

$$\begin{array}{ccc} Q & \xrightarrow{\theta} & Q' \\ {\scriptstyle\beta}\downarrow & \searrow^{\theta''} & \downarrow{\scriptstyle\theta'} \\ \mathcal{M} & \xleftarrow{\beta''} & Q'' \end{array} \qquad \begin{array}{ccc} Q^{log} & \xrightarrow{\theta^{log}} & Q'^{log} \\ {\scriptstyle\beta^{log}}\downarrow & \searrow^{\theta''^{log}} & \downarrow{\scriptstyle\theta'^{log}} \\ \mathcal{M} & \xleftarrow{\beta''^{log}} & Q''^{log}. \end{array}$$

Then $\overline{Q}''$ is generated by the elements $\overline{q}'_i = \overline{\theta}(q_i)$, and so $\overline{\theta''} : \overline{Q} \to \overline{Q}''$ is surjective. Since the natural homomorphism $\overline{Q''} \to \overline{Q''^{log}}$ is necessarily surjective, it follows that $\overline{\theta''^{log}}$ is also surjective. By construction θ''^{log} is sharp, and it follows that θ''^{log} is surjective. But $\beta''^{log} \circ \theta''^{log} = \beta^{log}$, which is an isomorphism, so θ'^{log} is also injective, hence bijective. Then β''^{log} is also an isomorphism, and so $\beta'' : Q'' \to \mathcal{M}$ is again a chart. □

Corollary 2.2.2. *Let $\mathcal{M}$ be a coherent sheaf of monoids on X and let x be a point of X. Then the category of germs of coherent charts for $\mathcal{M}$ at x is filtering.*

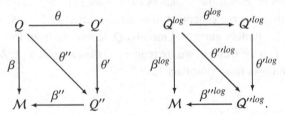

Proof We will verify the conditions F0′, F1′, and F2 (see Definition I.4.5.1 and the following discussion). Since $\mathcal{M}$ is coherent, the category of germs of coherent charts at x has at least one object. Let $\beta_i \colon Q_i \to \mathcal{M}_{|U_i}$ be finitely generated charts for the restrictions of $\mathcal{M}$ to neighborhoods U_i of x in X, for $i = 1, 2$, and let $U := U_1 \cap U_2$. Then $Q' := Q_1 \oplus Q_2$ is finitely generated and β_i factors through the map $\beta' \colon Q' \to \mathcal{M}_{|U}$ induced by β_1 and β_2. By Proposition 2.2.1, β' factors through a coherent chart $\beta'' \colon Q'' \to \mathcal{M}$ in some neighborhood of x, and so there is a commutative diagram:

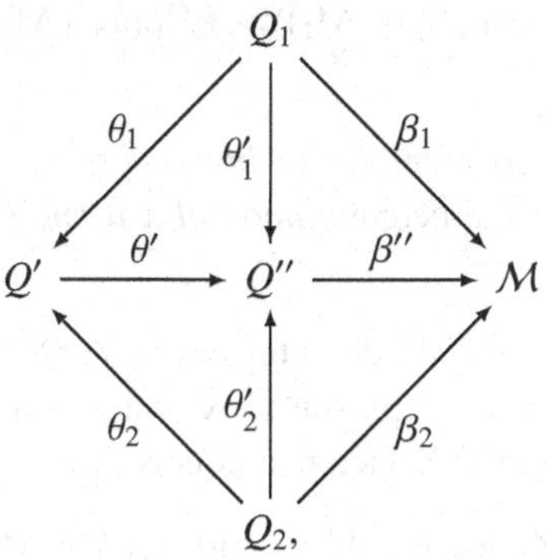

where β'' is a coherent chart for $\mathcal{M}$.

Similarly, if $\theta_1, \theta_2 \colon \beta \to \beta'$ is a pair of morphisms of coherent charts, the coequalizer Q'' of θ_1 and θ_2 is finitely generated, and there is a diagram

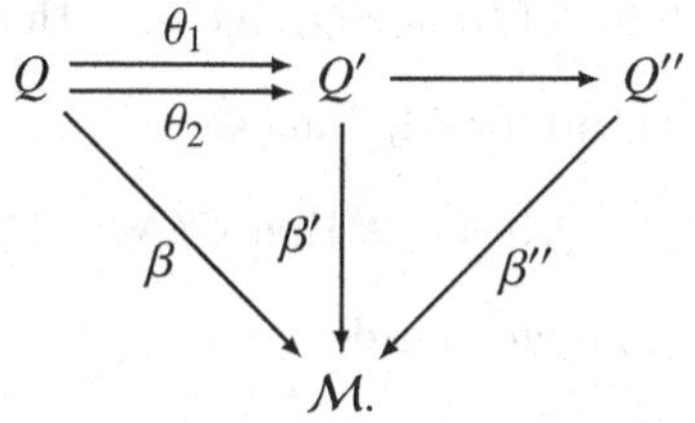

Then by Proposition 2.2.1, β'' factors through a coherent chart $Q''' \to \mathcal{M}$. □

The next result allows us to extend charts from a stalk to a neighborhood.

Proposition 2.2.3. *Let $\mathcal{M}$ be a coherent sheaf of monoids on X and let x be a point of X. Then the evident functor from the category of germs of coherent charts of $\mathcal{M}$ at x to the category of finitely generated charts of $\mathcal{M}_x$ is an equivalence.*

The proof of this proposition depends on some preliminary results.

Lemma 2.2.4. *Let $\mathcal{M}$ be a sheaf of monoids on X and let x be a point of X. If Q is a finitely generated monoid, the natural map*

$$(\mathrm{Hom}(Q, \mathcal{M}))_x \to \mathrm{Hom}(Q, \mathcal{M}_x)$$

is an isomorphism.

Proof By Theorem I.2.1.7, Q is of finite presentation, so the functor $\mathrm{Hom}(Q,\ )$ commutes with direct limits. □

Lemma 2.2.5. *Let $\mathcal{M}_1, \mathcal{M}_2$, and $\mathcal{N}$ be sheaves of monoids on X, let $\alpha_i\colon \mathcal{M}_i \to \mathcal{N}$ be logarithmic morphisms, and let x be a point of X.*

1. *If $\mathcal{M}_1$ is coherent, the natural map*

$$(\mathcal{H}om_{\mathcal{N}}(\mathcal{M}_1, \mathcal{M}_2))_x \to \mathrm{Hom}_{\mathcal{N}_x}(\mathcal{M}_{1_x}, \mathcal{M}_{2_x})$$

   is an isomorphism.

2. *If $\mathcal{M}_1$ and $\mathcal{M}_2$ are coherent, then a homomorphism $\theta\colon \mathcal{M}_1 \to \mathcal{M}_2$ over $\mathcal{N}$ is an isomorphism in a neighborhood of x if and only if its germ θ_x is an isomorphism.*

Proof Let $\beta_1\colon Q_1 \to \mathcal{M}_1$ be a chart for $\mathcal{M}_1$, with Q_1 finitely generated. Since α_1 and α_2 are logarithmic morphisms to $\mathcal{N}$ and β_1 is a chart for $\mathcal{M}_1$, any morphism from Q_1 to $\mathcal{M}_2$ over $\mathcal{N}$ factors uniquely through $\mathcal{M}_1$. That is,

$$\mathrm{Hom}_{\mathcal{N}}(Q_1, \mathcal{M}_2) \cong \mathrm{Hom}_{\mathcal{N}}(\mathcal{M}_1, \mathcal{M}_2).$$

This remains true on any neighborhood of x in X, so passing to the limit and applying Lemma 2.2.4 with $\mathcal{M} = \mathcal{M}_2$ and with $\mathcal{M} = \mathcal{N}$, we get

$$(\mathcal{H}om_{\mathcal{N}}(\mathcal{M}_1, \mathcal{M}_2))_x \cong (\mathcal{H}om_{\mathcal{N}}(Q_1, \mathcal{M}_2))_x \cong \mathrm{Hom}_{\mathcal{N}_x}(Q_1, \mathcal{M}_{2,x}).$$

But $Q_1 \to \mathcal{M}_{1_x}$ is also a chart for $\mathcal{M}_{1_x}$, and so

$$\mathrm{Hom}_{\mathcal{N}_x}(Q_1, \mathcal{M}_{2_x}) \cong \mathrm{Hom}_{\mathcal{N}_x}(\mathcal{M}_{1_x}, \mathcal{M}_{2_x}),$$

proving (1). Statement (2) is an immediate consequence. □

Lemma 2.2.6. *Let $\theta\colon \mathcal{M}_1 \to \mathcal{M}_2$ be a logarithmic homomorphism of coherent sheaves of monoids. If the germ of θ at a point x of X is an isomorphism, then θ is an isomorphism in some neighborhood of x.*

Proof This is an immediate consequence of (2) of Lemma 2.2.5, with $\alpha_1 = \theta$ and $\alpha_2 = \mathrm{id}_{\mathcal{M}_2}$. □

Proof of Proposition 2.2.3 Let us first prove that our "evident functor" is fully faithful. Let $\beta\colon Q \to \mathcal{M}$ and $\beta'\colon Q' \to \mathcal{M}$ be coherent charts for $\mathcal{M}$. Then $\beta_x\colon Q \to \mathcal{M}_x$ and $\beta'_x\colon Q' \to \mathcal{M}_x$ are charts for $\mathcal{M}_x$, and a morphism of charts $\beta_x \to \beta'_x$ is just a homomorphism $\theta\colon Q \to Q'$ such that $\beta'_x \circ \theta = \beta_x$. Then, for each $q \in Q$, the equality $\beta'(\theta(q)) = \beta(q)$ holds in some neighborhood of x and, since Q is finitely generated, in fact $\beta' \circ \theta = \beta$ in some neighborhood of x.

Thus the homomorphism θ defines a morphism of the germs of charts defined by β and β', proving that our functor is fully faithful. To show that it is essentially surjective, let β_x be a chart for $\mathcal{M}_x$. Then, by Lemma 2.2.4, β extends to a homomorphism from Q to $\mathcal{M}$ in some neighborhood of x. Moreover, β_x^{log} is an isomorphism and, since β^{log} is logarithmic, it follows from Lemma 2.2.6 that β^{log} is an isomorphism in some neighborhood U of x. Thus $\beta_{|_U}$ is a chart for $\mathcal{M}_{|_U}$. □

2.3 Exact and neat charts

It is often desirable to construct charts for a sheaf of monoids $\mathcal{M}$ that are as close as possible to the stalk of $\mathcal{M}$ at some given point x.

Definition 2.3.1. *Let $\mathcal{M}$ be a sheaf of integral monoids on a space X, let x be a point of X and let $\beta\colon Q \to \mathcal{M}(X)$ be an integral chart for $\mathcal{M}$.*

1. *β is exact at x if it satisfies the following equivalent conditions:*
   (a) *$\beta_x\colon Q \to \mathcal{M}_x$ is exact;*
   (b) *$\beta_x\colon Q \to \mathcal{M}_x$ is local;*
   (c) *$\overline{\beta}_x\colon \overline{Q} \to \overline{\mathcal{M}}_x$ is an isomorphism.*
2. *θ is neat at x if it satisfies the following equivalent conditions:* [3]
   (a) *Q is sharp and β is exact at x;*
   (b) *$\pi \circ \beta_x\colon Q \to \mathcal{M}_x \to \overline{\mathcal{M}}_x$ is an isomorphism;*
   (c) *$\pi^{gp} \circ \beta_x^{gp}\colon Q^{gp} \to \overline{\mathcal{M}}_x^{gp}$ is an isomorphism.*

The equivalence of the conditions in (1) follows immediately from Corollary 1.1.9. To check the equivalences in (2), note that if (2a) is true and Q is sharp, then (1a) and hence (1c) hold, and so does (2b). It is trivial that (2b) implies (2c). If (2c) is true, then $Q \to \overline{\mathcal{M}}_x$ is injective, so Q is sharp. Since β is a chart, $Q \to \overline{\mathcal{M}}_x$ is surjective, hence bijective, so β_x is exact by Proposition I.4.2.1. Thus (2c) implies (2a).

Remark 2.3.2. Let $\beta\colon Q \to \mathcal{M}$ be a chart for $\mathcal{M}$ that is subordinate to a fine monoid Q and let x be a point of X. Then $F := \beta^{-1}(\mathcal{M}_x^*)$ is a face of Q, and hence by (3) of Theorem I.2.1.17, there exists a $p \in F$ such that $\langle p \rangle = F$. Since $\beta(p)_x \in \mathcal{M}_x^*$, there exists a neighborhood U of x on which $\beta(p)$ is a unit, and so β factors through a map $\beta'\colon Q_F \to \mathcal{M}_{|_U}$. Then β'_x is local, hence exact. Thus, any fine chart for $\mathcal{M}$ locally factors through a chart that is exact at x.

We shall use the condition (2c) as a technique to construct exact charts.

[3] Some authors use the terminology "good chart" rather than "neat chart."

Definition 2.3.3. *A markup of an integral monoid M is a homomorphism ϕ from a finitely generated abelian group L to M^{gp} inducing a surjection $L \to \overline{M}^{gp}$. A morphism of markups of M is a homomorphism of abelian groups $\theta \colon L_1 \to L_2$ such that $\phi_2 \circ \theta = \phi_1$.*

Proposition 2.3.4. *Let M be an integral monoid such that $\overline{M}$ is fine.*

1. *The category of markups of M is filtering (and in particular not empty).*
2. *If $\phi \colon L \to M^{gp}$ is a markup of M, consider the induced map*

$$\beta \colon Q := L \times_{M^{gp}} M \to M.$$

Then the natural map $Q^{gp} \to L$ is an isomorphism, and β is a fine exact chart for M.

3. *Conversely, if $\beta \colon Q \to M$ is a fine exact chart for M, then $\beta^{gp} \colon Q^{gp} \to M^{gp}$ is a markup of M.*
4. *The correspondence $\phi \mapsto \beta$ gives a equivalence between the category of fine exact charts for M and the category of markups of M.*

Proof We will verify conditions F0′, F1′, and F2 described after Definition 4.5.1. If M is integral and $\overline{M}$ is fine, there exists a finitely generated free abelian group L together with a surjection $L \to \overline{M}^{gp}$. Since L is free and, by (1) of Proposition I.1.3.5, $\overline{M}^{gp} \cong M^{gp}/M^*$, this homomorphism lifts to a homomorphism $L \to M^{gp}$, which defines a markup of M. Let $\phi_1 : L_1 \to M^{gp}$ and $\phi_2 \colon L_2 \to M^{gp}$ be markups of M. Then $L' := L_1 \oplus L_2$ is a finitely generated group, the homomorphisms $\phi_i \colon L_i \to M^{gp}$ factor through a homomorphism $\phi' : L' \to M^{gp}$, and the induced map $L' \to \overline{M}^{gp}$ is necessarily surjective. Thus the markups ϕ_1 and ϕ_2 both map to the markup ϕ'. Finally, if θ and θ' are morphisms of markups $\phi_1 \to \phi_2$, then the induced map from the coequalizer of θ and θ' to M^{gp} is also a markup, and F2 is also satisfied.

Let ϕ be a markup of M and let $\beta \colon Q \to M$ be the map described in (2). We write, with an abuse of notation, $\overline{\phi}$ for the induced homomorphism $L \to \overline{M}^{gp}$. Note first that, since $M \to \overline{M}$ is exact,

$$Q := L \times_{M^{gp}} M = L \times_{M^{gp}} M^{gp} \times_{\overline{M}^{gp}} \overline{M} = L \times_{\overline{M}^{gp}} \overline{M}.$$

Since L and $\overline{M}$ are fine, it follows from (2) of Theorem I.2.1.17 that Q is fine. The integrality of M implies that $Q \subseteq L$ and hence $Q^{gp} \subseteq L$. To see the equality, choose $z \in L$ and then choose $\overline{m}_1, \overline{m}_2 \in \overline{M}$ such that $\overline{\phi}(z) = \overline{m}_1 - \overline{m}_2$. Since $L \to \overline{M}^{gp}$ is surjective, there exist $z_1, z_2 \in L$ such that $\overline{\phi}(z_i) = \overline{m}_i$, and then $\overline{\phi}(z - z_1 + z_2) = 0$. But $Q = L \times_{\overline{M}^{gp}} \overline{M}$, so z_1, z_2, and $z - z_1 + z_2$ all belong to Q. It follows that $z \in Q^{gp}$, and hence $Q^{gp} = L$. It is then clear from the definition of Q that β is exact, and so $\overline{\beta} \colon \overline{Q} \to \overline{M}$ is injective, by (5) of Proposition I.4.2.1.

Since ϕ is surjective, its pullback β is also surjective, hence $\overline{\beta}$ is surjective, hence an isomorphism. This proves that θ is an exact chart of M.

Conversely, if $\beta\colon Q \to M$ is a fine exact chart, then by condition (2c) of Definition 2.3.1, the homomorphism $\overline{\beta}\colon \overline{Q} \to \overline{M}$ is an isomorphism. Thus the map $\phi := \beta^{\mathrm{gp}}\colon Q^{\mathrm{gp}} \to M^{\mathrm{gp}}$ is a markup. It is immediate to verify that the correspondences $\phi \mapsto \beta$ and $\beta \mapsto \beta^{\mathrm{gp}}$ are functorial and form an equivalence of categories. □

Corollary 2.3.5. *Let $\mathcal{M}$ be a fine sheaf of monoids on X and let x be a point of X. Then the category of markups of $\mathcal{M}_x$ is equivalent to the category of germs at x of charts for $\mathcal{M}$ which are exact at x. In particular this category is filtering.*

Proof The proof follows immediately from Proposition 2.2.3 and statement (4) of Proposition 2.3.4. □

Corollary 2.3.6. *A sheaf of monoids $\mathcal{M}$ is fine (resp. fine and saturated) if and only if locally it admits a fine (resp. fine and saturated) chart.*

Proof Suppose that $\mathcal{M}$ is fine and let x be a point of X. Since $\mathcal{M}$ is integral, $\overline{\mathcal{M}}_x$ is integral. Since $\mathcal{M}$ admits a chart subordinate to a finitely generated monoid, by Proposition 1.1.8, $\overline{\mathcal{M}}_x$ is also finitely generated, hence fine. Then, by Proposition 2.3.4, $\mathcal{M}_x$ admits a markup, which in turn defines a chart for $\mathcal{M}_x$ subordinate to a fine monoid Q. Since $\mathcal{M}$ is by hypothesis coherent, this chart is in fact the germ of a chart for $\mathcal{M}$ at x by Proposition 2.2.3. Since $\overline{Q} \cong \overline{\mathcal{M}}_x$, Q is saturated if and only if $\mathcal{M}_x$ is. Thus $\mathcal{M}$ admits a chart subordinate to a fine (resp. fine and saturated) monoid, in some neighborhood of x. Conversely, if $\beta\colon Q \to \mathcal{M}$ is a chart for $\mathcal{M}$ and Q is fine (resp. fine and saturated) then, for each $x \in X$, Q maps surjectively to $\overline{\mathcal{M}}_x$, so $\overline{\mathcal{M}}_x$ is finitely generated, and $\mathcal{M} \cong Q^{log}$, so $\mathcal{M}$ is integral (resp. saturated). □

Proposition 2.3.7. *Let $\mathcal{M}$ be a fine sheaf of monoids on a topological space X and let x be a point of X. Then $\mathcal{M}$ admits a local chart that is neat at x if and only if the sequence*

$$0 \to \mathcal{M}_x^* \to \mathcal{M}_x^{\mathrm{gp}} \to \overline{\mathcal{M}}_x^{\mathrm{gp}} \to 0$$

splits. These conditions hold if $\overline{\mathcal{M}}^{\mathrm{gp}}$ is torsion free, and in particular if $\mathcal{M}$ is saturated.

Proof If the sequence splits, let $Q := \overline{\mathcal{M}}_x$. Then the splitting defines a homomorphism $\beta\colon Q \to Q^{\mathrm{gp}} \to \mathcal{M}_x^{\mathrm{gp}}$, which we claim factors through $\mathcal{M}_x$. Indeed, if $q \in Q$, then $\overline{\beta}(q) = q \in \overline{\mathcal{M}}_x$, and hence $\beta(q) = u + q$ for some $u \in \mathcal{M}_x^*$, so $\beta(q) \in \mathcal{M}_x$. It is clear that $\beta^{-1}(\mathcal{M}_x^*) = \overline{\beta}^{-1}(\{0\}) = \{0\}$, so β is local, and so $Q^{log} \to \mathcal{M}_x$ is strict and sharp, hence an isomorphism, by Proposition I.4.1.2.

Thus β is a chart for $\mathcal{M}_x$, which extends to a chart for $\mathcal{M}$ in a neighborhood of x by Proposition 2.2.3. It is neat at x by construction. If $\overline{\mathcal{M}}_x^{\mathrm{gp}}$ is torsion free, it is a torsion-free finitely generated group, hence free, and the sequence automatically splits. If $\mathcal{M}$ is saturated, we know $\overline{\mathcal{M}}_x^{\mathrm{gp}}$ is torsion free by (2) of Proposition I.1.3.5.

Conversely, if $\beta\colon Q \to \mathcal{M}$ is a chart for $\mathcal{M}$ that is neat at x, then by definition the map $\overline{\beta}_x\colon Q \to \overline{\mathcal{M}}_x$ is an isomorphism. Thus there is a splitting $\overline{\mathcal{M}}_x \to \mathcal{M}_x$ which induces a splitting $\overline{\mathcal{M}}_x^{\mathrm{gp}} \to \mathcal{M}_x^{\mathrm{gp}}$. □

Example 2.3.8. Let $n \geq 2$ be a natural number, let $\Gamma := \mathbf{Z} \oplus \mathbf{Z}/n^2\mathbf{Z}$, and let M be the submonoid of Γ generated by $(1,0), (1,\overline{1})$, and $(0,\overline{n})$. Then M^* is cyclic of order n and generated by $(0,\overline{n})$, the monoid $\overline{M}$ is given by generators m_1, m_2 satisfying the relation $nm_1 = nm_2$, and $\overline{M}^{\mathrm{gp}} \cong \mathbf{Z} \oplus \mathbf{Z}/n\mathbf{Z}$. The sequence

$$0 \to M^* \to M^{\mathrm{gp}} \to \overline{M}^{\mathrm{gp}} \to 0$$

does not split, and hence M does not admit a neat chart.

If $\beta\colon Q \to \mathcal{M}$ is a chart for $\mathcal{M}$ and $\gamma\colon Q \to \mathcal{M}^*$ is any homomorphism, then $\beta + \gamma$ is also a chart for $\mathcal{M}$. In fact it is almost true that, locally on X, any two charts can be compared in this way.

Proposition 2.3.9. *Let $\beta\colon Q \to \mathcal{M}$ and $\beta'\colon Q' \to \mathcal{M}$ be fine charts for a fine sheaf of monoids $\mathcal{M}$ on X. Let x be a point of x and suppose that either of the following condition holds:*

1. *β is neat at x; or*
2. *β is exact at x and Q'^{gp} is torsion free.*

Then in some neighborhood of x in X, there exist homomorphisms $\kappa\colon Q' \to Q$ and $\gamma\colon Q' \to \mathcal{M}^*$ such that $\beta' = \beta \circ \kappa + \gamma$.

Proof The solid arrows in the diagram

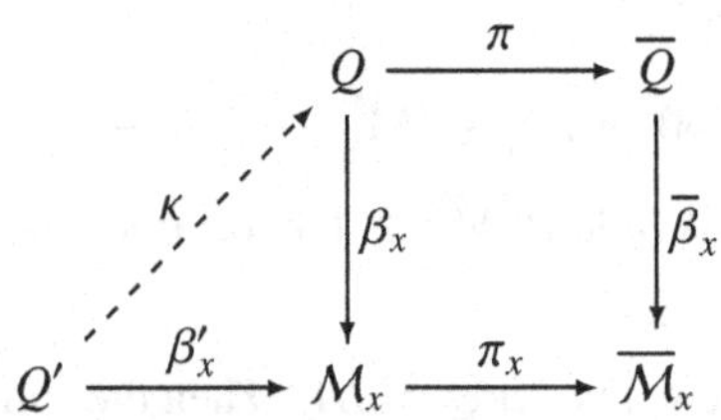

is given, and the square commutes. If β is neat at x, then the composition $\pi_x \circ \beta_x\colon Q \to \overline{\mathcal{M}}_x$ is an isomorphism, so there is a unique homomorphism $\kappa\colon Q' \to Q$ making the outer part of the diagram commute. If β is only exact at x, then $\overline{\beta}_x$ is an isomorphism, and we find a homomorphism $\overline{\kappa}\colon Q' \to \overline{Q}$

such that $\overline{\beta}_x \circ \overline{\kappa} = \pi_x \circ \beta'_x$. If Q'^{gp} is torsion free it is free and, since $\overline{\pi}^{\mathrm{gp}}$ is surjective, we can then choose a homomorphism $\kappa^{\mathrm{gp}} \colon Q'^{\mathrm{gp}} \to Q^{\mathrm{gp}}$ such that $\pi^{\mathrm{gp}} \circ \kappa^{\mathrm{gp}} = \overline{\kappa}^{\mathrm{gp}}$. Since π is exact, necessarily κ^{gp} sends Q' to Q. Thus we again find a homomorphism $\kappa \colon Q' \to Q$ making the outer diagram commute. The triangle may not commute, but, for each $q' \in Q$, $\pi_x(\beta'_x(q')) = \pi_x(\beta_x(\kappa(q')))$, so there is a unique $\gamma(q') \in \mathcal{M}^*_x$ such that $\beta'_x(q') = \beta_x(\kappa(q')) + \gamma(q')$. Evidently $\gamma \colon Q' \to \mathcal{M}^*_x$ is a homomorphism. Since Q' is of finite presentation, γ extends to a homomorphism $Q' \to \mathcal{M}^*$ in some neighborhood of x. □

2.4 Charts for morphisms

Definition 2.4.1. *Let $f \colon (X, \mathcal{M}_X) \to (Y, \mathcal{M}_Y)$ be a morphism of monoidal spaces. A chart for f subordinate to a homomorphism of monoids $\theta \colon P \to Q$ is a triple (α, θ, β) fitting into a commutative diagram*

$$\begin{array}{ccc} Q & \xrightarrow{\beta} & \Gamma(X, \mathcal{M}_X) \\ {\scriptstyle\theta}\uparrow & & \uparrow{\scriptstyle f^\flat} \\ P & \xrightarrow{\alpha} & \Gamma(Y, \mathcal{M}_Y), \end{array}$$

where α and β are charts for $\mathcal{M}_Y$ and $\mathcal{M}_X$, respectively.

Proposition 2.4.2. *Let $f \colon (X, \mathcal{M}_X) \to (Y, \mathcal{M}_Y)$ be a morphism of monoidal spaces and let $\alpha \colon P \to \Gamma(Y, \mathcal{M}_Y)$ be a chart for $\mathcal{M}_Y$, where P is finitely generated. Assume that $\mathcal{M}_X$ is coherent. Then, locally on X, α fits into a chart (α, θ, β) for f:*

$$\begin{array}{ccc} Q & \xrightarrow{\beta} & \Gamma(X, \mathcal{M}_X) \\ {\scriptstyle\theta}\uparrow & & \uparrow{\scriptstyle f^\flat} \\ P & \xrightarrow{\alpha} & \Gamma(Y, \mathcal{M}_Y), \end{array}$$

where Q is also finitely generated.

Proof Since the assertion is local on X, we may assume that $\mathcal{M}_X$ admits a chart $\beta \colon Q \to \mathcal{M}$, with Q finitely generated. Consider the commutative diagram

$$\begin{array}{ccccc} Q & \xrightarrow{\iota_Q} & P \oplus Q & \xrightarrow{\gamma} & \Gamma(X, \mathcal{M}_X) \\ & & \uparrow{\scriptstyle \iota_P} & & \uparrow{\scriptstyle f^\flat} \\ & & P & \xrightarrow{\alpha} & \Gamma(Y, \mathcal{M}_Y), \end{array}$$

where ι_P and ι_Q are the canonical inclusions and $\gamma(p,q) := f^\flat(\alpha(p)) + \beta(q)$. Since $\gamma \circ \iota_Q$ is a chart and $P \oplus Q$ is finitely generated, Proposition 2.2.1 implies that γ can be factored as $\gamma = \beta'\gamma'$, where $\beta' : Q' \to \mathcal{M}_X$ is a chart and Q' is finitely generated. Then $\theta := \gamma' \circ \iota_P$ fits into a chart for $f^\flat$ as desired. □

Remark 2.4.3. Suppose that, in the situation of Proposition 2.4.2, $\mathcal{M}_Y, \mathcal{M}_X$, and P are fine, and let x be a point of X. Then it follows from Remark 2.3.2 that β may be chosen to be exact at x.

The following more delicate theorem, due to K. Kato, shows how to construct charts that closely reflect the structure of a morphism of monoidal spaces. Recall from Definition 1.1.10 that if $f : X \to Y$ is a morphism of monoidal spaces, $\mathcal{M}_{X/Y}$ is the cokernel of $f^*_{log}(\mathcal{M}_Y) \to \mathcal{M}_X$.

Theorem 2.4.4 (Neat charts)**.** *Let $f : X \to Y$ be a morphism of fine monoidal spaces, let x be a point of X, and assume that* $\mathrm{Ext}^1(\mathcal{M}^{gp}_{X/Y,x}, \mathcal{M}^*_{X,x}) = 0$. *Let $\alpha : P \to \Gamma(Y, \mathcal{M}_Y)$ be a fine chart for $\mathcal{M}_Y$. Then, after X is replaced by some neighborhood of x, there exists a chart (α, θ, β) for f with the following properties.*

1. *Q is a fine monoid.*
2. *$\theta^{gp} : P^{gp} \to Q^{gp}$ is injective.*
3. *The homomorphism $Q^{gp}/P^{gp} \to \mathcal{M}^{gp}_{X/Y,x}$ induced by β is bijective.*

Moreover, the chart β can be chosen to be exact at x. A chart with properties (1)–(3) is said to be neat at x.

Proof Let $y := f(x)$, let N_x denote the image of $(f^*\mathcal{M}_Y)_x$ in $\mathcal{M}_{X,x}$ and let P' denote the image of P in $\mathcal{M}_{X,x}$. Because $P \to \mathcal{M}_{Y,y}$ is a chart, the map $P^{gp} \to \overline{\mathcal{M}}^{gp}_{Y,y}$ is surjective, and consequently N^{gp}_x is the subgroup of $\mathcal{M}^{gp}_{X,x}$ generated by $\mathcal{M}^*_{X,x}$ and P'^{gp}. Thus the map $\mathcal{M}^*_{X,x} \to N^{gp}_x/P'^{gp}$ is surjective. Since $\mathrm{Ext}^1(\mathcal{M}^{gp}_{X/Y,x}, \mathcal{M}^*_{X,x})$ vanishes by assumption and the functor $\mathrm{Ext}^1(\mathcal{M}^{gp}_{X/Y,x},\ )$ is right exact, it follows that $\mathrm{Ext}^1(\mathcal{M}^{gp}_{X/Y,x}, N^{gp}_x/P'^{gp})$ also vanishes. Then the map

$$\mathrm{Ext}^1(\mathcal{M}^{gp}_{X/Y,x}, P'^{gp}) \to \mathrm{Ext}^1(\mathcal{M}^{gp}_{X/Y,x}, N^{gp}_x)$$

is surjective, and the element of $\mathrm{Ext}^1(\mathcal{M}^{\mathrm{gp}}_{X/Y,x}, N^{\mathrm{gp}}_x)$ defined by the exact sequence

$$0 \to N^{\mathrm{gp}}_x \to \mathcal{M}^{\mathrm{gp}}_{X,x} \to \mathcal{M}^{\mathrm{gp}}_{X/Y,x} \to 0$$

lifts to a class in $\mathrm{Ext}^1(\mathcal{M}^{\mathrm{gp}}_{X/Y,x}, P'^{\mathrm{gp}})$. Since $P^{\mathrm{gp}} \to P'^{\mathrm{gp}}$ is surjective, it follows from the right exactness of $\mathrm{Ext}^1(\mathcal{M}^{\mathrm{gp}}_{X/Y,x}, \ )$ that this element lifts further to a class in $\mathrm{Ext}^1(\mathcal{M}^{\mathrm{gp}}_{X/Y,x}, P^{\mathrm{gp}})$. Thus, we have the following diagram,

$$\begin{array}{ccccccccc}
0 & \longrightarrow & P^{\mathrm{gp}} & \longrightarrow & L & \longrightarrow & \mathcal{M}^{\mathrm{gp}}_{X/Y,x} & \longrightarrow & 0 \\
 & & \downarrow & & \downarrow{\scriptstyle \phi} & & \downarrow{\scriptstyle \mathrm{id}} & & \\
0 & \longrightarrow & N^{\mathrm{gp}}_x & \longrightarrow & \mathcal{M}^{\mathrm{gp}}_{X,x} & \longrightarrow & \mathcal{M}^{\mathrm{gp}}_{X/Y,x} & \longrightarrow & 0,
\end{array}$$

whose rows are exact. Since $\mathcal{M}_{X/Y,x}$ is a quotient of the fine monoid $\overline{\mathcal{M}}_{X,x}$ it is also fine, and in particular $\mathcal{M}^{\mathrm{gp}}_{X/Y,x}$ is a finitely generated group. It follows that L is also a finitely generated group. Moreover, the map $P^{\mathrm{gp}} \to \overline{N}^{\mathrm{gp}}_x$ is surjective, and it follows from the diagram that $L \to \overline{\mathcal{M}}^{\mathrm{gp}}_{X,x}$ is also surjective. Thus ϕ is a markup of $\mathcal{M}_{X,x}$ in the sense of Definition 2.3.3. Let $\beta_x \colon Q := \phi^{-1}(\mathcal{M}_{X,x}) \to \mathcal{M}_{X,x}$ be the exact chart corresponding to ϕ, as explained in Corollary 2.3.5, and recall from Proposition 2.2.3 that β_x extends to a chart $\beta \colon Q \to \mathcal{M}_X$ in some neighborhood of x. Then $Q^{\mathrm{gp}} = L$, and it is clear from the diagram that β_x satisfies the conditions of the theorem. □

Remark 2.4.5. Let $\beta \colon Q \to \mathcal{M}_X$ be a chart for $\mathcal{M}_X$ that is subordinate to a fine monoid Q. Then $\alpha \colon 0 \to \mathcal{M}^*_X$ is a chart for $\mathcal{M}^*_X$ and one finds a chart $(\alpha, \beta, 0)$ for the canonical morphism $X \to \underline{X}$. This chart is neat at a point x of X if and only if the map $Q^{\mathrm{gp}} \to \overline{\mathcal{M}}^{\mathrm{gp}}_{X,x}$ is an isomorphism, that is, if and only if β is neat at x, by statement (2) of Definition 2.3.1. More generally, suppose we are given a morphism of monoidal spaces $f \colon X \to Y$ and a chart for f that is neat at a point x of f. Suppose further that $f^\flat_x$ induces an injection $\overline{\mathcal{M}}_{Y,f(x)} \to \overline{\mathcal{M}}_{X,x}$ and that $P \to \mathcal{M}_Y$ is neat at y. Then it follows that $Q \to \mathcal{M}_X$ is also neat at x. Indeed, in this case we have a commutative diagram with exact rows:

$$\begin{array}{ccccccccc}
0 & \longrightarrow & P^{\mathrm{gp}} & \longrightarrow & Q^{\mathrm{gp}} & \longrightarrow & \mathcal{M}^{\mathrm{gp}}_{X/Y,x} & \longrightarrow & 0 \\
 & & \downarrow{\scriptstyle \cong} & & \downarrow & & \downarrow{\scriptstyle \cong} & & \\
0 & \longrightarrow & \overline{\mathcal{M}}^{\mathrm{gp}}_{Y,f(x)} & \longrightarrow & \overline{\mathcal{M}}^{\mathrm{gp}}_{X,x} & \longrightarrow & \mathcal{M}^{\mathrm{gp}}_{X/Y,x} & \longrightarrow & 0.
\end{array}$$

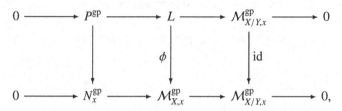

This diagram shows that $Q^{\mathrm{gp}} \to \overline{\mathcal{M}}^{\mathrm{gp}}_{X,x}$ is an isomorphism, so the chart $Q \to \mathcal{M}_{X,x}$ is neat at x.

Remark 2.4.6. More generally, let $f: X \to Y$ and $g: Y \to Z$ be morphisms of fine monoidal spaces, and let (β,ϕ,γ) and (α,θ,β) be charts for f and g respectively. Then $(\alpha,\phi\circ\theta,\gamma)$ is a chart for $g \circ f$. Let x be a point of X, and suppose that (β,ϕ,γ) is neat at x and that (α,θ,β) is neat at $y := f(x)$. Then it does not necessarily follow that $(\alpha,\phi\circ\theta,\gamma)$ is neat at x. Indeed if $\alpha: P \to \mathcal{M}_Z$, $\beta: Q \to \mathcal{M}_Y$, and $\gamma: R \to \mathcal{M}_X$, we have a commutative diagram with exact rows:

$$
\begin{array}{ccccccccc}
0 & \longrightarrow & Q^{\mathrm{gp}}/P^{\mathrm{gp}} & \longrightarrow & R^{\mathrm{gp}}/P^{\mathrm{gp}} & \longrightarrow & R^{\mathrm{gp}}/Q^{\mathrm{gp}} & \longrightarrow & 0 \\
 & & \downarrow \cong & & \downarrow \tilde{\gamma} & & \downarrow \cong & & \\
 & & \mathcal{M}^{\mathrm{gp}}_{Y/Z,f(x)} & \xrightarrow{a} & \mathcal{M}^{\mathrm{gp}}_{X/Z,x} & \longrightarrow & \mathcal{M}^{\mathrm{gp}}_{X/Y,x} & \longrightarrow & 0.
\end{array}
$$

The homomorphism $\tilde{\gamma}$ is necessarily surjective, but its kernel is isomorphic to the kernel of a. However, note that if $z := g(y)$, then $\mathcal{M}^{\mathrm{gp}}_{Y/Z,y}$ is the cokernel of the map $\overline{\mathcal{M}}_{Z,z} \to \overline{\mathcal{M}}_{Y,y}$ and $\mathcal{M}^{\mathrm{gp}}_{X/Z,x}$ is the cokernel of the map $\overline{\mathcal{M}}^{\mathrm{gp}}_{Z,z} \to \overline{\mathcal{M}}^{\mathrm{gp}}_{X,x}$. Thus a is injective if $\overline{\mathcal{M}}_{Y,y} \to \overline{\mathcal{M}}_{X,x}$ is injective, and we can conclude that the composed chart is indeed neat in this case. This fact is worth recording for future reference.

Corollary 2.4.7. *Let $f: X \to Y$ and $g: Y \to Z$ be morphisms of fine monoidal spaces, with charts (β,ϕ,γ) and (α,θ,β) for f and g, respectively. Let x be a point of X, and suppose that f is s-injective at x, that (β,ϕ,γ) is neat at x, and that (α,θ,β) is neat at $f(x)$. Then $(\alpha,\phi\circ\theta,\gamma)$ is neat at x.*

2.5 Constructibility and coherence

It is possible to give a fairly explicit description of what it means for a sheaf of integral monoids to be coherent.

Definition 2.5.1. *Let $\mathcal{E}$ be a sheaf of sets on a topological space X. A trivializing stratification for $\mathcal{E}$ is a locally finite partition Σ of X such that:*

1. *each element of Σ is connected and locally closed;*
2. *if S_1 and S_2 are elements of Σ and $S_1 \cap \overline{S}_2 \neq \emptyset$, then $S_1 \subseteq \overline{S}_2$;*
3. *the restriction of $\mathcal{E}$ to each S in Σ is constant.*

We say that a sheaf $\mathcal{E}$ on X is quasi-constructible if X has a trivializing stratification for $\mathcal{E}$.

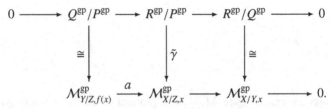

We will call a partition satisfying the conditions in Definition 2.5.1, with the exception of connectedness, a *quasi-trivializing partition.* If the sets occurring in such a partition are locally connected with finitely many components, then the partition can be refined into a trivializing partition. This is always the case if X is noetherian, for example, or, in the local complex analytic context, if the sets are locally defined by analytic functions.

Remark 2.5.2. Recall that a *Kolmogoroff space* is a topological space X such that given any two distinct points x and y of X, either x does not belong to the closure of y or y does not belong to the closure of x. For example, if Q is a monoid, spec(Q) is a Kolmogoroff space; it is finite if Q is finitely generated. Every point of a finite Kolmogoroff space is locally closed, and hence every sheaf on such a space is quasi-constructible. Furthermore, if $f\colon X \to Y$ is a continuous map and Σ is a trivializing partition for $\mathcal{E}$ on Y, then $f^{-1}(\Sigma)$ is a quasi-trivializing stratification for $f^{-1}(\mathcal{E})$ on X,

If X is noetherian, then every locally closed subset S of X is a union of finitely many irreducible components, and if $\mathcal{E}$ is constant on S it is constant on each of these. Thus we could require the elements of Σ to be irreducible (instead of connected) without significant lost of generality. If X is also sober [31, (0.2.1.1)], each of these irreducible subsets S contains a unique generic point σ, and for any $s \in S$, the cospecialization map $\mathcal{E}_s \to \mathcal{E}_\sigma$ is an isomorphism. Since we also want to work in the complex analytic context, we shall explain a different point of view.

If Σ is a trivializing stratification for $\mathcal{E}$ and $s \in S \in \Sigma$, then since $\mathcal{E}_{|S}$ is constant and S is connected, the natural map $\mathcal{E}(S) \to \mathcal{E}_s$ is an isomorphism. We write $\mathcal{E}_S$ for $\mathcal{E}(S)$ to emphasize this. If t belongs to the closure of S, then every neighborhood V of t contains a point s of $S \cap V$, and we have a natural map:

$$\mathcal{E}(V) \to \mathcal{E}_s \cong \mathcal{E}_S.$$

Taking the limit over all V, we find the *cospecialization map*,

$$\mathrm{cosp}_{t,S}\colon \mathcal{E}_t \to \mathcal{E}_S.$$

If T is the element of Σ containing t, and if now s is any point of S, then $T \subseteq S^-$, and we find a commutative diagram:

$$\begin{array}{ccc} \mathcal{E}_T & \xrightarrow{\operatorname{cosp}_{T,S}} & \mathcal{E}_S \\ \cong\big\downarrow & \nearrow^{\operatorname{cosp}_{t,S}} & \big\downarrow\cong \\ \mathcal{E}_t & \xrightarrow[\operatorname{cosp}_{t,s}]{} & \mathcal{E}_s. \end{array}$$

These cospecialization maps satisfy the following cocycle conditions: $\operatorname{cosp}_{S,S} = \mathrm{id}$ and, if $T \subseteq S^- \subseteq R^-$, then $\operatorname{cosp}_{T,R} = \operatorname{cosp}_{T,S} \circ \operatorname{cosp}_{S,R}$.

We shall say that a point x of X is a *central point* for Σ if x belongs to the closure of every element of Σ. Any point x of X has a neighborhood U such that x is a central point for $\Sigma_{|_U}$: it suffices to take a neighborhood U of x that meets only finitely many strata and then remove the closures of all the strata whose closures do not contain x.

Proposition 2.5.3. *Let $\mathcal{E}$ be a quasi-constructible sheaf on a topological space X and let x be a point of X.*

- *If x is a central point for a trivializing partition of $\mathcal{E}$, the natural map $\mathcal{E}(X) \to \mathcal{E}_x$ is an isomorphism.*
- *For all sufficiently small neighborhoods U of x in X, the natural map $\mathcal{E}(U) \to \mathcal{E}_x$ is an isomorphism.*

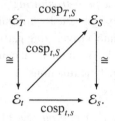

Proof It is clear that (1) implies (2), since any x has a neighborhood U with the property that it is central in any sub-neighborhood of U. So let us assume that x is a central point for a trivializing partition Σ. First we prove that the natural map $\mathcal{E}(X) \to \mathcal{E}_x$ is injective. The central point x belongs to the closure of the stratum containing every point y, so the map $\mathcal{E}(X) \to \mathcal{E}_y$ factors through the cospecialization map $\operatorname{cosp}_{x,y}$. Thus if two elements e, e' of $\mathcal{E}(X)$ have the same image in $\mathcal{E}_x$, they have the same image in $\mathcal{E}_y$ for every y, and hence must be equal. For the surjectivity, let e_x be an element of $\mathcal{E}_x$, let U be a neighborhood of x in X, and let e be an element of $\mathcal{E}(U)$ whose germ at x is e_x. Since x is a central point for $\mathcal{E}_{|_U}$, in fact e is unique. The Hausdorff maximality principle implies that there is a maximal open neighborhood of x to which e extends, and we may assume that U is such a maximal neighborhood. We claim that then $U = X$. Indeed, if y is any point of X, let $e'_y := \operatorname{cosp}_{x,y}(e_x)$, and let V be a neighborhood of y and e' be a section of $\mathcal{E}$ on V whose stalk at y is e'_y. After replacing V by a possibly smaller neighborhood of y, we may assume that y is central for $\mathcal{E}_{|_{V'}}$. If z is any point of $U \cap V'$, the closure of $S(z)$ contains both x and y, and $\operatorname{cosp}_{x,z} = \operatorname{cosp}_{y,z} \circ \operatorname{cosp}_{x,y}$. It follows that the stalk of e at z agrees with the stalk of e' at z. Since this is true at every point of $U \cap V'$, the restrictions of e' and of e to $U \cap V'$ agree. But then e and e' extend to $U \cup V'$, so in fact $y \in U$. □

Theorem 2.5.4. *An integral sheaf of monoids $\mathcal{M}$ is fine if it satisfies the following three conditions.*

1. *X admits an open covering on which $\overline{\mathcal{M}}$ is quasi-constructible.*
2. *For each $x \in X$, $\overline{\mathcal{M}}_x$ is finitely generated.*
3. *Whenever x and ξ are points of X with $x \in \overline{\xi}$, the cospecialization map $\operatorname{cosp}_{x,\xi} \colon \overline{\mathcal{M}}_x \to \overline{\mathcal{M}}_\xi$ identifies $\overline{\mathcal{M}}_\xi$ with the quotient of $\overline{\mathcal{M}}_x$ by a face.*

Conversely, any fine sheaf of monoids $\mathcal{M}$ satisfies conditions (2) and (3); it also satisfies (1) with "quasi-trivializing partition" in place of "trivializing partition." In fact, in the case of locally noetherian schemes, $\mathcal{M}$ satisfies (1).

Proof Suppose that $\mathcal{M}$ is fine. Properties (1) through (3) are local on X, so by Corollary 2.3.6 we may assume that $\mathcal{M}$ admits a fine chart $P \to \mathcal{M}$. Let $h \colon X \to S := \operatorname{spec}(Q)$ be the corresponding map of locally monoidal spaces. Then by Proposition 2.1.4, $\overline{\mathcal{M}} \cong h^{-1}(\overline{\mathcal{M}}_S)$. Since S is a finite Kolmogoroff space, $\overline{\mathcal{M}}_S$ is quasi-constructible and, by Remark 2.5.2, $\mathcal{M}$ admits a quasi-trivializing partition. Furthermore, properties (2) and (3) hold for $\overline{\mathcal{M}}_S$, and hence also for $\overline{\mathcal{M}}$.

Now suppose that $\mathcal{M}$ satisfies the conditions (1) through (3) and let x be a point of X. Without loss of generality we may assume that x is a central point for the trivializing partition Σ. Since $\overline{\mathcal{M}}_x$ is finitely generated, $\mathcal{M}_x$ admits a markup $L \to \mathcal{M}_x^{\mathrm{gp}}$ and hence an exact chart $\beta_x \colon Q \to \mathcal{M}_x$, as explained in in Proposition 2.3.4. Since Q is finitely generated, Lemma 2.2.4 tells us that, after replacing X by an open neighborhood of x, we can find a homomorphism $\beta \colon Q \to \mathcal{M}$ whose stalk at x is β_x. Then β is a chart for $\mathcal{M}$ at x, which, by Proposition 2.1.4, means that the induced map $Q/F_x \to \overline{\mathcal{M}}_x$ is an isomorphism, where $F_x := \beta_x^{-1}(\mathcal{M}_x^*)$. We claim that the same is true for every point y of X. For any such point we have a commutative diagram

$$\begin{array}{ccc}
Q/F_x & \xrightarrow{\;\overline{\beta}_x\;} & \overline{\mathcal{M}}_x \\
{\scriptstyle \pi_{x,y}}\downarrow & & \downarrow{\scriptstyle \operatorname{cosp}_{x,y}} \\
Q/F_y & \xrightarrow{\;\overline{\beta}_y\;} & \overline{\mathcal{M}}_y .
\end{array}$$

By condition (3), $\operatorname{cosp}_{x,y}$ is the quotient of $\overline{\mathcal{M}}_x$ by a face G, which can only be $\text{cosp}_{x,y}^{-1}(0)$. Since F_y is by definition the inverse image of 0 in Q, it follows that $\pi_{x,y}$ is the quotient of Q/F_x by $\overline{\beta}_x^{-1}(G)$, and hence that $\overline{\beta}_y$ is also an isomorphism. □

Proposition 2.5.5. *If $\mathcal{M}$ is a fine sheaf of monoids on a noetherian and sober space X, then $\Gamma(U, \overline{\mathcal{M}}_X)$ is fine, for every open set U of X,*

Proof It suffices to treat the case $U = X$. By Theorem 2.5.4 and Proposition 2.5.3, every point x admits an open neighborhood U_x such that the map $\overline{\mathcal{M}}_X(U_x) \to \overline{\mathcal{M}}_{X,x}$ is an isomorphism. In particular, $\overline{\mathcal{M}}_X(U_x)$ is a fine monoid. Since X is quasi-compact, there exists a finite set $\{U_{x_1}, \ldots, U_{x_n}\}$ of these neighborhoods that covers X. We prove that $\Gamma(U_m, \overline{\mathcal{M}}_X)$ is fine by induction on m, where $U_m := \cup\{U_{x_i} : i \leq m\}$. In fact, $\Gamma(U_m, \overline{\mathcal{M}}_X)$ is the fiber product of $\Gamma(U_{m-1}, \overline{\mathcal{M}}_X)$ and $\Gamma(U_{x_m}, \overline{\mathcal{M}}_X)$ over the integral monoid $\Gamma(U_m \cap U_{x_m}, \overline{\mathcal{M}}_X)$, so it is fine by statement (6) of Theorem I.2.1.17.

□

2.6 Coherent sheaves of ideals and faces

Let $\beta\colon Q \to \mathcal{M}$ be a homomorphism from a constant monoid Q to a sheaf of monoids $\mathcal{M}$ on a space X and let K be an ideal of Q. We denote by $\tilde{K}$ the sheaf associated to the presheaf taking an open set U to the ideal of $\mathcal{M}(U)$ generated by $\beta_U(K)$. If $\mathcal{K}$ is a sheaf of ideals in $\mathcal{M}$, if $\beta\colon Q \to \mathcal{M}$ is a chart for $\mathcal{M}$, and if $\mathcal{K} \cong \tilde{K}$, then we say that (Q, K) *is a chart for* $(\mathcal{M}, \mathcal{K})$.

Proposition 2.6.1. *Let $\mathcal{K}$ be a sheaf of ideals in a coherent sheaf of monoids $\mathcal{M}$ on a space X. Then the following conditions are equivalent.*

1. *Locally on X, $\mathcal{K}$ is generated by a finite set of global sections.*
2. *Locally on X, there exists a chart for $(\mathcal{M}, \mathcal{K})$.*

If these conditions are satisfied, $\mathcal{K}$ is said to be a coherent sheaf of ideals in $\mathcal{M}$. If this is the case and if $Q \to \mathcal{M}$ is any chart for $\mathcal{M}$ and x is a point of X, then there is an ideal K of Q such that (Q, K) is a chart for $(\mathcal{M}, \mathcal{K})$ in some neighborhood of x.

Proof To prove that (1) implies (2), we may and shall assume that $\mathcal{K}$ is generated by global sections $k_1, \ldots, k_n$ on all of X. Let x be a point of X. Since $\mathcal{M}$ is coherent, we may, after replacing X by a neighborhood of x, assume that there exist a finitely generated monoid Q and a chart $\beta\colon Q \to \Gamma(X, \mathcal{M})$. We shall show that there is an ideal K of Q such that (Q, K) is a chart for $(\mathcal{M}, \mathcal{K})$ in some neighborhood of x. Since β is a chart for $\mathcal{M}$, it induces a surjection $Q \to \overline{\mathcal{M}}_x$, by Proposition 2.1.4. For each i, choose a $q_i \in Q$ mapping to the germ of $\overline{k}_i$ at x. Replacing X by a suitable neighborhood of x, we may assume that each $\overline{\beta}(q_i) = \overline{k}_i$ in $\Gamma(X, \overline{\mathcal{M}})$. After a further shrinking of X, it follows that $\beta(q_i)$ and $\overline{k}_i$ differ by an element of $\Gamma(X, \mathcal{M}^*)$. Thus without loss of generality we may assume that $k_i = \beta(q_i)$ for all i. Let K be the ideal of Q generated by $q_1, \ldots, q_n$. Then it is clear that $\tilde{K} = \mathcal{K}$.

Conversely, if $\beta\colon (Q, K) \to (\mathcal{M}, \mathcal{K})$ is a chart with Q finitely generated, then by Theorem I.2.1.7, K is finitely generated as an ideal. If $(k_1, \ldots, k_n)$ generates K, then $(\beta(k_1), \ldots, \beta(k_n))$ generates $\mathcal{K}$. □

Remark 2.6.2. Let $\mathcal{K}$ be a coherent sheaf of ideals in a coherent sheaf of monoids on X. Suppose that $\beta\colon Q \to \mathcal{M}$ is a chart for $\mathcal{M}$ and x is a point of X. Let K be the inverse image in Q of $\overline{\mathcal{K}}_x$. Then the proof of Proposition 2.6.1 shows that in some neighborhood of x, K generates $\mathcal{K}$.

Proposition 2.6.3. *Let $\mathcal{M}$ be a fine sheaf of monoids on a topological space X. For each natural number i there exists a unique sheaf of ideals $\mathcal{K}_i$ of $\mathcal{M}$ such that, for each $x \in X$, $\mathcal{K}_{i,x}$ is the intersection of all the prime ideals $\mathfrak{p}$ of $\mathcal{M}_x$ of height i. Each of these ideals $\mathcal{K}_i$ is coherent. The stalk $\mathcal{K}_{i,x}$ of $\mathcal{K}_i$ at a point x of X is a proper ideal of $\mathcal{M}_x$ if and only if $\dim \mathcal{M}_x \geq i$. In particular, $\mathcal{I}_{\mathcal{M}} := \mathcal{K}_1$ is the interior ideal of $\mathcal{M}$.*

Proof The uniqueness is clear, since a sheaf of ideals is determined by the collection of its stalks. Because of this uniqueness, it will suffice to prove this proposition when $\mathcal{M}$ admits a chart β subordinate to a fine monoid Q. By Proposition 2.1.4, β induces an isomorphism $Q/F_x \to \overline{\mathcal{M}}_x$, where $F_x := \beta_x^{-1}(\mathcal{M}_X^*)$, for each $x \in X$. Then by Corollary I.2.3.9, the ideal K_i of Q discussed there corresponds to the intersection of the height-i primes of $\overline{\mathcal{M}}_{X,x}$. Thus the ideal of $\mathcal{M}_X$ generated by K_i has the desired properties. In particular, statement (4) of this corollary implies that the construction is compatible with further localization, so that these properties hold on all of X. The last statement also follows from Corollary I.2.3.9. □

It is also sometimes useful to work with sheaves of faces. For example, let $(X, \mathcal{M}_X)$ be a monoidal space and let U be an open subset of X. Then the subsheaf $\mathcal{F}$ of $\mathcal{M}_X$ consisting of those sections whose restriction to U are units is a sheaf of faces of $\mathcal{M}_X$.

Definition 2.6.4. *Let $\mathcal{M}$ be a sheaf of monoids on X and $\mathcal{F} \subseteq \mathcal{M}$ a sheaf of faces of $\mathcal{M}$. Then a* relative chart *for $\mathcal{F}$ is a chart $Q \to \mathcal{M}$ together with a face $F \subseteq Q$ such that $\mathcal{F}$ is the smallest sheaf of faces of $\mathcal{M}$ containing F. If, locally on X, such a chart exists then $\mathcal{F}$ is said to be* relatively coherent *in $\mathcal{M}$.*

The proof of the following result is so similar to that of Proposition 2.6.1

that we will omit it. (Recall from Theorem I.2.1.17 that, in a fine monoid, every face is monogenic.)

Proposition 2.6.5. *Let $\mathcal{F}$ be a sheaf of faces in a fine sheaf of monoids $\mathcal{M}$. Then the following conditions are equivalent.*

1. *Locally on X, $\mathcal{F}$ is generated as a sheaf of faces of $\mathcal{M}$ by a single global section.*
2. *Locally on X, $\mathcal{F}$ is generated as a sheaf of faces by a finite set of global sections.*
3. *$\mathcal{F}$ is relatively coherent in $\mathcal{M}$.*

If $\mathcal{F}$ is a relatively coherent sheaf of faces in $\mathcal{M}$, if $\beta\colon Q \to \mathcal{M}$ is a fine chart for $\mathcal{M}$, and if x is a point of X, then there is a face F of Q such that (Q, F) is a relative chart for $(\mathcal{M}, \mathcal{F})$ in some neighborhood of x. □

A relatively coherent sheaf of faces in a coherent sheaf of monoids need not be coherent as a sheaf of monoids. For a simple example, consider the monoid P given by generators a, b, c and the relation $a + b = 2c$. Let F be the face of P generated by $a = 2c - b$ and let $\mathfrak{p}$ be the complement of the face of P generated by b. Then the stalk of $\tilde{F}$ at $\mathfrak{p}$ is the face of P_b generated by a, which is the monoid generated by c, b, and $-b$. Thus $\tilde{F}_{\mathfrak{p}}/\tilde{F}^*_{\mathfrak{p}} \cong \mathbf{N}$, generated by the class of c. On the other hand, at the closed point $\mathfrak{m} := P^+$, $\tilde{F}_{\mathfrak{m}}$ is the free monoid generated by a. Thus the map $\tilde{F}_{\mathfrak{m}} \to \tilde{F}_{\mathfrak{p}}/\tilde{F}^*_{\mathfrak{p}}$ identifies with the homomorphism $\mathbf{N} \xrightarrow{\cdot 2} \mathbf{N}$. This map is not the quotient by a face, and so Theorem 2.5.4 shows that $\tilde{F}$ is not coherent. Other examples can be constructed from the non-simplicial monoid given by a, b, c, d with $a + b = c + d$. In fact, the next proposition shows that the coherence of $\tilde{F}$ is quite unusual.

Proposition 2.6.6. *Let F be a face of a fine sharp monoid P and let $X :=$ spec(P). Then the (relatively coherent) sheaf of faces $\mathcal{F}$ of $\mathcal{M}_X$ generated by F is coherent as a sheaf of monoids if and only if F is a direct summand of P.*

Proof By Theorem 2.5.4, $\mathcal{F}$ is coherent if and only if, for each pair of points (x, ξ) with $x \in \xi^-$, the map $\mathrm{cosp}_{x,\xi}\colon \mathcal{F}_x \to \mathcal{F}_\xi$ identifies $\overline{\mathcal{F}}_\xi$ with the quotient of $\mathcal{F}_x$ by the face $\mathrm{cosp}^{-1}_{x,\xi}(\mathcal{F}^*_\xi)$. It suffices to check this conditions if x is the unique closed point of X. For each $x \in X$, let G_x denote the face of P consisting of those elements that map to a unit in $\mathcal{M}_X$, i.e., the complement of the prime ideal of P corresponding to x. Then $\overline{\mathcal{M}}_x \cong P/G_x$ and $\overline{\mathcal{F}}_x \subseteq \overline{\mathcal{M}}_x$ corresponds to the face of P/G_x generated by $F/F \cap G_x$. Thus $\mathcal{F}$ is coherent if and only if for every face G of P, $F/F \cap G$ is a face of P/G. By Proposition I.4.7.10, this is the case if and only if F is a direct summand of P. □

Remark 2.6.7. If $\mathcal{M}$ is a fine sheaf of monoids on a topological space X, then the function $x \mapsto \operatorname{rk} \overline{\mathcal{M}}_x^{\mathrm{gp}}$ is upper semicontinuous, as follows easily from Theorem 2.5.4. This is not true for a relatively coherent sheaf of faces $\mathcal{F}$ in $\mathcal{M}$. However, it is immediate to verify that the set of points of X where $\overline{\mathcal{F}}_x = 0$, which we denote by $X^*(\mathcal{F})$ or $X^*_{\mathcal{F}}$, is open.

III

Logarithmic Schemes

1 Log structures and log schemes

Although the concepts of logarithmic geometry apply potentially to a wide range of situations, we shall not attempt to develop a language to carry this out in great generality here. We restrict ourselves to the case of algebraic geometry using the language of schemes, leaving to the future the task of building a foundation for logarithmic algebraic spaces, algebraic stacks, analytic varieties, and other types of geometry.

In the context of schemes, log structures in the étale and Zariski topologies each have their own advantages and disadvantages, so here we shall consider both. We refer the reader to chapter I of [15] for an introduction to the étale topology. If x is a scheme-theoretic point of a scheme X, we shall write $\overline{x}$ for a geometric point lying over x, i.e., a separably closed field extension of the residue field $k(x)$. The stalk of O_X at such a point is a strict Henselization of $O_{X,x}$, with residue field the separable closure of $k(x)$ in $k(\overline{x})$.

1.1 Log and prelog structures

Let (X, O_X) be a scheme, and let $\mathbf{Mon}_X$ denote the category of sheaves of (commutative) monoids on $X_{ét}$ or X_{zar}. The structure sheaf O_X is a sheaf of rings; when we refer to it as a sheaf of monoids we always will use its multiplicative structure.

Definition 1.1.1. *A prelogarithmic structure on a scheme X is a homomorphism of sheaves of monoids $\alpha\colon \mathcal{P} \to O_X$ on $X_{ét}$ (or X_{zar}). A logarithmic structure is a prelogarithmic structure such that the induced homomorphism $\alpha^{-1}(O_X^*) \to O_X^*$ is an isomorphism. If $\alpha\colon \mathcal{M} \to O_X$ and $\alpha'\colon \mathcal{M}' \to O_X$ are prelogarithmic or logarithmic structures, a morphism $\alpha \to \alpha'$ is a homomorphism $\theta\colon \mathcal{M} \to \mathcal{M}'$ such that $\alpha' \circ \theta = \alpha$.*

We denote by $\mathbf{Log}_X$ (resp. $\mathbf{Plog}_X$) the category of logarithmic (resp. prelogarithmic) structures on X. To save space and time, we may write "log" instead of "logarithmic."

Remark 1.1.2. A log structure α_X induces an isomorphism $\mathcal{M}_X^* \to \mathcal{O}_X^*$, and it is common practice to identify $\mathcal{O}_X^*$ and $\mathcal{M}_X^*$. Doing so requires the use of multiplicative notation for the monoid law on $\mathcal{M}_X$. When using additive notation for $\mathcal{M}_X$, we shall write λ_X for the mapping $\mathcal{O}_X^* \to \mathcal{M}_X$ induced by the inverse of α_X. Then $\lambda_X(uv) = \lambda_X(u) + \lambda_X(v)$, and $\lambda_X(u)$ can be thought of as the logarithm of the invertible function u. For any section f of $\mathcal{O}_X$, $\alpha_X^{-1}(f)$ is then the (possibly empty) set of logarithms of the function f.

Proposition 1.1.3. *Let X be a scheme.*

1. *The inclusion functor $\mathbf{Log}_X \to \mathbf{Plog}_X$ admits a left adjoint:*
$$\mathcal{P} \to \mathcal{O}_X \quad \mapsto \quad \mathcal{P}^{log} \to \mathcal{O}_X.$$
2. *The identity map $\mathcal{O}_X \to \mathcal{O}_X$ is the final object of $\mathbf{Plog}_X$ and of $\mathbf{Log}_X$. The initial object of $\mathbf{Plog}_X$ is the inclusion $\{1\} \to \mathcal{O}_X$, and the initial object of $\mathbf{Log}_X$ is the inclusion $\mathcal{O}_X^* \to \mathcal{O}_X$.*
3. *The categories $\mathbf{Plog}_X$ and $\mathbf{Log}_X$ admit pushouts, and their formation is compatible with the inclusion functor $\mathbf{Log}_X \to \mathbf{Plog}_X$.*

Proof The functor in (1) is nothing but the formation of the associated logarithmic morphism, which we already discussed in Proposition II.1.1.5. The verification of its adjointness to the inclusion functor is immediate, as is the verification of (2). For (3), suppose that we are given morphisms of (pre)log structures $\theta_i \colon \alpha_0 \to \alpha_i$ for $i = 1, 2$, where $\alpha_i \colon \mathcal{P}_i \to \mathcal{O}_X$ for $i = 0, 1, 2$. Let $\mathcal{P} := \mathcal{P}_1 \oplus_{\mathcal{P}_0} \mathcal{P}_2$ in the category $\mathbf{Mon}_X$. Then by the universal mapping property of the pushout, there is a unique morphism $\alpha \colon \mathcal{P} \to \mathcal{O}_X$ which is compatible with α_1 and α_2. It is clear that α satisfies the universal mapping property of a pushout in the category $\mathbf{Plog}_X$. It remains only to verify that α is a log structure if the α_i are. It suffices to work at the stalk at each point $\overline{x}$ of X. Suppose that m is a section of $\mathcal{P}_{\overline{x}}$ and $\alpha(m) \in \mathcal{O}_{X,\overline{x}}^*$. Write $m = \phi_1(m_1) + \phi_2(m_2)$, where $m_i \in \mathcal{P}_{i,\overline{x}}$ and $\phi_i \colon \mathcal{P}_i \to \mathcal{P}$ is the canonical map. Then $\alpha_1(m_1)\alpha_2(m_2) = \alpha(m) \in \mathcal{O}_{X,\overline{x}}^*$, so each $\alpha_i(m_i)$ is a unit. Since each α_i is a local homomorphism, each $m_i \in \mathcal{P}_{i,\overline{x}}^*$ and hence $m \in \mathcal{P}_{\overline{x}}^*$. Thus α is local. It follows from Proposition I.4.1.5 that it is also strict, hence a log structure. □

The inclusion $\mathcal{O}_X^* \to \mathcal{O}_X$ (the initial object of $\mathbf{Log}_X$) is called the *trivial log structure* on X. The final object, which is rarely used, can be called the *empty log structure.*

Remark 1.1.4. Formation of the log structure $\mathcal{P}^{log} \to \mathcal{O}_X$ associated to a prelog structure $\beta\colon \mathcal{P} \to \mathcal{O}_X$ involves a pushout in the category of sheaves of monoids: this is the sheaf associated to the presheaf sending each open set to the pushout in the category of monoids. We shall see later that, if $\mathcal{P}$ is integral, then this sheafification yields the same result when carried out in the Zariski or in the étale topology. More precisely, for each étale $f\colon X' \to X$, let $\mathcal{P}^{log}_{X'}$ denote the log structure on X_{zar} associated to $\mathcal{P} \to f^{-1}(\mathcal{O}_X) \to \mathcal{O}_{X'}$. We will see in in Proposition 1.4.1 that $X' \mapsto \mathcal{P}^{log}_{X'}(X')$ defines a sheaf on $X_{ét}$, and it follows that this sheaf is the étale log structure associated to β.

Definition 1.1.5. *Let $f\colon X' \to X$ be a morphism of schemes.*

1. *If $\alpha'\colon \mathcal{M}' \to \mathcal{O}_{X'}$ is a log structure on X', form the cartesian diagram*

$$\begin{array}{ccc} f_*^{log}(\mathcal{M}') & \xrightarrow{f_*^{log}(\alpha')} & \mathcal{O}_X \\ \downarrow & & \downarrow{\scriptstyle f^\sharp} \\ f_*(\mathcal{M}') & \xrightarrow{f_*(\alpha')} & f_*(\mathcal{O}_{X'}). \end{array}$$

   Then $f_*^{log}(\alpha')$ is a log structure on X, and

$$f_*^{log}\colon \mathbf{Log}_{X'} \to \mathbf{Log}_X$$

   is a functor from the category of log structures on X' to the category of log structures on X.

2. *If $\alpha\colon \mathcal{M} \to \mathcal{O}_X$ is a log structure on X, let*

$$f^*_{log}(\alpha)\colon f^*_{log}(\mathcal{M}) \to \mathcal{O}_X$$

   be the log structure associated to the prelog structure

$$f^{-1}(\mathcal{M}) \to f^{-1}(\mathcal{O}_X) \to \mathcal{O}_{X'}$$

   Then

$$f^*_{log}\colon \mathbf{Log}_X \to \mathbf{Log}_{X'}$$

   is a functor from the category of log structures on X to the category of log structures on X'.

3. *The functor f^*_{log} is left adjoint to the functor f_*^{log}.*

In this context, $f_*^{log}(\alpha')$ is called the *direct image* of α', and $f^*_{log}(\alpha)$ is called the *inverse image* of α. The sub- and superscripts "*log*" in the notation above are often omitted. The verifications of the statements in the definition are immediate.

Remark 1.1.6. If $f\colon X \to Y$ is a morphism of schemes and $\alpha_Y\colon \mathcal{M}_Y \to \mathcal{O}_Y$ is a log structure on Y, then the homomorphisms $f^{-1}(\mathcal{M}_Y) \to f^{-1}(\mathcal{O}_Y)$ and $f^{-1}(\mathcal{O}_Y) \to \mathcal{O}_X$ are both local, and hence so is the composite $f^{-1}(\mathcal{M}_Y) \to \mathcal{O}_X$. It follows that the construction of the associated log structure $f^*_{log}(\mathcal{M}_Y) \to \mathcal{O}_X$ is accomplished just by sharpening. As a consequence, the homomorphism $f^{-1}(\overline{\mathcal{M}}_Y) \to \overline{f^*_{log}(\mathcal{M}_Y)}$ is an isomorphism.

A log structure $\alpha\colon \mathcal{M} \to \mathcal{O}_X$ is said to be *coherent* if the sheaf of monoids $\mathcal{M}_X$ is coherent, that is, if locally on X there exist a finitely generated monoid Q and a chart $\gamma\colon Q \to \Gamma(X, \mathcal{M})$. Recall from Definition II.2.1.1 that this means that the associated logarithmic homomorphism $\gamma^{log}\colon Q^{log} \to \mathcal{M}$ is an isomorphism. Note that $\beta := \alpha\circ\gamma$ is a prelog structure on X, and since α is logarithmic, Q^{log} is the same whether computed with respect to γ or β, by Remark II.1.1.7. Thus, if γ is a chart for $\mathcal{M}$, then α is isomorphic to the log structure associated to the prelog structure $Q \to \mathcal{O}_X$, and to give a chart for $\mathcal{M}$ is equivalent to giving a prelog structure $\beta\colon Q \to \mathcal{O}_X$ together with an isomorphism $Q^{log} \cong \mathcal{M}_X$.

To relate log structures to divisors, we introduce the following notation. If X is a scheme and U is an open subset of X, let $\mathcal{Z}_X(U)$ be the set of closed subschemes Z of U, or equivalently, the set of quasi-coherent sheaves of ideals in $\mathcal{O}_U$. We endow $\mathcal{Z}_X(U)$ with the structure of a commutative monoid, where $Z_1 + Z_2$ is the (scheme-theoretic) union of Z_1 and Z_2, defined by the product of the ideals $\mathcal{I}_1$ and $\mathcal{I}_2$ defining Z_1 and Z_2. (This is the ideal sheaf generated by the products of sections of $\mathcal{I}_1$ and $\mathcal{I}_2$.) The identity element of this monoid is the empty subscheme, corresponding to the unit ideal $\mathcal{O}_X$. In fact we will mostly be interested in the submonoid corresponding to the ideals that are locally monogenic (including the zero ideal). The natural restriction maps make $\mathcal{Z}_X$ into a presheaf, and this presheaf is in fact a sheaf. If a is a section of $\mathcal{O}_X$ over a set U, let $Z(a)$ be the subscheme corresponding to the ideal sheaf generated by a. If a is a nonzero divisor of $\mathcal{O}_X$, then $Z(a)$ is an effective Cartier divisor, but this will not be true in general. Note also that the ideal sheaf $\mathcal{I}_Z$ is the *inverse* of the invertible sheaf typically assigned to the effective Cartier divisor Z. If $\alpha\colon \mathcal{M} \to \mathcal{O}_X$ is a prelog structure on X and m is a section of $\mathcal{M}$, the subscheme $Z(\alpha(m))$ depends only on the image of m in $\overline{\mathcal{M}} := \mathcal{M}/\mathcal{M}^*$, so there is a commutative diagram:

$$\begin{array}{ccc} \mathcal{M} & \xrightarrow{\ \alpha\ } & \mathcal{O}_X \\ \Big\downarrow{\scriptstyle \pi} & & \Big\downarrow{\scriptstyle Z} \\ \overline{\mathcal{M}} & \xrightarrow{\ \zeta\ } & \mathcal{Z}_X. \end{array} \qquad (1.1.1)$$

Remark 1.1.7. Let $\alpha\colon \mathcal{M} \to \mathcal{O}_X$ be a u-integral log structure on X, so that there is an exact sequence

$$0 \longrightarrow \mathcal{O}_X^* \xrightarrow{\ \lambda\ } \mathcal{M}_X^{\mathrm{gp}} \xrightarrow{\ \pi\ } \overline{\mathcal{M}}_X^{\mathrm{gp}} \longrightarrow 0,$$

in the sense that $\overline{\mathcal{M}}_X$ is the sheaf of orbits by the free action of $\mathcal{O}_X^*$ on $\mathcal{M}_X$. If q is a global section of $\overline{\mathcal{M}}_X$, the sheaf $\mathcal{L}_q^* := \pi_X^{-1}(q)$ is an $\mathcal{O}_X^*$-torsor on X, and hence defines an invertible sheaf $\mathcal{L}_q$ on X. The isomorphism class $[\mathcal{L}_q]$ of $\mathcal{L}_q$ is an element of the Picard group of X, and the induced map

$$\Gamma(X, \overline{\mathcal{M}}_X^{\mathrm{gp}}) \to \mathrm{Pic}(X) : q \mapsto [\mathcal{L}_q]$$

is nothing but the boundary map associated to the short exact sequence

$$0 \to \mathcal{O}_X^* \to \mathcal{M}_X^{\mathrm{gp}} \to \overline{\mathcal{M}}_X^{\mathrm{gp}} \to 0.$$

The image by α_X of $\mathcal{L}_q$ is a quasi-coherent and locally monogenic sheaf of ideals $\mathcal{I}_q$ in $\mathcal{O}_X$, and the subscheme defined by $\mathcal{I}_q$ is $\zeta(q)$ in the notation of diagram 1.1.1.

1.2 Log schemes and their charts

Definition 1.2.1. *A log scheme is a scheme X endowed with a log structure $\alpha_X\colon \mathcal{M}_X \to \mathcal{O}_X$. A morphism of log schemes is a morphism $f\colon X \to Y$ of the underlying schemes together with a homomorphism $f^\flat\colon \mathcal{M}_Y \to f_*(\mathcal{M}_X)$ such that the diagram*

$$\begin{array}{ccc} \mathcal{M}_Y & \xrightarrow{\ f^\flat\ } & f_*(\mathcal{M}_X) \\ {\scriptstyle \alpha_Y}\downarrow & & \downarrow{\scriptstyle f_*(\alpha_X)} \\ \mathcal{O}_Y & \xrightarrow{\ f^\sharp\ } & f_*(\mathcal{O}_X) \end{array}$$

commutes.

By adjunction, to give the homomorphism $f^\flat$ in this diagram is equivalent to giving a homomorphism $f^{-1}(\mathcal{M}_Y) \to \mathcal{M}_X$, which in turn is equivalent to giving a homomorphism $f_{log}^*(\mathcal{M}_Y) \to \mathcal{M}_X$. We will allow ourselves to denote any of these data by $f^\flat$. The commutative diagram in the definition implies that for each $x \in X$, the morphism $f_x^\flat\colon \mathcal{M}_{Y,f(x)} \to \mathcal{M}_{X,x}$ is automatically local, so f induces a morphism of monoidal spaces $(X, \mathcal{M}_X) \to (Y, \mathcal{M}_Y)$.

If X is a scheme then X, endowed with the trivial log structure $\mathcal{O}_X^* \subseteq \mathcal{O}_X$, is a log scheme, and the functor $X \mapsto (X, \mathcal{O}_X^*)$ is fully faithful. In this way we can and shall regard the category of schemes as a full subcategory of the category of log schemes.

The following proposition needs no proof.

Proposition 1.2.2. *If X is a log scheme, denote by $\underline{X}$ its underlying scheme or (equivalently) the same underlying scheme X endowed with the trivial log structure. Then the inclusion $\mathcal{O}_X^* \to \mathcal{M}_X$ underlies a morphism of log schemes $p_X : X \to \underline{X}$, which in fact is the universal morphism from X to a log scheme with trivial log structure.* □

Thus if Y is a (log) scheme with trivial log structure, there is a natural bijection $\mathrm{Mor}(X, Y) \cong \mathrm{Mor}(\underline{X}, Y)$, and the functor $X \to \underline{X}$ is left adjoint to the inclusion of the category of schemes into the category of log schemes.

Definition 1.2.3. *A log ring is a homomorphism β from a monoid P to the multiplicative monoid of a ring A. A local log ring is a local homomorphism from a monoid P into the multiplicative monoid of a local ring. If $P \to A$ is a log ring,* $\mathrm{Spec}(P \to A)$ *is the log scheme whose underlying scheme is $X :=$* $\mathrm{Spec}\, A$ *with the log structure associated to the prelog structure $P \to \mathcal{O}_X$ induced by the map $P \to A$.*

If P is a monoid and $e\colon P \to \mathbf{Z}[P]$ is its monoid algebra (see Section I.3.1), one can form the log scheme

$$\mathsf{A}_P := \mathrm{Spec}(e\colon P \to \mathbf{Z}[P]).$$

More generally, if S is a monoscheme, we denote by A_S the log scheme whose underlying scheme is $\underline{\mathsf{A}}_S$ (defined in Proposition II.1.9.1) with the log structure associated to the prelog structure $\pi^{-1}(\mathcal{M}_S) \to \mathcal{O}_{\underline{\mathsf{A}}_S}$, where $\pi\colon \underline{\mathsf{A}}_S \to S$ is the natural map.

A log scheme X gives rise to a monoidal space $(X, \mathcal{M}_X)$, to which the definitions and results of Chapter II apply. For example, a morphism $f\colon X \to Y$ is said to be *strict* if the induced homomorphism $f^*_{log}(\mathcal{M}_Y) \to \mathcal{M}_X$ is an isomorphism. By the same token, a *chart* for X is just a chart for its underlying sheaf of monoids. Such charts can be interpreted in a variety of ways, as the following proposition (which requires no proof) illustrates.

Proposition 1.2.4. *Let X be a log scheme, let Q be a monoid, and let S denote the monoidal space* $\mathrm{spec}(Q)$. *Then the following sets of data are equivalent.*

1. *A monoid homomorphism $\alpha\colon Q \to \Gamma(X, \mathcal{M}_X)$ for $\mathcal{M}_X$.*
2. *A morphism of locally monoidal spaces $a\colon (X, \mathcal{M}_X) \to (S, \mathcal{M}_S)$.*
3. *A morphism of log schemes $b\colon X \to \mathsf{A}_Q$.*

Then α is a chart for $\mathcal{M}_X$ if and only if a is strict if and only if b is strict. In this case, any one of these data is called a chart for X subordinate to Q. □

More generally, let S be a monoscheme and let A_S be the associated log scheme. Then the following sets of data are equivalent.

1. A morphism of locally monoidal spaces $a\colon (X, \mathcal{M}_X) \to (S, \mathcal{M}_S)$.
2. A morphism of log schemes $b\colon X \to \mathsf{A}_S$.

Then a is a strict if and only if b is, and in this case either of these sets of data is called a *chart for X subordinate to S.*

In general, a morphism $f\colon (X, \alpha_X) \to (Y, \alpha_Y)$ of log schemes has a *canonical factorization*

$$(X, \alpha_X) \xrightarrow{\ i\ } X_Y := (X, f^*_{log}(\alpha_Y)) \xrightarrow{\ f^s\ } (Y, \alpha_Y). \tag{1.2.1}$$

This factorization is uniquely determined by the fact that i is the identity on the underlying schemes and f^s is strict. There is an analogous factorization through Y equipped with the direct image log structure $f^{log}_*(\alpha_X)$, and in fact f fits into the commutative diagram

$$\begin{array}{ccc} (X, \alpha_X) & \xrightarrow{\ i\ } & (X, f^*_{log}(\alpha_Y)) \\ \downarrow & & \downarrow{\scriptstyle f^s} \\ (Y, f^{log}_*(\alpha_X)) & \xrightarrow{\ j\ } & (Y, \alpha_Y), \end{array}$$

where i and j are each the identity map on the underlying schemes. In some sense, $f^{log}_*(\alpha_X)$ is the log structure on Y that makes it as close as possible to X, and $f^*_{log}(\alpha_Y)$ is the log structure on X that makes it as close as possible to Y.

Proposition 1.2.5. *A morphism of log schemes $f\colon X \to Y$ is strict if and only if the diagram*

$$\begin{array}{ccc} X & \xrightarrow{\ f\ } & Y \\ {\scriptstyle p_X}\downarrow & & \downarrow{\scriptstyle p_Y} \\ \underline{X} & \xrightarrow{\ \underline{f}\ } & \underline{Y} \end{array}$$

is cartesian.

Proof Suppose that f is strict and let $h\colon T \to Y$ and $g\colon T \to \underline{X}$ be morphisms such that $p_Y \circ h = \underline{f} \circ g$. Then the morphism h includes the data of a homomorphism

$$h^\flat : \underline{h}^*_{log}(\mathcal{M}_Y) \to \mathcal{M}_T.$$

But $\underline{h}^*_{log}(\mathcal{M}_Y) \cong \underline{g}^*_{log}(\underline{f}^*_{log}(\mathcal{M}_Y))$, and since f is strict, $\underline{f}^*_{log}(\mathcal{M}_Y) \cong \mathcal{M}_X$. Thus $h^\flat$ amounts to a homomorphism $\underline{g}^*_{log}(\mathcal{M}_X) \to \mathcal{M}_T$. All these homomorphisms are compatible with α_T, and thus we have a morphism of log schemes $T \to X$ making the diagram commute. This morphism is uniquely determined by its underlying morphism of schemes, given by its composition with p_X, together with $h^\flat$, given by its composition with $f^\flat$, and hence is unique.

Suppose on the other hand that the diagram is cartesian. Then we have a commutative diagram

$$\begin{array}{ccccc} X & \xrightarrow{i} & X_Y & \xrightarrow{f^s} & Y \\ & \searrow^{p_X} & \downarrow & & \downarrow p_Y \\ & & \underline{X} & \xrightarrow{\underline{f}} & \underline{Y}, \end{array} \qquad (1.2.2)$$

where the top row is the canonical factorization (1.2.1) of f, with $\underline{i} = \mathrm{id}_{\underline{X}}$ and f^s strict. Since f^s is strict, the square is cartesian, and so it follows from our hypothesis that i is an isomorphism. Then $f = f^s \circ i$ is strict. □

Remark 1.2.6. It is also useful to reinterpret the definition (II.2.4.1) of a chart for a morphism of monoidal spaces in the context of log schemes. Thus if $f: X \to Y$ is a morphism of log schemes and $\theta: P \to Q$ is a homomorphism of monoids, a *chart for f subordinate to θ* can be interpreted as a commutative diagram of log schemes

$$(a, \theta, b) \qquad \begin{array}{ccc} X & \xrightarrow{b} & \mathsf{A}_Q \\ f \downarrow & & \downarrow \mathsf{A}_\theta \\ Y & \xrightarrow{a} & \mathsf{A}_P \end{array}$$

in which a and b are strict. In this context, we let Y_θ denote the log scheme whose underlying scheme is the fiber product $\underline{Y}_\theta := \underline{Y} \times_{\mathsf{A}_P} \underline{\mathsf{A}}_Q$, endowed with the log structure induced from the morphism $\underline{Y}_\theta \to \underline{\mathsf{A}}_Q$. Thus we have the commutative diagram

$$\begin{array}{ccccc} X & \xrightarrow{b_\theta} & Y_\theta & \xrightarrow{b} & \mathsf{A}_Q \\ & \searrow^{f} & \downarrow f_\theta & & \downarrow \mathsf{A}_\theta \\ & & Y & \xrightarrow{a} & \mathsf{A}_P, \end{array} \qquad (1.2.3)$$

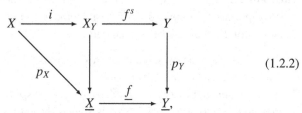

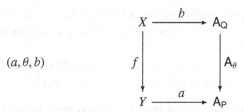

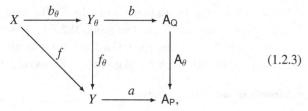

in which the square is cartesian and the horizontal maps are all strict. Note that in this diagram, the group scheme $\mathsf{A}^*_{Q/P} := \underline{\mathsf{A}}_{Q^{gp}/P^{gp}} = \underline{\mathsf{A}}^*_{Q/P}$ acts on A_Q over A_P and hence by base change also on Y_θ over Y.

More generally, we can define a chart for a morphism of log schemes subordinate to a morphism of monoschemes $T \to S$ in obvious ways.

The construction of exact and neat charts (Definition II.2.3.1) in the context of log schemes is worth describing explicitly.

Theorem 1.2.7. *Let $f\colon X \to Y$ be a morphism of fine log schemes for the étale (resp. Zariski) topology, let $\overline{x}$ be a geometric point of X (resp. a scheme theoretic point), and let $\overline{y}$ be its image in Y.*

1. *After Y is replaced by some étale (resp. Zariski) neighborhood of $\overline{y}$, Y admits a fine chart which is exact at $\overline{y}$.*
2. *If the sequence $0 \to \mathcal{O}^*_{Y,\overline{y}} \to \mathcal{M}^{gp}_{Y,y} \to \overline{\mathcal{M}}^{gp}_{Y,y} \to 0$ splits, then the chart in (1) may be taken to be neat at $\overline{y}$. This is always the case if $\mathcal{M}_{Y,\overline{y}}$ is saturated, or more generally if $\overline{\mathcal{M}}^{gp}_{Y,\overline{y}}$ is torsion free. In the étale case, it is true provided that the order of the torsion subgroup of $\overline{\mathcal{M}}^{gp}_{Y,\overline{y}}$ is invertible in $k(\overline{y})$. A neat chart always exists in some fppf[1] neighborhood of $\overline{y}$.*
3. *Suppose a fine chart $\alpha\colon P \to \mathcal{M}_Y$ is given. Then after X is replaced by some étale (resp. Zariski) neighborhood of $\overline{x}$, the morphism f admits a chart (α, θ, β), where $\beta\colon Q \to \mathcal{M}_X$ is exact at $\overline{x}$.*
4. *Suppose that $\mathrm{Ext}^1(\mathcal{M}^{gp}_{X/Y,\overline{x}}, \mathcal{O}^*_{X,\overline{x}}) = 0$. Then the chart in (3) may be taken to be neat at $\overline{x}$. This is always the case if $\mathcal{M}^{gp}_{X/Y,\overline{x}}$ is torsion free. In the étale case, it is true provided that the order of the torsion subgroup of $\mathcal{M}^{gp}_{X/Y,\overline{x}}$ is invertible in $k(\overline{x})$. A neat chart always exists in some fppf neighborhood of $\overline{x}$.*
5. *Suppose that f is s-injective at $\overline{x}$ and that (α, θ, β) is a chart for f that is neat at $\overline{x}$. Then β is automatically neat at $\overline{x}$ if α is neat at $\overline{y}$.*

Proof By hypothesis, $\mathcal{M}_Y$ is integral and coherent. In the Zariski topology Corollary II.2.3.6 therefore implies that the sheaf of monoids $\mathcal{M}_Y$ admits a fine chart $\alpha\colon P \to \mathcal{M}_Y$ in some Zariski neighborhood of $\overline{x}$. Let $F := \alpha^{-1}(\mathcal{M}^*_{Y,y})$, a face of P. Then by (3) of Theorem I.2.1.17 there is an $f \in F$ that generates F as a face of P. If we replace Y by the open subset on which $\alpha(f)$ is invertible, α factors through the localization P_F of P by F. The resulting chart $P_F \to \mathcal{M}_Y$ is local, hence exact, at $\overline{y}$ (see Definition II.2.3.1). If we are working in the étale topology, then we can apply Corollary 1.4.4 to replace Y by some étale neighborhood of $\overline{y}$ on which $\mathcal{M}_Y$ is given by a Zariski log structure. This proves

[1] "faithfully flat and of finite presentation"

(1). To prove (2), we can reduce as above to the case of Zariski log structures. Then Proposition II.2.3.7 shows that we can choose α to be neat provided that the sequence shown splits. Let n be the order of the torsion subgroup of $\overline{\mathcal{M}}^{\mathrm{gp}}_{Y,\overline{y}}$. If n is invertible in $k(\overline{y})$, then multiplication by n on $\mathcal{O}^*_{Y,\overline{y}}$ is surjective, since $\mathcal{O}_{Y,\overline{y}}$ is strictly Henselian. Then $\mathrm{Ext}^1(\overline{\mathcal{M}}^{\mathrm{gp}}_{Y,\overline{y}}, \mathcal{O}^*_{Y,\overline{y}}) = 0$ and the sequence splits. Since multiplication by n on $\mathcal{O}^*_{Y,\overline{y}}$ is always surjective in the flat topology, any class in this finite Ext group can be killed after a flat localization, and then the sequence will split. Statement (2) follows. Statement (3) is proved in the same way as (1), starting with Proposition II.2.4.2, and (4) is proved in the same was as (2), using Theorem II.2.4.4. Statement (5) follows from Remark II.2.4.5. □

Recall from Proposition 1.2.2 that the inclusion functor from the category of schemes to the category of log schemes has a left adjoint $X \mapsto \underline{X}$. This functor has a right adjoint as well, if we restrict our attention to coherent log structures.

Proposition 1.2.8. *If X is a log scheme, let*

$$X^* := \{x \in X : \overline{\mathcal{M}}_x = 0\}.$$

1. *If X^* is open in X and T is a log scheme with trivial log structure, then every morphism $T \to X$ factors (uniquely) through X^*.*
2. *If $\mathcal{M}_X$ is coherent, X^* is open in X.*

Thus the functor $X \mapsto X^*$ from the category of coherent log schemes to the category of schemes is right adjoint to the inclusion functor from schemes to coherent log schemes.

Proof If $f : T \to X$ is a morphism of log schemes and $t \in T$, then the homomorphism $f^\flat_t : \mathcal{M}_{X,f(t)} \to \mathcal{M}_{T,t}$ is local. Hence if $\mathcal{M}_{T,t}$ is dull, the same is true of $\mathcal{M}_{X,f(t)}$; that is, $f(t) \in X^*$. Thus if X^* is open in X, the map f factors through the open immersion $X^* \to X$. If $\mathcal{M}_X$ is coherent then Corollary II.2.1.6 implies that X^* is open in X. □

Let P be a monoid and let $\alpha_P : \mathcal{M}_P \to \mathcal{O}_{\mathsf{A}_P}$ be the log structure of A_P. The construction of $\mathcal{M}_P$ shows that there is a natural homomorphism

$$e : P \to \Gamma(\mathsf{A}_P, \mathcal{M}_P).$$

Proposition 1.2.9. *Let T be a log scheme and P a monoid. For each morphism $f : T \to \mathsf{A}_P$ of log schemes, consider the composition*

$$e_f : P \xrightarrow{e} \Gamma(\mathsf{A}_P, \mathcal{M}_P) \xrightarrow{f^\flat} \Gamma(T, \mathcal{M}_T).$$

Then $f \mapsto e_f$ defines a bijection

$$\mathrm{Mor}(T, \mathsf{A}_P) \xrightarrow{\cong} \mathrm{Hom}(P, \Gamma(T, \mathcal{M}_T)).$$

Proof Let $\theta\colon P \to \Gamma(T, \mathcal{M}_T)$ be a monoid homomorphism. The monoid homomorphism $\Gamma(T, \alpha_T) \circ \theta\colon P \to \Gamma(T, \mathcal{O}_T)$ induces a ring homomorphism $\mathbf{Z}[P] \to \Gamma(T, \mathcal{O}_T)$, and hence a morphism of schemes $\underline{f}\colon \underline{T} \to \underline{\mathsf{A}}_P$. Then the composition

$$P \to \Gamma(T, \mathcal{M}_T) = \Gamma(\underline{\mathsf{A}}_P, \underline{f}_*(\mathcal{M}_T)) \to \underline{f}_*(\mathcal{M}_T)$$

is compatible with the homomorphisms $P \to \mathcal{O}_{\mathsf{A}_P}$ and $\underline{f}_*(\mathcal{M}_T) \to \underline{f}_*(\mathcal{O}_T)$, and hence defines a homomorphism $P \to f_*^{log}(\mathcal{M}_T)$. Since $f_*^{log}(\alpha_T)$ is a log structure on $\underline{\mathsf{A}}_P$ and P is a chart for $\mathcal{M}_{\mathsf{A}_P}$, this homomorphism factors uniquely through a homomorphism $\mathcal{M}_{\mathsf{A}_P} \to f_*^{log}(\mathcal{M}_T)$. This in turns induces a morphism of log schemes $T \to \mathsf{A}_P$. We leave it to the reader to check that this construction is inverse to the assignment $f \mapsto e_f$. □

Corollary 1.2.10. *Let T be a scheme with trivial log structure and let P be a monoid. Then every morphism of log schemes $T \to \mathsf{A}_P$ factors uniquely through the inclusion $\mathsf{A}_P^* \to \mathsf{A}_P$, and in fact*

$$\mathsf{A}_P(T) \cong \mathsf{A}_{P^{gp}}(T) \cong \mathsf{A}_P^*(T).$$

Proof Using the previous proposition and the universal mapping property of $P \to P^{gp}$, we find:

$$\mathsf{A}_P(T) := \mathrm{Mor}(T, \mathsf{A}_P) \cong \mathrm{Hom}(P, \Gamma(T, \mathcal{O}_T^*)) \cong \mathrm{Hom}(P^{gp}, \Gamma(T, \mathcal{O}_T^*))$$

$$\cong \mathrm{Hom}(P^{gp}, \Gamma(T, \mathcal{O}_T)) \cong \mathrm{Mor}(T, \mathsf{A}_{P^{gp}}) := \mathsf{A}_P^*(T).$$ □

If P is fine, $\mathsf{A}_P^* = \underline{\mathsf{A}}_P^*$ and is the largest open subscheme of A_P on which the log structure is trivial, and the corollary says that the set of T-valued points of A_P is the same as the set of T-valued points of A_P^*.

The following result is an immediate consequence of Remark 1.1.6 and Proposition I.4.1.2.

Corollary 1.2.11. *Let $f\colon X \to Y$ be a morphism of log schemes. If f is strict, the induced map $f^{-1}(\overline{\mathcal{M}}_Y) \to \overline{\mathcal{M}}_X$ is an isomorphism, and the converse holds if $\mathcal{M}_X$ is u-integral.* □

1.3 Idealized log schemes

It is sometimes convenient to study closed subschemes of log schemes that are cut out by sheaves of ideals in the underlying monoid. In this context it is helpful to keep track of this sheaf of ideals. The notion of idealized log schemes provides a framework for this study. To avoid overburdening the exposition, we shall limit ourselves to explaining the main definitions and concepts.

Definition 1.3.1. *An idealized log scheme is a log scheme $(X, \mathcal{M}_X)$ equipped with a sheaf of ideals $\mathcal{K}_X \subseteq \mathcal{M}_X$ such that $\mathcal{K}_X \subseteq \alpha_X^{-1}(0)$. A morphism of idealized log schemes $f\colon X \to Y$ is a morphism of log schemes such that $f^\flat$ maps $f^{-1}(\mathcal{K}_Y)$ into $\mathcal{K}_X$.*

The functor which endows a log scheme X with the empty sheaf of ideals in $\mathcal{M}_X$ defines a fully faithful functor from the category of log schemes to the category of idealized log schemes. This functor is left adjoint to the functor from idealized log schemes to log schemes that forgets the ideal. On the other hand, endowing X with the sheaf of ideals $\alpha_X^{-1}(0)$ gives the right adjoint to the "forget the ideal" functor.

Let (P, K) be an acceptably idealized monoid (see Definition I.1.5.1 and the discussion which follows it) and let $\mathbf{Z}[P, K]$ be the quotient of the monoid algebra $\mathbf{Z}[P]$ by the ideal generated by the image of K. The map $P \to \mathbf{Z}[P, K]$ sends the elements of K to zero. We denote by $\mathsf{A}_{P,K}$ the idealized log scheme whose underlying scheme is Spec $\mathbf{Z}[P, K]$, with log structure associated to the prelog structure coming from the map $P \to \mathbf{Z}[P, K]$, and with the sheaf of ideas $\mathcal{K}$ in $\mathcal{M}_P$ generated by the image of K. If T is any idealized log scheme, then we can argue as in Proposition 1.2.9 to see that the set of morphisms $T \to \mathsf{A}_{P,K}$ can be identified with the set of monoid homomorphisms $P \to \Gamma(T, \mathcal{M}_T)$ sending K to $\Gamma(T, \mathcal{K}_T)$.

Definition 1.3.2. *A morphism of idealized log schemes f is ideally strict if $\mathcal{K}_X$ is generated by $f^{-1}(\mathcal{K}_Y)$.*

If X is a fine log scheme, the inverse image in $\mathcal{M}_X$ of the zero ideal of $\mathcal{O}_X$ need be coherent. For example, let $X :=$ Spec $(\mathbf{N} \to k[X, Y]/(XY))$, where n is sent to x^n. Then the stalk of $\alpha_X^{-1}(0)$ at the origin is empty, but the stalk at any other point on the y-axis is not. Hence $\alpha_X^{-1}(0)$ cannot be coherent. The following result shows that this cannot happen in some good situations.

Proposition 1.3.3. *Let K be an ideal in a fine monoid P, and let $X := \mathsf{A}_{P,K}$ and let $\tilde{K}$ be the sheaf of ideals of $\mathcal{M}_K$ generated by K. Then $\alpha_X^{-1}(0) = \tilde{K}$ and thus is a coherent sheaf of ideals in $\mathcal{M}_X$.*

Proof We work over an arbitrary base ring R. If x is a point of X, let β_x be the map $P \to \mathcal{O}_{X,x}$, and let $F_x := \beta_x^{-1}(\mathcal{O}_{X,x}^*)$. Evidently $\tilde{K}$ maps injectively to $\alpha_X^{-1}(0)$; to prove that the map is an isomorphism, suppose that $m \in \mathcal{M}_{X,x}$ and $\alpha_{X,x}(m) = 0$. Since $P \to \overline{\mathcal{M}}_{X,x}$ is surjective, there exists a $p \in P$ mapping to $\overline{m}$. Let $\mathfrak{m}_x \subseteq R[P]$ be the prime ideal corresponding to the point x. Then $\mathcal{O}_{X,x}$ is the localization of $R[P]/R[K]$ at $\mathfrak{m}_x$, and since e^p maps to zero in $\mathcal{O}_{X,x}$, there exists an $f \in A[P] \setminus \mathfrak{m}_x$ such that $fe^p \in R[K]$. Write $f := \sum a_q e^q$; then since $f \notin \mathfrak{m}_x$, there exists some $q \in F_x$ such that $a_q \neq 0$. Since $fe^p \in R[K]$, it follows that $q + p \in K$. Since $q \in F_x$ and since $\mathcal{K}_x$ is the ideal of $\mathcal{M}_{X,x}$ generated by K_{F_x}, we can conclude that $p \in \tilde{\mathcal{K}}_x$. □

Proposition 1.3.4. *Let X be a coherent log scheme, let $\mathcal{K}$ be a coherent sheaf of ideals in $\mathcal{M}_X$, and let $\mathcal{I}$ be the sheaf of ideals in $\mathcal{O}_X$ generated by $\alpha_X(\mathcal{K})$. Then $\mathcal{I}$ is a quasi-coherent sheaf of ideals in $\mathcal{O}_X$. Let X' denote the closed subscheme of X defined by $\mathcal{I}$, endowed with the log structure induced from X, and let $\mathcal{K}' \subseteq \mathcal{M}_{X'}$ be the sheaf of of ideals in $\mathcal{M}_{X'}$ generated by the image of $\mathcal{K}$. Then $(X', \mathcal{K}')$ is an idealized log scheme, and there is a natural morphism of idealized log schemes $(X', \mathcal{K}') \to (X, \emptyset)$.*

Proof The quasi-coherence of an étale sheaf of ideals in $\mathcal{O}_X$ can be verified étale locally, so we may without loss of generality assume that the log structure of X admits a chart. We may also assume without loss of generality that X is affine, and hence isomorphic to the spectrum of a log ring $\alpha\colon P \to A$. Since $\mathcal{K}$ is coherent, Proposition II.2.6.1 implies that, possibly after a further shrinking, $\mathcal{K}$ admits a chart $K \subseteq P$. Let I be the ideal of A generated by $\alpha(K)$. Then for any $a \in A$, $P \to A_a$ is a chart for the special affine subset of X defined by a, and K is still a chart for the corresponding sheaf of ideals of its sheaf of monoids. It follows that $\mathcal{I}$ is the sheaf of ideals of $\mathcal{O}_X$ generated by I. Since $\mathcal{I}$ is then quasi-coherent, it defines a closed subscheme X' of X, and since $\mathcal{K}'$ maps to zero in $\mathcal{O}_{X'}$, the pair $(X', \mathcal{K}')$ is an idealized log scheme, which maps to $(X, \emptyset)$. □

1.4 Zariski and étale log structures

In this section we will attempt to compare and connect the theories of log structures in the étale and Zariski topologies. Examples illustrating the differences will be given in Remark 1.6.4. The point is that it is often important for a log structure to account for all the branches of a divisor, which can sometimes only be distinguished in the étale topology.

Proposition 1.4.1. *Let X be a scheme, and let $\eta\colon X_{\acute{e}t} \to X_{\mathrm{zar}}$ be the canonical map from the étale topos of X to its Zariski topos.*

1. *Let $\alpha\colon \mathcal{M}_{\mathrm{zar}} \to \mathcal{O}_X$ be a u-integral log structure on X, and for each étale $f\colon X' \to X$, let $\mathcal{M}_{X'} := f^*_{log}(\mathcal{M}_{\mathrm{zar}})$. Then the assignment $X' \mapsto \mathcal{M}_{X'}(X')$ defines a sheaf $f^*_{log}(\mathcal{M}_{\mathrm{zar}})$ in the étale topology of X, and*

$$f^*_{log}(\alpha)\colon f^*_{log}(\mathcal{M}_{\mathrm{zar}}) \to \mathcal{O}_{X_{\acute{e}t}}$$

is a log structure $\eta^*_{log}(\mathcal{M}_{\text{zar}})$ *on* $X_{\acute{e}t}$*.*

2. *Let* $\alpha\colon \mathcal{M}_{\acute{e}t} \to \mathcal{O}_X$ *be a u-integral log structure on* $X_{\acute{e}t}$*, and suppose that* $\mathcal{M}_{\acute{e}t}$ *admits a global chart* $\gamma\colon P \to \Gamma(X, \mathcal{M}_{\acute{e}t})$*. Then* γ *factors through a global chart for* $\mathcal{M}_{\text{zar}} := f^{log}_*(\mathcal{M}_{\acute{e}t})$*, and the natural map* $\eta^*_{log}(\mathcal{M}_{\text{zar}}) \to \mathcal{M}_{\acute{e}t}$ *is an isomorphism.*

Proof To prepare for the proof of (1), let us consider the following situation. If $f\colon X' \to X$ is a morphism, let $X'' := X' \times_X X'$, let $p_i\colon X'' \to X'$, $i = 1, 2$, be the two projections, and let $g := f \circ p_1 = f \circ p_2 : X'' \to X$. If $\alpha_X\colon \mathcal{M}_X \to \mathcal{O}_X$ is a log structure on X, let $\mathcal{M}_{X'} := f^*_{log}(\mathcal{M}_X)$ and let $\mathcal{M}_{X''} := g^*_{log}(\mathcal{M}_X)$. Then there are canonical isomorphisms $\mathcal{M}_{X''} \cong p^*_{i,log}(\mathcal{M}_{X'})$, and hence each of the maps p_i induces a morphism of sheaves of monoids $f_*(\mathcal{M}_{X'}) \to g_*(\mathcal{M}_{X''})$. Then statement (1) follows immediately from the following lemma.

Lemma 1.4.2. *In the situation above, assume that* $f\colon X' \to X$ *is a faithfully flat and quasi-compact morphism of schemes. Let* $\alpha_X\colon \mathcal{M}_X \to \mathcal{O}_X$ *be a u-integral log structure in the Zariski topology of* X*. Then the natural maps*

$$\overline{\mathcal{M}}_X \to \mathrm{Eq}\left(f_*\overline{\mathcal{M}}_{X'} \rightrightarrows g_*\overline{\mathcal{M}}_{X''}\right)$$
$$\mathcal{M}_X \to \mathrm{Eq}\left(f_*\mathcal{M}_{X'} \rightrightarrows g_*\mathcal{M}_{X''}\right)$$

are isomorphisms.

Proof This lemma is a simple consequence of faithfully flat descent. First we prove the statement about $\overline{\mathcal{M}}$. Since $f^*_{log}(\overline{\mathcal{M}}) = f^{-1}(\overline{\mathcal{M}})$, this reduces to the following elementary statement about sheaves of sets.

Claim 1.4.3. Let S be a sheaf of sets on X. Then the natural map

$$S \to \mathrm{Eq}\left(f_*f^{-1}S \rightrightarrows g_*g^{-1}S\right)$$

is an isomorphism.

Proof The injectivity of $S \to f_*f^{-1}(S)$ is clear from the surjectivity of f. For the surjectivity, recall that since f is faithfully flat and quasi-compact, the underlying map on topological spaces is open and surjective [27, 2.4.6]. Let s' be a section of $f_*f^{-1}(S)$ such that $p_1^*(s') = p_2^*(s')$ in $g_*g^{-1}(S)$. For any point x of X there is at least one point x' of X' such that $f(x') = x$, and for any pair (x_1', x_2') of such points, there is a point x'' of X'' such that $p_i(x'') = x_i'$. The natural maps $F_x \to (f^{-1}(S))_{x_i'} \to (g^{-1}(S))_{x''}$, are isomorphisms, and because $p_1^*(s') = p_2^*(s')$, the stalks of s' at x_1' and x_2' correspond to the same element of F_x, which we denote by $s(x)$. Thus $x \mapsto s(x) \in \prod_x F_x$ is a "discontinuous section" of F such that $s(x) = s_{x'}$ whenever $f(x') = x$. It remains only to prove that s is in fact continuous. If $x \in X$, there exist a neighborhood U of x in X

and a section t of S over U whose stalk at x is $s(x)$. Choose a point x' of $f^{-1}(U)$ mapping to x. Then the stalk of s' at x' agrees with the stalk of $f^*(t)$ at x', and hence there is a neighborhood U' of x' in X' such that $f^*(t)_{|U'} = s'_{|U'}$. Then if $y' \in U'$, the stalk of t at $f(y')$ equals the stalk of s' at y', so $t_{f(y')} = s(f(y'))$. In other words, $t_y = s(y)$ for all y in the image of U'. Since f is open, this image contains a neighborhood of x, and so s is continuous, as required. □

Now to prove the lemma for $\mathcal{M}$, note that since $\mathcal{M}_X$ is u-integral, it is an $\mathcal{O}^*_X$-torsor over $\overline{\mathcal{M}}_X$, and similarly $\mathcal{M}_{X'}$ (resp. $\mathcal{M}_{X''}$) is an $\mathcal{O}^*_{X'}$-torsor (resp., an $\mathcal{O}_{X''}$-torsor) over $\overline{\mathcal{M}}_{X'}$ (resp., $\overline{\mathcal{M}}_{X''}$). Consequently the rows of the diagram

$$\begin{array}{ccccccccc} 0 & \longrightarrow & \mathcal{O}^*_X & \longrightarrow & \mathcal{M}_X & \longrightarrow & \overline{\mathcal{M}}_X & \longrightarrow & 0 \\ & & \downarrow & & \downarrow & & \downarrow & & \\ 0 & \longrightarrow & f_*(\mathcal{O}^*_{X'}) & \longrightarrow & f_*(\mathcal{M}_{X'}) & \longrightarrow & f_*(\overline{\mathcal{M}}_{X'}) & & \\ & & \downdownarrows & & \downdownarrows & & \downdownarrows & & \\ 0 & \longrightarrow & g_*(\mathcal{O}^*_{X''}) & \longrightarrow & g_*(\mathcal{M}_{X''}) & \longrightarrow & g_*(\overline{\mathcal{M}}_{X''}) & & \end{array}$$

are exact. The column on the left is an equalizer diagram because $\mathcal{O}^*_X$ is a sheaf in the fpqc topology. The argument of the previous paragraph shows that the column on the right is also an equalizer diagram, because $\overline{\mathcal{M}}_{X'} \cong f^{-1}(\overline{\mathcal{M}}_X)$ and $\overline{\mathcal{M}}_{X''} \cong g^{-1}(\overline{\mathcal{M}}_X)$. Now the exactness of the middle column follows by chasing the diagram (locally in the Zariski topology on X). □

Let us now turn to statement (2). Since the map $\mathcal{O}_{X_{zar}} \to \eta_*(\mathcal{O}_{X_{\acute{e}t}})$ is an isomorphism, $\eta_*(\mathcal{M}_{\acute{e}t}) = \eta_*^{log}(\mathcal{M}_{\acute{e}t})$. Let $\overline{x}$ be a geometric point of X lying over $x \in X$, and consider the following commutative diagram:

$$\begin{array}{ccccc} \mathcal{O}^*_{X,x} & \longrightarrow & \eta_*(\mathcal{M}_{\acute{e}t})_x & \longrightarrow & \eta_*(\overline{\mathcal{M}}_{\acute{e}t})_x \\ \downarrow & & \downarrow & & \downarrow {\scriptstyle \eta_{\overline{x}}} \\ \mathcal{O}^*_{X,\overline{x}} & \longrightarrow & \mathcal{M}_{\acute{e}t,\overline{x}} & \longrightarrow & \overline{\mathcal{M}}_{\acute{e}t,\overline{x}}. \end{array}$$

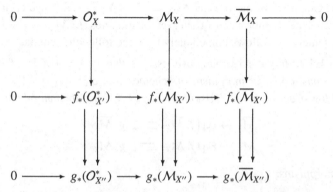

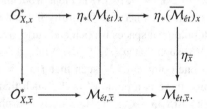

The bottom row is short exact, in the sense that the first map is injective and $\overline{\mathcal{M}}_{\acute{e}t,\overline{x}}$ is the quotient of $\mathcal{O}^*_{X,\overline{x}}$ acting on $\mathcal{M}_{\acute{e}t,\overline{x}}$. In fact the same is true of the top row, by Hilbert's Theorem 90. It follows that the vertical arrow on the right is injective. Indeed, if m_x and m'_x are two elements of $\eta_*(\mathcal{M}_{\acute{e}t})_x$ with the same image in $\overline{\mathcal{M}}_{\acute{e}t,\overline{x}}$, then their respective images $m_{\overline{x}}$ and $m'_{\overline{x}}$ in $\mathcal{M}_{\acute{e}t,\overline{x}}$ differ by the action of some $u \in \mathcal{O}^*_{X,\overline{x}}$. In fact this u must lie in $\mathcal{O}^*_{X,x}$, by descent, since $\mathcal{O}^*_X$ acts freely on $\mathcal{M}_{\acute{e}t}$. It follows that m_x and m'_x have the same image in $\eta_*(\overline{\mathcal{M}}_{\acute{e}t,\overline{x}})$, and the injectivity of $\eta_{\overline{x}}$ follows.

The globally given homomorphism γ induces a factorization of the homomorphism $\gamma_{\overline{x}}\colon P \to \mathcal{M}_{\acute{e}t,\overline{x}}$ through a map $\gamma_x\colon P \to \mathcal{M}_{\acute{e}t,x}$. Let $F_x := \gamma_{\overline{x}}^{-1}(\mathcal{O}^*_{X_{\overline{x}}}) = \gamma_x^{-1}(\mathcal{O}^*_{X,x})$. Then the composite

$$P/F_x \xrightarrow{\overline{\gamma}_x} \eta_*(\overline{\mathcal{M}}_{\acute{e}t})_x \xrightarrow{\eta_{\overline{x}}} \overline{\mathcal{M}}_{\acute{e}t,\overline{x}},$$

is an isomorphism. Since $\eta_{\overline{x}}$ is injective, it follows that it and $\overline{\gamma}_x$ are also isomorphisms. This implies that γ_x is a chart for $\eta_*(\mathcal{M}_{\acute{e}t})_x$, and the rest of the corollary follows immediately. □

Corollary 1.4.4. *Let X be a scheme and let $\alpha_X\colon \mathcal{M}_X \to \mathcal{O}_X$ be a u-integral log structure on $X_{\acute{e}t}$. Then $\mathcal{M}_X$ is coherent if and only if there exist an étale covering $\tilde{X} \to X$ and a coherent log structure $\tilde{\mathcal{M}}_X$ on $\tilde{X}_{zar}$ whose pullback to $\tilde{X}_{\acute{e}t}$ agrees with the pullback of $\mathcal{M}_X$.* □

Corollary 1.4.5. *If X is a log scheme, then the functor on the category of u-integral log schemes sending T to the set of morphisms $T \to X$ forms a sheaf in the Zariski (resp. étale, fpqc) topology.*

1.5 Log points and dashes

Definition 1.5.1. *A log point is a log scheme whose underlying scheme is the spectrum of a field. A log dash is a log scheme whose underlying scheme is the spectrum of a discrete valuation ring.*

Example 1.5.2. If P is sharp monoid, and k is a field, the homomorphism

$$k^* \oplus P \to k : (u, p) \mapsto \begin{cases} u & \text{if } p = 0 \\ 0 & \text{otherwise} \end{cases}$$

defines a log structure on $S := \operatorname{Spec} k$, and we denote the corresponding log point by S_P. Since P is sharp, $\overline{\mathcal{M}}_S \cong P$, and the given homomorphism $P \to \mathcal{M}_S$ is a chart for $\mathcal{M}_S$ and a splitting of the canonical map $\mathcal{M}_S \to \overline{\mathcal{M}}_S$. We therefore call S_P the *split log point* defined by k and P. In particular, $S_{\mathbf{N}}$ is sometimes called the *standard log point* over k.

Remark 1.5.3. Let S be a u-integral Zariski log point, whose underlying scheme is $\operatorname{Spec} k$. Since S has only one point, every sheaf on S is constant, and we identify it with its set of global sections. The exact sequence of abelian groups

$$0 \to k^* \to \mathcal{M}_S^{\mathrm{gp}} \to \overline{\mathcal{M}}_S^{\mathrm{gp}} \to 0 \tag{1.5.1}$$

defines an element of $\mathrm{Ext}^1(\overline{\mathcal{M}}_S^{\mathrm{gp}}, k^*)$. If k is algebraically closed, the group k^* is divisible, and if $\overline{\mathcal{M}}_S^{\mathrm{gp}}$ is finitely generated and torsion free, it is free. In either case, the extension group vanishes, and hence the sequence can be split; note that the set of splittings is naturally a pseudo-torsor under the action of $\mathrm{Hom}(\overline{\mathcal{M}}_S^{\mathrm{gp}}, k^*)$. A choice of splitting defines an isomorphism $S \to S_P$, where $P := \overline{\mathcal{M}}_S$.

If we are working in the étale topology, then the sheaves in (1.5.1) are not necessarily constant. We choose a separable closure k^s of k, defining a geometric point $\overline{x}$ of S, and look at the sequence of G-sets

$$0 \to k^{s*} \to \mathcal{M}_{\overline{x}}^{\mathrm{gp}} \to \overline{\mathcal{M}}_{\overline{x}}^{\mathrm{gp}} \to 0, \tag{1.5.2}$$

where G is the Galois group of k^s/k. If $\mathcal{M}$ is fine, then $\overline{\mathcal{M}}_{\overline{x}}$ is a fine sharp monoid, and hence by Corollary I.2.1.3 its automorphism group is finite. Then G operates through a finite quotient and there is a finite extension k'/k such that $\overline{\mathcal{M}}_{x'} = \overline{\mathcal{M}}_{\overline{x}}$, where $x' := \mathrm{Spec}(k')$. It follows from Hilbert's Theorem 90 and the spectral sequence with $E_2^{p,q} = H^p(G, \mathrm{Ext}^q(\overline{\mathcal{M}}_{x^s}^{\mathrm{gp}}, k^{s*}))$ that the obstruction to splitting the sequence of G-modules (1.5.2) again lies in $\mathrm{Ext}^1(\overline{\mathcal{M}}_{x'}^{\mathrm{gp}}, k^{s*})$.

A log point can be endowed with many idealized structures: one can choose any ideal $\mathcal{K} \subseteq \mathcal{M}_S^+$.

Proposition 1.5.4. *Let P be a fine sharp monoid and let S_P be the corresponding split log point over a field k. Then the automorphism group $\mathrm{Aut}(S_P/\underline{S}_P)$ of S_P relative to $\underline{S}_P$ is naturally identified with the opposite of the semidirect product $\mathsf{A}_P^*(k) \rtimes \mathrm{Aut}(P)$.*

Proof We can identify an element γ of $\mathrm{Aut}(S_P/\underline{S}_P)^{\mathrm{op}}$ with its action on $\mathcal{M}_S$. Such a γ acts on $P = \overline{\mathcal{M}}_S$, and the map $\gamma \to \overline{\gamma}$ defines a homomorphism of groups $\mathrm{Aut}(S_P/\underline{S}_P)^{\mathrm{op}} \to \mathrm{Aut}(P)$. This homomorphism is split by the map $\mathrm{Aut}(P) \to \mathrm{Aut}(S_P/\underline{S}_P)^{\mathrm{op}}$ sending an automorphism g of P to $(\mathrm{id}_{k^*} \oplus g)$. On the other hand, if $x \in \mathsf{A}_P^*(k) = \mathrm{Hom}(P, k^*)$, we can define an automorphism γ_x of $(k^* \oplus P)$ by $\gamma_x(u, p) := (x(p)u, p)$. Since $\alpha(u, p) = 0$ unless $p = 0$, this automorphism is also compatible with α and defines an automorphism of the log scheme $S/\underline{S}$, acting trivially on P. If γ is any such automorphism, its compatibility with α forces the equality $\gamma(u, 0) = (u, 0)$ and, since $\overline{\gamma}$ is the identity, $\gamma(u, q) = (x(q)u, q)$ for some homomorphism $x \colon P \to k^*$. Thus $\mathsf{A}_P^*(k)$ becomes identified with the kernel of the homomorphism $\mathrm{Aut}(S_P/\underline{S}_P)^{\mathrm{op}} \to \mathrm{Aut}(P)$. □

The next result shows that every fine log point is dominated by a standard log point.

Proposition 1.5.5. *If S is any fine log point, there exist an algebraically closed field k and a morphism from the standard log point over k to S.*

Proof Suppose that $\underline{S} = \operatorname{Spec}(k)$, where k is a field. Choose an algebraic closure $\overline{k}$ of k and endow $\operatorname{Spec}(\overline{k})$ with the log structure obtained by pulling back the log structure of S via the morphism $\operatorname{Spec}(\overline{k}) \to \operatorname{Spec}(k)$. Thus there is a strict morphism of log points $\overline{S} \to S$, and so we may assume without loss of generality that k is algebraically closed. Then the log structure on S is split, so that there exists an isomorphism $\mathcal{M}_S \cong k^* \oplus \overline{\mathcal{M}}_S$. Since $\overline{\mathcal{M}}_S$ is a fine monoid, there exists a local homomorphism $h\colon \overline{\mathcal{M}}_S \to \mathbf{N}$, by Proposition I.2.2.1. Since α_S sends $\overline{\mathcal{M}}_S^+$ to 0 and h is local, the homomorphism h defines a morphism of log schemes $S_{\mathbf{N}} \to S$. □

Example 1.5.6. A *standard log dash* is a log scheme of the form

$$T := \operatorname{Spec}(V' \subseteq V),$$

where V is a discrete valuation ring and $V' := V \setminus \{0\}$, viewed as a multiplicative monoid.

The terminology "dash" for the spectrum of a discrete valuation is a translation of the French usage "trait." Such a spectrum T has just two points, the closed point t and the generic point τ, which is an open subset. In contrast to the case of a standard log point, the sheaf of monoids $\mathcal{M}_T$ on a standard log dash is not constant. It has the following form:

$$\begin{array}{ccccc}
V' & \xrightarrow{\;=\;} & \mathcal{M}_t & \xrightarrow{\;\alpha_t\;} & \mathcal{O}_{T,t} = V \\
\downarrow & & \downarrow & & \downarrow \\
K^* & \xrightarrow{\;=\;} & \mathcal{M}_\tau & \xrightarrow{\;\alpha_\tau\;} & \mathcal{O}_{T,\tau} = K
\end{array}$$

where α_t is the inclusion $V' \to V$. The homomorphism of sheaves $\mathcal{M}_T \to \overline{\mathcal{M}}_T$ cannot split, since $\overline{\mathcal{M}}_t = V'/V^* \cong \mathbf{N}$ and $\overline{\mathcal{M}}_\tau = 0$, while $\mathcal{M}_\tau = K^*$. Splittings of the stalk $\mathcal{M}_t \to \overline{\mathcal{M}}_t$ do exist, and amount to choices of a uniformizing parameter of V.

Since the quotient V'/V^* is isomorphic to $\mathbf{N}$, the restriction of the standard log structure on Spec V to the spectrum of its residue field k is isomorphic to the standard log structure on Spec k. However, this isomorphism is not canonical: it depends on the choice of a uniformizing parameter π of V. Two such choices

π and π' define the same isomorphism if and only if $\pi' = u\pi$, where $u \in V^*$ is congruent to 1 modulo the maximal ideal of V.

There are other interesting log structures on dashes. For example, one has the *hollow log structure*

$$\begin{array}{ccccc} V' & \xrightarrow{=} & \mathcal{M}_t & \xrightarrow{\alpha_t} & \mathcal{O}_{T,t} = V \\ \downarrow & & \downarrow & & \downarrow \\ V' \oplus_{V^*} K^* & \xrightarrow{=} & \mathcal{M}_\tau & \xrightarrow{\alpha_\tau} & \mathcal{O}_{T,\tau} = K \end{array}$$

where now

$$\alpha_t \colon V' \mapsto V : v \mapsto \begin{cases} v & \text{if } v \in V^* \\ 0 & \text{otherwise} \end{cases}$$

and α_τ is the obvious map. The pullback of the hollow log structure to $\operatorname{Spec} k$ is also non-canonically isomorphic to the standard log structure on $\operatorname{Spec} k$. In between these extremes, for each n one can form the log structure associated to the prelog structure

$$\mathbf{N} \to V : 1 \mapsto \pi^n.$$

For $n = 1$, the corresponding log structure is isomorphic to the standard one, and as n approaches infinity, the induced log structure approaches (in some sense) the hollow one; see for example [60].

Remark 1.5.7. If T is a standard log dash, then $\operatorname{Aut}(T/\overline{T})$ is trivial, since $\mathcal{M}_{T,t} \subseteq \mathcal{O}_T$. On the other hand, if T has the hollow log structure, one finds as in the case of log points that $\operatorname{Aut}(T/\overline{T})$ is a semidirect product of $\operatorname{Aut}(Q)$ and $\mathsf{A}^*_{\mathsf{Q}}(V)$.

1.6 Compactifying log structures

One of the most important applications of log geometry is the study of compactifications and, more generally, open embeddings $U \to X$.

Definition 1.6.1. *Let U be a nonempty Zariski open subset of a scheme X and let $j \colon U \to X$ be the inclusion. The direct image log structure (Definition 1.1.5)*

$$\alpha_{U/X} \colon \mathcal{M}_{U/X} := j_*^{log}(\mathcal{O}_U^*) \to \mathcal{O}_X$$

is called the compactifying log structure associated to the open immersion j. A log structure $\alpha \to \mathcal{O}_X$ on X said to be compactifying if its subset of triviality X^* is open and the natural map $\alpha \to \alpha_{X^*/X}$ is an isomorphism.

The sheaf $\mathcal{M}_{U/X}$ is the inverse image of $j_*(\mathcal{O}_U^*) \subseteq j_*(\mathcal{O}_U)$ via the natural map $\mathcal{O}_X \to j_*(\mathcal{O}_U)$. This is just the sheaf of sections of $\mathcal{O}_X$ whose restriction to U is invertible. Thus $\alpha_{U/X}\colon j_*^{log}(\mathcal{O}_U^*) \to \mathcal{O}_X$ is injective, and its image is a sheaf of faces in the monoid $\mathcal{O}_X$. For example, $\alpha_{U/X}$ is the trivial log structure if $U = X$ and is the empty log structure if $U = \emptyset$. The log structure on standard log dash T is the compactifying log structure associated with the open immersion $\{\tau\} \to T$.

The construction of compactifying log structures is functorial, in the following sense. If $f\colon X \to Y$ is a morphism of schemes taking an open subset U of X to an open subset V of Y, then f induces a morphism of compactifying log schemes $f\colon (X, \alpha_{U/X}) \to (Y, \alpha_{V/Y})$.

The following proposition illustrates one sense in which the log scheme X, with the compactifying log structure associated with an open embedding $U \to X$, is closer to U than to the underlying scheme $\underline{X}$. We will see further illustrations later; see for example Corollary V.1.3.2.

Proposition 1.6.2. *Let X be a log scheme and let U be a Zariski open subset. Suppose that the log structure of X is the compactifying log structure of the open immersion $U \to X$. Then, for every log scheme Y, the natural map*

$$\mathrm{Mor}(X, Y) \to \{g \in \mathrm{Mor}(\underline{X}, \underline{Y}) : g(U) \subseteq Y^*\}$$

is bijective.

Proof Let $f\colon X \to Y$ be a morphism of log schemes and let $\underline{f}\colon \underline{X} \to \underline{Y}$ be the corresponding map of underlying schemes. Since $U \subseteq X^*$, it follows from Proposition 1.2.8 that $\underline{f}(U) \subseteq Y^*$. Moreover, since $\alpha_{U/X}$ is injective, the homomorphism $f^\flat\colon \mathcal{M}_Y \to \underline{f}_*(\mathcal{M}_X)$ is uniquely determined by $\underline{f}$, and so the correspondence $f \mapsto \underline{f}$ is injective. Suppose on the other hand that $g\colon \underline{X} \to \underline{Y}$ maps U to Y^*, and that m is a local section of $\mathcal{M}_Y$. Then $g^\sharp(\alpha_Y(m))$ is a local section of $g_*(\mathcal{O}_X)$ whose restriction to U is a unit and hence defines a section of $g_*(\mathcal{M}_X)$. The resulting homomorphism $\mathcal{M}_Y \to g_*(\mathcal{M}_X)$ defines the required homomorphism $X \to Y$. □

Proposition 1.6.3. *Let X be a locally noetherian scheme and let U be an open subset containing all the associated points of X. Denote by $\underline{\Gamma}_Y(Div_X^+)$ the sheaf of effective Cartier divisors on X with support in $Y := X \setminus U$.*

1. *The compactifying log structure $\mathcal{M}_{U/X}$ is integral, and the homomorphism*

ζ of diagram (1.1.1) fits into a commutative diagram

$$\begin{array}{ccc} \overline{\mathcal{M}}_{U/X} & \xrightarrow{\cong} & \underline{\Gamma}_Y(\mathit{Div}_X^+) \\ & \searrow^{\zeta} & \downarrow \\ & & \mathcal{Z}_X. \end{array}$$

Here the top arrow takes a local section q of $\overline{\mathcal{M}}_X$ to the effective Cartier divisor defined by the inverse of the invertible sheaf of ideals generated by any local section m of $\mathcal{M}_X$ mapping to q.

2. *If q is a section of $\overline{\mathcal{M}}_{U/X}$ corresponding to a section D of $\underline{\Gamma}_Y(\mathit{Div}_X^+)$, then $\mathcal{L}_q$ is the ideal sheaf $\mathcal{I}_D$ of the effective divisor D, and is equal to $\zeta(q)$ in the notation of Remark 1.1.7.*
3. *If $U' \subseteq U$ is another open subset containing all the associated points of X and if, for each $x \in U \setminus U'$, the height of $\mathcal{O}_{X,x}$ is at least two, then the natural map $\mathcal{M}_{U/X} \to \mathcal{M}_{U'/X}$ is an isomorphism.*

Proof If m is a local section of $\mathcal{M}_{U/X}$, then the restriction of $\alpha_{U/X}(m)$ to U is a unit and hence does not vanish at any associated point of X. This implies that the germ of $\alpha_{U/X}(m)$ at every point x of X is a nonzero divisor of $\mathcal{O}_{X,x}$. Thus, $\alpha_{U/X}(m)$ lies in the sheaf $\mathcal{O}'_X$ of nonzero divisors of X. Since $\alpha_{U/X}$ is an injective homomorphism of monoids, it follows that $\mathcal{M}_{U/X}$ is integral. Since $\mathcal{M}^*_{U/X} \cong \mathcal{O}^*_X$, $\alpha_{U/X}$ induces an injection $\overline{\alpha}\colon \overline{\mathcal{M}}_{U/X} \to \mathcal{O}'_X/\mathcal{O}^*_X$. The map sending an element of $\mathcal{O}'_X$ to the inverse of the invertible ideal that it generates induces an isomorphism $\mathcal{O}'_X/\mathcal{O}^*_X \to \mathit{Div}^+_X$. Since each $\alpha_{U/X}(m)$ restricts to a unit on U, the divisor it defines has support in Y. Conversely, if D is an effective Cartier divisor, then locally D can be expressed as the class of an element f of $\mathcal{O}'_X$. If D has support in Y, then $f_{|_U}$ is a unit, i.e., $f \in j^{log}_*(\mathcal{O}^*_U)$. This proves the first statement, and the second follows immediately from the definitions. For the third, it will suffice to prove that the map $\underline{\Gamma}_Y(\mathit{Div}^+_X) \to \underline{\Gamma}_{Y'}(\mathit{Div}^+_X)$ is an isomorphism, where $Y' := X \setminus U'$. Injectivity is clear. Let x be a point of X and let D be an effective Cartier divisor of $\mathcal{O}_{X,x}$ with support in Y', defined by a nonzero divisor a of $\mathcal{O}_{X,x}$. We have to prove that if $x \in Y' \setminus Y$, then a is a unit. If not, then $\mathcal{O}_{X,x}/(a)$ is a noetherian local ring that necessarily contains a minimal prime $\mathfrak{p}$, corresponding to a point z of X generizing x. By the Hauptidelsatz, z has height one. Then $z \notin Y$, since x is a specialization of z and $x \notin Y$. On the other hand, $z \in D$ and D has support in Y', so $z \in Y'$. This contradicts our hypothesis that every point of $Y' \setminus Y$ has height at least two. Here we have been working in the Zariski topology; the étale argument is the same. □

Remark 1.6.4. The formation of $\mathcal{M}_{U/X}$ can depend on whether we use the étale or the Zariski topology. Indeed, if $f\colon X' \to X$ is étale, the natural map $f^{-1}(Div_X^+) \to Div_{X'}^+$ is not an isomorphism in general. For an explicit example, take $X := \mathrm{Spec}(\mathbf{C}[x, y])$, let Y be the closed subset defined by $y^2 - (x^2 - x^3)$, let $U := X \setminus Y$, and let s denote the origin. Then, in the Zariski topology, the stalk of $\overline{\mathcal{M}}_{U/X} \cong \Gamma_Y(Div_X^+)$ at s has rank one, generated by the class of Y, since Y is irreducible. However, in the étale topology, the stalk at $\overline{s}$ has rank two, generated by the two branches of Y. To make this explicit, let $X' :=$ Spec $\mathbf{C}[y, t, t^{-1}]$, and let $f\colon X' \to X$ be the map defined by $x \mapsto 1 - t^2$. Then $X' \to X$ is étale, and the inverse image Y' of Y in X' is defined by $y^2-(1-t^2)^2t^2$, which clearly has two branches. See Figure 1.6.1.

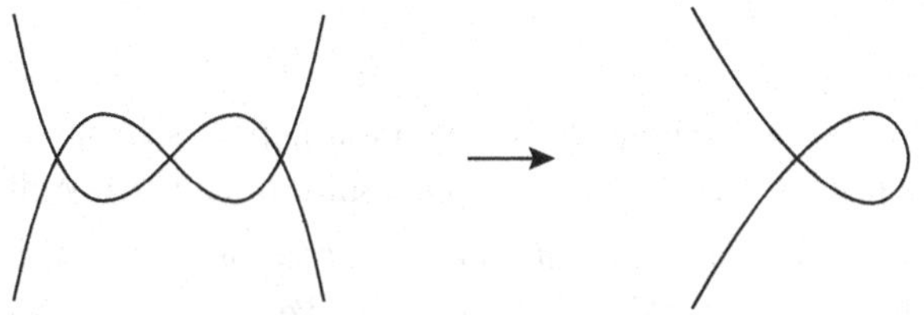

Figure 1.6.1 Étale vs. Zariski branches.

As pointed out in [22, 12.2.12], this difficulty is amplified in higher dimensions. For example let $X := \mathbf{C}[x, y, z]$ and let Y be the closed subset defined by $zy^2 - zx^2 + x^3$. Then the Henselization of Y at the origin s has only one branch, but it has two branches at points s' where $y = x = 0$ and $z \neq 0$. Thus, if $\mathcal{M}$ is the compactifying log structure associated to the inclusion of $X \setminus Y$ in X computed in the étale topology, $\overline{\mathcal{M}}_s$ has rank one but $\overline{\mathcal{M}}_{s'}$ has rank two. Then it follows from Corollary II.2.1.6 that $\mathcal{M}$ cannot be coherent. See Figure 1.6.2.

The following proposition allows us to avoid, with a suitable hypothesis, the difficulties illustrated by the examples in the previous remark.

Proposition 1.6.5. *Let X be a locally noetherian and normal scheme and let U be an open subset of X containing all the associated points of X. Assume that the complement Y of U has pure codimension one and that each of its irreducible components is geometrically unibranch [27, 6.15.1]. Then the map*

$$\eta^*(\mathcal{M}_{U/X_{zar}}) \to \mathcal{M}_{U/X_{\acute{e}t}}$$

is an isomorphism, where $\eta\colon X_{\acute{e}t} \to X_{zar}$ is the projection mapping.

Figure 1.6.2 Collapsing branches.

Proof The following lemma shows that on a normal scheme, the effective Cartier divisors are exactly embedded in the sheaf of effective Weil divisors.

Lemma 1.6.6. *Let X be a normal scheme and let Div_X^+ and $\mathcal{W}_X^+$ denote, respectively, the sheaves of effective Cartier and Weil divisors on X. Then there is a natural injective and exact homomorphism*

$$\nu\colon Div_X^+ \to \mathcal{W}_X^+.$$

Proof Since X is normal, each of its local rings is an integral domain, and so the sheaf $\mathcal{O}'_X$ of nonzero divisors consists of those sections f of $\mathcal{O}_X$ whose stalk at every point is not zero. As we have seen, this is a sheaf of submonoids of $\mathcal{O}_X$, and Div_X^+ can be identified with the quotient $\mathcal{O}'_X/\mathcal{O}^*_X$. The sheaf $\mathcal{W}_X^+$ is the sheaf associated to the presheaf that to every open U assigns the free monoid on the set of points $\eta \in U$ such that $\mathcal{O}_{U,\eta}$ has dimension one. Since X is regular in codimension one, each $\mathcal{O}_{U,\eta}$ is a discrete valuation ring, and the valuation maps induce a monoid homomorphism $\nu\colon \mathcal{O}'_X \to \mathcal{W}_X^+$ [34, II §6]. The normality of X implies that, for each $x \in X$, the local ring $\mathcal{O}_{X,x}$ is the intersection, in the fraction field $K_{X,x}$ of $\mathcal{O}_{X,x}$, of its localizations at height one primes. It follows that $\mathcal{O}'_{X,x}$ is the set of sections f of $K_{X,x}$ such that $\nu^{\mathrm{gp}}(f) \in \mathcal{W}_X^+$. Hence the homomorphism $\nu_x\colon \mathcal{O}'_{X,x} \to \mathcal{W}_{X,x}^+$ is exact. It follows that $Div_{X,x}^+ := \overline{\mathcal{O}'_{X,x}}$ is an exact submonoid of $\mathcal{W}_{X,x}^+$. □

Lemma 1.6.7. *Let $f\colon X' \to X$ be an étale morphism of normal and let $Y \subseteq X$ be a closed subscheme each of whose irreducible components is purely of codimension one and geometrically unibranch [27, 6.15.1]. Then, if $Y' := f^{-1}(Y)$,* *the natural maps*

$$f^{-1}(\underline{\Gamma}_Y(\mathcal{W}_X^+)) \to \underline{\Gamma}_{Y'}(\mathcal{W}_{X'}^+) \quad \textit{and} \quad f^{-1}(\underline{\Gamma}_Y(Div_X^+)) \to \underline{\Gamma}_{Y'}(Div_{X'}^+)$$

are isomorphisms.

Proof Let x' be a point in X' and let $x := f(x) \in X$. Since $X' \to X$ is étale, Y' is purely of codimension one in X'. The stalk of $\underline{\Gamma}_{Y'}(\mathcal{W}_{X'}^+)$ at x' is the free monoid generated by the generic points of Y' that generize x'. If ζ' is such a point, its image ζ in X is a generic point of Y generizing x. Since f is étale and Z is geometrically unibranch, then by [29, 18.10.7] $f^{-1}(Z)$ has a unique generic point ζ'' whose closure contains x', and so $\zeta'' = \zeta'$. Thus the first of our maps, for Weil divisors, is an isomorphism.

Since X and X' are normal, the vertical maps in the following diagram are injections:

$$\begin{array}{ccc} f^{-1}(\underline{\Gamma}_Y(Div_X^+)) & \longrightarrow & \underline{\Gamma}_{Y'}(Div_{X'}^+) \\ \big\downarrow & & \big\downarrow \\ f^{-1}(\underline{\Gamma}_Y(\mathcal{W}_X^+)) & \longrightarrow & \underline{\Gamma}_{Y'}(\mathcal{W}_{X'}^+). \end{array}$$

Since f is faithfully flat, a Weil divisor on X is Cartier if and only if its pullback to X' is Cartier. Thus the square is cartesian and, since the bottom arrow is an isomorphism, so is the top one. □

Proposition 1.6.5 follows from Lemma 1.6.7 and Propositions 1.4.1 and 1.6.3. □

1.7 DF log structures

The notion of a log structure that we have been discussing was formulated by Fontaine and Illusie and further developed by Kato. A similar but less flexible theory was suggested independently by Deligne and Faltings. Since it is somewhat simpler and still useful, we discuss it briefly.

Definition 1.7.1. *A DF structure on a scheme X is a finite sequence $\gamma.$ of homomorphisms $\gamma_i \colon \mathcal{L}_i \to \mathcal{O}_X$, where each $\mathcal{L}_i$ is an invertible sheaf on X.* [2]

Let us indicate how to pass from a DF structure to a log structure, following Kato's explanation in [48, Complement (1)]. Given a DF structure $\gamma.$, consider the sheaf $\mathcal{P}$ whose sections on a connected open set U consist of the set of

[2] In fact Deligne and Faltings take the equivalent but dual point of view: for them γ_i is an invertible sheaf $\mathcal{L}_i$ endowed with a global section s_i.

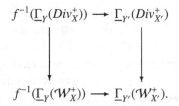

pairs (a, I), where $I := (I_1, \ldots, I_n) \in \mathbf{N}^n$ is a multi-index and a is a generator of the invertible sheaf $\mathcal{L}^I := \mathcal{L}_1^{I_1} \otimes \cdots \otimes \mathcal{L}_n^{I_n}$. The sheaf $\mathcal{P}$ has a natural structure of a sheaf of integral monoids. Let $\gamma \colon \mathcal{P} \to \mathcal{O}_X$ be the homomorphism sending (a, I) to $\gamma^I(a)$, where $\gamma^I := \gamma_1^{I_1} \otimes \cdots \otimes \gamma_n^{I_n} \colon \mathcal{L}^I \to \mathcal{O}_X$. The prelog structure γ defines a log structure $\alpha \colon \mathcal{M} \to \mathcal{O}_X$, with $\gamma = \alpha \circ \beta$, where $\beta \colon \mathcal{P} \to \mathcal{M}$ is a chart for $\mathcal{M}$. Not every fine log structure comes from a DF structure; see Remark 1.12.2 for a characterization. We refer to work of Bourne and Vistoli [8] for a more general notion of a DF structure, using the language of stacks.

If $\gamma_\cdot$ is a DF structure on X and $f \colon X' \to X$ is any morphism, then each γ_i induces a morphism $f^*(\mathcal{L}_i) \to \mathcal{O}_{X'}$, and thus one obtains a DF structure $f^*(\gamma_\cdot)$ on X'. The log structure associated to this DF structure is the pullback (Definition 1.1.5) of the log structure corresponding to $\gamma_\cdot$, as is easy to verify.

Example 1.7.2. Let $Y_1, \ldots, Y_n$ be a sequence of effective Cartier divisors in X, defined by invertible sheaves of ideals $\mathcal{I}_1, \ldots, \mathcal{I}_n$. Then the family of inclusions $\mathcal{I}_i \to \mathcal{O}_X, i = 1, \ldots, n$, defines a DF structure on X. More generally, given closed immersions $Y_i \subseteq X' \subseteq X''$, let $\mathcal{I}_i''$ be the ideal sheaf of Y_i in X'', and suppose that the restriction of $\mathcal{I}_i$ to X' is an invertible sheaf $\mathcal{L}_i$ on X'. Then the inclusion $\mathcal{I}_i \to \mathcal{O}_X$ induces a (not necessarily injective) map $\gamma_i \colon \mathcal{L}_i \to \mathcal{O}_{X'}$, and the collection $\gamma_\cdot$ defines a DF structure on X'.

Let $\gamma_\cdot$ be a DF structure and let $\alpha \colon \mathcal{M} \to \mathcal{O}_X$ be the corresponding log structure. For each i, the image $\mathcal{I}_i$ of γ_i is a quasi-coherent sheaf of ideals in $\mathcal{O}_X$, defining a closed subscheme Y_i of X. Note that, since each $\mathcal{L}_i$ is invertible, each $\mathcal{I}_i$ is locally monogenic but it need not be invertible and could, for example, be the zero ideal. In any case, the codimension of Y_i in X is everywhere less than or equal to one. The homomorphism $\mathcal{O}_X^* \to \mathcal{P} : u \mapsto (u, 0)$ identifies $\mathcal{O}_X^*$ with the sheaf of units $\mathcal{P}^*$ of $\mathcal{P}$. Let $\theta \colon \mathcal{P} \to \mathbf{N}^n$ be the homomorphism sending (a, I) to I. On any open set U on which all the sheaves $\mathcal{L}_i$ are trivial, a choice of generators $(a_1, \ldots, a_n)$ defines a splitting of θ sending I to $(a_1^{I_1} a_2^{I_2} \cdots a_n^{I_n}, I)$. It follows that we have an exact sequence: $0 \to \mathcal{O}_X^* \to \mathcal{P} \to \mathbf{N}^n \to 0$, so that $\overline{\theta}$ identifies $\overline{\mathcal{P}}$ with $\mathbf{N}^n$ and the chart $\beta \colon \mathcal{P} \to \mathcal{M}$ defines an isomorphism $\overline{\beta} \colon \mathbf{N}^n \to \overline{\mathcal{M}}$. Let q_i denote the global section of $\overline{\mathcal{M}}$ corresponding to the ith standard basis vector of $\mathbf{N}^n$ via this isomorphism. Then $Y_i = \zeta(q_i)$, in the notation of diagram 1.1.1.

Proposition 1.7.3. *Let $\alpha \colon \mathcal{M} \to \mathcal{O}_X$ be the log structure associated to a DF structure $\gamma_\cdot$, let $q_i \in \Gamma(X, \overline{\mathcal{M}})$ correspond to the ith standard basic vector of $\mathbf{N}^n$ as explained above.*

1. *The sheaf of monoids $\mathcal{M}$ is integral and, for each i, there is a natural iso-*

morphism $\mathcal{L}_i^* \cong \pi^{-1}(q_i)$, where $\pi\colon \mathcal{M} \to \overline{\mathcal{M}}$ is the natural map and $\mathcal{L}_i^*$ is the sheaf of local generators for $\mathcal{L}_i$.

2. Locally on X, the homomorphism $\overline{\beta}\colon \mathbf{N}^n \to \overline{\mathcal{M}}$ lifts to a chart $\mathbf{N}^n \to \mathcal{M}$. For each $x \in X$, the stalk of $\overline{\mathcal{M}}_x$ is freely generated by the images of those q_i such that $\overline{\gamma}_i(q_i)$ vanishes in $k(x)$.

3. Let $Y_i = \zeta(q_i)$, in the notation of diagram 1.1.1 of Remark 1.1.6. and let $Y := \cup\{Y_i : i = 1, \ldots, n\}$. Then the restriction of $\mathcal{M}$ to $U := X \setminus Y$ is trivial. Hence there is a natural homomorphism of log structures from α to the compactifying log structure $\alpha_{U/X}$ associated to $U \subseteq X$.

4. Suppose that X is normal, that each γ_i is injective, and that each Y_i is reduced and irreducible (resp. and geometrically unibranch). Then in the Zariski (resp. étale) topology, the map $\alpha \to \alpha_{U/X}$ is an isomorphism. In particular, the homomorphism $\zeta\colon \overline{\mathcal{M}} \to \underline{\Gamma}_Y(\mathcal{W}_X^+)$ sending $q_i \in \Gamma(X, \overline{\mathcal{M}})$ to $Y_i \in \underline{\Gamma}_Y(\mathcal{W}_X^+)$ is an isomorphism, as is the inclusion $\underline{\Gamma}_Y(Div_X^+) \to \underline{\Gamma}_Y(\mathcal{W}_X^+)$.

Proof Since $\mathcal{P}$ is a sheaf of integral monoids and is a chart for $\mathcal{M}$, it follows from Proposition II.1.1.8 that $\mathcal{M}$ is also integral. The map $\beta_i\colon \mathcal{L}_i^* \to \mathcal{M}$ factors through $\pi^{-1}(q_i)$, and the resulting morphism $\mathcal{L}_i^* \to \pi^{-1}(q_i)$ is a morphism of $\mathcal{O}_X^*$-torsors, hence an isomorphism. This proves (1). To prove statement (2), we may assume that all the invertible sheaves $\mathcal{L}_i$ are trivial, so that there is a splitting $\sigma\colon \mathbf{N}^n \to \mathcal{P}$ as explained in the previous paragraph. Then $\beta \circ \sigma\colon \mathbf{N}^n \to \mathcal{M}$ lifts $\overline{\beta}$ and hence is a chart for $\mathcal{M}$. Now, for each $x \in X$, the stalk $\overline{\mathcal{M}}_x$ of $\mathcal{M}$ at x is the quotient of $\mathbf{N}^n$ by the face F_x consisting of the set of elements $I \in \mathbf{N}^n$ that map to a unit in $\mathcal{M}_x$, or equivalently in $\mathcal{O}_{X,x}$. Since $\mathbf{N}^n$ is free with basis $(e_1, \ldots, e_n)$, statement (2) of Examples I.1.4.8 tells us that F_x is generated by the set of e_i that map to a unit in $\mathcal{O}_{X,x}$, or equivalently, the set of i such that q_i does not vanish in $k(x)$. Then $\mathbf{N}^n/F_x$ is freely generated by the remaining e_i's. Statement (2) of the proposition follows.

To prove statement (3), let m_i be a local section of $\mathcal{M}$ lifting q_i. Then $\alpha(m_i)$ is a unit outside Y_i, and hence m_i restricts to a unit of $\mathcal{M}$ outside Y_i. It follows that $\overline{\mathcal{M}}$ is trivial on U, and the existence of the map $\mathcal{M} \to \mathcal{M}_{U/X}$ follows from Proposition 1.6.2.

Now suppose the hypotheses in (4) are satisfied. Since each γ_i is injective and X is integral, $\mathcal{I}_i$ is an invertible sheaf of ideals and each Y_i is an effective Cartier divisor in X. It follows that U is a nonempty subset of X and, since X is normal, (1) of Proposition 1.6.3 exhibits an isomorphism $\overline{\mathcal{M}}_{U/X} \cong \underline{\Gamma}_Y(Div_X^+)$. Since the Y_i are the reduced irreducible components of Y, their images in the Zariski stalks of $\Gamma_Y(\mathcal{W}_X^+)$ form a basis at each point. Then we have a commutative diagram

$$\begin{array}{ccccc} \mathbf{N}^n & \xrightarrow{\cong} & \overline{\mathcal{M}} & \longrightarrow & \overline{\mathcal{M}}_{U/X} \\ & \searrow & \downarrow \zeta & & \downarrow \cong \\ & & \underline{\Gamma}_Y(\mathcal{W}^+_X) & \xleftarrow{\delta} & \underline{\Gamma}_Y(\mathcal{D}iv^+_X). \end{array}$$

We have just seen that ζ is an isomorphism, and it follows that δ is surjective. The normality of X implies that δ is also injective, hence an isomorphism. It follows that all the arrows are isomorphisms. When the Y_i are geometrically unibranch, the same is true in the étale context, by Proposition 1.6.5. □

Corollary 1.7.4. *Let $U \subseteq X$ be a dense open subset of a locally noetherian and locally factorial scheme X, let $Y := X \setminus U$, and let $\mathcal{M}_{U/X}$ be the Zariski compactifying log structure associated with $U \to X$. Then $\mathcal{M}_{U/X}$ is coherent, and $\overline{\mathcal{M}}_{U/X} \cong \Gamma_Y(\mathcal{W}^+_X)$. Moreover, if q is a global section of $\overline{\mathcal{M}}_{U/X}$ corresponding to an irreducible component Y_i of Y, then $\mathcal{L}_q \cong \mathcal{I}_{Y_i}$.*

Proof Let Z be the closure of the set of points y of Y such that $\dim \mathcal{O}_{X,y}$ is at least two, let Y' be the closure of $Y \setminus Z$, and let $U' := X \setminus Y'$. Then all the points of $U' \setminus U$ have height at least two, and so (3) of Proposition 1.6.3 implies that the natural map $\mathcal{M}_{U'/X} \to \mathcal{M}_{U/X}$ is an isomorphism. Thus without loss of generality we may replace U by U', that is, we may assume that all the irreducible components of Y have codimension one. Since X is locally factorial, each such component Y_i is a Cartier divisor, and its ideal sheaf $\mathcal{I}_i$ is invertible. The collection of inclusions $\mathcal{I}_i \to \mathcal{O}_X$ defines, as we have seen, a DF log structure $\mathcal{M}$, and Proposition 1.7.3 tells us that $\mathcal{M}_{U/X}$ is the associated log structure. □

Let $i\colon Y \to X$ be a reduced and irreducible Cartier divisor in a normal scheme X and let $j\colon U \to X$ be its complement. Proposition 1.7.3 shows that $\alpha_{U/X}\colon \mathcal{M}_{U/X} \to \mathcal{O}_X$ is the log structure associated to the DF structure $\mathcal{I}_Y \to \mathcal{O}_X$. It follows that $i^*_{log}(\alpha_{U/X})$ is the log structure associated to the DF structure $(\mathcal{I}_Y/\mathcal{I}^2_Y, \gamma)$, where γ is the zero map $\mathcal{I}_Y/\mathcal{I}^2_Y \to \mathcal{O}_X$. Thus the information that this log structure contains is no more or less than the conormal sheaf of Y in X. In particular, it depends only on the first infinitesimal neighborhood of Y in X. The next proposition generalizes this fact.

Proposition 1.7.5. *Let X be a normal and locally noetherian scheme, and let Y be a closed subset of codimension one, each of whose irreducible components Y_i is a Cartier divisor. For each natural number m, let X_m be the subscheme of X*

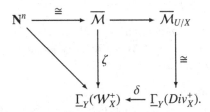

defined by $\mathcal{I}_Y^{m+1}$. Let $U := X\setminus Y$ and let $\alpha_{U/X}\colon \mathcal{M}_{U/X} \to X$ be its compactifying log structure.

1. *The restriction of the ideal of Y_i in X_m to X_{m-1} is an invertible sheaf $\mathcal{L}_i$ of $\mathcal{O}_{X_{m-1}}$-modules. Thus, by the construction of Example 1.7.2, there is a DF structure $\gamma_\cdot^{(m)}\colon \mathcal{L}_\cdot \to \mathcal{O}_{X_{m-1}}$ associated to the family of embeddings $Y_i \to X_m$.*
2. *The log structure associated to the DF structure $\mathcal{L}_\cdot \to \mathcal{O}_{X_{m-1}}$ is the restriction of $\alpha_{U/X}$ to X_{m-1}.*
3. *The restriction of $\alpha_{U/X}$ to X_{m-1} depends only on the scheme X_m and the embedding $Y \to X_m$.*

Proof We will work in the Zariski topology; in the étale topology one must assume that each Y_i is geometrically unibranch. Let $\mathcal{I}$ denote the ideal of Y in X, so that $\mathcal{I} \subseteq \mathcal{I}_i$ for all i. Then the ideal $\mathcal{I}_i''$ of Y_i in X_m is $\mathcal{I}_i/\mathcal{I}_i \cap \mathcal{I}^{m+1} = \mathcal{I}_i/\mathcal{I}^{m+1}$, and its restriction to X_{m-1} is

$$\mathcal{I}_i''/\mathcal{I}^m\mathcal{I}_i'' = \mathcal{I}_i/(\mathcal{I}^{m+1} + \mathcal{I}_i\mathcal{I}^m) = \mathcal{I}_i/\mathcal{I}_i\mathcal{I}^m \cong \mathcal{O}_{X_{m-1}} \otimes \mathcal{I}_i,$$

an invertible sheaf on X_{m-1}. This proves (1), and in fact we see that the DF structure $\gamma_\cdot^{(m)}$ on X_{m-1} is the restriction of the DF structure $\gamma_\cdot$ on X defined by the set of embeddings $Y_i \to X$. By (4) of the previous proposition, the latter is the compactifying log structure $\alpha_{U/X}$. Since the formation of the log structure associated to a DF structure is compatible with pullbacks, statement (2) follows, and (3) is an immediate consequence. □

1.8 Normal crossings and semistable reduction

Let us look more closely at the case of divisors with normal crossings, which has long been studied in other formulations. We review the fundamental results of [20], which were updated in the language of log geometry in [45] and [74]. If Z is a closed subscheme of X, we write $i_{Z/X}$ for the closed immersion $Z \to X$ and $\mathcal{I}_{Z/X}$ for the ideal sheaf of Z in X.

Definition 1.8.1. *Let X be a regular scheme. A divisor with (strict) normal crossings in X is a closed subscheme Y such that the intersection of every set of irreducible components of Y is also regular. A scheme Y is said to be a (strict) normal crossing scheme if, for every point y of Y, there exist a (Zariski) neighborhood U of y and a closed immersion identifying U with a divisor with strict normal crossings in a regular scheme X.*

We sometimes write DNC for "divisor with normal crossings." If U is an open subset of a DNC scheme, we shall call an embedding $U \to X$ as in the definition a *DNC coordinate system* on U. If Y is a closed subscheme of X, the condition that Y be a strict divisor with normal crossings X is local in the Zariski topology of X. More generally, a closed subscheme Y of a regular scheme X is said to be a divisor with (not necessarily strict) normal crossings if étale locally it is a strict DNC in X. It is standard, and easy to verify, that Y is a strict DNC in X if and only if, for each point x in X, there exists a regular sequence $(t_1, \ldots, t_m)$ generating the maximal ideal of $\mathcal{O}_{X,x}$ and a natural number r such that $t_1 \cdots t_r$ generates the ideal of Y in $\mathcal{O}_{X,x}$.

Let $Y \to X$ be a divisor with strict normal crossings in X, with irreducible components Y_i defined by ideals $\mathcal{I}_i$. Since X is regular, it is locally factorial and each $\mathcal{I}_i$ is an invertible sheaf of ideals. The family of inclusions $\gamma_i \colon \mathcal{I}_i \to \mathcal{O}_X$ thus defines a DF structure on X. The following result is then a straightforward consequence of Proposition 1.7.3; we omit the proof.

Proposition 1.8.2. *Let $Y \to X$ be a strict DNC in a regular scheme X, let $\gamma_\cdot \colon \mathcal{I}_\cdot \to \mathcal{O}_X$ be the DF structure defined by the ideals of the irreducible components of Y in X, and let $\alpha_X \colon \mathcal{M}_X \to \mathcal{O}_X$ denote the corresponding log structure.*

1. *The log structure α_X is compactifying.*
2. *Let x be a point of X and let $\{Y_1, \ldots, Y_r\}$ be the set of irreducible components of Y containing x. Then in a neighborhood of x there exists a chart $\beta \colon \mathbf{N}^r \to \mathcal{M}_X$ such that $\alpha_X(\beta(e_i))$ is a generator of $\mathcal{I}_i$ for each $i = 1, \ldots, r$.* □

As we shall see later, the log scheme associated to a divisor with strict normal crossing behaves very well: in Example 1.11.9 we will see that it is "log regular," and in Example IV.3.1.14 we will see under what circumstances it is "log smooth." If Y is a divisor with (not necessarily) strict normal crossing in X, one can still consider the compactifying log structure α_X on X corresponding to the complement of Y in X; in this case it is important to work with étale log structures. Then α_X is coherent and étale locally admits charts as in statement (2) of the proposition.

Normal crossing divisors often arise as a degenerate fiber of a smooth family. This situation has been much studied and was one of the founding motivations for log geometry.

Definition 1.8.3. *Let $f \colon X \to S$ be a flat morphism of schemes, where X is regular and S is the spectrum of a discrete valuation ring. One says that X/S has (strict) semistable reduction if its special fiber $\underline{Y}$ is a (strict) divisor with normal crossings in $\underline{X}$.*

Suppose that X is regular, that S is the spectrum of a DVR, that X/S is flat, and that π is a uniformizing parameter of X. Then, if x is a closed point of the closed fiber of X/S, the morphism has strict semistable reduction at x if and only if there is a regular sequence $(t_1, \ldots, t_n)$ generating the maximal ideal of $\mathcal{O}_{X,x}$ such that $\pi = t_1 \cdots t_r$ for some r.

Let X/S be a morphism with semistable reduction, and let η and s be the generic and closed points of S, respectively. For the sake of exposition, we assume the semistable reduction is strict and work in the Zariski topology. Thus the fiber Y over s is a union of regular irreducible components $Y_1, \ldots, Y_n$. The compactifying log structures associated to the open immersions $X_\eta \to \underline{X}$ and $\eta \to \underline{S}$ define log structures $\alpha_{X_\eta/X}$ and $\alpha_{\eta/S}$ on $\underline{X}$ and $\underline{S}$, respectively, and the morphism $\underline{f}$ underlies a morphism $f \colon X \to S$ of the corresponding log schemes. By Proposition 1.8.2, the compactifying log structures on X and S are induced from the DF structures corresponding to the family of maps $\mathcal{I}_{Y_i} \to \mathcal{O}_X$ and $\mathcal{I}_s \to \mathcal{O}_S$. Let x be a point of X, and restrict attention to a neighborhood on which each ideal sheaf $\mathcal{I}_{Y_i}$ is trivial. A choice of a generator t_i for each $\mathcal{I}_{Y_i}$ and π for $\mathcal{I}_s$ defines charts $\mathbf{N}^n \to \mathcal{M}_X$ and $\mathbf{N} \to \mathcal{M}_S$. Necessarily $f^*(\pi)$ is a unit times $t_1 t_2 \cdots t_r$, and we may adjust the choices so that the unit is 1. Thus we see that there is a chart for f subordinate to the diagonal homomorphism $\mathbf{N} \to \mathbf{N}^r$. One finds a commutative diagram

$$\begin{array}{ccccc} X & \xrightarrow{g} & S \times_{\mathsf{A}_{\mathbf{N}}} \mathsf{A}_{\mathbf{N}^r} & \longrightarrow & \mathsf{A}_{\mathbf{N}^r} \\ & \searrow & \downarrow & & \downarrow \\ & & S & \longrightarrow & \mathsf{A}_{\mathbf{N}}. \end{array}$$

The horizontal arrows in this diagram are strict, and one sees easily that the underlying morphism of g is flat, and that its fiber over the closed point of S is the closed subscheme of X defined by $(t_1, \ldots, t_r)$, which is regular. Thus, if X/S is locally of finite type, the map $\underline{g}$ is smooth. As we shall see in Chapter IV, and in particular in Corollary IV.3.1.18, the morphism $X \to S$ is an example of a smooth morphism of log schemes.

Let us summarize our conclusions for future reference.

Proposition 1.8.4. *Let $\underline{X}/\underline{S}$ be a semistable reduction morphism over the spectrum of a DVR. Then the étale log structures $\alpha_{X_\eta/X}$ and $\alpha_{\eta/S}$ are fine. Let X/S be the corresponding morphism of log schemes and let x be a point of X. Then, after restricting to some neighborhood of x in X, there exists a chart for X/S subordinate to the diagonal homomorphism $\mathbf{N} \to \mathbf{N}^r$. If X/S is of finite type, then the underlying morphism of the map $X \to S \times_{\mathsf{A}_{\mathbf{N}}} \mathsf{A}_{\mathbf{N}^r}$ is smooth.* □

The log structures associated to divisors with a normal crossings divisor Y in

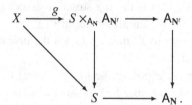

X can be restricted to Y; this restricted log structure "remembers" certain key aspects of the embedding. More generally, let $i\colon Y \to X$ be a closed immersion, let $\alpha\colon \mathcal{M} \to \mathcal{O}_X$ be the compactifying log structure associated to the complementary open embedding $U \to X$, and let $i^*_{log}(\alpha)$ be the restriction of α to Y. In some cases one can say explicitly, in terms of familiar data on Y, what this log structure is, and when such structures exist, without a priori knowledge of X. For example, as we saw in the discussion preceding Proposition 1.7.5, if Y is an irreducible Cartier divisor in a normal X, then $\mathcal{M}$ is the log structure associated to the DF structure $\mathcal{I}_Y \to \mathcal{O}_X$, and $i^*_{log}(\alpha)$ is just the data of the conormal sheaf $\mathcal{I}_Y/\mathcal{I}_Y^2$: it is the DF structure associated to the zero map $\mathcal{I}_Y/\mathcal{I}_Y^2 \to \mathcal{O}_Y$.

More generally, suppose that X is normal and that each irreducible component Y_i of Y is a reduced Cartier divisor (and is unibranch, if we are working in the étale topology). Then, as we saw in Proposition 1.7.3, the compactifying log structure $\mathcal{M}$ is the log structure attached to the DF structure defined by the family of ideals $\mathcal{I}_{Y_i} \to \mathcal{O}_X$, and $\tilde{\zeta}$ induces an isomorphism

$$\overline{\mathcal{M}} \cong \epsilon_*(\mathbf{N}_{\tilde{Y}}), \tag{1.8.1}$$

where $\epsilon\colon \tilde{Y} \to Y$ is the normalization of Y. Note that the sheaves in this isomorphism vanish outside Y, and that in fact the same isomorphism holds if we replace α by $i^*_Y(\alpha)$. This is the log structure associated to the restriction of $\{\mathcal{L}_i \to \mathcal{O}_X\}$ to Y, i.e., the family of maps: $i^*(\mathcal{L}_{Y_i}) \to \mathcal{O}_Y$. In particular, if $q_i \in \Gamma(X, \overline{\mathcal{M}})$ corresponds to Y_i, then $i^*_Y(\mathcal{L}_{Y_i})$ is the invertible sheaf $\mathcal{L}_{q_i}$ in the notation of Remark 1.1.7.

Let us now attempt to explicate the log structures that arise naturally on a strict normal crossing scheme. If Y is such a scheme, let $\epsilon\colon \tilde{Y} \to Y$ be its normalization. Then $\tilde{Y}$ is the disjoint union of the irreducible components $Y_1, \ldots, Y_n$ of Y, each of which is regular. For each j, let

$$\begin{aligned} Y^j &:= \cup\{Y_k : k \neq j\} = \overline{Y \setminus Y_j}, \\ D_j &:= Y_j \cap Y^j, \\ D &:= D_1 \cup \cdots \cup D_n \\ &= Y^1 \cap \cdots \cap Y^n \\ &= \cup\{Y_j \cap Y_k : j < k\} \\ &= Sing(Y). \end{aligned}$$

The subscheme $D \subseteq Y$ is often called the *double locus* of Y.

Lemma 1.8.5. *Let Y be a strict normal crossing scheme, with normalization $\tilde{Y} = Y_1 \sqcup \cdots \sqcup Y_n$.*

1. *D_j is a Cartier divisor in Y_j and in Y^j.*

2. The ideal $\mathcal{I}_{Y_j/Y}$ of Y_j in Y is annihilated by the ideal $\mathcal{I}_{Y^j/Y}$ of Y^j in Y, and in fact $\mathcal{I}_{Y_j/Y} \cong i_{Y^j/Y*}(\mathcal{I}_{D_j/Y^j})$, and $\mathcal{I}_{D_j/Y^j}$ is an invertible sheaf on Y^j.
3. The ideal $\mathcal{I}_{Y^j/Y}$ of Y^j in Y is annihilated by the ideal $\mathcal{I}_{Y_j/Y}$ of Y_j in Y, and in fact $\mathcal{I}_{Y^j/Y} \cong i_{Y_j/Y*}(\mathcal{I}_{D_j/Y_j})$, and $\mathcal{I}_{D_j/Y_j}$ is an invertible sheaf on Y_j.
4. Each $i^*_{D/Y}(\mathcal{I}_{Y_j/Y})$ is an invertible sheaf of $\mathcal{O}_D$-modules, as is

$$\begin{aligned}\mathcal{O}_D(-Y) &:= i^*_{D/Y}(\mathcal{I}_{Y_1/Y}) \otimes \cdots \otimes i^*_{D/Y}(\mathcal{I}_{Y_n/Y})\\ &\cong i^*_{D/Y_1}(\mathcal{I}_{D_1/Y^1}) \otimes \cdots \otimes i^*_{D/Y_n}(\mathcal{I}_{D_n/Y^n}).\end{aligned}$$

   For every j, we have

$$\mathcal{O}_D(-Y) \cong i^*_{D/Y}(\mathcal{I}_{Y^j/Y}) \otimes i^*_{D/Y}(\mathcal{I}_{Y_j/Y}).$$

5. If $i \colon Y \to X$ is a DNC coordinate system on Y and $\mathcal{I}_Y$ is the ideal of Y in X, then there is a natural isomorphism

$$i^*_{D/Y}(i^*(\mathcal{I}_Y)) \cong \mathcal{O}_D(-Y).$$

Proof All these statements can be proved under the assumption that there exists a DNC coordinate system $i \colon Y \to X$. We work in the local ring of a point y of Y and with a regular sequence $(t_1, \ldots, t_n)$ as in Definition 1.8.1. Then t_j generates the ideal of Y_j in X and $s_j := t_1 \cdots \hat{t}_j \cdots t_r$ generates the ideal of Y^j in X. The ideal $\mathcal{I}_{Y_j/X}$ of Y_j in X is an invertible sheaf of ideals; note that it becomes the unit ideal in $\mathcal{O}_{X,y}$ if $j > r$. The ideal of D_j in Y_j is generated by the image of s_j in $\mathcal{O}_X/(t_j)$ and the ideal of D_j in Y^j is generated by the image of t_j in $\mathcal{O}_X/(s_j)$. It follows from the fact that $(t_1, \ldots, t_r)$ is a regular sequence that (t_j, s_j) and (s_j, t_j) are regular sequences, and hence t_j is a nonzero divisor mod (s_j) and s_j is a nonzero divisor mod (t_j). This proves (1). Writing $\bar{t}$ for the image in $\mathcal{O}_{Y,y}$ of an element t of $\mathcal{O}_{X,x}$, we see that $\bar{t}_j\bar{s}_j = 0$. This proves (2) and (3). For (4) and (5), note that the inclusion $i_{D/Y} \colon D \to Y$ factors through Y^j and, by (2), $i^*_{D/Y}(\mathcal{I}_{Y_j/Y}) \cong i^*_{D/Y^j} i^*_{Y^j/Y}(\mathcal{I}_{Y_j/Y}) \cong i^*_{D/Y_j}(\mathcal{I}_{D_j/Y^j})$ and hence is invertible on D. For each j, we have a natural surjection $i^*_{Y/X}(\mathcal{I}_{Y_j/X}) \to \mathcal{I}_{Y_j/Y}$ and hence also a surjection $i^*_{D/Y} i^*_{Y/X}(\mathcal{I}_{Y_j/X}) \to i^*_{D/Y}(\mathcal{I}_{Y_j/Y})$. Since each of these is an invertible sheaf of $\mathcal{O}_D$-modules, the latter surjection is an isomorphism. Since $\mathcal{I}_{Y/X} = \mathcal{I}_{Y_1/X} \otimes \cdots \otimes \mathcal{I}_{Y_n/X}$, we find that

$$\begin{aligned}i^*_{D/Y} i^*_{Y/X}(\mathcal{I}_{Y/X}) &= i^*_{D/Y} i^*_{Y/X}(\mathcal{I}_{Y_1/X}) \otimes \cdots \otimes i^*_{D/Y} i^*_{Y/X}(\mathcal{I}_{Y_n/X})\\ &= i^*_{D/Y}(\mathcal{I}_{Y_1/Y}) \otimes \cdots \otimes i^*_{D/Y}(\mathcal{I}_{Y_n/Y})\\ &= \mathcal{O}_D(-Y).\end{aligned}$$

Furthermore, the ideal of Y^j in X is defined by the tensor product of all $\mathcal{I}_{Y_i/X}$ such that $i \neq j$, and hence $\mathcal{O}_D(-Y) \cong i^*_{D/Y}(\mathcal{I}_{Y^j/Y}) \otimes i^*_{D/Y}(\mathcal{I}_{Y_j/Y})$. □

Theorem 1.8.6. *Let Y be a strict normal crossings scheme, with irreducible components $Y_1, \ldots, Y_n$ and normalization $\epsilon\colon \tilde{Y} \to Y$. Then the following sets of data are naturally equivalent:*

1. *a log structure $\alpha\colon \mathcal{M} \to \mathcal{O}_Y$ such that the map $\zeta\colon \overline{\mathcal{M}} \to \mathcal{Z}_Y$ induced by α in turn induces an isomorphism $\overline{\mathcal{M}} \to \epsilon_*(\mathbf{N}_{\tilde{Y}})$;*
2. *for each j, an invertible sheaf of $\mathcal{O}_Y$-modules $\mathcal{L}_j$ and an isomorphism of $\mathcal{O}_{Y^j}$-modules*
$$\alpha_j\colon i^*_{Y^j/Y}(\mathcal{L}_j) \cong \mathcal{I}_{D_j/Y^j};$$
3. *for each j, an invertible sheaf of $\mathcal{O}_{Y_j}$-modules $\tilde{\mathcal{L}}_j$ and an isomorphism of $\mathcal{O}_{D_j}$-modules*
$$\beta_j\colon i^*_{D_j/Y_j}(\tilde{\mathcal{L}}_j) \cong i^*_{D_j/Y^j}(\mathcal{I}_{D_j/Y^j});$$
4. *an invertible sheaf of $\mathcal{O}_Y$-modules $\mathcal{L}$ and an isomorphism of $\mathcal{O}_D$-modules*
$$\beta\colon i^*_{D/Y}(\mathcal{L}) \cong \mathcal{O}_D(-Y).$$

If $i\colon Y \to X$ is a DNC embedding, then these data arise as follows: α is the restriction to Y of the compactifying log structure associated to $U \subseteq X$; $\mathcal{L}_j$ is the restriction $i^*_{Y/X}(\mathcal{I}_{Y_j/X})$ to Y of the ideal sheaf of Y_j in X; $\tilde{\mathcal{L}}_j$ is the restriction $i^*_{Y_j/X}(\mathcal{I}_{Y_j/X})$ to Y_j of the ideal sheaf of Y^j in X to Y_j; and $\mathcal{L}$ is the conormal sheaf of Y in X.

Proof For each j, the data in (1) give us a global section of $\overline{\mathcal{M}}$ defined by Y_j, whose inverse image in $\mathcal{M}$ is an $\mathcal{O}^*_Y$-torsor. Let $\mathcal{L}_j$ be the invertible sheaf corresponding to this global section, as explained in Remark 1.1.7. Then α maps $\mathcal{L}_j$ surjectively to the ideal $\mathcal{I}_j \subseteq \mathcal{O}_Y$. The restriction of this map to Y^j defines a surjection

$$\alpha_j\colon i^*_{Y^j/Y}(\mathcal{L}_j) \to i^*_{Y^j/Y}(\mathcal{I}_j) = \mathcal{I}_{D_j/Y_j}.$$

Since $\mathcal{I}_{D_j/Y_j}$ is invertible on Y^j, α_j is an isomorphism. This gives us the data in (2). On the other hand, thanks to statement (2) of Lemma 1.8.5, the data (2) yield homomorphisms

$$\mathcal{L}_j \to i_{Y^j/Y_*} i^*_{Y^j/Y}(\mathcal{L}_j) \cong i_{Y_j/Y*}(\mathcal{I}_{D_j/Y^j}) \cong \mathcal{I}_{Y_j/Y} \to \mathcal{O}_Y,$$

and thus define a DF structure $\gamma.$ on Y. Let $\mathcal{M}$ be the sheaf of monoids of the associated log structure and let $\zeta\colon \overline{\mathcal{M}} \to \mathcal{Z}_Y$ be the associated map. Since the image of γ_i is precisely $\mathcal{I}_{Y_i/Y}$, the map ζ sends $q_i \in \overline{\mathcal{M}}$ to $Y_i \in \mathcal{Z}_Y$. Thus the data in (1) and (2) are naturally equivalent. Furthermore, the restriction of α_j to $D \subseteq Y^j$ defines an isomorphism $i^*_{D/Y}(\mathcal{L}_j) \cong i^*_{D/Y^j} i^*_{Y^j/Y}(\mathcal{L}_j) \cong i^*_{D/Y^j}(\mathcal{I}_{D_j/Y^j})$.

Thus if $\mathcal{L} := \mathcal{L}_1 \otimes \cdots \otimes \mathcal{L}_n$, the tensor product of these isomorphisms yields an isomorphism

$$\begin{aligned}\beta\colon i^*_{D/Y}(\mathcal{L}) &\cong i^*_{D/Y^1}(\mathcal{I}_{D_1/Y^1}) \otimes \cdots \otimes i^*_{D/Y^n}(\mathcal{I}_{D_n/Y^n}) \\ &\cong i^*_{D/Y}(\mathcal{I}_1) \otimes \cdots \otimes i^*_{D/Y}(\mathcal{I}_n) \\ &\cong \mathcal{O}_D(-Y),\end{aligned}$$

and thus the data (4).

By (3) of Lemma 1.8.5 the ideal $\mathcal{I}_{D_j/Y_j}$ is an invertible sheaf of $\mathcal{O}_{Y_j}$-modules. Thus, given the data in (4), we can set

$$\tilde{\mathcal{L}}_j := \mathcal{I}^{-1}_{D_j/Y_j} \otimes i^*_{Y_j/Y}(\mathcal{L}),$$

so that we have isomorphisms:

$$\begin{aligned}i^*_{D_j/Y_j}(\tilde{\mathcal{L}}_j) &\cong i^*_{D_j/Y_j}(\mathcal{I}_{D_j/Y_j})^{-1} \otimes i^*_{D_j/Y_j} i^*_{Y_j/Y}(\mathcal{L}) \\ &\cong i^*_{D_j/Y_j}(\mathcal{I}_{D_j/Y_j})^{-1} \otimes i^*_{D_j/D} i^*_{D/Y}(\mathcal{L}) \\ &\cong i^*_{D_j/Y_j}(\mathcal{I}_{D_j/Y_j})^{-1} \otimes i^*_{D_j/D}(\mathcal{O}_D(-Y)) \\ &\cong i^*_{D_j/Y_j}(\mathcal{I}_{D_j/Y_j})^{-1} \otimes i^*_{D_j/Y_j}(\mathcal{I}_{D_j/Y_j}) \otimes i^*_{D_j/Y^j}(\mathcal{I}_{D_j/Y^j}). \\ &\cong i^*_{D_j/Y^j}(\mathcal{I}_{D_j/Y^j}).\end{aligned}$$

This gives the data of (3).

Finally we show how to pass from (3) to (2). Note first that for each j there are exact sequences:

$$0 \longrightarrow \mathcal{O}_Y \longrightarrow \mathcal{O}_{Y_j} \oplus \mathcal{O}_{Y^j} \longrightarrow \mathcal{O}_{D_j} \longrightarrow 0,$$

$$1 \longrightarrow \mathcal{O}^*_Y \longrightarrow \mathcal{O}^*_{Y_j} \oplus \mathcal{O}^*_{Y^j} \longrightarrow \mathcal{O}^*_{D_j} \longrightarrow 1,$$

$$\to H^0(\mathcal{O}^*_{D_j}) \to \mathrm{Pic}(Y) \to \mathrm{Pic}(Y_j) \oplus \mathrm{Pic}(Y^j) \to \mathrm{Pic}(D_j).$$

We make the construction suggested by the last of these explicit as follows. Given $(\tilde{\mathcal{L}}_j, \beta_j)$ as in (3), let $\mathcal{L}_j$ be the sheaf of pairs $(\tilde{\ell}, x) \in \tilde{\mathcal{L}}_j \times \mathcal{I}_{D_j/Y^j}$ such that $\beta_j(i^*_{D_j/Y_j}(\tilde{\ell})) = i^*_{D_j/Y^j}(x)$. Then $\mathcal{L}_j$ is an invertible sheaf of $\mathcal{O}_Y$-modules, and there is an induced isomorphism $i^*_{Y^j/Y}(\mathcal{L}_j) \cong \mathcal{I}_{D_j/Y^j}$. □

Remark 1.8.7. The assertion of the theorem that the sets of data (1)–(4) are "naturally equivalent" means that the categories (which we view as groupoids) in question are equivalent. We hope that the reader can imagine the definitions of the morphisms in these categories, which we have not made explicit. For example, a morphism in (2) means a collection of isomorphisms $\theta_j\colon \mathcal{L}_j \to \mathcal{L}'_j$ that are compatible with the isomorphisms α_j and α'_j, and a morphism in (4) means an isomorphism $\theta\colon \mathcal{L} \to \mathcal{L}'$ that is compatible with the isomorphisms

β and β'. We have not verified the compatibilities of these equivalences, a task we leave to the reader.

More generally, a scheme Y is said to be a *normal crossings scheme* if it admits an étale covering by a strict normal crossings scheme. A nodal curve is an example of a (non-strict) normal crossings scheme. For such a Y one can still consider the sheaf $\epsilon_*(\mathbf{N}_{\tilde{Y}})$ in the étale topology, the sheaf of *branches of* Y. Then if $\overline{y} \to \tilde{Y}$ is a geometric point, the strict Henselization of Y at $\overline{y}$ is a union of irreducible components as in Lemma 1.8.5, naturally indexed by the germs of the sheaf of branches at $\overline{y}$. This sheaf is no longer generated by its global sections, but the tensor product defining $O_D(-Y)$ is independent of the ordering of the branches, and hence this construction defines an invertible sheaf on D. In Theorem 1.8.6, statements (1) and (4) make sense in the étale topology, although (2) and (3) have no global meaning. Furthermore, the equivalence of the data of (1) and (4) is still valid: morphisms between the objects form an étale sheaf, and objects can be glued. (That is, the categories in questions form *stacks*.) Although the passage from (1) to (4) passes through (2) and (3), the result is independent of the ordering of the branches.

Now suppose that $f \colon X \to S$ is a morphism of log schemes coming from a semistable reduction situation. This morphism restricts to a morphism of log schemes $Y \to s$, where Y is the fiber over the closed point s of S. Moreover, the map $\overline{f}^{\flat} : \mathbf{N} \to \overline{M}_Y \cong \eta_*(\mathbf{N}_{\tilde{Y}})$ sends 1 to $\sum 1_{Y_i}$, where the Y_i are the branches of Y. The following proposition makes explicit the extra information provided by the log structures.

Proposition 1.8.8. *Let k be a field, let Y be a normal crossing scheme over k. and let $s_{\mathbf{N}}$:= Spec($\mathbf{N} \to k$), the standard log point over k. Then the following sets of data are equivalent:*

1. *a pair (α, f), where α is a log structure on Y satisfying condition (1) of Theorem 1.8.6 and f is a morphism of log schemes $f \colon Y \to s_{\mathbf{N}}$ whose underlying morphism $\underline{f}$ is the given k-scheme structure of Y and such that $\overline{f}^{\flat}$ sends 1 to $\sum 1_{Y_i}$;*
2. *an isomorphism $O_D \to O_D(-Y)$.*

Proof Suppose we are given a pair (α, f). By Theorem 1.8.6, we find an invertible sheaf $\mathcal{L}$ on Y and an isomorphism $i^*_{D/Y}(\mathcal{L}) \cong O_D(-Y)$ depending only on α. It follows from the construction that in fact $\mathcal{L} = \mathcal{L}_y$. Since by assumption $f^{\flat}(1) = y$, the morphism $f^{\flat}$ maps $\mathcal{L}^*_1$ to $\mathcal{L}^*_y$, and hence induces a trivialization of $\mathcal{L}$ and hence also of $O_D(-Y)$. Conversely, given a trivialization of $O_D(-Y)$, we let $\mathcal{L} := \mathcal{O}_Y$, and endow Y with the corresponding log structure. Then the global section 1 of $\mathcal{O}_Y = \mathcal{L} = \mathcal{L}_1 \otimes \cdots \otimes \mathcal{L}_n$ induces the morphism f. □

Remark 1.8.9. The origins of Proposition 1.8.8 trace back to the work of Friedman [20], who showed that the triviality of $\mathcal{O}_D(-Y)$ is a necessary condition for a DNC scheme Y to be the special fiber of a semistable degeneration. In his widely adopted terminology, a normal crossing scheme Y such that $\mathcal{O}_D(-Y)$ is said to be *d-semistable* if $\mathcal{O}_D(-Y)$ is trivial.

1.9 Coherence of compactifying log structures

In this section we take a closer look at the compactifying log structure associated with an open immersion $j\colon U \to X$, and, particular, we investigate when and in what senses it is coherent.

Proposition 1.9.1. *Let U be a dense open subset of a normal and locally noetherian scheme X and let $Y := X \setminus U$ be a closed subset purely of codimension one.*

1. *The stalks of $\overline{\mathcal{M}}_{U/X}$ are finitely generated monoids and are free if X is locally factorial.*
2. *If X is locally factorial and we are working in the Zariski topology, $\mathcal{M}_{U/X}$ is fine. The same is true in the étale topology provided X has an étale covering $f\colon X' \to X$ such that each irreducible component of $f^{-1}(Y)$ is unibranch.*

Proof Recall that from Proposition 1.6.3 we have

$$\overline{\mathcal{M}}_{U/X,x} \cong \underline{\Gamma}_Y(\mathcal{D}iv_X^+) \subseteq \underline{\Gamma}_Y(\mathcal{W}_X^+).$$

Since the inclusion is an equality if X is locally factorial, the first statement of the proposition will follow from the following lemma. The same argument works in the étale setting if one works in a strict Henselization of $\mathcal{O}_{X,x}$.

Lemma 1.9.2. *With the hypotheses of Proposition 1.9.1, the stalks of the sheaf $\underline{\Gamma}_Y(\mathcal{D}iv_X^+)$ (resp. of $\underline{\Gamma}_Y(\mathcal{W}_X^+)$) are fine (resp. fine and free) monoids.*

Proof According to Lemma 1.6.6, $\mathcal{D}iv_{X,x}^+$ is an exact submonoid of $\mathcal{W}_{X,x}^+$, and it follows that the same is true if we restrict to sections with supports in Y. But $\underline{\Gamma}_Y(\mathcal{W}_X^+)_x$ is just the free monoid on the set of prime ideals of height one in the local ring $\mathcal{O}_{X,x}$ that are contained in Y. Since Y is a proper closed subset of X, each of these is a minimal prime of the noetherian local ring $\mathcal{O}_{Y,x}$, and hence there only finitely many such primes. Thus $\underline{\Gamma}_{Y,x}(\mathcal{W}_X^+)$ is a fine monoid, and by (2) of Theorem I.2.1.17, the same is true of its exact submonoid $\underline{\Gamma}_Y(\mathcal{D}iv_X^+)_x$. □

Now suppose that X is locally factorial and that we are working in the Zariski topology. Each irreducible component Y_i of Y is defined by an invertible sheaf of ideals $\mathcal{I}_i$, and the family of inclusions $\mathcal{I}_i \to \mathcal{O}_X$ is a DF structure on X. Let $\mathcal{M}$ be the associated log structure. As we saw in Proposition 1.7.3, this log structure is fine and the natural map $\mathcal{M} \to \mathcal{M}_{U/X}$ is an isomorphism. Statement (2), in the case of the Zariski topology, follows. In the étale topology, one must first pass to an étale cover $\eta\colon \tilde{X} \to X$ such that the irreducible components of $\tilde{Y} := \eta^{-1}(Y)$ are unibranch. Then we can apply the same construction, thanks to Proposition 1.6.5. □

Remark 1.9.3. To see that the normality hypothesis in Lemma 1.9.2 is not superfluous, let X be the spectrum of the subring R of $\mathbf{C}[t]$ consisting of those polynomials f such that $f'(0) = 0$. This is a curve with a cusp at the origin x. Let $Y := \{x\}$, and for any complex number a, let D_a be the class of $t^2 - at^3$ in $Div^+_{X,x} = \mathcal{O}'_{X,x}/\mathcal{O}^*_{X,x}$. In the fraction field of X, we have

$$(t^2 - at^3)(t^2 - bt^3)^{-1} = (1 - at)(1 - bt)^{-1} = 1 + (b - a)t + \cdots,$$

which does not belong to $\mathcal{O}^*_{X,x}$ if $a \neq b$. Thus $D_a \neq D_b$ in $\underline{\Gamma}_{Y,x}(Div^+_X)$. It follows that $\underline{\Gamma}_{Y,x}(Div^+_X)$ is uncountable and hence is not finitely generated. Similar examples can be constructed with nodal curves.

The following result computes the compactifying log structure corresponding to a special affine open subset U of an affine toric variety $\underline{\mathsf{A}}_P$, when U is defined by element p of the defining monoid P. An important special case, Corollary 1.9.5, is obtained by taking p to be an element of P not contained in any proper face, and is a theorem of Kato [49, 11.6].

Theorem 1.9.4. *Let R be a normal integral domain, let A_P denote the log scheme $\operatorname{Spec}(P \to R[P])$, and let $\eta\colon X \to \mathsf{A}_P$ be a strict morphism of log schemes whose underlying morphism $\underline{\eta}$ of schemes is étale. Let p be an element of P and let $U := \eta^{-1}(D(e^p))$, where $D(e^p) = \operatorname{Spec} R[P_p] \subseteq \operatorname{Spec} R[P]$. Finally, let F be the the face of P generated by p, and let $\mathcal{F}$ be the sheaf of faces of $\mathcal{M}_X$ generated by F. Then the natural map $F \to \Gamma(X, \mathcal{M}_X)$ induces an isomorphism*

$$\gamma\colon \mathcal{F} \to \mathcal{M}_{U/X} := j^{log}_*(\mathcal{O}^*_U).$$

Proof If f is any element of F, f maps to a unit in $k[P_p]$, and consequently f also maps to a unit in $\Gamma(U, \mathcal{O}_U)$. Since $\mathcal{M}_{U/X} \subseteq \mathcal{O}_X$ is a sheaf of faces in the multiplicative monoid $\mathcal{O}_X$, the map $F \to \mathcal{O}_X$ induces a homomorphism of sheaves of monoids $\gamma\colon \mathcal{F} \to \mathcal{M}_{U/X}$. The homomorphism γ is sharp and $\mathcal{M}_{U/X}$ is integral, so, by Proposition I.4.1.2, it will suffice to prove that $\overline{\gamma}$ is an isomorphism. Since P is torsion free, $\underline{X}$ is integral and hence, by Proposition 1.6.3, $\overline{\mathcal{M}}_{U/X} \cong \underline{\Gamma}_Y(Div^+_{\underline{X}})$, where $Y := X \setminus U$.

First suppose that η is the identity map. If x is a point of X, the stalk of $\overline{\mathcal{F}}_x$ at x is the face F_x of P/G_x generated by p, where G_x is the set of elements in P such that $e^p(x) \neq 0$. Theorem I.3.5.8 says that the the natural homomorphism $F_x \to \underline{\Gamma}_Y(Div^+_{\underline{X}})_x$ is an isomorphism, proving the theorem in this case. More generally, note that Proposition I.3.5.2 asserts that each irreducible component of the closed subscheme Z of $\underline{\mathsf{A}}_P$ defined by e^p is unibranch. Hence by Lemma 1.6.7 the map $\eta^*(\underline{\Gamma}_Z(Div^+_{\underline{\mathsf{A}}_P})) \to \underline{\Gamma}_Y(Div^+_{\underline{X}})$ is an isomorphism, and the result follows. □

Corollary 1.9.5. *With the notation of Theorem 1.9.4, there is a natural isomorphism: $\mathcal{M}_X \to \mathcal{M}_{X^*/X}$. In particular, the natural homomorphism $P \to \mathcal{M}_{X^*/X}$ is a chart for $\mathcal{M}_{X^*/X}$.*

1.10 Hollow and solid log structures

Recall from Section 1.5 that the spectrum of a valuation ring V has two particularly interesting log structures: the standard (compactifying) log structure, where $\alpha_t\colon \mathcal{M}_t = V' \to V$ is the inclusion, and the hollow log structure, where $\alpha_t\colon \mathcal{M}_t = V' \to V$ sends v to 0 if $v \in \mathfrak{m}_V$. The standard log structure is what we shall call "solid," a notion that we will define and study in this section. This notion is related both to exactness and to the logarithmic analog of regularity, which we discuss in Section 1.11.

Definition 1.10.1. *A log ring $\beta\colon Q \to A$ is:*

1. *hollow if $\beta(Q^+) \subseteq \{0\}$,*
2. *solid if* $\mathrm{Spec}(\beta)\colon \mathrm{Spec}(A) \to \mathrm{Spec}(Q)$ *is locally surjective (Remark I.4.2.14),*
3. *very solid if for every prime ideal $\mathfrak{q}$ of Q, the ideal $\mathfrak{q}A$ of A generated by $\beta(\mathfrak{q})$ is a prime ideal of A such that $\beta^{-1}(\mathfrak{q}A) = \mathfrak{q}$.*

A prelog structure $\alpha\colon Q \to \mathcal{O}_X$ is hollow (resp. solid, resp. very solid) if for every geometric point $\overline{x}$ of X, $\beta_{\overline{x}}\colon Q_{\overline{x}} \to \mathcal{O}_{X,\overline{x}}$ is hollow (resp. solid, resp. very solid).

If Q is a Zariski log structure, then can one work with ordinary points instead of geometric points. It is clear that a log ring $\beta\colon Q \to A$ is solid if and only if its localization $Q_{\beta^{-1}(A^*)} \to A$ is solid. One checks immediately that a prelog structure $\beta\colon Q \to \mathcal{O}_X$ is hollow (resp. solid) if and only if the associated log structure α is. It is almost as immediate to verify that the log structure associated to a very solid prelog structure is very solid.

Remark 1.10.2. Let $\beta\colon Q \to A$ be a log ring, let $\mathfrak{q}$ be a prime of Q, and let $F := Q\setminus\mathfrak{q}$. Then the fiber of $\mathrm{Spec}(\beta)$ over $\mathfrak{q}$ is $\mathrm{Spec}(A/\mathfrak{q}A)_F$, which is nonempty if and only if $\beta(F) \cap \mathfrak{q}A = \emptyset$, or equivalently, if and only if $\mathfrak{q} = \beta^{-1}(\mathfrak{q}A)$. Thus β is solid if and only if, for every prime ideal $\mathfrak{p}$ of A and every prime ideal $\mathfrak{q}$ of Q contained in $\beta^{-1}(\mathfrak{p})$, $\mathfrak{q} = \beta_{\mathfrak{p}}^{-1}(\mathfrak{q}A_{\mathfrak{p}})$. In particular, β is solid if it is very solid. Indeed, suppose that $\mathfrak{p} \in \mathrm{Spec}(A)$ and $\mathfrak{q} \in \mathrm{Spec}(Q)$, with $\mathfrak{q} \subseteq \beta^{-1}(\mathfrak{p})$. If β is very solid, $\mathfrak{q}A$ is a prime ideal of A and $\beta^{-1}(\mathfrak{q}A) = \mathfrak{q}$. Since $\mathfrak{q}A \subseteq \mathfrak{p}$, $\mathfrak{q}A_{\mathfrak{p}}$ is a prime ideal of $A_{\mathfrak{p}}$, and $\beta_{\mathfrak{p}}^{-1}(\mathfrak{q}A_{\mathfrak{p}}) = \mathfrak{q}$. On the other hand, a solid log ring need not be very solid: the local log ring

$$\mathbf{N} \to \mathbf{C}[[x, y]]\colon n \mapsto x^n y^n$$

is solid but not very solid.

If $\beta\colon Q \to A$ is a solid local log ring, then Q^+ is in the image of $\mathrm{Spec}(\beta)$ since β is local, and since $\mathrm{Spec}(\beta)$ is locally surjective, it must also be surjective. The converse is not true: the local log ring

$$\beta\colon \mathbf{N} \oplus \mathbf{N} \to k[[x, y, z, w]]/(xy - zw) : (m, n) \mapsto x^m z^n$$

is not solid, although $\mathrm{Spec}(\beta)$ is surjective.

If $\beta\colon Q \to A$ is a hollow log ring, then $\beta^{-1}(\mathfrak{p}) = Q^+$ for every $\mathfrak{p} \in \mathrm{Spec}(A)$, so the map $\mathrm{Spec}(\beta)\colon \mathrm{Spec}(A) \to \mathrm{Spec}(Q)$ is constant. Thus if β is solid, local, and hollow, then $\mathrm{Spec}(Q) = \{Q^+\}$, and hence $Q^+ = \emptyset$ and Q is dull. In particular, the log structure associated to a solid and hollow local log ring is trivial.

Proposition 1.10.3. *Let $\beta\colon Q \to A$ be a local log ring, let $\theta\colon A \to A'$ be a local homomorphism of local rings, and let $\beta' := \theta \circ \beta$.*

1. *If β is hollow, then the same is true of β', and the converse holds if θ is injective (for example, if θ is flat).*
2. *If θ is flat and β' is very solid, then β is also very solid.*
3. *If θ is flat, then β is solid if and only if β' is solid. In particular, if A is noetherian and $\theta\colon A \to \hat{A}$ is its completion, then β is solid if and only if β' is solid, and if β' is very solid, then so is β.*

Proof The first statement is clear. If θ is flat, it is faithfully flat, since θ is a local homomorphism of local rings. Suppose that β' is very solid and that $\mathfrak{q}$ is a prime ideal of Q. Then the induced homomorphism $A/\mathfrak{q}A \to A'/\mathfrak{q}A'$ is faithfully flat, and since $A'/\mathfrak{q}A'$ is an integral domain, so is $A/\mathfrak{q}A$. Thus $\mathfrak{q}A$ is prime. Since $\beta'^{-1}(\mathfrak{q}A') = \mathfrak{q}$, it follows that $\beta^{-1}(\mathfrak{q}A) = \mathfrak{q}$. Thus β is also very solid. Since θ is faithfully flat, $\mathrm{Spec}(\theta)$ is locally surjective and surjective, and it follows that β is solid if and only if β' is solid. □

Proposition 1.10.4. *Let X be a fine log scheme whose underlying scheme $\underline{X}$ is reduced. Then the following conditions are equivalent*

1. *The log structure on X is hollow.*
2. *The sheaf of monoids $\overline{\mathcal{M}}_X$ on $X_{\acute{e}t}$ is locally constant.*
3. *The sheaf of abelian groups $\overline{\mathcal{M}}_X^{gp}$ on $X_{\acute{e}t}$ is locally constant.*
4. *The function $X \to \mathbf{N}$ sending $x \in X$ to $\mathrm{rk}(\overline{\mathcal{M}}^{gp}_{X,\overline{x}})$ is locally constant.*

Proof Suppose (1) holds. Since we are working locally on $X_{\acute{e}t}$, we may choose a point x of X, replace X by an étale neighborhood of x, and assume that there is a fine chart $\beta\colon P \to \mathcal{O}_X$ which is local at x. Since α_X is hollow, it follows that $\beta_x(p) = 0$ for every $p \in P^+$, i.e., that each $\beta(p)$ vanishes in some neighborhood of x. Since P^+ is finitely generated as an ideal of P, we may replace X by a neighborhood of x so that $\beta(P^+) = 0$. Then, for every $x' \in X$, $\beta^{-1}(\mathcal{O}^*_{X,x'}) = P^*$, and hence β induces an isomorphism $\overline{P} \to \overline{\mathcal{M}}_{X,x'}$. Thus $\overline{\mathcal{M}}_X$ is constant, proving that (2) holds. It is obvious that (2) implies (3) and that (3) implies (4). Suppose that (4) holds. Again we may assume that α_X admits a chart $\beta\colon P \to \mathcal{M}_X$ and that x is a point of X such that β_x is local. Let ξ be a point of $\mathcal{O}_{X,x}$ corresponding to a generization of x in X, which we denote by the same letter. Then $\overline{\mathcal{M}}_{X,\xi} \cong P/F_\xi$, where $F_\xi := \beta_x^{-1}(\mathcal{O}^*_{X,\xi})$. Then (4) implies that $\mathrm{rk}(P/F_\xi)^{gp} = \mathrm{rk}(\overline{P}^{gp})$ and, as we saw in Proposition I.2.3.5, this implies that $F_\xi = P^*$. It follows that β maps P^+ to the maximal ideal of $\mathcal{O}_{X,\xi}$, i.e., that $\beta(P^+)$ is contained in the prime ideal of $\mathcal{O}_{X,x}$ defined by ξ. Since this is true for every ξ, in fact $\beta(P^+)$ is contained in every prime ideal of $\mathcal{O}_{X,x}$ and hence in the nilradical of $\mathcal{O}_{X,x}$. When X is reduced, this implies that $\beta(P^+) = \{0\}$, so α_X is hollow. □

Corollary 1.10.5. *If X is a fine log scheme,*

$$X_i := \{x \in X : \mathrm{rk}(\overline{\mathcal{M}}^{gp}_{X,\overline{x}} = i\}$$

is a locally closed subset of X, and the restriction of α to its reduced subscheme $X_{i,red}$ is hollow.

Proof Recall from Corollary II.2.1.6 that each $X^{(n)} := \{x : \mathrm{rk}\,\overline{\mathcal{M}}^{gp}_{X,\overline{x}} \leq n\}$ is an open subset of X, and it follows that each X_i is locally closed. Thus $X_{i,red}$ has a natural structure of a fine log scheme, and the rank of $\overline{\mathcal{M}}^{gp}_X$ at each of its points is i. Since $\mathcal{M}_X$ is coherent, Theorem II.2.5.4 implies that the cospecialization maps $\overline{\mathcal{M}}^{gp}_{X,\overline{x}} \to \overline{\mathcal{M}}^{gp}_{X,\overline{x}}$ are surjective, hence bijective. Proposition 1.10.4 then implies that the log structure is constant. □

Corollary 1.10.6. *Let $\alpha\colon \mathcal{M}_X \to \mathcal{O}_X$ be a fine log structure on X. Then the following conditions are equivalent:*

1. *For every geometric point $\overline{x} \in X$ and every $m \in \mathcal{M}^+_{X,\overline{x}}$, $\alpha_{\overline{x}}(m)$ is nilpotent.*
2. *The sheaf of monoids $\overline{\mathcal{M}}_X$ is locally constant.*

Proof A sheaf of sets on the étale or Zariski topology of X is locally constant if and only if it is so on its reduced subscheme. Thus we may replace X by X_{red}, and the corollary follows from Proposition 1.10.4. □

Remark 1.10.7. It is not difficult to describe the structure of fine saturated and hollow log structures fairly explicitly, at least when $\mathcal{M}_X$ is constant. Given such a log structure, consider the exact sequence of abelian sheaves

$$0 \to \mathcal{O}^*_X \to \mathcal{M}^{gp}_X \to \overline{\mathcal{M}}^{gp}_X \to 0,$$

and the associated boundary homomorphism

$$H^0(X, \overline{\mathcal{M}}^{gp}_X) \to H^1(X, \mathcal{O}^*_X) = \mathrm{Pic}(X).$$

Let $P := H^0(X, \overline{\mathcal{M}}_X)$, a fine sharp monoid; note that P^{gp} is a finitely generated free abelian group because $\mathcal{M}_X$ is fine and saturated. Then the above homomorphism amounts to the data of a homomorphism $P \to \mathrm{Pic}(X)$. This homomorphism determines the log structure up to isomorphism. Indeed, let P be a saturated and sharp monoid and let $\theta\colon P \to \mathrm{Pic}(X)$ be a homomorphism of monoids. Since P^{gp} is free,

$$\mathrm{Hom}(P^{gp}, H^1(X, \mathcal{O}^*_X)) \cong H^1(X, \underline{\mathrm{Hom}}(P^{gp}, \mathcal{O}^*_X)) \cong \mathrm{Ext}^1(P^{gp}, \mathcal{O}^*_X).$$

The image of θ^{gp} in this Ext group is the isomorphism class of an extension of P^{gp} by $\mathcal{O}^*_X$. Choose a representative of this class; its pullback to P gives an exact sequence of monoids $0 \to \mathcal{O}^*_X \to \mathcal{M} \to P \to 0$. If m is a section of $\mathcal{M}$ over an open set U, its image in P is constant, and we let $\mathcal{M}^+(U)$ denote the set of all $m \in \mathcal{M}(U)$ whose image is not a unit. Then $\mathcal{M}^+$ forms a sheaf of ideals in $\mathcal{M}$, and in fact $\mathcal{M}^+(U)$ is the complement of $\mathcal{O}^*_X(U)$ for every U. The homomorphism $\mathcal{M} \to \mathcal{O}_X$ that is the inclusion on $\mathcal{O}^*_X$ and that sends $\mathcal{M}^+$ to 0 is the unique hollow log structure $\mathcal{M} \to \mathcal{O}_X$. It is clear that if $\theta\colon P \to \mathrm{Pic}(X)$ comes from $\mathcal{M}_X$ as above, then $\mathcal{M}$ is isomorphic to $\mathcal{M}_X$, compatibly with the log structures. Note, however, that our construction required choosing a representative extension from its extension class and it is thus not functorial.

Proposition 1.10.8. *A coherent log structure $\alpha\colon \mathcal{M}_X \to \mathcal{O}_X$ is solid if and only if, for every $\overline{x} \in X$, the map* $\mathrm{Spec}(\alpha_{\overline{x}})\colon \mathrm{Spec}(\mathcal{O}_{X,\overline{x}}) \to \mathrm{Spec}(\mathcal{M}_{X,\overline{x}})$ *is surjective.*

Proof Since the statement is local for the étale topology, we may assume assume that α admits a local chart and then work with Zariski log structures. Then, assuming that every $\mathrm{Spec}(\alpha_x)$ is surjective, we will show that in fact every $\mathrm{Spec}(\alpha_x)$ is locally surjective. This latter condition means that if $\xi \in$

$\operatorname{Spec}(\mathcal{O}_{X,x})$ and $\eta := \alpha_x^{-1}(\xi) \in \operatorname{Spec}(\mathcal{M}_{X,x})$, then the induced map $\operatorname{Spec}(\mathcal{O}_{X,\xi}) \to \operatorname{Spec}((\mathcal{M}_{X,x})_\eta)$ is surjective. The point ξ corresponds to a point of X to which we assign the same letter. Then we have a commutative diagram

$$\begin{array}{ccc} \mathcal{O}_{X,x} & \longrightarrow & \mathcal{O}_{X,\xi} \\ \alpha_x \uparrow & & \uparrow \alpha_\xi \\ \mathcal{M}_{X,x} & \longrightarrow & \mathcal{M}_{X,\xi}. \end{array}$$

The top arrow in this diagram can be identified with the localization of $\mathcal{O}_{X,x}$ at the prime ideal $\mathfrak{p}_\xi$ of of $\mathcal{O}_{X,x}$ corresponding to ξ. Because $\mathcal{M}_X$ is coherent, the bottom arrow can be identified with the localization of $\mathcal{M}_{X,x}$ at the prime ideal $\alpha_x^{-1}(\mathfrak{p}_\xi)$. By hypothesis $\operatorname{Spec}(\alpha_\xi)$ is surjective, proving the claim. The converse is obvious. □

Proposition 1.10.9. *Let $\beta\colon Q \to A$ be a solid (resp. very solid) log ring, let $\mathfrak{q}$ be a prime an ideal of $\mathfrak{q}$ and let F be its complementary face. Assume that Q is fine. Then $\beta_F\colon Q_F \to A_F$ and $\beta_\mathfrak{q}\colon F \to A/\mathfrak{q}A$ are also solid (resp. very solid).*

Proof Recall that $\operatorname{Spec}(Q_F) \to \operatorname{Spec}(Q)$ is an open immersion whose image is the set of primes $\mathfrak{p}$ of Q such that $\mathfrak{p} \cap F = \emptyset$, and that

$$\operatorname{Spec}(F) \cong \operatorname{Spec}(Q, \mathfrak{q}) \subseteq \operatorname{Spec}(Q)$$

is a closed subset, the set of primes $\mathfrak{p}$ containing $\mathfrak{q}$. Furthermore, the following diagrams are cartesian in the category of topological spaces:

$$\begin{array}{ccc} \operatorname{Spec}(A_F) & \longrightarrow & \operatorname{Spec}(A) \\ \downarrow & & \downarrow \\ \operatorname{Spec}(Q_F) & \longrightarrow & \operatorname{Spec}(Q) \end{array} \qquad \begin{array}{ccc} \operatorname{Spec}(A/\mathfrak{q}A) & \longrightarrow & \operatorname{Spec}(A) \\ \downarrow & & \downarrow \\ \operatorname{Spec}(F) & \longrightarrow & \operatorname{Spec}(Q). \end{array}$$

Thus if β is solid, the following lemma shows that β_F and $\beta_\mathfrak{q}$ are also solid.

Lemma 1.10.10. *Let $f\colon X \to Y$ and $g\colon Y' \to Y$ be continuous maps of sober topological spaces. If f is locally surjective, so is the induced map f' from $X' := X \times_Y Y'$ to Y'.*

Proof Let x' be a point of X', let x (resp. y', resp. y) be its image in X (resp. Y', resp. Y). Then $X'_{x'} \to Y'_{y'}$ can be identified with induced map $X_x \times_{Y_y} Y'_{y'} \to Y'_y$. Since $X_x \to Y_y$ is surjective, so is this induced map. □

Now suppose that β is very solid. Every prime of Q_F is of the form $\mathfrak{p}_F$, where

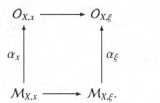

$\mathfrak{p}$ is a prime of Q not meeting F. Then $\mathfrak{p}A$ is a prime of A not meeting $\beta(F)$, and hence $\mathfrak{p}A$ is the inverse image in A of $\mathfrak{p}A_F = \mathfrak{p}_F A_F$. Since $\mathfrak{p} = \beta^{-1}(\mathfrak{p}A)$, it follows that $\mathfrak{p}_F = \beta_F^{-1}(\mathfrak{p}_F A_F)$. Since $\mathfrak{p}A_F$ is prime, we can conclude that β_F is very solid. If $\mathfrak{p}$ is a prime of F, then $\mathfrak{p} \cup \mathfrak{q}$ is the prime of $(Q, \mathfrak{q})$ corresponding to $\mathfrak{p}$. Since β is solid, $(\mathfrak{p} \cup \mathfrak{q})A$ is a prime of A and, since $(\mathfrak{p} \cup \mathfrak{q})A$ contains $\mathfrak{q}A$, the ideal $\mathfrak{p}A/\mathfrak{q}A \cong (\mathfrak{p} \cup \mathfrak{q})A/\mathfrak{q}A$ is prime. Since also $\mathfrak{p} \cup \mathfrak{q} = \beta^{-1}((\mathfrak{p} \cup \mathfrak{q})A)$, it follows that $\mathfrak{p} = \beta_{\mathfrak{q}}^{-1}(\mathfrak{p}A/\mathfrak{q}A)$. □

Proposition 1.10.11. *Let X be a fine Zariski log scheme. Then the set of points of X at which the log structure is very solid is open in X.*

Proof Suppose that the log structure $\alpha\colon \mathcal{M}_X \to \mathcal{O}_X$ is very solid at a point x of X, i.e., that $\alpha_x\colon \mathcal{M}_{X,x} \to \mathcal{O}_{X,x}$ is very solid. Since the statement is local on X, we may and shall assume that $X = \operatorname{Spec}(\beta\colon Q \to A)$, where Q is fine and A is noetherian. As usual, we may also assume that β is local at x. Let $\mathfrak{p}$ be the prime ideal of A corresponding to x, so that the homomorphism $Q \to A_{\mathfrak{p}}$ is local. Then the map $\operatorname{Spec}(\mathcal{M}_{X,x}) \to \operatorname{Spec}(Q)$ is bijective, and hence $\beta_{\mathfrak{p}}\colon Q \to A_{\mathfrak{p}}$ is very solid. Let $\lambda\colon A \to A_{\mathfrak{p}}$ be the localization homomorphism. For each $\mathfrak{q} \in \operatorname{Spec}(Q)$, the ideal $\mathfrak{q}A_{\mathfrak{p}}$ is prime and $\beta_{\mathfrak{p}}^{-1}(\mathfrak{q}A_{\mathfrak{p}}) = \mathfrak{q}$, since $\beta_{\mathfrak{p}}$ is very solid. Then $\lambda^{-1}(\mathfrak{q}A_{\mathfrak{p}})$ is a prime ideal $\mathfrak{p}'$ of A containing $\mathfrak{q}A$, and $\beta^{-1}(\mathfrak{p}') = \mathfrak{q}$. Since $(\mathfrak{q}A)_{\mathfrak{p}} = \mathfrak{p}'_{\mathfrak{p}} = \mathfrak{q}A_{\mathfrak{p}}$ and these ideals are finitely generated, there exists an element $a \in A \setminus \mathfrak{p}$ such that $(\mathfrak{q}A)_a = \mathfrak{p}'_a$ as ideals in A_a. In particular $\mathfrak{q}A_a$ is prime. Replacing A by A_a we see that $\mathfrak{q}A$ is prime and $\beta^{-1}(\mathfrak{q}A) = \mathfrak{q}$. Note that further localization by any $a' \in A \setminus \mathfrak{p}$ will preserve both of these conditions. Since $\operatorname{Spec}(Q)$ is finite, we may repeat this process for each prime of Q: there is an $a' \in A \setminus \mathfrak{p}$ such that, for every $\mathfrak{q} \in Q$, $\mathfrak{q}A_{a'}$ is prime and its inverse image in Q is $\mathfrak{q}$. Then $Q \to A_{a'}$ is very solid, and it follows that the associated log structure is also very solid. □

We next discuss some examples of solid log rings.

Proposition 1.10.12. *Let Q be an integral monoid such that Q^{gp} is torsion free and let R be an integral domain.*

1. *The canonical homomorphism $e\colon Q \to R[Q]$ is injective and very solid.*
2. *Suppose in addition that Q is finitely generated and sharp and that R is local. Then $R[[Q]]$ is a local integral domain, and the natural homomorphism $Q \to R[[Q]]$ is injective, local, and very solid.*

Proof Recall first from Proposition I.3.4.1 that, since R is an integral domain and Q^{gp} is torsion free, $R[Q]$ is an integral domain. Next note that if K is an ideal of Q, then $R[K]$ is the ideal of $R[Q]$ generated by the image of K, and $e^{-1}(R[K]) = K$. Furthermore, if K is prime and F is the complementary face, then $R[Q]/R[K] \cong R[F]$, which is again an integral domain, since F^{gp} is also torsion free. Thus $R[K]$ is prime, and hence $e\colon Q \to R[Q]$ is very solid. The injectivity of e is obvious.

If Q is finitely generated and sharp and R is local then, by Proposition I.3.6.1, $R[[Q]]$ is a local domain and can be identified with the $R[Q^+]$-adic completion of $R[Q]$. Furthermore if K is an ideal of Q, it is finitely generated, and $q+k=k$ implies $q=0$, so by Proposition I.3.6.5, $R[[K]]$ can be viewed as the $R[Q^+]$-adic completion of $R[K]$ and also as the ideal of $R[[Q]]$ generated by K. If K is prime and F is its complement, the quotient $R[[Q]]/R[[K]]$ is isomorphic to the integral domain $R[[F]]$, so $R[[K]]$ is prime. It is clear that $\hat{e}^{-1}(R[[K]]) = K$, so $\hat{e}\colon Q \to R[[Q]]$ is very solid. □

The following proposition gives a more subtle construction of very solid local log rings. Recall that an element of a commutative ring is called "prime" if it is a nonzero divisor and generates a prime ideal.

Proposition 1.10.13. *Let R be a noetherian local ring, let Q be a fine sharp monoid such that Q^{gp} is torsion free, and let f be an element of $R[[Q]]$ whose constant term is a prime element of R.*

1. *The quotient $A := R[[Q]]/(f)$ is a local integral domain.*
2. *The homomorphism $\beta\colon Q \to A$ is injective, local, and very solid.*

The main technical work in the proof will be accomplished in the following lemma.

Lemma 1.10.14. *With the notation above, the follow results hold.*

1. *For every ideal K of Q, multiplication by f on $R[[Q,K]] := R[[Q]]/R[[K]]$ is injective.*
2. *More generally, let S be a finitely generated Q-set and let T be a Q-subset. Assume that for every $s \in S$, $qs = s$ only if $q = 0$. Then multiplication by f is injective on the quotient $R[[S,T]]$ of $R[[S]]$ by $R[[T]]$.*
3. *If $Q \to Q'$ is an injective homomorphism of fine sharp monoids, the corresponding homomorphism $R[[Q]]/(f) \to R[[Q']]/(f)$ is injective.*
4. *If K is any ideal of Q, then $\beta^{-1}(KA) = K$. In particular, $\hat{e}^{-1}((f)) = \emptyset$.*
5. *If $\mathfrak{q} \subseteq Q$ is a prime ideal, then $\mathfrak{q}A$ is also prime.*

Proof Write

$$f = \pi + \sum\{a_q e^q : q \in Q^+\}.$$

Statement (1) is a special case of (2), so we just prove the latter. Suppose that $g := \sum\{b_s e^s : s \in S\} \in R[[S]]$ and $fg \in R[[T]]$. We claim that $g \in R[[T]]$, i.e., that the support $\sigma(g) := \{s \in S : b_s \neq 0\}$ is contained in T. If this is not the case, then by Proposition I.2.1.5, the nonempty set $\sigma(g) \setminus T$ has a minimal element r. Since $fg \in R[[T]]$, the coefficient of e^r in the product vanishes, so

$$\pi b_r + \sum \{a_q b_s : q + s = r, q \in Q^+, s \in S\} = 0.$$

Since π is a nonzero divisor, πb_r is not zero, and it follows that the sum is not empty. That is, there exists a pair (q, s) with $r = q + s$ and $a_q, b_s \neq 0$. In particular, $s \in \sigma(g)$, and $s \notin T$, because $r \notin T$. But $q \in Q^+$ and $r = q + s$, contradicting the minimality of r.

The proof of (3) is similar. To simplify the notation in its proof, we write P instead of Q' and assume that Q is a submonoid of P. Suppose that $g \in R[[Q]]$ and $h := \sum b_p e^p \in R[[P]]$ with $g = fh$. We claim that in fact $h \in R[[Q]]$, that is, that $\Phi := \{p \in P \setminus Q : b_p \neq 0\}$ is empty. If this is not the case, then, by Proposition I.2.1.5 applied to the fine monoid P, Φ has a minimal element r. Since $r \notin Q$, the coefficient of e^r in g is zero, so in fact

$$0 = \pi b_r + \sum \{a_q b_p : p + q = r, q \in Q^+, p \in P\}.$$

Necessarily the sum is not empty, which means that there exist $p \in P$ and $q \in Q^+$ such that $r = p + q$. But $p \notin Q$, contradicting the minimality of r in Φ.

For (4), let K be an ideal of Q and let q be an element of Q with $\beta(q) \in KA$. Then there exist $g \in R[[Q]]$ and $h \in R[[K]]$ such that $e^q = fg + h$. Then $fg = e^q - h$ belongs to $R[[K']]$, where K' is the ideal of Q generated by q and K. By statement (1), it follows that $g \in R[[K']]$. Since $R[[K'] = (e^q) + R[[K]]$, we can write $g = g'e^q + h'$, where $g' \in R[[Q]]$ and $h' \in R[[K]]$. Then $e^q(1 - fg') = fh' + h \in R[[K]]$ and, since $1 - fg'$ is a unit of $R[[Q]]$, it follows that $e^q \in R[[K]]$, hence $q \in K$. If we apply this statement with $K = \emptyset$, we find that $\beta^{-1}(0) = \emptyset$, hence $\hat{e}^{-1}((f)) \cap Q = \emptyset$.

First we prove (5) when $\mathfrak{q}$ is the empty ideal; in this case we need to show that A is an integral domain. Thanks to (3) and Corollary I.2.2.7, it suffices to prove this when $Q = \mathbf{N}^r$. To prove that A is a domain, it suffices to prove that $\mathrm{Gr}_I A$ is a domain, where I is the ideal of A generated by Q^+, because A is I-adically separated. But $\mathrm{Gr}_I(A) \cong \mathrm{Gr}_{R[Q^+]} R[Q]/\mathrm{in}(f) \cong R/\pi[\mathbf{N}^r]$, since the initial term, $\mathrm{in}(f)$, of f in $\mathrm{Gr}_I(A)$ is just π. This last ring is an integral domain because π is prime. More generally, if $\mathfrak{q}$ is a prime ideal of Q, let $F := Q \setminus \mathfrak{q}$, and let $\overline{f}$ be the image of f in $R[[Q]]/R[[\mathfrak{q}]] \cong R[[F]]$. Then F is a fine sharp monoid with F^{gp} torsion free, and the constant term of $\overline{f}$ is again π. Then $A/\mathfrak{q}A \cong R[[F]]/(\overline{f})$, which is an integral domain, as we have just seen. □

Proof of Proposition 1.10.13 The fact that A is an integral domain follows from (5) of Lemma 1.10.14, and the fact that A and β are local follows from the fact that $R[[Q]]$ and $Q \to R[[Q]]$ are local (Proposition I.3.6.1). The fact that β is very solid follows from (4) and (5) of the lemma. □

The next proposition investigates the behavior of solidity under base change.

Proposition 1.10.15. *Let $\alpha\colon P \to A$ be a log ring, let $Y := \operatorname{Spec}(\alpha)$, and let $\theta\colon P \to Q$ be an injective local homomorphism of fine monoids. Let x be a point of $Y_Q := Y \times_{\mathsf{A}_P} \mathsf{A}_Q$ lying over a point y of Y. If α is solid at y, then the induced homomorphism $\beta\colon Q \to Y_Q$ is solid at x. The converse holds if θ is locally exact.*

Proof We need the following lemma.

Lemma 1.10.16. *Suppose that $\theta\colon P \to Q$ is an injective homomorphism of fine monoids and that $\phi\colon \mathbf{Z}[P] \to \mathbf{Z}[Q]$ is the corresponding homomorphism of monoid algebras. Then the obvious map from $\underline{\mathsf{A}}_Q$ to the set-theoretic fiber product $\underline{\mathsf{A}}_P \times_{\mathbf{a}_P} \mathbf{a}_Q$ is surjective.*

Proof Let $\mathfrak{q}$ be a prime ideal of Q, let $\mathfrak{p} := \theta^{-1}(\mathfrak{q})$ and let J be a prime ideal of A such that $J \cap P = \mathfrak{p}$. The lemma asserts that there exists a prime ideal K of $\mathbf{Z}[Q]$ such that $K \cap Q = \mathfrak{q}$ and $\phi^{-1}(K) = J$.

We check a sequence of special cases. If P and Q are groups, $\mathbf{a}_Q$ and $\mathbf{a}_P$ are singletons, and the lemma just asserts that $\operatorname{Spec}(\phi)$ is surjective.

Case 1: P and Q are groups and the cokernel of θ is torsion free. Then θ admits a section, hence so does ϕ, and hence $\operatorname{Spec}(\phi)$ is surjective.

Case 2: P and Q are groups and there exists an n such that $nQ \subseteq P$. In this case the morphism ϕ is injective and finite, so $\operatorname{Spec}(\phi)$ is surjective.

Case 3: P and Q are groups. Let $P' := \{q \in Q : nq \in P \text{ for some } n > 0\}$. Then Case 1 implies that $\operatorname{Spec}(\mathbf{Z}[Q] \to \mathbf{Z}[P'])$ is surjective and Case 2 implies that $\operatorname{Spec}(\mathbf{Z}[P'] \to \mathbf{Z}[P])$ is surjective. Thus $\operatorname{Spec}(\mathbf{Z}[Q] \to \mathbf{Z}[P])$ is surjective.

Case 4: $\mathfrak{q} = \emptyset$. Then also $\mathfrak{p} = \emptyset$. Let k be the fraction field of the integral domain $\mathbf{Z}[P]/J$. Since $J \cap P = \emptyset$, every element of P maps to a unit in k, the homomorphism $P \to k$ factors through P^{gp}, and the homomorphism $\mathbf{Z}[P] \to k$ factors through $\mathbf{Z}[P^{\mathrm{gp}}]$. Thus J lifts to a prime ideal J' of $\mathbf{Z}[P^{\mathrm{gp}}]$. By Case 3, J' lifts to a prime ideal K' of $\mathbf{Z}[Q^{\mathrm{gp}}]$, whose intersection with $\mathbf{Z}[Q]$ is the desired prime ideal lifting J and $\mathfrak{q}$.

Case 5: In general, let $G := Q \setminus \mathfrak{q}$ and let $P := P \setminus \mathfrak{p}$. Since J contains $\mathfrak{p}$ it corresponds to a prime ideal J' of $\mathbf{Z}[P, \mathfrak{p}] \cong \mathbf{Z}[F]$, and, since $J \cap P = \mathfrak{p}$, $J' \cap F = \emptyset$. Furthermore, the homomorphism ϕ induces a homomorphism $\phi'\colon \mathbf{Z}[P, \mathfrak{p}] \to \mathbf{Z}[Q, \mathfrak{q}]$, which can be identified with the homomorphism $\mathbf{Z}[F] \to \mathbf{Z}[G]$ induced by $\theta'\colon F \to G$. Then $\mathfrak{q}' := G \cap \mathfrak{q} = \emptyset$, so $\theta'^{-1}(\mathfrak{q}') = \emptyset = J' \cap F$. By Case 4, there exists a prime K' of $\mathbf{Z}[G]$ such that

$K' \cap G = \mathfrak{q}'$ and $\phi'^{-1}(K') = J'$. The corresponding prime K of $\mathbf{Z}[Q]$ fulfills the requirements of the lemma. □

To prove the proposition, we begin with the commutative diagram

$$\begin{array}{ccccc} X := Y_Q & \xrightarrow{b'} & \mathsf{A}_{\mathsf{Q}} & \xrightarrow{a'} & \mathbf{a}_Q \\ \Big\downarrow f & & \Big\downarrow \mathsf{A}_\theta & & \Big\downarrow \mathbf{a}_\theta \\ Y & \xrightarrow{b} & \mathsf{A}_{\mathsf{P}} & \xrightarrow{a} & \mathbf{a}_P, \end{array}$$

in which the left square is cartesian.

Let us prove that if α is solid at y, then β is solid at x, i.e., that the map $X_{(x)} \to \mathbf{a}_Q$ is locally surjective. We may and shall replace Q by its localization at $\mathfrak{q}$ and P by its localization at $\mathfrak{p} := \theta^{-1}(\mathfrak{q})$. As we saw in Remark 1.10.2, we must show that, for every $x' \in X_{(x)}$ lying over $\mathfrak{q} := a'(b'(x))$, the map $X_{(x')} \to \mathbf{a}_Q$ is surjective. Since $y' := f(x')$ is a generization of x and $Y_{(y)} \to \mathbf{a}_P$ is locally surjective, $Y_{(y')} \to \mathbf{a}_P$ is also locally surjective, so we may as well assume that $x' = x$ and $y' = y$. Let $\tau \in \mathbf{a}_Q$ be a generization of the image t of x in $\mathbf{a}_Q$. Then the image σ of τ in $\mathbf{a}_P$ is a generization of the image s of y. Since $Y_{(y)} \to \mathbf{a}_P$ is locally surjective, there exists a point $\eta \in Y$ such that $a(b(\eta)) = \sigma$. By the lemma, there exists a point η' of A_{Q} such that $a'(\eta') = \tau$ and $\mathsf{A}_\theta(\eta') = b(\eta)$. Since the left square is cartesian, there exists a point ζ of X such that $b'(\zeta) = \eta'$, and hence $a'(\beta'(\zeta)) = \tau$.

For the converse, suppose that θ is locally exact and that β is solid. Since θ is locally exact and local, $\mathbf{a}_\theta$ is locally surjective and surjective, by Proposition I.4.2.2. Then it follows from the lemma that A_θ is surjective and, since the square is cartesian, that f is surjective. Since β is solid, $a'b'$ is locally surjective and hence so is $\mathbf{a}_\theta a'b' = abf$. Since f is surjective, it follows that ab is locally surjective, as desired. □

The next proposition gives a geometric consequence of solidity that will play an important role in the characterization of log regularity.

Proposition 1.10.17. *Let $\beta\colon Q \to A$ be a log ring, where A is a noetherian local ring and Q is a toric monoid, and let I_β be the ideal of A generated by $\beta(Q^+)$. Then*

$$\dim(A) \le \dim(A/I_\beta) + \dim(Q),$$

and equality holds if β is solid.

Proof Let $\hat{A}$ be the completion of A and let $\hat{\beta}\colon Q \to \hat{A}$ be the induced homomorphism. Recall from Proposition 1.10.3 that $\hat{\beta}$ is solid if β is. Furthermore

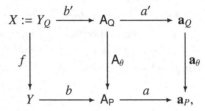

$\dim(A) = \dim(\hat{A})$ and $\dim A/I_\beta) = \dim(\hat{A}/I_\beta)$, so we may and shall assume, without loss of generality, that A is complete.

Let $F := \beta^{-1}(A^*)$, and let $\beta_F \colon Q_F \to A$ be the induced local homomorphism. Since $\dim(Q_F) \le \dim(Q)$, and since $\operatorname{Spec}(\beta_F)$ is locally surjective if and only if $\operatorname{Spec}(\beta)$ is, we may without loss of generality replace β by β_F. Thus we may and shall assume β is local. Since Q is saturated, $Q \cong Q^* \oplus \overline{Q}$, so we may also assume that Q is sharp. Let r be the dimension of A/I_β, and choose a sequence $(t_1, \ldots, t_r)$ in A whose image in A/I_β is a system of parameters. The sequence $(t_1, \ldots, t_r)$ defines a homomorphism $\mathbf{N}^r \to A$, and hence a homomorphism $\gamma \colon P := Q \oplus \mathbf{N}^r \to A$. Since the image of $(t_1, \ldots, t_r)$ in A/I_β contains a system of parameters, the quotient A/Q^+A has finite length.

To prove the inequality, assume first that A contains a field. Then it also contains a copy of its residue field k [17, 7.7]. Since γ takes Q^+ into the it induces a homomorphism $\psi \colon k[[Q]] \to A$, and since A/Q^+A has finite length and A is complete, it follows that $k[[Q]] \to A$ is a finite homomorphism of local rings. Thus $\dim(A) \le \dim(k[[Q]])$, and Proposition I.3.4.2 implies that $\dim k[[Q]] = r + \dim(Q)$.

If A does not contain a field, it does contain a Cohen ring V of its residue field [26, 19.8.8], which necessarily has positive characteristic p. Proceeding as before, we obtain a finite homomorphism $\phi \colon V[[Q]] \to A$ such that A/Q^+A has finite length. Let $J := \operatorname{Ker}(\phi)$, so that ϕ factors through an injective and finite homomorphism $V[[Q]]/J \to A$, and the image of $\operatorname{Spec}(\phi)$ is the set of prime ideals containing J. We claim that J is not zero, and in fact that it contains an element with a nonzero constant term. In the contrary case, $J \subseteq Q^+$, and so there is a prime ideal $\mathfrak{p}$ of A such that $\phi^{-1}(\mathfrak{p}) = V[[Q^+]]$. But then $\mathfrak{p}$ contains P^+A and since A/Q^+A has finite length, necessarily $\mathfrak{p}$ is the maximal ideal $\mathfrak{m}$ of A. This is a contradiction, since $p \in \phi^{-1}(\mathfrak{m})$ and $p \notin V[[Q^+]]$, so indeed $J \neq 0$. Since $V[[Q]]$ is an integral domain of dimension $1 + \dim Q + r$, its quotient by J has strictly smaller dimension, and hence the dimension of A is at most $\dim Q + r$.

Now suppose that β is solid. Let $\mathfrak{q}_0 \supset \cdots\cdots \supset q_d$ be a chain of prime ideals in Q with $d = \dim(Q)$, and let $\mathfrak{p}_{-r} \supset \cdots \supset \mathfrak{p}_0$ be a chain of prime ideals in A/I_β with $r = \dim A/I_\beta$. By construction, the homomorphism $Q \to A/I_\beta$ is hollow, and consequently $\beta^{-1}(\mathfrak{p}_0) = Q^+$. Since $\operatorname{Spec}(\beta)$ is locally surjective, there exists a chain $\mathfrak{p}_0 \supset \cdots \supset \mathfrak{p}_d$ in $\operatorname{Spec}(A)$ with $\beta^{-1}(\mathfrak{p}_i) = \mathfrak{q}_i$. Thus $\mathfrak{p}_{-r} \supset \cdots \supset \mathfrak{p}_d$ is a chain of distinct prime ideals in A, showing that $\dim(A) \ge r + d$. □

Variant 1.10.18. An idealized log ring $\beta \colon (Q, K) \to A$ is solid if the map $\operatorname{Spec}(\beta) \colon \operatorname{Spec}(A) \to \operatorname{Spec}(Q, K)$ is locally surjective. We claim that, for every idealized log local ring, $\dim(A) \le \dim(A/I_\beta) + \dim(Q, K)$, and equality holds if β is solid.

To check this, we may assume without loss of generality that K is reduced and hence is the intersection of prime ideals $\mathfrak{p}_1, \ldots, \mathfrak{p}_r$. Then $\mathrm{Spec}(Q, K)$ is the union of the closed sets $\mathrm{Spec}(Q, \mathfrak{p}_i)$, and each $\mathrm{Spec}(A/\mathfrak{p}_i A)$ is the inverse image of each $\mathrm{Spec}(Q, \mathfrak{p}_i)$ under the map $\mathrm{Spec}(\beta)$. Thus $\mathrm{Spec}(A)$ is set-theoretically the union of closed subschemes $\mathrm{Spec}(A/\mathfrak{p}_i A)$, and hence its dimension is the maximum of the dimensions of these closed subschemes. Suppose that $\mathfrak{p}$ is the prime $\mathfrak{p}_i$ that achieves this maximum and let $A' := A/\mathfrak{p}$. Then $\dim(Q, K) \ge \dim(Q, \mathfrak{p}) = \dim(G)$, where $G := Q \setminus \mathfrak{p}$. Consider the log ring $\beta' \colon G \to A'$ induced by β. Note that $A/I_\beta \cong A'/I_{\beta'}$, since $Q^+ = \mathfrak{p} \cup G^+$ and $\mathfrak{p}$ maps to zero in A'. By Proposition 1.10.17, we have

$$\dim(A) = \dim(A') \le \dim(A'/I_{\beta'}) + \dim(G) \le \dim(A/I_\beta) + \dim(Q, K),$$

as required. On the other hand, if β is solid, we can extend a chain of prime ideals of A/I_β to A, just as in the non-idealized case, to prove the reverse inequality.

1.11 Log regularity

We now discuss a logarithmic analog of the notion of regularity, due to K. Kato [49] and further developed by Gabber and Ramero in [22]. Roughly speaking, a local log ring is regular if its singularity is completely accounted for by its log structure.

Theorem 1.11.1. *Let $\beta \colon Q \to A$ be a local log ring, where A is noetherian and $\overline{Q}$ is fine and saturated. Let I_β be the ideal of A generated by $\beta(Q^+)$. Suppose that*

1. *the quotient ring A/I_β is regular, in the usual sense.*

Then the following conditions are equivalent:

2. *β is very solid;*
3. *β is solid;*
4. $\dim(A) = \dim(A/I_\beta) + \dim(Q)$.

If these conditions are verified, β is said to be (log) regular. A fine saturated log scheme X is regular at $x \in X$ if the local log ring $\mathcal{M}_{X,x} \to \mathcal{O}_{X,x}$ is regular, and X is regular if it is so at each of its points.

We have already seen, in Remark 1.10.2 and Proposition 1.10.17, that condition (2) implies (3) and that (3) implies (4). The equivalence of (2)–(4) in the presence of condition (1) will be proved in the course of the next theorem, which also gives a structure theorem for regular local log rings. More equivalent conditions appear in Proposition 1.11.5 below.

Since we are assuming that $\overline{Q}$ is fine and saturated, the group $\overline{Q}^{\mathrm{gp}}$ is finitely generated and free, so the natural projection $Q \to \overline{Q}$ splits, and β is regular if and only if the induced homomorphism $\overline{Q} \to A$ is regular. Thus we are reduced to the case in which Q is sharp.

Theorem 1.11.2. *Let $\beta\colon Q \to A$ be a local log ring, where A is a noetherian local ring and Q is fine, sharp, and saturated. Let k be the residue field of A and let I_β be the ideal of A generated by $\beta(Q^+)$.*

1. *If A contains a field, then β is regular if and only if there exists a commutative diagram of the form*

$$\begin{array}{ccc} Q & \longrightarrow & k[[Q \oplus \mathbf{N}^r]] \\ {\scriptstyle\beta}\Big\downarrow & & \Big\downarrow{\scriptstyle\psi}\,\cong \\ A & \longrightarrow & \hat{A}, \end{array}$$

where the top arrow is the obvious one and $\hat{A}$ is the completion of A.

2. *In the mixed characteristic case, let V be a Cohen ring of k and let $p > 0$ be the characteristic of k. Then β is regular if and only if there exists a commutative diagram of the form*

$$\begin{array}{ccc} Q & \longrightarrow & V[[Q \oplus \mathbf{N}^r]] \\ {\scriptstyle\beta}\Big\downarrow & & \Big\downarrow{\scriptstyle\phi} \\ A & \longrightarrow & \hat{A}, \end{array}$$

where ϕ is a surjection whose kernel is a principal ideal generated by an element f of $V[[Q \oplus \mathbf{N}^r]]$ whose constant term is p.

Proof Suppose that we have a diagram as in (1) above. Then $\hat{A}/I_\beta\hat{A} \cong k[[\mathbf{N}^r]]$, which is regular, and it follows that A/I_β is also regular. Furthermore, by Proposition 1.10.12, the homomorphism $Q \oplus \mathbf{N}^r \to \hat{A}$ is very solid. It follows that the same is true for the homomorphisms $Q \to \hat{A}$ and $\beta\colon Q \to A$, by Proposition 1.10.3.

Next suppose that we have a diagram as in (2). Then $\hat{A}/I_\beta \cong V[[\mathbf{N}^r]]/(g)$, where g is the image of f in $V[[\mathbf{N}^r]$. Since g belongs to the maximal $\mathfrak{m}$ of this ring but not to $\mathfrak{m}^2$, this quotient is a regular local ring. Furthermore, by

Proposition 1.10.13, the homomorphism $Q \oplus \mathbf{N}^r \to \hat{A}$ is very solid, and hence so is $\beta\colon Q \to A$.

We have thus proved that the existence of a diagram as in (1) or (2) implies that $\hat{A}/I_\beta\hat{A}$ is regular and that β is very solid, so that conditions (2)–(4) of Theorem 1.11.1 hold.

To finish the proofs of Theorems 1.11.2 and 1.11.1 we need to show that if A/I_β is regular and $\dim(A) = \dim(A/I_\beta) + \dim(Q)$ then we have a diagram as in statement (1) or (2) of the theorem. We may as well replace A by $\hat{A}$ in this argument.

Suppose first that A contains a field. Then, by [17, 7.7], A contains a copy of its residue field k, and we will use a more precise version of the argument in the proof of Proposition 1.10.17. Since A/I_β is a regular local ring, there exists a sequence $(t_1, \ldots, t_r)$ of elements of A whose image in A/I_β is a regular sequence of parameters. Then the corresponding homomorphism $k[[\mathbf{N}^r]] \to A/I_\beta$ is an isomorphism. Since β is a local homomorphism, the map $k[Q \oplus \mathbf{N}^r] \to A$ induced by β and $(t_1, \ldots, t_r)$ extends to a morphism $\psi\colon k[[Q \oplus \mathbf{N}^r]] \to A$, which is necessarily surjective, since it is so modulo I_β and Q generates I_β. Since $k[[Q \oplus \mathbf{N}^r]]$ is an integral domain of dimension $\dim(Q) + \dim(A/I_\beta) = \dim(A)$, in fact ψ must be an isomorphism. This shows that we have the diagram in statement (1).

If the residue field k has characteristic zero, then A contains a field, so we may and shall suppose that k has characteristic p. Again we let $(t_1, \ldots, t_r)$ be a sequence in A whose image in the regular local ring A/I_β is a regular sequence of parameters. Then if $P := Q \oplus \mathbf{N}^r$, we obtain a surjective homomorphism $\phi\colon V[[P]] \to A$; let J be its kernel. By construction, P^+A is the maximal ideal of A, which contains p, so $p \in J + V[[P^+]]$. That is, there exists an element $f \in J$ whose constant term is p. By Lemma 1.10.14, $V[[P]]/(f)$ is an integral domain of dimension $\dim(Q) + r = \dim A$, and hence $(f) = J$. Thus (2) implies the existence of the diagram in statement (2). □

Corollary 1.11.3. *Let Q be a fine sharp and saturated monoid and let $\beta\colon Q \to A$ be a regular log local ring. Assume that A contains its residue field k. Then the homomorphism $k[Q] \to A$ is flat.*

Proof According to statement (1) of Theorem 1.11.2, the homomorphism β fits into an isomorphism of the form $k[[Q \oplus \mathbf{N}^r]] \to \hat{A}$. Since the homomorphism $k[Q] \to k[[Q]] \to k[[Q \oplus \mathbf{N}^r]]$ is flat and since $A \to \hat{A}$ is faithfully flat, it follows that $k[Q] \to A$ is also flat. □

Corollary 1.11.4. *The underlying local ring of a regular local log ring is Cohen–Macaulay.*

Proof Suppose $\beta: Q \to A$ is a regular local log ring. Since a noetherian local ring is Cohen–Macaulay if and only if its completion is so [26, 0,16.5.2], we may and shall assume that A is complete. If R is a DVR or a field and Q is a fine saturated and sharp monoid, it follows from Theorem I.3.4.3 that $R[[Q]]]$ is Cohen–Macaulay, and hence so is the quotient of $R[[Q]]$ by any nonzero divisor [26, 0,16.5.5]. Thus the corollary follows from the structure theorem 1.11.2. □

A noetherian local ring R with residue field k is regular if and only if it has finite projective dimension, which is true if and only if $\mathrm{Tor}_i^R(k,k)$ vanishes for i sufficiently large. The closest analogy of this result currently known is the following.

Proposition 1.11.5. *Let $\beta: Q \to A$ be a local log ring, where Q is saturated, A is noetherian, and A/I_β is a regular local ring. Then the following conditions are equivalent.*

1. *β is regular.*
2. *$\mathrm{Tor}_i^{\mathbf{Z}[Q]}(\mathbf{Z}[S,T],A) = 0$ for all $i > 0$, whenever T is a sub-Q-set of a finitely generated Q-set S such that $qs \neq s$ for all $q \in Q^+$ and all $s \in S$.*
3. *$\mathrm{Tor}_1^{\mathbf{Z}[Q]}(\mathbf{Z}[Q]/\mathbf{Z}[Q^+],A) = 0$.*
4. *$(\mathrm{Gr}_{\mathbf{Z}[Q^+]}\mathbf{Z}[Q]) \otimes_{\mathbf{Z}} A/I_\beta \to \mathrm{Gr}_{I_\beta} A$ is an isomorphism.*

Proof We may assume without loss of generality that A is complete. Suppose that β is regular and that A contains a field, so that we have the diagram in statement (1) of Theorem 1.11.2. Let $F^\cdot$ be a resolution of $\mathbf{Z}[S,T]$ by flat $\mathbf{Z}[Q]$-modules. Since $\mathbf{Z}[S,T]$ is flat over $\mathbf{Z}$,

$$F_k^\cdot := F^\cdot \otimes_{\mathbf{Z}[Q]} k[Q] \cong F^\cdot \otimes_{\mathbf{Z}} k$$

is a resolution of $k[S,T]$ by flat $k[Q]$-modules. Since $A \cong k[[Q \oplus \mathbf{N}^r]]$, it is flat over $k[Q]$, and so

$$F^\cdot \otimes_{\mathbf{Z}[Q]} A \cong F_k^\cdot \otimes_{k[Q]} A$$

is still acyclic. Thus $\mathrm{Tor}_i^{\mathbf{Z}[Q]}(\mathbf{Z}[S,T],A)$ vanishes for $i > 0$.

Now suppose that we have a diagram of the form (2) in Theorem 1.11.2, and let $R := V[[\mathbf{N}^r]]$. Since $R[[Q]]$ is flat over $\mathbf{Z}[Q]$, and A is an $R[[Q]]$-module,

$$\begin{aligned}\mathrm{Tor}_i^{\mathbf{Z}[Q]}(\mathbf{Z}[S,T],A) &\cong \mathrm{Tor}_i^{R[[Q]]}(\mathbf{Z}[S,T] \otimes_{\mathbf{Z}[Q]} R[[Q]],A)\\ &\cong \mathrm{Tor}_i^{R[[Q]]}(R[[S,T]],A).\end{aligned}$$

Now $R[[Q]] \xrightarrow{f} R[[Q]]$ is a resolution of $R[[Q]]/(f)$ by flat $R[[Q]]$-modules, so these Tor groups are computed as the cohomology of the complex

$$V[[S,T]] \xrightarrow{f} V[[S,T]].$$

The homomorphism $R[[Q]] \to A$ induces a surjection $R \to A/I_\beta$ whose kernel is generated by the constant term g of f in R. Since A/I_β is an integral domain, g is a prime element of R, and so (2) of Lemma 1.10.14 tells us that multiplication by f is injective on $V[[S,T]]$, whence the vanishing of the Tor_i.

It is trivial that (2) implies (3). Suppose (3) holds. Let $J := \mathbf{Z}[Q^+]$, and note that the $\mathbf{Z}[Q]$-module J^n/J^{n+1} is a direct sum of copies of $\mathbf{Z}[Q]/\mathbf{Z}[Q^+]$. Thus assumption (3) implies that $\mathrm{Tor}_1^{\mathbf{Z}[Q]}(J^n/J^{n+1}, A) = 0$ for all n. Consider the following diagram, in which the rows are exact and the vertical arrows are surjective:

$$\begin{array}{ccccccccc}
 & & J^{n+1}\otimes A & \xrightarrow{a_n} & J^n \otimes A & \longrightarrow & J^n/J^{n+1}\otimes A & \longrightarrow & 0 \\
 & & \downarrow{\scriptstyle b_{n+1}} & & \downarrow{\scriptstyle b_n} & & \downarrow{\scriptstyle c_n} & & \\
0 & \longrightarrow & I_\beta{}^{n+1} & \xrightarrow{i_n} & I_\beta{}^n & \longrightarrow & I_\beta{}^n/I_\beta{}^{n+1} & \longrightarrow & 0.
\end{array}$$

The vanishing of the Tor_1 implies that a_n is injective for all n, and since $b_0 = \mathrm{id}$ is injective, it follows by induction that b_n is injective for all n. But then it also follows that c_n is bijective for all n, so the homomorphism

$$(\mathrm{Gr}_J \mathbf{Z}[Q]) \otimes_{\mathbf{Z}[Q]} A \to \mathrm{Gr}_{I_\beta} A$$

is an isomorphism. Since $\mathrm{Gr}_J \mathbf{Z}[Q]$ is annihilated by J, statement (4) follows.

Let us prove that (4) implies the existence of a diagram (1) or (2); we write out only the latter case. We assume without loss of generality that A is complete. As in the proof of Theorem 1.11.2, we construct a surjective homomorphism $V[[\mathbf{N}^r \oplus Q]] \to A$ whose kernel contains an element f whose constant term is p and such that the induced map $R := V[[\mathbf{N}^r]] \to A/I_\beta$ corresponds to a regular system of parameters for A/I_β. It follows that the induced map $R/(g) \to A/I_\beta$ is an isomorphism, where g is the image of f in R. The proof will be finished if we can show that the homomorphism $R[[Q]]/(f) \to A$ is an isomorphism. It will suffice to prove that the associated graded map $\mathrm{Gr}_J (R[[Q]]/(f)) \to \mathrm{Gr}_{I_\beta} A$ is an isomorphism. By (2) of Lemma 1.10.14, multiplication by f on $\mathrm{Gr}_J R[[Q]]$ is injective, so that $(f) \cap J^n \subseteq fJ^n + J^{n+1}$. Therefore, since g is the initial form of f,

$$\begin{aligned}
\mathrm{Gr}_J (R[[Q]]/(f)) &\cong (\mathrm{Gr}_J R[[Q]])/(g) \cong (R/(g)) \otimes (\mathrm{Gr}_J \mathbf{Z}[Q]) \\
&\cong A/I_\beta \otimes \mathrm{Gr}_J \mathbf{Z}[Q],
\end{aligned}$$

so our claim follows from (4). □

Theorem 1.11.6. *Let X be a fine saturated log scheme, regular at a point x of X. Then the underlying local ring $\mathcal{O}_{X,x}$ is regular if and only if $\overline{\mathcal{M}}_{X,x}$ is a free monoid. In particular, this is the case if* $\dim \mathcal{M}_{X,x} \leq 1$.

Proof Since $\mathcal{M}_{X,x}$ is saturated, we can choose a section of the projection $\mathcal{M}_{X,x} \to \overline{\mathcal{M}}_{X,x}$, and the induced homomorphism

$$\beta \colon Q := \overline{\mathcal{M}}_{X,x} \to A := \mathcal{O}_{X,x}$$

is a chart. Then β is a regular log local ring, and in particular A/I_β is a regular local ring. Then its maximal ideal can be generated by $m := \dim(A/I_\beta)$ elements, which we may lift to a sequence $(a_1, \ldots, a_m)$ in A. Since β is regular, $\dim(A) = m + d$, where $d := \dim(Q)$. If Q is free, it can be generated by d elements $(q_1, \ldots, q_d)$, and then $(a_1, \ldots, a_m, \beta(q_1), \ldots, \beta(q_d))$ is a sequence of generators for the maximal ideal of A. Since $\dim(A) = m + d$, it follows that A is regular.

Suppose conversely that A is a regular local ring. Then its completion $\hat{A}$ is again regular and $Q \to \hat{A}$ is also regular, so we may as well assume that A is complete. If A contains a field then, according to Theorem 1.11.2, A is isomorphic to $k[[P]]$, where $P \cong Q \oplus \mathbf{N}^r$ for some $r \geq 0$. Then Lemma 1.11.7 below implies that P is a free monoid, and Corollary I.4.7.13 allows us to conclude that Q is also free. If A has mixed characteristic p, Theorem 1.11.2 says that $A \cong V[[P]]/(f)$ where $f \in V[[P]]$ has p as its constant term. If A is regular, necessarily $V[[P]]$ is regular, and then $k[[P]]$ is also regular, and so again it follows that P and Q are free. □

Lemma 1.11.7. *If P is a fine sharp monoid and k is a field, then $k[[P]]$ is regular if and only if P is free.*

Proof If P is free, $k[[P]]$ is a formal power series ring, hence regular. Suppose conversely that $A := k[[P]]$ is regular. Then $\mathfrak{m}_A/\mathfrak{m}_A^2$ is a vector space of dimension $d := \dim(A) = \operatorname{rank}(P^{\mathrm{gp}})$. On the other hand, it follows from Proposition I.3.6.5 that $\mathfrak{m}_A/\mathfrak{m}_A^2 \cong k[P^+, P_2^+]$, where P_2^+ is the square of the ideal P^+. The set $P^+ \setminus P_2^+$ is a basis for this vector space and therefore has cardinality d. On the other hand, by Proposition I.2.1.2, this set generates P. Thus we have a surjective homomorphism $\mathbf{N}^d \to P$, which induces a surjection $\mathbf{Z}^d \to P^{\mathrm{gp}}$. Since d is the rank of P^{gp}, this surjection is an isomorphism, and hence $\mathbf{N}^d \to P$ is also an isomorphism. □

The following result is useful amplification of Theorem 1.11.6.

Theorem 1.11.8. *Let $\alpha \colon P \to A$ and $\beta \colon Q \to A$ be local log rings, where P*

and Q are fine, sharp, and saturated, and let $\gamma\colon P \oplus Q \to A$ be the log ring induced by α and β. If γ is regular, the following conclusions hold.

1. *The local log ring $\alpha'\colon P \to A/I_\beta$ is regular, and $\dim(A/I_\beta) = \dim(A/I_\gamma)$.*
2. *The local log ring $\beta\colon Q \to A$ is regular if and only if P is free.*

Proof Let d be the rank of Q^{gp}, let r be the rank of P^{gp}, and let $A' := A/I_\beta$. Then $A/I_\gamma \cong A'/I'_\alpha$, where I'_α is the image of I_α in A'. Since γ is regular, $A'' := A/I_\gamma$ is regular of dimension $\dim(A) - (d + r)$. On the other hand, Proposition 1.10.17 applied to β implies that $\dim(A) \le d + \dim(A')$, and hence

$$\dim(A') \ge \dim(A) - d = \dim(A'') + r.$$

But the same proposition applied to α' says that

$$\dim(A') \le \dim(A'') + r,$$

and so in fact equality holds, and it follows that α' is regular.

If P is free, I'_α can be generated by r elements, say $(a_1, \ldots, a_r)$. Since the quotient A'' of A' is regular, its maximal ideal can be generated by e elements, where $e := \dim(A'') = \dim(A') - r$. If $(a_{r+1}, \ldots, a_{r+e})$ is a sequence in A' lifting such a set of generators, then $(a_1, \ldots, a_{r+e})$ is a sequence of generators for the maximal ideal of A' of length $\dim(A')$. It follows that A' is a regular local ring and then that β is also regular. Conversely, if β is regular, then A' is also a regular ring, and it follows from Theorem 1.11.6 that P is free. □

Example 1.11.9. Let Y be a divisor with strict normal crossings in a regular scheme $\underline{X}$, let α be the compactifying log structure associated to the complement of Y, and let X be the log scheme $(\underline{X}, \alpha)$. Then X is a regular log scheme. Indeed, if $x \in X$, it follows from Proposition 1.8.2 that α is the log structure associated to the DF structure defined by the set of irreducible components $Y_1, \ldots, Y_r$ containing x and that $\overline{\mathcal{M}}_{X,x} \cong \mathbf{N}^r$. The ideal of $\mathcal{O}_{X,x}$ generated by $\mathcal{M}_{X,x}$ is the ideal of $Y_1 \cap \cdots \cap Y_r$, which is regular and of dimension $\dim(\mathcal{O}_{X,x}) - r$. Thus X satisfies condition (4) of Theorem 1.11.1.

If $\beta\colon Q \to A$ is a regular local log ring and $\mathfrak{q}$ is a prime of Q, then $\mathfrak{q}A$ is a prime ideal of A. The induced homomorphism $F \to A/\mathfrak{q}$ is then a local log ring and the next proposition shows that this log ring is again regular.

Proposition 1.11.10. *Let $\beta\colon Q \to A$ be a regular local log ring. For each prime ideal $\mathfrak{q}$ of Q, the local log ring $\beta_\mathfrak{q}\colon F := Q \setminus \mathfrak{q} \to A/\mathfrak{q}A$ is again regular.*

Proof Proposition 1.10.9 tells us that $\beta_\mathfrak{q}\colon F \to A/\mathfrak{q}A$ is solid. Furthermore, $Q^+ = F^+ \cup \mathfrak{q}$, and hence $(A/\mathfrak{q}A)/F^+(A/\mathfrak{q}A) \cong A/Q^+A$, which is by hypothesis a regular ring. □

The following result globalizes the previous one and shows that the solidity and regularity of log structures are preserved on passage to suitably defined subschemes.

Proposition 1.11.11. *Let X be a fine log scheme, let $\mathcal{K} \subseteq \mathcal{M}_X$ be a coherent sheaf of ideals, and let Y be the closed subscheme of X defined by $\mathcal{K}$. Suppose that $\mathcal{K}_y$ is prime for every y of Y, and let $\mathcal{F}$ be the subsheaf of $\mathcal{M}_Y$ consisting of the sections whose stalks at every point do not belong to $\mathcal{K}_y$.*

1. *$\mathcal{F}$ is a sheaf of faces in $\mathcal{M}_Y$ and, for each $y \in Y$, its stalk is $\mathcal{M}_{Y,y} \setminus \mathcal{K}_y$. Furthermore, $\mathcal{F}$ is coherent as a sheaf of monoids, and the induced map $\alpha_Y : \mathcal{F} \to \mathcal{O}_Y$ is a fine log structure.*
2. *If α_X is solid (resp. very solid, resp. regular), then the same is true of α_Y.*

Proof Since $\mathcal{M}_{X,x} \setminus \mathcal{K}_x$ is face of $\mathcal{M}_{X,x}$, it is clear that $\mathcal{F}$ is a sheaf of faces in $\mathcal{M}_X$; the description of the stalks of $\mathcal{F}$ is also immediate to check. To see that $\mathcal{F}$ is coherent is a local question around a point y of Y, so we may assume that there is a chart $\beta \colon Q \to \mathcal{M}_X$ and that $K := \beta^{-1}(\mathcal{K}_y)$ generates $\mathcal{K}$. Then K is prime, so its complement is a face F of Q. We claim that the induced map $F \to \mathcal{F}$ is a chart for $\mathcal{F}$. Let y' be any point of Y and let $G := \beta^{-1}(\mathcal{O}^*_{Y,y'})$. Then $G \subseteq F$, since K maps to zero in $\mathcal{O}_{Y,y'}$. It follows that F_G is the complement of K_G in Q_G, and hence that the map $F_G/G \to \overline{\mathcal{F}}_{y'}$ is an isomorphism, as required.

Suppose that β is solid. The subscheme Y of X is defined by the ideal of $\mathcal{O}_X$ generated by $\mathcal{K}$. Hence, for each point y of X, there is a cartesian diagram

$$\begin{array}{ccccc} & & Y_{(y)} & \longrightarrow & X_{(x)} \\ & & \downarrow & & \downarrow \\ \operatorname{Spec}(\mathcal{F}_y) & \xrightarrow{\cong} & \operatorname{Spec}(\mathcal{M}_{X,y}, \mathcal{K}_y) & \longrightarrow & \operatorname{Spec}(\mathcal{M}_{X,x}). \end{array}$$

Since the diagram is set-theoretically cartesian and the right vertical arrow is locally surjective, Lemma 1.10.10 shows that the left vertical arrow is also locally surjective. If β is very solid and $\mathfrak{p}$ is a prime ideal of $\mathcal{F}_y$, then $\mathfrak{p} \cup K_y$ is a prime ideal of $\mathcal{M}_{X,y}$, hence $(\mathfrak{p} \cup K_y)\mathcal{O}_{X,y}$ is prime, and hence $\mathfrak{p}\mathcal{O}_{Y,y}$ is prime. If X is regular, Proposition 1.11.10 implies that Y is also regular. □

Theorem 1.11.12. *The log structure of a regular log scheme is compactifying; that is, the natural map $\mathcal{M}_X \to \mathcal{M}_{X^*/X}$ is an isomorphism. More generally, let $\mathcal{F}$ be a relatively coherent sheaf of faces of $\mathcal{M}_X$ and let $X_{\mathcal{F}}$ be the log scheme $\mathcal{F} \to \mathcal{O}_X$. Then $X^*_{\mathcal{F}}$ is open in X, and the log structure of $X_{\mathcal{F}}$ is compactifying.*

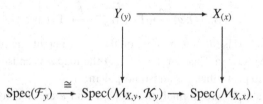

Proof We may assume that there is a chart β for α_X subordinate to a fine saturated monoid Q and that $\mathcal{F}$ is the sheaf of faces of $\mathcal{M}_X$ generated by an element q of Q. Then $X_{\mathcal{F}}$ is the special affine open subset of X defined by $\beta(q)$, and its complement Y is the closed subset defined by $\beta(q)$. We must show that the natural map $\overline{\mathcal{F}} \to \underline{\Gamma}_Y(Div_X^+)$ is an isomorphism. Working at the stalks, we may assume that X is the spectrum of a regular local log ring $Q \to A$, and that Q is sharp. Then if F is the face of Q generated by q, the natural map $F \to \overline{\mathcal{F}}_x$ is an isomorphism. Let $\hat{A}$ be the completion of A and let $\hat{X}$ be the corresponding local log ring. Then we have maps

$$F \to \Gamma_Y(Div_X^+) \to \Gamma_{\hat{Y}}(Div_{\hat{X}}^+),$$

and the second of these is injective, since $A \to \hat{A}$ is faithfully flat. Hence if the theorem is true for $\hat{A}$, it is also true for A, so we may assume without loss of generality that A is complete.

By Theorem 1.11.2, we may assume that β has the form $Q \to R[[Q]]/(f)$, where R is a normal noetherian local ring and and the constant term of f is a prime element of R. Let X' be the localization of $\mathrm{Spec}(Q \to R[Q])$ at the ideal $\mathfrak{m}[Q^+]$, where $\mathfrak{m}$ is the maximal ideal of R. Then we have a strict morphism $X \to X'$ and a diagram

$$\begin{array}{ccccc}
F & \xrightarrow{a} & \Gamma_{Y'}(Div_{X'}^+) & \xrightarrow{b} & \Gamma_Y(Div_X^+) \\
\Big\downarrow{\scriptstyle c} & & \Big\downarrow{\scriptstyle d} & & \Big\downarrow{\scriptstyle e} \\
\mathcal{W}^+(Q,q) & \xrightarrow{a'} & \Gamma_{Y'}(\mathcal{W}_{X'}^+) & \xrightarrow{b'} & \Gamma_Y(\mathcal{W}_X^+),
\end{array}$$

where $\mathcal{W}^+(Q,q)$ is the free monoid on the set of height one primes of Q containing q. We saw in Theorem I.3.5.8 that the map a is an isomorphism, so it will suffice to prove that b is an isomorphism.

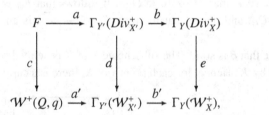

Since X and X' are normal, the maps d and e are injective. By Proposition I.3.5.2, the irreducible components of $\mathrm{Spec}(R[Q]/(q))$ are defined by the height one prime ideals $\mathfrak{p}$ of Q containing q, i.e., the map a' is an isomorphism. Furthermore since X is very solid, each $\mathfrak{p}$ remains prime in $R[[Q]]/(f)$. It follows that the irreducible components of $Y := \mathrm{Spec}(R[[Q]](q))$ are defined by the same set of prime ideals. Thus the map b' is also an isomorphism. It follows that b is injective. To prove its surjectivity, let D be an element of $\Gamma_Y(Div_X^+)$; there is a unique $Z' = \sum n_{\mathfrak{p}}\mathfrak{p} \in \mathcal{W}^+(Q,q)$ such that $e(D) = b'(a'(Z))$. Let K be the ideal of Q defined by Z', i.e., $K = \{q' \in Q : v_{\mathfrak{p}}(q') \geq n_{\mathfrak{p}} \text{ for all } \mathfrak{p}\}$. By statement (1) of Lemma 1.10.14, $(f) \cap R[[K]] = fR[[K]]$, so $R[[K]]/(fR[[K]])$ is isomorphic to the ideal of D in A, a principal ideal. Thus $R[[K]]/(fR[[K]])$ is monogenic as an A-module, hence $R[[K]]$ is monogenic as an $R[[Q]]$-module, by Nakayama's lemma. This implies that $a'(Z)$ is in the image of d and concludes the proof. □

It is natural to ask if a localization of a regular log local ring is regular, and if, under a suitable excellence hypothesis, the regular locus of a log scheme is open. We shall see that both these questions have affirmative answers. However we defer the discussion until Section 3.5, after the discussion of logarithmic smoothness.

1.12 Frames for log structures

The following definition is modeled after a notion introduced by K. Kato and T. Saito [51, 4.1].

Definition 1.12.1. *Let $(X, \mathcal{M}_X)$ be a monoidal space.*

1. *If Q is a monoid, a frame for X subordinate to Q is a homomorphism $\phi\colon \overline{Q} \to \Gamma(X, \overline{\mathcal{M}}_X)$, which locally on X lifts to a chart $Q \to \mathcal{M}_X$.*
2. *More generally, if T is a monoscheme, a frame for X subordinate to T is a morphism of monoidal spaces $(X, \overline{\mathcal{M}}_X) \to (T, \overline{\mathcal{M}}_T)$ which, locally on X, lifts to a chart $X \to T$.*
3. *If S is an s-fan (Remark II.1.9.4), an s-frame for X subordinate to S is a morphism of monoidal spaces $f\colon (X, \overline{\mathcal{M}}_X) \to (S, \mathcal{M}_S)$ such that, locally on X and S, there exist a monoscheme T and an isomorphism $j\colon (S, \mathcal{M}_S) \to (T, \overline{\mathcal{M}}_T)$ such that $j \circ f$ is a frame.*

Remark 1.12.2. Let $\alpha\colon \mathcal{M} \to \mathcal{O}_X$ be the log structure on X associated to a DF structure. Then the map $\mathbf{N}^r \to \overline{\mathcal{M}}$ constructed in Proposition 1.7.3 is a frame. Conversely, suppose that a log structure α on X admits a frame $\mathbf{N}^r \to \mathcal{M}$. Then for each basis e_i of $\mathbf{N}^r$, the $\mathcal{O}_X^*$-torsor associated to image e_i in $\mathcal{M}$ defines an invertible sheaf $\mathcal{L}_i$, and α defines a DF structure $\{\gamma_i\colon \mathcal{L}_i \to \mathcal{O}_X\}$. Then one can check that α is the log structure associated to $\gamma.$, using the fact that $\mathbf{N}^r \to \mathcal{M}_X$ is a frame. Thus a log structure comes from a DF structure if and only if it admits a frame subordinate to a finitely generated free monoid.

The next result is an adaptation of a construction of Kato [49, 10.1]. It implies that every regular log scheme (in the Zariski topology) has a canonical s-frame.

Proposition 1.12.3. *Let X be a fine very solid log scheme for the Zariski topology. Then there exist an s-fan $S_X = (S_X, \mathcal{M}_S)$ and morphisms*

$$i\colon (S_X, \mathcal{M}_S) \to (X, \overline{\mathcal{M}}_X), \quad s\colon (X, \overline{\mathcal{M}}_X) \to (S_X, \mathcal{M}_S),$$

with $s \circ i = \mathrm{id}$*, such that* s *is universal: any morphism from* $(X, \overline{\mathcal{M}}_X)$ *to an s-fan factors uniquely through* s*. Moreover, the morphism* s *is open and is an s-frame for* X*.*

Proof Since α_X is very solid, for each point x of X, the ideal $\mathcal{M}^+_{X,x}\mathcal{O}_{X,x}$ is prime and hence corresponds to a generization $s(x)$ of x. Then the inverse image of $\mathfrak{m}_{X,s(x)}$ in $\mathcal{O}_{X,x}$ is $\mathcal{M}^+_{X,x}\mathcal{O}_{X,x}$, and it follows that the cospecialization homomorphism $\mathcal{M}_{X,x} \to \mathcal{M}_{X,s(x)}$ is local and then that the map $\overline{\mathcal{M}}_{X,x} \to \overline{\mathcal{M}}_{X,s(x)}$ is an isomorphism.

Let us say that a point x of X is an "s-point" if $x = s(x)$. Let S_X be the set of s-points of X, and let $i\colon S_X \to X$ be the inclusion. Endow S_X with the topology induced from its embedding in X and let $\mathcal{M}_S$ be the restriction of $\overline{\mathcal{M}}_X$ to S_X. Since $s(x)$ is a generization of x, it belongs to every neighborhood of x, and consequently formation of the spaces S_X and the maps i and s is compatible with passage to open subsets of X. Thus, for every open subset U of X, $s(U) = S_X \cap U$. It follows that s is a continuous open mapping. Since x is a specialization of $s(x)$, there is a cospecialization homomorphism $\mathcal{M}_{S_X,s(x)} = \overline{\mathcal{M}}_{X,s(x)} \to \overline{\mathcal{M}}_{X,x}$, and the integrality of $\mathcal{M}_X$ makes it easy to check that these maps fit into a homomorphism $\mathcal{M}_S \to s_*(\overline{\mathcal{M}}_X)$. We have seen that each homomorphism $\mathcal{M}_{S,s(x)} \to \overline{\mathcal{M}}_{X,x}$ is local, so that s becomes a morphism of locally monoidal spaces $(X, \overline{\mathcal{M}}_X) \to (S, \mathcal{M}_S)$.

Let $g\colon (X, \overline{\mathcal{M}}_X) \to T$ be a morphism, where T is an s-fan; we claim that there is a unique $g'\colon S_X \to T$ such that $g' \circ s = g$. Since $s \circ i = \mathrm{id}$, necessarily $g' = g \circ i$, and we must check that, with this definition of g', in fact $g' \circ s = g$. We can verify this equality locally on X, and hence may assume that T is affine, so that the map g is given by a monoid homomorphism $\beta\colon Q \to \overline{\mathcal{M}}_X$. If x is a point of X, then $g(x) = \beta^{-1}(\overline{\mathcal{M}}^+_{X,x}) \in \mathrm{Spec}(Q)$, and $g'(s(x)) = g(s(x)) = \beta^{-1}(\overline{\mathcal{M}}^+_{X,s(x)})$. This coincides with $g(x)$ because the homomorphism $\overline{\mathcal{M}}_{X,x} \to \overline{\mathcal{M}}_{X,s(x)}$ is local. Thus $g = g' \circ s$ on the level of sets. We must also verify that $g^\flat = s^\flat \circ g'^\flat$, i.e., that for each point x of X, the upper left triangle in the following diagram commutes:

$$\begin{array}{ccc}
\mathcal{M}_{T,g(x)} & \xrightarrow{\;g^\flat_x\;} & \overline{\mathcal{M}}_{X,x} \\
{\scriptstyle g'^\flat_{s(x)}}\big\downarrow & \nearrow{\scriptstyle s^\flat_x} & \big\downarrow{\scriptstyle \mathrm{cosp}} \\
\mathcal{M}_{S,s(x)} & \xleftarrow{\;i^\flat_{s(x)}\;} & \overline{\mathcal{M}}_{X,s(x)}.
\end{array}$$

Recall that $g(x) = g(s(x))$, so $\mathrm{cosp} \circ g^\flat_x = g^\flat_{s(x)}$, and so

$$g'^\flat_{s(x)} := i^\flat_{s(x)} \circ g^\flat_{s(x)} = i^\flat_{s(x)} \circ \mathrm{cosp} \circ g^\flat_x,$$

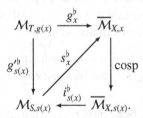

so the square commutes. Since all the arrows in the lower commutative right triangle are isomorphisms, it follows that the upper left triangle commutes as well.

It remains to prove that S_X is an s-fan and that s is an s-frame. These questions are also local on X, so we may and shall assume that X is affine, say $X = \operatorname{Spec}(\beta\colon Q \to A)$, where Q is a fine monoid. Choose a point x of X; after localizing Q, we may assume that $\beta_x\colon Q \to \mathcal{O}_{X,x}$ is local. Then the map $\overline{Q} \to \overline{\mathcal{M}}_{X,x}$ is an isomorphism, and since α_X is very solid it follows that β_x is very solid. As we saw in the proof of Proposition 1.10.11, we may replace X by some neighborhood such that $\beta\colon Q \to A$ is very solid. Assuming this is the case, let S be the s-fan defined by Q and let $g\colon (X, \overline{\mathcal{M}}_X) \to S$ be the map defined by β. Then $g\colon X \to S$ is locally surjective, hence open and, since its image contains the closed point of S_Q, g is surjective. Then $g' := g \circ i$ is also surjective and open. If $\mathfrak{p}$ is the prime ideal of A corresponding to a point y of X, then $s(y)$ corresponds to the prime ideal $\mathfrak{q}A$, where $\mathfrak{q} = \beta^{-1}(\mathfrak{p})$ is the prime ideal of Q corresponding to $g(y)$. If y is an s-point, it follows that y is uniquely determined by $g(y)$, and hence g' is injective. Since g' is continuous and open and bijective, it is a homeomorphism. Finally, the map $g'^{\flat}\colon \overline{\mathcal{M}}_{S_Q} \to \mathcal{M}_S$ is an isomorphism because β is a chart. Thus $S_X \cong S_Q$, and hence is an s-fan and s is an s-frame. □

It follows from the universal property of $s\colon X \to S_X$ that its formation is functorial: if $f\colon X \to Y$ is a morphism of fine and very solid schemes, there is an induced map $S_f\colon S_X \to S_Y$ such that $s_Y \circ f = S_f \circ s_X$. Such functoriality does not hold for the map $i\colon S_X \to X$. For example, if $X = \mathsf{A}_{\mathbf{N}}$, S_X consists of the vertex (origin) and the generic point, but, if $Y = \underline{\mathsf{A}}_{\mathbf{N}}$, S_Y consists of just the generic point, and the canonical map $X \to Y$ is the identity on underlying schemes and does not map S_X to S_Y.

Remark 1.12.4. Let X be a fine and very solid log scheme as in Proposition 1.12.3. The proposition shows that, for every $x \in X$, there is a point ξ (namely, the point $is(x)$), with the following properties.

1. The point x is a specialization of the point ξ.
2. The maximal ideal of $\mathcal{O}_{X,\xi}$ is generated by the image of the maximal ideal of $\mathcal{M}_{X,\xi}$.
3. $\dim \mathcal{O}_{X,\xi} = \dim \mathcal{M}_{X,\xi}$.

The first two properties of ξ follow from the construction, and property (3) follows from (2) and the equality implied by Proposition 1.10.17 in the solid case.

2 Morphisms of log schemes

In this section we discuss some of the important properties of morphisms of log schemes. We begin with a discussion of fiber products, whose formation is delicate in the categories of integral and separated log schemes. We then discuss some properties of morphisms of log schemes that are related to the corresponding homomorphism of sheaves of monoids. The important notion of smooth morphisms of log schemes will be covered in a later chapter. Some readers may prefer to skip ahead to that topic, deferring their study of some of the material here.

Before embarking on the technical discussion of morphisms of log schemes, let us first describe the set of morphisms from a log point into a log scheme. It is important to remember that the passage from a log scheme X to its underlying scheme $\underline{X}$ forgets a good deal of the geometry. In some contexts, it is helpful to picture the set of "log point" valued points of a log scheme.

Proposition 2.0.1. *Let X be a fine log scheme and let $p_X \colon X \to \underline{X}$ be the canonical map from X to the corresponding scheme with trivial log structure. Let S be an integral log point whose underlying scheme is the spectrum of an algebraically closed field k, and let $P := \overline{\mathcal{M}}_S$. Given a k-valued point $\overline{x} \in \underline{X}(k)$, let $Q_{\overline{x}}$ be the stalk of $\overline{\mathcal{M}}_X$ at $\overline{x}$ and let $p_X^{-1}(\overline{x})$ denote $\{g \colon S \to X : \underline{g} = \overline{x}\}$. Then there exist maps*

$$h \colon p_X^{-1}(\overline{x}) \to \operatorname{Hom}_{loc}(Q_{\overline{x}}, P), \quad \mathsf{A}^*_{Q_{\overline{x}}}(k) \times p_X^{-1}(\overline{x}) \to p_X^{-1}(\overline{x})$$

making $p_X^{-1}(\overline{x})$ an $\mathsf{A}^*_{Q_{\overline{x}}}(k)$-torsor over h. In particular:

1. *$p_X^{-1}(\overline{x})$ is empty if and only if $P = 0$ and $\overline{x} \notin X^*(k)$;*
2. *if $P = \mathbf{N}$, then $p_X^{-1}(\overline{x})$ is an $\mathsf{A}^*_{Q_{\overline{x}}}(k)$-torsor over the set of interior lattice points in the dual of $Q_{\overline{x}}$.*

Proof The point $\overline{x}$ defines a homomorphism $\overline{x}^\sharp \colon \mathcal{O}_{X,x} \to k$. Then an element g of $p_X^{-1}(\overline{x})$ amounts to a local homomorphism of monoids $g^\flat \colon \mathcal{M}_{X,\overline{x}} \to \mathcal{M}_S$ whose restriction to $\mathcal{O}^*_{X_{\overline{x}}}$ is given by $\overline{x}^\sharp$. In particular, $h(g) := \overline{g}^\flat$ is a local homomorphism $Q_{\overline{x}} \to P$. A point of $\mathsf{A}^*_{Q_{\overline{x}}}(k)$ can be viewed as a homomorphism $\gamma \colon \mathcal{M}_{X_{\overline{x}}} \to k^*$ which annihilates $\mathcal{O}^*_{X,x}$. Then $\gamma g^\flat \colon \mathcal{M}_{X,\overline{x}} \to k^*$ defines an element γg of $p_X^{-1}(\overline{x})$ with $h(\gamma g) = h(g)$. Moreover, if $g' \in p_X^{-1}(\overline{x})$ and $h(g') = h(g)$, then there is a unique γ such that $g' = \gamma g$. To show that this action makes $p_X^{-1}(\overline{x})$ a torsor over h, it remains only to show that h is surjective.

Since P is sharp, $\operatorname{Hom}_{loc}(\mathcal{M}_{X,\overline{x}}, P) = \operatorname{Hom}_{loc}(Q_{\overline{x}}, P)$. There is an exact sequence

$$\operatorname{Hom}(\mathcal{M}^{\mathrm{gp}}, k^*) \to \operatorname{Hom}(\mathcal{M}^{\mathrm{gp}}_{X,\overline{x}}, \mathcal{M}^{\mathrm{gp}}_S) \to \operatorname{Hom}(\mathcal{M}^{\mathrm{gp}}_{X,\overline{x}}, P^{\mathrm{gp}}) \to \operatorname{Ext}^1(\mathcal{M}^{\mathrm{gp}}_{X,\overline{x}}, k^*)$$

and, since k is algebraically closed, k^* is divisible and the Ext^1 group vanishes. Thus given any local homomorphism $\nu\colon \overline{\mathcal{M}}_{X,\overline{x}} \to P$, there is a homomorphism $\rho\colon \mathcal{M}^{gp}_{X,\overline{x}} \to \mathcal{M}^{gp}_S$ such that $\overline{\rho}^{gp} = \nu^{gp}$. Since $\mathcal{M}_S \to P$ is exact, it follows that ρ maps $\mathcal{M}_{X,\overline{x}}$ to $\mathcal{M}_S$, and the resulting homomorphism is local because ν is local.

If $P = 0$ and $\overline{x} \notin X^*(k)$, then $Q_{\overline{x}} \neq 0$ and there are no local homomorphisms $Q_{\overline{x}} \to P$. Hence $p_X^{-1}(\overline{x})$ is empty in this case. If $P = \mathbf{N}$, then the set of local homomorphisms $\overline{Q}_{\overline{x}} \to \mathbf{N}$ can be identified with the interior lattice points of the dual to $Q_{\overline{x}}$, by Remark I.2.2.8. More generally, if $P \neq 0$, choose $p \in P^+$ and note that the corresponding homomorphism $\mathbf{N} \to P$ is local. Proposition I.2.2.1 shows that there is a local homomorphism $\theta\colon Q \to \mathbf{N}$, and so the homomorphism $Q \to P$ sending q to $\theta(q)p$ is also local. It follows that $p_X^{-1}(\overline{x})$ is not empty. □

2.1 Fibered products of log schemes

Just as in the case of ordinary schemes, the existence of products in the category of log schemes has deep consequences and many subtleties.

Proposition 2.1.1. *Let X be a scheme. Then the category of prelog (resp. log) structures on X admits colimits. The colimit of a finite family of coherent log structures is coherent.*

Proof Let $\mathcal{M}_\cdot := \{\alpha_i\colon \mathcal{M}_i \to \mathcal{O}_X : i \in I\}$ be a family of prelog structures on X and let $\mathcal{M}$ be the colimit of the family $\mathcal{M}_\cdot$ in the category of sheaves of monoids on X. Then the maps $\{\alpha_i : i \in I\}$ induce a map $\beta\colon \mathcal{M} \to \mathcal{O}_X$, and β is the colimit of $\{\alpha_i : i \in I\}$ in the category of prelog structures on X. If each α_i is in fact a log structure, then the log structure $\alpha := \beta^a$ associated to β is the limit of $\{\alpha_i : i \in I\}$ in the category of log structures on X.

To prove that finite colimits of coherent log structures are coherent, it will suffice to check that the direct sum of two coherent log structures is coherent and that the coequalizer of two maps between coherent log structures is coherent. Suppose that, for $i = 1, 2$, each $\alpha_i\colon \mathcal{M}_i \to \mathcal{O}_X$ is a coherent log structure. Localizing on X, we may assume that each $\mathcal{M}_i$ admits a chart $\gamma_i\colon P_i \to \mathcal{M}_i$, with P_i finitely generated. Then each $\alpha_i \circ \gamma_i\colon P_i \to \mathcal{O}_X$ is a prelog structure, and the induced map $P := P_1 \oplus P_2 \to \mathcal{O}_X$ is their direct sum in the category of prelog structures. Since the formation of associated log structures is a left adjoint, it commutes with colimits, and hence the associated prelog structure $P^a \to \mathcal{O}_X$ is the direct sum of the log structures $\mathcal{M}_i$. Since P is finitely generated, $\mathcal{M}$ is coherent. Now suppose that $\alpha\colon \mathcal{M} \to \mathcal{O}_X$ and $\beta\colon \mathcal{N} \to \mathcal{O}_X$ are log structures and ϕ_1 and ϕ_2 are morphisms $\alpha \to \beta$. Choose a finitely generated chart $P \to \mathcal{M}$ for α. By Proposition II.2.4.2, we may find, locally on X, homomorphisms $\theta_i \colon P \to Q_i$ of finitely generated monoids and charts $Q_i \to \mathcal{N}$. By Corollary II.2.2.2, we may find a finitely generated Q and a chart $Q \to \mathcal{N}$ through which these charts factor. Thus we may assume that $Q_1 = Q_2 = Q$. Let R be the coequalizer of the two maps $\theta_i \colon P \to Q$ in the category of monoids. Then R is finitely generated, and the map $Q \to \mathcal{O}_X$ factors through R. Then the log structure $R^a \to \mathcal{O}_X$ is coherent and is the coequalizer of ϕ_1 and ϕ_2 in the category of log structures.

The most important case of colimits of log structures is that of coproducts. If α_0 is a log structure and $\theta_i \colon \alpha_0 \to \alpha_i$, with $i = 1, 2$ is a pair of morphisms log structures, then the coproduct $\alpha \colon \mathcal{M} \to \mathcal{O}_X$ is formed by taking the log structure associated to the prelog structure $\mathcal{M}_1 \oplus_{\mathcal{M}_0} \mathcal{M}_2 \to \mathcal{O}_X$. If all these log structures are coherent, then locally on X one can find coherent charts $P_i \to \mathcal{M}_i$ and compatible morphisms $P_0 \to P_i$. Then the induced map $P_1 \oplus_{P_0} P_2 \to \mathcal{M}$ is a chart for $\mathcal{M}$. □

As we shall see, the construction of fibered products in the category of log schemes is straightforward, and in particular the underlying scheme of such a product is the fiber product of the underlying schemes. This is not the case for fibered products in the categories of integral and saturated log schemes, which we discuss later.

Proposition 2.1.2. *The category of log schemes admits fibered products, and the functor $X \to \underline{X}$ taking a log scheme to its underlying scheme commutes with fibered products. The fibered product of coherent log schemes is coherent.*

Proof Let $f \colon X \to Z$ and $g \colon Y \to Z$ be morphisms of log schemes and let $\underline{X}'$ denote the fiber product $\underline{X} \times_{\underline{Z}} \underline{Y}$ in the category of schemes, with projection maps $g' \colon \underline{X}' \to \underline{X}$ and $f' \colon \underline{X}' \to \underline{Y}$, and $h := f \circ g' = g \circ f'$. Then on $\underline{X}'$ we find morphisms of log structures $h^*_{log}(\alpha_Z) \to g'^*_{log}(\alpha_X)$ and $h^*_{log}(\alpha_Z) \to f'^*_{log}(\alpha_Y)$. It is straightforward to verify that if $\alpha_{X'}$ is the coproduct of these log structures, then the log scheme $X' := (\underline{X}', \alpha_{X'})$, together with its maps $X' \to X$ and $X' \to Y$, is the fibered product $X \times_Z Y$. If the log structures α_X, α_Y, and α_Z are coherent, so are their pullbacks to X', and hence by Proposition 2.1.1, so is $\alpha_{X'}$. By construction, the scheme underlying $(X', \alpha_{X'})$ is the fiber product $\underline{X} \times_{\underline{Z}} \underline{Y}$. □

Remark 2.1.3. Suppose that the morphism $f \colon X \to Z$ is strict, i.e., that $f^*_{log}(\alpha_Z) \cong \alpha_X$ on X. Then on X', $h^*_{log}(\alpha_Z) \cong g'^*_{log}(\alpha_X)$. It follows that

$$f'^*_{log}(\alpha_Y) \oplus_{h^*_{log}(\alpha_Z)} g'^*_{log}(\alpha_X) \cong f'^*_{log}(\alpha_Y).$$

Thus, the morphism f' is also strict. In other words, the family of strict maps is stable under base change.

Since the amalgamated sum of integral (resp. saturated) monoids need not be integral (resp. saturated), the construction of fibered products in the category of fine (resp. fine and saturated) log schemes is more delicate, and in fact involves some of the main technical difficulties of logarithmic algebraic geometry.

The following result—just a reformulation of descent theory—is a basic tool in the construction of log schemes, following the pattern we sketched for monoschemes in Proposition II.1.3.2. It applies, for example, to the category of schemes, log schemes, fine log schemes, etc.

Proposition 2.1.4. *Let* $\mathbf{LS}'$ *be a full subcategory of the category of log schemes which is locally defined, i.e., such that any log scheme* X *admitting an open covering whose objects are in* $\mathbf{LS}'$ *also belongs to* $\mathbf{LS}'$. *Let* S *be a scheme and denote by* $\mathbf{LS}'_S$ *the category of pairs* (X, f) *where* X *is an object of* $\mathbf{LS}$ *and* f *is a morphism* $\underline{X} \to S$. *Finally, let* F *be a functor from* $\mathbf{LS}'_S$ *to the category of sets.*

1. *If* F *admits a Zariski open cover by representable functors, then* F *is also representable.*
2. *If* S *admits a Zariski open covering* $\{S_i \to S\}$ *such that the restriction of* F *to each* S_i *is representable, then* F *is representable.*
3. *If* S *admits an étale open covering* $\{S_i \to S\}$ *such that the restriction of* F *to each* S_i *is representable by an object in* $\mathbf{LS}'_{S_i}$ *whose underlying scheme is affine over* S_i, *then* F *is representable by an object whose underlying scheme is affine over* S.

Proof Statement (1) follows from the fact that topological spaces and sheaves thereon can be glued along open coverings, as in Proposition II.1.3.2, thanks to the locality of the subcategory $\mathbf{LS}'$. Statement (2) is a direct consequence, since $\{F_i := F \times_S S_i\}$ is a Zariski open cover of F. Since gluing schemes along étale open subsets can yield algebraic spaces that are not schemes, we need the affineness condition in (3). Specifically, suppose that F_i is represented by an object X_i of $\mathbf{LS}'_S$ such that $\underline{X}_i$ is affine over S_i. Since F_i is the restriction of F to S_i and X_i represents F_i, we have isomorphisms between the restriction of X_i to $S_i \cap S_j$ and the restriction of X_j to $S_i \times_S S_j$, and these isomorphisms satisfy the cocycle condition over $S_i \times_S S_j \times_S S_k$. This collection of isomorphisms forms "descent data" with respect to the given covering of S. Since $\underline{X}_i$ is affine over S_i, it corresponds to a quasi-coherent sheaf $\mathcal{A}_i$ of $\mathcal{O}_{S_i}$-algebras. The descent data on the schemes $\underline{X}_i$ define define gluing data on these sheaves and hence a quasi-coherent sheaf of $\mathcal{O}_S$-algebras $\mathcal{A}$ on the étale topology of S. By Hilbert's Theorem 90 for quasi-coherent sheaves [15, I-2], this sheaf descends to the Zariski topology of S and hence defines an affine scheme X over S. Then the log structures α_i on the étale topology of each X_i glue to a log structure α on the étale topology of X, and define an object (X,α) of $\mathbf{LS}'_S$. Finally, the isomorphisms of functors $(X_i,\alpha_i) \to F_i$ glue to define an isomorphism $(X,\alpha) \to F$. □

Proposition 2.1.5. *Let* **LS** *(resp.* $\mathbf{LS}_{\mathrm{coh}}, \mathbf{LS}_{\mathrm{fin}}, \mathbf{LS}_{\mathrm{fs}}$*) denote the category of log schemes (resp. of coherent log schemes, fine log schemes, fine and saturated log schemes).*

1. *The inclusion functor* $\mathbf{LS}_{\mathrm{fin}} \to \mathbf{LS}_{\mathrm{coh}}$ *admits a right adjoint* $X \mapsto X^{\mathrm{int}}$. *For each object* X *of* $\mathbf{LS}_{\mathrm{coh}}$*, the map* $\underline{X}^{\mathrm{int}} \to \underline{X}$ *is a closed immersion.*
2. *The inclusion functor from* $\mathbf{LS}_{\mathrm{fin}} \to \mathbf{LS}_{\mathrm{fs}}$ *admits a right adjoint* $X \mapsto X^{\mathrm{sat}}$*, and for each object* X *of* $\mathbf{LS}_{\mathrm{fin}}$*, the corresponding morphism* $\underline{X}^{\mathrm{sat}} \to \underline{X}$ *is finite and surjective.*

Proof Let X be a coherent (resp. fine) log scheme, let h_X be its Yoneda functor on the category of coherent (resp. fine) log schemes and let h'_X denote its restriction to $\mathbf{LS}' := \mathbf{LS}^{\mathrm{fin}}$ (resp. $\mathbf{LS}^{\mathrm{fs}}$). We claim that h'_X is representable in $\mathbf{LS}'$. Let $\mathbf{LS}'_{\underline{X}}$ denote the category of fine (resp. fine and saturated) log schemes T endowed with a morphism $\underline{T} \to \underline{X}$ and, for any object (T,p) of $\mathbf{LS}'_{\underline{X}}$, let $F(T) := \{f\colon T \to X : \underline{f} = p\}$. Then h'_X will be representable in $\mathbf{LS}'$ if and only if F is representable in $\mathbf{LS}'_{\underline{X}}$, as is tautological to verify.

To prove the proposition, first suppose that $X = \mathsf{A}_P$, where P is a finitely generated (resp. fine) monoid. In this case let $P' := P^{\mathrm{int}}$ (resp. P^{sat}), and let $X' := \mathsf{A}_{P'}$. Then if T is any fine (resp. fine and saturated) log scheme,

$$h'_X(T) = \mathrm{Hom}(P, \mathcal{M}_T) = \mathrm{Hom}(P', \mathcal{M}_T) = h_{X'}(T).$$

Let $g\colon X' \to X$ be the natural map, so that $(X',\underline{g})$ is an object of $\mathbf{LS}'_{\underline{X}}$. Then $(X',\underline{g})$ represents the functor F on $\mathbf{LS}_{\underline{X}}$. When P is finitely generated, the homomorphism $P \to P^{\mathrm{int}}$ is surjective and hence the morphism $\underline{X}' \to \underline{X}$ is a closed immersion. When P is fine, the homomorphism $P \to P^{\mathrm{sat}}$ is finite and injective, and hence the morphism $X' \to X$ is finite and surjective. Thus the result is proved in this case.

Somewhat more generally, suppose that X admits a chart subordinate to a finitely generated (resp. fine) monoid P. Then we find a strict morphism $f\colon X \to Y := \mathsf{A}_P$. By Proposition 1.2.5, the morphism $X \to \underline{X} \times_{\underline{Y}} Y$ is an isomorphism, and it follows that the map $h'_X \to \underline{X} \times_{\underline{Y}} h'_Y$ is also an isomorphism. As we saw in the previous paragraph, $h'_Y \cong h_{Y'}$; so $X' := \underline{X} \times_{\underline{Y}} Y'$, an object of $\mathbf{LS}'$, represents the functor h'_X.

The proposition now follows from Proposition 2.1.4. Indeed, if X is coherent (resp. fine), then $\underline{X}$ has an étale covering $\{\underline{X}_i \to \underline{X}\}$ such that each X_i admits a chart subordinate to a finitely generated (resp. fine) monoid. In each case, the functor F is representable by an object (X'_i, p_i), where $p_i \colon \underline{X}'_i \to \underline{X}$ is affine, and hence F is representable. □

Notice that the morphisms of topological spaces underlying the maps $X^{\rm int} \to X$ and $X^{\rm sat} \to X^{\rm int}$ are not in general homeomorphisms and, in particular, that we cannot identify $\mathcal{M}_{X^{\rm int}}$ with $\mathcal{M}_X^{\rm int}$ or $\mathcal{M}_{X^{\rm sat}}$ with $\mathcal{M}_X^{\rm sat}$, in general.

Corollary 2.1.6. *The category of fine log schemes (resp. of fine and saturated log schemes) admits finite projective limits. If X and Y are fine (resp. fine and saturated) log schemes over a fine (resp. fine and saturated) log scheme Z, then the natural map from the underlying scheme of the fibered product $X \times_Z Y$ to the fibered product of underlying schemes is a closed immersion (resp. a finite morphism).*

Proof If $X \to Z$ and $Y \to Z$ are morphisms of fine log schemes, then it follows from the universal mapping properties that $(X \times_Z Y)^{\rm int}$, together with its induced maps to X, Y, and Z, is the fibered product of X and Y over Z in the category of fine log schemes. Analogous constructions works for fine and saturated log schemes. □

Example 2.1.7. In classical geometry, the fiber product of points is never empty, because the tensor product of field extensions is never zero. However in the category of fine log schemes, the fiber product of morphisms of log points can be empty. This can happen because the pushout of a local homomorphism can fail to be local (see Example I.4.1.6). The following example is based on a different phenomenon involving the behavior of units.

Let P be the monoid with generators p_1, p_2 and relation $mp_1 = mp_2$ and let S_P be the split log point associated to P over some field k. If ζ is an mth root of unity in k, define a homomorphism of monoids $\theta_\zeta \colon P \to k^* \oplus \mathbf{N}$ by sending p_1 to $(1, 1)$ and p_2 to $(\zeta, 1)$. This homomorphism defines a morphism of log schemes $f_\zeta \colon S_{\mathbf{N}} \to S_P$. We claim that if $\zeta \neq \zeta'$, the fiber product of f_ζ and $f_{\zeta'}$ in the category of integral log schemes is empty. Indeed, if t were a point of the fiber product in the category of integral log schemes, there would be a commutative diagram

$$\begin{array}{ccc} P & \xrightarrow{\theta_\zeta} & k^* \oplus \mathbf{N} \\ {\scriptstyle \theta_{\zeta'}}\downarrow & & \downarrow{\scriptstyle \theta'} \\ k^* \oplus \mathbf{N} & \xrightarrow{\theta} & \mathcal{M}_{T,t} \end{array}$$

with $\mathcal{M}_{T,t}$ an integral monoid containing k^*. Let $q := \theta(1, 1)$ and $q' := \theta'(1, 1)$

in $\mathcal{M}_{T,t}$. Then $q' = \theta'\theta_\zeta(p_1) = \theta\theta_{\zeta'}(p_1) = q$, and $\zeta + q' = \theta'\theta_\zeta(p_2) = \theta\theta_{\zeta'}(p_2) = \zeta' + q$. We conclude that $\zeta + q = \zeta' + q$. The integrality of $\mathcal{M}_{T,t}$ then implies that $\zeta = \zeta'$. Note that, in this example, neither θ_ζ nor $\theta_{\zeta'}$ is exact. For ways to avoid these difficulties, see Proposition 2.2.3 and Lemma 2.4.9.

Fiber products also exist in the category of idealized log schemes. The construction is straightforward. Given morphisms of idealized log schemes $X \to Z$ and $Y \to Z$, one constructs the fiber product $X \times_Z Y$ in the category of sheaves, and endows its sheaf of monoids with the sheaf of ideals $\mathcal{K}$ generated by the sheaves of ideals $\mathcal{K}_X$ and $\mathcal{K}_Y$. Then $\mathcal{K}$ maps to zero in $\mathcal{O}_{X\times_Z Y}$ and it is easy to check that the universal mapping property of fiber products is satisfied. An important example is the notion of the idealized log fiber over a point.

Definition 2.1.8. *If y is a point of a log scheme Y, the* idealized log point at y *is the idealized log scheme*

$$y_{\mathcal{K}} := \operatorname{Spec}\left((i^*_{log}(\mathcal{M}_{Y,y}, \mathcal{M}^+_{Y,y})) \to k(y)\right),$$

where $i\colon y \to Y$ is the inclusion and $\mathcal{M}^+_{Y,y}$ is the maximal ideal of $\mathcal{M}_{Y,y}$. If $f\colon X \to Y$ is a morphism of fine log schemes, the idealized log fiber of f over y *is the fiber product*

$$X_{\mathcal{K}} := y_{\mathcal{K}} \times_Y X,$$

computed in the category of idealized log schemes.

2.2 Exact morphisms

Let $f\colon X \to Y$ be a morphism of integral log schemes. Following our conventions (see Definition 1.1.11), we say that f is exact at a point x of X if for every (equivalently, for some) geometric point $\overline{x}$ lying over x, the homomorphism $f^\flat_{\overline{x}}\colon f^*_{log}(\mathcal{M}_{Y,f(\overline{x})}) \to \mathcal{M}_{X,\overline{x}}$ is exact or, equivalently, if the corresponding homomorphisms $\overline{\mathcal{M}}_{Y,f(\overline{x})}) \to \overline{\mathcal{M}}_{X,\overline{x}}$ and $\mathcal{M}_{Y,f(\overline{x})} \to \mathcal{M}_{X,\overline{x}}$ are exact.

Our first result is an immediate consequence of Proposition I.4.2.1.

Proposition 2.2.1. *In the category of integral log schemes, the following statements hold.*

1. *Let $f\colon X \to Y$ and $g\colon Y \to Z$ be morphisms and let x be a point of X, with $y := f(x)$ and $z := g(y)$. If f is exact at x and g is exact at y, then $g \circ f$ is exact at x. If $g \circ f$ is exact at x, then g is exact at y. If $g \circ f$ is exact at x and $\overline{g}^\flat_z$ is surjective (or is small and $\mathcal{M}_{Y,y}$ is saturated) then f is exact at x.*
2. *If $f\colon X \to Y$ is exact at x, then $\overline{f}^\flat_{\overline{x}}\colon \overline{\mathcal{M}}_{Y_{f(\overline{x})}} \to \overline{\mathcal{M}}_{X,\overline{x}}$ is injective. Thus an exact morphism is s-injective.*

3. *If $f\colon X \to Y$ and $h\colon Y' \to Y$ are morphisms, let $X' := X' \times_Y X$, with projections $f'\colon X' \to Y'$ and $g\colon X' \to X$. If $x' \in X$ and f is exact at $g(x)$, then f' is exact at x'. In particular, the class of exact morphisms is stable under fiber products.* □

The following result relates exactness of morphisms and exactness of their charts.

Proposition 2.2.2. *Let $f\colon X \to Y$ be a morphism of fine log schemes, equipped with a fine chart (α, θ, β). Suppose that $x \in X$ and $y = f(x)$.*

1. *If β is exact at x and θ is exact, then α is exact at y and f is exact at x. If f is exact at x and α is exact at y, then θ is exact.*
2. *If θ is locally exact then f is exact. Conversely, suppose that f is exact at every point in the fiber X' over $y := f(x)$, that the idealized fiber $X'_{\mathcal{K}}$ is solid (in the sense of Variant 1.10.18), and that β is exact at x. Then θ is locally exact and hence* **Q***-integral.*

Proof Since we are assuming the existence of charts in this proposition, we do not need to distinguish between geometric and scheme-theoretic points. Suppose that $\beta_x\colon Q \to \mathcal{M}_{X,x}$ and θ are exact. Then the homomorphism $f^\flat_x \circ \alpha_y = \beta_x \circ \theta\colon P \to \mathcal{M}_{X,x}$ is exact. It follows that α_y is exact and hence by Definition II.2.3.1 that $\overline{\alpha}_y$ is an isomorphism. Since $\overline{\theta}$ is also exact and $\overline{\beta}_x$ is an isomorphism, it follows that $\overline{f}^\flat_x$ is exact, so f is exact at x. Conversely, if $\alpha_y\colon P \to \mathcal{M}_{Y,y}$ and $f^\flat_x\colon \mathcal{M}_{Y,y} \to \mathcal{M}_{X,x}$ are exact, then $\beta_x \circ \theta = f^\flat_x \circ \alpha_y$ is exact, and it follows that θ is exact.

Suppose that θ is locally exact. Let $G := \beta^{-1}(\mathcal{M}^*_{X,x})$ and let $F := \theta^{-1}(G) = \alpha^{-1}(\mathcal{M}^*_{Y,y})$. Since θ is locally exact, the induced homomorphism $P_F \to Q_G$ is exact. Replacing β by the chart $\beta_G\colon Q_G \to \mathcal{M}_X$ in some neighborhood of x, we see from part (1) that f is exact at x. For the converse, let $S := \operatorname{Spec}(Q)$, $T := \operatorname{Spec}(P)$, and $g := \operatorname{Spec}(\theta)$. Then α and β define morphisms of monoidal spaces $b\colon (Y, \mathcal{M}_Y) \to T$ and $a\colon (X, \mathcal{M}_X) \to S$. Let $s = a(x)$ and $t := b(y)$. According to Theorem I.4.7.7, it will suffice to prove that θ is critically exact, i.e., that $g^\flat_{s'}$ is exact for every $s' \in g^{-1}(t)$. Since β is exact at x, the homomorphism $Q \to \overline{\mathcal{M}}_{X,x}$ is an isomorphism and the map $\operatorname{Spec}(\mathcal{M}_{X,x}) \to S$ is a homeomorphism, and we identify these spaces. Then $s' \in \operatorname{Spec}(\mathcal{M}_{X,x}, \mathcal{K})$ and, since the idealized log fiber$X'_{\mathcal{K}}$ is solid, the map $a\colon X'_{\mathcal{K}} \to \operatorname{Spec}(\mathcal{M}_{X,x}, \mathcal{K})$ is locally surjective. Since $x \in X'_{\mathcal{K}}$ maps to s and s' is a generization of s, there is a generization x' of x such that $a(x') = s'$. Since β is a chart, the map $\overline{\mathcal{M}}_{S,s'} \to \overline{\mathcal{M}}_{X,x'}$ is an isomorphism. By hypothesis f is exact at x', so the composed homomorphism $\overline{\mathcal{M}}_{T,t} \cong \overline{\mathcal{M}}_{Y,y} \to \overline{\mathcal{M}}_{X,x'} \to \overline{\mathcal{M}}_{S,s'}$ is exact. □

The following result is sometimes called the "four point lemma" for log schemes. It shows how exactness helps overcome the difficulties entailed by the necessity to work with fiber products in the category of integral log schemes.

Proposition 2.2.3. *Let $f_1 : X_1 \to Y$ and $f_2 : X_2 \to Y$ be morphisms of fine log schemes, and let X' be their fiber product in the category of integral log schemes. Suppose that $x_1 \in X_1$ and $x_2 \in X_2$ are points such that $f_1(x_1) = f_2(x_2)$. Then if f_1 is exact at x_1 or f_2 is exact at x_2, there exists a point $x' \in X'$ such that $p_1(x') = x_1$ and $p_2(x') = x_2$.*

Proof The statement is étale local around the points, so we may assume that there exists an exact chart $P \to \mathcal{M}_{Y,y}$ for $\mathcal{M}_{Y,y}$, with P fine. By Proposition II.2.4.2, we may also assume that there exists a chart for each f_i subordinate to a homomorphism of fine monoids $\theta_i : P \to Q_i$, and we may also assume that the homomorphisms $Q_i \to \mathcal{M}_{X_i,x}$ are local, equivalently, exact. Suppose that f_1 is exact at x_1. Then, by Proposition 2.2.2, θ_1 is exact. Let $Q := Q_1 \oplus_P Q_2$, formed in the category of integral monoids. By Proposition I.4.2.5, the homomorphisms $Q_i \to Q$ are local and the homomorphism $Q_1^* \oplus_{P^*} Q_2^* \to Q^*$ is an isomorphism.

Let k be the residue field of the point $y := f(x_i)$ of Y and let k_i be the residue field of x_i. Let $\gamma := \alpha_{Y,y} \circ \theta : P \to k$ and $\beta_i := \alpha_{X_i,x_i} \circ \theta_i : Q_i \to k_i$; since these homomorphisms are local, they send the maximal ideals of P^+ (resp. Q_i^+) to 0. There are then morphisms of log schemes $S := \mathrm{Spec}(\beta) \to Y$ and $T_i := \mathrm{Spec}(\beta_i) \to X_i$ sending the unique point of S to y and the unique point of T_i to x_i.

The k-algebra $k_1 \otimes_k k_2$ is not zero and hence admits a homomorphism to a field k'. The homomorphisms $Q_1^* \to k_1$ and $Q_2^* \to k_2$ define a homomorphism

$$\beta^* : Q^* \cong Q_1^* \oplus_{P^*} Q_2^* \to k_1 \otimes_k k_2 \to k'.$$

Extend this to a homomorphism $\beta : Q \to k'$ by sending Q^+ to zero, and let $T := \mathrm{Spec}(\beta)$. Since the homomorphisms $Q_i \to Q$ are local, we have morphisms $T \to T_i$, which agree when composed with the map to S. We find morphisms $T \to X_i$, which agree when composed with the map to Y, and since T is an integral log scheme, a morphism $T \to X_1 \times_Y X_2$. The image of the unique point of T has the desired properties. □

Corollary 2.2.4. *Let $f_1 : X_1 \to Y$ and $f_2 : X_2 \to Y$ be morphisms of fine log schemes and let X' (resp. X'') be their fiber product in the category of coherent (resp. fine) log schemes. If f_1 or f_2 is exact, the morphism $i : X'' \to X'$ is a nil immersion.*

Proof By (1) of Proposition 2.1.5, the morphism i is a closed immersion, and

Proposition 2.2.3 implies that i is surjective. These two facts imply that i is a nil immersion. □

Corollary 2.2.5. *A surjective and exact morphism of fine (resp. fine saturated) log schemes is universally surjective in the category of fine log schemes. If $f: X \to Y$ is a morphism of fine log schemes that is universally bijective in the category of fine log schemes and Y is saturated, then f is exact.*

Proof The first statement is a direct consequence of Proposition 2.2.3, together with the fact that the map $X^{\text{sat}} \to X$ is surjective for the saturated case. We write the proof of the converse for fine log schemes, leaving the saturated case to the reader. Suppose that f is universally bijective, i.e., that for every $Y' \to Y$, the underlying morphism of topological spaces $\underline{X}' \to \underline{Y}'$ is bijective, where $X' := X \times_Y Y'$ in the category of fine log schemes. Let $\overline{x}$ be a geometric point of X and let $\overline{y}$ be its image in Y. To prove that $f^\flat_{\overline{x}}$ is exact, we may replace Y by the log point $\overline{y}$, since the induced map $X_{\overline{y}} \to \overline{y}$ is again universally bijective. Thus we may assume without loss of generality that Y is a log point and even that $\underline{Y}$ is the spectrum of an separably closed field $k = k(\overline{y}) = k(\overline{x})$. Then $\underline{X}$ has a single point x and, since k is separably closed, $\mathcal{M}_{X,x} \cong \mathcal{M}_{X,\overline{x}}$. Choose a splitting $\alpha: P := \overline{\mathcal{M}}_{Y,y} \to \mathcal{M}_{Y,y}$ and a fine chart (α, θ, β) for f, with $\beta: Q \to \mathcal{M}_{X,x}$ exact. It will suffice to prove that θ is exact. Let $\phi: P \to P'$ be a local homomorphism of fine monoids, and let $\theta': P' \to Q'$ be the integral pushout of θ. The homomorphism $P' \to k$ sending P'^+ to 0 and P'^* to 1 defines a prelog structure on k, and hence a log point Y'. Since ϕ is local and P is sharp, we find a morphism of log points $Y' \to Y$ for which ϕ is a chart. Then $\theta': P' \to Q'$ is a chart for $X' \to Y'$, and by assumption there is a point $x' \in X'$ mapping to the unique point of Y'. It follows that there is a prime ideal of Q' mapping to the maximal ideal of P', and consequently that θ' is local. Thus θ is universally local, in the sense of Proposition I.4.2.3. Since P is saturated, it follows from this proposition that θ is exact. □

Example 2.2.6. To show that Corollary 2.2.5 is not trivial, here is an example of a morphism that is surjective but not universally so. Consider the following homomorphisms of monoids:

$$\sigma: \mathbf{N} \oplus \mathbf{N} \to \mathbf{N} : (a, b) \mapsto a + b,$$
$$\sigma': \mathbf{N} \oplus \mathbf{N} \to \mathbf{N} : (c, d) \mapsto c,$$
$$\tau: \mathbf{N} \oplus \mathbf{N} \to \mathbf{N} \oplus \mathbf{N} : (a, b) \mapsto (a + b, b),$$

leading to the following diagrams of monoids and log schemes:

$$\begin{array}{ccc} & & \mathbf{N}\oplus\mathbf{N} \\ & \swarrow^{\sigma'} & \uparrow \tau \\ \mathbf{N} & \xleftarrow{\ \sigma\ } & \mathbf{N}\oplus\mathbf{N}\end{array} \qquad\qquad \begin{array}{ccc} & & Z' \\ & \nearrow^{i'} & \downarrow f \\ X & \xrightarrow{\ i\ } & Z.\end{array}$$

Then f is a monomorphism in the category of fine log schemes, as follows from Proposition 1.2.4 and the fact that τ^{gp} is surjective. Let $X \to Y \to Z$ be the canonical factorization (1.2.1) of i, with $j\colon X \to Y$ the identity on topological spaces and $Y \to Z$ strict. Then j is surjective but not exact, and we shall see that it is not universally surjective.

The log scheme Y is $\operatorname{Spec}(\alpha\colon \mathbf{N}\oplus\mathbf{N} \to \mathbf{Z}[t])$, where α sends (a,b) to t^{a+b}. Let $Y' := Y \times_Z Z'$. Then the morphism $Y' \to Y$ is also a monomorphism of fine log schemes. Since $Y \to Z$ is strict, the underlying scheme of Y' is

$$\underline{Y} \times_{\underline{Z}} \underline{Z}' \cong \operatorname{Spec}(\mathbf{Z}[t,s]/(ts = t)).$$

The maps i' and j define a map $j'\colon X \to Y'$. Since $Y' \to Y$ is a monomorphism, the map $(\mathrm{id}, j')\colon X \to X \times_Y Y'$, is an isomorphism, but j' is not surjective.

We saw in Section I.4.2 that the condition of exactness becomes much more powerful when it holds at sufficiently many nearby points. The next theorem makes use of the notion of the solidity of the idealized fiber (see Variant 1.10.18) to guarantee that there are enough such points, in the context of log schemes.

Theorem 2.2.7. *Let $f\colon X \to Y$ be a morphism of fine saturated log schemes, let x be a point of X, and let $y := f(x)$, with the log structure inherited from Y. Let $X' := X \times_Y y$ be the fiber of f over y and let $X'_{\mathcal{K}} := (\underline{X}', \alpha_{X'}, \mathcal{K}_{X'})$ be its idealized version. If the idealized log fiber $X'_{\mathcal{K}}$ is solid at x, the following conditions are equivalent and are also equivalent to the existence, in some étale neighborhood of x, of a chart for f subordinate to a $\mathbf{Q}$-integral homomorphism of fine monoids.*

1. *f is $\mathbf{Q}$-integral in some neighborhood of x.*
2. *f is exact in some neighborhood of x.*
3. *f is s-injective in some neighborhood of x.*
4. *f is $\mathbf{Q}$-integral at x.*
5. *$f'\colon X' \to y$ is exact in some neighborhood of x.*
6. *$f'\colon X' \to y$ is s-injective in some neighborhood of x.*
7. $\dim(\mathcal{O}_{X',\bar{x}}/(\mathcal{M}^+_{X,\bar{x}}\mathcal{O}_{X',xx})) + \dim(\mathcal{M}_{X,\bar{x}}) = \dim(\mathcal{M}_{Y,\bar{y}}) + \dim(\mathcal{O}_{X',\bar{x}})$.

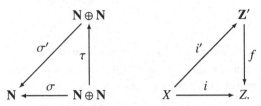

8. $\dim(\mathcal{M}_{X,\overline{x}}) = \dim(\mathcal{M}_{Y,\overline{y}}) + \dim(\mathcal{M}_{X,\overline{x}}, \mathcal{K}_{X',\overline{x}})$.

In any case, the implications denoted by the solid arrows in the figure

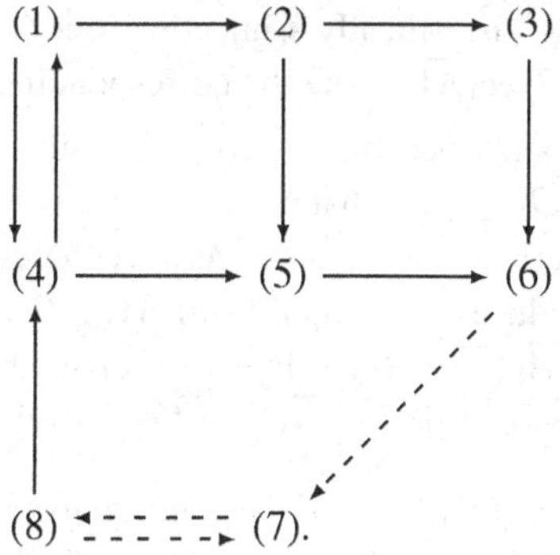

hold.

Proof Since the theorem is a local statement, we may replace X and Y by Zariski neighborhoods of x and y respectively. In fact, if $U \to X$ and $V \to Y$ are étale neighborhoods of x and y with U mapping to V, and if the theorem is true for the restriction of f to U, then it is true for f as well, because the image of U in X is Zariski open. Thus we may assume that the morphism f admits a chart $(\alpha, \theta\beta)$, where $\theta \colon P \to Q$ is a homomorphism of fine monoids. Replacing P and Q by suitable localizations, we may also assume that $P \to \mathcal{M}_{Y,y}$ and $Q \to \mathcal{M}_{X,x}$ are local, so that $\overline{P} \cong \overline{\mathcal{M}}_{Y,y} \cong \overline{\mathcal{M}}_{Y,\overline{y}}$ and $\overline{Q} \cong \overline{\mathcal{M}}_{X,x} \cong \overline{\mathcal{M}}_{X,\overline{x}}$. Let K_θ be the ideal of Q generated by $\theta(P^+)$. Then $(Q, K_\theta) \to (\mathcal{M}_{X',x}, \mathcal{K}_{X'})$ is a chart for the idealized log scheme $X'_{\mathcal{K}}$.

The implications indicated by the downward arrows in the first row of the figure are trivial. Statement (1) implies (2) because a **Q**-integral homomorphism of saturated log schemes is exact, by Theorem I.4.7.7. Statement (2) implies (3) and statement (5) implies (6) because an exact homomorphism is s-injective, by Proposition I.4.2.1. We next claim that (4) implies (1). Since a homomorphism θ is **Q**-integral if and only if $\overline{\theta}$ is, hypothesis (4) implies that $\theta \colon P \to Q$ is **Q**-integral. It follows from Corollary I.4.6.4 that, for every face F of P mapping into a face G of Q, the induced homomorphism $P_F \to Q_G$ is also **Q**-integral. For every $x' \in X$, the homomorphism $f^\flat_{x'} \colon \mathcal{M}_{Y_{f(x')}} \to \mathcal{M}_{X,x'}$ is charted by such a homomorphism, and hence is again **Q**-integral. Thus f is **Q**-integral on all of X. Finally, (8) implies (4) by Theorem I.4.7.7. We have proved all the implications indicated by solid arrows.

Now suppose that $X'_{\mathcal{K}}$ is solid at x. Then, by Variant 1.10.18,

$$\dim(\mathcal{O}_{X',x}) = \dim((\mathcal{O}_{X',x}/\mathcal{M}^+_{X,x}\mathcal{O}_{X',x})) + \dim(\mathcal{M}_{X,x}, \mathcal{K}_x).$$

Thus condition (7) is equivalent to (8).

It remains to prove that (6) implies (8) when $X'_{\mathcal{K}}$ is solid at x. By Theorem I.4.7.7, it will suffice to show that (6) implies that the homomorphism $\overline{\mathcal{M}}_{Y,y} \to \overline{\mathcal{M}}_{X,x} = \overline{\mathcal{M}}_{X',x}$ is critically s-injective. Let G be an $\overline{f}^{\flat}_{x}$-critical face of $\overline{\mathcal{M}}_{X',x}$, and let $\mathfrak{q} \in \operatorname{Spec}(\overline{\mathcal{M}}_{X',x})$ be the corresponding prime ideal. Since G is $\overline{f}^{\flat}_{x}$-critical, $\mathfrak{q}$ belongs to $\operatorname{Spec}(\mathcal{M}_{X',x}, \mathcal{K}_{X',x})$ and, since $X'_{\mathcal{K}}$ is solid, there is a prime ideal $\mathfrak{p}$ of $\operatorname{Spec}(\mathcal{O}_{X',x})$ such that $\alpha^{-1}_{X',x}(\mathfrak{p}) = \mathfrak{q}$. Let ξ be the point of X' corresponding to $\mathfrak{p}$, a generization of x. Since $\mathcal{M}_X$ is coherent, the cospecialization map $\mathcal{M}_{X,x} \to \mathcal{M}_{X,\xi}$ induces an isomorphism $\overline{\mathcal{M}}_{X,x}/G \cong \overline{\mathcal{M}}_{X,\xi}$. Statement (6) implies that the map $\overline{\mathcal{M}}_{Y,y} \to \overline{\mathcal{M}}_{X,\xi}$ is injective. Since this is true for every $f^{\flat}_{x}$-critical face G, the homomorphism $\overline{\mathcal{M}}_{Y,y} \to \overline{\mathcal{M}}_{X,x}$ is critically s-injective. □

Proposition 2.2.8. *Let $f \colon X \to Y$ be a morphism of fine saturated log schemes. Assume that the underlying morphism of schemes $\underline{f} \colon \underline{X} \to \underline{Y}$ is locally surjective. Then f is exact if Y is solid, and f is $\mathbf{Q}$-integral if X is also solid.*

Proof Choose $x \in X$ and let $y := f(x)$. Replacing x and y by étale neighborhoods, we may assume that $\overline{\mathcal{M}}_{X,x} \cong \overline{\mathcal{M}}_{X,\overline{x}}$ and similarly for y. We have a commutative square

$$\begin{array}{ccc} X_{(x)} & \xrightarrow{\;a\;} & \operatorname{Spec}(\mathcal{M}_{X,x}) \\ {\scriptstyle f_{(x)}}\downarrow & & \downarrow{\scriptstyle c} \\ Y_{(y)} & \xrightarrow{\;b\;} & \operatorname{Spec}(\mathcal{M}_{Y,y}). \end{array}$$

If Y is solid, the map b is surjective and, since $\underline{f}$ is locally surjective, the map $f_{(x)}$ is surjective. It follows that c is surjective, and then Proposition I.4.2.2 implies that $f^{\flat}_{x}$ is exact. If X is also solid, the map a is surjective, and hence its restriction to the fiber over the closed point of $\operatorname{Spec}(\mathcal{M}_{Y,y})$ is also surjective. Thus the idealized log fiber $X'_{\mathcal{K}}$ is also solid, and so it follows from the previous theorem that f is $\mathbf{Q}$-integral. □

This next result illustrates a connection between exactness and flatness that the reader may have come to suspect.

Corollary 2.2.9. *Let $f \colon X \to Y$ be morphism of solid locally noetherian log schemes such that $\underline{f}$ is flat. Then f is exact and $\mathbf{Q}$-integral.*

Proof A flat morphism of locally noetherian schemes is locally surjective, and so the corollary follows from the previous proposition. □

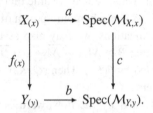

2.3 Immersions and small morphisms

Definition 2.3.1. *A morphism $f\colon X \to Y$ of log schemes is an immersion (resp. closed immersion) if the underlying morphism of schemes $\underline{f}$ is an immersion (resp. closed immersion) and the homomorphism $f^\flat : f^*_{log}(M_Y) \to M_X$ is surjective.*

If $i\colon X \to Y$ is a closed immersion, let

$$\mathcal{I}_{X/Y} := \operatorname{Ker}\Big(i^{-1}(\mathcal{O}_Y) \to \mathcal{O}_X\Big) \quad \text{and} \quad \mathcal{K}_{X/Y} := \operatorname{Ker}\Big((i^{-1}(\mathcal{M}_Y^{\mathrm{gp}}) \to \mathcal{M}_X^{\mathrm{gp}}\Big).$$

There is an exact sequence

$$1 \to (1 + i^{-1}\mathcal{I}_{X/Y}) \to i^{-1}(\mathcal{O}_Y^*) \to \mathcal{O}_X^* \to 1,$$

leading to a zero-sequence

$$0 \to (1 + \mathcal{I}_{X/Y}) \to i^{-1}(\mathcal{M}_Y) \to i^*_{log}(\mathcal{M}_Y) \to 0,$$

which is exact if Y is u-integral. Thus there is a homomorphism of abelian sheaves

$$\Big(1 + i^{-1}(\mathcal{I}_{X/Y})\Big) \to \mathcal{K}_{X/Y}, \tag{2.3.1}$$

which is injective if $\mathcal{M}_Y$ is u-integral and surjective if i is strict. Conversely, if $\mathcal{M}_Y$ is integral and (2.3.1) is an isomorphism, then i is strict. An exact immersion of integral log schemes is strict, since necessarily f is s-injective, by (2) of Proposition I.4.2.1.

Definition 2.3.2. *Let $f\colon X \to Y$ be a morphism of fine log schemes and let x be a point of X. Then f is:*

1. *small at x if, for every (equivalently, for one) geometric point $\overline{x}$ over x, the homomorphism $f^\flat_{\overline{x}}\colon (f^*_{log,\overline{x}}\mathcal{M}_Y)_{\overline{x}} \to \mathcal{M}_{X,\overline{x}}$ is small;*
2. *s-finite at x if, for every (equivalently, for one) geometric point $\overline{x}$ over x, the homomorphism $\overline{f}^\flat_{\overline{x}}\colon \overline{\mathcal{M}}_{Y,f(\overline{x})} \to \overline{\mathcal{M}}_{X,\overline{x}}$ makes $\overline{\mathcal{M}}_{X,\overline{x}}$ a finitely generated $\overline{\mathcal{M}}_{Y,f(\overline{x})}$-set;*
3. *Kummer at x if, for every (equivalently. for one) geometric point $\overline{x}$ over X, the homomorphism $\overline{f}^\flat_{\overline{x}}\colon \overline{\mathcal{M}}_{Y,f(\overline{x})} \to \overline{\mathcal{M}}_{X,\overline{x}}$ is Kummer.*

See Definition I.4.3.1 for these notions for homomorphisms of monoids.

Proposition 2.3.3. *If $f\colon X \to Y$ is a morphism of fine log schemes and x is a point of X, the following conditions are equivalent.*

1. *f is small at x.*
2. *For every (equivalently, for one) geometric point $\overline{x}$ over x, the homomorphism $\overline{f}^\flat_{\overline{x}}\colon \overline{\mathcal{M}}_{Y,f(\overline{x})} \to \overline{\mathcal{M}}_{X,\overline{x}}$ is small.*

3. *For every (equivalently, for one) geometric point $\overline{x}$ over X, the group $\mathcal{M}^{gp}_{X/Y,\overline{x}}$ is finite.*
4. *f is small in a neighborhood of x.*

Proof The equivalence of the first three conditions is straightforward. If they are satisfied, we may, after an étale shrinking of X and Y, assume that f admits a chart which is exact at x and at $f(x)$, subordinate to a homomorphism of fine monoids $\theta\colon P \to Q$. Then $\overline{f}^{\flat}_{\overline{x}}$ coincides with $\overline{\theta}$, and hence $\overline{\theta}$ is small. By Proposition I.4.3.3 it follows that, for every face G of Q, the induced map $\overline{P} \to Q/G$ is small, and hence that f is small. □

Proposition 2.3.4. *If $f\colon X \to Y$ is a morphism of fine log schemes and x is a point of X, the following conditions are equivalent.*

1. *f is s-finite at x.*
2. *For every (equivalently for one) geometric point $\overline{x}$ over x, the homomorphism $\overline{f}^{\flat}_{\overline{x}}\colon \overline{\mathcal{M}}_{Y,f(\overline{x})} \to \overline{\mathcal{M}}_{X,\overline{x}}$ is $\mathbf{Q}$-surjective.*

If f is s-finite at x, then it is s-finite in some neighborhood of x.

Proof Proposition I.4.3.6 implies that conditions (1) and (2) are equivalent. Suppose that they are verified and that f admits a fine chart (α, θ, β), where α is exact at $f(x)$ and β is exact at x. Then $\overline{\theta}\colon \overline{P} \to \overline{Q}$ identifies with $\overline{f}^{\flat}_{x}\colon \overline{\mathcal{M}}_{Y,f(x)} \to \overline{\mathcal{M}}_{X,x}$ and, consequently, is $\mathbf{Q}$-surjective. If x' is any point of X, the homomorphism $\overline{f}^{\flat}_{x'}\colon \overline{\mathcal{M}}_{Y,f(x')} \to \overline{\mathcal{M}}_{X,x'}$ identifies with a homomorphism $P/F \to Q/G$, where F and G are suitable faces of P and Q, and consequently is also $\mathbf{Q}$-surjective. □

It follows from the results of Section I.4.3 that a morphism is Kummer if and only if it is s-injective and s-finite. If Y is saturated, then by Corollary I.4.3.10 f is Kummer if and only if it is exact and small. An s-finite morphism is small, and an immersion is s-finite. If X is a coherent log scheme, the map $X^{\text{int}} \to X$ is a closed immersion and the map $X^{\text{sat}} \to X^{\text{int}}$ is s-finite. If $f\colon X \to Y$ and $g\colon Y \to Z$ are small, then $g \circ f$ is small if f and g are small, and if $g \circ f$ is small, g is small.

Proposition 2.3.5. *Let $f\colon X \to Y$ be a small morphism of fine log schemes. Then, locally on X and Y, f admits a factorization $f = g \circ f'$, where $f'\colon X \to Y'$ is exact and Kummer and $g\colon Y' \to Y$ admits a fine chart (α, ϕ, α'), where ϕ^{gp} is an isomorphism. If f is a closed immersion, then in fact f' is a strict closed immersion.*

Proof Since our assertion is local on X and Y, we may assume that we are working in the Zariski topology. Let x be a point of X and let y be its image in Y. We may assume that f admits a chart subordinate to a homomorphism of fine monoids $\theta\colon P \to Q$. By Remarks II.2.3.2 and II.2.4.3, we may also assume that $\alpha_y\colon P \to \mathcal{M}_{Y,y}$ and $\beta_x\colon Q \to \mathcal{M}_{X,x}$ are exact. Then the maps $\overline{P} \to \overline{\mathcal{M}}_{Y,y}$ and $\overline{Q} \to \overline{\mathcal{M}}_{Y,y}$ are isomorphisms. Since $\overline{f}^{\flat}_x\colon \overline{\mathcal{M}}_{Y,y} \to \overline{\mathcal{M}}_{X,x}$ is small, it follows that $\overline{\theta}\colon \overline{P} \to \overline{Q}$ is small.

Now consider the factorization $P \xrightarrow{\tilde{\theta}} P^{\theta} \xrightarrow{\theta^e} Q$ described in Proposition I.4.2.17. The homomorphism $\tilde{\theta}^{\mathrm{gp}}$ is an isomorphism and the homomorphism θ^e is exact. Moreover, since $\overline{\theta}$ is small, it follows from Corollary I.4.3.12 that $P^{\theta} \to Q$ is locally exact, and hence that $\mathsf{A}_Q \to \mathsf{A}_{P^\theta}$ is exact. Let $Y' := Y \times_{\mathsf{A}_P} \mathsf{A}_{P^\theta}$ and let $\alpha'\colon P' \to \mathcal{M}_{Y'}$ be the natural map. Then $(\alpha, \tilde{\theta}, \alpha')$ is a chart for g' with the desired properties. Since $X \to \mathsf{A}_Q$ is strict, it follows that $f\colon X \to Y'$ is also exact and small, hence Kummer. If f is a closed immersion it is proper, and since g is affine it is separated, and hence f' is also proper. Necessarily f' is an immersion, hence a closed immersion, and since it is exact, it must also be strict, since an exact morphism is s-injective. □

Example 2.3.6. Let $f\colon X \to Y$ be a morphism of fine log schemes and let $X \times_Y X$ denote the fiber product, computed in the category of integral log schemes. Then the diagonal morphism $\Delta_{X/Y}\colon X \to X \times_Y X$ is an immersion. A chart (a, θ, b) for f will induce a chart for $\Delta_{X/Y}$ that we can use to give an explicit description of the construction in Proposition 2.3.5, based on the exactification construction discussed in Proposition I.4.2.19.

Suppose that $\theta\colon P \to Q$ is a homomorphism of integral monoids. Recall that $(Q/P)^{\mathrm{gp}} \cong Q^{\mathrm{gp}}/P^{\mathrm{gp}}$, where we write $Q^{\mathrm{gp}}/P^{\mathrm{gp}}$ for the cokernel of θ^{gp}. Carrying over the discussion in Remark I.3.3.6 to the category of log schemes, we see that the homomorphism of monoids

$$\theta\colon Q \to Q \oplus Q^{\mathrm{gp}}/P^{\mathrm{gp}} : q \mapsto (q, [q])$$

induces a morphism of log schemes

$$m_{Q/P}\colon \mathsf{A}_Q \times \mathsf{A}^*_{Q/P} \to \mathsf{A}_Q, \tag{2.3.2}$$

defining a right action of the group scheme $\mathsf{A}^*_{Q/P}$ on the log scheme A_Q, over the log scheme A_P. The homomorphism

$$\phi\colon Q \oplus_P Q \to Q \oplus Q^{\mathrm{gp}}/P^{\mathrm{gp}} : [(q_1, q_2)] \mapsto (q_1 + q_2, [q_1]]$$

corresponds to the morphism of log schemes

$$h_{Q/P}\colon \mathsf{A}_Q \times \mathsf{A}^*_{Q/P} \to \mathsf{A}_Q \times_{\mathsf{A}_P} \mathsf{A}_Q : (\alpha, \gamma) \mapsto (\alpha, \alpha\gamma), \tag{2.3.3}$$

and the summation homomorphism $\sigma\colon Q \oplus_P Q \to Q$ corresponds to the diagonal morphism $\mathsf{A}_Q \to \mathsf{A}_Q \times_{\mathsf{A}_P} \mathsf{A}_Q$. As explained in Proposition I.4.2.19, the homomorphism ϕ^{gp} is an isomorphism and ϕ identifies $Q \oplus Q^{\mathrm{gp}}/P^{\mathrm{gp}} \to Q$ with the exactification $(Q \oplus Q)^{\sigma} \to Q$ of σ.

Now, if (a, θ, b) is a chart for f, we find a chart $(b, \sigma, b \times b)$ for $\Delta_{X/Y}$. Then the construction of Proposition 2.3.5 gives a factorization

$$\Delta_{X/Y} : X \xrightarrow{\Delta'_{X/Y}} (X \times_Y X)' \xrightarrow{h_{X/Y}} X,$$

where $(X \times_Y X)' \to X$ is the projection

$$(X \times_Y X) \times_{(\mathsf{A}_Q \times_{\mathsf{A}_P} \mathsf{A}_Q)} (\mathsf{A}_Q \times \mathsf{A}^*_{Q/P}) \to X.$$

The following diagram, in which all the squares are cartesian, may be helpful.

$$\begin{array}{ccccc}
 & & (X \times_Y X)' & \xrightarrow{h_{X/Y}} & X \times_Y X \\
 & \nearrow^{\Delta'_{X/Y}} & \downarrow & & \downarrow{\scriptstyle \mathrm{id} \times b} \\
X & \xrightarrow{(\mathrm{id}, 1)} & X \times \mathsf{A}^*_{Q/P} & \xrightarrow{h_X} & X \times_{\mathsf{A}_P} \mathsf{A}_Q \\
{\scriptstyle b}\downarrow & & \downarrow{\scriptstyle b \times \mathrm{id}} & & \downarrow{\scriptstyle b \times \mathrm{id}} \\
\mathsf{A}_Q & \xrightarrow{(\mathrm{id}, 1)} & \mathsf{A}_Q \times \mathsf{A}^*_{Q/P} & \xrightarrow{h_{Q/P}} & \mathsf{A}_Q \times_{\mathsf{A}_P} \mathsf{A}_Q\,.
\end{array} \qquad (2.3.4)$$

2.4 Inseparable morphisms and Frobenius

Definition 2.4.1. *Let X and Y be log schemes in characteristic p.*

1. *The* absolute Frobenius endomorphism *of X is the morphism $F_X\colon X \to X$ that is the identity on the underlying topological space of X and for which $F_X^{\flat}\colon \mathcal{M}_X \to \mathcal{M}_X$ is multiplication by p and $F_X^{\sharp}\colon \mathcal{O}_X \to \mathcal{O}_X$ is the pth-power map.*
2. *If $f\colon X \to Y$ is a morphism of log schemes, then X' is the fiber product $X \times_{F_Y} Y$, and the* relative Frobenius morphism *$F_{X/Y}\colon X \to X'$ is the morphism whose composition with the projection $X' \to X$ is F_X and whose composition with the projection $X' \to Y$ is f.*

The use of the relative Frobenius morphism requires some caution, since the fiber product $X \times_{F_Y} Y$ need not be integral or saturated even if X and Y are. Thus its construction may depend on the category in which this fiber product is taken. One way to avoid this difficulty is to require that the morphism $f\colon X \to$

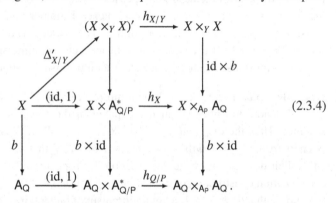

Y be *saturated*; see Section 2.5; in this case all three constructions agree, and furthermore the morphism $F_{X/Y}$ is exact. Kato's seminal article [48] defines a morphism f to be of *Cartier type* if it is integral and if the the relative Frobenius morphism $F_{X/Y}$ is exact. As we shall see in Corollary 2.5.4, it turns out that a morphism f of fine saturated log schemes is of Cartier type if and only if it is saturated.

The Frobenius endomorphism F_X of a fine log scheme X is s-finite, hence small, and it is exact if X is saturated. If $f: X \to Y$ is a morphism of fine log schemes and X' is formed in the category of integral (resp. saturated) log schemes, then $F_{X/Y}$ is is small (resp. small and exact; see Proposition I.4.4.2). In the category of integral log schemes, the exactification construction of the morphism $F_{X/Y}$ explained in Proposition 2.3.5 is canonical, and hence globalizes, as we shall see in Corollary V.3.3.8.

Frobenius morphisms are special cases of "weakly inseparable morphisms," a notion due to Kato. Our definition is slightly stronger than the original one he proposed [48, 4.9]; we have added the requirement that $\underline{f}$ be radicial. As Theorem 2.4.3 shows, this amounts to replacing "homeomorphism" by "universal homeomorphism."

Definition 2.4.2. *A morphism of integral log schemes $f: X \to Y$ in characteristic p is weakly inseparable if it satisfies the equivalent conditions of Theorem 2.4.3 below. A morphism f is inseparable if it is exact and weakly inseparable.*

Theorem 2.4.3. *Let $f: X \to Y$ be a morphism of fine log schemes in characteristic $p > 0$. Then the conditions (1)–(3) below are equivalent. Furthermore, they imply that $\underline{f}$ is a universal homeomorphism.* [3]

1. (a) *For every $\overline{x} \in X$, the homomorphism $f^{\flat}_{\overline{x}}: \mathcal{M}_{Y,f(\overline{x})} \to \mathcal{M}_{X,\overline{x}}$ is a p-isogeny (Definition I.4.4.5).*
   (b) *The underlying morphism $\underline{f}: \underline{X} \to \underline{Y}$ is radicial and a homeomorphism.*
2. (a) *Condition (1a) holds.*
   (b) *The morphism $\underline{f}$ is affine and surjective.*
   (c) *For every local section b of $f_*\mathcal{O}_X$, locally there exists $r > 0$ such that b^{p^r} is in the image of $f^\sharp: \mathcal{O}_Y \to f_*(\mathcal{O}_X)$.*
3. (a) *For each $\overline{x} \in X$, the homomorphism $\overline{f}^{\flat}_{\overline{x}}: \overline{\mathcal{M}}_{Y,f(\overline{x})} \to \overline{\mathcal{M}}_{X,\overline{x}}$ is a p-isogeny in the sense of Definition I.4.4.5.*
   (b) *The morphism $\underline{f}$ is affine, and the kernel of $\mathcal{O}_Y \to f_*(\mathcal{O}_X)$ is a nil ideal.*
   (c) *Condition (2c) holds.*

Proof The proof will depend on the following lemmas.

[3] We thank O. Gabber for correcting some shortcomings in our first proof of this theorem.

Lemma 2.4.4. *Let $\theta\colon A \to B$ be a local homomorphism of local rings in characteristic $p > 0$. Then θ induces a p-isogeny $A^* \to B^*$ if and only if the following two conditions are verified.*

1. *For every $a \in \operatorname{Ker}(\theta)$, there is an $r > 0$ such that $a^{p^r} = 0$.*
2. *For every $b \in B$, there exists an $r > 0$ such that b^{p^r} is in the image of θ.*

If these conditions are satisfied, then θ is a universal homeomorphism and in particular is surjective and radicial.

Proof Suppose that θ induces a p-isogeny $\theta^*\colon A^* \to B^*$. If $a \in \operatorname{Ker}(\theta)$, then a is not a unit of A, hence $1 + a$ is a unit, and $\theta^*(1 + a) = 1$. Since θ^* is a p-isogeny, it follows that $(1 + a)^{p^r} = 1$ for some $r > 0$, and then $a^{p^r} = 0$. If $b \in B$ and b is a unit, then since θ^* is a p-isogeny there exist $r > 0$ and $a \in A$ such that $\theta(a) = b^{p^r}$. If b is not a unit, then $1 + b$ is a unit, and there exist $u \in A^*$ and $r > 0$ such that $(1 + b)^{p^r} = \theta(u)$. Then $\theta(u - 1) = b^{p^r}$. We leave the proof of the converse, which is similar, to the reader.

Condition (2) implies that $\operatorname{Spec}(\theta)$ is injective and, since it is stable under base change, it implies that $\operatorname{Spec}(\theta)$ is radicial. Conditions (1) and (2) together imply that $\operatorname{Spec}(\theta)$ is surjective and integral. Thus by [29, 18.12.11], $\operatorname{Spec}(\theta)$ is a universal homeomorphism. □

Lemma 2.4.5. *Let $\theta\colon A \to B$ be a local homomorphism of local rings of characteristic $p > 0$, and let $\theta'\colon A' \to B'$ be the corresponding homomorphism of their strict henselizations. Assume that* Spec(θ) *is surjective and radicial and that θ' satisfies conditions (1) and (2) of Lemma 2.4.4. Then θ also satisfies these conditions.*

Proof Consider the diagram

$$\begin{array}{ccccc}
A' & \xrightarrow{\theta''} & A' \otimes_A B & \xrightarrow{\phi} & B' \\
\uparrow{\scriptstyle\alpha} & & \uparrow{\scriptstyle\gamma} & \nearrow{\scriptstyle\beta} & \\
A & \xrightarrow{\theta} & B. & &
\end{array}$$

The maps α and β are ind-étale and hence the same is true of γ and ϕ [4, tag 097H]. Moreover $\operatorname{Spec}(\theta)$ is universally bijective, and hence $\operatorname{Spec}(\theta'')$ is bijective. Since θ' satisfies the conditions of Lemma 2.4.4, $\operatorname{Spec}(\theta') = \operatorname{Spec}(\phi) \circ \operatorname{Spec}(\theta'')$ is bijective and, since $\operatorname{Spec}(\theta')$ is bijective, $\operatorname{Spec}(\phi)$ is also bijective. Since ϕ is flat it is faithfully flat, hence injective.

Now, if $a \in \operatorname{Ker}(\theta)$, then $\alpha(a) \in \operatorname{Ker}(\theta'')$, and hence $\alpha(a^{p^r}) = 0$ for some $r > 0$. Since α is injective, in fact $a^{p^r} = 0$. If $b \in B$, there exist $r > 0$ and $a' \in A'$

such that $\theta'(a') = \beta(b)^{p^r}$. Since ϕ is injective, we can conclude that $\theta''(a') = \gamma(b^{p^r})$. In the following commutative diagram of A-modules, the bottom row is exact by construction and the top row is consequently exact, since α is flat:

$$\begin{array}{ccccc} A' & \xrightarrow{\theta''} & A' \otimes_A B & \xrightarrow{\pi'} & A' \otimes_A (B/A) \\ \Big\uparrow{\scriptstyle \alpha} & & \Big\uparrow{\scriptstyle \gamma} & & \Big\uparrow{\scriptstyle \delta} \\ A & \xrightarrow{\theta} & B & \xrightarrow{\pi} & B/A. \end{array}$$

In fact α is faithfully flat, so, for every A-module M, the natural map $M \to A' \otimes_A M$ is injective. We have seen that $\pi'(\gamma(b^{p^r})) = 0$, hence $\pi(b^{p^r}) = 0$, so there is an $a \in A$ such that $\theta(a) = b^{p^r}$. □

Now suppose that condition (1) of Theorem 2.4.3 holds. Since $\underline{f}$ is a homeomorphism, it is easily seen to be affine. (For example, one could observe that f is quasi-compact and apply Serre's criterion.) Thus conditions (2a) and (2b) hold. To prove (2c), begin by observing that, if y is the image of a point x of X, then the natural map $f_*(\mathcal{O}_X)_x \to \mathcal{O}_{X,x}$ is an isomorphism, since $\underline{f}$ is a homeomorphism. Thus it will suffice to show that, for every $b \in B := \mathcal{O}_{X,x}$, there exists an $r > 0$ such that b^{p^r} is in the image of $A := \mathcal{O}_{Y,y}$. Let $\overline{x}$ be a geometric point of X lying over x, let $\overline{y}$ be its image in Y, and let $A' := \mathcal{O}_{Y,\overline{y}}$ and $B' := \mathcal{O}_{X,\overline{x}}$, the strict Henselizations of A and B, respectively. By statement (5) of Proposition I.4.4.8, condition (1a) implies that the group homomorphism $\mathcal{O}^*_{Y,\overline{y}} \to \mathcal{O}^*_{X,\overline{x}}$ is a p-isogeny. It follows that the homomorphism $A' \to B'$ satisfies conditions (1) and (2) of Lemma 2.4.4. By Lemma 2.4.5, the same is true of the homomorphism $A \to B$. Thus (2c) is satisfied.

Suppose that (2) of Theorem 2.4.3 holds and let $\overline{x}$ be a geometric point of X mapping to $\overline{y} \in Y$. Then, by statement (5) of Proposition I.4.4.8, the homomorphisms $\overline{f}^{\flat}_{\overline{x}}$ and $f^{\flat *}_{\overline{x}}$ are p-isogenies. In particular, (3a) is satisfied. To prove (3b), choose a point y of Y; since f is surjective there is an $x \in X$ with $f(x) = y$. Furthermore, Lemmas 2.4.4 and 2.4.5 imply that the kernel of $f^\sharp$ is a nil ideal. Thus (3b) is also satisfied.

Conditions (3b) and (3c) imply that $\underline{f}$ is integral, radicial, and surjective, hence a universal homeomorphism by [29, 18.12.11]. Moreover, these conditions remain true after a strict étale localization on Y. Since $\underline{f}$ is a universal homeomorphism, it is also a homeomorphism for the étale topology [3, VIII,1.1]. Let $\overline{x}$ be a geometric point of X mapping to $\overline{y}$ in Y. Since every étale neighborhood of x in X is pulled back from an étale neighborhood of y in Y, it follows that the homomorphism $f^\sharp_{\overline{x}} \colon \mathcal{O}_{Y,\overline{y}} \to \mathcal{O}_{X,\overline{x}}$ still satisfies conditions (3b) and (3c). Then Lemma 2.4.4 implies that $f^\sharp_{\overline{x}}$ induces a p-isogeny $\mathcal{O}^*_{Y,\overline{y}} \to \mathcal{O}^*_{X,\overline{x}}$. Condition (3a) asserts that $\overline{f}^\flat_{\overline{x}}$ is a p-isogeny, and so (5) of Proposition I.4.4.8 implies that $f^\flat_{\overline{x}}$ is also a p-isogeny. □

It will be useful to know that weakly inseparable morphisms admit weakly inseparable charts.

Proposition 2.4.6. *Let $f\colon X \to Y$ be a weakly inseparable morphism of fine (resp. fine saturated) log schemes. Then, étale locally on Y, there exists a chart (α, θ, β) for f such that θ is a weakly inseparable homomorphism of fine (resp. fine saturated) monoids.*

Proof Since $\underline{f}$ is a universal homeomorphism, it is also a homeomorphism for the étale topology [32, I, §11], and so localizing on X and Y is the same. In particular, we may assume that the log structures on X and Y are Zariski. Let x be a point of X and let y be its image in Y. Applying Lemma 2.4.7 below to the p-isogeny $f^\flat_x\colon \mathcal{M}_{Y,y} \to \mathcal{M}_{X,x}$, we find a fine chart (α, θ, β) for $f^\flat_x$ with θ a p-isogeny of fine (resp. fine saturated) monoids. By Proposition II.2.2.3, this chart extends to a chart for f in neighborhoods of x and y. □

Lemma 2.4.7. *Suppose that $\phi\colon M_y \to M_x$ is a local homomorphism of integral monoids, where $\overline{M}_y$ and $\overline{M}_x$ are finitely generated and ϕ (resp. ϕ^{gp}) is a p-isogeny. Then there exists a fine chart (α, θ, β) for ϕ such that θ (resp. θ^{gp}) is a p-isogeny, where θ is a homomorphism of fine (resp. fine saturated) monoids.*

Proof Suppose that ϕ is a p-isogeny. Choose a finitely generated submonoid P of M_y such that the induced map $P \to \overline{M}_y$ is surjective, and then choose a finitely generated submonoid Q of M_x containing $\phi(P)$ whose image surjects to $\overline{M}_x$. Since ϕ is a p-isogeny and Q is finitely generated, there exists some $r > 0$ such that $p^r Q \subseteq \operatorname{Im} \phi$. Hence there is a finitely generated submonoid A of M_y such that $\phi(A) = p^r Q$. Replacing P by $P + A$, we may assume that $p^r Q \subseteq \phi(P)$ and $\phi(P) \subseteq Q$. Now let $P' := Q \times_{M_x} P$, with its natural maps $\alpha'\colon P' \to M_y$ and $\theta\colon P' \to Q$. This monoid is finitely generated by (6) of Theorem I.2.1.17. We claim that α' induces a surjection to $\overline{M}_y$. Indeed, if $m \in M_y$, there exists $a \in P$ mapping to $\overline{m}$, and then $(\phi(a), a) \in P'$. Moreover, if $b \in Q$, there exist $r > 0$ and $a \in P$ such that $p^r b = \phi(a)$, and then $(p^r b, a) \in P'$ with $\theta(p^r b, a) = p^r b$. Finally, suppose that $(b_1, a_1), (b_2, a_2) \in P'$ and $\theta(b_1, a_1) = \theta(b_2, a_2)$. Then $b_1 = b_2$ and $\phi(a_1) = \phi(a_2)$, and it follows that $p^m a_1 = p^m a_2$ for some m. Thus θ is a p-isogeny. If M_y and M_x are saturated, we may replace θ by its saturation θ^{sat}, which is again a p-isogeny by Proposition I.4.4.8.

Suppose instead that ϕ^{gp} is a p-isogeny. Then the same construction applied to ϕ^{gp} produces finitely generated markups (Definition II.2.3.3)

$b\colon L_x \to M_x$, $a\colon L_y \to M_y$ and a p-isogeny $\theta\colon L_y \to L_x$ such that $\phi \circ a = b \circ \theta$. Then the corresponding homomorphism of charts $P := a^{-1}(M_y) \to Q := b^{-1}(M_x)$ fits into a chart for ϕ and has the desired properties. □

Proposition 2.4.8. *In the category of fine log schemes in positive characteristic, the following statements hold.*

1. *If $f\colon X \to Y$ is weakly inseparable and Y is saturated, then f is inseparable.*
2. *If f and g are two composable morphisms and $h = g \circ f$, then if any two of f, g, and $g \circ f$ are weakly inseparable, so is the third.*
3. *If f and g are inseparable, so is $g \circ f$. If $g \circ f$ is inseparable then f is inseparable if g is, and the converse holds if Y is saturated.*
4. *The class of weakly inseparable (resp. of inseparable) morphisms is closed under base change in the category of fine log schemes and also in the category of fine saturated log schemes.*

Proof Statement (1) is proved in the same way as Proposition I.4.4.7. It is easy to verify that if any two of $\underline{f}$, $\underline{g}$ and $\underline{g} \circ \underline{f}$ are radicial homeomorphisms, then so is the third. Proposition I.4.4.8 shows that the analogous result holds for p-isogenies of integral monoids, and statement (2) follows. Since f is exact if $g \circ f$ is exact and since weakly inseparable morphisms to a saturated target are necessarily inseparable, (3) also holds.

Now suppose that $f\colon X \to Y$ and $g\colon Y' \to Y$ are morphisms of fine log schemes and that f is weakly inseparable, and let $f'\colon X' \to Y'$ be the pullback of f along g, in the category of integral monoids. Recall that the underlying scheme $\underline{X}'$ of X' is a closed subscheme of the fiber product $X'' := \underline{Y}' \times_{\underline{Y}} \underline{X}$. If $\overline{x}'$ is a point of X' mapping to $\overline{x} \in X$, the morphism $\overline{f}'^{\flat}_{\overline{x}'}$ is the integral pushout of the morphism $\overline{f}^{\flat}_{\overline{x}}$, and hence is again a p-isogeny, by Proposition I.4.4.8. The morphisms $\underline{X}' \to \underline{X}''$ and $\underline{X}' \to \underline{Y}'$ are affine and satisfy condition (2c) of Theorem 2.4.3, and hence the same is true for their composition $\underline{f}'$. It remains to prove that the kernel of the map $\mathcal{O}_{Y'} \to f_*(\mathcal{O}'_X)$ is a nil ideal, equivalently, that X' maps surjectively to Y'. Since the map $X'' \to Y'$ is surjective, it will suffice to prove that $X' \to X''$ is surjective. When f is exact, the "four-point lemma," Proposition 2.2.3, suffices; fortunately, the analogous result also holds for weakly inseparable morphisms, as Lemma 2.4.9 below shows. This completes the proof in the category of fine log schemes.

To finish the proof in the category of fine saturated log schemes, it remains to prove that the map $\underline{X}'^{\mathrm{sat}} \to \underline{Y}$ is a radicial homeomorphism. Since this is true for $X' \to Y$ and since $\underline{X}''^{\mathrm{sat}} \to \underline{X}'$ is surjective, we need only show that $\underline{X}'^{\mathrm{sat}} \to \underline{X}'$ is radicial. This is a local statement, which we can check in the presence of a weakly inseparable chart (α, θ, β) for f as Proposition 2.4.6. We may also assume that there is a chart (α, ϕ, γ) for g, where $\phi\colon P \to P'$ is a homomorphism of fine saturated monoids. Since $P \to Q$ is a p-isogeny and P' is saturated, the homomorphism $Q' \to Q'^{\mathrm{sat}}$ is a p-isogeny (see statement (3) of Proposition I.4.4.8). Then the corresponding morphism of schemes $\mathsf{A}_{Q'^{\mathrm{sat}}} \to \mathsf{A}_{Q'}$ is radicial. Since $\underline{X}'^{\mathrm{sat}} \to \underline{X}'$ is obtained by base change from this map, it is also radicial.

Lemma 2.4.9. *Let $f\colon X \to Y$ be a morphism of fine log schemes, and let $f'\colon X' \to Y'$ be its pullback along a morphism $Y' \to Y$, in the category of fine log schemes. Suppose that $\overline{x}$ is a geometric point of X such that the morphism $f^{\flat}_{\overline{x}}\colon \mathcal{M}_{Y,f(\overline{x})} \to \mathcal{M}_{X,\overline{x}}$ is a p-isogeny, where p is the characteristic exponent of $k(\overline{x})$. Let $\overline{y}'$ be a point of Y' mapping to $\overline{y} := f(\overline{x})$. Then there is a point $\overline{x}'$ of X' mapping to $\overline{x}$ and to $\overline{y}'$.*

Without loss of generality we may replace X, Y, and Y' by the log points defined by $\overline{x}, \overline{y}$, and $\overline{y}'$, respectively. In particular, $\underline{X} = \underline{Y} = \operatorname{Spec} k$, where k is an algebraically closed field, and $\underline{Y} = \operatorname{Spec} k'$, where $k \to k'$ is a field extension and k' is also algebraically closed. Then we can identify y with $\overline{y}$, x with $\overline{x}$, and y' with $\overline{y}'$. Then $\mathcal{M}^*_{Y,y} \to \mathcal{M}^*_{X,x}$ is the identity map of k^*, $\mathcal{M}^*_{Y,y} \to \mathcal{M}'^*_{Y',y'}$ is the inclusion $k^* \to k'^*$, and the morphisms of log schemes are given by local homomorphisms $\theta\colon \mathcal{M}_{Y,y} \to \mathcal{M}_{X,x}$ and $\phi\colon \mathcal{M}_{Y,y} \to \mathcal{M}'_{Y,y}$ extending these homomorphisms. Let $\mathcal{M}' := \mathcal{M}'_{Y',y'} \oplus_{\mathcal{M}_{Y,y}} \mathcal{M}_{X,x}$ in the category of integral monoids. By Proposition I.4.4.9, the homomorphisms $\phi'\colon \mathcal{M}_{X,x} \to Q'$ and $\theta'\colon \mathcal{M}'_{Y',y'} \to Q'$ are local and the kernel K of $\mathcal{M}^*_{Y',y'} \oplus_{\mathcal{M}^*_{Y,y}} \mathcal{M}^*_{X,x} \to Q'^*$ is p^∞-torsion. But $\mathcal{M}^*_{Y',y'} \oplus_{\mathcal{M}^*_{Y,y}} \mathcal{M}^*_{X,x} = k'^* \oplus_{k^*} k^* \cong k'^*$ and, since k' is a field of characteristic $p > 0$, in fact $K = \{1\}$, and hence the map $k'^* \to Q'^*$ is injective. Since k' is algebraically closed, the group k'^* is divisible, and hence there is a homomorphism $\sigma\colon Q'^* \to k'^*$ splitting the inclusions $k'^* \to Q'^*$. Extend σ to a homomorphism $Q' \to k'$ by sending Q'^+ to zero to obtain a local log ring $X'' := \operatorname{Spec}(Q' \to k')$. Then the homomorphisms $\theta'\colon \mathcal{M}'_{Y',y'} \to Q'$ and $\phi'\colon \mathcal{M}_{X,x} \to Q'$ define morphisms of log schemes $X'' \to Y'$ and $X'' \to X$, which agree when projected to Y, and hence a morphism $X'' \to X'$. The image of the closed point of X'' has the desired property. □

For more about inseparable morphisms of log schemes, we refer to the important paper of I. Vidal [77].

2.5 Integral and saturated morphisms

The following definitions satisfy the conventions established in II.1.1.11.

Definition 2.5.1. *Let $f\colon X \to Y$ be a morphism of integral log schemes, let x*

be a point of X, and let **P** *be one of the following properties: integral (Definition I.4.6.2), saturated, or n-saturated for some integer n (Definition I.4.8.2). Then f is said to satisfy the property* **P** *at x if for every geometric point $\overline{x}$ lying over x, the homomorphism $f^\flat_{\overline{x}}\colon \mathcal{M}_{Y,f(\overline{x})} \to \mathcal{M}_{X,\overline{x}}$ has* **P**. *The morphism f is said to satisfy* **P** *if it does so at every point of X.*

It is clear that if $f^\flat_{\overline{x}}$ has one of the properties **P** for some geometric point lying over x, then it does so for every such geometric point. Note that, as a consequence of statement (2) of Proposition I.4.6.3 (resp. of statement (4) of Proposition I.4.8.5), the homomorphism $f^\flat_{\overline{x}}\colon \mathcal{M}_{Y,f(\overline{x})} \to \mathcal{M}_{X,\overline{x}}$ has **P** if and only if $\overline{f}^\flat_{\overline{x}}\colon \overline{\mathcal{M}}_{Y,f(\overline{x})} \to \overline{\mathcal{M}}_{X,\overline{x}}$ does.

Proposition 2.5.2. *Let $f\colon X \to Y$ be a morphism of fine log schemes, let x be a point of X, and let n be a natural number.*

1. *If f is saturated or just n-saturated at x, then it is integral at x. If f is integral at x, it is exact at x.*
2. *If f n-saturated at x for some $n > 1$ and if X is saturated at x and Y is saturated at $f(x)$, then f is saturated at x.*
3. *Let (α, θ, β) be a fine chart for f that is neat at x and such that α is neat at y. Then f has one of the properties* **P** *listed in Definition 2.5.1 if and only if θ has the corresponding property* **P**.
4. *If f has property* **P** *at x, then it does so in a neighborhood of x.*

Proof Saturated and n-saturated homomorphisms are integral by their very definition. If f is integral at x, the homomorphism $f^\flat_{\overline{x}}$ is local and integral, hence exact, by statement (4) of Proposition I.4.6.3. This proves statement (1). Statement (2) follows from Corollary I.4.8.16. Let (α, θ, β) be a chart for f that is neat at x and such that α is neat at y. Since $f^\flat_{\overline{x}}$ is exact, it is s-injective, and hence, by Remark II.2.4.5, the chart β is also neat at x. Thus the monoids P and Q are sharp and the homomorphisms $P \to \overline{\mathcal{M}}_{Y,f(\overline{x})}$ and $Q \to \overline{\mathcal{M}}_{X,\overline{x}}$ are isomorphisms, and statement (3) becomes clear. Suppose f has **P** at x. To verify that the same is true in a neighborhood of x, it is harmless to replace X by an fppf neighborhood of x. Thus, by Theorem 1.2.7, we may assume that there exists a chart (α, θ, β) for f which is neat at x and such that α is neat at y. Then $\theta\colon P \to Q$ has property **P**. If G is a face of Q and $F := \theta^{-1}(G)$, it follows from Corollary I.4.6.4 and Proposition I.4.8.8 that the induced homomorphism $P/F \to Q/G$ still has **P**. But (α, θ, β) is a chart for f on all of X, so, for every $x' \in X$, the homomorphism $\overline{f}^\flat_{\overline{x}'}\colon \overline{\mathcal{M}}_{Y,f(\overline{x}')} \to \overline{\mathcal{M}}_{X,f(\overline{x}')}$ identifies with such a homomorphism, and hence also has **P**. □

Proposition 2.5.3. *Let $f\colon X \to Y$ and $g\colon Y \to Z$ be morphisms of fine log* *schemes, let x be a point of x, let $y := f(x)$ and $z := g(y)$, and let* **P** *be one of these properties: integral, saturated, or n-saturated.*

1. *If f is has* **P** *at x and g has* **P** *at y, then $g \circ f$ has* **P** *at x. If $g \circ f$ has* **P** *at x and f has* **P** *at x, then g has* **P** *at y. If $g \circ f$ has* **P** *at x and $\overline{g}^{\flat}_{\overline{y}} \colon \overline{\mathcal{M}}_{Z,\overline{z}} \to \overline{\mathcal{M}}_{Y,\overline{y}}$ is surjective for some geometric point $\overline{y}$ over y, then f has* **P** *at x.*
2. *The family of morphisms satisfying* **P** *is stable under base change in the category of fine log schemes. In particular, if $X \to Y$ and $Y' \to Y$ are morphisms of fine (resp. fine saturated) log schemes and f is integral (resp. saturated) then $X' := X \times_Y Y'$ is an integral (resp. saturated) log scheme and the morphism $f' \colon X' \to Y'$ is integral (resp. saturated).*
3. *If $f \colon X \to Y$ is a morphism of fine log schemes and the rank of $\overline{\mathcal{M}}^{\mathrm{gp}}_{Y,f(\overline{x})}$ is less than or equal to one, then f is integral at x.*

Proof When **P** is integral, statement (1) follows from statement (1) of Proposition I.4.6.3, and statement (3) follows from (5) of that proposition. Statement (2) follows from statements (1) of the same proposition and (3) of Definition I.4.6.2. When **P** is saturated or n-saturated, statement (1) follows from Proposition I.4.8.5, and statement (2) follows from Proposition I.4.8.9. Statement (3) follows from (5) of Proposition I.4.6.3. □

Proposition 2.5.3 shows that the family of integral morphisms of fine log schemes is stable under base change in the categories of log schemes and in the category of integral log schemes. However, the analogous statement in the category of *saturated* log schemes fails, as Example I.4.6.5 shows.

Corollary 2.5.4. *Let $f \colon X \to Y$ be an integral morphism of fine saturated log schemes in characteristic p. Then f is saturated if and only if the relative Frobenius morphism $F_{X/S} \colon X \to X^{(p)}$ (Definition 2.4.1) is exact.*

Proof Let $\overline{x}$ be a geometric point of X, and let $\overline{x}' := F_{X/Y}(\overline{x})$ and $\overline{y} := f(\overline{x})$. Since f is integral, the fiber product $X^{(p)} := X \times_{F_Y} Y$ is an integral log scheme, and $\overline{\mathcal{M}}_{X^{(p)},\overline{x}'}$ is the pushout of $\overline{\mathcal{M}}_{X,\overline{x}}$ and $\overline{\mathcal{M}}_{Y,\overline{y}}$ over the pth toric Frobenius endomorphism of $\overline{\mathcal{M}}_{Y,\overline{y}}$. Since $\overline{\mathcal{M}}_{X,\overline{x}}$ is saturated and $F_{X/Y}$ is exact, it follows that $\overline{\mathcal{M}}_{X^{(p)},\overline{x}'}$ is also saturated. The exactness of $F_{X/Y}$ at $\overline{x}$ means that the homomorphism $f^{\flat}_{\overline{x}}$ is p-quasi-saturated at $\overline{x}$ (Definition I.4.8.2) and, since f is integral, it is in fact p-saturated at $\overline{x}$. Since $f^{\flat}_{\overline{x}}$ is integral, it is locally exact and, by Corollary I.4.8.16, it follows that $f^{\flat}_{\overline{x}}$ is saturated. □

Theorem 2.5.5. *Let $f \colon X \to Y$ be a morphism of fine saturated log schemes. Consider the following conditions.*

1. *The morphism f is saturated.*

2. *For every geometric point $\overline{x}$ of X such that* $\dim \mathcal{M}_{X,\overline{x}} = \dim \mathcal{M}_{Y,f(\overline{x})}$, *the morphism $f^{\flat}_{\overline{x}}$ is strict.*
3. *For every geometric point $\overline{x}$ of X such that* $\dim \mathcal{M}_{X,\overline{x}} = \dim \mathcal{M}_{Y,f(\overline{x})} = 1$, *the homomorphism $f^{\flat}_{\overline{x}}$ is strict.*
4. *For every geometric point $\overline{x}$ of X, the group $\mathcal{M}^{\mathrm{gp}}_{X/Y,\overline{x}}$ is torsion free.*

Then the following implications hold.

1. *Condition (1) implies conditions (2), (3), and (4), and condition (2) implies condition (3).*
2. *If f is $\mathbf{Q}$-integral (resp. integral) and X is solid, then (2) implies (1) (resp. (3) implies (1)).*
3. *If f is $\mathbf{Q}$-integral and the idealized fibers of f are solid, then (4) implies (1).*

Proof Since a saturated homomorphism is by definition integral, the fact that (1) implies (3) and (4) follows from Theorem I.4.8.14; the fact that (2) implies (3) is trivial. Suppose that f is $\mathbf{Q}$-integral (resp. integral) and that X is solid. Let $\overline{x}$ be a geometric point of X and let $\mathfrak{q}$ be a prime ideal of $\overline{\mathcal{M}}_{X,\overline{x}}$ whose height is the same as the height of $f^{\flat -1}_{\overline{x}}(\mathfrak{q})$ (resp. and such that this height is one) and let G be the corresponding face of $\overline{\mathcal{M}}_{X,\overline{x}}$. Since X is solid, there exists a generization $\overline{x}'$ of $\overline{x}$ such that the cospecialization map identifies $\overline{\mathcal{M}}_{X,\overline{x}'}$ with $\overline{\mathcal{M}}_{X,\overline{x}}/G$. Then condition (2) (resp. condition (3)) implies that $f^{\flat}_{\overline{x}'}$ is strict. Since this is true for every such $\mathfrak{q}$, Theorem I.4.8.14 implies that $f^{\flat}_{\overline{x}}$ is saturated. Finally, suppose that f is $\mathbf{Q}$-integral and that its idealized fibers are solid. Let G be an $f^{\flat}_{\overline{x}}$-critical face of $\mathcal{M}_{X,\overline{x}}$, and let $\mathfrak{q}$ be the corresponding prime ideal of $\mathcal{M}_{X,\overline{x}}$. Since the idealized fiber of f over $\overline{y}$ is solid, it contains a point $\overline{x}'$ such that the cospecialization map identifies $\overline{\mathcal{M}}_{X\overline{x}'}$ with the quotient of $\overline{\mathcal{M}}_{X,\overline{x}}$ by G. Condition (4) asserts that the cokernel of the homomorphism $\overline{f}^{\mathrm{gp}}_{\overline{x}'}$ is torsion free, and since this is true for every such G, Theorem I.4.8.14 implies that $f^{\flat}_{\overline{x}}$ is saturated. □

For more about integral and saturated morphisms, we refer to the discussion in Section IV.4. In particular, if f is an integral morphism of fine log schemes that is smooth in the sense of log geometry (Definition IV.3.1.1), then the underlying morphism of schemes $\underline{f}$ is flat (see Theorem IV.4.3.5). Proposition IV.4.1.9 shows that the idealized fibers of a logarithmically smooth (or just flat) morphism are solid, so that (3) of Theorem 2.5.5 applies to such morphisms. Furthermore, Theorem IV.4.3.6 shows that a smooth and integral morphism f is saturated if and only if the fibers of $\underline{f}$ are reduced.

2.6 Log blowups

Blowing up is an important technique in algebraic geometry. It provides, for example, a universal way to render a coherent sheaf of ideals invertible. The logarithmic analog is equally important and admits an analogous description. Log blowups in the category of integral log schemes behave in surprising ways; in particular they are compatible with base change and are monomorphisms, despite the fact that they are not set-theoretically injective.[4]

Proposition 2.6.1. *Let X be a fine log scheme and let $\mathcal{K}$ be a sheaf of ideals in $\mathcal{M}_X$. Then the following conditions are equivalent.*

1. *Étale locally on X, $\mathcal{K}$ is generated as a sheaf of ideals by a single element.*
2. *Étale locally on X, $\mathcal{K}$ is isomorphic as a sheaf of $\mathcal{M}_X$-sets to $\mathcal{M}_X$.*
3. *The sheaf of ideals $\mathcal{K}$ is coherent and, for each geometric point $\overline{x}$ of X, its stalk $\mathcal{K}_{\overline{x}}$ is monogenic as an $\mathcal{M}_{\overline{x}}$-set.*

A sheaf of ideals with this property is said to be invertible.

4. *If $f\colon X \to Y$ is a morphism of fine log schemes and $\mathcal{K}$ is an invertible sheaf of ideals in $\mathcal{M}_Y$, then the ideal of $\mathcal{M}_X$ generated by $f^{-1}(\mathcal{K})$ is invertible.*
5. *If $\mathcal{I}$ and $\mathcal{J}$ are ideal sheaves in $\mathcal{M}_x$, then $\mathcal{I} + \mathcal{J}$ is invertible if and only if each of $\mathcal{I}$ and $\mathcal{J}$ is invertible.*

Proof If (1) holds, a choice of a local generator of $\mathcal{K}$ defines a surjection of $\mathcal{M}_X$-sets from $\mathcal{M}_X$ to $\mathcal{K}$. This surjection is necessarily injective because $\mathcal{M}_X$ is integral. Thus (1) implies (2), and it is clear that (2) implies (3). If (3) holds, observe that a local generator of $\mathcal{K}_{\overline{x}}$ will also generate $\mathcal{K}$ in some étale neighborhood of $\overline{x}$, because of the coherence of $\mathcal{K}$. Thus (3) implies (1). It is clear that condition that condition (1) for $\mathcal{K} \subseteq \mathcal{M}_Y$ implies condition (1) for the ideal of $\mathcal{M}_X$ generated by $f^{-1}(\mathcal{K})$, proving statement (4). Statement (5) is proved as in the case of monoschemes; see Lemma II.1.7.1. □

Definition 2.6.2. *A morphism $h\colon X' \to X$ of fine (resp. fine saturated) log schemes is a* log blowup *if there exists a coherent sheaf of ideals $\mathcal{K}$ in $\mathcal{M}_X$ with the following properties.*

1. *The ideal of $\mathcal{M}_{X'}$ generated by $h^{-1}(\mathcal{K})$ is invertible.*
2. *If $h'\colon T \to X$ is any morphism of fine (resp. fine saturated) log schemes such that the ideal of $\mathcal{M}_T$ generated by $h'^{-1}(\mathcal{K})$ is invertible, then h' factors uniquely through h.*

In this case we say that $h\colon X' \to X$ is a log blowup of X along $\mathcal{K}$.

[4] We have already seen this phenomenon in the category of monoschemes, in Section II.1.7

The universal property (2) in the definition ensures that if a log blowup of X along $\mathcal{K}$ exists, it is unique up to unique isomorphism. Thus we may, and shall often, write $X_{\mathcal{K}} \to X$ for any such map, and refer to it as *the* (log) blowup of X along $\mathcal{K}$. Note that, because of statement (4) of Proposition 2.6.1 (which holds only in the category of integral log schemes), the universal mapping property for log blowups is *stronger* than the universal property for blowups in the category of schemes [34, II,7.14]. (In the latter case, morphisms to the blowup of a coherent sheaf of ideals in $\mathcal{O}_X$ may not preserve invertibility of the pullback ideal.)

Theorem 2.6.3. *In the category of fine (resp. fine saturated) log schemes, the following statements hold.*

1. *The base change of a log blowup is again a log blowup. Specifically, if $h\colon X_{\mathcal{K}} \to X$ is a log blowup of X along $\mathcal{K}$ then, for any $g\colon X' \to X$, the fiber product $X' \times_X X_{\mathcal{K}}$ (computed in the appropriate category) is a log blowup of the ideal of $\mathcal{M}_{X'}$ generated by $g^{-1}(\mathcal{K})$.*
2. *The log blowup of X along any coherent sheaf of ideals of $\mathcal{M}_X$ always exists.*
3. *Every log blowup is proper, a monomorphism, and universally surjective.*
4. *If $\mathcal{I}$ and $\mathcal{J}$ are coherent sheaves of ideals in $\mathcal{M}_X$, then there are unique isomorphisms of log schemes over X*

$$X_{\mathcal{I}+\mathcal{J}} \cong X_{\mathcal{I}} \times_X X_{\mathcal{J}} \cong (X_{\mathcal{I}})_{\mathcal{J}} \cong (X_{\mathcal{J}})_{\mathcal{I}}.$$

Proof Let $\mathcal{K}$ be a coherent sheaf of ideals in $\mathcal{M}_X$, let $h\colon X_{\mathcal{K}} \to X$ be a log blowup of X along $\mathcal{K}$, and let T be any fine (resp. fine saturated) log scheme. The universal mapping property of the log blowup $X_{\mathcal{K}} \to X$ allows us to identify the set of morphisms $T \to X_{\mathcal{K}}$ as the subset of all morphisms $T \to X$ rendering $\mathcal{K}$ invertible. In particular, the map $X_{\mathcal{K}} \to X$ is a monomorphism. The universal mapping property also implies that the category of log blowups is stable under base change. Indeed, let $g\colon X' \to X$ be a morphism, and let $h'\colon X'' := X' \times_X X_{\mathcal{K}} \to X_{\mathcal{K}}$ be the pullback of h along g. Let $g'\colon X'' \to X_{\mathcal{K}}$ be the natural projection. Then by (4) of Proposition 2.6.1, the ideal of $\mathcal{M}_{X''}$ generated by $h'^{-1}(\mathcal{K}') = g'^{-1}(h^{-1}(\mathcal{K}))$ is invertible. Moreover, if $f\colon T \to X'$ is any morphism rendering $\mathcal{K}'$ invertible, the composition $g \circ f$ renders $\mathcal{K}$ invertible, and hence factors uniquely through $X_{\mathcal{K}}$. Thus the map f factors uniquely through X'', and X'' satisfies the universal property of a log blowup of the ideal $\mathcal{K}'$.

We prove the existence of log blowups in several steps. We first do the construction in the category of fine log schemes. To obtain the blowup in the saturated category it will suffice to replace the fine log blowup by its saturation.

Lemma 2.6.4. *Let X be a fine log scheme and let $\mathcal{K}$ be a coherent sheaf of ideals in $\mathcal{M}_X$. Suppose that X admits a chart a subordinate to a fine monoid P, or more generally to a fine monoscheme S. Suppose also that there exists a coherent sheaf of ideals $\mathcal{I}$ in $\mathcal{M}_S$ such that $a^{-1}(\mathcal{I})$ generates $\mathcal{K}$. Then there is a commutative diagram of locally monoidal spaces*

$$\begin{array}{ccc} X_{\mathcal{K}} & \xrightarrow{a_{\mathcal{K}}} & S_{\mathcal{I}} \\ {\scriptstyle h}\downarrow & & \downarrow{\scriptstyle g} \\ X & \xrightarrow{a} & S, \end{array}$$

where g is the log blowup of S along $\mathcal{I}$ (Theorem II.1.7.2), where h is a log blowup of X along $\mathcal{K}$, and where $a_{\mathcal{K}}$ is a chart for $X_{\mathcal{K}}$. In particular, the triple $(a, g, a_{\mathcal{K}})$ is a chart for h.

Proof First suppose that $X = \mathsf{A}_S$, the log scheme associated to the monoscheme A_S (see Proposition 1.2.4). Let $h\colon X_{\mathcal{K}} := \mathsf{A}_{S_{\mathcal{I}}} \to \mathsf{A}_S$ be the morphism of log schemes associated to g. Then the map of locally monoidal spaces $a_{\mathcal{K}}\colon X_{\mathcal{K}} \to S_{\mathcal{I}}$ is strict and the ideal of $X_{\mathcal{K}}$ generated by $\mathcal{K}$ is the same as the ideal generated by $\mathcal{I}$ and hence is invertible. To check that h is universal, suppose that $f\colon T \to X$ is a morphism of integral log schemes such that the ideal $\mathcal{K}_T$ of $\mathcal{M}_T$ generated by $\mathcal{K}$ is invertible. Then $a \circ f\colon T \to S$ is a morphism of integral monoidal spaces such that the ideal of $\mathcal{M}_T$ generated by $\mathcal{I}$ is invertible, and hence $a \circ f$ factors uniquely through g, by the universal property of $S_{\mathcal{I}}$. Since T is a log scheme, the resulting morphism $T \to S_{\mathcal{I}}$ factors uniquely through a map $\tilde{f}\colon T \to \mathsf{A}_{S_{\mathcal{I}}}$, and necessarily $h \circ \tilde{f} = f$. Thus h is a log blowup and the lemma is proved in this case.

For the general case, observe that the morphism a factors uniquely through a morphism of log schemes $X \to \mathsf{A}_S$. Then by (1) of Theorem 2.6.3, the morphism $X_{\mathcal{K}} := X \times_{\mathsf{A}_S} S_{\mathcal{I}} \to X$ is a log blowup. Furthermore, $X_{\mathcal{K}} \to \mathsf{A}_{S_{\mathcal{I}}}$ is strict, and hence so is the morphism $X_{\mathcal{K}} \to S_{\mathcal{I}}$. □

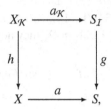

If the log structure of X is Zariski, then we can cover X by open sets U admitting charts, and on each U we can construct the blowup $U_{\mathcal{K}}$. For any two open sets U and V in this covering, there is a unique isomorphism $U_{\mathcal{K}} \times_X V \cong V_{\mathcal{K}} \times_X U$, and these isomorphisms automatically satisfy the cocycle condition. It then follows by standard methods that these local constructions glue together to form a global object $X_{\mathcal{K}} \to X$ satisfying the requisite universal mapping property. On the other hand, if the log structure on X only admits charts étale locally, this global construction could a priori yield an algebraic space rather than a scheme. The following additional argument [5] overcomes this difficulty.

Lemma 2.6.5. *Let $h\colon X_{\mathcal{K}} \to X$ be a log blowup in the category of fine (resp. fine saturated) log schemes. Then the invertible sheaf of $\mathcal{O}_{X_{\mathcal{K}}}$-modules on $X_{\mathcal{K}}$ corresponding to the torsor of local generators of $h^{-1}(\mathcal{K})$ is relatively ample for the morphism h. In particular, the morphism $X_{\mathcal{K}} \to X$ is projective.*

Proof We can verify this statement locally on X, and in particular we may assume that there is a fine chart $\alpha\colon (P, K) \to (\mathcal{M}_X, \mathcal{K})$. Since all our constructions and the desired conclusions are compatible with base change, we are reduced to the case in which $X = \mathsf{A}_P$ and $\mathcal{K}$ is the ideal generated by K. Then $X_{\mathcal{K}}$ is the log scheme associated to monoscheme A_{P_K} obtained by blowing up A_P along the ideal K. As explained in Theorem II.1.7.2, the invertible sheaf of ideals on $\underline{\mathsf{A}}_{P_K}$ generated by K is ample, so Proposition II.1.9.5 implies that the corresponding invertible sheaf of $\mathcal{O}_{X_{\mathcal{K}}}$-modules is also ample. □

As we have seen, to prove the existence of log blowups in general, it will suffice to consider the case in which X is affine and admits an affine étale covering $\eta\colon \underline{\tilde{X}} \to \underline{X}$ such that the induced log structure on $\tilde{X}$ is Zariski. If $\mathcal{K}$ is a coherent sheaf of ideals in $\mathcal{M}_X$, we have seen that the log blowup $\tilde{X}_{\mathcal{K}}$ of $\tilde{X}$ along the ideal of $\mathcal{M}_{\tilde{X}}$ generated by $\mathcal{K}$ exists as a log scheme. The universal mapping property of the blowup endows $\tilde{X}_{\mathcal{K}}$ with descent data for the covering $\tilde{X} \to X$, but, because $\tilde{X}_{\mathcal{K}}$ is not affine over $\tilde{X}$, these data are not a priori effective. However, the descent data also apply to the invertible sheaf $\tilde{\mathcal{L}}$ on $\tilde{X}_{\mathcal{K}}$ constructed in Lemma 2.6.5. Then as explained in [32, VIII, 7.8], one can descend the sheaf of graded algebras associated to $\mathcal{L}$, and then construct a scheme $\underline{X}_{\mathcal{K}}$ and an isomorphism $\underline{X}_{\mathcal{K}} \times_{\underline{X}} \underline{\tilde{X}} \cong \underline{\tilde{X}}_{\mathcal{K}}$ compatible with these data. Then the sheaf of monoids $\mathcal{M}_{\tilde{X}_{\mathcal{K}}}$ descends to an (étale) sheaf of monoids on $X_{\mathcal{K}}$. The ideal sheaf of $\mathcal{M}_{X_{\mathcal{K}}}$ generated by $\mathcal{K}$ is invertible because it is so on $\tilde{X}_{\mathcal{K}}$, and the universal mapping property is easy to verify.

We have already seen that a log blowup is a monomorphism. Properness can be checked locally on X and is preserved by base change, and follows from the fact that $X \to X_{\mathcal{K}}$ is projective, as we observed in Lemma 2.6.5. Statement (4) follows from the universal mapping property of log blowups and statement (5) of Proposition 2.6.1.

It remains only to prove that log blowups are universally surjective. Since the base change of a log blowup is again a log blowup, the surjectivity of log blowups will prove their universal surjectivity. Moreover, since the saturation map $X_{\mathcal{K}}^{\mathrm{sat}} \to X_{\mathcal{K}}$ is always surjective, it suffices to treat the surjectivity in

[5] due to Martin Olsson

the category of integral log schemes. Thus the theorem will follow from the stronger statement provided by the following proposition. □

Proposition 2.6.6. *Let X be a fine (resp. fine and saturated) log scheme, let $h\colon X' \to X$ be a log blowup, and let $\overline{x}$ be a geometric point of X. Then there exists a point $\overline{x}'$ of X' such that $h(\overline{x}') = \overline{x}$ and such that the induced map*

$$h^{\flat}_{\overline{x}'}\colon \overline{\mathcal{M}}^{\mathrm{gp}}_{X,\overline{x}} \to \overline{\mathcal{M}}^{\mathrm{gp}}_{X',\overline{x}'}$$

is surjective with finite kernel (resp. is an isomorphism).

Proof This statement is local on X, so we may assume that there is a chart $\alpha\colon P \to \mathcal{M}_X$ and that h is the log blowup of X along the ideal of $\mathcal{M}_X$ generated by K. Then the morphism $X \to \mathsf{A}_P$ is strict, hence exact, and therefore satisfies the four point lemma, Proposition 2.2.3. Thus we are reduced to proving the statement when $X = \mathsf{A}_P$. In this case the lemma follows immediately from the analogous statement for monoschemes, Proposition II.1.7.7. We should remark that the saturation hypothesis could be weakened: it is enough to require that the sheaf $\overline{\mathcal{M}}^{\mathrm{gp}}_X$ be torsion free. □

The following important application of log blowups is due to [40].

Theorem 2.6.7. *Let $f\colon X \to Y$ be a quasi-compact morphism of fine log schemes. Then, locally on Y, there exists a log blowup $Y' \to Y$ such that the base-changed map $f'\colon X' \to Y'$ is* **Q***-integral.*

Proof First suppose that f admits a chart (a, θ, b), where $\theta\colon P \to Q$ is a homomorphism of fine monoids, or, more generally, a morphism $g\colon T \to S$ of fine and quasi-compact monoschemes. By Theorem II.1.8.1, there exists a coherent sheaf of ideals $\mathcal{I}$ in $\mathcal{M}_S$ such that the induced morphism $g_{\mathcal{I}}\colon T_{\mathcal{I}} \to S_{\mathcal{I}}$ is exact. By Theorem II.1.8.7, this morphism is in fact **Q**-integral. Let $\mathcal{K}$ be the ideal of $\mathcal{M}_Y$ defined by $\mathcal{I}$ and let $f_{\mathcal{K}}\colon X_{\mathcal{K}} \to Y_{\mathcal{K}}$ be the blowup of f along $\mathcal{K}$. Then by Lemma 2.6.4, we have morphisms of locally monoidal spaces $X_{\mathcal{K}} \to T_{\mathcal{I}}$ and $Y_{\mathcal{K}} \to S_{\mathcal{I}}$, fitting into a chart for $f_{\mathcal{K}}$. It follows that $f_{\mathcal{K}}$ is also **Q**-integral.

In general, there exists a finite collection of étale maps $X_i \to X$ over which such charts exists, and for each of these there exists an ideal $\mathcal{K}_i$ of $\mathcal{M}_Y$ such that $X_{i,\mathcal{K}_i} \to Y$ is **Q**-integral. Let $\mathcal{K} := \sum \mathcal{K}_i$, and recall from Theorem 2.6.3 that $X_{\mathcal{K}}$ is the fiber product of the $X_{\mathcal{K}_i}$ over X. It follows from Proposition I.4.6.3 that **Q**-integrality is preserved by base change. Hence each $X_{i\mathcal{K}} \to Y_{\mathcal{K}}$ is **Q**-integral, and so $X_{\mathcal{K}} \to Y_{\mathcal{K}}$ is also **Q**-integral. □

IV

Differentials and Smoothness

To motivate the definitions to come, let us recall the classical notion of "differentials with log poles" along a divisor with normal crossings, which served as a model for some of the original notions of log geometry. Let X/Z be a smooth morphism and let $Y \subseteq X$ be a divisor with strict normal crossings in X relative to Z. This means that, locally on X, there exist a sequence $(t_1, \ldots, t_n)$ of sections of $\mathcal{O}_X$ such that the corresponding map $X \to \mathbf{A}^n_Z$ is étale and an integer $r \geq 0$ such that the ideal $\mathcal{I}$ of Y in X is defined by $(t_1 \cdots t_r)$. The sheaf of differentials of X/Z is the target of the universal derivation $d\colon \mathcal{O}_X \to \Omega^1_{X/Z}$ and, in the presence of coordinates $(t_1, \ldots, t_n)$, is the free $\mathcal{O}_X$-module generated by $dt_1, \ldots, dt_n$. Let $j\colon U \to X$ be the open immersion complementary to the closed immersion $Y \to X$. Then the sheaf $\Omega^1_{X/S}(\log Y)$ of *differentials with log poles along* Y [13] is the subsheaf of $j_* j^*(\Omega^1_{X/Z})$ generated by $(\operatorname{dlog} t_1, \ldots, \operatorname{dlog} t_r, dt_{r+1}, \ldots, dt_n)$, where $\operatorname{dlog} t := t^{-1}dt$. Its dual is the subsheaf of $\mathcal{D}er_{X/Z}(\mathcal{O}_X)$ generated by $(t_1\partial/\partial t_1, \ldots, t_r\partial/\partial t_r, \partial/\partial t_{r+1}, \ldots, \partial/\partial t_n)$. This is the sheaf of derivations of $\mathcal{O}_X$ that preserve the ideal $\mathcal{I}$. Let $\alpha\colon \mathcal{M}_{U/X} \to \mathcal{O}_X$ be the compactifying log structure associated with $U \to X$, that is, the sheaf of sections of $\mathcal{O}_X$ whose restriction to U is invertible. Then each t_i defines a section m_i of $\mathcal{M}_{U/X}$, and the map $\beta\colon \mathbf{N}^r \to \mathcal{M}$ defined by $(m_1, \ldots, m_r)$ is a chart for $\mathcal{M}_{U/X}$. There is a unique homomorphism

$$\mathbf{N}^r \to \Omega^1_{X/Z}(\log Y)$$

sending the ith basis element e_i to dt_i/t_i that agrees with the homomorphism

$$\mathcal{O}^*_X \to \Omega^1_{X/Z} \subseteq \Omega^1_{X/Z}(\log Y) : u \mapsto \operatorname{dlog} u$$

on $\beta^{-1}(\mathcal{O}^*_X)$. It then follows from the construction of $\mathcal{M}$ that there is a unique homomorphism

$$\operatorname{dlog}\colon \mathcal{M} \to \Omega^1_{X/Z}$$

such that $\alpha(m)\,\mathrm{dlog}(m) = d\alpha(m)$ for all sections m of $\mathcal{M}$, and $\Omega^1_{X/Z}$ is the target of the universal such homomorphism. Note that in particular, $t_i\,\mathrm{dlog}(m_i) = dt_i$.

We hope that this familiar example motivates the formal algebraic definitions of log derivations and log differentials, which we explain in the forthcoming sections. These definitions turn out to fit beautifully into the deformation theory of log schemes, in the general geometric framework envisioned by Grothendieck, thus providing one of the main justifications for the foundations of log geometry.

1 Derivations and differentials

1.1 Derivations and differentials of log rings

For the sake of simplicity and concreteness, let us begin by discussing derivations and differentials for log rings. Recall from Definition III.1.2.3 that a log ring is a homomorphism $\alpha\colon P \to A$ from a monoid P into the underlying multiplicative monoid of a ring A, and that if α and β are log rings, a homomorphism $\theta\colon \alpha \to \beta$ is a commutative diagram

$$\begin{array}{ccc} Q & \xrightarrow{\beta} & B \\ \uparrow{\scriptstyle\theta^\flat} & & \uparrow{\scriptstyle\theta^\sharp} \\ P & \xrightarrow{\alpha} & A. \end{array}$$

Definition 1.1.1. *Let θ be a homomorphism of log rings as given above and let E be a B-module. Then a (log) derivation of θ with values in E is a pair (D,δ), where $D\colon B \to E$ is an A-linear homomorphism and $\delta\colon Q \to E$ is a homomorphism of monoids, such that:*

1. *$D(bb') = bD(b') + b'D(b)$ for every pair of elements b, b' of B;*
2. *$D(\beta(q)) = \beta(q)\delta(q)$ for every $q \in Q$;*
3. *$\delta(\theta^\flat(p)) = 0$ for every $p \in P$.*

We denote by $\mathrm{Der}_\theta(E)$ the set of all derivations of θ with values in E. If (D_1,δ_1) and (D_2,δ_2) are derivations of θ and b_1, b_2 are elements of B, then $(b_1D_1 + b_2D_2, b_1\delta_1 + b_2\delta_2)$ is also a derivation, so that $\mathrm{Der}_\theta(E)$ has a natural structure of a B-module. Furthermore, if $h\colon E \to E'$ is a homomorphism of B-modules and $(D,\delta) \in \mathrm{Der}_\theta(E)$, then $(h \circ D, h \circ \delta) \in \mathrm{Der}_\theta(E')$. so that Der_θ becomes a functor from the category of B-modules into itself. Recall that if b is an element of B that acts bijectively on E, then any derivation D of B/A with values in E extends uniquely to the localization $\lambda\colon B \to B_b$. Thus if β_b is the log ring $Q \to B \to B_b$, the natural map $\mathrm{Der}_{\lambda\circ\theta}(E) \to \mathrm{Der}_\theta(E)$ is an isomorphism. Finally, note that if $q \in \beta^{-1}(B^*)$, then (2) implies that $\delta(q) = \beta(q)^{-1}D(\beta(q))$, i.e., $\delta(q)$ is the "logarithmic derivative of $\beta(q)$ with respect to D." It follows that δ factors uniquely through $Q \to Q^{log}$, where $\beta^{log}\colon Q^{log} \to B$ is the logarithmic homomorphism associated to β, in the sense of Proposition II.1.1.5. Thus the natural map $\mathrm{Der}_{\theta^{log}}(E) \to \mathrm{Der}_\theta(E)$ is an isomorphism, where θ^{log} is the associated homomorphism of log rings with Q^{log} in place of Q. Furthermore, if Q is dull and $D \in \mathrm{Der}_{B/A}(E)$, then $\delta(q) := \beta(-q)D(\beta(q))$ is the unique homomorphism $Q \to E$ such that $(D,\delta) \in \mathrm{Der}_\theta(E)$. Thus the natural map $\mathrm{Der}_\theta(E) \to \mathrm{Der}_{B/A}(E)$ is an isomorphism in this case.

It is straightforward to construct the universal derivation of a homomorphism of log rings; its target will be the module of logarithmic differentials.

Proposition 1.1.2. *Let $\theta\colon \alpha \to \beta$ be a homomorphism of log rings. Then the functor Der_θ is representable by a universal derivation*

$$(d,d)\colon (B,Q) \to \Omega^1_\theta,$$

whose target is called the module of (log) differentials of θ.

Proof There are of course many ways to construct Ω^1_θ by generators and relations. If one is willing to make use of the standard module of differentials $d\colon B \to \Omega^1_{B/A}$, one can take

$$\Omega^1_\theta := \left(\Omega^1_{B/A} \oplus (B \otimes Q^{gp}/P^{gp})\right)/R.$$

Here Q^{gp}/P^{gp} is the cokernel of the map $P^{gp} \to Q^{gp}$ and R is the submodule of $\Omega^1_{B/A} \oplus (B \otimes Q^{gp}/P^{gp})$ generated by elements of the form

$$(d\beta(q), -\beta(q) \otimes \pi(q)) \quad \text{for } q \in Q,$$

where $\pi\colon Q \to Q^{gp}/P^{gp}$ is the canonical map. Then $d\colon B \to \Omega^1_\theta$ sends b to the class of $(db, 0)$ and $d\colon Q \to \Omega^1_\theta$ sends q to the class of $(0, 1 \otimes \pi(q))$. We omit the straightforward verification that this construction has the desired universal property. □

The formation of the module of log differentials enjoys the familiar functoriality properties of its classical counterpart. Let us record the essential ones here.

Proposition 1.1.3. *Let*

$$\begin{array}{ccc} (\beta\colon Q \to B) & \xrightarrow{\phi} & (\beta'\colon Q' \to B') \\ \uparrow \theta & & \uparrow \theta' \\ (\alpha\colon P \to A) & \longrightarrow & (\alpha'\colon P' \to A') \end{array}$$

be a commutative diagram of log rings. Then there is a canonical homomorphism

$$B' \otimes_B \Omega^1_\theta \to \Omega^1_{\theta'}$$

compatible with the universal derivations. This homomorphism is an isomorphism if the square is cartesian, or if $\alpha = \alpha'$, $Q = Q'$, and $B \to B'$ is a localization.

Proof A B'-module E' gives rise to a B-module $\phi_*(E')$, and composition with $\theta^\sharp$ and $\theta^\flat$ induces a natural map

$$\mathrm{Der}_{\theta'}(E') \to \mathrm{Der}_\theta(\phi_*(E')). \tag{1.1.1}$$

Applying this map to the universal derivation with values in $\Omega^1_{\theta'}$, one finds a canonical element in $\mathrm{Der}_\theta(\theta_*(\Omega^1_{\theta'}))$ and hence a map

$$\Omega^1_\theta \to \phi_*(\Omega^1_{\theta'}).$$

The homomorphism in the proposition is deduced from this one by adjunction. Now suppose that the diagram is cartesian. Then $B' = A' \otimes_A B$ and $Q' = P' \oplus_P Q$, and it follows easily that, for any B'-module E', the map in equation (1.1.1) is an isomorphism. Consequently

$$\mathrm{Hom}_{B'}(B' \otimes_B \Omega^1_\theta, E') \cong \mathrm{Hom}_B(\Omega^1_\theta, \phi_*(E')) \cong \mathrm{Hom}_{B'}(\Omega^1_{\theta'}, E'),$$

and since this holds for every E', it follows that the map is an isomorphism. In the case of localizations, suppose that B' is the localization of B by b, and let (D, δ) be a derivation of θ with values in E'. Then multiplication by b acts bijectively on E' and, as we have observed, (1.1.1) is an isomorphism and consequently so is the homomorphism in the proposition. □

The module of log differentials has an especially simple description in the case of monoid algebras.

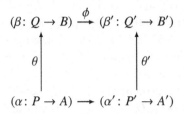

Proposition 1.1.4. *Let R be a commutative ring, let $\theta^\flat\colon P \to Q$ be a monoid*

homomorphism, and let $\theta^\sharp\colon R[P] \to R[Q]$ be the corresponding homomorphism of monoid algebras. Write $\theta := (\theta^\flat, \theta^\sharp)$ for the resulting homomorphism of log rings and, with some abuse of notation, $\pi\colon Q \to Q^{\mathrm{gp}}/P^{\mathrm{gp}}$ for the homomorphism from Q to the cokernel of $(\theta^\flat)^{\mathrm{gp}}$.

1. *For every $R[Q]$-module E, the natural map*

$$\mathrm{Der}_\theta(E) \to \mathrm{Hom}(Q^{\mathrm{gp}}/P^{\mathrm{gp}}, E)$$

is an isomorphism. Namely, given any $\delta \in \mathrm{Hom}(Q^{\mathrm{gp}}/P^{\mathrm{gp}}, E)$, there is a unique $D \in \mathrm{Der}_{R[Q]/R[P]}$ such that $(D,\delta) \in \mathrm{Der}_\theta(E)$.

2. *Define $D\colon R[Q] \to R[Q] \otimes Q^{\mathrm{gp}}/P^{\mathrm{gp}}$ and $\delta\colon Q \to R[Q] \otimes Q^{\mathrm{gp}}/P^{\mathrm{g}}$ by*

$$D(e^q) := e^q \otimes \pi(q) \quad \text{and} \quad \delta(q) := q \otimes \pi(q).$$

Then (D,δ) is a universal log derivation of θ, and hence induces a natural isomorphism

$$\Omega^1_\theta \to R[Q] \otimes Q^{\mathrm{gp}}/P^{\mathrm{gp}}.$$

In particular, if $Q^{\mathrm{gp}}/P^{\mathrm{gp}}$ is a finitely generated group the order of whose torsion is invertible in R, then Ω^1_θ is a free $R[Q]$-module.

Proof Note that the map in (1) is injective, since, for any $(D,\delta) \in \mathrm{Der}_\theta(E)$, $D(e^q) = e^q\delta(q)$. To prove the surjectivity, given $\delta \in \mathrm{Hom}(Q^{\mathrm{gp}}/P^{\mathrm{gp}}, E)$, let D be the unique R-linear map sending e^q to $e^q\delta(q)$. If q and q' are elements of Q,

$$\begin{aligned} D(e^q e^{q'}) &= D(e^{q+q'}) \\ &= e^{q+q'} \otimes \delta(q+q') \\ &= e^q e^{q'} \otimes (\delta(q) + \delta(q')) \\ &= e^q e^{q'} \otimes \delta(q') + e^q e^{q'} \otimes \delta(q) \\ &= e^q D(e^{q'}) + e^{q'} D(e^q). \end{aligned}$$

Furthermore,

$$D(e^{\theta^\flat(p)}) = e^{\theta^\flat(p)}\delta(\theta^\flat(p)) = 0.$$

Thus D is a derivation of $R[Q]$ over $R[P]$ and (D,δ) is the required element of $\mathrm{Der}_\theta(E)$.

Statement (1) implies statement (2), since, for every $R[Q]$-module E,

$$\begin{aligned} \mathrm{Hom}_{R[Q]}(\Omega^1_\theta, E) &\cong \mathrm{Der}_\theta(E) \cong \mathrm{Hom}_{\mathbf{Z}}(Q^{\mathrm{gp}}/P^{\mathrm{gp}}, E) \\ &\cong \mathrm{Hom}_{R[Q]}(R[Q] \otimes Q^{\mathrm{gp}}/P^{\mathrm{gp}}, E). \end{aligned}$$ □

Corollary 1.1.5. *If G is an abelian group, then there is a universal derivation*

$$R[G] \to R[G] \otimes G\colon e^g \mapsto e^g \otimes g.$$

Proof The previous result shows that this is the case when $R[G]$ is replaced by the log ring $G \to R[G]$. Since G is dull, the log differentials agree with the usual differentials, as we observed in the discussion immediately following Definition 1.1.1. □

Corollary 1.1.6. *Let $0 \to G' \to G \to G'' \to 0$ be an exact sequence of abelian groups and let $I \subseteq R[G]$ be the kernel of the corresponding homomorphism $R[G] \to R[G'']$. Then there is a unique isomorphism*

$$\eta\colon R[G''] \otimes_{\mathbf{Z}} G' \cong I/I^2$$

mapping $1 \otimes g'$ to $(e^{g'} - 1) \pmod{I^2}$ for all $g' \in G'$.

Proof If $g' \in G'$, then $e^{g'} - 1 \in I$. If also $h' \in G'$, then

$$\begin{aligned} e^{g'+h'} - 1 &= e^{g'} e^{h'} - 1 \\ &= (e^{g'} - 1)(e^{h'} - 1) + (e^{g'} - 1) + (e^{h'} - 1) \\ &= (e^{g'} - 1) + (e^{h'} - 1) \pmod{I^2}. \end{aligned}$$

Thus the map $G' \to I/I^2$ sending g' to the class of $e^{g'} - 1$ is a group homomorphism. Since I/I^2 is an $R[G]/I \cong R[G'']$-module, this homomorphism induces by adjunction an $R[G'']$-linear map $\eta\colon R[G''] \otimes G' \to I/I^2$, as in the statement of the corollary. Since I is generated by the set of all $e^{g'} - 1$ with $g' \in G'$, η is surjective.

On the other hand, we constructed in Corollary 1.1.5 a (universal) derivation $D\colon R[G] \to R[G] \otimes G$ sending each e^g to $e^g \otimes g$. Consider the composite

$$I \xrightarrow{D} R[G] \otimes G \to (R[G]/I) \otimes G \cong R[G''] \otimes G. \tag{1.1.2}$$

Since the last of these $R[G]$-modules is annihilated by I and since D is a derivation, the map (1.1.2) is in fact $R[G]$-linear and annihilates I^2. Furthermore, for any $g' \in G'$, $D(e^{g'} - 1) = e^{g'} \otimes g' = 1 \otimes g' \in R[G''] \otimes G$. Since I is generated by such elements, the map (1.1.2) factors through a homomorphism $\overline{d}\colon I/I^2 \to R[G''] \otimes G'$, sending the class of $e^{g'} - 1$ to $1 \otimes g'$. Then $\overline{d}\eta = \mathrm{id}$ and, since η is surjective, it must be an isomorphism. □

Corollary 1.1.7. *Let $\phi\colon P \to Q$ be a homomorphism of finitely generated abelian groups and let K be the kernel of ϕ. Then the corresponding homomorphism $R[P] \to R[Q]$ is flat if and only if $R \otimes K = 0$.*

Proof First observe that if ϕ is injective, Q is a free P-set, hence $R[Q]$ is a free $R[P]$-module, and in particular the homomorphism $R[P] \to R[Q]$ is faithfully flat. Now let P' denote the image of ϕ. Since $R[P'] \to R[Q]$ is faithfully flat, we see that $R[P] \to R[Q]$ is flat if and only if $R[P] \to R[P']$ is flat. Thus we are reduced to the case in which ϕ is surjective.

Let $I \subseteq R[P]$ be the kernel of $R[\phi]$. If $R[P] \to R[P]/I$ is flat, necessarily $I^2 = I$, and then Corollary 1.1.6 implies that $R \otimes K = 0$. Conversely, if $R \otimes K = 0$, then $I = I^2$. Since I is finitely generated, Nakayama's lemma implies that, at each point x of $\underline{\mathsf{A}}_{\mathsf{P}}$, either $I_x = \mathcal{O}_{X,x}$ or $I_x = 0$. Thus the map $\underline{\mathsf{A}}_{\mathsf{P}'} \to \underline{\mathsf{A}}_{\mathsf{P}}$ is an open immersion, hence flat. □

Remark 1.1.8. In the category of schemes, the sheaf of differentials $\Omega^1_{X/Y}$ can be identified with the conormal sheaf of the diagonal embedding $X \to X \times_Y X$. The logarithmic version of this useful interpretation is not straightforward, because in general the diagonal embedding is not strict, and the notion of the conormal sheaf in this case requires some preparation, as we shall see later in Section 3.4. We can, however, explain here how to think about the conormal sheaf of the diagonal in the case of monoid algebras.

If $\theta\colon P \to Q$ is a homomorphism of integral monoids and $\mathsf{A}_\theta\colon \mathsf{A}_\mathsf{Q} \to \mathsf{A}_\mathsf{P}$ is the corresponding morphism of log schemes, then the diagonal embedding $\mathsf{A}_\mathsf{Q} \to \mathsf{A}_\mathsf{Q} \times_{\mathsf{A}_\mathsf{P}} \mathsf{A}_\mathsf{Q}$ corresponds to the homomorphism $\sigma\colon Q \oplus_P Q \to Q$ given by summation. Since σ is surjective, to make it strict is the same as to make it exact, and Proposition I.4.2.19 allows us to identify the universal exactification of σ with the embedding

$$i\colon \mathsf{A}_\mathsf{Q} \to \mathsf{A}^*_{\mathsf{Q}/\mathsf{P}} \times \mathsf{A}_\mathsf{Q} : \alpha \mapsto (1, \alpha).$$

Let J (resp. I) be the kernel of the homomorphism $R[i^\#]$ (resp. of the augmentation homomorphism $R[Q^{\mathrm{gp}}/P^{\mathrm{gp}}] \to R$). Corollary 1.1.6 constructed an isomorphism: $I/I^2 \cong R \otimes Q^{\mathrm{gp}}/P^{\mathrm{gp}}$, and hence by base change we have $J/J^2 \cong R[Q] \otimes Q^{\mathrm{gp}}/P^{\mathrm{gp}}$. Thus the isomorphism of Proposition 1.1.4 identifies Ω^1_θ with the conormal J/J^2 of the exactified diagonal embedding.

The following result will have a geometric explanation in Remark 3.2.5, where we will see that the morphisms of log schemes corresponding to ϕ and ψ are étale.

Proposition 1.1.9. *Suppose that we are given a commutative diagram of log rings*

$$\begin{array}{ccc}
(Q \xrightarrow{\beta} B) & \xrightarrow{\psi} & (Q' \xrightarrow{\beta'} B') \\
\uparrow{\scriptstyle\theta} & \nearrow{\scriptstyle\theta'} & \uparrow{\scriptstyle\theta''} \\
(P \xrightarrow{\alpha} A) & \xrightarrow{\phi} & (P' \xrightarrow{\alpha'} A'),
\end{array}$$

where $\phi^{\flat \mathrm{gp}}\colon P^{\mathrm{gp}} \to P'^{\mathrm{gp}}$ and $\psi^{\flat \mathrm{gp}}\colon Q^{\mathrm{gp}} \to Q'^{\mathrm{gp}}$ are isomorphisms and where the natural maps $A \otimes_{\mathbf{Z}[P]} \mathbf{Z}[P'] \to A'$ and $B \otimes_{\mathbf{Z}[Q]} \mathbf{Z}[Q'] \to B'$ are isomorphisms. Then the natural maps

$$B' \otimes_B \Omega^1_{\theta} \to \Omega^1_{\theta'} \quad \text{and} \quad \Omega^1_{\theta'} \to \Omega^1_{\theta''}$$

are also isomorphisms.

Proof To prove that the second arrow above is an isomorphism, we must prove that for any B'-module E', the natural map $\mathrm{Der}_{\theta''}(E') \to \mathrm{Der}_{\theta'}(E')$ is an isomorphism. This map is obviously injective. Suppose that $(D', \delta') \in \mathrm{Der}_{\theta'}(E')$. Since ϕ^{gp} is an isomorphism, δ' annihilates the image of P'. Hence $D'(\alpha'(p')) = \alpha(p')\delta'(p') = 0$ for every $p' \in P'$, and it follows that $D'(a\alpha'(p')) = 0$ for every $a \in A$ and every $p' \in P'$. Since A' is generated by such products, D' annihilates A' and so (D', δ') belongs to $\mathrm{Der}_{\theta''}(E')$.

For the first arrow, we must prove that $\mathrm{Der}_{\theta'}(E') \to \mathrm{Der}_{\theta}(E')$ is an isomorphism for every E'. Injectivity follows from the facts that $\psi^{\flat \mathrm{gp}}$ is an isomorphism and that B' is generated by B and Q. To prove surjectivity, let ϵ (resp. ϵ') denote the log ring $Q \to \mathbf{Z}[Q]$ (resp. $Q' \to \mathbf{Z}[Q']$). Since $\psi^{\flat \mathrm{gp}}$ is an isomorphism, it induces an isomorphism $\mathrm{Hom}(Q', E') \to \mathrm{Hom}(Q, E')$. Then Proposition 1.1.4 implies that ψ induces an isomorphism $\mathrm{Der}_{\epsilon'}(E') \to \mathrm{Der}_{\epsilon}(E')$. Thus if $(D, \delta) \in \mathrm{Der}_{\theta}(E')$, there is a unique $(D', \delta') \in \mathrm{Der}_{\epsilon'}(E')$ such that $\delta' \circ \psi^{\mathrm{gp}} = \delta$. Necessarily $\delta'^{\mathrm{gp}} \circ \theta'^{\flat \mathrm{gp}} = 0$, and it follows that D' annihilates $\mathbf{Z}[P]$. Then $D' \in \mathrm{Der}_{\mathbf{Z}[Q']/\mathbf{Z}[P]}(E')$ and $D \in \mathrm{Der}_{B/A}$ restrict to the same element of $\mathrm{Der}_{\mathbf{Z}[Q]/\mathbf{Z}[P]}$. Applying Lemma 1.1.10 below with $R = \mathbf{Z}[P]$, $S = \mathbf{Z}[Q]$, and $S' = \mathbf{Z}[Q']$, we conclude that there is a $D'' \in \mathrm{Der}_{B'/A}$ restricting to D'. Then $(D'', \delta') \in \mathrm{Der}_{\theta'}(E')$ restricts to $(D, \delta) \in \mathrm{Der}_{\theta}(E')$, completing the proof. □

Lemma 1.1.10. *Suppose we are given a commutative diagram of commutative ring homomorphisms*

$$\begin{array}{ccccc} A & \longrightarrow & B & \longrightarrow & B' \\ \uparrow & & \uparrow & & \uparrow \\ R & \longrightarrow & S & \longrightarrow & S', \end{array}$$

in which the square on the right is cocartesian. Then if E' is a B'-module, the natural map

$$\mathrm{Der}_{B'/A}(E') \to \mathrm{Der}_{B/A}(E') \times_{\mathrm{Der}_{S/R}(E')} \mathrm{Der}_{S'/R}(E')$$

is an isomorphism. Consequently, the following sequence is exact:

$$B' \otimes_S \Omega^1_{S/R} \to B' \otimes_B \Omega_{B/A} \oplus B' \otimes_{S'} \Omega^1_{S'/R} \to \Omega^1_{B'/A} \to 0.$$

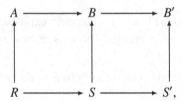

Proof The injectivity is clear from the fact that B' is generated as an algebra by the images of B and S'. For the surjectivity, suppose that we are given E'-valued derivations of B/A and of S'/R that agree on S, which we denote by the same letter D. Define a function $\langle\ ,\ \rangle\colon B \times S' \to E'$ by setting $\langle b, s'\rangle := s'D(b) + bD(s')$. This function is evidently $\mathbf{Z}$-bilinear. Moreover, if $s \in S$,

$$\begin{aligned}\langle sb, s'\rangle &= s'D(sb) + sbD(s')\\ &= s'sD(b) + s'bD(s) + sbD(s')\\ &= ss'D(b) + bD(ss')\\ &= \langle b, ss'\rangle.\end{aligned}$$

Even though D is not S-linear, this equality implies that $\langle\ ,\ \rangle$ factors through an R-linear map $D\colon B \otimes S' \to E'$. One checks immediately that D is a derivation of B'/A. □

1.2 Derivations and differentials of log schemes

Although log schemes are the focus of our study, it is convenient to define derivations and differentials for prelog schemes as well.

Definition 1.2.1. *Let $f\colon X \to Y$ be a morphism of prelog schemes and let $\mathcal{E}$ be a sheaf of $\mathcal{O}_X$-modules. A derivation (or, for emphasis, log derivation) of X/Y with values in $\mathcal{E}$ is a pair (D, δ), where $D\colon \mathcal{O}_X \to \mathcal{E}$ is a homomorphism of abelian sheaves and $\delta\colon \mathcal{M}_X \to \mathcal{E}$ is a homomorphism of sheaves of monoids such that the following conditions are satisfied:*

1. *$D(\alpha_X(m)) = \alpha_X(m)\delta(m)$ for every local section m of $\mathcal{M}_X$;*
2. *$\delta(f^\flat(n)) = 0$ for every local section n of $f^{-1}(\mathcal{M}_Y)$,*
3. *$D(ab) = aD(b) + bD(a)$ for every pair of local sections a, b of $\mathcal{O}_X$;*
4. *$D(f^\sharp(c)) = 0$ for every local section c of $f^{-1}(\mathcal{O}_Y)$.*

We denote by $\mathrm{Der}_{X/Y}(\mathcal{E})$ the set of all such derivations. Then the presheaf $\mathcal{D}er_{X/Y}(\mathcal{E})$ that to every open subset U of X assigns the set of derivations of U/Y with values in $\mathcal{E}_{|_U}$ is in fact a sheaf. Furthermore, if (D_1, δ_1) and (D_2, δ_2) are sections of $\mathcal{D}er_{X/Y}(\mathcal{E})$, so is $(D_1 + D_2, \delta_1 + \delta_2)$, and if a is a section of $\mathcal{O}_X$ and (D, δ) an section of $\mathcal{D}er_{X/Y}(\mathcal{E})$, then $(aD, a\delta)$ also belongs to $\mathrm{Der}_{X/Y}(\mathcal{E})$. Thus $\mathcal{D}er_{X/Y}(\mathcal{E})$ has a natural structure of a sheaf of $\mathcal{O}_X$-modules.

Variant 1.2.2. Derivations for idealized log schemes are defined in exactly the same way as in Definition 1.2.1. Thus, if $f\colon X \to Y$ is a morphism of idealized log schemes, and $(D, \delta) \in \mathrm{Der}_{X/Y}(\mathcal{E})$, we do not require that $\delta(k) = 0$ for $k \in \mathcal{K}_X$, and we have $\Omega^1_{X/Y} = \Omega^1_{(X,\emptyset)/(Y,\emptyset)}$. The reason for this definition will become apparent from the relationship between derivations and deformations explained in Theorem 2.2.2 and Variant 2.2.3.

The formation of $\mathrm{Der}_{X/Y}$ is functorial in $\mathcal{E}$: an $\mathcal{O}_X$-linear map $h\colon \mathcal{E} \to \mathcal{E}'$ induces a homomorphism

$$\mathrm{Der}_{X/Y}(h)\colon \mathrm{Der}_{X/Y}(\mathcal{E}) \to \mathrm{Der}_{X/Y}(\mathcal{E}') \quad (D, \delta) \mapsto (h \circ D, h \circ \delta).$$

The following proposition explains how Der is also functorial in X/Y.

Proposition 1.2.3. *Let*

$$\begin{array}{ccc} X' & \xrightarrow{g} & X \\ {\scriptstyle f'}\downarrow & & \downarrow{\scriptstyle f} \\ Y' & \xrightarrow{h} & Y \end{array}$$

be a commutative diagram of prelog schemes.

1. *Composition with $g^\sharp$ and $g^\flat$ induces a morphism of functors*

$$\mathrm{Der}_g\colon \mathrm{Der}_{X'/Y'} \to \mathrm{Der}_{X/Y} \circ g_*,$$

which for an $\mathcal{O}_{X'}$-module $\mathcal{E}'$ is the map

$$\mathrm{Der}_{X'/Y'}(\mathcal{E}') \to \mathrm{Der}_{X/Y}(g_*(\mathcal{E}'))\colon (D', \delta') \mapsto (D' \circ g^\sharp, \delta' \circ g^\flat).$$

2. *The functoriality morphism Der_g is an isomorphism in the following cases*
   (a) *f' is the morphism of log schemes associated to the morphism f of prelog schemes.*
   (b) *The diagram is cartesian in the category of prelog schemes.*
   (c) *The diagram is cartesian in the category of log schemes.*

Proof The verification that composition with $g^\sharp$ and $g^\flat$ takes derivations to derivations is immediate. To prove (2a), recall from Proposition II.1.1.5 and especially diagram (1.1.1) of its proof, that the log structure $\mathcal{M}_X^{log} \to \mathcal{O}_X$ associated to the prelog structure $\mathcal{M}_X \to \mathcal{O}_X$ is obtained from the cocartesian square in the following diagram:

$$\begin{array}{ccccc} \alpha_X^{-1}(\mathcal{O}_X^*) & \longrightarrow & \mathcal{O}_X^* & & \\ \downarrow & & \downarrow & \searrow & \\ \mathcal{M}_X & \longrightarrow & \mathcal{M}_X^{log} & \longrightarrow & \mathcal{O}_X. \end{array}$$

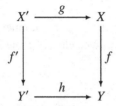

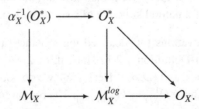

Thus the sheaf of monoids $\mathcal{M}_X^{log}$ is locally generated by the images of $\mathcal{M}_X$ and $\mathcal{O}_X^*$, and it follows that Der_g is injective. Conversely, if $(D,\delta) \in \mathrm{Der}_{X/Y}(\mathcal{E})$, define

$$\partial\colon \mathcal{O}_X^* \to \mathcal{E} : u \mapsto u^{-1}Du.$$

Then $\partial(uv) = \partial u + \partial v$, and ∂ is a homomorphism $\mathcal{O}_X^* \to \mathcal{E}$. Furthermore, if m is a section of $\alpha_X^{-1}(\mathcal{O}_X^*)$, then

$$\partial(\alpha_X(m)) = \alpha_X(m)^{-1}D(\alpha_X(m)) = \alpha_X(m)^{-1}\alpha_X(m)\delta(m) = \delta(m).$$

Since the diagram is cocartesian, it follows that there is a unique $\delta^{log}\colon \mathcal{M}_X^{log} \to \mathcal{E}$ which agrees with δ on $\mathcal{M}_X$ and with ∂ on $\mathcal{O}_X^*$. Since $\mathcal{M}_X^{log}$ is generated by $\mathcal{M}_X$ and $\mathcal{O}_X^*$, in fact $D\alpha_X(m) = \alpha_X(m)\delta^{log}(m)$ for every section m of $\mathcal{M}_X^{log}$. Furthermore, since $\mathcal{M}_Y^{log}$ is generated by $\mathcal{O}_Y^*$ and $\mathcal{M}_Y$, δ^{log} annihilates the image of $f^{-1}(\mathcal{M}_Y^{log})$. Thus (D,δ^{log}) is a section of $\mathrm{Der}_{X'/Y'}(\mathcal{E})$ that restricts to (D,δ). This shows that the functoriality map is also surjective and completes the proof of statement (2a). In the case (2b), let $p := h\circ f' = f\circ g$. If $(D,\delta) \in \mathrm{Der}_{X/Y}(g_*\mathcal{E}')$, then, by the classical version of (2b), there is a unique $D' \in \mathrm{Der}_{\underline{X}'/\underline{Y}'}(\mathcal{E}')$, which restricts to $D \in \mathrm{Der}_{\underline{X}/\underline{Y}}$. Since the diagram is cartesian in the category of prelog schemes, the map

$$f'^{-1}(\mathcal{M}_{Y'}) \oplus_{p^{-1}(\mathcal{M}_Y)} g^{-1}(\mathcal{M}_X) \to \mathcal{M}_{X'}$$

is also an isomorphism, and the map $\delta\colon \mathcal{M}_X \to g_*(\mathcal{E}')$ induces a unique map $\mathcal{M}_{X'} \to g_*(\mathcal{E}')$ that annihilates $f'^{-1}(\mathcal{M}_{Y'})$. It follows that (D',δ') satisfies conditions (1) and (2) of Definition 1.2.1 as well, and this completes the proof of (2b). Finally, we observe that (2c) is a consequence of (2a) and (2b), since the log structure of the fiber product in the category of log schemes is the log structure associated to the prelog structure of the fiber product in the category of prelog schemes. □

Theorem 1.2.4. *Let $f\colon X \to Y$ be a morphism of prelog schemes. Then the functor $\mathcal{E} \mapsto \mathrm{Der}_{X/Y}(\mathcal{E})$ is representable by a universal object*

$$\mathcal{O}_X \xrightarrow{d} \Omega^1_{X/Y}, \quad \mathcal{M}_X \xrightarrow{d} \Omega^1_{X/Y} \quad \left(\text{or} \quad \mathcal{M}_X \xrightarrow{dlog} \Omega^1_{X/Y}\right).$$

Proof There are of course many constructions possible. For example, we can take

$$\Omega^1_{X/Y} = \left(\Omega^1_{\underline{X}/\underline{Y}} \oplus (\mathcal{O}_X \otimes \mathcal{M}_X^{gp})\right)/\mathcal{R},$$

where $\mathcal{R}$ is the sub $\mathcal{O}_X$-module generated by sections of the form

$$(d\alpha_X(m), -\alpha_X(m)\otimes m) \text{ for } m \in \mathcal{M}_X, \quad (0, 1\otimes f^\flat(n)) \text{ for } n \in f^{-1}(\mathcal{M}_Y),$$

with the evident maps

$$d\colon \mathcal{O}_X \to \Omega^1_{X/Y}, \quad d\colon \mathcal{M}_X \to \Omega^1_{X/Y}.$$

Proposition 1.2.11 below gives an alternative construction that does not use the sheaf $\Omega^1_{\underline{X}/\underline{Y}}$ as an ingredient. □

Remark 1.2.5. When using additive notation for $\mathcal{M}_X$, it seems sensible to write d for the map $\mathcal{M}_X \to \Omega^1_{X/Y}$. Then $\alpha_X\colon \mathcal{M}_X \to \mathcal{O}_X$ behaves like an exponential map, which is consistent with the equation $d\alpha_X(m) = \alpha_X(m)dm$. In this case, the canonical injection $\mathcal{O}^*_X \to \mathcal{M}_X$ needs a symbol λ, which should be regarded as a logarithm map, and one has $d\lambda(u) = u^{-1}du$, as expected. When the monoid law on $\mathcal{M}_X$ is written multiplicatively and $\mathcal{O}^*_X$ is viewed as a submonoid of $\mathcal{M}_X$, it is more natural (and more usual) to write $dlog$ for the universal map $\mathcal{M}_X \to \Omega^1_{X/Y}$, since $dlog(mn) = dlog(m) + dlog(n)$ and $dlog(u) = u^{-1}du$ if $u \in \mathcal{O}^*_X \subseteq \mathcal{M}_X$.

The proofs of the next two corollaries are immediate and therefore omitted.

Corollary 1.2.6. *Let $f\colon X \to Y$ be a morphism of log schemes given by a morphism of log rings θ as in Definition 1.1.1. Then $\Omega^1_{X/Y}$ is the quasi-coherent sheaf associated to the A-module Ω^1_θ.* □

Corollary 1.2.7. *Let $f\colon X \to Y$ be a morphism of schemes with trivial log structure and let $\mathcal{E}$ be a sheaf of $\mathcal{O}_X$-modules. Then $\mathcal{D}er_{X/Y}(\mathcal{E})$ can be identified with the usual sheaf of derivations of $\underline{X}/\underline{Y}$ with values in $\mathcal{E}$, i.e., with the sheaf of homomorphisms of abelian groups $D\colon \mathcal{O}_X \to \mathcal{E}$ satisfying conditions (3) and (4) of Definition 1.2.1.* □

Corollary 1.2.8. *If $f\colon X \to Y$ is a morphism of coherent log schemes, $\Omega^1_{X/Y}$ is quasi-coherent, and it is of finite type (resp. of finite presentation) if $\underline{f}$ is of finite type (resp. of finite presentation).*

Proof This assertion is of a local nature on X, so we may assume that X and Y are affine, and by Proposition II.2.4.2, that f admits a coherent chart. Then f comes from a morphism of log rings whose underlying monoids P and Q are finitely generated. By Corollary 1.2.6, $\Omega^1_{X/Y}$ is quasi-coherent. Since $\Omega^1_{\underline{X}/\underline{Y}}$ is of finite type (resp. of finite presentation) if f is, and since Q^{gp} is a finitely generated abelian group, $\Omega^1_{X/Y}$ is of finite type (resp. of finite presentation) if $\underline{f}$ is. □

We shall see that a derivation (D, δ) of log schemes is uniquely determined by δ. This observation leads to a new construction of the sheaf of differentials, which is in some ways more convenient that the standard one.

Proposition 1.2.9. *Suppose that $f\colon X \to Y$ is a morphism of log schemes,* *that $\mathcal{E}$ is a sheaf of $\mathcal{O}_X$-modules, and that (D, δ) is a pair of homomorphisms of sheaves of monoids satisfying conditions (1) and (2) of Definition 1.2.1. Then D is uniquely determined by δ and necessarily satisfies conditions (3) and (4).*

Proof We shall need the following simple lemma.

Lemma 1.2.10. *If X is any scheme, the image of $\mathcal{O}_X^* \to \mathcal{O}_X$ generates $\mathcal{O}_X$ as sheaf of additive monoids. That is, any local section of $\mathcal{O}_X$ can locally be written as a sum of sections of $\mathcal{O}_X^*$. In particular, if X is a log scheme, $\mathcal{O}_X$ is generated, as a sheaf of additive monoids, by the image of $\alpha_X \colon \mathcal{M}_X \to \mathcal{O}_X$.*

Proof Let a be a local section of $\mathcal{O}_X$ and let x be a point of X. If a maps to a unit in the local ring $\mathcal{O}_{X,x}$, then a is a unit in some neighborhood of x, and hence a is locally in the image of $\mathcal{O}_X^*$. If a maps to an element of the maximal ideal of $\mathcal{O}_{X,x}$, then $a - 1$ maps to a unit, and so $a = 1 + (a - 1)$ is locally the sum of two units. □

The lemma evidently implies that D is uniquely determined by δ, and that condition (4) follows from condition (2). To check (3), observe that if m and n are sections of $\mathcal{M}_X$ and $a := \alpha_X(m)$ and $b := \alpha_X(n)$, then

$$\begin{aligned} D(ab) &= D(\alpha_X(m)\alpha_X(n)) \\ &= D(\alpha_X(m+n)) \\ &= \alpha_X(m+n)\delta(m+n) \\ &= \alpha_X(m)\alpha_X(n)(\delta(n)+\delta(m)) \\ &= \alpha_X(m)\alpha_X(n)\delta(n) + \alpha_X(m)\alpha_X(n)\delta(m) \\ &= aD(b) + bD(a). \end{aligned}$$

If $a_i = \alpha_X(m_i)$ and $a = a_1 + a_2$, then again

$$D(ab) = D(a_1 b + a_2 b) = a_1 D(b) + bD(a_1) + a_2 D(b) + bD(a_2) = aD(b) + bD(a).$$

A similar argument with b, together with an application of Lemma 1.2.10, shows that (3) holds for any sections a and b of $\mathcal{O}_X$. □

The following proposition gives an alternative construction of the sheaf of differentials of a morphism of log schemes, as a suitable quotient of the sheaf $\mathcal{O}_X \otimes \mathcal{M}_X^{\mathrm{gp}}$. We consider the subsheaves of $\mathcal{O}_X \otimes \mathcal{M}_X^{\mathrm{gp}}$ defined as follows:

- $\mathcal{R}_1 \subseteq \mathcal{O}_X \otimes \mathcal{M}_X^{\mathrm{gp}}$ is the subsheaf of sections locally of the form
$$\sum_i \alpha_X(m_i) \otimes m_i - \sum_i \alpha_X(m_i') \otimes m_i',$$
where $(m_1, \ldots, m_k)$ and $(m_1', \ldots, m_{k'}')$ are sequences of local sections of $\mathcal{M}_X$ such that $\sum_i \alpha_X(m_i) = \sum_i \alpha_X(m_i')$;

- $\mathcal{R}_2$ is the image of the map
$$\mathcal{O}_X \otimes f^{-1}(\mathcal{M}_Y^{\text{gp}}) \to \mathcal{O}_X \otimes \mathcal{M}_X^{\text{gp}}.$$

Proposition 1.2.11. *Let $f \colon X \to Y$ be a morphism of log schemes. Then there is a unique isomorphism of sheaves of $\mathcal{O}_X$-modules:*
$$\Omega^1_{X/Y} \cong \left(\mathcal{O}_X \otimes \mathcal{M}_X^{\text{gp}}\right) / (\mathcal{R}_1 + \mathcal{R}_2),$$
sending dm to the class of $1 \otimes m$ for every local section m of $\mathcal{M}_X$. In particular, $\Omega^1_{X/Y}$ is generated as a sheaf of $\mathcal{O}_X$-modules, by the image of $d \colon \mathcal{M}_X \to \Omega^1_{X/Y}$.

Proof The main difficulty is contained in the following lemma.

Lemma 1.2.12. *The sheaves $\mathcal{R}_1$ and $\mathcal{R}_2$ are sheaves of sub-$\mathcal{O}_X$-modules of $\mathcal{O}_X \otimes \mathcal{M}_X^{\text{gp}}$.*

Proof This result is clear for $\mathcal{R}_2$; the difficulty is $\mathcal{R}_1$. For a finite sequence $\mathbf{m} := (m_1, \ldots, m_k)$ of sections of $\mathcal{M}_X$, let
$$s(\mathbf{m}) := \sum_i \alpha_X(m_i) \in \mathcal{O}_X$$
$$r(\mathbf{m}) := \sum_i \alpha_X(m_i) \otimes m_i \in \mathcal{O}_X \otimes \mathcal{M}_X^{\text{gp}}.$$

Let S be the sheaf of pairs $(\mathbf{m}, \mathbf{m}')$ of finite sequences of sections of $\mathcal{M}_X$ such that $s(\mathbf{m}) = s(\mathbf{m}')$. Then $\mathcal{R}_1$ is the subsheaf of sections of $\mathcal{O}_X \otimes \mathcal{M}_X^{\text{gp}}$ locally of the form $r(\mathbf{m}) - r(\mathbf{m}')$ for some local section $(\mathbf{m}, \mathbf{m}')$ of S.

Note first that the pair $(0, 0)$ belongs to S and that $r(0) - r(0) = 0$, so that $0 \in \mathcal{R}_1$. Next, note that, since $(\mathbf{m}', \mathbf{m}) \in S$ if $(\mathbf{m}, \mathbf{m}') \in S$, it follows that $-r \in \mathcal{R}_1$ whenever $r \in \mathcal{R}_1$. Now if $(\mathbf{m}, \mathbf{m}')$ and $(\mathbf{n}, \mathbf{n}') \in S$, let $\mathbf{p}$ (resp. $\mathbf{p}'$) denote the concatenation of $\mathbf{m}$ and $\mathbf{n}$ (resp. of $\mathbf{m}'$ and $\mathbf{n}'$). Then $(\mathbf{p}, \mathbf{p}') \in S$ and
$$r(\mathbf{m}) - r(\mathbf{m}') + r(\mathbf{n}) - r(\mathbf{n}') = r(\mathbf{p}) - r(\mathbf{p}').$$

Thus $\mathcal{R}_1$ is an abelian subsheaf of $\mathcal{O}_X \otimes \mathcal{M}_X^{\text{gp}}$.

It remains to check that $\mathcal{R}_1$ is stable under multiplication by sections a of $\mathcal{O}_X$. Lemma 1.2.10 shows that it suffices to check this for $a = \alpha_X(m)$, with m a section of $\mathcal{M}_X$. Let us first observe that S is stable under the action of $\mathcal{M}_X$ by translation. Thus, if $\mathbf{m} = (m_1, \ldots, m_k)$ is a sequence of sections of $\mathcal{M}_X$ and m is any section of $\mathcal{M}_X$, let $\mathbf{m} + m := (m_1 + m, \ldots, m_k + m)$ and let $a := \alpha_X(m)$. Then
$$\begin{aligned} s(\mathbf{m} + m) &= \sum \alpha_X(m_i + m) \\ &= \alpha_X(m) \sum \alpha_X(m_i) \\ &= \alpha_X(m) s(\mathbf{m}) \end{aligned}$$

$$= as(\mathbf{m}).$$

Hence if $(\mathbf{m}, \mathbf{m}') \in S$,

$$s(\mathbf{m} + m) = \alpha_X(m)s(\mathbf{m}) = \alpha_X(m)s(\mathbf{m}') = s(\mathbf{m}' + m),$$

so that $(\mathbf{m} + m, \mathbf{m}' + m) \in S$. Next, we compute as follows:

$$\begin{aligned} r(\mathbf{m} + m) &= \sum \alpha_X(m_i + m) \otimes (m_i + m) \\ &= \sum \alpha_X(m_i)\alpha_X(m) \otimes m_i + \sum \alpha_X(m_i)\alpha_X(m) \otimes m \\ &= a \sum \alpha_X(m_i) \otimes m_i + a \sum \alpha_X(m_i) \otimes m \\ &= ar(\mathbf{m}) + as(\mathbf{m}) \otimes m. \end{aligned}$$

Hence, if $(\mathbf{m}, \mathbf{m}') \in S$, we have, since $s(\mathbf{m}) = s(\mathbf{m}')$,

$$\begin{aligned} a(r(\mathbf{m}) - r(\mathbf{m}')) &= ar(\mathbf{m}) - ar(\mathbf{m}') \\ &= ar(\mathbf{m}) + as(\mathbf{m}) \otimes m - ar(\mathbf{m}') - as(\mathbf{m}') \otimes m \\ &= r(\mathbf{m} + m) - r(\mathbf{m}' + m). \end{aligned}$$

Since $(\mathbf{m} + m, \mathbf{m}' + m) \in S$, we can conclude that $a(r(\mathbf{m}) - r(\mathbf{m}')) \in \mathcal{R}_1$. □

Now let us write Ω for the quotient of $\mathcal{O}_X \otimes \mathcal{M}_X^{gp}$ by $\mathcal{R}_1 + \mathcal{R}_2$ and let $d\colon \mathcal{M}_X \to \Omega$ be the map described in the statement. We claim that there is a unique derivation $d\colon \mathcal{O}_X \to \Omega$ such that $d\alpha_X(m) = \alpha_X(m)dm$ for all local sections m of $\mathcal{O}_X$. As we have explained, the uniqueness follows from Lemma 1.2.10. If a is any section of $\mathcal{O}_X$, choose a sequence $\mathbf{m}$ of local sections of $\mathcal{M}_X$ with $s(\mathbf{m}) = a$. Then it follows from the definition of $\mathcal{R}_1$ that the image of $r(\mathbf{m})$ in Ω is independent of the choice of $\mathbf{m}$. Let $d\colon \mathcal{O}_X \to \Omega$ be the map of abelian sheaves such that $ds(\mathbf{m})$ is the class of $r(\mathbf{m})$ for every sequence $\mathbf{m}$. In particular, if m is a section of $\mathcal{M}_X$ and $\mathbf{m} := (m)$, then $\alpha_X(m) = s(\mathbf{m})$ and so $d\alpha_X(m)$ is the class of $r(\mathbf{m}) = \alpha_X(m) \otimes m$. Thus, $d\alpha_X(m) = \alpha_X(m)dm$, and the pair (d, d) satisfies (1) and (2), hence also (3) and (4), of Definition 1.2.1.

To check that (d, d) is universal, suppose that $\mathcal{E}$ is a sheaf of $\mathcal{O}_X$-modules and $(D, \delta) \in \mathrm{Der}_{X/Y}(\mathcal{E})$. Since $\mathcal{E}$ is a sheaf of abelian groups, δ factors uniquely through $\mathcal{M}_X^{gp}$, and since $\mathcal{E}$ is a sheaf of $\mathcal{O}_X$-modules, it factors through a unique $\mathcal{O}_X$-linear map $\theta\colon \mathcal{O}_X \otimes \mathcal{M}_X^{gp} \to \mathcal{E}$. Property (2) of Definition 1.2.1 implies that

θ annihilates $\mathcal{R}_2$. If $\mathbf{m}$ is a sequence of sections of $\mathcal{M}_X$, then

$$\begin{aligned}\theta(r(\mathbf{m})) &= \theta\Big(\sum_i \alpha_X(m_i)\otimes m_i\Big)\\ &= \sum_i \alpha_X(m_i)\delta(m_i)\\ &= \sum_i D(\alpha_X(m_i))\\ &= D\Big(\sum_i \alpha_X(m_i)\Big)\\ &= D(s(\mathbf{m})).\end{aligned}$$

Consequently $\theta(r(\mathbf{m})) = \theta(r(\mathbf{m}'))$ whenever $(\mathbf{m},\mathbf{m}') \in S$, so θ factors uniquely through an $\mathcal{O}_X$-linear map $h\colon \Omega \to \mathcal{E}$. This is the unique homomorphism such that $h(d(m)) = \delta(m)$ for every local section m of $\mathcal{M}$. It follows that $h(d(a)) = D(a)$ for every local section a of $\mathcal{O}_X$. □

The difference between classical and log differentials is revealed by the *Poincaré residue* mapping, of which there are several versions. The next example explains a version for log rings, and the proposition which follows it is a geometric incarnation.

Example 1.2.13. Let $\theta\colon (\alpha\colon P \to A) \to (\beta\colon Q \to B)$ be a homomorphism of log rings, let F be a face of Q containing the image of P, and let I be the ideal of B generated by $\beta(\mathfrak{p}_F)$. Define $\delta\colon Q \to B/I \otimes (Q/F)^{\mathrm{gp}}$ to be the homomorphism sending q to $1 \otimes \pi_F(q)$, where $\pi_F(q)$ is the image of q in $(Q/F)^{\mathrm{gp}}$. Then $\beta(q)\delta(q) = 0$ for every $q \in Q$, so $(0,\delta)$ defines an element of $\mathrm{Der}_\theta(B/I \otimes (Q/F)^{\mathrm{gp}})$. The corresponding homomorphism

$$\rho_F\colon \Omega^1_\theta \to B/I \otimes (Q/F)^{\mathrm{gp}}$$

sends dq to $1 \otimes \pi_F(q)$ if $q \in Q$ and db to 0 if $b \in B$. In particular, if F is a facet of Q and Q is saturated, then $(Q/F)^{\mathrm{gp}} \cong \mathbf{Z}$, and ρ_F can be viewed as a map $\Omega^1_\theta \to B/I$.

Proposition 1.2.14. *Let $f\colon X \to Y$ be a morphism of coherent idealized log schemes. Suppose that for every $x \in X$, $\mathcal{K}_{X,x}$ is a prime ideal disjoint from the image of $f^{-1}(\mathcal{M}_Y)$. Let $\mathcal{F}$ be the subsheaf of sections of $\mathcal{M}_X$ whose stalks do not belong to $\mathcal{K}$ (see Proposition III.1.11.11), and define*

$$\delta\colon \mathcal{M}_X \to \mathcal{O}_X \otimes (\mathcal{M}_X/\mathcal{F})^{\mathrm{gp}} : m \mapsto 1 \otimes \pi_F(m),$$

where $\pi_F(m)$ is the image of m in $(\mathcal{M}_X/F)^{\mathrm{gp}}$. Then there is a unique $\mathcal{O}_X$-linear *map ρ making the following diagram commute:*

$$\begin{array}{ccc} \mathcal{M}_X & & \\ \downarrow{\scriptstyle dlog} & \searrow{\scriptstyle \delta} & \\ \Omega^1_{X/Y} & \xrightarrow{\rho} & O_X \otimes (\mathcal{M}_X/\mathcal{F})^{gp}, \end{array}$$

The map ρ is called the Poincaré residue mapping, and $\rho(da) = 0$ for every $a \in O_X$.

Proof Any $m \in \mathcal{M}_{X,x}$ belongs either to $\mathcal{K}_{X,x}$ or to $\mathcal{F}_x$. Since $\alpha_X(m) = 0$ if $m \in \mathcal{K}_{X,x}$ and $\pi_F(m) = 0$ if $m \in \mathcal{F}_x$ it follows that $\alpha_X(m) \otimes \delta(m) = 0$ for every section m of $\mathcal{M}_X$. Then $(0, \delta)$ is a log derivation of X/Y with values in $O_X \otimes (\mathcal{M}_X/\mathcal{F})^{gp}$. The existence and uniqueness of ρ follow from the universal mapping property of $\Omega^1_{X/Y}$. □

Next we discuss functoriality and base change for the sheaf of log differentials.

Proposition 1.2.15. *Let C be a commutative diagram*

$$\begin{array}{ccc} X' & \xrightarrow{g} & X \\ {\scriptstyle f'}\downarrow & & \downarrow{\scriptstyle f} \\ Y' & \xrightarrow{h} & Y \end{array}$$

of prelog schemes. Then there is a unique homomorphism

$$d_C \colon g^*(\Omega^1_{X/Y}) \to \Omega^1_{X'/Y'}$$

sending $1 \otimes da$ to $dg^\sharp(a)$ for every section a of $g^{-1}(O_X)$ and $1 \otimes dlog(m)$ to $dlog\, g^\flat(m)$ for every section m of $g^{-1}(\mathcal{M}_X)$. This morphism d_C is an isomorphism in the following cases:

1. *f' is the morphism of log schemes associated to to the morphism f of prelog schemes;*
2. *The diagram is cartesian in the category of prelog (resp. log, coherent, fine, fine and saturated log) schemes.*

In either of these cases, there is a commutative diagram of isomorphisms

$$\begin{array}{ccc} & \Omega^1_{X'/Y} & \\ \nearrow & & \searrow \\ f'^*(\Omega^1_{Y'/Y}) \oplus g^*(\Omega^1_{X/Y}) & \longrightarrow & \Omega^1_{X'/X} \oplus \Omega^1_{X'/Y'}. \end{array}$$

Proof Let $\mathcal{E}'$ be any sheaf of $\mathcal{O}_{X'}$-modules and let (D, δ) be an element of $\mathrm{Der}_{X'/Y'}(\mathcal{E}')$. As we saw in Proposition 1.2.3, there is a natural homomorphism

$$\mathrm{Der}_{X'/Y'}(\mathcal{E}') \to \mathrm{Der}_{X/Y}(g_*\mathcal{E}').$$

The existence of and uniqueness of the map on differentials d_C follows from their defining universal property. If f' is the morphism of log schemes corresponding to the morphism of prelog schemes f, then d_C is an isomorphism by (2a) of Proposition 1.2.3. If the diagram is cartesian in the category of prelog (resp. log schemes), d_C is an isomorphism by (2b) (resp. (2c)). Since the fiber product of coherent log schemes is the same as the fiber product in the category of log schemes, it follows that d_C is an isomorphism in this case as well. The proofs for fine and fine saturated log schemes follow from Lemma 1.2.16 below.

The last statement follows formally. Indeed, since the map $g^*(\Omega^1_{X/Y}) \to \Omega^1_{X'/Y'}$ is an isomorphism, the map $g^*(\Omega^1_{X/Y}) \to \Omega^1_{X'/Y}$ provides a splitting of the map $\Omega^1_{X'/Y} \to \Omega^1_{X'/Y'}$. By the same token, the map $f'^*(\Omega^1_{Y'/Y}) \to \Omega^1_{X'/X}$ is also an isomorphism, and the map $f'^*(\Omega^1_{Y'/Y}) \to \Omega^1_{X'/Y}$ provides a splitting of the map $\Omega^1_{X'/Y} \to \Omega^1_{X'/X}$. □

Lemma 1.2.16. *Let $f: X \to Y$ be a morphism of coherent (resp. fine) log schemes, and let $r: X' \to X$ be the universal map from X^{fin} (resp. X^{sat}) into X. Then the map $d_r: r^*(\Omega^1_{X'/Y}) \to \Omega^1_{X/Y}$ is an isomorphism. If $s: Y' \to Y$ is defined similarly, the map $d_s: \Omega^1_{X'/Y} \to \Omega^1_{X'/Y'}$ is an isomorphism.*

Proof To prove that these morphisms are isomorphisms is a local problem, so we may assume that f is given by a homomorphism of log rings. In this case the lemma follows from Proposition 1.1.9. □

Example 1.2.17. Associated to any morphism $f: X \to Y$ of prelog schemes is a commutative square, mapping f to the morphism $\underline{f}$ of underlying schemes. Thus there is a canonical map $\Omega^1_{\underline{X}/\underline{Y}} \to \Omega^1_{X/Y}$ sending (D, δ) to D. If f is strict, the diagram is cartesian, and this homomorphism is an isomorphism by Proposition 1.2.15. For example, if $X \to X_Y \to Y$ is the canonical factorization (equation III.1.2.1) of f, then the map $\Omega^1_{\underline{X}/\underline{Y}} \to \Omega^1_{X_Y/Y}$ is an isomorphism.

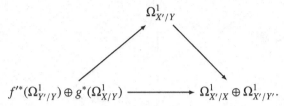

2 Thickenings and deformations

Deformation theory in log geometry works very nicely, as long as one works with deformations along strict closed immersions and with log structures that are u-integral or nearly so. Many of the ideas in this section were suggested by L. Illusie.

2.1 Thickenings and extensions

Definition 2.1.1. *A log thickening is a strict closed immersion $i\colon S \to T$ of log schemes such that:*

1. *the ideal $\mathcal{I}$ of S in T is a nil ideal;*
2. *the subgroup $1 + \mathcal{I}$ of $\mathcal{O}_T^* \cong \mathcal{M}_T^*$ operates freely on $\mathcal{M}_T$.*

A log thickening of order n is a log thickening such that $\mathcal{I}^{n+1} = 0$.

If T is u-integral, condition (2) in Definition 2.1.1 is automatic. A thickening $i\colon S \to T$ induces a homeomorphism of the underlying topological spaces of S and T, and it is common to identify them.

Proposition 2.1.2. *Let $i\colon S \to T$ be a log thickening, with ideal $\mathcal{I}$.*

1. *The commutative square*

$$\begin{array}{ccc} \mathcal{O}_T^* & \longrightarrow & \mathcal{M}_T \\ \big\downarrow & & \big\downarrow {\scriptstyle i^\flat} \\ \mathcal{O}_S^* & \longrightarrow & \mathcal{M}_S \end{array}$$

is cartesian and cocartesian (i.e., $\mathcal{O}_T^*$ is the inverse image of $\mathcal{O}_S^*$ in $\mathcal{O}_T$, and $\mathcal{M}_S$ is the amalgamated sum of $\mathcal{O}_S^*$ and $\mathcal{M}_T$).

2. $\mathrm{Ker}(\mathcal{O}_T^* \to \mathcal{O}_S^*) = \mathrm{Ker}(\mathcal{M}_T^{\mathrm{gp}} \to \mathcal{M}_S^{\mathrm{gp}}) = 1 + \mathcal{I}$.
3. *The action of $1 + \mathcal{I}$ on $\mathcal{M}_T$ (resp. on $\mathcal{M}_T^{\mathrm{gp}}$) makes it a torsor over $\mathcal{M}_S$ (resp. over $\mathcal{M}_S^{\mathrm{gp}}$). That is, the maps*

$$(1 + \mathcal{I}) \times \mathcal{M}_T \to \mathcal{M}_T \times_{\mathcal{M}_S} \mathcal{M}_T \quad \text{and} \quad (1 + \mathcal{I}) \times \mathcal{M}_T^{\mathrm{gp}} \to \mathcal{M}_T^{\mathrm{gp}} \times_{\mathcal{M}_S^{\mathrm{gp}}} \mathcal{M}_T^{\mathrm{gp}},$$

$$(u, m) \mapsto (m, um)$$

are isomorphisms.

4. *The map $\mathcal{M}_T \to \mathcal{M}_S \times_{\mathcal{M}_S^{\mathrm{gp}}} \mathcal{M}_T^{\mathrm{gp}}$ is an isomorphism.*

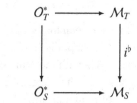

Proof The fact that the square in (1) is cartesian just amounts to the statement that the homomorphism $i^\flat$ is local. The fact that the diagram is cocartesian comes from the fact that i is strict, so that $\mathcal{M}_S$ is the log structure associated to the prelog structure $\mathcal{M}_T \to \mathcal{O}_S$. Since $\mathcal{I}$ is a nilideal, any local section a of $\mathcal{I}$ is locally nilpotent, and hence $1 + a$ is a unit of $\mathcal{O}_T$. It is clear that $1 + \mathcal{I}$ is exactly the kernel of the homomorphism $\mathcal{O}_T^* \to \mathcal{O}_S^*$. Since $\mathcal{M}_T \to \mathcal{O}_T$ is a log structure, $\mathcal{M}_T^* = \mathcal{O}_T^*$, and since the action of $1 + \mathcal{I}$ on $\mathcal{M}_T$ is free, the map $1 + \mathcal{I} \to \mathcal{M}_T^{\rm gp}$ is injective, and evidently its image is contained in the kernel of the map $\mathcal{M}_T^{\rm gp} \to \mathcal{M}_S^{\rm gp}$. Conversely, any local section x of $\mathcal{M}_T^{\rm gp}$ is the class of $m' - m$ for two sections of $\mathcal{M}_T$ and, if x maps to zero in $\mathcal{M}_S^{\rm gp}$, there exists a local section n of $\mathcal{M}_S$ such that $i^\flat(m') + n = i^\flat(m) + n$. Locally n lifts to a section m'' of $\mathcal{M}_T$, and the equation then becomes $i^\flat(m' + m'') = i^\flat(m + m'')$. Since i is strict, there then exists a $u \in 1 + \mathcal{I}$ such that $m' + m'' = u + m + m''$, and hence $m' - m = u$ in $\mathcal{M}_T^{\rm gp}$. This shows that $x \in 1 + \mathcal{I}$ and completes the proof of (2). These same arguments also prove (3).

When T and S are integral, statement (4) amounts to the fact that the morphism i is exact, a consequence of its strictness. Let us check it in the general case. Let (m, x) be a local section of $\mathcal{M}_S \times_{\mathcal{M}_S^{\rm gp}} \mathcal{M}_T^{\rm gp}$. We may locally write $m = i^\flat(m')$ for a local section of $\mathcal{M}_T$ and x as the class of $m_2 - m_1$ for local sections m_i of $\mathcal{M}_T$. Since m' and x have the same image in $\mathcal{M}_S^{\rm gp}$, there exists a local section m'' of $\mathcal{M}_T$ such that

$$i^\flat(m'') + i^\flat(m') + i^\flat(m_1) = i^\flat(m'') + i^\flat(m_2).$$

Then there is a local section u of $1 + \mathcal{I}$ such that $u + m'' + m' + m_1 = m'' + m_2$ in $\mathcal{M}_T$. Then $u + m'$ is a section of $\mathcal{M}_T$ mapping to (m, x). Suppose on the other hand that m and m' are sections of $\mathcal{M}_T$ with the same image in $\mathcal{M}_S \times_{\mathcal{M}_S^{\rm gp}} \mathcal{M}_T^{\rm gp}$. Since the images in $\mathcal{M}_S$ of m and m' are the same, $m' = u + m$ for some section u of $1 + \mathcal{I}$, and since the images of m and m' in $\mathcal{M}_T^{\rm gp}$ are the same, u maps to 0 in $\mathcal{M}_T^{\rm gp}$. Since $1 + \mathcal{I}$ acts freely on $\mathcal{M}_T$, this implies that $u = 0$, so $m' = m$, completing the proof. □

Proposition 2.1.3. *Let $i\colon S \to T$ be a log thickening.*

1. *T is coherent (resp. integral, resp. fine, resp. fine and saturated) if and only if S is.*
2. *Let $\beta\colon P \to \mathcal{M}_T$ be a homomorphism from a constant monoid P to $\mathcal{M}_T$. Then β is a chart for $\mathcal{M}_T$ if and only if $i^\flat \circ \beta$ is a chart for $\mathcal{M}_S$.*

Proof [1] Let us first prove statement (2). Let $\mathcal{N}_T \to \mathcal{M}_T$ (resp. $\mathcal{N}_S \to \mathcal{M}_S$) be

[1] This proof is due to O. Gabber

the logarithmic homomorphism associated to β (resp. to $i^\flat \circ \beta$). These fit into a commutative diagram

$$\begin{array}{ccc} \mathcal{N}_T & \longrightarrow & \mathcal{M}_T \\ \downarrow & & \downarrow \\ \mathcal{N}_S & \longrightarrow & \mathcal{M}_S. \end{array}$$

The two vertical arrows are the quotient mappings by the action of the group $(1 + \mathcal{I})$, and the action on $\mathcal{M}_T$ is free, since $S \to T$ is a log thickening. It follows that the action on $\mathcal{N}_T$ is also free. Then the top vertical arrow is an isomorphism if and only if the bottom one is. It follows that $\mathcal{M}_T$ is integral if and only if $\mathcal{M}_S$ is integral.

To prove (1), suppose that $\beta: P \to \mathcal{M}_S$ is a chart for $\mathcal{M}_S$ subordinate to a finitely generated (resp. fine, resp. fine and saturated) monoid P. Then $P^{\rm gp}$ is a finitely generated group, and we choose a surjection θ from a finitely generated free abelian group L to $P^{\rm gp}$. By statement (6) of Theorem I.2.1.17, the monoid $P' := P \times_{P^{\rm gp}} L$ is finitely generated (resp. fine, resp. fine and saturated). It is easy to see that the natural map $P'^{\rm gp} \to L$ is an isomorphism and that $P' \to \mathcal{M}_S$ is still a chart for $\mathcal{M}_S$. Replacing P by P', we may assume without loss of generality that $P^{\rm gp}$ is free. Then the homomorphism $P \to \mathcal{M}_S^{\rm gp}$ lifts locally to $\mathcal{M}_T^{\rm gp}$, and hence by (4) of Proposition 2.1.2, the homomorphism $\beta: P \to \mathcal{M}_S$ lifts locally to $\mathcal{M}_T$. As we have seen, any such lift is a chart for $\mathcal{M}_T$. □

Definition 2.1.4. *Let $f: X \to Y$ be a morphism of log schemes and let $\mathcal{E}$ be a quasi-coherent sheaf of $\mathcal{O}_X$-modules. A Y-extension of X by $\mathcal{E}$ is a commutative diagram*

$$\begin{array}{ccc} X & \xrightarrow{\;i\;} & T \\ {\scriptstyle f}\downarrow & \swarrow & \\ Y, & & \end{array}$$

where i is a log thickening of order one with $\mathcal{E} = \ker(i^\sharp)$. If $u: \mathcal{F} \to \mathcal{E}$ is a homomorphism of quasi-coherent sheaves of $\mathcal{O}_X$-modules and $i: X \to S$ (resp. $j: X \to T$) is a Y-extension of X by $\mathcal{E}$ (resp. by $\mathcal{F}$), then a morphism of Y-extensions over u *is a Y-morphism $g: S \to T$ such that $g \circ i = j$ and $g^\sharp(ax) = g^\sharp(a)u(x)$, for sections a of $\mathcal{O}_T$ and x of $\mathcal{F}$. When $\mathcal{E} = \mathcal{F}$ and $u = \mathrm{id}$, one says simply that g is a* morphism of Y-extensions.

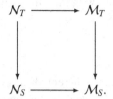

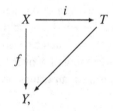

If T is a Y-extension of X by $\mathcal{E}$, then the map $i^\flat\colon \mathcal{M}_T \to \mathcal{M}_X$ and the action of $\mathcal{E} \subseteq \mathcal{M}_T^*$ on $\mathcal{M}_T$ make $\mathcal{M}_T$ into a $\mathcal{E}$-torsor over $\mathcal{M}_X$. The category of Y-extensions of X with a fixed $\mathcal{E}$ (with morphisms over $\mathrm{id}_{\mathcal{E}}$) is a groupoid: every morphism is an isomorphism.

Remark 2.1.5. If $\mathcal{E}$ is a quasi-coherent sheaf of $\mathcal{O}_X$-modules, the *trivial Y-extension of X by $\mathcal{E}$*, denoted $X \oplus \mathcal{E}$, is the log scheme T defined by $\mathcal{O}_T := \mathcal{O}_X \oplus \mathcal{E}$, where $(a, e)(a', e') := (aa', ae' + a'e)$, and by $\mathcal{M}_T := \mathcal{M}_X \oplus \mathcal{E}$, where $\alpha_T(m, e) := (\alpha_X(m), \alpha_X(m)e)$. The kernel of $\mathcal{O}_T \to \mathcal{O}_X$ is the ideal $(0, \mathcal{E}) \subseteq \mathcal{O}_T$, which acts freely on $\mathcal{M}_T$, so that (id_X, i) is a Y-extension of X by $\mathcal{E}$. Furthermore, we have an evident retraction $T \to X$. Conversely, a Y-extension is trivial (isomorphic to $X \oplus \mathcal{E}$) if and only if i admits a Y-retraction $r\colon T \to X$.

One can endow the set $\mathrm{Ext}_Y(X, \mathcal{E})$ of isomorphism classes of Y-extensions of X by $\mathcal{E}$ with an abelian group structure in a natural way. If $i\colon X \to S$ and $j\colon X \to T$ are Y-extensions of X by $\mathcal{E}$, then the sum of the classes of i and j in $\mathrm{Ext}_Y(X, \mathcal{E})$ is formed by first taking the Y-extension of X by $\mathcal{E} \oplus \mathcal{E}$ given by the fibered products $\mathcal{O}_S \times_{\mathcal{O}_X} \mathcal{O}_T$ and $\mathcal{M}_S \times_{\mathcal{M}_X} \mathcal{M}_T$, and then taking the class of the pushout along the additional law $\mathcal{E} \oplus \mathcal{E} \to \mathcal{E}$. The identity element of $\mathrm{Ext}_Y(X, \mathcal{E})$ is the class of $X \oplus \mathcal{E}$. If a is a section of $\mathcal{O}_X$ and T is an object of $\mathrm{Ext}_Y(X, \mathcal{E})$, then the pushout along the endomorphism of $\mathcal{E}$ defined by a defines the class of aT in $\mathrm{Ext}_Y(X, \mathcal{E})$.

Variant 2.1.6. An *idealized log thickening* is a log thickening $i\colon S \to T$ of idealized log schemes that is ideally strict as well as strict. Thus $\mathcal{K}_T$ is the unique sheaf of ideals in $\mathcal{M}_T$ whose restriction to $\mathcal{M}_S$ agrees with $\mathcal{K}_S$, and $\alpha_T(k) = 0$ for every section k of $\mathcal{K}_T$. If $f\colon X \to Y$ is a morphism of idealized log schemes and $\mathcal{E}$ is a quasi-coherent sheaf on X, then an *idealized Y-extension* of X by $\mathcal{E}$ is a first-order idealized log thickening of X over Y whose ideal sheaf is $\mathcal{E}$. The trivial Y-extension of X by $\mathcal{E}$ is such an extension, since $\alpha_T(k, e) = 0$ if k is a section of $\mathcal{K}_T$.

Example 2.1.7. Let $f\colon X \to Y$ and $g\colon Y \to Z$ be morphisms of log schemes such that the underlying morphism of schemes $\underline{f}$ is affine, and let $i\colon X \to S$ be a Z-extension of X by a quasi-coherent $\mathcal{O}_X$-module $\mathcal{E}$. Then $f_*\mathcal{E}$ is quasi-coherent on Y, and we can construct a Z-extension $f_*(i) := j\colon Y \to T$ of Y by

$f_*\mathcal{E}$ and a universal commutative diagram

$$\begin{array}{ccc} X & \xrightarrow{\ i\ } & S \\ {\scriptstyle f}\downarrow & & \downarrow \\ Y & \xrightarrow{\ j\ } & T, \end{array}$$

as follows. Since $\mathcal{E}$ is quasi-coherent and f is affine, there is an exact sequence of sheaves

$$0 \to f_*\mathcal{E} \to f_*\mathcal{O}_S \to f_*\mathcal{O}_X \to 0$$

on Y. Let $\mathcal{O}_T$ be the fiber product of $f_*\mathcal{O}_S$ and $\mathcal{O}_Y$ over $f_*\mathcal{O}_X$, which fits into an exact sequence

$$0 \to f_*\mathcal{E} \to \mathcal{O}_T \to \mathcal{O}_Y \to 0.$$

Since $\mathcal{F} := f_*\mathcal{E}$ is quasi-coherent, there is a closed immersion $j\colon Y \to T$ with square-zero ideal $\mathcal{F}$ corresponding to this exact sequence. Since i is a log thickening, $1+\mathcal{E}$ acts freely on $\mathcal{M}_S$, with quotient $\mathcal{M}_X$. Moreover, $\mathcal{E}$ is a square-zero ideal, so as an abelian sheaf $1+\mathcal{E} \cong \mathcal{E}$, and consequently $R^1 f_*(1+\mathcal{E}) = 0$. It follows that $1+\mathcal{F} = f_*(1+\mathcal{E})$ acts freely on $f_*(\mathcal{M}_S)$ with quotient $f_*(\mathcal{M}_X)$. Let $\mathcal{M}_T$ be the fiber product of $f_*(\mathcal{M}_S)$ and $\mathcal{M}_Y$ over $f_*(\mathcal{M}_X)$. Then $1+\mathcal{F}$ acts freely on $\mathcal{M}_T$ with quotient $\mathcal{M}_Y$, and the the map $\alpha_T\colon \mathcal{M}_T \to \mathcal{O}_T$ induced by α_S is a log structure. Then $j\colon Y \to T$ is an extension of Y by $\mathcal{F}$, and the square above is cocartesian in the category of log schemes.

2.2 Differentials and deformations

The geometric motivation for the definition of log derivations lies in the study of extensions of morphisms to thickenings.

Definition 2.2.1. *Let $f\colon X \to Y$ be a morphism of log schemes. A log thickening over X/Y is a commutative diagram consisting of the solid arrows in the diagram*

$$\begin{array}{ccc} S & \xrightarrow{\ i\ } & T \\ {\scriptstyle g}\downarrow & \overset{\tilde g}{\swarrow} & \downarrow{\scriptstyle h} \\ X & \xrightarrow[\ f\ ]{} & Y, \end{array}$$

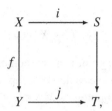

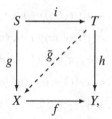

where i is a log thickening (Definition 2.1.1). A deformation of g to T is an element of

$$\mathrm{Def}_{X/Y}(g,T) := \{\tilde{g}\colon T \to X : \tilde{g} \circ i = g, f \circ \tilde{g} = h\}.$$

The morphism i in the diagram is a homeomorphism for the Zariski topology, and we have identified the underlying topological spaces of S and T. If i has finite order, the étale topologies of S and T can also be identified. Thus $\mathrm{Def}_{X/Y}(g,T)$ forms a sheaf on S, and we can identify $\tilde{g}_*$ with g_*. Then a deformation of g to T amounts to a pair of homomorphisms

$$\tilde{g}^\sharp\colon \mathcal{O}_X \to g_*\mathcal{O}_T \quad \text{and} \quad \tilde{g}^\flat\colon \mathcal{M}_X \to g_*\mathcal{M}_T,$$

such that $\alpha_T \circ g^\flat = g^\sharp \circ \alpha_X$ that are compatible with h and f.

The following key result reveals the geometric meaning of log derivations.

Theorem 2.2.2. *Let $i\colon S \to T$ be a first-order log thickening over X/Y. Then there is an action of $\mathrm{Der}_{X/Y}(g_*\mathcal{I}_T)$ on $g_*\mathrm{Def}_{X/Y}(g,T)$, with respect to which $g_*\mathrm{Def}_{X/Y}(g,T)$ becomes a pseudo-torsor under $\mathrm{Der}_{X/Y}(g_*\mathcal{I}_T)$. With multiplicative notation for the monoid law of $\mathcal{M}_T$, the action is given as follows:*

$$\begin{aligned}\mathrm{Der}_{X/Y}(g_*\mathcal{I}) \times g_*\mathrm{Def}_{X/Y}(g,T) &\to g_*\mathrm{Def}_{X/Y}(g,T)\\ ((D,\delta),g_1) &\mapsto g_2 := (g_1^\sharp + D, (1+\delta)g_1^\flat).\end{aligned}$$

Proof Explicitly, for $a \in \mathcal{O}_X$ and $m \in \mathcal{M}_X$,

$$g_2^\sharp(a) := g_1^\sharp(a) + Da \quad \text{and} \quad g_2^\flat(m) := g_1^\flat(m)(1+\delta(m)).$$

This makes sense because $Da \in g_*\mathcal{I}_T \subseteq g_*\mathcal{O}_T$ and $1+\delta(m) \in g_*(1+\mathcal{I}_T) \subseteq g_*\mathcal{O}_T^* \subseteq g_*\mathcal{M}_T$. We claim that g_2 is another deformation of g to T. It is standard and immediate to verify that $g_2^\sharp$ is a homomorphism of sheaves of $f^{-1}(\mathcal{O}_Y)$ algebras, because D is a derivation relative to Y and $\mathcal{I}_T^2 = 0$. Moreover, since $\mathcal{I}_T^2 = 0$, the map $\mathcal{I}_T \to \mathcal{O}_T^* \subseteq \mathcal{M}_T$ sending b to $1+b$ is a homomorphism of sheaves of monoids, and hence $g_2^\flat$ is also a homomorphism. Since $\delta \circ f^\flat = 0$, it still the case that $g_2^\flat \circ f^\flat = h^\flat$. Furthermore, if $m \in \mathcal{M}_X$,

$$\begin{aligned}g_2^\sharp(\alpha_X(m)) &= g_1^\sharp(\alpha_X(m)) + D\alpha_X(m)\\ &= g_1^\sharp(\alpha_X(m)) + g^\sharp(\alpha_X(m))\delta(m)\\ &= g_1^\sharp(\alpha_X(m))(1+\delta(m))\\ &= \alpha_T(g_1^\flat(m))(1+\delta(m))\\ &= \alpha_T\big((1+\delta(m))g_1^\flat(m)\big)\\ &= \alpha_T(g_2^\flat(m)).\end{aligned}$$

Thus g_2 really is a morphism of log schemes. Furthermore, $g_2 \circ i = g$ because D and δ map into $\mathcal{I}_T$.

It is immediate from the formulas that the mapping they define is an action of the group $\mathrm{Der}_{X/Y}(g_* \mathcal{I}_T)$ on $\mathrm{Def}_{X/Y}(g, T)$. The action is free because $1 + \mathcal{I}_T$ acts freely on $\mathcal{M}_T$. If g_1 and g_2 are deformations of g to T, $(g_1^\flat, g_2^\flat)$ defines a homomorphism of sheaves of monoids

$$g^{-1}\mathcal{M}_X \to \mathcal{M}_T \times_{\mathcal{M}_S} \mathcal{M}_T \xrightarrow{\epsilon} (1 + \mathcal{I}_T) \times \mathcal{M}_T \xrightarrow{pr} 1 + \mathcal{I}_T \to \mathcal{I}_T,$$

where ϵ is the inverse of the isomorphism $(u, m_1) \mapsto (m_1, um_2)$ in (3) of Proposition 2.1.2, and the last map is the first-order logarithm homomorphism $u \mapsto u - 1$. Let $\delta\colon \mathcal{M}_X \to g_* \mathcal{I}_T$ be the map obtained by adjunction. Since $f \circ g_2 = f \circ g_1$, it follows that δ annihilates the image of $\mathcal{M}_Y$. Moreover, $D\colon g_2^\sharp - g_1^\sharp$ defines a derivation $\mathcal{O}_X \to g_* \mathcal{I}_T$, and reversing the calculation in the previous paragraph shows that $g^\sharp(\alpha_X(m))\delta(m) = D(\alpha_X(m))$ for every $m \in \mathcal{M}_X$. Thus (D, δ) is a derivation of X/Y with values in $g_* \mathcal{I}_T$. We conclude that $\mathrm{Def}_{X/Y}(g, T)$ is a pseudo-torsor under the action of $\mathrm{Der}_{X/Y}(g_* \mathcal{I}_T)$. □

Variant 2.2.3. Let X/Y be a morphism of idealized log schemes and let $i\colon S \to T$ be a first-order idealized log thickening over X/Y. Then the formula given in Theorem 2.2.2 defines an action of $\mathrm{Der}_{X/Y}(g_* \mathcal{I}_T)$ on $\mathrm{Def}_{X/Y}(g, T)$. Indeed, if g_1 is a morphism $T_1 \to Y$, then it follows from the formulas in the theorem that the pair $(g_2^\sharp, g_2^\flat)$ defines a morphism of idealized log schemes.

Corollary 2.2.4. *If $i\colon X \to T$ is a Y-extension of the log scheme X with ideal $\mathcal{I}$, then* $\mathrm{Aut}(i) \cong \mathrm{Der}_{X/Y}(\mathcal{I})$.

2.3 Fundamental exact sequences

In most cases, standard arguments from classical algebraic geometry carry over to the logarithmic case to produce the familiar exact sequences exhibiting the behavior of differentials with respect to closed immersions and compositions.

Proposition 2.3.1. *Let $f\colon X \to Y$ and $g\colon Y \to Z$ be morphisms of logarithmic schemes. Then the functoriality maps defined in Proposition 1.2.15 fit into an exact sequence of sheaves of $\mathcal{O}_X$-modules*

$$f^* \Omega^1_{Y/Z} \to \Omega^1_{X/Z} \to \Omega^1_{X/Y} \to 0.$$

Proof This is proved just as in the classical case: the morphisms in the sequence are induced by the commutative squares formed by $g, f, g \circ f, \mathrm{id}_Z$ and $g \circ f, \mathrm{id}_X, f, g$. One checks from the definitions that, for any $\mathcal{O}_X$-module $\mathcal{E}$, the sequence

$$0 \to \mathrm{Der}_{X/Y}(\mathcal{E}) \to \mathrm{Der}_{X/Z}(\mathcal{E}) \to \mathrm{Der}_{Y/Z}(f_* \mathcal{E}) \tag{2.3.1}$$

is exact. The exactness of the sequence of differentials then follows from the universal properties. □

Proposition 2.3.2. *Let $i: X \to Y$ and $g: Y \to Z$ be morphisms of logarithmic schemes, where i is a strict closed immersion defined by an ideal sheaf $\mathcal{I}$. Then there is an exact sequence of sheaves of $\mathcal{O}_X$-modules*

$$\mathcal{I}/\mathcal{I}^2 \xrightarrow{\overline{d}} i^*(\Omega^1_{Y/Z}) \to \Omega^1_{X/Z} \to 0,$$

where the map $\overline{d}$ sends the class of an element a of $\mathcal{I}$ to the image of da in $i^*(\Omega^1_{Y/Z})$. If the first infinitesimal neighborhood T of X in Y admits a Z-retraction and $\mathcal{M}_Y$ is u-integral, then $\overline{d}$ is injective and split.

Proof One verifies immediately that the composition of $d: I \to \Omega^1_{Y/Z}$ with the map $\Omega^1_{Y/Z} \to i^*(\Omega^1_{Y/Z})$ is $\mathcal{O}_Y$-linear, and hence that this composition factors through a map $\overline{d}$ as claimed. To prove the exactness of the sequence, it suffices to prove that, for every sheaf $\mathcal{E}$ of $\mathcal{O}_X$-modules, the sequence obtained by applying $\mathrm{Hom}(\ ,\mathcal{E})$ is exact. By the universal mapping property of the sheaf of differentials, this amounts to verifying that the sequence

$$0 \to \mathrm{Der}_{X/Z}(\mathcal{E}) \to \mathrm{Der}_{Y/Z}(i_*\mathcal{E}) \to \mathrm{Hom}(\mathcal{I}, i_*\mathcal{E})$$

is exact. The injectivity of the map $\mathrm{Der}_{X/Z}(\mathcal{E}) \to \mathrm{Der}_{Y/Z}(i_*\mathcal{E})$ follows from the fact that $i^\flat: \mathcal{M}_Y \to \mathcal{M}_X$ is surjective. Let (D, δ) be a derivation of Y/Z with values in $i_*(\mathcal{E})$ such that $Da = 0$ for every section a of $\mathcal{I}$. Then D factors through $i_*\mathcal{O}_X$; we must also check that δ factors through $i_*(\mathcal{M}_X)$. Since i is strict, if m_1 and m_2 are two sections of $i^{-1}(\mathcal{M}_Y)$ with the same image in $\mathcal{M}_X$, then locally on X there exists some $u \in 1 + \mathcal{I}$ such that $m_2 = um_1$. Hence $\delta(m_2) = \delta(u) + \delta(m_1) = u^{-1}Du + \delta(m_1)$, and $Du = 0$ since D annihilates $\mathcal{I}$. Hence $\delta(m_2) = \delta(m_1)$, as required.

Let $j: T \to Y$ be the first infinitesimal neighborhood of X in Y, i.e., the strict closed subscheme of Y defined by $\mathcal{I}^2$. If $\mathcal{M}_Y$ is u-integral, $i^{-1}(1+\mathcal{I})$ acts freely on $i^{-1}(\mathcal{M}_T)$, and $i': X \to T$ is a first-order log thickening of X over Y/Z. Suppose that $r: T \to X$ is a morphism over Z such that $r \circ i' = \mathrm{id}_X$. Then j and $i \circ r$ are two deformations of i to T and, by Theorem 2.2.2, there is a unique $h: \Omega^1_{Y/Z} \to \mathcal{I}/\mathcal{I}^2$ such that $h(da) = j^\sharp(a) - (ir)^\sharp(a)$ for every local section a of $\mathcal{O}_Y$. If $a \in \mathcal{I}$, we see that $h(da) = j^\sharp(a)$, i.e., the image of a in $\mathcal{I}/\mathcal{I}^2$, and we have found a splitting of the map $\overline{d}$. □

Corollary 2.3.3. *Let $f: Y \to Z$ be a morphism of coherent log schemes, let $\mathcal{K} \subseteq \mathcal{M}_Y$ be a coherent sheaf of ideals, and let $i: X \to Y$ be the strict closed immersion of log schemes defined by $\mathcal{K}$ (see Proposition 1.3.4). Then there is a natural isomorphism $i^*(\Omega^1_{Y/Z}) \cong \Omega^1_{X/Z}$.*

Proof The ideal $\mathcal{I}$ of X in Y is generated by $\alpha_X(\mathcal{K})$ as a sheaf of ideals. In fact it is even generated by $\alpha_X(\mathcal{K})$ as an abelian sheaf, since $\alpha_X(\mathcal{K})$ is stable under the action of $\mathcal{O}_X^*$ and any section of $\mathcal{O}_X$ can be locally written as a sum of sections of $\mathcal{O}_X^*$. If k is a local section of $\mathcal{K}$, $d\alpha_X(k) = \alpha_X(k)dk$, which already maps to zero in $i^*(\Omega^1_{Y/Z})$. Thus the map $\overline{d}\colon \mathcal{I}/\mathcal{I}^2 \to i^*(\Omega^1_{Y/Z})$ in the exact sequence of Proposition 2.3.2 vanishes, and hence the map $i^*(\Omega^1_{Y/Z}) \to \Omega^1_{X/Z}$ is an isomorphism. □

Let $f\colon X \to Y$ and $g\colon Y \to Z$ be morphisms of log schemes. We can use the action (Theorem 2.2.2) of derivations on deformations and the group of isomorphism classes of extensions (Remark 2.1.5) to prolong the exact sequence (2.3.1). For a quasi-coherent sheaf of $\mathcal{O}_X$-modules $\mathcal{E}$, let $X \oplus \mathcal{E}$ be the trivial Y-extension of X by $\mathcal{E}$. The morphism $X \oplus \mathcal{E} \to Y$ is $r \circ f$, where $r\colon X \oplus \mathcal{E} \to X$ is the canonical retraction. Then, if $\partial \in \mathrm{Der}_{Y/Z}(f_*\mathcal{E})$, let $\tilde{f}_\partial := \partial + r \circ f$, again using the action of $\mathrm{Der}_{X/Y}(f_*\mathcal{E})$ defined in Theorem 2.2.2. Then the log scheme $X \oplus \mathcal{E}$ with the morphism $\tilde{f}_\partial$ defines a new thickening of X over Y, which we denote by $X \oplus_\partial \mathcal{E}$. Passing to isomorphism classes, we find a map $\mathrm{Der}_{X/Y}(f_*\mathcal{E}) \to \mathrm{Ext}_Y(X, \mathcal{E})$. We leave the proof of the following result to the reader, since it uses no further input particular to log geometry.

Proposition 2.3.4. *Let $f\colon X \to Y$ and $g\colon Y \to Z$ be morphisms of u-integral log schemes and let $\mathcal{E}$ be a quasi-coherent sheaf of $\mathcal{O}_X$-modules. Then, with the notation of the previous paragraph, the map*

$$\mathrm{Der}_{X/Y}(f_*\mathcal{E}) \to \mathrm{Ext}_Y(X, \mathcal{E}) : \partial \mapsto [X \oplus_\partial \mathcal{E}]$$

is a group homomorphism. Furthermore, the exact sequence (2.3.1) prolongs to an exact sequence

$$0 \to \mathrm{Der}_{X/Y}(\mathcal{E}) \to \mathrm{Der}_{X/Z}(\mathcal{E}) \to \mathrm{Der}_{Y/Z}(f_*\mathcal{E}) \to \mathrm{Ext}_Y(X, \mathcal{E}) \to \mathrm{Ext}_Z(X, \mathcal{E}).$$

If f is affine, the sequence prolongs further to an exact sequence including the sequence:

$$\ldots \mathrm{Der}_{Y/Z}(f_*\mathcal{E}) \longrightarrow \mathrm{Ext}_Y(X, \mathcal{E}) \to \mathrm{Ext}_Z(X, \mathcal{E}) \xrightarrow{f_*} \mathrm{Ext}_Z(Y, f_*\mathcal{E}),$$

where f_* is the map of extension classes induced by the construction iin Example 2.1.7. □

Proposition 2.3.5. *Let $f\colon X \to Y$ be a morphism of log schemes and $\overline{x}$ a* *geometric point of X. Then there is a commutative diagram*

$$\begin{array}{ccccc} & & \mathcal{M}^{\rm gp}_{X,\overline{x}} & \longrightarrow & \mathcal{M}^{\rm gp}_{X/Y,\overline{x}} & \\ & & {\scriptstyle dlog}\downarrow & \searrow^{\pi} & \downarrow{\scriptstyle \overline{\pi}} & \\ \Omega^1_{\underline{X}/\underline{Y}}(\overline{x}) & \longrightarrow & \Omega^1_{X/Y}(\overline{x}) & \xrightarrow{\rho_{X/Y,\overline{x}}} & k(\overline{x})\otimes\mathcal{M}^{\rm gp}_{X/Y,\overline{x}} & \to 0. \end{array}$$

Here $\overline{\pi}(m)$ sends a section m of $\mathcal{M}^{\rm gp}_{X/Y,\overline{x}}$ to $1\otimes m$, and the bottom row is exact.

The map $\rho_{X/Y,\overline{x}}$ is sometimes called the *Poincaré residue* at $\overline{x}$.

Proof Consider the canonical factorization $X \to X_Y \to Y$, where X_Y is the scheme $\underline{X}$ with the log strucure induced from Y. Then $X_Y \cong \underline{X}\times_{\underline{Y}} Y$, so the base-change formula for differentials induces an isomorphism $\Omega^1_{\underline{X}/\underline{Y}} \to \Omega^1_{X_Y/Y}$. We get from Proposition 2.3.1 an exact sequence:

$$\Omega^1_{\underline{X}/\underline{Y}} \to \Omega^1_{X/Y} \to \Omega^1_{X/X_Y} \to 0.$$

We shall prove that the composite map

$$\theta\colon \mathcal{M}^{\rm gp}_X \to \Omega^1_{X/Y} \to \Omega^1_{X/X_Y}$$

induces an isomorphism

$$\theta_{\overline{x}}\colon k(\overline{x})\otimes\mathcal{M}^{\rm gp}_{X/Y,\overline{x}} \xrightarrow{\cong} \Omega^1_{X/X_Y}(\overline{x}). \qquad (2.3.2)$$

The image of $\mathcal{M}_Y$ in $\Omega^1_{X/Y}$ is zero by definition, and the image of $\mathcal{O}^*_{X,\overline{x}}$ is zero in $\Omega^1_{X/X_Y,\overline{x}}$. Thus θ kills $f^*\mathcal{M}^{\rm gp}_Y$ and hence factors through $\mathcal{M}^{\rm gp}_{X/Y}$ and induces the map $\theta_{\overline{x}}$ in (2.3.2). We know from Lemma 1.2.10 that Ω^1_{X/X_Y} is generated by the image of $\mathcal{M}^{\rm gp}_{X/Y}$, so $\theta_{\overline{x}}$ is clearly surjective. If $m\in\mathcal{M}^*_{X,\overline{x}}$, then its image in $\mathcal{M}_{X/Y,\overline{x}}$ is zero, and hence $\pi(m)$ is zero, and if $m\in\mathcal{M}^+_{X,\overline{x}}$, then $\alpha_X(m)$ maps to zero in $k(\overline{x})$. Thus in any case $\alpha_X(m)\pi(m)=0$, and the pair

$$(0,\pi)\colon \mathcal{O}_X\oplus\mathcal{M}^{\rm gp}_{X,\overline{x}} \to k(\overline{x})\otimes\mathcal{M}^{\rm gp}_{X/Y,\overline{x}}$$

is a logarithmic derivation of X/Y. Thus there is a unique map $r\colon \Omega^1_{X/Y,\overline{x}} \to k(\overline{x})\otimes M^{g}_{X/Y,\overline{x}}$ such that $r(dm) = \pi(m)$ for all $m\in\mathcal{M}_{X,\overline{x}}$. Evidently r kills $d\mathcal{O}_{X,\overline{x}}$, hence also the image of $\Omega^1_{\underline{X},\overline{x}/\underline{Y}}$. Consequently r factors through a map $\overline{r}\colon \Omega^1_{X/X_Y,\overline{x}} \to k(\overline{x})\otimes\mathcal{M}^{\rm gp}_{X/Y,\overline{x}}$, which is inverse to $\theta_{\overline{x}}$. □

Corollary 2.3.6. *If X is a log point over a field k, the Poincaré residue map $\rho_{X/k}$ induces an isomorphism: $\Omega^1_{X/k,\overline{x}} \cong k\otimes\overline{\mathcal{M}}^{\rm gp}_{\overline{x}}$.*

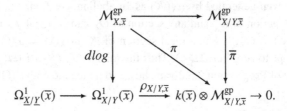

3 Logarithmic smoothness

The basic definitions and main properties of smoothness for log schemes follow closely Grothendieck's functorial and geometric approach to smoothness in algebraic geometry. Nevertheless, local models of smooth morphisms of log schemes are considerably more complicated than in the classical case.

3.1 Definitions and examples

Definition 3.1.1. *A morphism of log schemes $f: X \to Y$ is formally smooth (resp. formally unramified, resp. formally étale) if, for every n and every nth-order log thickening (Definition 2.1.1) $S \to T$ over X/Y, the given morphism $g: S \to X$ locally admits at least one (resp. at most one, resp. exactly one) deformation (Definition 2.2.1) to T. A morphism f is smooth (resp. étale) if it is formally smooth (resp. étale) and in addition $\mathcal{M}_X$ and $\mathcal{M}_Y$ are coherent and $\underline{f}$ is locally of finite presentation. A morphism f is unramified if it is formally unramified and $\underline{f}$ is locally of finite type.*

Since an nth-order log thickening can be written as a succession of first-order thickenings, it is enough to check the deformation conditions in Definition 3.1.1 when $n = 1$. In this case, the sheaf $\mathcal{D}ef_{X/Y}(g, T)$ of deformations of g is a pseudo-torsor under $\mathcal{D}er_{X/Y}(g_* I_T)$ by Theorem 2.2.2. Thus the formal smoothness condition says that this pseudo-torsor is locally nonempty, i.e., is in fact a torsor.

Remark 3.1.2. The family of formally smooth (resp. formally étale, resp. formally unramified) morphisms is stable under composition and base change in the category of log schemes. If $g: Y \to Z$ is formally étale, then a morphism $f: X \to Y$ is formally smooth if and only if $g \circ f$ is formally smooth. If $X \to Z$ and $Y \to Z$ are formally étale, then any Z-morphism $X \to Y$ is formally étale. If $g \circ f: X \to Z$ is formally unramified, then $f: X \to Y$ is also formally unramified. These properties follow immediately from the definitions.

Proposition 3.1.3. *A morphism $f: X \to Y$ of log schemes is formally unramified if $\Omega^1_{X/Y} = 0$. The converse holds if X and Y are coherent.*

Proof If $i: S \to T$ is a log thickening over X/Y, the sheaf of deformations of $g: S \to X$ to T is a pseudo-torsor under $\mathcal{D}er_{X/Y}(g_*\mathcal{I}) \cong \mathcal{H}om(\Omega^1_{X/Y}, g_*\mathcal{I})$, by Theorem 2.2.2. If $\Omega^1_{X/Y} = 0$, these groups vanish, and so deformations are unique when they exist and X/Y is formally unramified. If X and Y are coherent, the sheaf $\Omega^1_{X/Y}$ is quasi-coherent by Corollary 1.2.8. Then we can form the trivial extension T of X/Y by $\Omega^1_{X/Y}$ (see Remark 2.1.5), and the set of deformations of id_X is a torsor under $\mathcal{E}nd(\Omega^1_{X/Y})$. If X/Y is unramified, the retraction $T \to X$ is the unique such deformation, so $\Omega^1_{X/Y} = 0$. □

Proposition 3.1.4. *Let $f: X \to Y$ be a morphism of log schemes.*

1. *The morphism f is formally smooth if and only if for every open subset U of X and every quasi-coherent $\mathcal{O}_U$-module $\mathcal{E}$, every Y-extension of U by $\mathcal{E}$ is locally trivial.*
2. *If X and Y are coherent and $\underline{f}$ is locally of finite presentation, then f is smooth if and only if, for every finite-order log thickening $S \to T$ over X/Y with S-affine, the morphism $S \to X$ can be deformed to T.*

Proof It follows immediately from the definition that if f is formally smooth, every Y-extension $U \to T$ of an affine open subset U of X locally admits a section $T \to X$ and hence is locally trivial. Conversely, suppose that every such extension is locally trivial and that $(i: S \to T, g: S \to X)$ is a log thickening of order one over X/Y, with ideal $\mathcal{I}$. The thickening i defines an element of ξ of $\mathrm{Ext}_Y(S, \mathcal{I})$. Assuming without loss of generality that X and S are affine, we may form the direct image extension (Example 2.1.7) of X/Y by $g_*(\mathcal{I})$:

$$\begin{array}{ccc} S & \xrightarrow{\ i\ } & T \\ {\scriptstyle g}\downarrow & & \downarrow{\scriptstyle g'} \\ X & \xrightarrow{\ j\ } & g_*(T). \end{array}$$

By assumption, this extension is locally trivial, and hence locally on X there exists a retraction $r: g_*(T) \to X$. Then $r \circ g'$ is the desired deformation of g.

To prove (2), suppose that f is smooth and that $i: S \to T$ is log thickening of order n over X/Y. Working by induction, we may assume that $n = 1$. Since X and Y are coherent log schemes and f is locally of finite presentation, the sheaf $\Omega^1_{X/Y}$ is quasi-coherent and locally of finite presentation, by Corollary 1.2.8. Hence $\mathcal{D}er_{X/Y}(g_*(\mathcal{I})) \cong \mathcal{H}om(g^*\Omega^1_{X/Y}, \mathcal{I})$ is quasi-coherent and, since S is affine, $H^1(S, \mathcal{D}er_{X/Y}(g_*(\mathcal{I})) = 0$. Thus every torsor under this sheaf of groups has a global section, and in particular $\mathrm{Def}_{X/Y}(g, T)$ is not empty. □

Corollary 3.1.5. *In the definition of smooth (resp., unramified, étale) morphisms, it is sufficient to consider thickenings where $g: S \to X$ is an open immersion.* □

If $f: X \to Y$ is a morphism of schemes, and if X and Y are endowed with the trivial log structure, then f is formally smooth (resp. étale, unramified) if and only if $\underline{f}$ is. More generally:

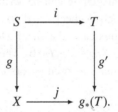

Proposition 3.1.6. *Let $f: X \to Y$ be a strict morphism of log schemes. If the underlying morphism of schemes $\underline{f}: \underline{X} \to \underline{Y}$ is formally smooth (resp. étale, unramified), then the same is true of f. The converse holds if the log structure on Y is u-integral.*

Proof If f is strict, the map $X \to \underline{X} \times_{\underline{Y}} Y$ is an isomorphism. Thus if $\underline{f}$ is formally smooth the same is true of f, since smoothness is preserved by base change. The same holds for étale and unramified morphisms. To prove the converse, suppose that f is smooth and that $\underline{S} \to \underline{T}$ is a thickening over $\underline{X}/\underline{Y}$. Endow $\underline{T}$ with the inverse image of the log structure on Y, and note that $\mathcal{M}_T$ is u-integral by (3) of Proposition II.1.1.8. Then $S \to T$ is a log thickening over X/Y. Since f is smooth, locally there exists a deformation $\tilde{g}: T \to X$ of $g: X \to X$, and then $\underline{\tilde{g}}$ is a deformation of $\underline{g}$. It follows that $\underline{f}$ is smooth. If f is unramified, so is $\underline{f}$, since $\Omega^1_{X/Y} \cong \Omega^1_{\underline{X}/\underline{Y}}$. □

The next set of results explains when the morphisms of log schemes modeled on monoid homomorphisms are unramified, smooth, or étale. This important result will be key to understanding the local structure of log smooth morphisms, which is more complicated than in the classical case.

Theorem 3.1.7. *Let $\theta: P \to Q$ be a homomorphism of finitely generated monoids and let $f: \mathsf{A}_Q \to \mathsf{A}_P$ be the corresponding morphism of log schemes over a base ring R. Then the following conditions are equivalent.*

1. *The order of the cokernel* Cok *of θ^{gp} is finite and invertible in R.*
2. *The morphism of log schemes $f: \mathsf{A}_Q \to \mathsf{A}_P$ is unramified over R.*
3. *The morphism of group schemes $f_{|\mathsf{A}^*_Q}: \mathsf{A}^*_Q \to \mathsf{A}^*_P$ is unramified over R.*

Proof If (1) holds, then $R \otimes \mathrm{Cok} = 0$. By Proposition 1.1.4, $\Omega^1_{\mathsf{A}_Q/\mathsf{A}_P}$ is the quasi-coherent sheaf associated to $R[Q] \otimes \mathrm{Cok}$, and hence vanishes. Then it follows from Proposition 3.1.3 that f is unramified. If f is unramified, so is its restriction $f_{|\mathsf{A}^*_Q}$ to the open subset A^*_Q. Finally, if $f_{|\mathsf{A}^*_Q}$ is unramified, Proposition 3.1.3 implies that $\Omega^1_{\mathsf{A}^*_Q/\mathsf{A}^*_P} = 0$, hence $R[Q^{gp}] \otimes \mathrm{Cok} = 0$. Since $R[Q^{gp}]$ is a nonzero free R-module, this vanishing also implies that $R \otimes \mathrm{Cok}$ vanishes. □

Theorem 3.1.8. *Let $\theta: P \to Q$ be a homomorphism of finitely generated monoids. and let $f: \mathsf{A}_Q \to \mathsf{A}_P$ be the corresponding morphism of log schemes over a base ring R. Then the following conditions are equivalent:*

1. *The kernel and the torsion part of the cokernel of θ^{gp} are finite groups whose order is invertible in R.*
2. *The morphism of log schemes $f: \mathsf{A}_Q \to \mathsf{A}_P$ is smooth over R.*
3. *The morphism of group schemes $f_{|\mathsf{A}^*_Q}: \mathsf{A}^*_Q \to \mathsf{A}^*_P$ is smooth over R.*

Proof Suppose that (1) holds. Recall from Proposition 1.2.9 that, for any log scheme T, the set of morphisms $T \to \mathsf{A}_Q$ is identified with the set of monoid homomorphisms $Q \to \Gamma(T, \mathcal{M}_T)$. Thus a log thickening $i\colon S \to T$ over f can be thought of as a solid commutative square of the form.

$$\begin{array}{ccc} P & \xrightarrow{\ \theta\ } & Q \\ {\scriptstyle \phi}\big\downarrow & \overset{\tilde{\psi}}{\swarrow} & \big\downarrow{\scriptstyle \psi} \\ \Gamma(T, \mathcal{M}_T) & \xrightarrow{\ \iota\ } & \Gamma(S, \mathcal{M}_S). \end{array}$$

We must show that, locally on T, there is a homomorphism $\tilde{\psi}$ as shown that makes both triangles commute. Recall from statement (4) of Proposition 2.1.2 that the homomorphism $\mathcal{M}_T \to \mathcal{M}_T^{\mathrm{gp}} \times_{\mathcal{M}_S} \mathcal{M}_S$ is an isomorphism. Thus it will suffice to find a map in the analogous diagram in which all the monoids are replaced by their corresponding groups.

Since the question is local on T, we may assume without loss of generality that T is affine, and we may also assume that i is a first-order thickening. By statement (2) of Proposition 2.1.2, the kernel of the surjection $\mathcal{M}_T^{\mathrm{gp}} \to \mathcal{M}_S^{\mathrm{gp}}$ is $1 + \mathcal{I}$ and, since $\mathcal{I}^2 = 0$, this sheaf of groups is isomorphic to $\mathcal{I}$. Since $\mathcal{I}$ is quasi-coherent, $H^1(T, \mathcal{I}) = 0$, and the sequence

$$0 \to \Gamma(\mathcal{I}) \to \Gamma(\mathcal{M}_T^{\mathrm{gp}}) \to \Gamma(S, \mathcal{M}_S^{\mathrm{gp}}) \to 0$$

is exact. Let $E := \Gamma(\mathcal{M}_T^{\mathrm{gp}}) \times_{\Gamma(\mathcal{M}_S^{\mathrm{gp}})} Q^{\mathrm{gp}}$, which fits into the following commutative diagram:

$$\begin{array}{ccccccccc} 0 & \longrightarrow & \mathrm{Ker}(\theta^{\mathrm{gp}}) & \longrightarrow & P^{\mathrm{gp}} & & & & \\ & & {\scriptstyle \phi''}\big\downarrow & & {\scriptstyle \phi'}\big\downarrow & \overset{\theta^{\mathrm{gp}}}{\searrow} & & & \\ 0 & \longrightarrow & \Gamma(\mathcal{I}) & \longrightarrow & E & \xrightarrow{\ \pi\ } & Q^{\mathrm{gp}} & \longrightarrow & 0 \\ & & {\scriptstyle \mathrm{id}}\big\downarrow & & \big\downarrow & & \big\downarrow{\scriptstyle \rho} & & \\ 0 & \dashrightarrow & \Gamma(\mathcal{I}) & \longrightarrow & \mathrm{Cok}(\phi') & \xrightarrow{\ \pi'\ } & \mathrm{Cok}(\theta^{\mathrm{gp}}) & \longrightarrow & 0. \end{array}$$

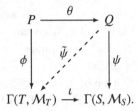

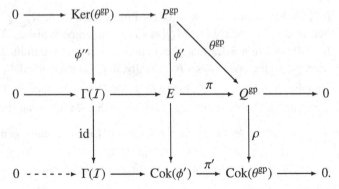

In this diagram, the columns and solid-arrowed rows are exact. Moreover, $\Gamma(\mathcal{I})$ is an R-module and $\mathrm{Ker}(\theta^{\mathrm{gp}})$ is a finite group whose order is invertible in R. It follows that ϕ'' vanishes, so the bottom row is also exact at $\Gamma(\mathcal{I})$. But then the middle row is the pullback of the bottom row, i.e., the square at the bottom right is cartesian. Observe that $\mathrm{Ext}^1(\mathrm{Cok}(\theta^{\mathrm{gp}}), \Gamma(\mathcal{I})) = 0$, since $\Gamma(\mathcal{I})$ is an R-module and the order of the torsion part of $\mathrm{Cok}(\theta^{\mathrm{gp}})$ is invertible in R. Thus the sequence on the bottom splits: there is a homomorphism $\sigma\colon \mathrm{Cok}(\theta^{\mathrm{gp}}) \to \mathrm{Cok}(\phi')$ such that $\pi'\sigma = \mathrm{id}$. Since the square at the bottom right is cartesian, such a splitting also defines a map $Q^{\mathrm{gp}} \to E$, which necessarily agrees with the given map $P^{\mathrm{gp}} \to E$. This map gives the desired deformation of g and completes the proof that (1) implies (2).

It is apparent from the definitions that the restriction of a smooth map to any open subset is smooth, and it follows that (2) implies (3). Thus it remains only to prove that (3) implies (1). For this implication we may as well replace P by P^{gp} and Q by Q^{gp}. Thus we may and shall assume that P and Q are finitely generated abelian groups. Let P' be the image of P in Q, so that the map θ factors as $\theta = P \xrightarrow{\phi} P' \xrightarrow{\theta'} Q$, where ϕ is surjective and θ' is injective. We get a corresponding factorization of maps of group schemes

$$f = \underline{\mathsf{A}}_{\mathsf{Q}} \xrightarrow{f'} \underline{\mathsf{A}}_{\mathsf{P}'} \xrightarrow{g} \underline{\mathsf{A}}_{\mathsf{P}},$$

where g is a closed immersion and f' is dominant. Observe that the group homomorphism θ' makes Q into a P'-set and, since θ' is injective, each P' orbit is isomorphic to P'. Thus Q is a free P'-set, and hence $R[Q]$ is a free $R[P']$-module and f' is faithfully flat. Let x be a point of $\underline{\mathsf{A}}_{\mathsf{Q}}$, let $y := f'(x) \in \underline{\mathsf{A}}_{\mathsf{P}'} \subseteq \underline{\mathsf{A}}_{\mathsf{P}}$, and let s be the image of x in $\operatorname{Spec} R$. Then y lies in the inverse image of s in

$$Y'_s := \operatorname{Spec} k(s) \times_{\operatorname{Spec} R} \underline{\mathsf{A}}_{\mathsf{P}'} = \operatorname{Spec} k(s)[P'],$$

and the fiber of y in $\underline{\mathsf{A}}_{\mathsf{Q}}$ can be identified with its fiber in

$$X_s := \operatorname{Spec} k(s) \times_{\operatorname{Spec} R} \underline{\mathsf{A}}_{\mathsf{Q}} = \operatorname{Spec} k(s)[Q].$$

Now the dimension of X_s is the rank of the abelian group Q, the dimension of Y_s is the rank of P', and the morphism $f'_s\colon X_s \to Y'_s$ is faithfully flat. It follows that all the fibers of f'_s have dimension equal to the rank of Q minus the rank of P', i.e., the rank r of Q/P. Since f is smooth, its sheaf of relative differentials is locally free, and its rank at any point x is the dimension of the fiber containing it [34, III.10.4], which we have just seen is r. By Proposition 1.1.4, we conclude that $R[Q] \otimes Q/P$ is locally free of rank r and, since $R[Q]$ is faithfully flat over R, it follows that $R \otimes Q/P$ is locally free of rank r. Write Q/P as a direct sum of a free abelian group F and a finite group T. Then F has rank r, and it follows that the map $R[Q/P] \to R[F]$ is an isomorphism and that $R \otimes T = 0$. This implies that the order of T is invertible in R.

It remains to prove that $\mathrm{Ker}(\theta)$ is a finite group whose order is invertible in

R. Since f is smooth, it is flat and since, f' is faithfully flat, it follows that the closed immersion g is flat. Then the result follows from Corollary 1.1.7. □

Corollary 3.1.9. *Let Q be a finitely generated monoid, let*

$$\mathsf{A}_Q := \operatorname{Spec} Q \to R[Q],$$

and let $S := \operatorname{Spec} R$ (with trivial log structure). Then the following conditions are equivalent.

1. *The order of the torsion subgroup of Q^{gp} is invertible in R.*
2. *The morphism of log schemes $\mathsf{A}_Q \to S$ is smooth.*
3. *The group scheme $\mathsf{A}^*_Q := \operatorname{Spec} R[Q^{\mathrm{gp}}]$ is smooth over S.* □

Corollary 3.1.10. *Let $\theta\colon P \to Q$ be a morphism of finitely generated monoids. and let $f\colon \mathsf{A}_Q \to \mathsf{A}_P$ be the corresponding morphism of log schemes over a base ring R. Then the following conditions are equivalent:*

1. *The kernel and cokernel of θ^{gp} are finite groups whose order is invertible in R.*
2. *The morphism of log schemes $f\colon \mathsf{A}_Q \to \mathsf{A}_P$ is étale over R.*
3. *The morphism of group schemes $f_{|\mathsf{A}^*_Q} : \mathsf{A}^*_Q \to \mathsf{A}^*_P$ is étale over R.*

In particular, if θ^{gp} is an isomorphism, then $f_{|\mathsf{A}^*_Q} : \mathsf{A}^*_Q \to \mathsf{A}^*_P$ is an isomorphism and $f\colon \mathsf{A}_Q \to \mathsf{A}_P$ is étale. □

Corollary 3.1.11. *If X is a coherent log scheme, the canonical maps $X^{\mathrm{int}} \to X$ and $X^{\mathrm{sat}} \to X^{\mathrm{int}}$ are étale.*

Proof Let Q be a finitely generated (resp. fine) monoid, and let Q' denote Q^{fin} (resp. Q^{sat}). It follows from Corollary 3.1.10 that the map $\mathsf{A}_{Q'} \to \mathsf{A}_Q$ is étale. If now X is a coherent (resp. fine) log scheme, we can verify that the map $X' \to X$ is étale locally on X, so we may assume that there exists a chart $X \to \mathsf{A}_Q$, where Q is finitely generated (resp. fine). Then, by the construction in Proposition III.2.1.5, X' is the fiber product of X and $\mathsf{A}_{Q'}$ over A_Q. Since étaleness is preserved by base change, it follows that $X' \to X$ is also étale. □

Corollary 3.1.12. *Let $f\colon X \to Y$ be a morphism of coherent log schemes with X fine. Then f is smooth if and only if the canonical factorization $\tilde{f}\colon X \to Y^{\mathrm{int}}$ is smooth, and the same holds with f^{sat} in place of f^{int}.*

Proof Let $\zeta\colon Y^{\mathrm{int}} \to Y$ be the canonical map, and consider the following diagram, in which the square is cartesian:

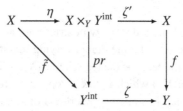

$$\begin{array}{ccccc} X & \xrightarrow{\eta} & X \times_Y Y^{\text{int}} & \xrightarrow{\zeta'} & X \\ & \underset{\tilde{f}}{\searrow} & \downarrow{\scriptstyle pr} & & \downarrow{\scriptstyle f} \\ & & Y^{\text{int}} & \xrightarrow{\zeta} & Y. \end{array}$$

Since ζ is smooth, so is ζ' and, since $\zeta' \circ \eta = \text{id}$ is étale, η is also étale. If f is smooth, pr is smooth, and hence $\tilde{f} = pr \circ \eta$ is also smooth. If $\tilde{f}$ is smooth, then $f = \zeta \circ \tilde{f}$ is also smooth. The proof for Y^{sat} is analogous. □

Proposition 3.1.13. *Let $f\colon X \to Y$ be the morphism of log schemes admitting a coherent chart $\theta\colon P \to Q$ and let x be a point of X, satisfying the following conditions.*

1. *The kernel of θ is a finite group whose order is invertible in $k(x)$.*
2. *The cokernel (resp. the torsion part of the cokernel) of θ is a finite group whose order is invertible in $k(x)$.*
3. *The map $\underline{b}_\theta\colon \underline{X} \to Y_\theta := \underline{Y} \times_{\underline{\mathsf{A}}_P} \underline{\mathsf{A}}_Q$ is étale (resp. smooth) in some neighborhood of x.*

Then $f\colon X \to Y$ is étale (resp. smooth) in some neighborhood of x.

Proof Consider the commutative diagram of log schemes constructed in Remark III.1.2.6:

$$\begin{array}{ccccc} X & \xrightarrow{b_\theta} & Y_\theta & \xrightarrow{b} & \mathsf{A}_Q \\ & \underset{f}{\searrow} & \downarrow{\scriptstyle f_\theta} & & \downarrow{\scriptstyle \mathsf{A}_\theta} \\ & & Y & \longrightarrow & \mathsf{A}_P, \end{array}$$

in which the square is cartesian. Let n be the order of the kernel of θ and let $y := f(x)$. Condition (1) implies that n is a unit in $k(x)$ and hence also in $k(y)$. It follows that n is a unit in the local ring of y in Y. The same argument for the order m of the torsion subgroup of the cokernel of θ shows that m is invertible in the local ring of Y at y. Thus, after a further localization, we may assume that Y is a scheme over a ring R in which m and n are invertible, and we view A_P and A_Q as schemes over R. Then, by Corollary 3.1.10) (resp. Theorem 3.1.8), the map A_θ is étale (resp. smooth). The same conclusion holds for f_θ by base change. Since $X \to \mathsf{A}_Q$ is a chart for X, the map b_θ is strict. Since $\underline{b}_\theta$ is étale (resp. smooth), it follows from Proposition 3.1.6 that the same is true for b_θ.

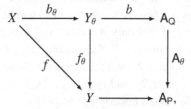

Since the family of smooth (resp. étale) maps is closed under composition, this completes the proof. □

Example 3.1.14. Let Y be a divisor with strict normal crossing in a smooth scheme $\underline{X}$ over field k, let α be the compactifying log structure associated to the complement of Y, and let X be the log scheme $(\underline{X}, \alpha)$. Then X/k is smooth at every closed point x of X at which $k(x)/k$ is separable. To see this, recall from Proposition III.1.8.2 that, in some neighborhood of x, there is a chart $a\colon X \to \mathsf{A}_{\mathbf{N}^r}$, where $Y_1, \ldots, Y_r$ is the set of irreducible components of Y passing through x. Then it suffices to see that the underlying morphism $\underline{X} \to \underline{\mathsf{A}}_{\mathbf{N}^r} = \mathbf{A}^r$ is smooth. This morphism is flat, since its defining sequence $(t_1, \ldots, t_r)$ is regular, so it suffices to prove that the fiber $Y_1 \cap \cdots \cap Y_r$ over the origin is smooth. Since $k(x)/k$ is separable, the map $m_x/m_x^2 \to \Omega^1_{X/k}(x)$ is an isomorphism. Since Y is a divisor with strict normal crossings in X, the image of the sequence $(t_1, \ldots, t_r)$ in m_x/m_x^2 is linearly independent, and it follows from the Jacobian criterion that $Y_1 \cap \cdots \cap Y_r$ is smooth. We note that the separability hypothesis is not superfluous. For example, if $k := \mathbf{F}_p(s)$, if $\underline{X} = k[t]$, and if Y is the closed subscheme defined by $t^p - s$, then the corresponding log scheme X is not smooth over k.

Example 3.1.15. Let n be an integer and let $\theta\colon \mathbf{N} \to \mathbf{N}$ be multiplication by n. Then the corresponding morphism $f\colon \mathsf{A}_{\mathbf{N}} \to \mathsf{A}_{\mathbf{N}}$ is étale if and only if n is invertible in the base ring R. The map $\underline{f}$ on underlying schemes is a finite covering, tamely and totally ramified over the origin. More generally, let $\theta\colon P \to Q$ be an injective and small homomorphism of fine monoids, so that $Q^{\mathrm{gp}}/P^{\mathrm{gp}}$ is a finite group. Then the corresponding morphism of log schemes is étale over a base ring R if and only if the order of this group is invertible in R. For another example, consider P, the monoid given by generators (a, b, c) satisfying the relation $a + b = 2c$, and the homomorphism $\theta\colon P \to \mathbf{N} \oplus \mathbf{N}$ given by $\theta(a) = (2, 0)$, $\theta(b) = (0, 2)$, and $\theta(c) = (1, 1)$. Then θ is small and even Kummer, so A_θ is étale but the underlying morphism of schemes is not flat.

Example 3.1.16. Let $\theta\colon P \to Q$ be a homomorphism of monoids such that θ^{gp} is an isomorphism. Then $\mathsf{A}_\theta\colon \mathsf{A}_Q \to \mathsf{A}_P$ is étale. For example, let n an integer greater than 1 and let

$$\theta\colon \mathbf{N}^n \to \mathbf{N}^n \quad \text{by} \quad (a_1, a_2, \ldots, a_n) \mapsto (a_1, a_2 + a_1, \ldots, a_n + a_1).$$

Then the corresponding map θ^{gp} is an isomorphism and A_θ is étale. However, the underlying map on schemes $\underline{\mathsf{A}}_\theta\colon \mathbf{A}^n \to \mathbf{A}^n$ is an affine piece of a blowup and is not even flat.

Example 3.1.17. Let r be a positive integer and let $\phi\colon \mathbf{N} \to \mathbf{N}^r$ be the diagonal map sending a natural number n to $(n, n, \ldots, n)$. Then the corresponding morphism of log schemes $\mathsf{A}_{\mathbf{N}^r} \to \mathsf{A}_{\mathbf{N}}$ is smooth. The map of underlying schemes sends a point $(x_1, \ldots, x_r)$ to the point $x_1 x_2 \cdots x_r$, and is the standard model of semistable reduction. Notice that there are commutative diagrams

$$
\begin{array}{ccc}
\mathbf{N}^r & \xleftarrow{\ \theta\ } & \mathbf{N}^r \\
 & {\scriptstyle \phi}\nwarrow & \uparrow{\scriptstyle \pi} \\
 & & \mathbf{N}
\end{array}
\qquad\qquad
\begin{array}{ccc}
\mathsf{A}_{\mathbf{N}^r} & \xrightarrow{\ \mathsf{A}_\theta\ } & \mathsf{A}_{\mathbf{N}^r} \\
 & {\scriptstyle \mathsf{A}_\phi}\searrow & \downarrow{\scriptstyle \mathsf{A}_\pi} \\
 & & \mathsf{A}_{\mathbf{N}},
\end{array}
$$

where θ is the map in Example 3.1.16 corresponding to a partial blowup and π is the map $n \mapsto (n, 0, \ldots, 0)$ corresponding to a projection. Thus in the log world, a stable reduction mapping can be factored as an étale map followed by a standard projection.

We can now easily see that semistable reduction morphisms, when endowed with their natural log structures, become (logarithmically) smooth.

Corollary 3.1.18. *A morphism of log schemes arising as in Proposition III.1.8.4 from a semistable reduction morphism of finite type is smooth.*

Proof Let $f: X \to S$ be such a morphism and let $\overline{x}$ be a geometric point of X. According to Proposition III.1.8.4, we may find, after restricting to some étale neighborhood of $\overline{x}$, a chart for f subordinate to a homomorphism $\theta: P \to Q$, where $P = \mathbf{N}$, $Q = \mathbf{N}^r$, and $\theta(n) = (n, n, \ldots, n)$, and such that the induced map $\underline{X} \to \underline{S}_\theta := \underline{S} \times_{\mathsf{A}_P} \mathsf{A}_Q$ is smooth. Then Proposition 3.1.13 implies that f is smooth. □

More generally, if $(m_1, m_2, \ldots, m_r)$ is a sequence of positive integers, the morphism of log schemes corresponding to the map

$$\mathbf{N} \to \mathbf{N}^r \quad \text{given by} \quad n \mapsto (m_1 n, m_2 n, \ldots, m_r n)$$

is smooth if and only if the greatest common divisor of $(m_1, m_2, \ldots m_r)$ is invertible in the base ring R. This follows immediately from Corollary 3.1.10. Notice that the multiplicities of the divisors of the special fiber, i.e., the integers m_i themselves, need not be invertible in R. For example, if p is prime, the map of log schemes corresponding to the map of monoids $\mathbf{N} \to \mathbf{N} \oplus \mathbf{N}$ sending n to (n, pn) is log smooth in any characteristic.

Let us explain how this situation arises in classical algebraic geometry. Let $\underline{S}$ be the spectrum of a discrete valuation ring V, and let $\underline{X}$ be a V-scheme of finite type. Let x be a point of $\underline{X}$ lying over the closed point s of $\underline{S}$, and assume that X is regular at x. Then the local ring $\mathcal{O}_{X,x}$ of X at x is a unique factorization domain, and hence a uniformizer π of V can be written as a product $\pi = ut_1^{m_1} \cdots t_r^{m_r}$, where each t_i is irreducible in $\mathcal{O}_{X,x}$. It follows from Corollary 1.7.4 that the compactifying log structure α associated to the generic fiber of $\underline{X}/\underline{S}$ is fine, and the sequence $(t_1, \ldots, t_r)$ provides a chart for α. Then $\underline{f}$ induces a morphism of log schemes $f\colon X \to S$, where S is $\underline{S}$ equipped with the compactifying log structure coming from the generic point. Suppose that the closed subscheme of X defined by the ideal $(t_1, \ldots, t_r)$ is of codimension r in X and smooth over the residue field k of V. If all $m_i = 1$, then $\underline{f}$ is a semistable reduction morphism, and so f is a smooth morphism of log schemes. In fact this smoothness can hold more generally, as the following proposition illustrates.

Proposition 3.1.19. *Let X/S be a morphism of finite type, where $\underline{X}$ is regular and $\underline{S}$ is the spectrum of a DVR, and that X and S are endowed with the compactifying log structures induced from the generic fiber over S. Let x be a closed point of the special fiber of X/S, and in the local ring $\mathcal{O}_{X,x}$ write the uniformizer π is as a product $ut_1^{m_1} \cdots t_r^{m_r}$, where u is a unit in $\mathcal{O}_{X,x}$ and each t_i is irreducible. Suppose that the closed subscheme of $\underline{X}$ defined by $(t_1, \ldots, t_r)$ is of codimension r and smooth over the residue field k of V and that at least one m_i is invertible in k. Then the morphism f is smooth at x.*

Proof If m is invertible in k and u is a unit in $\mathcal{O}_{X,x}$, then the equation $t^m - u$ defines a finite étale algebra over $\mathcal{O}_{X,x}$. Hence after an étale localization, we may assume that $u = v^m$. If m_i is invertible in k, we may then replace t_i by vt_i, and the equation for π becomes $\pi = t_1^{m_1} \cdots t_r^{m_r}$. Let $\alpha\colon \mathbf{N} \to V$ be $1 \mapsto \pi$, let $\theta\colon \mathbf{N} \to \mathbf{N}^r$ be $n \mapsto (m_1 n, m_2 n, \ldots, m_r n)$, and let $\beta\colon \mathbf{N}^r \to \mathcal{O}_X$ be $e_i \mapsto t_i$. Then (α, θ, β) is a chart for f. The map $X \to S_\theta := S \times_{\mathsf{A}_\mathbf{N}} \mathsf{A}_{\mathbf{N}^r}$ is strict, and its homomorphism of rings is given by

$$A := V[s_1, \ldots, s_r]/(s_1^{m_1} \cdots s_r^{m_r} - \pi) \to \mathcal{O}_{X,x} : s_i \mapsto t_i.$$

Since $\mathcal{O}_{X,x}$ is Cohen–Macaulay and the codimension of the quotient defined by $(t_1, \ldots, t_r)$ is r, the sequence $(t_1, \ldots, t_r)$ is regular. Since A is a regular local ring, it follows that the the morphism $X \to S_\theta$ is flat at x. Its fiber over the point of S_θ defined by $(s_1, \ldots, s_r)$ is by hypothesis smooth at x, so $\underline{X} \to \underline{S}_\theta$ is smooth at x. By Proposition 3.1.13, it follows that f is also smooth at x. □

Example 3.1.20. In the situation of Proposition 3.1.19, the morphism f can be smooth even if all m_i are divisible by p. Furthermore, a morphism of log schemes X/Y can be smooth even if $\mathcal{M}^{\mathrm{gp}}_{X/Y}$ is p-torsion. For example, the homomorphism $\mathbf{N} \to \mathbf{Z} \oplus \mathbf{N}$ sending n to (n, pn) defines a smooth morphism of log schemes. If we take the base change defined by sending $1 \in \mathbf{N}$ to $\pi \in V$, where V is a DVR of mixed characteristic p, the resulting log scheme X is smooth over the standard log dash associated to V, and its underlying ring homomorphism is given by $V \to V[u, u^{-1}, t]/(ut^p = \pi)$. A further base change to the standard log point gives an example of a smooth log scheme over the standard log point in which $\mathcal{M}_{X/S} \cong \mathbf{Z}/p\mathbf{Z}$.

The concepts of smooth, unramified, and étale morphisms of idealized log schemes are defined just as in Definition 3.1.1, using idealized log thickenings as test objects. The categorical and differential properties of such morphisms behave as expected. Let us here indicate some of the key points. An important family of étale morphisms of idealized log schemes can be constructed as follows.

Variant 3.1.21. Let $(X, \mathcal{K})$ be a coherent idealized log scheme, let $\mathcal{K}'$ be a coherent sheaf of ideals in $\mathcal{M}_X$ containing $\mathcal{K}$, and let $(X', \mathcal{K}_{X'})$ denote the idealized log subscheme of X defined by $\mathcal{K}'$ (Proposition III.1.3.4). Then the morphism of idealized log schemes $j\colon (X', \mathcal{K}_{X'}) \to (X, \mathcal{K}_X)$ is étale.

Proof It suffices to show the following. For every idealized log thickening i over $(X', \mathcal{K}_{X'})/(X, \mathcal{K}_X)$ as in the diagram

$$\begin{array}{ccc} (S, \mathcal{K}_S) & \xrightarrow{i} & (T, \mathcal{K}_T) \\ \downarrow g & & \downarrow h \\ (X', \mathcal{K}_{X'}) & \xrightarrow{j} & (X, \mathcal{K}_X), \end{array}$$

the morphism h factors through j. To see this, observe that since i is ideally strict, $\mathcal{K}_T \cong \mathcal{M}_T \times_{\mathcal{M}_S} \mathcal{K}_S$. The commutativity of the square implies that $i^\flat h^\flat(\mathcal{K}') \subseteq g^\flat(\mathcal{K}_{X'}) \subseteq \mathcal{K}_S$ and hence that $h^\flat(\mathcal{K}') \subseteq \mathcal{K}_T$. Since $\alpha_T(\mathcal{K}_T) = 0$, it follows that $h^\sharp \alpha_X(\mathcal{K}') = 0$ and hence that the morphism $\underline{h}$ factors through $\underline{j}$. Since $\mathcal{M}_{X'}$ is the quotient of $\mathcal{M}_X$ by $1 + \alpha_X(\mathcal{K}')$, it also follows that $h^\flat$ factors through $\mathcal{M}_{X'}$, and this factorization necessarily takes $\mathcal{K}_{X'}$ to $\mathcal{K}_T$. □

For example, if Q is a fine sharp toric monoid, then the log scheme $\mathsf{A}_Q := \operatorname{Spec}(Q \to \mathbf{Z}[Q])$ is smooth over $\mathbf{Z}$. The map of idealized log schemes $\mathsf{A}_{Q,Q^+} \to \mathsf{A}_{Q,\emptyset}$ is étale. Thus A_{Q,Q^+} is smooth over $\mathbf{Z}$ as an idealized log scheme although the underlying log scheme A_Q is not smooth over $\mathbf{Z}$.

Let us check the idealized version of Proposition 3.1.6.

Variant 3.1.22. An ideally strict morphism $f_{\mathcal{K}}\colon (X, \mathcal{K}_X) \to (Y, \mathcal{K}_Y)$ of fine idealized log schemes is smooth (resp. unramified, resp. étale) if and only if the corresponding map of underlying log schemes $f\colon X := (X,\emptyset) \to Y := (Y,\emptyset)$ is smooth (resp. unramified, resp. étale).

Proof If $f_{\mathcal{K}}$ is ideally strict, then the natural map $(X,\mathcal{K}_X) \to (Y,\mathcal{K}_Y) \times_Y X$ is an isomorphism. Hence if the map f is smooth (resp. unramified, resp. étale) then so is $f_{\mathcal{K}}$, because each of these respective properties is preserved by base change. Suppose conversely that $f_{\mathcal{K}}$ is smooth, and let $S \to T$ be an affine first-order thickening over X/Y, in the notation of Definition 2.2.1. Let $\mathcal{K}_T$ be the sheaf of ideals of $\mathcal{M}_T$ generated by the image of $h^{-1}(\mathcal{K}_Y)$, and note that $\alpha_T(\mathcal{K}_T) = 0$, because $\alpha_T \circ h^\flat = h^\sharp \circ \alpha_Y$ and $\alpha_Y(\mathcal{K}_Y) = 0$. Thus $(T,\mathcal{K}_T)$ is an idealized log scheme. Similarly, let $\mathcal{K}_S$ be the ideal of $\mathcal{M}_S$ generated by the image of $g^{-1}(\mathcal{M}_X)$, so that there is a morphism of idealized log schemes $i\colon (S,\mathcal{K}_S) \to (T,\mathcal{K}_T)$. This morphism is strict because f is strict, and hence is an idealized log thickening. Then the rest of the proof follows the pattern of the proof of Proposition 3.1.6. □

3.2 Differential criteria for smoothness

The next set of results follows the standard pattern from algebraic geometry.

Proposition 3.2.1. *If $f\colon X \to Y$ is a smooth morphism of idealized log schemes, then $\Omega^1_{X/Y}$ is locally free of finite type.*

Proof Without loss of generality we assume that X and Y are affine. If $\mathcal{E}$ is a quasi-coherent sheaf on X, let $X \oplus \mathcal{E}$ denote the trivial Y-extension (Remark 2.1.5) of X by $\mathcal{E}$. Then, by Theorem 2.2.2, the set of retractions $X \oplus \mathcal{E} \to X$ is canonically bijective with $\mathrm{Hom}(\Omega^1_{X/Y}, \mathcal{E})$. A surjective map of quasi-coherent $\mathcal{O}_X$-modules $\mathcal{E} \to \mathcal{E}'$ gives rise to a corresponding log thickening $X \oplus \mathcal{E}' \to X \oplus \mathcal{E}$ and, by the smoothness of X/Y, every retraction $X \oplus \mathcal{E}' \to X$ lifts to $X \oplus \mathcal{E}$, since X is affine. It follows that, for every surjective homomorphism $\mathcal{E} \to \mathcal{E}'$, the induced map $\mathrm{Hom}(\Omega^1_{X/Y}, \mathcal{E}) \to \mathrm{Hom}(\Omega^1_{X/Y}, \mathcal{E}')$ is also surjective. Since $\Omega^1_{X/Y}$ is of finite presentation, it follows that it is locally free. □

Theorem 3.2.2. *Let $g\colon Y \to Z$ be a smooth morphism of coherent and u-integral log schemes and let $i\colon X \to Y$ be a strict closed immersion defined by a finite type ideal $\mathcal{I}$ of $\mathcal{O}_Y$. Then $X \to Z$ is smooth if and only if the map $\overline{d}$ in the sequence of Proposition 2.3.2,*

$$\mathcal{I}/\mathcal{I}^2 \xrightarrow{\overline{d}} i^*(\Omega^1_{Y/Z}) \to \Omega^1_{X/Z} \to 0,$$

is injective and locally split.

Proof The proof is standard; we will recall the main outline for the convenience of the reader. Let $j\colon X \to T$ be the first infinitesimal neighborhood of X in Y. If X/Z is smooth, then locally on X there exists a retraction $T \to X$ and hence, by Proposition 2.3.2, the sequence shown is locally split. Conversely, suppose that the sequence is locally split, and let X' be a Z-extension of an affine open subset U of X by a quasi-coherent $\mathcal{E}$. By Proposition 3.1.4, it will suffice to prove that X' is trivial. Since Y/Z is smooth, there exists a deformation $\tilde{h}\colon X' \to Y$ of the inclusion $U \to Y$. Necessarily $\tilde{h}$ factors through T and induces a homomorphism $\tilde{h}^*\colon \mathcal{I}/\mathcal{I}^2$ to $i_*\mathcal{E}$. Since the map $\overline{d}$ is split, $\tilde{h}^*$ can be extended to a map $\Omega^1_{Y/Z} \to g_*(\mathcal{E})$. Such a map corresponds to a section $\xi = (D, \delta)$ of $\mathrm{Der}_{Y/Z}(g_*\mathcal{E})$. Then the deformation $\tilde{h}' := -\xi + \tilde{h}$ annihilates $\mathcal{I}/\mathcal{I}^2 \subseteq \mathcal{O}_T$ and $1 + \mathcal{I}/\mathcal{I}^2 \subseteq \mathcal{M}_T$, and hence factors through $i\colon X \to T$. This proves that X' is trivial and hence that X/Z is formally smooth. Since $\mathcal{I}$ is of finite type and Y/Z is locally of finite presentation, X/Z is also locally of finite presentation, hence smooth. □

Theorem 3.2.3. *Let $f\colon X \to Y$ and $g\colon Y \to Z$ be morphisms of coherent log schemes, and consider the exact sequence from Proposition 2.3.1*

$$f^*(\Omega^1_{Y/Z}) \xrightarrow{s} \Omega^1_{X/Z} \xrightarrow{t} \Omega^1_{X/Y} \to 0.$$

1. *If f is smooth, the map s is injective and locally split.*
2. *If $g \circ f$ is smooth and s is injective and locally split, then f is smooth.*

Proof Without loss of generality we assume that X is affine. Suppose that f is smooth, and let $T := X \oplus f^*(\Omega^1_{Y/Z})$ denote the trivial Y-extension (Remark 2.1.5) of X by $f^*(\Omega^1_{Y/Z})$. Thus $\tilde{f}\colon T \to Y$ is $f \circ r$, where $r\colon T \to X$ is the retraction. The adjunction map $\Omega^1_{Y/Z} \to f_*f^*(\Omega^1_{Y/Z})$ induces a derivation

$$D \in \mathrm{Der}_{Y/Z}(f_*f^*(\Omega^1_{Y/Z})),$$

and hence a deformation $\tilde{f}' := D + \tilde{f}$ of f to T. Since X/Y is smooth and X is affine, the Y-extension of X defined by $\tilde{f}'$ is also trivial, so there exists a retraction $r'\colon T \to X$ such that $f \circ r' = \tilde{f}'$. Then r' and r are both deformations of id_X to T over Z, and hence differ by an element of $\mathrm{Der}_{X/Z}(f^*(\Omega^1_{Y/Z})) \cong \mathrm{Hom}(\Omega^1_{X/Z}, f^*(\Omega^1_{Y/Z}))$. The homomorphism thus given is the desired splitting of s.

Suppose on the other hand that $h := g \circ f$ is smooth and that s is injective and locally split. To prove that X/Y is smooth, we shall show that every Y-extension $i\colon X \to T$ of X is locally split. If $\tilde{f}\colon T \to Y$ is the structure morphism of T/Y, then $\tilde{h} := g \circ \tilde{f}$ makes i into a Z-extension of X and, since X/Z is smooth, this extension is locally split. Thus we may assume that there is a Z-retraction

$\tilde{r}\colon T \to X$ such that $h \circ \tilde{r} = \tilde{h}$. This retraction will not be a splitting of T as a Y-extension unless $f \circ \tilde{r} = \tilde{f}$. In any case, $f \circ \tilde{r}$ and $\tilde{f}$ are two deformations of f to T, and hence differ by a homomorphism $\phi\colon f^*(\Omega^1_{Y/Z}) \to \mathcal{I}_T$. Let $\rho\colon \Omega^1_{X/Z} \to f^*(\Omega^1_{Y/Z})$ be a splitting of s. Then $\phi \circ \rho$ corresponds to a derivation D of X/Z with values in $\mathcal{I}_T$, and $\tilde{r}' := D + \tilde{r}$ is now a splitting of $X \to T$ that is compatible with the maps to Y. □

Corollary 3.2.4. *Let $f\colon X \to Y$ and $g\colon Y \to Z$ be morphisms of coherent log schemes.*

1. *If f is étale, then the natural map $f^*(\Omega^1_{Y/Z}) \to \Omega^1_{X/Z}$ is an isomorphism.*
2. *If g is unramified, the natural map $\Omega^1_{X/Z} \to \Omega^1_{X/Y}$ is an isomorphism.*

Proof If f is étale then it is smooth, so the map s in the exact sequence of Theorem 3.2.3 is injective. Since f is also unramified, $\Omega^1_{X/Y} = 0$, and hence the map $f^*(\Omega^1_{Y/Z}) \to \Omega^1_{X/Z}$ is an isomorphism. If g is unramified, $\Omega^1_{Y/Z} = 0$, and the same exact sequence shows that $\Omega^1_{X/Z} \to \Omega^1_{X/Y}$ is an isomorphism. □

Remark 3.2.5. We can now explain the geometry behind Proposition 1.1.9. The diagram of log rings appearing there becomes a diagram of log schemes:

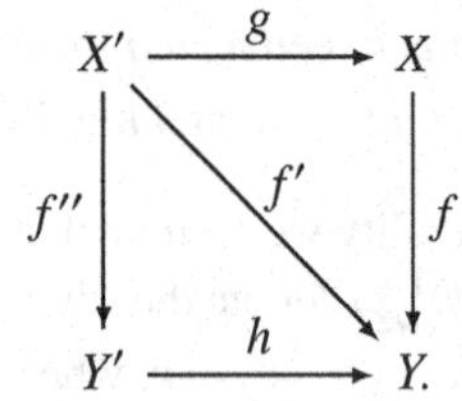

The homomorphisms $\phi\colon P \to P'$ and $\psi\colon Q \to Q'$ induce morphisms of log schemes $\mathsf{A}_{P'} \to \mathsf{A}_P$ and $\mathsf{A}_{Q'} \to \mathsf{A}_Q$. Since ϕ^{gp} and ψ^{gp} are isomorphisms, these morphisms are étale, by Corollary 3.1.10. Since $g\colon X' \to X$ (resp. $h\colon Y' \to Y$) is obtained by base change of A_ϕ (resp. A_ψ) via the map $\mathrm{Spec}(B) \to \mathsf{A}^*_Q$ (resp. $\mathrm{Spec}(A) \to \mathsf{A}^*_P$), the maps g and h are étale. Since g is étale, the map $g^*(\Omega^1_{X/Y}) \to \Omega^1_{X'/Y}$ is an isomorphism and, since h is étale, the map $\Omega^1_{X'/Y} \to \Omega^1_{X'/Y'}$ is an isomorphism by Corollary 3.2.4.

The following result is a familiar echo of the existence of local coordinates for smooth morphisms in classical algebraic geometry. However, it involves an unspecified *logarithmically* étale morphism, whose structure can, as we have seen, be quite complicated. We shall give more explicit local models of logarithmically smooth and étale morphisms in the next section.

Theorem 3.2.6. *Let $h\colon X \to Z$ be a smooth morphism of coherent log* *schemes and let $\overline{x}$ be a geometric point of X. Then in a strict étale neighborhood of $\overline{x}$, there exists a diagram*

$$\begin{array}{ccc} X & \xrightarrow{f} & Z \times \mathsf{A}_{\mathbf{N}^r} \\ & \searrow^{h} & \downarrow g \\ & & Z \end{array}$$

in which f is étale.

Proof Recall from Lemma 1.2.10 that the map $\mathcal{O}_X \otimes \mathcal{M}_X^{\mathrm{gp}} \to \Omega^1_{X/Z}$ is surjective. It follows that the fiber $\Omega^1_{X/Z}(\overline{x})$ of $\Omega^1_{X/Z}$ at $\overline{x}$ is spanned as a $k(\overline{x})$-vector space by the image of the map $\mathcal{M}_{X,\overline{x}} \to \Omega^1_{X/Z}(\overline{x})$. Thus there exists a finite sequence $(m_1, m_2, \ldots, m_r)$ of local sections of $\mathcal{M}_X$ whose images in the vector space $\Omega^1_{X/Z}(\overline{x})$ form a basis. Since X/Z is smooth, $\Omega^1_{X/Z}$ is locally free, and hence $(dm_1, \ldots, dm_r)$ give rise to a basis for its stalk at $\overline{x}$. Restricting to some (strict) étale neighborhood of $\overline{x}$, we may assume that the m_i are global sections and therefore define a map of log schemes $p\colon X \to \mathsf{A}_{\mathbf{N}^r}$. Let $Y := Z \times \mathsf{A}_{\mathbf{N}^r}$, let $f\colon X \to Y$ be the map (h, p), and let $g\colon Y \to Z$ be the projection mapping. Consider the exact sequence

$$\to f^*\Omega^1_{Y/Z} \xrightarrow{s} \Omega^1_{X/Z} \to \Omega^1_{X/Y} \to 0$$

from Proposition 2.3.1. The sequence $(dm_1, dm_2, \ldots, dm_r)$ also forms a basis for $\Omega^1_{Y/Z,\overline{x}}$, and s takes this sequence to our basis for $\Omega^1_{X/Z,\overline{x}}$. It follows that s induces an isomorphism on the stalks at $\overline{x}$, hence in some neighborhood of $\overline{x}$. Replacing X by such a neighborhood, we find that $\Omega^1_{X/Y} = 0$ and that s is an isomorphism. Then it follows from Theorem 3.2.3 that f is smooth. Since $\Omega^1_{X/Y} = 0$, it follows from Proposition 3.1.3 that f is also unramified, hence étale. □

3.3 Charts for smooth morphisms

The following theorem describes the local structure of a smooth morphism of log schemes.

Theorem 3.3.1. *Let $f\colon X \to Y$ be a smooth (resp. étale) morphism of fine log schemes, let $a\colon Y \to \mathsf{A}_P$ be a fine chart for Y, and let $\overline{x}$ be a geometric point of X. Then, after X and Y are replaced by étale neighborhoods of $\overline{x}$ and $f(\overline{x})$ respectively, a fits into a chart (a, θ, b) for f with the following properties.*

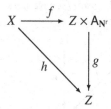

1. *The homomorphism* $\theta\colon P \to Q$ *is injective and the torsion subgroup of* $\mathrm{Cok}(\theta)^{\mathrm{gp}}$ *has order invertible in* $\mathcal{O}_X$ *(resp. is finite of order invertible in* $\mathcal{O}_X$*).*
2. *The morphism* b_θ *in the diagram (see Remark III.1.2.6)*

$$\begin{array}{ccccc} X & \xrightarrow{b_\theta} & Y_\theta & \xrightarrow{b} & \mathsf{A}_Q \\ & \searrow^{f} & \downarrow{\scriptstyle f_\theta} & & \downarrow{\scriptstyle \mathsf{A}_\theta} \\ & & Y & \xrightarrow{a} & \mathsf{A}_P \end{array}$$

   is étale and strict.
3. *The chart* b *is exact at* $\overline{x}$*. If* f *is étale, the chart* (a, θ, b) *can be chosen to be neat at* $\overline{x}$*. If* f *is smooth and the order of the torsion subgroup of* $\mathcal{M}^{\mathrm{gp}}_{X/Y,\overline{x}}$ *is invertible in* $k(x)$*, the chart* (a, θ, b) *can be chosen to be neat at* $\overline{x}$*, provided* b_θ *is allowed to be smooth (but not necessarily étale).*

Proof First note that $\Omega^1_{X/Y}(\overline{x}) = 0$ if f is étale, and so it follows from Proposition 2.3.5 that $k(\overline{x}) \otimes \mathcal{M}^{\mathrm{gp}}_{X/Y,\overline{x}} = 0$. Then $\mathcal{M}^{\mathrm{gp}}_{X/Y,\overline{x}}$ is a finite abelian group whose order is invertible in $k(\overline{x})$. Suppose more generally that the order of the torsion part of $\mathcal{M}^{\mathrm{gp}}_{X/Y,\overline{x}}$ is invertible in $k(\overline{x})$. Localizing X, we may assume that this order is invertible in $\mathcal{O}_X$. Then (4) of Theorem III.1.2.7 tells us that, after a further étale localization, a can be embedded in a chart for f that is neat at $\overline{x}$ and subordinate to a homomorphism $\theta\colon P \to Q$. In particular, θ^{gp} is injective, and the homomorphism $Q^{\mathrm{gp}}/P^{\mathrm{gp}} \to \mathcal{M}^{\mathrm{gp}}_{X/Y,\overline{x}}$ induced by b is bijective. Thus property (1) is certainly satisfied. By Corollary 3.1.9 (resp. Corollary 3.1.10), the morphism A_θ is smooth (resp. étale). Hence, by Remark 3.1.2, the base-changed map $f_\theta\colon Y_\theta \to Y$ is also smooth (resp. étale). If f is étale, it follows from Remark 3.1.2 that b_θ is also étale and, since b_θ is strict, Proposition 3.1.6 implies that $\underline{b}_\theta$ is also étale. If f is only smooth, consider the diagram

$$\begin{array}{ccccccc} \Omega^1_{Y_\theta/Y}(\overline{x}) & \longrightarrow & \Omega^1_{X/Y}(\overline{x}) & \longrightarrow & \Omega^1_{X/Y_\theta}(\overline{x}) & \longrightarrow & 0 \\ \uparrow{\scriptstyle \cong} & & \downarrow{\scriptstyle \rho_{X/Y,x}} & & & & \\ k(\overline{x}) \otimes Q^{\mathrm{gp}}/P^{\mathrm{gp}} & \longrightarrow & k(\overline{x}) \otimes \mathcal{M}^{\mathrm{gp}}_{X/Y,\overline{x}}, & & & & \end{array} \tag{3.3.1}$$

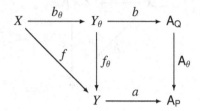

where the exact sequence along the top comes from Proposition 2.3.1, the map $\rho_{X/Y,x}$ from Proposition 2.3.5, and the left vertical isomorphism from Propositions 1.1.4 and 1.2.15. Since the chart is neat at x, the horizontal arrow at the bottom left is an isomorphism, and it follows that the top left horizontal arrow is injective. Then it follows from Theorem 3.2.3 that b_θ is smooth at x and, since b_θ is strict, that $\underline{b}_\theta$ is also smooth.

Now suppose that f is smooth, without any assumption on $\mathcal{M}_{X/Y}$. Let us apply Theorem 3.2.6 to find, after a localization, a diagram

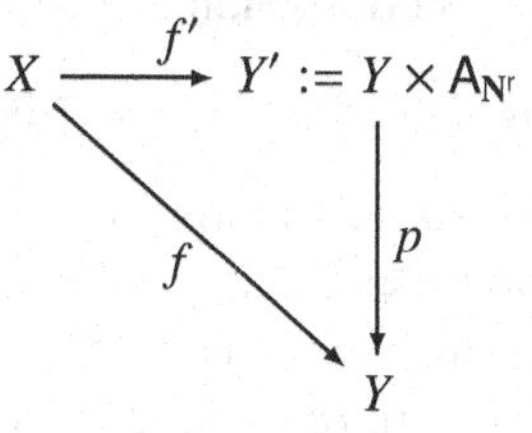

in which f' is étale. Since $\alpha := a^\flat$ is a chart for $\mathcal{M}_Y$,

$$\alpha' := \alpha \oplus \mathrm{id}\colon P \oplus \mathbf{N}^r \to \mathcal{M}_{Y'}$$

is a chart for Y'. Now let us apply the étale case we have already proved to find a chart for f' that is subordinate to a morphism $\theta'\colon P \oplus \mathbf{N}^r \to Q$ satisfying conditions (1) and (2). Let $\theta\colon P \to Q$ be the composite of θ' with the inclusion $P \to P \oplus \mathbf{N}^r$. Then θ^{gp} is injective, and there is an exact sequence

$$0 \to \mathbf{Z}^r \to Q^{\mathrm{gp}}/P^{\mathrm{gp}} \to Q^{\mathrm{gp}}/(\mathbf{Z}^g \oplus P^{\mathrm{gp}}) \to 0.$$

It follows that the torsion subgroup of Q^{gp}/P^g injects in the torsion subgroup of $Q^g/(\mathbf{Z}^g \oplus P^{\mathrm{gp}})$, and hence has order invertible in $\mathcal{O}_X$. Finally, observe that the two squares in the diagram

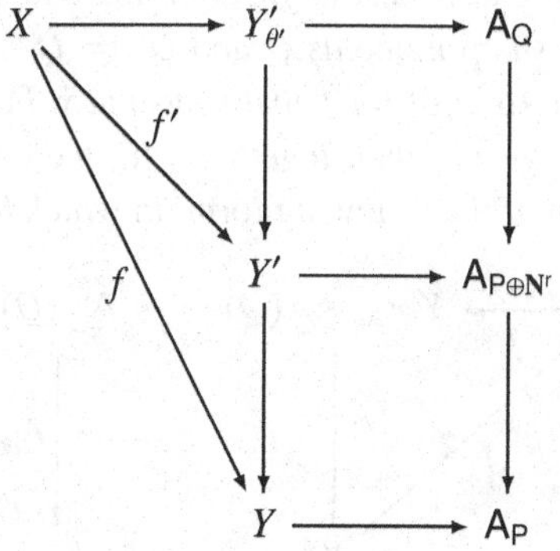

are cartesian, and hence so is the rectangle. Since $\underline{X} \to \underline{Y}'_{\theta'}$ is étale, (2) is also satisfied. Let F be the face of Q consisting of those elements of Q that map to units in $\mathcal{O}_{X,\overline{x}}$. After a replacement of X by a Zariski open neighborhood of $\overline{x}$, the map $X \to \mathsf{A}_Q$ will factor through A_{Q_F}, and the map $Q_F \to \mathcal{M}_{\overline{x}}$ will be local, hence exact. Then (3) will hold as well. □

Remark 3.3.2. The chart constructed in the smooth case of Theorem 3.3.1 may not be neat. Indeed, if f is smooth, it can happen that $\mathcal{M}^{\mathrm{gp}}_{X/Y}$ can have torsion which is not invertible in $\mathcal{O}_X$, and that a flat (not étale) localization can be required before a neat chart can be found. For example, let $f\colon X \to Y$ be the morphism of log schemes corresponding to the morphism of monoids

$$n \mapsto (n, pn) : \mathbf{N} \to \mathbf{Z} \oplus \mathbf{N}.$$

As we saw in Example 3.1.20, this morphism is smooth in characteristic p although the multiplicity of the special fiber is p. Note that the cokernel of the map $\overline{f}^{-1}(\mathcal{M}^{\mathrm{gp}}_{Y,x}) \to \overline{\mathcal{M}}^{\mathrm{gp}}_{X,x}$ at any point of the special fiber is $\mathbf{Z}/p\mathbf{Z}$, and thus the chart $\mathbf{N} \to \mathbf{Z} \oplus \mathbf{N}$ is not neat. In order to find a neat chart, one must take a flat cover $\tilde{X}$ of X given by taking a pth root of the function corresponding to $(1, p) \in \mathbf{Z} \oplus \mathbf{N}$, but then the morphism $\tilde{X} \to Y$ is no longer smooth.

It is in general impossible to construct charts that are both neat and étale—for example one cannot do this for schemes with trivial log structure. It will be possible, however, if we relax the notion of a chart somewhat. If Q' is a monoid and Q is a submonoid, we let $\underline{\mathsf{A}}_{Q'}(Q) := \operatorname{Spec}(Q \to \mathbf{Z}[Q'])$. We shall see that, when Q is sharp, the local behavior of $\underline{\mathsf{A}}_{Q'}(Q)$ near its vertex provides a good model for the local behavior of a smooth log scheme, or a smooth morphism of log schemes, rather generally.

Theorem 3.3.3. *Let $f\colon X \to Y$ be a smooth morphism of fine log schemes, and let $\overline{x} \in X$ be a geometric point and $\overline{y} := f(\overline{x})$. Suppose that f is s-injective at $\overline{x}$, that the order of the torsion subgroup of $\mathcal{M}^{\mathrm{gp}}_{X/Y,\overline{x}}$ is invertible in $k(\overline{x})$, and that α is a chart for Y that is neat at $\overline{y}$, subordinate to a fine sharp monoid P. Then there exist fine sharp monoids Q and $Q' := Q \oplus \mathbf{N}^r$, a homomorphism $\theta\colon P \to Q$, and a chart (α, θ, β) for f that is neat at $\overline{x}$. Furthermore, there exists a homomorphism $\mathbf{N}^r \to \mathcal{O}_X$ that, together with the chart (α, θ, β), fits into a commutative diagram of the following form, in which h is strict and étale:*

$$\begin{array}{ccccc} X & \xrightarrow{\;h\;} & Y \times_{\mathsf{A}_P} \underline{\mathsf{A}}_{Q'}(Q) & \longrightarrow & \underline{\mathsf{A}}_{Q'}(Q) \\ & \searrow^{f} & \Big\downarrow{\scriptstyle f'} & & \Big\downarrow{\scriptstyle \underline{\mathsf{A}}_{\theta\oplus 0}(Q)} \\ & & Y & \longrightarrow & \mathsf{A}_P\,. \end{array}$$

The morphism $X \to \underline{\mathsf{A}}_{Q'}(Q)$ (resp. $Y \to \mathsf{A}_P$) takes the point $\overline{x}$ (resp. $\overline{y}$) to the vertex of $\underline{\mathsf{A}}_{Q'}(Q)$ (resp. of A_P). If f is $\mathbf{Q}$-integral (resp. integral, resp. saturated), then the same is true for the homomorphisms $P \to Q$ and $P \to Q'$.

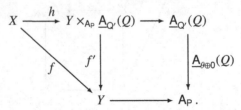

Proof Replacing X and Y by suitable étale neighborhoods of $\overline{x}$ and $\overline{y}$, we work with Zariski log structures and with schematic points in place of geometric points. Thanks to the assumption on $\mathcal{M}^{\mathrm{gp}}_{X/Y,x}$, we may apply Theorem III.1.2.7 to obtain a chart (α,θ,β) for f that is neat at x. Statement (5) of that theorem implies that the chart β is neat at x, since f is s-injective at x. Let $Y_\theta =: Y \times_{\mathsf{A}_P} \mathsf{A}_Q$. There is a commutative diagram in which the row and column are exact:

$$\begin{array}{ccccccc}
 & & \Omega^1_{\underline{X}/\underline{Y}}(x) & & & & \\
 & & \downarrow & \searrow^{a} & & & \\
\Omega^1_{Y_\theta/Y}(x) & \xrightarrow{c} & \Omega^1_{X/Y}(x) & \longrightarrow & \Omega^1_{X/Y_\theta}(x) & \longrightarrow & 0 \\
 & \searrow_{b} & \downarrow{\scriptstyle \rho_x} & & & & \\
 & & k(x) \otimes \mathcal{M}^{\mathrm{gp}}_{X/Y,x}. & & & &
\end{array}$$

The column comes from Proposition 2.3.5 and the row from Proposition 2.3.1. As we saw in the discussion of diagram 3.3.1 in the proof of Proposition 3.3.1, the map b is an isomorphism. Then it follows from this diagram that c is injective and and that a is surjective. Thus we can choose elements $(t_1,\ldots,t_r)$ in the maximal ideal of x such that $(a(dt_1),\ldots,a(dt_r))$ is a basis of $\Omega^1_{X/Y_\theta}(x)$. These elements define a homomorphism $\mathbf{N}^r \to \mathcal{O}_X$, possibly after a further shrinking of X. We get a map

$$X \to X' =: \mathbf{A}^r \times Y_\theta \cong Y \times_{\mathsf{A}_P} \underline{\mathsf{A}}_{Q'}(Q),$$

and the map $\Omega^1_{X'/Y}(x) \to \Omega^1_{X/Y}(x)$ is an isomorphism. By Theorem 3.2.3, this implies that $X \to X'$ is étale, and it is strict by construction. Since the chart $P \to \mathcal{M}_Y$ is neat at $\overline{y}$, the homomorphism $P \to \mathcal{M}_{Y,y}$ is local, and it follows that the morphism $Y \to \mathsf{A}_P$ takes $\overline{y}$ to the vertex of A_P. The same argument applies to x. If f is $\mathbf{Q}$-integral (resp. integral, resp. saturated), then the homomorphism $\overline{f}^\flat_x \colon \overline{\mathcal{M}}_{Y,y} \to \overline{\mathcal{M}}_{X,x}$ has the same property, and hence so does $\theta\colon P \to Q$. Since $Q \to Q \oplus \mathbf{N}^r$ is saturated, the composition $P \to Q'$ is also $\mathbf{Q}$-integral (resp. integral, resp. saturated). □

Corollary 3.3.4. *Let X be a fine log scheme, smooth over a field with trivial log structure, and let $\overline{x}$ be a geometric point of X such that the order of the torsion subgroup of $\overline{\mathcal{M}}^{\mathrm{gp}}_{X,\overline{x}}$ is invertible in $k(\overline{x})$. Then, in some étale neighborhood of $\overline{x}$, there exist a fine chart $\beta\colon Q \to \mathcal{M}_X$ that is neat at $\overline{x}$ and a homomorphism*

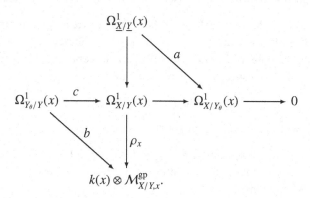

$\mathbf{N}^r \to \mathcal{O}_X$ *such that the induced map*

$$X \to \underline{\mathsf{A}}_{\mathsf{Q}\oplus\mathsf{N}^r}(Q)$$

is étale and strict and takes $\overline{x}$ *to the vertex of* $\underline{\mathsf{A}}_{\mathsf{Q}\oplus\mathsf{N}^r}(Q)$. □

The idealized analog of the local structure theorem for smooth and étale morphisms is proved in precisely the same way as the non-idealized version.

Variant 3.3.5. Let $f\colon X \to Y$ be a smooth (resp. étale) morphism of fine idealized log schemes, let $a\colon Y \to \mathsf{A}_{\mathsf{P,J}}$ be a fine chart for Y, and let $\overline{x}$ be a geometric point of X. Then, after X and Y are replaced by étale neighborhoods of $\overline{x}$ and $f(\overline{x})$ respectively, there is a chart

$$a\colon Y \to \mathsf{A}_{\mathsf{P,J}}, \quad b\colon X \to \mathsf{A}_{\mathsf{Q,K}}, \quad \theta\colon P \to Q$$

for f with the following properties.

1. The homomorphism $\theta\colon P \to Q$ is injective and the order of the torsion part of its cokernel is invertible in $\mathcal{O}_X$ (resp. and its cokernel is finite of order invertible in $\mathcal{O}_X$).
2. The morphism b_θ in the diagram (see Remark III.1.2.6)

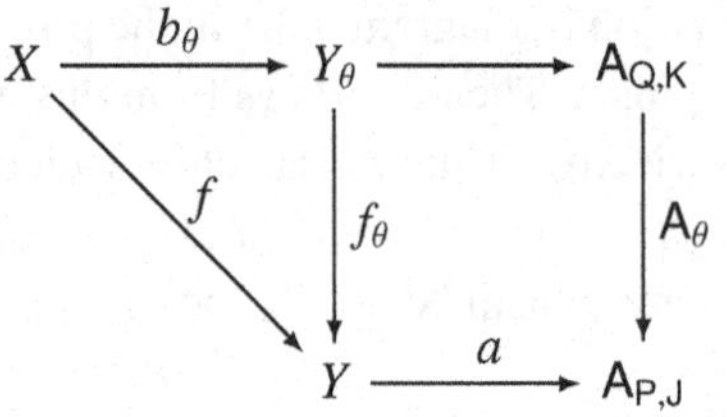

   is étale and strict.
3. The chart b is exact at $\overline{x}$. If f is étale, the chart (a, θ, b) can be chosen to be neat at $\overline{x}$.

Corollary 3.3.6. *Let* $f\colon X \to Y$ *be an étale morphism of fine idealized log schemes. Then, locally on* X *and* Y, f *admits a factorization* $f = f'' \circ f'$, *where* $f'\colon X \to X'$ *is a closed immersion defined by a coherent sheaf of ideals in* $\mathcal{M}_{X'}$ *and* $f''\colon X' \to Y$ *is étale and ideally strict.*

Proof We may suppose that there exists a chart for f of the form described in Variant 3.3.5. Let J' be the ideal of Q generated by the image of J and let $Y'_\theta := Y \times_{\mathsf{A}_{\mathsf{P,J}}} \mathsf{A}_{\mathsf{Q,J'}}$. Then $Y'_\theta \to Y$ is ideally strict and étale, $Y_\theta \to Y'_\theta$ is a closed immersion, and $X \to Y_\theta$ is ideally strict and étale. Thus $\underline{X} \to \underline{Y}_\theta$ is an étale morphism of schemes. Since $\underline{Y}_\theta \to \underline{Y}'_\theta$ is a closed immersion, we may, after a further Zariski localization if necessary, find an étale morphism $\underline{X}' \to \underline{Y}'_\theta$ such that $\underline{X}' \times_{\underline{Y}'_\theta} \underline{Y}_\theta \cong \underline{X}$. Endow X' with the idealized log structure induced from

Y'_θ, and let $\mathcal{K}$ be the sheaf of ideals of $\mathcal{M}_{X'}$ generated by K. Then $f'': X' \to Y$ is étale and ideally strict, and $f': X \to X'$ is the closed immersion defined by $\mathcal{K}$. □

Let $X \to Y$ be an étale morphism and let $i: S \to T$ be a (log) thickening. Then every Y-morphism $S \to X$ extends uniquely to T. In standard algebraic geometry, the same unique extension property holds more generally: it suffices for i to be an integral radicial homeomorphism [3, Exp. VIII, Thm 1.1]. The follow result is a logarithmic analog.

Proposition 3.3.7. *Let $f: X \to Y$ be an étale morphism of fine log schemes in characteristic p and let $i: S \to T$ be an inseparable morphism of fine log schemes over Y. Then every Y-morphism $S \to X$ lifts uniquely to a Y-morphism $T \to X$.*

Proof First let us verify this when $X \to Y$ is the morphism corresponding to an injective homomorphism of fine monoids $\theta: P \to Q$ such that $\mathrm{Cok}(\theta^{\mathrm{gp}})$ is a finite group whose order is prime to p. Using Proposition III.1.2.9, we see that the proposition in this case amounts to the assertion that the diagram

$$\begin{array}{ccc} \mathrm{Hom}(Q, \mathcal{M}_T) & \longrightarrow & \mathrm{Hom}(P, \mathcal{M}_T) \\ \big\downarrow & & \big\downarrow \\ \mathrm{Hom}(Q, \mathcal{M}_S) & \longrightarrow & \mathrm{Hom}(P, \mathcal{M}_S) \end{array}$$

is cartesian. Since $S \to T$ is by definition exact, we may replace all monoids by their corresponding group envelopes in this diagram. The localization of the diagram by the powers of p is cartesian because the homomorphism $P^{\mathrm{gp}} \to Q^{\mathrm{gp}}$ becomes an isomorphism, and its localization by the powers of the order n of $\mathrm{Cok}(\theta)$ is cartesian because $\mathcal{M}_T^{\mathrm{gp}} \to \mathcal{M}_S^{\mathrm{gp}}$ becomes an isomorphism. Since $(p, n) = 1$, it follows that the diagram itself is cartesian.

For the general case, observe that, because of the uniqueness, it suffices to prove the statement locally on S and X. Thus we may assume without loss of generality that f admits a chart as described in Theorem 3.3.1. The argument in the previous paragraph gives us a unique $T \to \mathsf{A}_Q$ over A_P, and hence a unique $T \to Y_\theta$ over Y. Since $\underline{X} \to \underline{Y}_\theta$ is étale and $\underline{S} \to \underline{T}$ is an integral radicial homeomorphism, the result in the classical case implies that there is a unique $\underline{T} \to \underline{X}$ compatible with the given maps $\underline{S} \to \underline{X}$ and $\underline{T} \to \underline{Y}_\theta$. Since $X \to Y_\theta$ is strict, the morphisms $T \to Y_\theta$ and $\underline{T} \to \underline{X}$ correspond to a unique $T \to X$, as desired. □

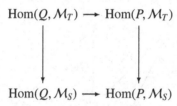

Corollary 3.3.8. *Let $f: X \to Y$ be a weakly inseparable morphism of fine log*

schemes in characteristic p. Then f admits a factorization $X \to Y' \to Y$, where $X \to Y'$ is inseparable and $Y' \to Y$ is étale, unique up to unique isomorphism. An inseparable and étale morphism of fine log schemes is an isomorphism.

Proof Since f is small, such a factorization exists étale locally on X, by Proposition III.2.3.5. Since $\underline{f}$ is an integral radicial homeomorphism, "étale locally on $\underline{X}$" is equivalent to "étale locally on $\underline{Y}$," by [3, Exp. VIII, Thm 1.1]. Let $X \to Y_1' \to Y$ and $X \to Y_2' \to Y$ be two such factorizations. Then, by Proposition 3.3.7, there is a unique map $Y_1' \to Y_2'$ compatible with these factorizations, and this morphism must be an isomorphism. The morphisms $\underline{Y}' \to \underline{Y}$ are necessarily weakly inseparable by Proposition III.2.4.8, hence affine, and hence the local constructions patch, by Proposition III.2.1.4. Suppose that f is inseparable and étale. Then $f = \mathrm{id}_Y \circ f$ and $f = f \circ \mathrm{id}_X$ are two factorizations of f as the composition of a separable and an étale map. The uniqueness of such a factorization implies that f is an isomorphism. □

Example 3.3.9. Let $f\colon X \to Y$ be a morphism of fine log schemes in characteristic p. We can form a commutative diagram

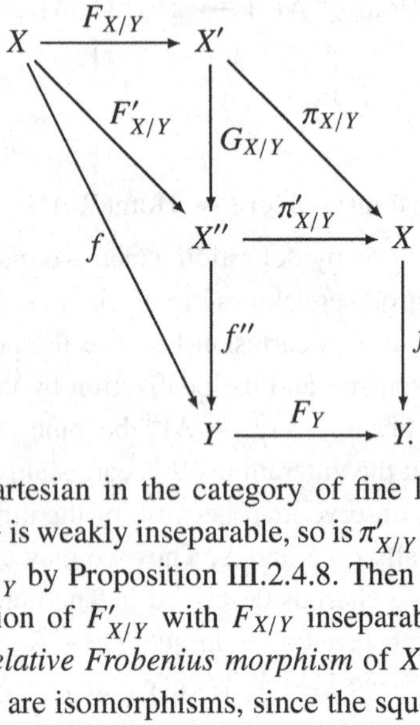

Here the square is cartesian in the category of fine log schemes and $F_X = \pi'_{X/Y} \circ F'_{X/Y}$. Since F_Y is weakly inseparable, so is $\pi'_{X/Y}$ and, since F_X is weakly inseparable, so is $F'_{X/Y}$ by Proposition III.2.4.8. Then $F'_{X/Y} = G_{X/Y} \circ F_{X/Y}$ is the unique factorization of $F'_{X/Y}$ with $F_{X/Y}$ inseparable and $G_{X/Y}$ étale. We call $F_{X/Y}$ the *exact relative Frobenius morphism* of X/Y. Note that the maps $\pi'^{*}_{X/Y}(\Omega^1_{X/Y}) \to \Omega^1_{X''/Y}$ are isomorphisms, since the square is cartesian, and the maps $G^*_{X/Y}(\Omega^1_{X''/Y}) \to \Omega^1_{X'/Y}$ are isomorphisms because $G_{X/Y}$ is étale. It follows that the maps

$$\pi^*_{X/Y}(\Omega^1_{X/Y}) \to \Omega^1_{X'/Y}$$

are also isomorphisms.

Proposition 3.3.10. *Formation of the exact relative Frobenius morphism in the category of fine log schemes is compatible with strict étale localization and with base change. Thus if $f: X \to Y$ is a morphism of fine log schemes and if $g: \tilde{X} \to X$ is strict and étale, then the squares in the diagram*

$$\begin{array}{ccccc} \tilde{X} & \xrightarrow{F_{\tilde{X}/Y}} & \tilde{X}' & \xrightarrow{\pi_{\tilde{X}/Y}} & \tilde{X} \\ \Big\downarrow g & & \Big\downarrow g' & & \Big\downarrow g \\ X & \xrightarrow{F_{X/Y}} & X' & \xrightarrow{\pi_{X/Y}} & X \end{array}$$

are cartesian, and the morphism g' is strict and étale. If $\tilde{Y} \to Y$ is a morphism of fine log schemes and $\tilde{X} := \tilde{X} \times_{\tilde{Y}} X$, then the squares in the diagram are cartesian and the vertical maps are the projections.

Proof If g is strict and étale, consider the commutative diagram

$$\begin{array}{ccccccccc} \tilde{X} & \xrightarrow{a} & X \times_{X''} \tilde{X}'' & \xrightarrow{b} & X' \times_{X''} \tilde{X}'' & \xrightarrow{\pi''} & \tilde{X}'' & \xrightarrow{\pi'_{\tilde{X}/Y}} & \tilde{X} \\ & {\scriptstyle g}\searrow & \Big\downarrow \pi & & \Big\downarrow \pi' & & \Big\downarrow g'' & & \Big\downarrow g \\ & & X & \xrightarrow{F_{X/Y}} & X' & \xrightarrow{G_{X/Y}} & X'' & \xrightarrow{\pi'_{X/Y}} & X. \end{array}$$

Since $X'' := X \times_{F_Y} Y$ and $\tilde{X}' := \tilde{X} \times_{F_Y} Y$, the square on the far right is cartesian. The remaining two squares are cartesian by construction, and hence so are the rectangles. The composition along the top is the absolute Frobenius endomorphism of $\tilde{X}$, which is weakly inseparable, and hence by Proposition III.2.4.8, all the horizontal morphisms in the top row are weakly inseparable. The morphisms g and π are strict and étale, hence a is strict and étale. It then follows from Corollary 3.3.8 that a is an isomorphism. We can thus identify the composite $\pi'' \circ b$ with the morphism $F'_{\tilde{X}/Y}$. The morphism $G_{X/Y}$ is étale, hence so is π'', and the morphism $F_{X/Y}$ is exact, hence so is b. Thus the factorization $b \circ \pi''$ of $F'_{\tilde{X}/Y}$ is the canonical one, i.e., b identifies with $F_{\tilde{X}/Y}$ and π'' with $G_{\tilde{X}/Y}$. Then $\pi'_{\tilde{X}/Y} \circ p''$ identifies with $\pi_{\tilde{X}/Y}$ and the morphism π' with g'. This completes the proof of the proposition when g is étale. The proof for base-change maps is similar but simpler. □

3.4 Unramified morphisms and the conormal sheaf

The main ideas in this section are due to L. Illusie and W. Bauer [5], and of course K. Kato.

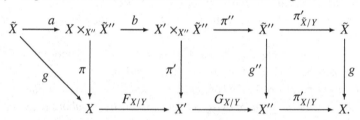

Proposition 3.4.1. *Let $f: X \to Y$ be an unramified morphism of log schemes. Then f is small, and $\mathcal{M}^{gp}_{X/Y}$ is, locally on X, annihilated by an integer invertible in $\mathcal{O}_X$.*

Proof Proposition 2.3.5 implies that $k(\overline{x}) \otimes \mathcal{M}^{gp}_{X/Y,\overline{x}}$ is a quotient of $\Omega^1_{X/Y}(\overline{x})$ for every geometric point $\overline{x}$ of X. Since X/Y is unramified, $\Omega^1_{X/Y}$ vanishes, and hence the same is true of $k(\overline{x}) \otimes \mathcal{M}^{gp}_{X/Y,\overline{x}}$. It follows that $\mathcal{M}^{gp}_{X/Y}$, a finitely generated group, must in fact be finite of order prime to the characteristic of $k(\overline{x})$. Then Proposition III.2.3.3 implies that f is small in some neighborhood of $\overline{x}$. □

Theorem 3.4.2. *Let $f: X \to Y$ be a morphism of fine log schemes. Assume that $\underline{f}$ is locally of of finite type. Then the following conditions are equivalent.*

1. *The morphism f is unramified.*
2. *The sheaf $\Omega^1_{X/Y}$ vanishes.*
3. *The diagonal morphism $\Delta_{X/Y}: X \to X \times_Y X$ is étale.*
4. *For every $Y' \to Y$, every section $Y' \to X \times_Y Y'$ of the projection $X \times_Y Y' \to Y'$ is étale.*
5. *For every log point (resp. standard log point) S and every morphism $S \to Y$, the base-changed map $X \times_Y S \to S$ is unramified.*
6. *For every point y of Y, the base changed map $X_y \to y$ is unramified.*
7. *Étale locally on $\underline{X}$, there exists a chart (a, θ, b) for f subordinate to an injective homomorphism of fine monoids $\theta: P \to Q$ such that* $\mathrm{Cok}(\theta^{gp})$ *is finite of order invertible in $\mathcal{O}_X$ and such that the induced map $\underline{b}_\theta: \underline{X} \to \underline{Y}_\theta$ is unramified. In fact, for any point x of X, one can arrange for (a, θ, b) to be neat at x.*
8. *Étale locally on $\underline{X}$, there exists a factorization $f = g \circ i$ where g is étale and i is a strict closed immersion.*

In particular, any immersion of fine log schemes is unramified.

Proof We have already seen in Proposition 3.1.3 that a morphism of fine log schemes is formally unramified if and only if $\Omega^1_{X/Y} = 0$. Since we are also assuming that $\underline{f}$ is locally of finite type, the equivalence of (1) and (2) follows. In particular, suppose that $f: X \to Y$ is an immersion of fine log schemes. Then $\underline{f}$ is an immersion, and hence locally of finite type and unramified, so $\Omega^1_{\underline{X}/\underline{Y}} = 0$. Furthermore, the map $f^*_{log}(\mathcal{M}_Y) \to \mathcal{M}_X$ is surjective, so $\mathcal{M}_{X/Y} = 0$. Then it follows from the exact sequence in Proposition 2.3.5 that $\Omega^1_{X/Y} = 0$, so f is unramified.

Suppose that f is unramified. Since f is of finite type, the diagonal morphism $\Delta_{X/Y}$ is locally of finite presentation. Since $\Delta_{X/Y}$ is an immersion, it is unramified, and to see that it is formally étale, it is enough to check that it is formally smooth. Let $i\colon S \to T$ be a first-order log thickening over $\Delta_{X/Y}$. The given map $h\colon T \to X \times_Y X$ defines a pair of maps $h_1, h_2\colon T \to X$ with the same projection k to Y and the same restriction g to X. Then (i, g, k) defines a log thickening over f, and h_1, h_2 are two deformations of g to X. Since f is unramified, these two deformations agree, so h factors uniquely through $\Delta_{X/Y}$ and defines a deformation of g to X. Conversely, if $\Delta_{X/Y}$ is étale and $S \to T$ is a log thickening over f, then any two deformations $\tilde{g}_1, \tilde{g}_2$, of g to T define a log thickening (i, g, k) over $\Delta_{X/Y}$. Since $\Delta_{X/Y}$ is étale, the map g deforms uniquely to T and defines a factorization of k through $\Delta_{X/Y}$. It follows that $\tilde{g}_1 = \tilde{g}_2$ and hence that f is formally unramified. Thus conditions (2) and (3) are equivalent.

It is clear that condition (4) implies condition (3): take Y' to be X. Suppose conversely that (3) holds, that $Y' \to Y$ is a morphism, and that $s\colon Y' \to Y' \times_Y X$ is a section of the map $f'\colon Y' \times_Y X \to Y'$. Then the diagram

$$\begin{array}{ccc} Y' & \xrightarrow{\;s\;} & Y' \times_Y X \\ \downarrow & & \downarrow \\ X & \xrightarrow{\;\Delta_{X/Y}\;} & X \times_Y X \end{array}$$

is cartesian. Since $\Delta_{X/Y}$ is étale, so is s.

Since the family of unramified maps is stable under base change, (1) implies (5) and (6). Suppose that (5) holds, for all standard log points S. Let x be a point of X and let y be its image in Y. According to Proposition III.1.5.5, there exist a standard log point S and a morphism of log schemes $S \to x$, where x is endowed with the log structure induced from X. Let $S \to Y$ be the morphism obtained by composing $S \to x$ with the morphisms $x \to X$ and $X \to Y$. Then the morphism $f'\colon X' := X \times_Y S \to S$ admits a section s taking the unique point of S to a point x' of X' whose image via the map $g\colon X' \to X$ is x. If (5) holds, the morphism f' is unramified, and hence $\Omega^1_{X'/S} = 0$. By Proposition 1.2.15, $\Omega^1_{X'/S} \cong g^*(\Omega^1_{X/Y})$, where $g\colon X' \to X$ is the projection, and it follows that the fiber $g^*(\Omega^1_{X/Y})(x')$ of $g^*(\Omega^1_{X/Y})$ at x' vanishes. Since $g(x') = x$, we conclude that $\Omega^1_{X/Y}(x)$ also vanishes. Since X/Y is of finite type, the module $\Omega^1_{X/Y}$ is of finite type and hence, by Nakayama's lemma, it follows that $\Omega^1_{X/Y,x}$ vanishes as well. Since x was arbitrary, the sheaf $\Omega^1_{X/Y}$ vanishes and so f is unramified. Suppose (6) holds and $S \to Y$ is a morphism, where S is a log point. Let y be the image of the unique point of S, and endow it with the log structure induced from Y. Then the morphism $S \to Y$ factors through the morphism $y \to Y$. Since $X_y \to y$ is unramified, so is the morphism $X \times_Y S \to S$, and so (6) implies (5).

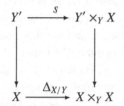

Suppose that f admits a chart as described in condition (7). Theorem 3.1.7 implies that the map $\mathsf{A}_\theta \colon \mathsf{A}_Q \to \mathsf{A}_P$ is unramified, and hence so is the map $h \colon Y_\theta \to Y$ induced by base change. Since $b_\theta \colon X \to Y_\theta$ is strict and $\underline{b}_\theta$ is unramified, it follows from Proposition 3.1.6 that b_θ is also unramified. Then $f = h \circ b_\theta$ is also unramified. Supposing conversely that f is unramified, we follow the method of the proof of the structure theorem 3.3.1 of smooth morphisms. Since our statement is local, we may assume that we are working with Zariski log structures. Let x be a point of X and let $y := f(x)$. It follows from Proposition 3.4.1 that $\mathcal{M}^{\mathrm{gp}}_{X/Y,x}$ has finite order, invertible in $k(x)$. Hence by Theorem III.1.2.7, after a further localization of X, f admits a chart (a, θ, b) which is neat at x. In particular, $Q^{\mathrm{gp}}/P^{\mathrm{gp}} \cong \mathcal{M}^{\mathrm{gp}}_{X/Y,x}$, and hence $Q^{\mathrm{gp}}/P^{\mathrm{gp}}$ is a finite group whose order is invertible in $k(x)$, hence also in some neighborhood of x. Let $Y_\theta := Y \times_{\mathsf{A}_P} \mathsf{A}_Q$, let $b_\theta \colon X \to Y_\theta$ be the morphism induced by f and b, and let $x' := b_\theta(x)$. Since $X \to Y$ is unramified, the same is true of the map $b_\theta \colon X \to Y_\theta$ and, since b_θ is also strict, $\underline{b}_\theta$ is unramified by Proposition 3.1.6.

Since immersions and étale morphisms are unramified, and since the family of unramified morphisms is stable under composition, it is clear that (8) implies (1). On the other hand, if f is unramified, then étale locally on $\underline{X}$ there exists a chart as in (7). Note that, by Corollary 3.1.10, the map $f_\theta \colon Y_\theta \to Y$ is étale. By the structure theorem for unramified morphism of schemes [32, Corollaire 7.8], $\underline{b}_\theta$ can, locally on $\underline{X}$, be written as a composite of a closed immersion $i \colon \underline{X} \to Y'$ and an étale map $g' \colon Y' \to \underline{Y}_\theta$. Endow Y' with the log structure induced from Y_θ. Then i becomes a strict closed immersion and g' a strict étale morphism of log schemes. We now have $f = f_\theta \circ g' \circ i$, where i is a strict closed immersion and $f_\theta \circ g'$ is an étale morphism. □

The previous result can be used to construct "strict infinitesimal neighborhoods" of a closed immersion, or, more generally, of an unramified morphism.

Theorem 3.4.3. *Let $f \colon X \to Y$ be a morphism of fine log schemes and, for each natural number n, let $\mathsf{Thick}_n(X/Y)$ denote the category whose objects are log thickenings $i \colon S \to T$ of order less than or equal to n over X/Y(Definition 2.2.1) and whose morphisms are commutative squares over $X \to Y$. Then f is unramified if and only if, for every n, the category $\mathsf{Thick}_n(X/Y)$ has a final object i_n of the form*

$$\begin{array}{ccc} X & \xrightarrow{\;i_n\;} & Y_n \\ {\scriptstyle \mathrm{id}_X}\downarrow & & \downarrow{\scriptstyle f_n} \\ X & \xrightarrow{\;f\;} & Y. \end{array}$$

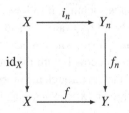

Furthermore, the formation of i_n, when it exists, is compatible with base change $Y' \to Y$. The thickening i_n is called the nth strict infinitesimal neighborhood of X in Y.

Proof Let us first prove that the existence of such a universal object implies that f is unramified. Suppose that $i: S \to T$ is a log thickening over X/Y of order n, given by the maps $g: S \to X$ and $h: T \to Y$. Since i_n is a final object in $\mathsf{Thick}_n(X/Y)$, there is a unique morphism $h_n: T \to Y_n$ such that $h_n \circ i = i_n \circ g$ and $f_n \circ h_n = h$. If $\tilde{g}: T \to X$ is a deformation of g to T, then $i_n \circ \tilde{g} \circ i = i_n \circ g$ and furthermore $f_n \circ i_n \circ \tilde{g} = f \circ \tilde{g} = h$. It follows from the uniqueness of h_n that $i_n \circ \tilde{g} = h_n$. Hence $i_n \circ \tilde{g}' = i_n \circ \tilde{g}$ for any other deformation $\tilde{g}'$ of g to T. But i_n is a closed immersion, hence a monomorphism, and it follows that $\tilde{g} = \tilde{g}'$. Thus deformations of g, when they exist, are unique, and hence f is unramified.

Before embarking on the proof of the converse, let us check that the formation of i_n is compatible with base change. Thus we suppose that $i_n: X \to Y_n$ is a final object of $\mathsf{Thick}_n(X/Y)$ and that $q: Y' \to Y$ is a morphism of fine log schemes. Let $p: X' := X \times_Y Y' \to X$ and $i'_n: X' \to Y'_n := Y_n \times_Y Y'$ be the induced maps. Then i'_n is an object of $\mathsf{Thick}_n(X'/Y')$. If $i': S' \to T'$ is any other such object, with maps $g': S' \to X$ and $h': T' \to Y'$, then $p \circ g': S' \to X$ and $q \circ h': T' \to Y$ make i' into an object of $\mathsf{Thick}_n(X/Y)$, which then maps uniquely to $i_n: X \to Y_n$. The induced map $T' \to Y'_n$ then defines the (necessarily unique) morphism $i' \to i'_n$ in $\mathsf{Thick}_n(X'/Y')$.

To prove the existence of i_n, let us first assume that f admits a factorization $f = f' \circ i$, where $f': Y' \to Y$ is étale and $i: X \to Y'$ is a strict closed immersion. Then the nth infinitesimal neighborhood $i'_n: X \to Y'_n$ of X in Y' is a log thickening of order n, and the map $Y'_n \to Y' \to Y$ allows us to view it as an nth-order log thickening of X in Y. We claim that this object, which we now denote by i_n, is the final object in the category $\mathsf{Thick}_n(X/Y)$. Indeed, if $i: S \to T$ is any object of $\mathsf{Thick}_n(X/Y)$, the map $h: T \to Y$ factors uniquely through a map $h': T \to Y'$, because $Y' \to Y$ is étale. The ideal $\mathcal{I}_{X/Y'}$ of X in Y' then maps to the ideal $\mathcal{I}_{S/T}$ of S in T, and hence $\mathcal{I}^{n+1}_{X/Y}$ maps to zero in $\mathcal{O}_T$. Thus the map $\mathcal{O}_{Y'} \to h'_*(\mathcal{O}_T)$ factors uniquely through $f_{n*}(\mathcal{O}_{Y_n})$. Since i_n is strict, it follows also that the map $\mathcal{M}_{Y'} \to h'_*(\mathcal{M}_T)$ factors uniquely through $\mathcal{M}_{Y_n}$. Thus the map $T \to Y$ factors uniquely through Y_n.

In the general case, Theorem 3.4.2 implies that any unramified f admits such a factorization étale locally on $\underline{X}$, and hence the nth strict infinitesimal neighborhood of X in Y exists étale locally. It remains to show that these locally constructed objects glue. Our argument shows that there is a strict étale covering $\tilde{Y} \to Y$ such that $\mathsf{Thick}_n(\tilde{X}/\tilde{Y})$ admits a final object $\tilde{i}_n$. By the compatibility of its formation with base change, we see that there is a unique isomorphism between the pullbacks of $\tilde{i}_n$ via the two projections $\tilde{Y} \times_Y \tilde{Y} \to Y$, and that this isomorphism satisfies the cocycle condition on $\tilde{Y} \times_Y \tilde{Y} \times_Y \tilde{Y}$. Thus we have descent data for the embedding $\tilde{i}_n$, with respect to the étale morphism $\tilde{Y} \to Y$. The uniqueness of the construction will give descent data, but since the maps $Y_n \to Y$ need not be affine in general, the resulting global object Y_n could, a priori, be an algebraic space rather than a scheme. One can conclude by observing that $Y_{n,red} = X_{red}$ is a scheme and then referring to the literature (for example [53, Theorem III,3.3]) to see that Y_n is consequently also a scheme. Let us roughly sketch an alternative argument showing how to overcome this problem, without resorting to the language of algebraic spaces.

Our argument shows that there is a strict étale covering $\tilde{Y} \to Y$ such that $\mathsf{Thick}_n(\tilde{X}/\tilde{Y})$ admits a final object $\tilde{i}_n$. By the compatibility of its formation with base change, we see that there is a unique isomorphism between the pullbacks of $\tilde{i}_n$ via the two projections $\tilde{Y} \times_Y \tilde{Y} \to Y$, and that this isomorphism satisfies the cocycle condition on $\tilde{Y} \times_Y \tilde{Y} \times_Y \tilde{Y}$. Thus we have descent data for the embedding $\tilde{i}_n$, with respect to the étale morphism $\tilde{Y} \to Y$. If $\tilde{U} \to \tilde{X}$ is an étale map, let $\tilde{Y}_{n,\tilde{U}}$ be the restriction of $\tilde{Y}_n$ to $\tilde{U}$, and let $\mathcal{A}(\tilde{U}) := \Gamma(\tilde{Y}_{n,\tilde{U}}, \mathcal{O}_{Y_{n,\tilde{U}}})$. The descent data imply that $\mathcal{A}(\tilde{U})$ does not depend the map $\tilde{U} \to \tilde{X}$ and that $\mathcal{A}$ defines a sheaf of rings (not a sheaf of $\mathcal{O}_X$-algebras) on $X_{ét}$.

Let $\eta\colon \underline{X}_{ét} \to \underline{X}_{\mathrm{zar}}$ be the canonical map from the étale to the Zariski topology. We shall show by induction on n that the natural map $\eta^*\eta_*(\mathcal{A}) \to \mathcal{A}$ is an isomorphism. This statement is just Hilbert's Theorem 90 when $n = 0$. Suppose its truth for $n-1$, let $\mathcal{A}'$ be the structure sheaf of Y_{n-1}, viewed as a sheaf on $X_{ét}$, and look at the exact sequence $0 \to \mathcal{I} \to \mathcal{A} \to \mathcal{A}' \to 0$. Then $\mathcal{I}$ is in fact a quasi-coherent sheaf of $\mathcal{O}_X$-modules on $X_{ét}$. By étale descent, $R^1\eta_*(\mathcal{I}) = 0$ and the map $\eta^*\eta_*(\mathcal{I}) \to \mathcal{I}$ is an isomorphism. It follows that $\eta^*\eta_*\mathcal{A} \to \mathcal{A}$ is also an isomorphism. Then $\eta_*(\mathcal{A})$ forms a quasi-coherent sheaf of rings on $\underline{X}_{\mathrm{zar}}$ and hence defines a scheme Y_n, and $\eta^*(\mathcal{O}_{Y_n}) \cong \mathcal{A}$. Thus the pullback of Y_n to $\tilde{X}$ is $\tilde{Y}_n$, and the map $\tilde{Y}_n \to Y$ descends to a map $Y_n \to Y$. The fact that $X \to Y_n$ is a log thickening of order n over X/Y can be checked étale locally on $\underline{X}$, as can the fact that it is the final such object. This concludes the argument. □

Corollary 3.4.4. *Let* Lognet *denote the category of unramified morphisms f of fine log schemes, with morphisms $f' \to f$ given by commutative squares. For $n \in \mathbf{N}$, let* Thick_n *be the full subcategory of* Lognet *whose objects are the log thickenings of order less than or equal to n. Then the inclusion functor* $\mathsf{Thick}_n \to$ Lognet *admits a right adjoint $(f\colon X \to Y) \mapsto (T_n(f)\colon X \to Y_n)$ (so that f and $T_n(f)$ have the same source).*

Proof If $f\colon X \to Y$ is an unramified morphism of fine log schemes, let $T_n(f)\colon X \to Y_n$ be the final object of $\mathsf{Thick}_n(X/Y)$ constructed in Theorem 3.4.3 above. If $f' \to f$ is a morphism in Lognet given by the maps $p\colon X' \to X$ and $q\colon Y' \to Y$, then composition with p and q define a functor $\mathsf{Thick}_n(X'/Y') \to \mathsf{Thick}_n(X/Y)$. Then the final object i'_n of $\mathsf{Thick}_n(X'/Y')$ can be viewed as an object of $\mathsf{Thick}_n(X/Y)$ that maps uniquely to the final object i_n of $\mathsf{Thick}_n(X/Y)$. Thus we find a morphism $i'_n \to i_n$ in Thick_n, and T_n defines a functor $\mathsf{Lognet} \to \mathsf{Thick}_n$. If $i\colon S \to T$ is an object of Thick_n and $f\colon X \to T$ an object of Lognet, a morphism $i \to f$ in Lognet makes i into an object of $\mathsf{Thick}_n(X/Y)$, which then maps uniquely to the final object $T_n(f)$. Thus T_n is right adjoint to the inclusion functor $\mathsf{Thick}_n \to \mathsf{Lognet}$. □

Let $i\colon X \to Y$ be a log unramified morphism of fine log schemes, and let $i_1\colon X \to Y_1$ be its first strict infinitesimal neighborhood. The ideal of X in Y_1 is a square-zero ideal, hence an O_X-module, called the *conormal sheaf* of X in Y and denoted by $C_{X/Y}$. It depends functorially on i: a morphism from $i'\colon X' \to Y'$ to $i\colon X \to Y$ given by $f\colon X' \to X$ and $g\colon Y' \to Y$ induces a morphism of thickenings $i'_1 \to i_1$ and hence a morphism $f^*(C_{X/Y}) \to C_{X'/Y'}$.

Remark 3.4.5. Let $f\colon X \to S$ be a morphism of fine log schemes for which $\underline{f}$ is locally of finite presentation, and let $i_1\colon X \to P^1_{X/S}$ be the first strict infinitesimal neighborhood of the diagonal morphism $\Delta_{X/S}\colon X \to X \times_S X$. Thus the sheaf of ideals defining i_1 is the conormal sheaf C_Δ of $\Delta_{X/S}$. According to Theorem 2.2.2, the two maps $p_i\colon P^1_{X/S} \to X$ differ by a homomorphism

$$\Omega^1_{X/S} \to C_\Delta.$$

In fact, this homomorphism is an isomorphism. This may be checked geometrically as follows. Given any quasi-coherent $\mathcal{E}$, Theorem 2.2.2 identifies the set of deformations of id_X to the trivial extension $X \oplus \mathcal{E}$ with the set of homomorphisms $\Omega^1_{X/S} \to \mathcal{E}$. On the other hand, if $\tilde{g}$ is such a deformation, and if r is the retraction $X \oplus \mathcal{E} \to X$, then (r, g) defines a morphism $X \oplus \mathcal{E} \to P^1_{X/S}$ and hence a morphism $C_\Delta \to \mathcal{E}$. In particular, taking $\mathcal{E}$ to be $\Omega^1_{X/S}$, one finds a homomorphism $C_\Delta \to \Omega^1_{X/S}$, which is easily seen to be inverse to the map defined above.

3.5 Smoothness and regularity

Our main goal in this section is to explore the relationship between regularity and smoothness in the context of log geometry, echoing the main results in the classical setting. In particular, we prove the openness of the regular locus for log schemes satisfying a suitable excellence condition. These ideas will be further developed in the section on logarithmic flatness.

Theorem 3.5.1. *Let X/S be a fine saturated log scheme over the spectrum S of a perfect field k with trivial log structure, and let x be a point of X. Assume that the underlying morphism $\underline{X}/\underline{S}$ is of finite type. Then X/S is smooth in a neighborhood of x if and only if X is regular at x.*

Proof This statement can be verified étale locally on X, so we may assume that we are working in the Zariski topology.

Suppose that X is regular at x. Since X is saturated, statement (2) of Theorem 1.2.7 tells us that, after passing to an étale neighborhood of x, we may find a chart $\beta\colon Q \to \mathcal{M}_X$ that induces an isomorphism $Q \to \overline{\mathcal{M}}_{X,x}$. Then β induces a strict morphism $b\colon X \to \mathsf{A}_Q := \operatorname{Spec}(Q \to k[Q])$. By Corollary III.1.11.3, the homomorphism $k[Q] \to \mathcal{O}_{X,x}$ is flat, and hence $\underline{b}$ is flat in a neighborhood of x. Let s be the vertex of A_Q, that is, the k-valued point defined by Q^+. Since X is regular at x, the fiber $\underline{X}_s$ of b over s is regular at x. Since k is perfect and $\underline{X}_s$ is a regular k-scheme of finite type, the morphism $\underline{X}_s/k$ is smooth at x [29, 17.15.1]. Since $\underline{b}$ is flat at x and since its fiber over s is smooth at x, it follows from [32, 2.1] that $\underline{b}$ is also smooth in a neighborhood of x.

Conversely, assume that X/S is smooth. By Theorem 3.3.1, there exist a fine monoid Q such that the order of the torsion part of Q^{gp} is invertible in $\mathcal{O}_X$ and a chart $b\colon X \to \mathsf{A}_Q$ such that $\underline{b}$ is étale. As we saw in Proposition I.3.3.1, the closed subscheme $\underline{\mathsf{A}}_{Q,Q^+}$ of $\underline{\mathsf{A}}_Q$ defined by Q^+ is isomorphic to $\operatorname{Spec} k[Q^*]$. Since Q^* is a finitely generated group whose torsion subgroup has order invertible in k, the scheme $\underline{\mathsf{A}}_{Q^*}$ is smooth over k, and hence regular. Since $\underline{b}$ is étale, $\dim(\mathcal{O}_{X,x}) = \dim(\underline{\mathsf{A}}_Q)$ and $\dim(\mathcal{O}_{X_s,x}) = \dim(\underline{\mathsf{A}}_{Q,Q^+})$, where $X_s := \operatorname{Spec}(\mathcal{O}_{X,x}/Q^+\mathcal{O}_{X,x})$. Then

$$\dim(\mathcal{O}_{X,x}) = \dim(\underline{\mathsf{A}}_Q) = \operatorname{rk}(Q) = \operatorname{rk}(\overline{Q}) + \operatorname{rk}(Q^*) = \dim(Q) + \dim(\mathcal{O}_{X_s,x}),$$

so X is regular at x. □

Theorem 3.5.2. *Let X be a fine saturated and locally noetherian log scheme.*

1. *The set* Reg(X) *of points of X at which X is regular is stable under generization.*
2. *If the scheme underlying X is quasi-excellent [4, Tag 07QS], then* Reg(X) *is open.*

The proof of this theorem will require some preliminary results, of independent interest. The main difficulty is (1). This result was first formulated by Kato [49, 7.1], but his proof is incomplete.[2] We follow the methods of Gabber and Ramero in [22, 12.5.47] and Gabber in [42, X,1.1.1(b)].

[2] His argument only shows regularity at generizations of the generic points of the log strata.

Proposition 3.5.3. *Let $f: X \to Y$ be a smooth morphism of fine saturated log schemes whose underlying schemes are locally noetherian. Then X is regular at a geometric point $\overline{x}$ if and only if Y is regular at $f(\overline{x})$.*

Proof Replacing X and Y by étale neighborhoods of $\overline{x}$ and $f(\overline{x})$, we may assume that X and Y have charts, and we may therefore work in the Zariski topology and replace geometric points by scheme-theoretic points. We may also assume that there exists a chart (a, θ, b) for f of the form specified in Theorem 3.3.1, and we use the notation established there. Since $X \to Y_\theta$ is étale and strict, the log scheme X is regular at x if and only if Y_θ is regular at $b_\theta(x)$. Thus we may assume that $X = Y_\theta$. Since our statement is local, we may and shall assume that X and Y are affine, say $Y = \mathrm{Spec}(\alpha: P \to A)$ and $X = \mathrm{Spec}(\beta: Q \to B)$. We may also assume that A is a local ring. Since now $X = Y_\theta$, we have $B = A \otimes_{\mathbf{Z}[P]} \mathbf{Z}[Q]$. Let I_β be the ideal of B generated by Q^+, let I_α be the ideal of A generated by P^+, and let $X^+ \subseteq X$ and $Y^+ \subseteq Y$ be the corresponding closed log subschemes.

Lemma 3.5.4. *In the situation described in the previous paragraph, the following statements hold.*

1. *The restriction $X^* \to Y^*$ of f to the log trivial locus is smooth and of relative dimension equal to* $\mathrm{rank}(Q^{\mathrm{gp}}/P^{\mathrm{gp}})$.
2. *The morphism $\underline{X}^+ \to \underline{Y}^+$ induced by f is smooth at x and of relative dimension equal to* $\mathrm{rank}(Q^*/P^*)$.

Proof Let q (resp. p) be an element of the interior of Q (resp. P), so that the localization of Q by q is Q^{gp} (resp. the localization of P by p is P^{gp}). Then, if $b := \beta(b)$ and $a := \alpha(q)$,

$$B_b \cong A \otimes_{\mathbf{Z}[P]} \mathbf{Z}[Q^{\mathrm{gp}}] \cong A_a \otimes_{\mathbf{Z}[P^{\mathrm{gp}}]} \mathbf{Z}[Q^{\mathrm{gp}}].$$

Since $P^{\mathrm{gp}} \to Q^{\mathrm{gp}}$ is injective, and since the order of the torsion subgroup of its cokernel is invertible in A, the corresponding morphism of group schemes $\mathsf{A}_{Q^{\mathrm{gp}}} \to \mathsf{A}_{P^{\mathrm{gp}}}$ is smooth and of relative dimension $\mathrm{rank}(Q^{\mathrm{gp}}/P^{\mathrm{gp}})$, when pulled back to $\mathrm{Spec}(A_a)$. This proves (1).

For (2), we have

$$\begin{aligned} B/I_\beta &= B \otimes_{\mathbf{Z}[Q]} \mathbf{Z}[Q, Q^+] \\ &\cong A \otimes_{\mathbf{Z}[P]} \otimes \mathbf{Z}[Q] \otimes_{\mathbf{Z}[Q]} \mathbf{Z}[Q, Q^+] \\ &\cong A \otimes_{\mathbf{Z}[P]} \mathbf{Z}[P, P^+] \otimes_{\mathbf{Z}[P,P^+]} \mathbf{Z}[Q, Q^+] \\ &\cong A/I_\alpha \otimes_{\mathbf{Z}[P,P^+]} \mathbf{Z}[Q, Q^+] \\ &\cong A/I_\alpha \otimes_{\mathbf{Z}[P^*]} \mathbf{Z}[Q^*]. \end{aligned}$$

The homomorphism $P^* \to Q^*$ is injective; let us note also that the order of the torsion subgroup of its cokernel is invertible in A. Indeed, if $p > 0$ is the residue characteristic of A and $u \in Q^*$ with $pu \in P^*$ then, since Q^{gp}/P^{gp} is p-torsion free, necessarily $u \in P^{gp}$. Since P is saturated, it follows that $u \in P$ and, since θ is local, that $u \in P^*$. Hence u maps to zero in Q^*/P^*. The corresponding morphism of group schemes $\mathsf{A}_{Q^*} \to \mathsf{A}_{P^*}$ therefore becomes smooth when pulled back to $\operatorname{Spec}(A/I_\alpha)$; its relative dimension is the rank of Q^*/P^*. Thus, the morphism $\operatorname{Spec}(B/I_\beta) \to \operatorname{Spec}(A/I_\alpha)$ is smooth of the same relative dimension. □

Now suppose that α is regular at y. Then the underlying local ring A/I_α is regular and thus (2) of Lemma 3.5.4 implies that the ring B/I_β is regular at x. Furthermore, α is solid at y, so it follows from Proposition III.1.10.15 that β is solid at x. Thus β is also regular at x, by the equivalence of the conditions in Theorem III.1.11.1.

The proof of the converse, made possible by help from O. Gabber, is more difficult. It will use the following commutative algebra result from [27, 5.5.8 and 5.6.1].

Theorem 3.5.5. *Let $f\colon X' \to Y'$ be a dominant and finite type morphism of integral noetherian schemes, with generic points ξ and η, respectively, and let x be a point in X with image y in Y'. Then*

$$\dim_y(Y') + \text{tr.deg.}k(\xi)/k(\eta) \geq \dim_x(X') + \text{tr.deg.}k(x)/k(y),$$

with equality if Y' is universally catenary. □

To complete the proof of Proposition 3.5.3, suppose that X is regular at x. Then $\mathcal{O}_{X,x}$ is an integral domain, and we may assume that X is also integral. Let $\overline{A}$ be the image of A in B, which is also integral, and let $\overline{Y}$ be $\operatorname{Spec}\overline{A}$. Applying Theorem 3.5.5 with $Y' := \underline{\overline{Y}}$ and $X' := \underline{X}$, we see that

$$\dim_y Y \geq \dim_y \overline{Y} + \text{tr.deg.}k(\xi)/k(\eta) \geq \dim_x X + \text{tr.deg.}k(x)/k(y).$$

By (1) of Lemma 3.5.4, we see that $\text{tr.deg.}k(\xi)/k(\eta) = \text{rk}(Q^{gp}/P^{g})$, so this inequality becomes:

$$\dim_y Y \geq \dim_x X + \text{tr.deg.}k(x)/k(y) - \text{rk}(Q^{gp}) + \text{rk}(P^{gp}).$$

Let X^+ be the closed subscheme of X defined by the ideal generated by Q^+, with similar notation for Y. Since X is regular at x, $\dim_x X = \dim_x X^+ + \text{rk}(\overline{Q}^{gp})$, so we get

$$\dim_y Y \geq \dim_x X^+ + \text{tr.deg.}k(x)/k(y) - \text{rk}(Q^*) + \text{rk}(P^{gp}). \tag{3.5.1}$$

We also know that $\underline{X}^+$ is regular at x, and hence by (2) of Lemma 3.5.4, $\underline{Y}^+$

is regular at y, and in particular is universally catenary. Then Theorem 3.5.5 applied to the morphism $X^+ \to Y^+$ gives the equality:

$$\dim_y Y^+ + \text{tr.deg}.k(\xi^+)/k(\eta^+) = \dim_x X^+ + \text{tr.deg}.k(x)/k(y).$$

Statement (2) of Lemma 3.5.4 also implies that

$$\text{tr.deg}.k(\xi^+)/k(\eta^+) = \text{rk}(Q^*/P^*).$$

Thus

$$\dim_x X^+ = \dim_y Y^+ + \text{rk}(Q^*) - \text{rk}(P^*) - \text{tr.deg}.k(x)/k(y).$$

Substituting this equation into the inequality (3.5.1) yields

$$\dim_y Y \geq \dim_y Y^+ + \text{rk}(P^{\text{gp}}) - \text{rk}(P^*) = \dim_y Y^+ + \text{rk}(\overline{P}^{\text{gp}}),$$

as required. □

The next result will allow us to reduce the rank of a log structure by a carefully constructed modification.

Proposition 3.5.6. *Let X be a fine locally noetherian log scheme and let x, ξ be two points of X, where x is a specialization of ξ. Then there exist an étale morphism of fine log schemes $f: X' \to X$ and points x', ξ' of X' lying over x and ξ respectively, such that x' is a specialization of ξ' such that $\dim(\mathcal{M}_{X',x'}) \leq 1$, and such that $X'^* \to X^*$ is an isomorphism.*

Proof Without loss of generality we may assume that $X = \text{Spec}(Q \to A)$, and that x corresponds to a prime $\mathfrak{p}$ of A and ξ to a prime $\mathfrak{q}$ of A contained in $\mathfrak{p}$. We may also assume that $Q \to A_{\mathfrak{p}}$ is local and that A is noetherian. Then, by [34, Exercise II,4.11], there exists a discrete valuation ring V with fraction field K endowed with local homomorphisms $A_{\mathfrak{p}} \to V$ and $A_{\mathfrak{q}} \to K$. Let us identify V/V^* with $\mathbf{N}$, so that the homomorphism $Q \to A_{\mathfrak{p}} \to V \to V/V^*$ defines a local homomorphism $h: Q \to \mathbf{N}$. Let $Q' := Q^{\text{gp}} \times_{\mathbf{Z}} \mathbf{N}$, which is a fine monoid, by Corollary I.2.1.21. The homomorphism $\theta: Q \to Q'$ is local, because its composition with $Q' \to \mathbf{N}$ is the local homomorphism h. Moreover, $Q'^{\text{gp}} = Q^{\text{gp}}$ and $Q' \to \mathbf{N}$ is exact. By Proposition I.4.2.1, the induced homomorphism $\overline{Q'} \to \mathbf{N}$ is injective and hence $\dim(Q') \leq 1$. We have a homomorphism of log rings $(Q \to A) \longrightarrow (Q' \to V)$. Thus if T is the log dash $\text{Spec}(Q' \to V)$, there is a commutative diagram of log schemes

$$\begin{array}{ccc} T & \longrightarrow & \mathsf{A}_{Q'} \\ \downarrow & & \downarrow \mathsf{A}_{\theta} \\ X & \longrightarrow & \mathsf{A}_{Q}. \end{array}$$

Since θ^{gp} is an isomorphism, the morphism of log schemes A_θ is étale and induces an isomorphism on the log trivial loci. Then the same is true of the induced morphism

$$X' := X \times_{\mathsf{A}_Q} \mathsf{A}_{Q'} \to X.$$

The commutative diagram above defines a lifting $T \to X'$ of $T \to X$. Let x' (resp. ξ') be the image in X' of the closed (resp. generic) point of T. Then ξ' is a specialization of x'. Furthermore, since $X \to \mathsf{A}_Q$ is strict, so is $X' \to \mathsf{A}_{Q'}$, and it follows that the dimension of $\mathcal{M}_{X',x'}$ is at most one. □

Proof of Theorem 3.5.2 Let x be a point of $\mathrm{Reg}(X)$ and let ξ be a generization of x. Let A be the local ring of X at x, and assume we are given a fine and saturated chart $\beta\colon Q \to A$ of X, which we may assume is local. Then β is a regular local log ring. The point ξ corresponds to a prime ideal $\mathfrak{p}$ of A; let $\mathfrak{q} := \beta^{-1}(\mathfrak{p})$ and let $F := Q \setminus \mathfrak{q}$. We claim that the local log ring $\beta_F\colon Q_F \to A_\mathfrak{p}$ is again regular. Since $\beta\colon Q \to A$ is (very) solid, Proposition III.1.10.9 implies that the same is true of β_F. Thus by the equivalence of the conditions in Theorem III.1.11.1, it will suffice to prove that $A_\mathfrak{p}/\mathfrak{q}A_\mathfrak{p}$ is a regular ring. We know from Proposition III.1.11.10 that $\beta_\mathfrak{q}\colon F \to A/\mathfrak{q}A$ is a regular local log ring, and our target Theorem 3.5.2 asserts that that the local log ring $(F \to A_\mathfrak{p}/\mathfrak{q}A_\mathfrak{p})$ is again regular. However, the log structure of $F \to A_\mathfrak{p}$ is trivial, since $\beta(F) \subseteq A^*_\mathfrak{p}$, so log regularity implies standard regularity in this case. Thus we are reduced to proving (1) of Theorem 3.5.2 when $\mathfrak{p}$ lies in the log trivial locus X^*.

Let us write X for $\mathrm{Spec}(\beta)$. We must show that X is regular at ξ if $\xi \in X^*$. Let us apply Proposition 3.5.6 to find a morphism $f\colon X' \to X$ as described there. Then X' is regular and saturated at x' and $\dim(\mathcal{M}_{X',x'}) \le 1$, so by Theorem III.1.11.6, the underlying scheme X' is regular at x'. Since X' is quasi-excellent, it is also regular at ξ' [26, 0,17.3.2]. But since the log structure of X is trivial at ξ, X' is isomorphic to X in a neighborhood of ξ', and hence X is regular at ξ. This completes the proof of (1).

Since X is locally noetherian, a subset R of X is open if and only if it is closed under generization and every point ξ of R has an open neighborhood U such that $\{\xi\}^- \cap U$ is contained in R [27, 6.11.6.1]. Thus the following lemma will complete the proof of Theorem 3.5.2. □

Lemma 3.5.7. *With the hypotheses of Theorem 3.5.2, suppose that ξ is a point of X at which X is regular. Then there is a nonempty open subset of $\{\xi\}^-$ at every point of which X is regular.*

Proof Since X is regular at ξ it is very solid at ξ. Hence Proposition III.1.10.11 implies that X is very solid in some neighborhood of ξ, which we may assume to be all of X. Without loss of generality we may also assume that $X = \mathrm{Spec}(\beta\colon Q \to A)$, where Q is a fine and saturated monoid and where A is a quasi-excellent ring. Let X^+ be the closed subscheme of X defined by I_β. Since X is quasi-excellent, the same is true of X^+, and in particular $\mathrm{Reg}(X^+)$ is open. Since X^+ is regular at ξ, there is an open neighborhood of ξ in X^+ that is regular and we may assume that this neighborhood is all of X^+. Note that, for every point $x \in X^+$, $I_\beta \mathcal{O}_{X,x} = \mathcal{M}^+_{X,x}\mathcal{O}_{X,x}$. Thus $\mathcal{O}_{X,x}/\mathcal{M}^+_{X,x}\mathcal{O}_{X,x}$ is regular and, by the equivalence of the conditions in Theorem III.1.11.1, it follows that X is regular at x. □

4 Logarithmic flatness

4.1 Definition and basic properties

As we have seen, a morphism of log schemes is smooth (resp. étale) if and only if it locally admits a model resembling a morphism of monoid schemes $\underline{\mathsf{A}}_\mathsf{Q} \to \underline{\mathsf{A}}_\mathsf{P}$ whose corresponding morphism of tori $\mathsf{A}^*_\mathsf{Q} \to \mathsf{A}^*_\mathsf{P}$ is smooth (resp. étale). Following K. Kato's original idea, as explained and developed by W. Bauer [5], we use a similar method to define the notion of flatness for morphisms of log schemes. Note that if $\theta\colon P \to Q$ is an injective homomorphism of integral monoids, then $P^{\mathrm{gp}} \to Q^{\mathrm{gp}}$ is also injective, and so Q^{gp} is a free P^{gp}-set and the corresponding homomorphism $\mathbf{Z}[P^{\mathrm{gp}}] \to \mathbf{Z}[Q^{\mathrm{gp}}]$ is flat. Thus the corresponding morphism of log schemes $\mathsf{A}_\mathsf{Q} \to \mathsf{A}_\mathsf{P}$ will be an example of a (log) flat morphism.

In this section we work in the category of integral or fine log schemes, and in particular all fibered products are taken in the category of integral log schemes. We omit any discussion of idealized log schemes.

Definition 4.1.1. *A morphism of fine log schemes $f\colon X \to Y$ is (log) flat if, fppf locally on X and Y, there exists a chart (a, θ, b) for f such that $\theta\colon P \to Q$ is an injective homomorphism of fine monoids and such that the morphism of schemes $\underline{b}_\theta\colon \underline{X} \to \underline{Y}_\theta = \underline{Y} \times_{\underline{\mathsf{A}}_\mathsf{P}} \underline{\mathsf{A}}_\mathsf{Q}$ is flat.*

Before investigating this definition further, it will be useful to establish some basic properties of the class of flat morphisms. Let us say that a chart (a, θ, b) for a morphism f is flat if θ is injective and the map $\underline{b}_\theta\colon \underline{X} \to \underline{Y}_\theta$ is flat, as in Definition 4.1.1. Thus a morphism is flat if and only if, fppf locally on X and Y, it admits a flat chart.

Proposition 4.1.2. *In the category of fine log schemes, the following statements hold.*

1. *Smooth morphisms are flat.*
2. *A strict morphism of fine log schemes $f: X \to Y$ is flat if and only if the underlying morphism of schemes $\underline{f}: \underline{X} \to \underline{Y}$ is flat.*
3. *If $f: X \to Y$ is flat and $Y' \to Y$ is a morphism, then the base-changed morphism $f': X \times_Y Y' \to Y'$ is also flat. The same result holds in the category of fine saturated log schemes.*
4. *If $f: X \to Y$ and $g: Y \to Z$ are flat, then the composed morphism $g \circ f$ is also flat.*

Proof The fact that smooth morphisms are flat follows from Theorem 3.3.1, which asserts that a smooth morphism locally admits a chart subordinate to an injective homomorphism θ such that $\underline{b}_\theta$ is smooth (and hence flat).

Suppose that f is strict and that $\underline{f}$ is flat. Locally on Y, choose a chart $a: Y \to \mathsf{A}_P$ for Y. Then $b := a \circ f$ is a chart for X and (a, id_P, b) is a chart for f. Then $Y_\theta = Y$ and $\underline{b}_\theta = \underline{f}$, which is flat. It follows that f is flat. Conversely, suppose that f is flat and strict. To prove that $\underline{f}$ is flat, we may work fppf locally on X and Y, and so we may assume that f admits a flat chart (a, θ, b). Choose a point x of X mapping to a point y of Y. Then the homomorphism $\alpha := a^\flat: P \to \mathcal{M}_{Y,y}$ factors through its localization P_F, where $F := \alpha^{-1}(\mathcal{M}^*_{Y,y})$, and similarly $\beta := b^\flat: Q \to \mathcal{M}_{X,x}$ factors through its localization by $G := \beta^{-1}(\mathcal{M}^*_{X,x})$. The homomorphism $\theta': P_F \to Q_G$ induced by θ is still injective, and we obtain a chart (a', θ', b') for f. The morphism $\mathsf{A}_{Q_G} \to \mathsf{A}_Q$ is an open immersion, and hence so is the morphism $j: Y_{\theta'} \to Y_\theta$. Since $b_\theta = j \circ b'_{\theta'}$ and $\underline{b}_\theta$ is flat, it follows that $\underline{b}'_{\theta'}$ is also flat. Thus, changing notation, we may assume that α and β are local and hence induce isomorphisms $\overline{P} \to \overline{\mathcal{M}}_{Y,y}$ and $\overline{Q} \to \overline{\mathcal{M}}_{X,x}$, respectively. Since f is strict, the map $\overline{\mathcal{M}}_{Y,y} \to \overline{\mathcal{M}}_{X,x}$ is an isomorphism, and hence so is $\overline{\theta}$. Then $\overline{\theta}$ is integral and, by (2) of Proposition I.4.6.3, the homomorphism $\theta: P \to Q$ is also integral. Since θ is injective and integral, Proposition I.4.6.7 tells us that $\underline{\mathsf{A}}_\theta$ is flat. Since $\underline{f} = \underline{\mathsf{A}}_\theta \circ \underline{b}_\theta$, it is also flat.

To prove statement (3), let $f: X \to Y$ be a flat morphism of fine log schemes, let $g: Y' \to Y$ be an arbitrary morphism of fine log schemes, and let $f': X' \to Y'$ be the base change of f along g. The assertion that f' is flat may be checked fppf locally on X', and hence also fppf locally on X and Y'. Thus we may assume that f admits a chart (a, θ, b) as in Definition 4.1.1. Thanks to Proposition II.2.4.2, we may also assume that the given chart a for Y fits into a chart (a, ϕ, a') for the morphism g, where $\phi: P \to P'$ is a homomorphism of fine monoids. Let $Q' := Q \oplus_P P'$, computed in the category of integral monoids. Then the homomorphism $\theta': P' \to Q' := Q \oplus_P P'$ is again injective, the morphism

$$b': X' = Y' \times_Y X \to \mathsf{A}_{P'} \times_{\mathsf{A}_P} \mathsf{A}_Q = \mathsf{A}_{Q'}$$

is a chart for X', and (a', θ', b') is a chart for $f' \colon X' \to Y'$. The morphism

$$b'_{\theta'} \colon X' = Y' \times_Y X \to Y'_{\theta'} := Y' \times_{\mathsf{A}_{P'}} \mathsf{A}_{Q'} \cong Y' \times_{\mathsf{A}_P} \mathsf{A}_Q$$

can be identified with $\mathrm{id}_{Y'} \times_Y b_\theta$ and, since $\underline{b}_\theta$ is flat, so is the morphism $\underline{b}_{\theta'}$. Thus the morphism $f' \colon X' \to Y'$ is again flat. This proves statement (3) in the category of fine log schemes.

To prove (3) in the category of fine saturated log schemes, recall that if X, Y, and Y' are saturated, then X' may not be, and the fiber product in the category of saturated log schemes is formed by replacing Q' by its saturation Q'' and X' by $X'' := X' \times_{\mathsf{A}_{Q'}} \mathsf{A}_{Q''}$ (formed in the category of integral log schemes). Since the homomorphism $Q' \to Q''$ is injective, the morphism $\mathsf{A}_{Q''} \to \mathsf{A}_{Q'}$ is flat, and hence so is $X'' \to X'$. Then it will follow from statement (4) that the composition $X'' \to X' \to Y'$ is also flat.

The proof of statement (4) relies on the following lemma, due to O. Gabber.

Lemma 4.1.3. *Let $f \colon X \to Y$ be a flat morphism of fine log schemes and let $a \colon Y \to \mathsf{A}_P$ be a fine chart for Y. Then, fppf locally on X and Y, there exists a flat chart (a, θ, b) for f.*

Proof Let x be a point of X, let $y := f(x)$, and let $\alpha := a^\flat \colon P \to \mathcal{M}_Y$. Then $\alpha_y \colon P \to \mathcal{M}_{Y,y}$ factors through its localization $\alpha^{loc} \colon P^{loc} \to \mathcal{M}_{Y,y}$, and, after replacing X and Y by suitable open subsets, we may further assume that α^{loc} factors through $\Gamma(Y, \mathcal{M}_Y)$. Then a factors through the open immersion $\mathsf{A}_{P^{loc}} \to \mathsf{A}_P$. Since this open immersion is flat, it suffices to prove the lemma with α^{loc} in place of α. Thus we may also assume that a is exact at y.

Since f is flat and the assertion is fppf local, we may assume that there exists a flat chart (a', θ', b') for f, corresponding to monoid homomorphisms $\alpha' \colon P' \to \mathcal{M}_Y, \beta' \colon Q' \to \mathcal{M}_X$, and $\theta' \colon P' \to Q'$. After further shrinking of X and Y if necessary, we may also assume that a' is exact at y and that b' is exact at x. We proceed with a series of reductions.

Choose a surjection ϕ from a finitely generated free abelian group L to Q'^{gp} and let $Q'' := \phi^{-1}(Q')$ and $P'' := Q'' \times_{Q'} P'$. Then $\mathrm{Ker}(\phi)$ is contained in Q''^* and $Q' = Q''/\mathrm{Ker}(\phi)$. Since $\mathrm{Ker}(\phi)$ is also contained in P'', it follows that the map $Q'' \oplus_{P''} P' \to Q'$ is an isomorphism. The homomorphisms $P'' \to P'$ and $Q'' \to Q'$ are exact and surjective, and hence the induced morphisms $a'' \colon Y \to \mathsf{A}_{P''}$, $b'' \colon X \to \mathsf{A}_{Q''}$, and $\theta'' \colon P'' \to Q''$ form a chart for f, with θ'' injective. Since $\mathsf{A}_{Q'} \cong \mathsf{A}_{Q''} \times_{\mathsf{A}_{P''}} \mathsf{A}_{P'}$, the map $Y_{\theta'} \to Y_{\theta''}$ is an isomorphism and, since $b_{\theta'}$ is flat, $b''_{\theta''}$ is also flat. Thus (a'', θ'', b'') is a flat chart for f. Moreover $Q''^{\mathrm{gp}} \cong L$ and in particular is torsion free. Replacing (a', θ', b') by (a'', θ'', b''), we may therefore assume that P'^{gp} and Q'^{gp} are torsion free.

Since a is exact and y and P'^{gp} is torsion free, Proposition II.2.3.9 implies that there exist homomorphisms $\kappa\colon P' \to P$ and $\gamma\colon P' \to \mathcal{M}^*_{Y,y}$ such that $\alpha' = \alpha \circ \kappa + \gamma$. After shrinking X and Y, we may assume that γ defines a morphism $g\colon Y \to \mathsf{A}^*_{P'}$, and after a finite flat covering, we may assume that g lifts to a morphism $h\colon Y \to \mathsf{A}^*_{Q'}$. This h corresponds to a homomorphism $\delta\colon Q' \to \mathcal{O}^*_Y$ such that $\delta \circ \theta' = \gamma$. Let $\alpha'' := \alpha' - \gamma$ and $\beta'' := \beta' - \delta$, defining morphisms $a''\colon Y \to \mathsf{A}_{P'}$ and $b''\colon X \to \mathsf{A}_{Q'}$ respectively, and let $\theta'' = \theta'$. Then (a'', θ'', b'') is again a chart for f. Furthermore, there is a commutative diagram

$$\begin{array}{ccccc} X & \xrightarrow{b'_{\theta'}} & Y_{\theta'} & \xrightarrow{=} & Y \times_{a'} \mathsf{A}_{Q'} \\ & \searrow^{b''_{\theta''}} & \downarrow k & & \downarrow k \\ & & Y_{\theta''} & \xrightarrow{=} & Y \times_{a''} \mathsf{A}_{Q'}, \end{array}$$

where $k(y, z) = (y, h(y)z)$. Then $b''_{\theta''}$ is flat, because k is an isomorphism and $b'_{\theta'}$ is flat. Thus (a'', θ'', b'') is also a flat chart for f. Changing notation again, we may assume that $\alpha' = \alpha \circ \kappa$.

Since α' and α are exact at y, the map $\overline{\kappa}$ is an isomorphism, and it follows that $P \cong P' \oplus_{P'^*} P^*$. Let $\theta\colon P \to Q := Q' \oplus_{P'} P$ be the pushout of θ'. Then $Q \cong Q' \oplus_{P'^*} P$ and it follows that the induced homomorphism $\beta\colon Q \to \mathcal{M}_X$ is also a chart. The diagram

$$\begin{array}{ccc} X & \xrightarrow{b'_{\theta'}} & Y \times_{\mathsf{A}_{P'}} \mathsf{A}_{Q'} \\ & \searrow^{b_\theta} & \downarrow \cong \\ & & Y \times_{\mathsf{A}_P} \mathsf{A}_Q \end{array}$$

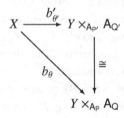

shows that (a, θ, b) is a flat chart for f, as desired. □

Now, to prove statement (4) of the proposition, suppose that $f\colon X \to Y$ and $g\colon Y \to Z$, are flat morphisms Then, fppf locally on Y and Z, there exists a flat and fine chart (c, ϕ, a) for g, where c is subordinate to M and a to P. The lemma implies that, after a further fppf localization on X and Y, there exists a flat chart (a, θ, b) for f. Then $\psi := \theta \circ \phi$ is injective and (b, ψ, c) is a chart for

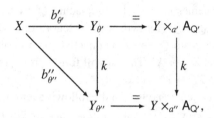

$g \circ f$, so it will suffice to check that $\underline{b}_\psi$ is flat. Consider the diagram

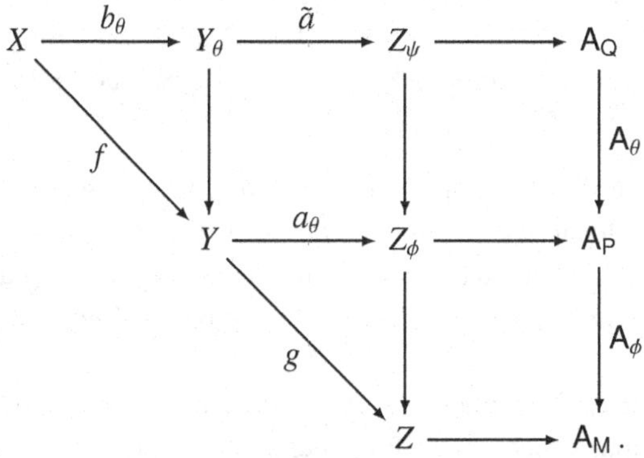

All the squares are cartesian and, by hypothesis the morphisms $\underline{b}_\theta$ and $\underline{a}_\theta$ are flat. It follows by base change that $\underline{\tilde{a}}$ is flat and hence that the composition $\underline{b}_\psi = \underline{\tilde{a}} \circ \underline{b}_\theta$. is flat. This completes the proof of the proposition. □

This definition of log flatness is not well adapted to proving that a morphism of log schemes is *not* flat: a flat morphism can (trivially) admit charts such that $\underline{b}_\theta$ is not flat. Theorem 4.1.4 and Corollary 4.1.6 will remedy this weakness by giving more intrinsic characterizations. It is inspired by the discussion of flatness due to Martin Olsson using the classifying stack of log structures [62], as recast in [51, Lemma 4.3.1]. It makes crucial use of the action $m_{Q/P}\colon \mathsf{A}_Q \times \mathsf{A}^*_{Q/P} \to \mathsf{A}_Q$ of the group scheme $\mathsf{A}^*_{Q/P}$ on the log scheme A_Q and the corresponding morphism $h_{Q/P}\colon \mathsf{A}_Q \times \mathsf{A}^*_{Q/P} \to \mathsf{A}_Q \times_{\mathsf{A}_P} \mathsf{A}_Q$, described in Example III.2.3.6.

Theorem 4.1.4. *Let $f\colon X \to Y$ be a morphism of fine log schemes. Then, in the category of fine (resp. fine saturated) log schemes, the following conditions are equivalent.*

1. *The morphism f is flat.*
2. *Fppf locally on X and Y, the following holds. For every diagram of the form*

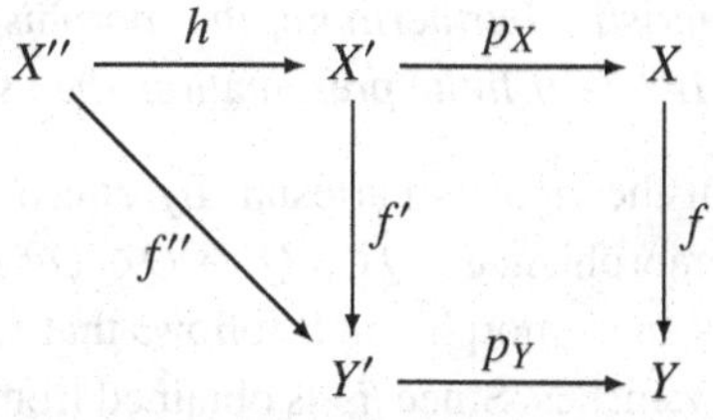

in which the square is cartesian, h is étale, and f'' is strict, the underlying morphism $\underline{f}''$ is flat.

3. *The same as (2), for diagrams with the additional assumptions that p_Y is flat and that the relative characteristic $\mathcal{M}_{X''/X'}$ of the morphism h vanishes.*
4. *Fppf locally on X and Y, for every fine chart (a, θ, b) for f in which θ is injective, the morphism $\underline{m}_\theta \colon \underline{X} \times \mathsf{A}^*_{\mathsf{Q/P}} \to \underline{Y}_\theta$ defined by $\underline{b}_\theta$ and the action $m_{Q/P}$ of $\mathsf{A}^*_{\mathsf{Q/P}}$ on $\underline{\mathsf{A}}_{\mathsf{Q}}$ is flat.*

Proof Suppose that f is flat and that we are given a diagram as in (2) (after a possible fppf localization). Statement (3) of Proposition 4.1.2 implies that the morphism f' is flat, statement (1) implies that h is flat, and statement (4) implies that f'' is flat. Since f'' is strict, statement (2) implies that $\underline{f}''$ is flat, as required.

It is trivial that (2) implies (3). The implication of (4) by (3) will use the following lemma, which shows how a chart for f produces a diagram of the type described in (3).

Lemma 4.1.5. *Let (a, θ, b) be a chart for a morphism of integral log schemes $f \colon X \to Y$, where $\theta \colon P \to Q$ is an injective homomorphism of fine monoids. Let*

$$m_Y \colon Y_\theta \times \mathsf{A}^*_{\mathsf{Q/P}} \to Y_\theta$$

be the map induced from $m_{Q/P}$ by base change along a, and let

$$h_X \colon X \times \mathsf{A}^*_{\mathsf{Q/P}} \to X \times_{\mathsf{A}_{\mathsf{P}}} \mathsf{A}_{\mathsf{Q}}$$

be the map induced from $h_{Q/P}$ by base change along b. Consider the commutative diagram

$$\begin{array}{ccccc} X \times \mathsf{A}^*_{\mathsf{Q/P}} & \xrightarrow{h_X} & X \times_{\mathsf{A}_{\mathsf{P}}} \mathsf{A}_{\mathsf{Q}} & \xrightarrow{pr_X} & X \\ {\scriptstyle b_\theta \times \mathrm{id}} \downarrow & \searrow^{m_\theta} & \downarrow {\scriptstyle f \times \mathrm{id}} & & \downarrow {\scriptstyle f} \\ Y_\theta \times \mathsf{A}^*_{\mathsf{Q/P}} & \xrightarrow{m_Y} & Y_\theta & \xrightarrow{pr_Y} & Y. \end{array} \qquad (4.1.1)$$

Then the right square is cartesian, and the morphism h_X is étale with vanishing relative characteristic. Furthermore, the morphism m_θ is strict and the morphism pr_Y is flat. If $\underline{f}$ is of finite presentation, then so are $\underline{m}_\theta$ and $b_\theta \otimes \mathrm{id}$.

Proof The square on the right is cartesian by construction. The map $h_{Q/P}$ comes from the homomorphism $\phi \colon Q \oplus_P Q \to Q \oplus Q^{\mathrm{gp}}/P^{\mathrm{gp}}$ defined in Example III.2.3.6, and ϕ^{gp} is an isomorphism. It follows that $h_{Q/P}$ is étale and that its relative characteristic vanishes. Since h_X is obtained from $h_{Q/P}$ by base change, it shares the same properties. The morphisms m_Y and $b_\theta \times \mathrm{id}$ are strict, and it follows that m_θ is also strict. Since θ is injective, the morphism $\mathsf{A}_{\mathsf{Q}} \to \mathsf{A}_{\mathsf{P}}$ is flat, and hence so is the morphism $pr_Y\colon Y_\theta \to Y$. Suppose that $\underline{f}$ is locally of finite presentation. Since $\underline{m}_Y$ is also locally of finite presentation, so are $\underline{h}_X$ and $\underline{f}\times\underline{\mathrm{id}}$, and it follows that $\underline{m}_\theta$ is locally of finite presentation. Since $\underline{m}_Y$ is locally of finite type, it also follows that $\underline{b}_\theta \times \mathrm{id}$ is locally of finite presentation. □

The lemma shows that if (a, θ, b) is any chart for f in which θ is injective, then m_θ fits as the morphism f'' in a diagram satisfying the conditions in statement (3). Thus if f satisfies the hypotheses of (3), the morphism $\underline{m}_\theta$ is flat, so (4) is also satisfied.

Now suppose that f satisfies (4). Since flatness can be checked fppf locally on X and Y, we may assume that f admits a chart (a, θ, b), where $\theta\colon P \to Q$ is a homomorphism of fine monoids. Let $Q' := P^{\mathrm{gp}} \oplus Q$, let $\theta'\colon P \to Q'$ be (inc, θ), and let $\beta'\colon Q' \to \mathcal{M}_X$ be $(0, \beta)$. Then $(\alpha, \theta', \beta')$ is a chart for f and θ' is injective. Changing notation, we may assume that θ was already injective. Since $\mathsf{A}^*_{Q/P}$ is faithfully flat and of finite presentation over $\mathrm{Spec}(\mathbf{Z})$ and since its log structure is trivial, the projection $\pi\colon \tilde{X} := X \times \mathsf{A}^*_{Q/P} \to X$ is an fppf cover of X, and $(a, \theta, b\circ\pi)$ is a chart for $\tilde{f} := f\circ\pi$. The map $\tilde{b} := m_{Q/P}\circ(b\times\mathrm{id})\colon \tilde{X} \to \mathsf{A}_Q$ is also a chart for $\tilde{f}$, because $m_{Q/P}$ and $b \times \mathrm{id}$ are strict. Since $\tilde{b}_\theta = m_\theta$ and $\underline{m}_\theta$ is flat by hypothesis, $(a, \theta, \tilde{b})$ is a flat chart for f, and hence f is flat. □

Corollary 4.1.6. *A morphism of fine log schemes is flat if, fppf locally on X and Y, it admits a fine chart (a, θ, b) in which θ is injective and the map $\underline{m}_\theta\colon X \times \mathsf{A}^*_{Q/P} \to Y_\theta$ is flat. Conversely, if f is flat, and (a, θ, b) is any local chart in which θ is injective, then $\underline{m}_\theta$ is flat.* □

The following result, due to K. Kato, shows that neat charts for flat morphisms are flat.

Theorem 4.1.7. *Let $f\colon X \to Y$ be a flat morphism of fine log schemes such that $\underline{f}$ is locally of finite presentation. Suppose that (a, θ, b) is a chart for f that is neat at point x of X. Then $\underline{b}_\theta$ is flat in some neighborhood of x.*

Proof Let us consider the diagram of Lemma 4.1.5 formed from the chart (a, θ, b). We proceed in several steps.

The morphism $\underline{m}_\theta$ is flat.
Since our chart (a, θ, b) is neat, the homomorphism $\theta\colon P \to Q$ is injective. Then statement (4) of Theorem 4.1.4 tells us that $\underline{m}_\theta$ is flat.

The action of $\mathsf{A}^*_{Q/P}$ on $\underline{Y}^+_\theta := \underline{Y}_\theta \times_{\underline{\mathsf{A}}_Q} \underline{\mathsf{A}}_{(Q,Q^+)}$ is trivial.
Since the chart (a, θ, b) is neat at x, the homomorphism $b^\flat$ induces an isomorphism $Q^{\mathrm{gp}}/P^{\mathrm{gp}} \to \mathcal{M}^{\mathrm{gp}}_{X/Y,\overline{x}}$. Note that $b^\flat$ maps Q^* into $\mathcal{O}^*_{X,\overline{x}}$, which maps to zero

in $\mathcal{M}^{\mathrm{gp}}_{X/Y,x}$. It follows that Q^* is contained in $\theta(P^{\mathrm{gp}})$. Thus $\mathsf{A}^*_{Q/P} = \mathsf{A}^*_{Q/(P+Q^*)}$, which acts trivially on $\mathsf{A}^*_{(Q,Q^+)}$ by Proposition I.3.3.5.

The morphism $\underline{b}^+_\theta \colon \underline{X}^+ := b_\theta^{-1}(Y^+_\theta) \to \underline{Y}^+_\theta$ is flat.
Consider the following diagram:

$$\begin{array}{ccc} \underline{X}^+ \times \mathsf{A}^*_{Q/P} & \xrightarrow{\underline{pr}_{X^+}} & \underline{X}^+ \\ {\scriptstyle \underline{b}^+_\theta \times \mathrm{id}}\downarrow & \searrow{\scriptstyle \underline{m}_\theta{}^{+}} & \downarrow{\scriptstyle \underline{b}^+_\theta} \\ \underline{Y}^+_\theta \times \mathsf{A}^*_{Q/P} & \xrightarrow{\underline{m}^+_Y} & \underline{Y}^+_\theta . \end{array}$$

The lower left triangle of this diagram comes from the lower left corner of diagram (4.1.1), and the square commutes because the action of $\mathsf{A}^*_{Q/P}$ on $\underline{Y}^+_\theta$ is trivial. It follows that the upper right triangle commutes. Since $\underline{m}_\theta$ is flat, so is $\underline{m}_\theta{}''^{+}$ and, since $\underline{pr}_{X^+}$ is faithfully flat, it follows that $\underline{b}^+_\theta$ is flat.

The morphism $\underline{b}_\theta$ is flat.
The lower left corner of diagram (4.1.1) gives the following diagram:

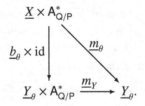

$$\begin{array}{ccc} \underline{X} \times \mathsf{A}^*_{Q/P} & & \\ {\scriptstyle \underline{b}_\theta \times \mathrm{id}}\downarrow & \searrow{\scriptstyle \underline{m}_\theta} & \\ \underline{Y}_\theta \times \mathsf{A}^*_{Q/P} & \xrightarrow{\underline{m}_Y} & \underline{Y}_\theta . \end{array}$$

Since β is neat at x, it is local at x, so $b(x)$ lies in $\underline{\mathsf{A}}_{(Q,Q^+)}$ and hence x belongs to $\underline{X}^+$, the fiber of $\underline{X}$ over $\underline{Y}^+_\theta$. We have seen that $\underline{m}_\theta$ and $\underline{b}^+_\theta$ are flat and that $\underline{b}_\theta$ is locally of finite presentation. Thus the criterion of flatness along the fibers [30, 11.3.10] implies that $\underline{b}_\theta \times \mathrm{id}$ is flat in a neighborhood of x. Since $\mathsf{A}^*_{Q/P}$ is faithfully flat over Spec($\mathbf{Z}$), the flatness of $\underline{b}_\theta$ follows. □

Remark 4.1.8. Recall that a morphism of schemes $f \colon X \to Y$ is said to be flat at a point x of X if the corresponding local homomorphism of local rings $\mathcal{O}_{Y,f(x)} \to \mathcal{O}_{X,x}$ is flat, and that a morphism f is flat if and only if it is flat at each point of its domain. Furthermore, if f is flat at x and is locally of finite presentation, then it is flat in some neighborhood of x [30, 11.3.10]. Let us use the analogous terminology for morphisms of fine log schemes. Then it is easy to see from Theorem 4.1.4 that a morphism of fine log schemes is flat if and only if it is flat at each point of its domain and that if f is flat at x and $\underline{f}$ is locally of finite presentation, then f is flat in some neighborhood of x.

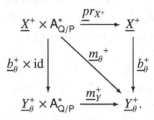

The next result shows that the idealized fibers of a flat morphism of fine log schemes are solid. The important corollary that follows it relates the notions of exactness, s-injectivity, and **Q**-integrality for smooth and flat morphisms.

Proposition 4.1.9. *Let $f\colon X \to Y$ be a flat morphism of fine log schemes. Assume that the underlying morphism of schemes is locally of finite presentation. Then the idealized fibers of f are solid. In particular, the idealized fibers of a smooth morphism of fine log schemes are solid.*

Proof We may assume without loss of generality that Y is a log point and that we are working in the Zariski topology. Let $\mathcal{K}$ be the sheaf of ideals of $\mathcal{M}_X$ generated by $\mathcal{M}_Y^+$. The proposition asserts that the idealized log scheme $(X, \mathcal{K})$ is solid, i.e., that for every point x of X, the map $\operatorname{Spec}(\mathcal{O}_{X,x}) \to \operatorname{Spec}(\mathcal{M}_{X,x}, \mathcal{K}_x)$ is locally surjective. Suppose that $X' \to X$ is a strict and ideally strict locally surjective morphism of fine idealized log schemes. Thus X will be solid if X' is. We may therefore assume without loss of generality that the field k is algebraically closed and furthermore we may work fppf locally on $\underline{X}$. Then, by Remark III.1.5.3, we may assume that Y is split, so that it admits a neat chart a subordinate to a fine sharp monoid P. By Theorem III.1.2.7, we may assume that the chart $P \to \mathcal{M}_Y$ extends to a chart (a, θ, b), neat at x, with b subordinate to a fine monoid Q. Then, by Theorem 4.1.7, the morphism $\underline{b}_\theta\colon \underline{X} \to \underline{Y}_\theta = \operatorname{Spec}(k[Q, J_\theta^+])$ is flat, where J_θ is the ideal of Q generated by $\theta(P^+)$. It follows that $\underline{b}_\theta$ is locally surjective. Since $Y_\theta \to \operatorname{Spec}(Q, J_\theta)$ is locally surjective the same is true of the map $X \to \operatorname{Spec}(Q, J_\theta)$. □

The following result follows immediately from the previous proposition and Theorem III.2.2.7. See also Proposition 4.3.1, which does not require the saturation assumption.

Corollary 4.1.10. *Let $f\colon X \to Y$ be a morphism of fine saturated log schemes. Assume that f is smooth or, more generally, that it is flat and that its underlying morphism of schemes is locally of finite presentation. Then conditions (1)–(8) of Theorem III.2.2.7 are equivalent. In particular, f is exact if and only if it is s-injective if and only if it is* **Q***-injective.* □

4.2 Flatness and smoothness

In classical algebraic geometry, a morphism is étale (smooth and unramified) if and only if it is flat and unramified. (In fact the latter is used as the definition of "étale" in [32].) We will see that the analogous result holds in the log case.

Corollary 4.2.1. *Let $f\colon X \to Y$ be a morphism of fine log schemes, and as-*

sume that $\underline{f}$ is locally of finite presentation. Then f is étale if and only if it is flat and unramified.

Proof It is clear that an étale map is flat and unramified. Suppose conversely that f is flat and unramified. Let x be a point of X; since f is unramified, we can, after some étale localization, find a chart (a, θ, b) for f that is neat at x, as we saw in Theorem 3.4.2. By Theorem 4.1.7, the associated morphism $\underline{b}_\theta \colon \underline{X} \to \underline{Y}_\theta$ is flat in a neighborhood of x. Since $f = g \circ b_\theta$ is unramified, b_θ is also unramified and, since it is strict, it follows that $\underline{b}_\theta$ is unramified. Since $\underline{b}_\theta$ is flat and unramified, it is étale [32] and, since b_θ is strict, it too is étale. Since θ is injective and $\mathrm{Cok}(\theta^{\mathrm{gp}})$ is finite of order invertible in $k(x)$, it follows from Corollary 3.1.10 that A_θ is étale, hence g is étale, and hence $f = g \circ b_\theta$ is also étale. □

The following result is a logarithmic version of the theorem on"flatness along the fibers."

Theorem 4.2.2. *Let $f \colon X \to Y$ and $g \colon Y \to Z$ be morphisms of fine log schemes, where $\underline{f}$ and $\underline{g}$ are locally of finite presentation. Let x be a point of X, let $z := f(g(x))$, and assume that both $g \circ f$ and the base-changed morphism $f_z \colon X_z \to Y_z$ are flat in some neighborhood of x. Then $f \colon X \to Y$ is flat in some neighborhood of x.*

Proof Let $\tilde{f} \colon \tilde{X} \to \tilde{Y}$ be an fppf localization of f, and let $\tilde{g} \colon \tilde{Y} \to Z$ be the composition $\tilde{Y} \to Y \to Z$. Then, for any $S \to Z$, the morphism $\tilde{f}_S \colon \tilde{X}_S \to \tilde{Y}_S$ is an fppf localization of f_S and hence is again flat. Thus our hypothesis on f and g is stable under fppf localization.

First assume that f is strict. Choose, after an fppf localization, a chart (a, θ, b) for g that is neat at y, where $\theta \colon P \to Q$ is a homomorphism of fine monoids. Since f is strict, the composite $c := b \circ f$ is a chart for X, and (a, θ, c) is a chart for $g \circ f$ that is neat at x. Since $g \circ f$ is flat, it follows from Theorem 4.1.7 that $\underline{c}_\theta \colon X \to Z_\theta$ is flat in some neighborhood of x. Let $z' := b_\theta(y) \in Z_\theta$ and $z := g(y) \in Z$, and let S' (resp. S) be the log point obtained by endowing z' (resp. z) with the log structure of Z_θ (resp. of Z). By assumption, the morphism $f_S \colon X \times_Z S \to Y \times_Z S$ is flat. Then the base-changed morphism

$$f_{S'} \colon (X \times_Z S) \times_{(Y \times_Z S)} (Y \times_{Z_\theta} S') \to Y' \times_{Z_\theta} S'$$

is also flat. But

$$(X \times_Z S) \times_{(Y \times_Z S)} (Y \times_{Z_\theta} S') \cong X \times_{Z_\theta} S',$$

so $f_{S'}$ can be viewed as the fiber of f along $S' \to Z_\theta$. Since f is strict, so is $f_{S'}$, and we conclude that the morphism $\underline{f}\colon \underline{X}_{z'} \to \underline{Y}_{z'}$ is flat. Since $\underline{c}_\theta\colon \underline{X} \to \underline{Z}_\theta$ is flat, it follows from the criterion of flatness along the fibers [30, 11.3.10] that $\underline{f}$ is flat in some neighborhood of x, as required.

We shall use Theorem 4.1.4 to reduce to the case in which f is strict. Suppose we are given a diagram as in (3) of that theorem. We must show that the (strict) morphism $f''\colon X'' \to Y'$ is flat. Since p_Y is flat, so is its base change p_X, and since $g \circ f$ is flat, so is $g' := g \circ f \circ p_X$. The morphism $h\colon X'' \to X'$ is étale, hence flat, and hence $g'' := g' \circ h\colon X'' \to Z$ is flat. By assumption, $f_z\colon X_z \to Y_z$ is flat, and it follows that $f'_z\colon X'_z \to Y'_z$ is also flat. Since h is étale, we conclude that $h_z\colon X''_z \to Y'_z$ is flat and hence that $f''_z = f'_z \circ h_z\colon X''_z \to Y'_z$ is flat. Since f'' is strict, the previous paragraph shows that $\underline{f}''$ is flat, concluding the proof. □

Corollary 4.2.3. *Let $f\colon X \to Y$ be a morphism of fine log schemes whose underlying morphism of schemes is locally of finite presentation. Let x be a point of X and let $y := f(x)$. Then f is smooth in a neighborhood of x if only if, in some neighborhood of x, f is flat and the induced morphism $f_y\colon X_y := f^{-1}(y) \to y$ is smooth.*

Proof Theorem 3.2.6 shows that, after replacing X and Y by étale neighborhoods of x and y, we may factor the smooth morphism $f_y\colon X_y \to y$ through an étale morphism $f'_y\colon X_y \to y \times \mathsf{A}_{\mathbf{N}^r}$. The morphism $X_y \to \mathsf{A}_{\mathbf{N}^r}$ is defined by global sections $m_1, \ldots, m_r$ of the sheaf $\mathcal{M}_{X_y}$. Since the map of sheaves $\mathcal{M}_X \to \mathcal{M}_{X_y}$ is surjective, these sections can be lifted to sections of $\mathcal{M}_X$, in some neighborhood of x. Then f'_y extends to a morphism $f'\colon X \to Y \times \mathsf{A}_{\mathbf{N}^r}$, whose fiber over y is f'_y. Since f'_y is unramified at x, Theorem 3.4.2 implies that f' is also unramified at x. Since f'_y is étale, it is also flat. The morphism f is the composition of $f'\colon X \to Y \times \mathsf{A}_{\mathbf{N}^r}$ with the projection $g\colon Y \times \mathsf{A}_{\mathbf{N}^r} \to Y$, and by assumption f is flat. Moreover $\underline{f}'$ and $\underline{g}$ are locally of finite presentation, so, by Theorem 4.2.2, it follows that f' is also flat. Thus f' is flat and unramified, and so, by Corollary 4.2.1, f' is in fact étale. Since the projection g is smooth, we can conclude that $f = g \circ f'$ is smooth. □

The following result explores the relationship between the flatness of a morphism of log schemes and the flatness of its underlying morphism of schemes. We will return to this question in the context of integral morphisms of log schemes, in Theorem 4.3.5.

Proposition 4.2.4. *Let $f\colon X \to Y$ be a morphism of fine saturated log schemes whose underlying morphism of schemes is locally of finite presentation and flat. Let x be a point of X, let $y := f(x)$, and suppose that the fiber $X_y \to y$ is* *smooth (resp. flat) in a neighborhood of x. Then f is smooth (resp. flat) in a neighborhood of x.*

Proof Thanks to Corollary 4.2.3 it suffices to prove this result for the flat case. Flatness can be checked fppf locally on X and Y, so, by Theorem III.1.2.7, we may assume that we have a chart (α, θ, β) for f that is neat at x. Then $f = f_\theta \circ b_\theta$, where $f_\theta \colon Y \times_{\mathsf{A}_P} \times \mathsf{A}_Q \to Y$ is the projection to Y, and it will suffice to prove that $\underline{b}_\theta$ is flat. We have charts $a_y \colon y \to \mathsf{A}_P$ and $b_y \colon X_y \to \mathsf{A}_Q$, and (a_y, θ, b_y) is a chart for f_y that is neat at x. Since $f_y = f_{\theta,y} \circ b_{\theta,y}$ is flat, it follows from Theorem 4.1.7 that $\underline{b}_{\theta,y}$ is flat. Since $\underline{f}$ is locally of finite presentation and $\underline{f}_\theta$ is locally of finite type, it follows that $\underline{b}_\theta$ is locally of finite presentation. Since $\underline{f}$ and $\underline{b}_{\theta,y}$ are flat, the criterion of flatness along the fibers [30, 11.3.10] implies that $\underline{b}_\theta$ is flat, as required. □

A variant of Theorem 4.1.4 holds for other kinds of morphisms as well. We content ourselves with the following statement; for a more thorough and general discussion, we refer to [51, Lemma 4.3.1].

Theorem 4.2.5. *Let* **P** *be one of the following properties of morphisms of schemes: flat, smooth, étale. Let $f \colon X \to Y$ be a morphism of fine log schemes, with $\underline{f}$ locally of finite presentation. Then f is* **P** *if and only if, for every diagram as in (2) or (3) of Theorem 4.1.4, the morphism $\underline{f}''$ is* **P**.

Proof Suppose that f is smooth. Then in any diagram as in (2) or (3) of Theorem 4.1.4, the morphism f'' is also smooth, and by hypothesis is strict. By Proposition 3.1.6 a strict morphism is smooth if and only if its underlying morphism of schemes is smooth, so $\underline{f}'$ is indeed smooth. A similar argument works for étale morphisms. Suppose conversely that $\underline{f}''$ in every such diagram is smooth. Let $S \to T$ be a log thickening over Y, together with a Y-morphism $S \to X$. We must show that this morphism admits a local extension $T \to X$. According to Corollary 3.1.5, we may assume that $S \to X$ is an open immersion, and in particular that it is strict. The assertion is local, so we may assume without loss of generality that $S = X$, that X is affine, and that the morphism $X \to Y$ admits a chart subordinate to an injective homomorphism of fine monoids $\theta \colon P \to Q$ as in Theorem 3.3.1. Since A_θ is smooth and $X \to T$ is a log thickening, the morphism $X \to \mathsf{A}_Q$ extends to a morphism $T \to \mathsf{A}_Q$ over A_P and, since $X \to T$ is strict, the morphism $T \to \mathsf{A}_Q$ is a chart for T. Form diagrams as in Lemma 4.1.5 for X and for T. Then $\underline{T}''$ is a thickening of $\underline{X}''$ over $\underline{Y}_\theta$ and, since $\underline{f}''$ is smooth, there is a retraction $\underline{T}'' \to X''$ over $\underline{Y}_\theta$. Since the morphisms $T'' \to Y_\theta$ and $X'' \to Y_\theta$ are strict, this retraction is compatible with the log structures, and defines a retraction $r \colon T'' \to X''$. Then the morphism $T \to T'' = \mathsf{A}^*_{Q/P} \times T : t \mapsto (1, t)$, followed by $r \colon T'' \to X'' = \mathsf{A}^*_{Q/P} \times X \to X$

gives the desired morphism $T \to X$. The argument for étale morphisms is left to the reader. □

4.3 Flatness, exactness, and integrality

The following result is due to C. Nakayama [56].

Proposition 4.3.1. *A flat and exact morphism of fine log schemes is* **Q**-*integral.*

Proof Suppose that $f\colon X \to Y$ is flat and exact and that $\overline{x}$ is a geometric point of X and $\overline{y}$ is its image in Y. Our claim is that the homomorphism $f^\flat_{\overline{x}}\colon \mathcal{M}_{Y,\overline{y}} \to \mathcal{M}_{\overline{x}}$ is $\mathbf{Q}$-integral. According to Theorem I.4.7.7, it will suffice to show that, for every face G of $\mathcal{M}_{\overline{x}}$ lying over $\mathcal{M}^*_{\overline{y}}$, the induced homomorphism $\mathcal{M}_{\overline{y}} \to (\mathcal{M}_{\overline{x}})_G$ is exact. Working fppf locally on X and Y, we may assume that f admits a chart (a, θ, b) such that the morphism $\underline{b}_\theta\colon \underline{X} \to \underline{Y}_\theta$ is flat. Then we may replace $\overline{x}$ and $\overline{y}$ by their corresponding scheme-theoretic points x and y, and X and Y by their corresponding local log schemes $\operatorname{Spec}(Q \to \mathcal{O}_{X,x})$ and $\operatorname{Spec}(P \to \mathcal{O}_{Y,y})$. We may also assume that the charts a and b are exact at x and y, respectively. The faces G and Q^* define prime ideals $\mathfrak{q}'$ and $\mathfrak{q}$ of $\mathbf{Z}[Q]$ lying over $\mathbf{Z}[P^+]$, with $\mathfrak{q}' \subseteq \mathfrak{q}$, giving rise to points η' and η of $Y_\theta := Y \times_{\mathsf{A}_P} \mathsf{A}_Q$ lying over y, with η a specialization of η'. Since b is exact at x, it follows that $b_\theta(x) = \eta$. Since $\underline{b}_\theta$ is a flat morphism, it is locally surjective, and hence there is a point x' of X such that $b_\theta(x') = \eta'$. Since f is exact, the homomorphism $f^\flat_{x'}\colon \mathcal{M}_y \to \mathcal{M}_{x'} = (\mathcal{M}_{\overline{x}})_G$ is exact, as required. □

In classical algebraic geometry, a surjective and flat morphism is faithfully flat, and many properties of schemes and morphisms of schemes can checked after passing to a faithfully flat cover. The situation in log geometry is more complicated and seems to require an additional hypothesis of exactness. We explain here the main ideas, referring to [41] for more details.

Theorem 4.3.2. *Let $f\colon X \to Y$ and $g\colon Y \to Z$ be morphisms of fine saturated log schemes. Assume that $\underline{f}$ and $\underline{g}$ are locally of finite presentation. Suppose that f is exact and flat at x. Then $g \circ f$ is flat x if and only if g is flat at $f(x)$.*

Proof If f is flat at x and g is flat at $f(x)$, then $g \circ f$ is flat at $f(x)$, by Proposition 4.1.2. Assume that f is exact and flat at x and that $g \circ f$ is flat at x. To prove that g is flat at $f(x)$ we may work fppf locally around $y := f(x)$. By Theorem III.1.2.7, we may assume that there exists a chart (a, θ, b) for g that is neat at y, and then that there is a chart (b, ϕ, c) for f that is neat at x. We may also assume that a is exact at z, that b is exact at y, and that c is exact at x. Since f is exact at x, it is s-injective, and it follows from Corollary II.2.4.7 that

$(a, \theta\phi, c)$ is also neat at x. Suppose that the charts a, b, and c are subordinate to the monoids P,Q, and R, respectively. Then we have a commutative diagram

$$\begin{array}{ccccc} X & \xrightarrow{c_\phi} & Y \times_{\mathsf{A}_Q} \mathsf{A}_R & \longrightarrow & Y \\ & \searrow^{c_{\theta\phi}} & \Big\downarrow b' & & \Big\downarrow b_\theta \\ & & Z \times_{\mathsf{A}_P} \mathsf{A}_R & \xrightarrow{\mathrm{id} \times \underline{\mathsf{A}}_\phi} & Z \times_{\mathsf{A}_P} \mathsf{A}_Q, \end{array}$$

in which the square is cartesian. Since $g \circ f$ and f are flat and the charts are neat, it follows from Theorem 4.1.7 that $\underline{c}_{\theta\phi}$ and $\underline{c}_\phi$ are flat at x and hence that the map $\underline{b}'$ is flat at y. Since f is exact at x and b is exact at y, it follows from (2) of Proposition I.4.2.1 that $\phi\colon Q \to R$ is also exact. Since it is injective by construction, we see from Corollary I.4.2.8 that the map $\underline{\mathsf{A}}_\phi$ universally descends flatness. Thus $\underline{b}_\theta$ is flat at y, and it follows that the morphism g is also flat at y. □

Corollary 4.3.3. *Let $f\colon X \to Y$ and $h\colon Y' \to Y$ be morphisms of fine log schemes, with $\underline{f}$ and $\underline{h}$ locally of finite presentation. Suppose that h is flat, exact, and surjective. Then f is flat if and only if the induced map $f'\colon X \times_Y Y' \to Y$ is flat.*

Proof If f is flat, then Proposition 4.1.2 implies that f' is flat. If h is exact and surjective, it is universally surjective, by Corollary III.2.2.5. Thus the projection $p\colon X \times_Y Y' \to X$ is surjective and flat. If f' is flat, then $f \circ p = h \circ f'$ is flat, so it follows from Theorem 4.3.2 that f is flat. □

Examples 4.3.4. Here are two examples which illustrate that the hypothesis of exactitude in Theorem 4.3.2 is not superfluous.

The first example is a flat and Kummer morphism that does not descend flatness. Let P be the submonoid of $\mathbf{N}$ generated by 2 and 3. This monoid is not 2-saturated, so multiplication by 2 on P is Kummer but not exact. As we saw in Proposition I.4.2.7, it follows that the corresponding endomorphism θ of $\mathbf{Z}[P]$ is not universally injective. Let us make this explicit. The ring $\mathbf{Z}[P]$ can be identified with the ring $A := \mathbf{Z}[x, y]/(y^2 - x^3)$. Let I be the ideal of A generated by y and let J be the ideal generated by y and x^2, let $Z := \operatorname{Spec}(P \to A/I)$, and let $Y := \operatorname{Spec}(P \to A/J)$. The inclusion $i\colon Y \to Z$ is strict and is not flat. The endomorphism $\phi_2\colon \mathsf{A}_P \to \mathsf{A}_P$ induced by multiplication by 2 is flat as a morphism of log schemes and Kummer, but not exact. Then the map $Z' \to Z$ obtained by base change of ϕ_2 along $Z \to \mathsf{A}_P$ is also flat and Kummer. The base change $i'\colon Y' \to Z'$ of i along $Z' \to Z$ is an isomorphism, hence flat, although i is not flat.

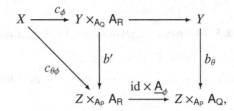

The second example, due to C. Nakayama, is a universally surjective and flat morphism that does not descend flatness. Let $P := \mathbf{N} \oplus \mathbf{N}$, and identify $\mathbf{Z}[P]$ with $\mathbf{Z}[x, y]$ in the obvious way. Let $Z := \operatorname{Spec}(P \to \mathbf{Z}[x, y]/(x^2, y^2))$ and $Y := \operatorname{Spec}(P \to \mathbf{Z}[x, y]/(x^2, xy, y^2))$, and let $i\colon Y \to Z$ be the obvious inclusion. The log blowup $Z' \to Z$ of Z at the ideal P^+ of P is flat and universally surjective, and the base-changed map $Y' \to Z'$ is an isomorphism, but i is not flat.

The following result relates the integrality of a morphism f of log schemes to the flatness of f and of its underlying morphism $\underline{f}$.

Theorem 4.3.5. *Let $f\colon X \to Y$ be a morphism of locally noetherian fine log schemes, where $\underline{f}$ is locally of finite presentation.*

1. *If f is flat and integral, then $\underline{f}$ is also flat.*
2. *If f is smooth and integral, then the morphisms f and $\underline{f}$ are flat, and if in addition X and Y are saturated, then the fibers of $\underline{f}$ are Cohen–Macaulay.* [3]
3. *If f and $\underline{f}$ are flat and Y is regular, then f is integral.*

Proof To prove the first statement, assume that f is flat and integral, let x be a point of X, and let y be its image in Y. The flatness of $\underline{f}$ can be verified fppf locally on $\underline{X}$ and Y, so we can replace X and Y by fppf neighborhoods of x and y as is convenient during the course of the proof. By (2) of Theorem III.1.2.7, there exists, after an fppf localization of $\underline{Y}$, a chart $a\colon Y \to \mathsf{A}_P$ for Y that is neat at y, where P is a fine sharp monoid. Statement (4) of the same theorem asserts that, after a further localization, there exists a chart (a, θ, b) for f that is neat and exact at x. Then Theorem 4.1.7 implies that the morphism $\underline{b}_\theta\colon \underline{X} \to \underline{Y} \times_{\underline{\mathsf{A}}_P} \underline{\mathsf{A}}_Q$ is flat. Since f is integral, the homomorphism $\overline{f}^\flat_{x}\colon \overline{\mathcal{M}}_{Y,y} \to \overline{\mathcal{M}}_{X,x}$ is integral. Since a is neat at y, the homomorphism $P \to \overline{\mathcal{M}}_{Y,y}$ is an isomorphism, and hence the induced homomorphism $P \to \overline{\mathcal{M}}_{X,x}$ is integral. Since $Q \to \mathcal{M}_{X,x}$ is an exact chart, the homomorphism $\overline{Q} \to \overline{\mathcal{M}}_{X,x}$ is also an isomorphism, and it follows that $\theta\colon P \to Q$ is also integral, by statement (2) of Proposition I.4.6.3. Since P is sharp and θ is integral, Proposition I.4.6.7 implies that the map of schemes $\underline{\mathsf{A}}_Q \to \underline{\mathsf{A}}_P$ is flat, and hence so is the projection $\underline{Y} \times_{\underline{\mathsf{A}}_P} \underline{\mathsf{A}}_Q \to \underline{Y}$. Since $\underline{f}$ is the composition of this projection with the flat morphism $\underline{b}_\theta$, it too is flat.

If f is smooth then it is also flat, and hence if f is integral, it follows from (1) that $\underline{f}$ is also flat. Suppose in addition that X and Y are saturated. Statement (2) of Theorem III.1.2.7 asserts that, after a harmless étale localization, there exists a chart $a\colon Y \to \mathsf{A}_P$ that is neat at y. After a further étale localization, we can fit a into a chart (a, θ, b) for f as in Theorem 3.3.1. Since $\underline{b}_\theta\colon X \to Y_\theta$ is étale, it will suffice to show that the fiber of $f_\theta\colon Y_\theta \to Y$ over y is Cohen–Macaulay.

[3] This result is due to T. Tsuji [76, II.4.1].

Since f_θ is obtained by base change from A_θ, it will suffice to show that the fiber of $\underline{\mathsf{A}}_\theta$ over the vertex of $\underline{\mathsf{A}}_P$ is Cohen–Macaulay. The homomorphism $P \to \overline{\mathcal{M}}_{Y,y}$ is an isomorphism because a is neat at y and the homomorphism $\overline{Q} \to \overline{\mathcal{M}}_{X,x}$ is an isomorphism because b is exact at x. Since f is integral, it follows that $\overline{\theta}\colon P \to \overline{Q}$ is integral, and hence, by Proposition I.4.6.3, the homomorphism $\theta\colon P \to Q$ is also integral. Since θ is local and P is sharp, Proposition 4.6.7 implies that $\underline{\mathsf{A}}_\theta$ is flat. Since P and Q are saturated, the schemes $\underline{\mathsf{A}}_P$ and $\underline{\mathsf{A}}_Q$ are Cohen–Macaulay (Theorem II.3.4.3). Then our conclusion follows from the fact that the fibers of a flat morphism of Cohen–Macaulay schemes are again Cohen–Macaulay [27, 6.5].

The proof of statement (3) is more involved. Suppose that f and $\underline{f}$ are flat and that Y is regular. Then Y is locally noetherian, and consequently so is X. Since Y is regular, it is solid, and $\underline{f}$ is locally surjective because it is flat. It follows from Proposition III.2.2.8 that f is exact and hence s-injective.

Let $\overline{x}$ be a geometric point of X and let $\overline{y}$ be its image in Y. We claim that the map $\overline{f}^\flat_{\overline{x}}\colon \overline{\mathcal{M}}_{\overline{y}} \to \overline{\mathcal{M}}_{\overline{x}}$ is integral. Since Y is regular, it is saturated. By Theorem III.1.2.7, we see that, after an étale localization of Y and an fppf localization of X, there is a chart a for Y that is neat at $\overline{y}$ and which fits into a chart (a, θ, b) for f that is neat at $\overline{x}$. Furthermore, since f is s-injective, b is also neat at $\overline{x}$. We can and will work with scheme-theoretic points from now on.

We use the notation of diagram III.1.2.3 . Since f is flat and the chart is neat at x, it follows from Theorem 4.1.7 that $\underline{b}_\theta$ is flat at x. Let $x' := b_\theta(x)$. Then the homomorphism of local rings $b^\sharp_{\theta,x}\colon \mathcal{O}_{Y_\theta,x'} \to \mathcal{O}_{X,x}$ is flat, hence faithfully flat. Since $f^\sharp_{\underline{x}}$ is also flat, it follows that $f^\sharp_{\theta,y}\colon \mathcal{O}_{Y,y} \to \mathcal{O}_{Y_\theta,x'}$ is flat. Thus we have reduced the theorem to the case in which $X = Y_\theta$. We may also replace Y and X by the spectra of the completions of their respective local rings.

Suppose first that the local ring $\mathcal{O}_{Y,y}$ contains a field k, which we may assume is its residue field. Then we may replace $\mathsf{A}_P := \operatorname{Spec}(P \to \mathbf{Z}[P])$ by $\operatorname{Spec}(P \to k[P])$ and A_Q by $\operatorname{Spec}(Q \to k[Q])$. Since Y is regular, the homomorphism $k[P] \to \mathcal{O}_{Y,y}$ is flat, as we saw in Corollary III.1.11.3. Then $\underline{Y} \to \underline{\mathsf{A}}_P$ is flat and hence the same is true for the map $X \to \mathsf{A}_Q$, by base change. We then have a commutative diagram of local homomorphisms of local rings:

$$
\begin{array}{ccc}
k[[Q]] & \xrightarrow{\ \beta\ } & \mathcal{O}_{X,x} \\
\uparrow{\scriptstyle \theta} & & \uparrow{\scriptstyle \phi} \\
k[[P]] & \xrightarrow{\ \alpha\ } & \mathcal{O}_{Y,y}.
\end{array}
$$

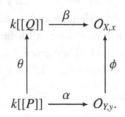

The regularity of Y implies that α is flat and, since β is obtained from α by base change and completion, it too is flat, hence faithfully flat. Since ϕ is flat by assumption, we can conclude that θ is flat. Then Proposition I.4.6.8 implies that $P \to Q$ is integral.

The mixed-characteristic case is more complicated. Write A for the local ring of Y at y and B for the local ring of X at x, and let V be a Cohen ring of the residue field k of A. By Theorem III.1.11.2, we can write $A = V[[P \oplus \mathbf{N}^r]]/(f)$, where $f \in V[[P \oplus \mathbf{N}^r]]$ has constant term p. Then $B \cong V[[Q \oplus \mathbf{N}^r]]/(g)$, where g is the image of f. Our assumption on f implies that $A \to B$ is flat, and we need to prove that $V[[P]] \to V[[Q]]$ is flat.

To simplify the notation, we let $P' := P \oplus \mathbf{N}^r$ and $Q' := Q \oplus \mathbf{N}^r$. Since $V[Q] \to V[Q']$ and $V[P] \to V[P']$ are faithfully flat, it is enough to prove that $V[P'] \to V[Q']$ is flat, and so we may as well replace P by P' and Q by Q'.

Let J be an ideal of P and let K be the ideal of Q generated by J. We have an exact sequence

$$0 \to V[[J]] \longrightarrow V[[P]] \longrightarrow V[[P, J]] \to 0. \tag{4.3.1}$$

Statement (1) of Lemma III.1.10.14, applied to the ideal J of P and the element f of $V[[P]]$, implies that multiplication by f is injective on $V[[P, J]]$. Since $A = V[[P]]/(f)$, we conclude that $\mathrm{Tor}_1^{V[[P]]}(V[[P, J]], A) = 0$. Therefore, tensoring the sequence (4.3.1) with A yields the exact sequence

$$0 \to V[[J]] \otimes_{V[[P]]} A \longrightarrow A \to A/JA \longrightarrow 0.$$

Since $A \to B$ is flat, the sequence

$$0 \to V[[J]] \otimes_{V[[P]]} B \xrightarrow{\gamma} B \longrightarrow B/JB \to 0 \tag{4.3.2}$$

is also exact.

On the other hand, tensoring the sequence (4.3.1) with $V[[Q]]$ yields the exact sequence

$$0 \to T \longrightarrow V[[J]] \otimes_{V[[P]]} V[[Q]] \longrightarrow V[[Q]] \longrightarrow V[[Q, K]] \to 0,$$

where $T := \mathrm{Tor}_1^{V[[P]]}(V[[P, J]], V[[Q]])$. Thus we find the exact sequence

$$0 \to T \longrightarrow V[[J]] \otimes_{V[[P]]} V[[Q]] \longrightarrow V[[K]] \to 0. \tag{4.3.3}$$

Statement (2) of Lemma III.1.10.14, applied to the pair of Q-sets $(K, \emptyset)$ and the element g of $V[[Q]]$, implies that multiplication by g on $V[[K]]$ is injective. Since $B = V[[Q]]/(g)$, we conclude that $\mathrm{Tor}_1^{V[[Q]]}(V[[K]], B) = 0$, and hence that the sequence

$$0 \to T \otimes_{V[[Q]]} B \longrightarrow V[[J]] \otimes_{V[[P]]} B \xrightarrow{\delta} V[[K]] \otimes_{V[[Q]]} B \to 0 \tag{4.3.4}$$

is exact. But we saw in equation (4.3.2) that the map $\gamma\colon V[[J]] \otimes_{V[[P]]} B \to B$ is injective and, since γ factors through δ, it follows that δ is also injective and hence that $T \otimes_{V[[Q]]} B = 0$. Since T is finitely generated over $V[[Q]]$ and g belongs to the maximal ideal of $V[[Q]]$, Nakayama's lemma implies that $T = 0$.

Taking $J = P^+$, we conclude that $\mathrm{Tor}_1^{V[[P]]}(V[[P, P^+], V[[Q]]) = 0$, so we have an exact sequence

$$0 \to V[[P^+]] \otimes_{V[[P]]} V[[Q]] \to V[[Q]] \to V[[Q, K]],$$

where K is the ideal of Q generated by P^+. Since $V[[Q, K]]$ is p-torsion free, the sequence

$$0 \to k[[P^+]] \otimes_{k[[P]]} k[[Q]] \to k[[Q]] \to k[[Q, K]]$$

is exact, so $\mathrm{Tor}_1^{k[[P]]}(k[[P, P^+], k[[Q]]) = 0$. Then, by Proposition I.4.6.8, the homomorphism $\theta\colon P \to Q$ is integral, completing the proof. □

The next result, due to T. Tsuji [76, II.4.2, II.4.11], gives a purely geometric criterion for a smooth and integral morphism to be saturated.

Theorem 4.3.6. *Let $f\colon X \to Y$ be a smooth and integral morphism of fine saturated log schemes. Then the following conditions are equivalent.*

1. *The morphism f is saturated.*
2. *The fibers of f are reduced.*
3. *The fibers of f are generically reduced (Serre's condition R_0).*

If Y is regular, it is enough to check that (3) holds for the fiber over every geometric point $\overline{y}$ of Y such that $\dim \mathcal{O}_{Y,\overline{y}} = \dim \mathcal{M}_{Y,\overline{y}} = 1$.

Proof Let x be a point of X and let $y := f(x)$, regarded as a log scheme with the log structure induced from Y. Then the base-changed map $f_y\colon X_y \to y$ is again smooth, and each of the conditions (1)–(3) is true for f if and only if it is true for f_y. Thus we may assume that Y is a log point, without loss of generality.

The theorem can be verified locally in a neighborhood of each point x of X, so we may assume that f admits a chart as described in Theorem 3.3.1. Moreover, since Y is saturated, we may assume that the chart $P \to \mathcal{M}_{Y,y}$ is neat, so P is sharp. In this situation the homomorphisms $P \to \overline{\mathcal{M}}_{Y,y}$ and $\overline{Q} \to \overline{\mathcal{M}}_{X,x}$ are isomorphisms, so $\overline{f}^\flat_x$ is saturated if and only if θ is. Let k be the residue field of y and let J_θ be the ideal of $k[Q]$ generated by $\theta(P^+)$. By construction, the morphism

$$\underline{b}_\theta\colon \underline{X} \to \underline{Y}_\theta = \operatorname{Spec} k[Q, J_\theta]$$

is étale and the torsion part of $\text{Cok}(\theta^{\text{gp}})$ has order n invertible in k.

Assume that f is saturated. Then θ and $\overline{\theta}$ are also saturated, and hence by Corollary I.4.8.17 the ring $k[Q, J_\theta]$ is reduced. Since b_θ is étale, it follows that X is also reduced, proving that (1) implies (2).

The implication of (3) by (2) is trivial. Suppose that (3) holds, i.e., that X is generically reduced. Since the map $\underline{b}_\theta\colon \underline{X} \to Y_\theta$ is étale, its image contains an open neighborhood V of the point $b_\theta(x)$, i.e., the point y' of Y_θ defined by the maximal ideal Q^+ of Q. By Remark I.4.2.11, y' is in the closure of every generic point of Y_θ. Since X is generically reduced and b_θ is étale, V is also generically reduced and, since V contains all the generic points of Y_θ, it follows that Y_θ satisfies R_0. Then Corollary I.4.8.17, implies that θ is saturated.

For the last statement, we assume that Y is regular. By Proposition 3.5.3, X is also regular, hence solid. Working locally, we may assume that our log structures are in the Zariski topology. By Theorem III.2.5.5, it will suffice to prove that f is strict at every point x of X such that $\dim \mathcal{M}_{X,x} = \dim \mathcal{M}_{Y,y} = 1$. Let x be such a point and let $y := f(x)$. Since X is regular, the ideal of $\mathcal{O}_{X,x}$ generated by $\mathcal{M}^+_{X,x}$ is prime and hence defines a generization x' of x. Let $y' := f(x')$, a generization of y. As we observed in Remark III.1.12.4, $\dim \mathcal{O}_{X,x'} = \dim \mathcal{M}_{X,x'} = 1$. Since X is saturated, $\overline{\mathcal{M}}_{X,x'} \cong \mathbf{N}$, and since X is regular, it follows from Theorem III.1.11.6 that $\mathcal{O}_{X,x'}$ is also regular, hence a DVR. Furthermore, the log structure on $\mathcal{O}_{X,x'}$ defined by $\mathcal{M}_{X,x'}$ is the standard one. Since f is integral and smooth, the underlying morphism $\underline{f}$ is flat, by Theorem 4.3.5, hence the map $\mathcal{O}_{Y,y'} \to \mathcal{O}_{X,x'}$ is flat, and hence $\dim \mathcal{O}_{Y,y'} \leq \dim \mathcal{O}_{X,x'} = 1$. On the other hand, since Y is regular, the ideal $\mathcal{M}^+_{Y,y'}$ generates a prime ideal $\mathfrak{p}$ of $\mathcal{O}_{Y,y'}$, and $\dim \mathcal{O}_{Y,y'} = \dim \mathcal{M}_{Y,y'} + \dim \mathcal{O}_{Y,y'}/\mathfrak{p}\mathcal{O}_{Y,y'}$. Since $\dim \mathcal{M}_{Y,y'} = 1$ and $\dim \mathcal{O}_{Y,y'} \leq 1$, it follows that $\dim \mathcal{O}_{Y,y'} = 1$ and that $\mathfrak{p}$ is the maximal ideal of $\mathcal{O}_{Y,y'}$. Then $\mathcal{O}_{Y,y'}$ is also a DVR with the standard log structure. The assumption (3) applies to the fiber of f over y', and in particular we can conclude that the ring $\mathcal{O}_{X,x'}/f^\sharp_{x'} m_{Y,y'}$ is reduced, i.e., that $m_{Y,y'}$ generates the maximal ideal of $\mathcal{O}_{X,x'}$. Since the log structures on the DVRs $\mathcal{O}_{X,x'}$ and $\mathcal{O}_{Y,y'}$ are the standard ones, we can conclude that the maximal ideal of $\overline{\mathcal{M}}_{Y,y'}$ generates the maximal ideal of $\overline{\mathcal{M}}_{X,x'}$. Thus $\overline{f}^\flat_{x'}\colon \overline{\mathcal{M}}_{Y,y'} \to \overline{\mathcal{M}}_{X,x'}$ is an isomorphism, and f is strict at x'. Since the cospecialization maps $\overline{\mathcal{M}}_{X,x} \to \overline{\mathcal{M}}_{X,x'}$ and $\overline{\mathcal{M}}_{Y,y} \to \overline{\mathcal{M}}_{Y,y'}$ are isomorphisms, it follows that f is also strict at x, as required. □

V

Betti and de Rham Cohomology

In this chapter we concentrate on geometry over the field **C** of complex numbers. Let $X/\mathbf{C}$ be a log scheme whose underlying scheme $\underline{X}/\mathbf{C}$ is locally of finite type. It is a straightforward matter to associate to $X/\mathbf{C}$ a log analytic space X_{an} (see Definition 1.1.4). A more subtle construction, due to K. Kato and C. Nakayama [50], is the ringed space $(X_{log}, \mathcal{O}_X^{log})$, which maps canonically to the ringed space $(X_{an}, \mathcal{O}_X^{an})$ and which gives a very good geometric picture of the meaning of the log structure on X. The space X_{log} fits into a commutative diagram of topological spaces

$$\begin{array}{ccc} & & X_{log} \\ & \overset{j_{log}}{\nearrow} & \downarrow{\scriptstyle \tau_X} \\ X^*_{an} & \xrightarrow{\ j\ } & X_{an}, \end{array}$$

in which τ_X is proper and is in some sense a "real blowup" of X_{an} determined by its log structure. If $X/\mathbf{C}$ is (logarithmically) smooth, then X_{log} is a topological manifold with boundary $X_{log} \setminus X^*_{an}$. This diagram expresses the sense in which $X_{log} \to X_{an}$ is a compactification of the open immersion $j\colon X^* \to \underline{X}$ that preserves its homotopy-theoretic properties. We call X_{log} the *Betti realization*[1] of X and define the Betti cohomology of X to be the singular cohomology of X_{log}.

[1] The space X_{log} is sometimes referred to as the *Kato–Nakayama space* of X.

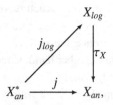

1 Betti realizations of log schemes

1.1 The space X_{an}

Let X be a scheme of finite type over $\mathbf{C}$. The set $X(\mathbf{C})$ of $\mathbf{C}$-valued points of X is classically endowed with the topology induced from the classical (strong) topology on $\mathbf{C}$. This is the weakest topology with the property that for every Zariski open subset U of X and every section f of $\mathcal{O}_X(U)$, the function $U(\mathbf{C}) \to \mathbf{C}$ given by f is continuous. We should remark that one gets the same result if one uses étale open sets $U \to X$ instead of Zariski open sets. This follows from the implicit function theorem in complex analysis, which says that if $U \to X$ is étale, then every point of $U(\mathbf{C})$ has a strong neighborhood basis of open sets V such that the restriction $V \to X(\mathbf{C})$ is an open embedding. If U is affine and $(f_1, \ldots, f_n)$ is a set of generators for $\mathcal{O}_X(U)$ over $\mathbf{C}$, the topology on $U(\mathbf{C})$ is also the weakest topology such that each f_i is continuous, and it is the topology induced from $\mathbf{C}^n$ via the closed immersion $U(\mathbf{C}) \to \mathbf{C}^n$ given by $(f_1, \ldots, f_n)$. We denote by X_{an} or X^{an} the topological space $X(\mathbf{C})$ with this topology.

When Q is a fine monoid and $X = \underline{\mathsf{A}}_Q$, the underlying set of X_{an} can be identified with $\mathsf{X}_Q = \operatorname{Hom}(Q, \mathbf{C})$, and its topology has the following useful description.

Proposition 1.1.1. *Let Q be a fine monoid, let $X := \underline{\mathsf{A}}_Q$, and let $x_0 \colon Q \to \mathbf{C}$ be an element of X_{an}.*

1. *Let S be a set of generators for Q, and for each $\delta > 0$, let*

$$U_{\delta, x_0} := \{x \in X_{an} : |x(s) - x_0(s)| < \delta \text{ for all } s \in S.\}$$

Then the set of all such U_δ forms a neighborhood basis for x_0 in X_{an}.

2. *In particular, if Q is sharp and S is the set of irreducible elements of Q, then the set of all sets of the form*

$$U_\delta := \{x : |x(s)| < \delta \text{ for all } x \in S\}$$

forms a neighborhood basis for the vertex of X_{an}. □

Proof If S is a set of generators for Q, then the corresponding set of elements of $\mathbf{C}[Q]$ generates the algebra $\mathbf{C}[Q] = \mathcal{O}_X(X)$. Hence X_{an} has the weakest topology such that, for all $s \in S$, the function $e_s \colon X \to \mathbf{C}$ given by evaluation is s is continuous. Statement (1) follows immediately from this fact. Statement (2) follows from (1) along with Proposition I.2.1.2. □

The neighborhood bases described in Proposition 1.1.1 allow us to construct a useful local version of the deformation retracts associated to a face of Q (see Proposition I.3.3.1).

Proposition 1.1.2. *Let Q be a fine monoid, let F be a face of Q, and let x_0 be a point of $\underline{\mathsf{A}}_{\mathsf{F}}$, viewed as an element of X_{Q} via the closed embedding i_F (Proposition I.3.3.1). Then x_0 has a neighborhood basis of open sets that are stable under the retraction $r_F\colon \mathsf{X}_{\mathsf{Q}} \to \mathsf{X}_{\mathsf{F}}$ and under the homotopy making i_F a strong deformation retract.*

Proof Let S be any finite set of generators for Q, and for each $\delta > 0$ let U_δ be the open neighborhood of x_0 defined in Proposition 1.1.1. Recall that if $x \in \mathsf{X}_{\mathsf{Q}}$, then $i_F r_F(x) \in \mathsf{X}_{\mathsf{Q}}$ is the homomorphism $Q \to \mathbf{C}$ sending q to $0 = x_0(q)$ if $q \notin F$ and to $x(q)$ if $q \in F$. Thus $i_F r_F$ maps U_{δ,x_0} to U_{δ,x_0}. Let us verify that the homotopies $f\colon \mathsf{X}_{\mathsf{N}} \times \mathsf{X}_{\mathsf{Q}} \to \mathsf{X}_{\mathsf{Q}}$ between id and $i_F \circ r_F$ constructed in Proposition I.3.3.1 induce a map $[0,1] \times U_{\delta,x_0} \to U_{\delta,x_0}$. Let $h\colon Q \to \mathbf{N}$ be a homomorphism with $h^{-1}(0) = F$. If $t \in [0,1]$ and $x \in U_{\delta,x_0}$, then $y := f(t,x)$ is by definition the homomorphism sending each $q \in Q$ to $t^{h(q)}x$. If $q \notin F$, then $x_0(q) = 0$ and $|y(q) - x_0(q)| = |t^{h(q)}x(q)| \le |x(q)| < \delta$. On other hand, if $q \in F$, then $h(q) = 0$ and $|y(q) - x_0(q)| = |t^0 x(q) - x_0(q)| = |x(q) - x_0(q)| < \delta$. □

So far we have discussed only the topological space X_{an} associated to a scheme of finite type over $\mathbf{C}$. To truly pass into the realm of analytic geometry, we need to introduce the sheaf $\mathcal{O}_X^{an}$ of analytic functions on X_{an}. We refer to [34, Appendix B] for precise definitions. Let us note here the following explicit description of the ring of germs of analytic functions at the vertex of a monoid scheme over $\mathbf{C}$.

Proposition 1.1.3. *Let Q be a fine sharp monoid, let x_0 be the vertex of X_{Q} and let h be any local homomorphism $Q \to \mathbf{N}$. Then a formal power series $\alpha := \sum_q a_q e^q \in \mathbf{C}[[Q]]$ converges in some neighborhood of x_0 if and only if the set of (extended) real numbers*

$$\left\{ \frac{\log |a_q|}{h(q)} : q \in Q^+ \right\}$$

is bounded above.

Proof We let T be the set of irreducible elements of Q, and use the notation of Proposition 1.1.1. Suppose that $\alpha = \sum_q a_q e^q$, and that $b \in \mathbf{R}$ is an upper bound for the set of all $h(q)^{-1} \log |a_q|$ with $q \in Q^+$. Choose $\epsilon > 0$, let $\lambda_t := -(b+\epsilon)h(t)$ for each $t \in T$, and choose a positive number δ such that $\delta < e^{\lambda_t}$ for all t. Then U_δ is an open neighborhood of x_0 in X_{Q}, and $\log |x(t)| < \lambda_t$ for all $t \in T$ and all

$x \in U_\delta$. Any $q \in Q$ can be written as $q = \sum n_t t$ with $n_t \in \mathbf{N}$. Hence if $x \in U_\delta$,

$$\begin{aligned}
\log |a_q x(q)| &= \log |x(q)| + \log |a_q| \\
&\leq \log |x(q)| + bh(q) \\
&\leq \sum_t (n_t \log |x(t)| + bn_t h(t)) \\
&\leq \sum_t n_t (\lambda_t + bh(t)) \\
&\leq \sum_t n_t(-\epsilon h(t)) \\
&\leq -\epsilon h(q)
\end{aligned}$$

Thus $|a_q x(q)| \leq r^{h(q)}$, where $r := e^{-\epsilon} < 1$. By Proposition I.2.2.9, there exist C and m such that $\{q : h(q) = i\}$ has cardinality less than Ci^m for all i. Then the set of partial sums of the series $\sum_q |a_q x(p)|$ is bounded by the set of partial sums of the series $\sum_i Ci^m r^i$. Since this latter series converges, so does the former.

Suppose on the other hand that $\alpha := \sum a_q e^q$ and that $\{h(q)^{-1} \log |a_q| : q \in Q^+\}$ is unbounded. For $c \in \mathbf{R}^+$, define $x_c \colon Q \to \mathbf{C}$ by $x_c(q) := c^{-h(q)}$. Then $x_c \in \mathsf{X}_Q$ and, if $\delta > 0$ and c is chosen large enough that $\log c > (h(t))^{-1}(-\log \delta)$ for all $t \in T$, then $x_c \in U_\delta$. For every such c, there are infinitely many $q \in Q^+$ such that $|a_q| > (1 + c)^{h(q)}$. For such a q,

$$|a_q x_c(q)| \geq (1 + c)^{h(q)} c^{-h(q)} = (1/c + 1)^{h(q)} \geq 1,$$

so the series $\sum_q a_q x_c(q)$ cannot converge. □

If $X/\mathbf{C}$ is a scheme of finite type, we let X_{an} denote the associated analytic space. Then there is a morphism of locally ringed spaces $a \colon X_{an} \to X$.

Definition 1.1.4. *A log analytic space is an analytic space X together with a logarithmic homomorphism of sheaves of commutative monoids $\alpha_X \colon \mathcal{M}_X \to \mathcal{O}_X$. If $X/\mathbf{C}$ is a log scheme whose underlying scheme $\underline{X}/\mathbf{C}$ is locally of finite type over $\mathbf{C}$, then the associated log analytic space X_{an} is $\underline{X}_{an}$, with the induced log structure $\alpha_X^{an} \colon \mathcal{M}_X^{an} := a_{log}^*(\mathcal{M}_X) \to \mathcal{O}_X^{an}$.*

We shall not develop the theory of smooth morphisms of log analytic spaces in detail here. We hope it will suffice to say that a morphism of fine log analytic spaces spaces is smooth if it admits local charts of the form described in Theorem IV.3.3.1, and that the analogies of the related results for schemes go through without difficulty. In particular, smooth log analytic spaces admit the following useful local description.

Proposition 1.1.5. *If Q and Q' are fine monoids with $Q \subseteq Q'$, let $\mathsf{X}_{Q'}(Q)$ denote the log analytic space associated to* $\operatorname{Spec}(Q \to \mathbf{C}[Q'])$.

1. *If r is a natural number and Q is a fine monoid, the log analytic space $X_{\mathbf{N}^r \oplus Q}(Q)$ is smooth over $\mathbf{C}$.*

2. *Let x be a point of a fine and smooth log analytic space X. Then there exist a neighborhood U of x in X, a natural number r, a fine sharp monoid Q, and an isomorphism from U to an open subset of $X_{\mathbf{N}^r \oplus Q}(Q)$ taking x to the vertex of $X_{\mathbf{N}^r \oplus Q}(Q)$.*

Proof Since the log scheme $\operatorname{Spec}(Q \to \mathbf{C}[\mathbf{N}^r \oplus Q])$ is smooth over $\mathbf{C}$, so is its associated log analytic space. Conversely, if x is a point of a smooth and fine analytic space X, then the analytic analog of Corollary IV.3.3.4 says that, after replacing X by an étale neighborhood of x, there exists a strict étale map $f\colon X \to X_{\mathbf{N}^r \oplus Q}(Q)$ taking x to the vertex of the latter. The restriction of f to a small neighborhood of x is then an isomorphism onto its image. □

The results and statements in this section have obvious idealized analogs, which we leave to the reader.

1.2 The space X_{log}

Let us begin with a simple example, the log scheme $A_{\mathbf{N}}$. The associated complex analytic space $X_{\mathbf{N}}$ is just the set of complex numbers $\mathbf{C}$ with its usual topology. The space $X_{\mathbf{N}}^{\log}$ turns out to be the real blowup of $\mathbf{C}$ at the origin, which is effectuated by the introduction of polar coordinates. Consider the following monoids (with monoid law given by multiplication) and monoid homomorphisms:

$$\begin{aligned}
\mathbf{S}^1 &:= \{z \in \mathbf{C} : |z| = 1\}\\
\mathbf{R}_{\geq} &:= \{r \in \mathbf{R} : r \geq 0\},\\
\mathbf{Z}(1) &:= \{2\pi i n : n \in \mathbf{Z}\} \subseteq \mathbf{C}\\
\mathbf{R}(1) &:= \{ri : r \in \mathbf{R}\} \subseteq \mathbf{C}\\
\arg\colon \mathbf{C}^* &\to \mathbf{S}^1 : u \mapsto u/|u|\\
\operatorname{abs}\colon \mathbf{C} &\to \mathbf{R}_{\geq} : z \mapsto |z|\\
\tau\colon \mathbf{S}^1 \times \mathbf{R}_{\geq} &\to \mathbf{C} : (\zeta, \rho) \mapsto \zeta\rho\\
\exp\colon \mathbf{R}(1) &\to \mathbf{S}^1 : \theta \mapsto \exp(\theta) := e^{\theta}.
\end{aligned}$$

There is a commutative diagram:

$$\begin{array}{ccccc}
\mathbf{R}(1)\times\mathbf{R}_{>} & \hookrightarrow & \mathbf{R}(1)\times\mathbf{R}_{\geq} & =: & \tilde{\mathsf{X}}^{\log}_{\mathbf{N}} \\
\Big\downarrow{\scriptstyle \exp\times\mathrm{id}} & & \Big\downarrow{\scriptstyle \exp\times\mathrm{id}} & & \Big\downarrow{\scriptstyle\eta} \\
\mathbf{S}^1\times\mathbf{R}_{>} & \hookrightarrow & \mathbf{S}^1\times\mathbf{R}_{\geq} & =: & \mathsf{X}^{\log}_{\mathbf{N}} \\
\Big\downarrow{\scriptstyle\tau'} & & \Big\downarrow{\scriptstyle\tau} & \swarrow{\scriptstyle\tau} & \\
\mathbf{C}^* & \hookrightarrow & \mathbf{C}. & &
\end{array}$$

The map η in this diagram is a universal covering of $\mathsf{X}^{\log}_{\mathbf{N}}$, with covering group $\mathbf{Z}(1)$. The map τ' is a homeomorphism, with inverse

$$\mathbf{C}^* \to \mathbf{S}^1\times\mathbf{R}_{>} : z \mapsto (\arg(z), \mathrm{abs}(z)).$$

Then

$$\tau^{-1}(z) = \begin{cases} \{z\} & \text{if } z \neq 0 \\ \mathbf{S}^1 & \text{if } z = 0, \end{cases}$$

and, as the diagram shows, τ restricts to an isomorphism on $\tau^{-1}(\mathbf{C}^*)$. Thus τ replaces the origin of $\mathbf{C}$ by $\mathbf{S}^1$, which can be identified with the set of rays through the origin. This is the sense in which τ is the real blowup of $\mathbf{C}$ at the origin, as shown in Figure 1.2.1.

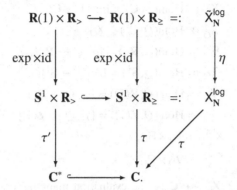

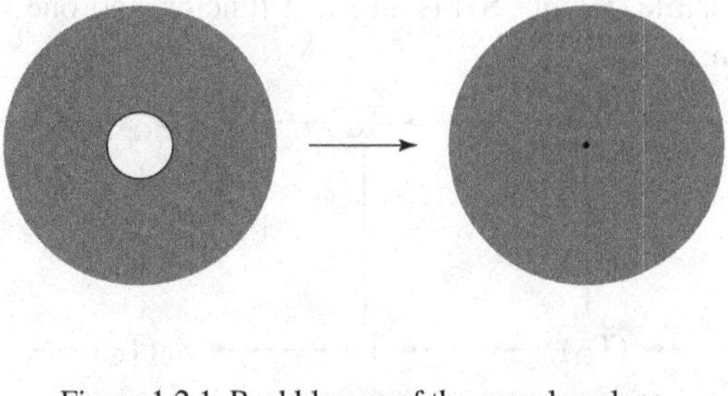

Figure 1.2.1 Real blowup of the complex plane.

A similar construction can be carried out with $\mathbf{N}$ replaced by any finitely generated monoid Q. We use the following notation:

$$\begin{aligned}
\mathsf{X}_\mathsf{Q} &:= \operatorname{Hom}(Q, \mathbf{C}) = \{z\colon Q \to \mathbf{C}\},\\
\mathsf{X}^*_\mathsf{Q} &: \operatorname{Hom}(Q, \mathbf{C}^*) \subseteq \mathsf{X}_\mathsf{Q},\\
\mathsf{R}_\mathsf{Q} &:= \operatorname{Hom}(Q, \mathbf{R}_\geq) = \{\rho\colon Q \to \mathbf{R}_\geq\},\\
\mathsf{T}_\mathsf{Q} &:= \operatorname{Hom}(Q, \mathbf{S}^1) = \{\zeta\colon Q \to \mathbf{S}^1\},\\
\mathsf{V}_\mathsf{Q} &:= \operatorname{Hom}(Q, \mathbf{R}(1)) = \{\theta\colon Q \to \mathbf{R}(1)\},\\
\mathsf{I}_\mathsf{Q} &:= \operatorname{Hom}(Q, \mathbf{Z}(1)) = \{\gamma\colon Q \to \mathbf{Z}(1)\},\\
\mathsf{X}^{\log}_\mathsf{Q} &:= \mathsf{T}_\mathsf{Q} \times \mathsf{R}_\mathsf{Q},\\
\tilde{\mathsf{X}}^{\log}_\mathsf{Q} &:= \mathsf{V}_\mathsf{Q} \times \mathsf{R}_\mathsf{Q}\,.
\end{aligned}$$

If $q \in Q$, let $e_q\colon \mathsf{X}_\mathsf{Q} \to \mathbf{C}$ be the evaluation mapping $x \mapsto x(q)$, and use the same notation for the similarly defined maps $\mathsf{T}_\mathsf{Q} \to \mathbf{S}^1$ and $\mathsf{R}_\mathsf{Q} \to \mathbf{R}$. Endow each of the sets X_Q, T_Q, and X_Q with the weak (product) topology induced by the standard topology on the target and the functions e_q for q belonging to any set of generators for Q. Then X_Q is identified with the topological space underlying the analytic space associated to $\underline{\mathsf{A}}_\mathsf{Q}$. The multiplication map $\mathbf{S}^1 \times \mathbf{R}_\geq \to \mathbf{C}$ induces a map

$$\tau_Q\colon \mathsf{T}_\mathsf{Q} \times \mathsf{R}_\mathsf{Q} \to \mathsf{X}_\mathsf{Q} : (\zeta, \rho) \mapsto \zeta\rho. \tag{1.2.1}$$

If $x = \tau_Q(\zeta, \rho)$, then (ζ, ρ) can be viewed as a set of polar coordinates for x.

Remark 1.2.1. Since $\mathbf{S}^1$ is compact, T_Q is a compact topological group. Let Q^{gp}_t denote the torsion subgroup of Q^{gp} and let Q^{gp}_f denote the quotient of Q^{gp} by Q^{gp}_t. Then Q^{g}_t is a finite abelian group, Q^{gp}_f is a finitely generated free abelian group, and there is an exact sequence

$$0 \to Q^{\mathrm{gp}}_t \to Q^{\mathrm{gp}} \to Q^{\mathrm{gp}}_f \to 0.$$

Since $\mathbf{S}^1$ is divisible, $\operatorname{Hom}(\ ,\mathbf{S}^1)$ is an exact functor, and one obtains the following diagram:

$$\begin{array}{ccccccccc}
0 & \longrightarrow & \mathsf{T}_{Q^{\mathrm{gp}}_f} & \longrightarrow & \mathsf{T}_\mathsf{Q} & \longrightarrow & \mathsf{T}_{Q^{\mathrm{gp}}_t} & \longrightarrow & 0\\
 & & \downarrow{\scriptstyle =} & & \downarrow{\scriptstyle =} & & \downarrow{\scriptstyle =} & & \\
0 & \longrightarrow & (\mathsf{T}_\mathsf{Q})^0 & \longrightarrow & \mathsf{T}_\mathsf{Q} & \longrightarrow & \pi_0(\mathsf{T}_\mathsf{Q}) & \longrightarrow & 0.
\end{array}$$

Here $(\mathsf{T}_\mathsf{Q})^0$ is the connected component of T_Q containing the identity and $\pi_0(\mathsf{T}_\mathsf{Q})$ is the group of connected components of T_Q (a finite group). When Q is toric, $Q^{\mathrm{gp}}_t = 0$ and T_Q is already connected.

The exponential map $\mathbf{R}(1) \to \mathbf{S}^1$ induces a map

$$\eta_Q \colon \tilde{\mathsf{X}}_{\mathsf{Q}}^{\log} \to \mathsf{X}_{\mathsf{Q}}^{\log}, \tag{1.2.2}$$

which maps surjectively onto $(\mathsf{T}_{\mathsf{Q}})^0$. This map is an even covering, and $\tilde{\mathsf{X}}_{\mathsf{Q}}^{\log}$ is contractible. In particular, if Q is toric, the map η_Q is a universal covering of X_{Q}. The natural (right) action of I_{Q} on $\tilde{\mathsf{X}}_{\mathsf{Q}}^{\log}$, given by

$$((\theta,\rho)\gamma)(q) := (\theta(q)+\gamma(q),\rho(q)), \tag{1.2.3}$$

makes $\tilde{\mathsf{X}}_{\mathsf{Q}}^{\log}$ into a torsor over $\mathsf{X}_{\mathsf{Q}}^{\log}$. It follows that I_{Q} identifies with the group of covering transformations of η_Q. Thus I_{Q} can be viewed as the fundamental group of $\mathsf{X}_{\mathsf{Q}}^{\log}$; since the group is abelian, the choice of base point is not crucial.

There is an obvious closed embedding $\mathbf{R}_{\geq} \to \mathbf{C}$, admitting a retraction $\text{abs}\colon \mathbf{C} \to \mathbf{R}_{\geq}$, both of which are group homomorphisms. There are corresponding morphisms $\mathsf{R}_{\mathsf{Q}} \subseteq \mathsf{X}_{\mathsf{Q}}$ and $\mathsf{X}_{\mathsf{Q}} \to \mathsf{R}_{\mathsf{Q}}$, so that R_{Q} is a retract X_{Q}. There is also a closed embedding of topological groups $\mathbf{S}^1 \to \mathbf{C}^*$, with a retraction $\arg\colon \mathbf{C}^* \to \mathbf{S}^1$, giving a closed embedding $\mathsf{T}_{\mathsf{Q}} \to \mathsf{X}_{\mathsf{Q}}^*$ with a retraction $\mathsf{X}_{\mathsf{Q}}^* \to \mathsf{T}_{\mathsf{Q}}$. We find a commutative diagram:

$$\begin{array}{ccc} \mathsf{T}_{\mathsf{Q}}\times\mathsf{R}_{\mathsf{Q}} & \longrightarrow & \mathsf{X}_{\mathsf{Q}}^*\times\mathsf{X}_{\mathsf{Q}} \\ \Big\downarrow{\scriptstyle pr_2} & & \Big\downarrow \\ \mathsf{R}_{\mathsf{Q}} & \xleftarrow{\ \text{abs}\ } & \mathsf{X}_{\mathsf{Q}}, \end{array}$$

where the right vertical arrow is the action of $\mathsf{X}_{\mathsf{Q}}^*$ on X_{Q} by multiplication.

Proposition 1.2.2. *If Q is an integral monoid, then two points x and x' of X_{Q} have the same image in R_{Q} if and only they are in the same orbit under the action of T_{Q}.*

Proof If $\zeta \in \mathsf{T}_{\mathsf{Q}}$ and $x' = \zeta x$, it is clear that $\text{abs}(x') = \text{abs}(x)$. Conversely suppose that $\text{abs}(x') = \text{abs}(x)$. Then $F := x^{-1}(\mathbf{C}^*) = x'^{-1}(\mathbf{C}^*)$, and x and x' induce homomorphisms $F^{\text{gp}} \to \mathbf{C}^*$. Furthermore, x'/x is a homomorphism $F^{\text{gp}} \to \mathbf{S}^1$. Since $\mathbf{S}^1$ is a divisible group, this homomorphism extends to a homomorphism $\zeta\colon Q^{\text{gp}} \to \mathbf{S}^1$. Then $x'(q) = \zeta(q)x(q)$ for all $q \in F$ and, since both sides are zero for $q \notin F$, in fact the same holds for all q. Thus $x' = \zeta x$. □

Proposition 1.2.3. *Let Q be a finitely generated monoid. Then τ_Q is proper. If Q is fine, then τ_Q is surjective and, for each $x \in \mathsf{X}_{\mathsf{Q}}$, $\tau_Q^{-1}(x)$ is naturally a torsor under $\mathsf{T}_{\mathsf{Q}/\mathsf{F}_x}$, where $F_x := x^{-1}(\mathbf{C}^*)$.*

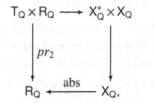

Proof Consider the commutative diagram

$$\begin{array}{ccc} \mathsf{T_Q} \times \mathsf{R_Q} & \xrightarrow{\mathrm{id} \times i} & \mathsf{T_Q} \times \mathsf{X_Q} \\ \downarrow{\scriptstyle \tau_Q} & & \downarrow{\scriptstyle \lambda} \\ \mathsf{X_Q} & \xleftarrow{pr_2} & \mathsf{T_Q} \times \mathsf{X_Q} . \end{array}$$

Here i is the inclusion $\mathsf{R_Q} \to \mathsf{X_Q}$, and $\lambda(\zeta, x) := (\zeta, \zeta x)$. The closed immersion $\mathrm{id} \times i$ is continuous and proper and λ is a homeomorphism, with inverse $(\zeta, x) \mapsto (\zeta, \zeta^{-1} x)$. Thus λ is continuous and proper and, since $\mathsf{T_Q}$ is compact, the projection mapping pr_2 is also continuous and proper. It follows that τ_Q is continuous and proper.

Now suppose that Q is integral. Any $x \in \mathsf{X_Q}$ defines a map $F_x^{\mathrm{gp}} \to \mathbf{C}^*$, and $x(q) = \arg(x(q))\mathrm{abs}(x(q))$ for every $q \in F_x$. Since $\mathbf{S}^1$ is divisible and $F_x^{\mathrm{gp}} \subseteq Q^{\mathrm{gp}}$, the homomorphism $\arg \circ x \colon F_x^{\mathrm{gp}} \to \mathbf{S}^1$ extends to Q^{gp} and, for any such extension ζ, we have $\tau_Q(\zeta, \mathrm{abs} \circ x) = x$. Furthermore, the set of such extensions is a torsor under $\mathrm{Hom}(Q/F_x, \mathbf{S}^1)$. Finally, note that if $\tau_Q(\zeta, \rho) = \tau_Q(\zeta', \rho') = x$, then $\rho = \mathrm{abs}(x) = \mathrm{abs}(x')$ and $\zeta(q) = \zeta'(q) = \arg(x(q))$ for all $q \in F$. Therefore $\zeta' \zeta^{-1}$ lies in $\mathrm{Hom}(Q/F_x, \mathbf{S}^1)$. □

Let us now turn to the global construction of X_{log}. We alert the reader that there seems to be no possible construction of a space $\tilde{X}_{log}$ globalizing the space $\tilde{\mathsf{X}}_{\mathsf{Q}}^{\mathsf{log}}$. Proposition 1.4.1 will discuss a sheaf-theoretic substitute.

Definition 1.2.4. *Let X be a log analytic space.*

1. *X_{log} is the set of pairs (x, σ), where x is a point of X and $\sigma \colon \mathcal{M}_{X,x} \to \mathbf{S}^1$ is a homomorphism of monoids fitting into the commutative diagram*

$$\begin{array}{ccc} \mathcal{M}_{X,x} & \xrightarrow{\sigma} & \mathbf{S}^1 \\ \uparrow{\scriptstyle \alpha_{X,x}} & & \uparrow{\scriptstyle \arg} \\ \mathcal{O}_{X,x}^* & \xrightarrow{x^*} & \mathbf{C}^*, \end{array}$$

   and $\tau_X \colon X_{log} \to X$ is the map sending (x, σ) to x.
2. *For each section m of $\mathcal{M}_X$ on an open subset U of X, $\arg(m)$ is the function*

$$\arg(m) \colon \tau_X^{-1}(U) \to \mathbf{S}^1 : (x, \sigma) \mapsto \sigma(m_x).$$

3. *X_{log} is given the weak topology defined by the map $\tau_X \colon X_{log} \to X$ and the family of maps $\{\arg(m) : U \subseteq X, m \in \mathcal{M}_X(U)\}$.*

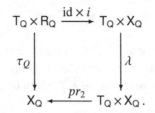

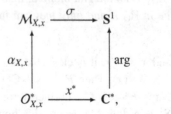

Proposition 1.2.5. *Let X be a log analytic space and X_{log} its Betti realization.*

1. *The map $\tau_X \colon X_{log} \to X$ is a continuous and separated, and is a homeomorphism if the log structure on X is trivial.*
2. *Let x be a point of x and let*

$$\mathsf{T}_x := \mathrm{Hom}(\overline{\mathcal{M}}_{X,x}, \mathbf{S}^1).$$

   Then T_x acts naturally on the fiber $\tau_X^{-1}(x)$, and this action defines a torsor if $\mathcal{M}_X$ is u-integral. In particular, τ_X is surjective if $\mathcal{M}_X$ is u-integral.
3. *The construction of X_{log} is functorial in X, and the morphism τ_X is natural. That is, a morphism of log schemes $f \colon X \to Y$ induces the commutative diagram*

$$\begin{array}{ccc} X_{log} & \xrightarrow{f_{log}} & Y_{log} \\ \downarrow{\scriptstyle \tau_X} & & \downarrow{\scriptstyle \tau_Y} \\ X_{an} & \xrightarrow{f_{an}} & Y_{an}. \end{array}$$

   This diagram is cartesian if f is strict.

Proof The continuity of τ_X is built into the definition of the topology on X_{log}. To see that τ_X is separated, suppose that (x_1, σ_1) and (x_2, σ_2) are distinct points of X_{log} with the same image in X. Then $x := x_1 = x_2$, so $\sigma_1 \neq \sigma_2$. Hence there is an $m_x \in \mathcal{M}_{X,x}$ such that $\sigma_1(m_x) \neq \sigma_2(m_x)$ in $\mathbf{S}^1$. Choose an open neighborhood U of x in X and a section m of $\mathcal{M}_X(U)$ whose germ at x is m_x. Since $\mathbf{S}^1$ is Hausdorff, there exist disjoint open neighborhoods V_1, V_2 of $\sigma_1(m_x), \sigma_2(x)$. Then $\arg(m)^{-1}(V_1)$ and $\arg(m)^{-1}(V_2)$ are disjoint neighborhoods of (x, σ_1) and (s, σ_2) in X_{log}. It is clear from the definition that τ_X is bijective if the log structure on X is trivial. Moreover, if u is a section of $\mathcal{M}_X = \mathcal{O}_X^*$, then $\arg(u)$ is already continuous on X, and hence τ_X is a homeomorphism.

Suppose $(x, \sigma) \in X_{log}$ and $\zeta \in \mathsf{T}_x$. Define $\zeta\sigma \colon \mathcal{M}_{X,x} \to \mathbf{S}^1$ by the formula $(\zeta\sigma)(m) := \zeta(\overline{m})\sigma(m)$. Since $\zeta(u) = 1$ for $u \in \mathcal{O}_{X,x}^*$, it follows that $(x, \zeta\sigma) \in X_{log}$. Thus we have defined an action of T_x on $\tau_X^{-1}(x)$.

Since $\mathcal{M}_X$ is u-integral, the sequence

$$1 \to \mathcal{O}_{X,x}^* \to \mathcal{M}_{X,x}^{\mathrm{gp}} \to \overline{\mathcal{M}}_{X,x}^{\mathrm{gp}} \to 0$$

is exact, and the divisibility of $\mathbf{S}^1$ implies that the sequence

$$0 \to \mathrm{Hom}(\overline{\mathcal{M}}_{X,x}^{\mathrm{gp}}, \mathbf{S}^1) \to \mathrm{Hom}(\mathcal{M}_{X,x}^{\mathrm{gp}}, \mathbf{S}^1) \to \mathrm{Hom}(\mathcal{O}_{X,x}^*, \mathbf{S}^1) \to 0$$

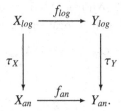

is also exact. It follows that $\arg \circ x^* \colon \mathcal{O}_{X,x}^* \to \mathbf{S}^1$ can be extended to $\mathcal{M}_{X,x}^{\mathrm{gp}}$, and hence that τ_X is surjective. Thanks to this exact sequence, the set of all $\sigma \in \mathrm{Hom}(\mathcal{M}^{gp}_{X,x}, \mathbf{S}^1)$ mapping to $\arg \in \mathrm{Hom}(\mathcal{O}^*_{X,x}, \mathbf{S}^1)$ is naturally a torsor under $\mathrm{Hom}(\overline{\mathcal{M}}^{gp}_{X,x}, \mathbf{S}^1)$. The set of all such σ is precisely $\tau_X^{-1}(x)$.

Let $f\colon X \to Y$ be a morphism of log schemes. If $(x, \sigma) \in X_{log}$, let $f(\sigma)$ be the composition

$$f(\sigma) := \mathcal{M}_{Y,f(x)} \xrightarrow{f^\flat_x} \mathcal{M}_{X,x} \xrightarrow{\sigma} \mathbf{S}^1.$$

Then $(f(x), f(\sigma)) \in Y_{log}$, and $(x, \sigma) \mapsto (f(x), f(\sigma))$ defines a map $f_{log}\colon X_{log} \to Y_{log}$. For each open subset V of Y and each section m of $\mathcal{M}_Y(V)$, the function $\arg(m) \circ f_{log} = \arg(f^\flat(m))$ is continuous on $f_{log}^{-1}(\tau_Y^{-1}(V))$, and it follows that f_{log} is continuous. The commutativity of the diagram is immediate from the definitions. If f is strict, $f^*_{log}(\mathcal{M}_Y) \to \mathcal{M}_X$ is an isomorphism and, since the homomorphism $f^{-1}(\mathcal{M}_Y) \to \mathcal{M}_X$ is local, the diagram

$$\begin{array}{ccc} \mathcal{O}^*_{X,x} & \longrightarrow & \mathcal{M}_{X,x} \\ \uparrow & & \uparrow \\ \mathcal{O}^*_{Y,f(x)} & \longrightarrow & \mathcal{M}_{Y,f(x)} \end{array}$$

is cocartesian, for each x in X. It follows that f_{log} maps $\tau_X^{-1}(x)$ bijectively to $\tau_Y^{-1}(f(x))$. Thus the diagram in the proposition is set-theoretically cartesian. Since f is strict, every section of $\mathcal{M}_X$ is locally the product of a section of $\mathcal{O}^*_X$ and a section of $f^{-1}(\mathcal{M}_Y)$, and it follows that that the topology on X_{log} is the weak topology defined by τ_X and f_{log}. That is, X_{log} has the product topology induces from X and Y_{log}, and the diagram is cartesian in the category of topological spaces. □

Remark 1.2.6. If T is any topological space, we let $\mathbf{S}^1_T$ (or $\underline{\mathbf{S}^1}$) denote the sheaf of continuous functions from T to $\mathbf{S}^1$. The construction of X_{log} includes a homomorphism of abelian sheaves $\arg\colon \tau_X^{-1}(\mathcal{M}_X) \to \mathbf{S}^1_{X_{log}}$ with the property that its composition with the inclusion $\tau_X^{-1}(\mathcal{O}^*_X) \to \tau_X^{-1}(\mathcal{M}^{gp}_X)$ agrees with the usual arg function. It follows easily from the construction that this homomorphism is universal. That is, if $\tau\colon T \to X$ is a continuous mapping of topological spaces and $a\colon \tau^{-1}(\mathcal{M}_X) \to \mathbf{S}^1_T$ is an extension of arg, there is a unique map $\tau'\colon T \to X_{log}$ such that $\tau_X \circ \tau' = \tau$ and such that a is the composition of $\tau'^{-1}(\arg)$ with the natural map $\tau'^{-1}(\mathbf{S}^1_{X_{log}}) \to \mathbf{S}^1_T$.

Before proceeding, we verify that the general construction given in Definition 1.2.4 agrees with the previous construction $\mathsf{X}^{\mathsf{log}}_{\mathsf{Q}}$ for monoid schemes.

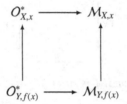

Proposition 1.2.7. *Let Q be a finitely generated monoid, let A_Q be the associated log scheme, and let $\mathsf{A}_\mathsf{Q}^{\log}$ be its Betti realization. Then there is a commutative diagram*

$$\begin{array}{ccc} \mathsf{A}_\mathsf{Q}^{\log} & \xrightarrow{\;\xi^{log}\;} & \mathsf{X}_\mathsf{Q}^{\log} \\ {\scriptstyle \tau_{\mathsf{A}_\mathsf{Q}^{\log}}}\big\downarrow & & \big\downarrow{\scriptstyle \tau_Q} \\ \mathsf{A}_\mathsf{Q}^{\mathrm{an}} & \xrightarrow{\;\xi\;} & \mathsf{X}_\mathsf{Q}, \end{array}$$

in which the horizontal maps are homeomorphisms.

Proof Let $X := \mathsf{A}_\mathsf{Q}$. By definition, $X = \operatorname{Spec}(Q \to \mathbf{C}[Q])$, and so is equipped with a chart $\beta\colon Q \to \Gamma(X, \mathcal{M}_X)$ and a homomorphism $\gamma\colon Q \to \Gamma(X, \mathcal{O}_X)$. The map ξ in the diagram takes a homomorphism $x\colon \mathbf{C}[Q] \to \mathbf{C}$ to $x \circ \gamma\colon Q \to \mathbf{C}$ and is, as we have already observed, a homeomorphism. We will henceforth identify x with $x \circ \gamma$.

If $(x, \sigma) \in X_{log}$, the composition $\zeta\colon Q \xrightarrow{\beta_x} \mathcal{M}_{X,x} \xrightarrow{\sigma} \mathbf{S}^1$ is a point of T_Q, and $\rho\colon Q \xrightarrow{\gamma_x} \mathcal{O}_{X,x} \xrightarrow{x^*} \mathbf{C} \xrightarrow{\text{abs}} \mathbf{R}_\geq$ is a point of R_Q. Then

$$\xi^{log}(x, \sigma) := (\zeta, \rho) \in \mathsf{X}_\mathsf{Q}^{\log} := \mathsf{T}_\mathsf{Q} \times \mathsf{R}_\mathsf{Q}\,.$$

Note that $\rho(q) = |x(q)|$ for all $q \in Q$, so $F_x := x^{-1}(\mathbf{C}^*) = \rho^{-1}(\mathbf{R}_>)$. Moreover, $\gamma_x(q) \in \mathcal{O}_X^*$ if $q \in F_x$, hence $\zeta(q) := \sigma(\beta_x(q)) = \arg(x(\gamma_x(q)))$ and $x(q) = \zeta(q)\rho(q)$. On other hand, $x(q) = \zeta(q)\rho(q) = 0$ if $q \notin F_x$. This shows that $\xi(\tau_X(x, \sigma)) = \tau_Q(\zeta, \rho)$, so that the diagram commutes.

Conversely, suppose that $(\zeta, \rho) \in \mathsf{X}_\mathsf{Q}^{\log}$. Let $x := \zeta\rho \in X_Q$ and let $F_x := x^{-1}(\mathbf{C}^*) = \rho^{-1}(\mathbf{R}_>)$. Since $x(q) = \zeta(q)\rho(q)$ and $\rho(q) \in \mathbf{R}_>$ for $q \in F_x$, in fact $\zeta(q) = \arg \circ x(q)$ for all $q \in F_x$. Since β is a chart for $\mathcal{M}_X$, the square in the diagram

$$\begin{array}{ccccc} F_x & \xrightarrow{\;i_x\;} & Q & & \\ {\scriptstyle \gamma'_x}\big\downarrow & & \big\downarrow & \overset{\zeta}{\searrow} & \\ \mathcal{O}^*_{X,x} & \longrightarrow & \mathcal{M}_{X,x} & \overset{\sigma}{\dashrightarrow} & \mathbf{S}^1 \end{array}$$

is cocartesian. Since $\arg \circ \gamma'_x = \zeta \circ i_x$ on F_x, there is a unique σ as shown making the diagram commute. Then $(x, \sigma) \in X_{log}$ is the unique point of X_{log} mapping to $x \in X$ and to $(\zeta, \rho) \in \mathsf{X}_\mathsf{Q}^{\log}$. This shows that the map ξ^{log} is bijective.

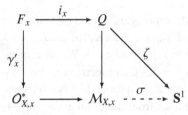

Finally we must show that ξ^{log} is a homeomorphism. For each $q \in Q$, we have evaluation maps $e_q\colon \mathsf{X}_{\mathsf{Q}}^{\mathsf{log}} \to \mathbf{S}^1$ and $\epsilon_q\colon \mathsf{X}_{\mathsf{Q}}^{\mathsf{log}} \to \mathbf{R}_{\geq}$, and $\mathsf{X}_{\mathsf{Q}}^{\mathsf{log}}$ has the weak topology defined by these maps. Thus it will suffice to show that X_{log} has the weak topology $\mathcal{T}'$ defined by the maps $e_q \circ \xi^{log}$ and $\epsilon_q \circ \xi^{log}$. If $q \in Q$ and $(x,\sigma) \in X_{log}$, then $\epsilon_q(\xi(x)) = |x(q)|$, so $\epsilon_q \circ \xi$ is continuous on X. Similarly, $e_q(\xi^{log}(x,\sigma)) = \sigma(\beta_x(q)) = \arg(\beta(q))(x,\sigma)$, so $e_q \circ \xi^{log}$ is also continuous. Thus the topology $\mathcal{T}$ of X_{log} is at least as strong as $\mathcal{T}'$.

Since $\xi\colon X \to \mathsf{X}_{\mathsf{Q}}$ is a homeomorphism and τ_Q is continuous, it follows that τ_X is $\mathcal{T}'$-continuous. Let U be an open subset of X and m a section of $\mathcal{M}_X(U)$. We claim that that $\arg(m)\colon \tau_X^{-1}(U) \to \mathbf{S}^1$ is $\mathcal{T}'$-continuous. This can be verified locally over U. Since β is a chart for $\mathcal{M}_X$, locally on U there exist a section u of $\mathcal{O}_X^*$ and an element q of Q such that $m = u + \beta(q)$. Then

$$\arg(m) = \arg(u \circ \tau_X)\arg(\beta(q) = \arg(u \circ \tau_X)e_q \circ \xi^{log}.$$

Since $\arg(u \circ \tau_X)$ and $e_q \circ \xi^{log}$ are continuous, so is $\arg(m)$. □

Corollary 1.2.8. *If X is a fine log analytic space, then $\tau_X\colon X_{log} \to X$ is proper.*

Proof If Q is a fine monoid and $X = \mathsf{A}_{\mathsf{Q}}^{\mathsf{an}}$, the properness of τ_X follows from Propositions 1.2.7 and 1.2.3. For the general case, we may verify the properness of τ_X locally on X, and hence we may assume without loss of generality that there is a chart $X \to \mathsf{A}_{\mathsf{Q}}^{\mathsf{an}}$, where Q is a fine monoid. Then the cartesian square in (3) of Proposition 1.2.5, with $Y = \mathsf{A}_{\mathsf{Q}}^{\mathsf{an}}$, shows that τ_X is proper. □

Corollary 1.2.9. *Let $f\colon X \to Y$ be a morphism of u-integral log analytic spaces such that $\underline{f}\colon \underline{X} \to \underline{Y}$ is an isomorphism.*

1. *If $\mathcal{M}_X$ is fine, then f_{log} is proper.*
2. *If $\mathcal{M}_X$ is fine and f is s-injective, the map $f_{log}\colon X_{log} \to Y_{log}$ is surjective, and τ_Y is also proper.*
3. *If $\mathcal{M}_X$ is fine and f is an immersion, then f_{log} is a closed embedding.*

Proof We identify the topological spaces of X and Y and hence $\tau_Y \circ f_{log}$ with τ_X. The properness of f_{log} can be verified locally on $\underline{X} = \underline{Y}$, so we may as well assume that this space is Hausdorff. We have seen in Corollary 1.2.8 that the coherence of $\mathcal{M}_X$ implies the properness of $\tau_X = \tau_Y \circ f_{log}$, and since τ_Y is separated, it follows that f_{log} is proper. If f is s-injective, the map $\overline{\mathcal{M}}_Y \to \overline{\mathcal{M}}_X$ is injective, and it follows from the u-integrality of $\mathcal{M}_X$ that the map $\mathcal{M}_Y \to \mathcal{M}_X$ is also injective. Then it follows from the divisibility of $\mathbf{S}^1$ that f_{log} is surjective, and the properness of τ_X implies that of τ_Y. If f is an immersion, $f^\flat$ is surjective, hence evidently f_{log} is injective and, since it is proper, it is a closed embedding. □

Corollary 1.2.10. *If the log structure on X is fine, the map $X^{\mathrm{sat}}_{log} \to X_{log}$ is a homeomorphism.*

Proof This statement can be verified locally on X, so we may assume that $\mathcal{M}_X$ admits a chart subordinate to a fine monoid Q, which we view as a strict morphism $X \to \mathsf{A}_Q$. Then $X^{\mathrm{sat}} \cong X \times_{\mathsf{A}_Q} \mathsf{A}_{Q^{\mathrm{sat}}}$, and hence, by (3) of Proposition 1.2.5, $X^{\mathrm{sat}}_{log} \cong X_{log} \times_{\mathsf{A}^{\mathrm{log}}_Q} \mathsf{A}^{\mathrm{log}}_{Q^{\mathrm{sat}}}$. Therefore it will suffice to prove that the map $\mathsf{A}^{\mathrm{log}}_{Q^{\mathrm{sat}}} \to \mathsf{A}^{\mathrm{log}}_Q$ is a homeomorphism. According to Proposition 1.2.7, this map can be identified with the map $\mathsf{X}^{\mathrm{log}}_{Q^{\mathrm{sat}}} \to \mathsf{X}^{\mathrm{log}}_Q$, so it will suffice to prove that $\mathsf{T}_{Q^{\mathrm{sat}}} \to \mathsf{T}_Q$ and $\mathsf{R}_{Q^{\mathrm{sat}}} \to \mathsf{R}_Q$ are homeomorphisms. Since $\mathsf{T}_Q := \mathrm{Hom}(Q, \mathbf{S}^1)$ and $\mathbf{S}^1$ is saturated, the map $\mathsf{T}_{Q^{\mathrm{sat}}} \to \mathsf{T}_Q$ is bijective, and a hence a homeomorphism since it is proper and continuous. The set R_Q (resp. $\mathsf{R}_{Q^{\mathrm{sat}}}$) is the disjoint union of its subsets $\mathsf{R}_{F^{\mathrm{gp}}}$ as F ranges over the faces of Q (resp. of Q^{sat}), and the maps $Q \to Q^{\mathrm{sat}}$, $Q \to C_{\mathbf{R}}(Q)$, and $Q^{\mathrm{sat}} \to C_{\mathbf{R}}(Q^{\mathrm{sat}}) = C_{\mathbf{R}}(Q)$ induce bijections between the respective sets of faces, by Corollary I.2.3.8. For each face F of Q (resp. of Q^{sat}),

$$\mathsf{R}_{F^{\mathrm{gp}}} := \mathrm{Hom}(F^{\mathrm{gp}}, \mathbf{R}_{\geq}) \cong \mathrm{Hom}(F^{\mathrm{gp}}, \mathbf{R}_{>}) \cong \mathrm{Hom}_{\mathbf{R}}(C_{\mathbf{R}}(F), \mathbf{R}),$$

where $C_{\mathbf{R}}(F)$ is the corresponding face of the cone $C_{\mathbf{R}}(Q)$. It follows that the natural map $\mathsf{R}_{Q^{\mathrm{sat}}} \to \mathsf{R}_Q$ is also bijective. Since $\mathsf{R}_{Q^{\mathrm{sat}}}$ has the weak topology generated by the evaluation functions e_x for $x \in Q^{\mathrm{sat}}$, and since each such x belongs to Q^{gp}, this is the same as the weak topology generated by the functions x_q for $q \in Q$. It follows that the map $\mathsf{R}_{Q^{\mathrm{sat}}} \to \mathsf{R}_Q$ is also a homeomorphism. □

In applying Corollary 1.2.10, be cautious of the fact that, although the map $X^{\mathrm{sat}}_{log} \to X_{log}$ is a homeomorphism, the map $X^{\mathrm{sat}} \to X$ need not be, and if it is not, the map from the fibers of $\tau_{X^{\mathrm{sat}}}$ to the fibers of τ_X is not bijective.

Proposition 1.2.11. *Let $f\colon X \to Z$ and $g\colon Y \to Z$ be morphisms of fine log analytic spaces. Then in the category of log analytic spaces, the natural map*

$$(X \times_Z Y)_{log} \to X_{log} \times_{Z_{log}} Y_{log}$$

is a homeomorphism. The same statement holds in the category of fine (resp. fine saturated) log analytic spaces provided that f or g is exact.

Proof Let $X' := X \times_Z Y$ in the category of log analytic spaces. Then the underlying topological space of X' is the cartesian product of the corresponding topological spaces. Since the conclusions of the proposition may be verified locally, we may assume that there exist charts for the morphisms $X \to Z$ and $Y \to Z$, subordinate to homomorphisms of monoids $R \to Q$ and $R \to P$, respectively. These charts induce a chart $X \times_Z Y \to \mathsf{X}_Q \times_{\mathsf{X}_R} \mathsf{X}_P$. Then, using (3) of

Proposition 1.2.5, we reduce to showing that the map $\mathsf{X}^{\log}_{\mathsf{Q}\oplus_{\mathsf{R}}\mathsf{P}} \to \mathsf{X}^{\log}_{\mathsf{Q}} \times_{\mathsf{X}^{\log}_{\mathsf{R}}} \mathsf{X}^{\log}_{\mathsf{P}}$ is an isomorphism. This follows immediately from the definitions and the universal property of the pushout monoid.

Now suppose that the spaces X, Y, and Z are fine and that f is exact. Let $X'' := X'^{\text{int}}$. The analog of Corollary III.2.2.4 holds in the category of log analytic spaces, so the map $X'' \to X' = X \times_Z Y$ is a nil immersion and its underlying map of topological spaces is a homeomorphism. Since the map $X''_{log} \to X_{log} \times_{Z_{log}} Y_{log}$ is a continuous morphism between spaces that are proper over X'', it is also proper and hence will be a homeomorphism if it is bijective. Thus it will suffice to show that the induced map on fibers over X'' is bijective.

Let x' be a point of X'' mapping to x in X, to $y \in Y$, and to $z \in Z$. Endow x, x', y, and z with the log structures induced from their ambient spaces. Since the map $x \to X$ is strict, (3) of Proposition 1.2.5 implies that the map $x_{log} \to \tau_X^{-1}(x)$ is a homeomorphism, and similarly for y, z, and x'. Thus we are reduced to the case in which all our spaces are log points. Choose charts for the morphisms $X \to Z$ and $Y \to Z$ that are subordinate to homomorphisms of fine monoids $R \to Q$ and $R \to P$, respectively. Replacing these charts by their localizations, we may assume that the homomorphisms $Q \to \mathcal{M}_X$, $P \to \mathcal{M}_Y$, and $R \to \mathcal{M}_Z$ are local and hence exact. Assuming as we may that f is exact, it follows that the homomorphism $R \to Q$ is also exact. We find a chart $Q' \to \mathcal{M}_{X'}$, where Q' is the integral pushout Q' of $R \to Q$ and $R \to P$. Then by (5) of Proposition I.4.2.5, the map $\overline{Q} \oplus_{\overline{R}} \overline{P} \to \overline{Q}'$ is an isomorphism, and hence $\mathsf{T}_{\overline{Q}'} \cong \mathsf{T}_{\overline{Q}} \times_{\mathsf{T}_{\overline{R}}} \mathsf{T}_{\overline{P}}$. Since $X_{log} \cong \mathsf{T}_{\overline{Q}}$, and similarly for Y, Z, and X', our result is proved, in the category of fine log analytic spaces. The saturated case follows from this and Corollary 1.2.10. □

Variant 1.2.12. If (Q, K) is an idealized monoid, we let

$$\begin{aligned}
\mathsf{T}_{\mathsf{Q,K}} &:= \mathsf{T}_{\mathsf{Q}},\\
\mathsf{V}_{\mathsf{Q,K}} &:= \mathsf{V}_{\mathsf{Q}},\\
\mathsf{R}_{\mathsf{Q,K}} &:= \{\rho \in \mathsf{R}_{\mathsf{Q}} : \rho(k) = 0 \text{ for all } k \in K\},\\
\mathsf{X}^{\log}_{\mathsf{Q,K}} &:= \mathsf{T}_{\mathsf{Q,K}} \times \mathsf{R}_{\mathsf{Q,K}},\\
\tilde{\mathsf{X}}^{\log}_{\mathsf{Q,K}} &:= \mathsf{V}_{\mathsf{Q,K}} \times \mathsf{R}_{\mathsf{Q,K}}.
\end{aligned}$$

If X is an idealized log analytic space, the space X_{log} is defined as in Definition 1.2.4, and the evident analogs of the results in this section are straightforward. We should also remark that the construction of the space X_{log} can be useful for some log structures that are not coherent; see Variant 1.3.5.

1.3 Local topology of X_{log}

The following result illustrates the geometric meaning of the construction X_{log}. It shows that if X is a smooth scheme over $\mathbf{C}$, then X_{log} is a topological manifold with boundary, with interior X^*.

Theorem 1.3.1. *Let X be a fine and smooth log analytic space, let x' be a point of X_{log}, and let n be the dimension of $\underline{X}$ at the point $\tau_X(x')$. Then there exist an open neighborhood U of x' in X_{log}, an open subset W of $\mathbf{R}_\geq \times \mathbf{R}^{2n-1}$, and a homeomorphism $U \to W$ taking $U \cap X^*$ to $W \cap (\mathbf{R}_> \times \mathbf{R}^{2n-1})$.*

Proof If $x := \tau_X(x')$ belongs to X^*, the underlying analytic space $\underline{X}$ is smooth at x and the map τ_X is an isomorphism in a neighborhood of x'. Hence X_{log} is a smooth manifold of real dimension $2n$ in a neighborhood of x', by classical results. If the log structure of X is not trivial at x, choose a chart as described in Proposition 1.1.5. Then Q is a sharp fine monoid of dimension $d := n - r > 0$, and we are reduced to proving the result when x is the vertex of $\mathsf{X}_{\mathbf{N}^r \oplus \mathsf{Q}}(Q)$. Then

$$X_{log} = \mathsf{X}^{\log}_{\mathbf{N}^r \oplus \mathsf{Q}}(Q) \cong \mathbf{C}^r \times \mathsf{X}^{\log}_{\mathsf{Q}} \cong \mathbf{C}^r \times \mathsf{T}_\mathsf{Q} \times \mathsf{R}_\mathsf{Q},$$

with x' mapping to the vertex of R_Q. By Corollary II.1.10.6, there is a homeomorphism $(\mathsf{R}_\mathsf{Q}, \mathsf{R}^*_\mathsf{Q}) \cong (\mathbf{R}_\geq \times \mathbf{R}^{d-1}, \mathbf{R}_> \times \mathbf{R}^{d-1})$. Since $\mathbf{C}^r \cong \mathbf{R}^{2r}$ and T_Q is locally homeomorphic to $\mathbf{R}^{2r}$, we find a neighborhood U of x' in X_{log}, an open subset W of $\mathbf{R}^{2r+d} \times \mathbf{R}_\geq \times \mathbf{R}^{d-1} \cong \mathbf{R}_\geq \times \mathbf{R}^{2n-1}$, and a homeomorphism $U \to W$ taking $U \cap X^*$ to $W \cap (\mathbf{R}_> \times \mathbf{R}^{2n-1})$. □

Corollary 1.3.2. *If X is a fine and smooth log analytic space, the map $j_{log} \colon X^*_{an} \to X_{log}$ is locally aspheric. Consequently:*

1. *The functors j_{log*} and j^*_{log} induce an equivalence between the categories of locally constant sheaves on X^* and on X_{log}. In particular, if $\mathcal{E}$ is a locally constant abelian sheaf on X^*, then $j_{log*}(\mathcal{E})$ is locally constant on X_{log} and $R^i j_{log*}(\mathcal{E}) = 0$ for $i > 0$.*
2. *If $\mathcal{E}$ is locally constant on X^*, then the natural map $R\tau_{X*}(j_{log*}\mathcal{E}) \to Rj_*(\mathcal{E})$ is an isomorphism.*

Proof Recall that a morphism of locally compact Hausdorff spaces $f \colon Y \to Z$ is said to be "locally aspheric" if every point of Z admits a basis B_z of neighborhoods U such that each $f^{-1}(U)$ is (nonempty and) contractible. Suppose that this is the case and that $\mathcal{F}$ is a locally constant sheaf on Y. Then, if $z \in Z$ and $U \in B_z$, the restriction of $\mathcal{F}$ to $f^{-1}(U)$ is constant and, since $f^{-1}(U)$ is connected and not empty, the natural map $f_*(\mathcal{F})(U) \to \mathcal{F}_y$ is an isomorphism for every $y \in f^{-1}(U)$. Thus the map $f^* f_*(\mathcal{E}) \to \mathcal{E}$ is an isomorphism. It follows that f induces an equivalence between the categories of locally constant sheaves on Y and on Z. Furthermore, if $\mathcal{F}$ is abelian, then $H^q(f^{-1}(U), \mathcal{E})$ vanishes if $q > 0$, since $f^{-1}(U)$ is contractible and $\mathcal{E}$ is constant on $f^{-1}(U)$. Consequently $R^q f_*(\mathcal{E}) = 0$ for $q > 0$, the natural map $H^q(Z, f_*(\mathcal{E})) \to H^q(Y, \mathcal{E})$ is an isomorphism, and in fact $f_*(\mathcal{E}) \cong Rf_*(\mathcal{E})$ in the derived category. Thus the asphericity of j_{log} implies statement (1), and statement (2) will then follow.

To prove that the map $j_{log}\colon X^* \to X_{log}$ is locally aspheric is a local question on X_{log}. Let x' be a point of X_{log} mapping to $x \in X$. If $x \in X^*$, Theorem 1.3.1 tells us that j_{log} looks locally like the identity map from $\mathbf{R}^{2n}$ to itself, and since every point of $\mathbf{R}^{2n}$ has a basis of contractible neighborhoods, the result is clear. If $x \in X \setminus X^*$, then j_{log} looks locally like the inclusion of $\mathbf{R}_> \times \mathbf{R}^{2n-1}$ in $\mathbf{R}_\geq \times \mathbf{R}^{2n-1}$. Assuming without loss of generality that x' is the origin, we observe that the family of sets

$$U_\epsilon = \{(r_1, r_2, \ldots, r_n) \in \mathbf{R}_\geq \times \mathbf{R}^{2n-1} : |r_i| < \epsilon\}, \epsilon > 0\}$$

is a neighborhood basis for x'. Since each

$$U_\epsilon^* = \{(r_1, r_2, \ldots, r_n) : 0 < r_1 < \epsilon, |r_i| < \epsilon, \text{ for } i = 2, \cdots n\}$$

is contractible, the result is again clear. □

The following theorem is a relative version of Theorem 1.3.1, for morphisms of log analytic spaces. The proof is somewhat more complicated; for it we refer the reader to the original source [57, 3.5]. This result also holds for some log structures that are only relatively coherent, as does Corollary 1.3.2. (See [57, 3.7] and Variant 1.3.5.)

Theorem 1.3.3. *Let $f\colon X \to Y$ be an exact and smooth morphism of fine log analytic spaces. Then $f_{log}\colon X_{log} \to Y_{log}$ is a topological submersion, whose fibers are topological manifolds with boundary, where the boundary consists of those points of X_{log} where the log structure is not vertical.* □

The following proposition gives more information about the topology of the morphism $\tau_X\colon X_{log} \to X$.

Proposition 1.3.4. *Let X be a fine saturated log analytic space. Then every point of X_{log} has a neighborhood basis consisting of compact sets K such that the fibers of $\tau_{X|_K}$ are either empty or contractible.* [2]

Proof We may and shall assume that X admits a chart $X \to \mathsf{A}_P$, where P is a

[2] The proof of this result is due to Piotr Achinger.

toric monoid. Then the diagram

$$\begin{array}{ccc} X_{log} & \longrightarrow & \mathsf{X}_{\mathsf{P}}^{\mathsf{log}} \\ {\scriptstyle \tau_X}\downarrow & & \downarrow{\scriptstyle \tau} \\ X_{an} & \longrightarrow & \mathsf{X}_{\mathsf{P}} \end{array}$$

is cartesian. The restriction of $X_{log} \to \mathsf{X}_{\mathsf{P}}^{\mathsf{log}}$ to a compact neighborhood of any point is proper. Thus we are reduced to proving the proposition when $X = \mathsf{X}_{\mathsf{P}}$, which we henceforth assume.

To linearize the computations, we use the covering $\eta\colon \tilde{\mathsf{X}}_{\mathsf{P}}^{\mathsf{log}} \to \mathsf{X}_{\mathsf{P}}^{\mathsf{log}}$, (1.2.2). This morphism is a local homeomorphism, and in particular $\mathsf{X}_{\mathsf{P}}^{\mathsf{log}}$ has the quotient topology induced by η. It therefore suffices to prove our claim with $\tilde{\tau} := \tau \circ \eta$ in place of τ.

Let (ρ, θ) be a point of $\tilde{\mathsf{X}}_{\mathsf{P}}^{\mathsf{log}}$ and let $x := e^{\theta}\rho$ be its image in X_{P}. Let $F_x := \rho^{-1}(\mathbf{R}^+)$, a face of P. Then if (ρ', θ') is another point in $\tilde{\tau}^{-1}(x)$, necessarily $\rho = \rho' = |x|$, and $x = e^{\theta(p)}\rho(p) = e^{\theta'(p)}\rho(p)$ for all $p \in P$. Since $\rho(f) \neq 0$ for $f \in F_x$, it follows that $\theta'(f) - \theta(f) \in \mathbf{Z}(1)$ for every such f. In fact $\tilde{\tau}^{-1}(x)$ is a torsor under

$$V''_x := \{\theta \in \mathsf{V}_{\mathsf{P}} : \theta(F_x^{\mathrm{gp}}) \subseteq \mathbf{Z}(1)\}.$$

The connected component of this space containing zero is

$$V'_x := \{\theta \in \mathsf{V}_{\mathsf{P}} : \theta(F_x^{\mathrm{gp}}) = 0\} \cong \mathsf{V}_{\mathsf{P}/\mathsf{F}_x},$$

and we have an exact sequence

$$0 \to V'_x \to V''_x \to \mathsf{I}_{\mathsf{F}_x} \to 0.$$

For each face F of P, let S_F be a finite set of generators for F^{gp}, let $S := \cup S_F$, and let T be a finite set of generators for P. Note that S is finite because P has only finitely many faces. Choose some $r \in (0, \pi/3)$, let U be the set of points ρ' in R_{P} such that $|\rho'(t) - \rho(t)| \leq r$ for all $t \in T$, and let W be the set of points θ' of V_{P} such that $|\theta'(s) - \theta(s)| \leq r$ for all $s \in S$. Then $K := U \times W$ is a compact neighborhood of (ρ, θ), and the set of all such K forms a neighborhood basis of (ρ, θ). We claim that the nonempty fibers of the restriction of $\tilde{\tau}$ to K are contractible.

Fix some (ρ', θ') in K and, for each $s \in S$, let

$$D_{s,\theta'} := \{\delta \in \mathsf{V}_{\mathsf{P}} : |\theta'(s) + \delta(s) - \theta(s)| \leq r\}.$$

Then $D_{s,\theta'}$ is a convex neighborhood of 0 in V_{P}, and hence so is

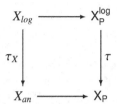

$D_{\theta'} := \cap\{D_{s,\theta'} : s \in S\}$. For $\delta \in \mathsf{V}_P$, $\theta' + \delta \in W$ if and only if $\delta \in D_{\theta'}$. If this is the case,

$$|\delta(s)| \le |\theta'(s) + \delta(s) - \theta(s)| + |\theta(s) - \theta'(s)| \le 2r < \pi$$

for every s in S. Furthermore, $(\rho', \theta' + \delta) \in \tilde{\tau}^{-1}(x)$ if and only if $\delta \in V''_x$. If also $\delta \in D_{\theta'}$, then $\delta(F_{x'}) \subseteq \mathbf{Z}(1)$ and, since S contains a set of generators for $F^{gp}_{x'}$, the inequality above implies that $\delta(F^{gp}_x) = 0$. Thus $D_{\theta'} \cap V''_x = D_{\theta'} \cap V'_x$, a convex set, and $(\rho', \theta' + \delta) \in \tilde{\tau}^{-1}(x)$ if and only if $\delta \in D'_\theta \cap V'_x$. It follows that $\tilde{\tau}^{-1}(x) \cap K$ is convex, hence contractible, as claimed. □

Variant 1.3.5. Let X be a fine log analytic space, let $\mathcal{F}$ be a relatively coherent sheaf of faces in X and let $\underline{X}(\mathcal{F})$ be the log analytic space whose underlying analytic space is $\underline{X}$ and with the log structure defined by the map $\mathcal{F} \to \mathcal{O}_X$. Then the morphisms of log analytic spaces $X \to \underline{X}(\mathcal{F}) \to \underline{X}$ define a factorization

$$\tau_X \colon X_{log} \xrightarrow{\tau_{X/\mathcal{F}}} \underline{X}(\mathcal{F})_{log} \xrightarrow{\tau_{\underline{X}(\mathcal{F})}} X.$$

As we saw in Proposition 1.2.5 and Corollary 1.2.8, the morphism $\tau_{\underline{X}(\mathcal{F})}$ is separated and the morphism τ_X is proper, and it follows that $\tau_{X/\mathcal{F}}$ is also proper. Moreover, the argument of Proposition 1.2.5 shows that $\tau_{X/\mathcal{F}}$ is surjective, and that the fiber over a point (x, σ) of $\underline{X}(\mathcal{F})_{log}$ is a torsor under $\mathcal{M}_{X,x}/\mathcal{F}_x$. Since $\tau_{X/\mathcal{F}}$ is surjective and $\tau_X = \tau_{\underline{X}(\mathcal{F})} \circ \tau_{X/\mathcal{F}}$ is proper, it also follows that $\tau_{\underline{X}(\mathcal{F})}$ is proper. However, this map does not necessarily satisfy the conclusion of Proposition 1.3.4. Finally, we point out that it is proved in [57, 5.1] that if X is smooth and the stalks of $\mathcal{M}_X/\mathcal{F}$ are free, then $\underline{X}(\mathcal{F})_{log}$ is again a manifold with boundary, whose interior is $\underline{X}^*(\mathcal{F}) := \{x \in X : \overline{\mathcal{F}}_x = 0\}$. Without the freeness hypothesis, the argument there shows that the pair $\left(\underline{X}(\mathcal{F})_{log}, \underline{X}^*(\mathcal{F})\right)$ is locally isomorphic to a product of the form $(\mathbf{R}_{\geq} \times \mathsf{X}_Q, \mathbf{R}_{>0}, \times \mathsf{X}_Q)$ for some fine sharp monoid Q. By Corollary I.3.3.2, the space X_Q is locally contractible, and it follows that the inclusion $\underline{X}^*(\mathcal{F}) \to \underline{X}(\mathcal{F})_{log}$ is still locally aspheric, and Corollary 1.3.2 also applies to the morphism $j_{log} \colon \underline{X}^*(\mathcal{F}) \to \underline{X}(\mathcal{F})_{log}$. (See Remark II.2.6.7.)

1.4 $\mathcal{O}_X^{log}$ and the exponential map

If Y and Z are topological spaces, let us write $\underline{Z}_Y$ for the sheaf that to every open set V of Y assigns the set of continuous functions $V \to Z$. (We sometimes omit the subscript and/or the underline if no confusion seems likely to result, and indeed we have already used this notation several times.)

On an analytic space X, one has the fundamental exact sequence of abelian sheaves

$$0 \longrightarrow \underline{\mathbf{Z}}(1) \longrightarrow \mathcal{O}_X \xrightarrow{\exp} \mathcal{O}_X^* \longrightarrow 1,$$

where $\exp(f) := \sum_{n=0}^{\infty} f^n/n!$ and $\mathbf{Z}(1)$ is the subgroup of $\mathbf{C}$ generated by $2\pi i$. If X is endowed with a log structure, then on X_{log} one can define an analogous sequence in which $\mathcal{M}_X^{gp}$ replaces $\mathcal{O}_X^*$. In order to do so, we must also replace $\mathcal{O}_X$ by a larger sheaf $\mathcal{L}_X$ on X_{log} whose sections correspond to logarithms of sections of $\mathcal{M}_X^{gp}$, which would be multivalued on X.

Recall again that the universal covering of $\mathbf{S}^1$ is given by the exponential map

$$\exp\colon \mathbf{R}(1) \to \mathbf{S}^1 : z \mapsto \exp(z) = \sum_0^{\infty} z^n/n!,$$

where $\mathbf{R}(1)$ is the set of purely imaginary numbers. Locally on $\mathbf{S}^1$, the map exp has a section. Then on any topological space X there is an exact sequence of abelian sheaves on X

$$0 \to \underline{\mathbf{Z}}(1) \to \underline{\mathbf{R}}(1) \to \underline{\mathbf{S}}^1 \to 0.$$

Proposition 1.4.1. *If X is a log analytic space, let $\mathcal{L}_X$ be the fiber product, in the category of abelian sheaves on X_{log}, in the following diagram:*

$$\begin{array}{ccc}
\mathcal{L}_X & \xrightarrow{\exp} & \tau_X^{-1}(\mathcal{M}_X^{gp}) \\
\Big\downarrow{\scriptstyle \theta} & & \Big\downarrow{\scriptstyle \arg} \\
\underline{\mathbf{R}}(1) & \xrightarrow{\exp} & \underline{\mathbf{S}}^1.
\end{array}$$

1. *There is a commutative diagram of abelian sheaves on X_{log}, in which the squares on the right are cartesian and the rows are exact,*

$$\begin{array}{ccccccccc}
0 & \longrightarrow & \underline{\mathbf{Z}}(1) & \longrightarrow & \tau_X^{-1}(\mathcal{O}_X) & \xrightarrow{\exp} & \tau_X^{-1}(\mathcal{O}_X^*) & \longrightarrow & 0 \\
 & & \Big\downarrow{\scriptstyle \mathrm{id}} & & \Big\downarrow{\scriptstyle \epsilon} & & \Big\downarrow{\scriptstyle \lambda} & & \\
0 & \longrightarrow & \underline{\mathbf{Z}}(1) & \longrightarrow & \mathcal{L}_X & \xrightarrow{\exp} & \tau_X^{-1}(\mathcal{M}_X^{gp}) & \longrightarrow & 0 \\
 & & \Big\downarrow{\scriptstyle \mathrm{id}} & & \Big\downarrow{\scriptstyle \theta} & & \Big\downarrow{\scriptstyle \arg} & & \\
0 & \longrightarrow & \underline{\mathbf{Z}}(1) & \longrightarrow & \underline{\mathbf{R}}(1) & \xrightarrow{\exp} & \underline{\mathbf{S}}^1 & \longrightarrow & 0,
\end{array}$$

and where $\theta \circ \epsilon(f)$ is the imaginary part $\mathrm{Im}(f)$ *of* f.

2. *There is an exact sequence*

$$0 \longrightarrow \tau_X^{-1}(\mathcal{O}_X) \xrightarrow{\ \epsilon\ } \mathcal{L}_X \xrightarrow{\ \overline{\exp}\ } \tau_X^{-1}(\overline{\mathcal{M}}_X^{\mathrm{gp}}) \longrightarrow 0.$$

Proof The middle row of the diagram is by definition the pullback of the bottom row via the map arg: $\tau_X^{-1}(\mathcal{M}_X^{\mathrm{gp}}) \to \mathbf{S}^1$. The top row is just the standard exponential sequence on X, pulled back to X_{log}. The homomorphism ϵ in statement (2) exists because $\arg(\lambda(\exp f)) = \exp(\mathrm{Im}(f))$ for every section f of $\mathcal{O}_X$, Then the top row is necessarily the pull-back of the middle row along λ, and it follows that the right upper square is cartesian. Chasing the diagram gives the exact sequence in (2). □

Sections of $\mathcal{L}_X$ may be thought of as logarithms of sections of $\mathcal{M}_X$. Such functions and their powers are often introduced in an ad hoc manner in the study of differential equations with regular singular points. Here they arise quite naturally, and we shall also want to consider the sheaf of $\tau_X^{-1}(\mathcal{O}_X)$-algebras that they generate. This sheaf plays an important role in the logarithmic Riemann–Hilbert correspondence ([50], [61], and [40]).

Definition 1.4.2. *Let X be a u-integral log analytic space. Then $\mathcal{O}_X^{log}$ is the universal $\tau^{-1}(\mathcal{O}_X)$-algebra equipped with a map $\mathcal{L}_X \to \mathcal{O}_X^{log}$ such that the diagram*

$$\begin{array}{ccc} \tau^{-1}(\mathcal{O}_X) & \xrightarrow{\ \epsilon\ } & \mathcal{L}_X \\ & \searrow & \downarrow \\ & & \mathcal{O}_X^{log} \end{array}$$

commutes.

The algebra $\mathcal{O}_X^{log}$ may be constructed explicitly by taking the quotient of $\tau_X^{-1}(\mathcal{O}_X) \otimes_{\mathbf{Z}} S^{\cdot}(\mathcal{L}_X)$ by the ideal generated by the sections of the form $1 \otimes \epsilon(f) - f \otimes 1$, for f a local section of $\tau_X^{-1}(\mathcal{O}_X)$. Alternatively, tensor the exact sequence of statement (2) of Proposition 1.4.1 with the (torsion-free) abelian sheaf $\tau^{-1}(\mathcal{O}_X)$ to obtain the top row of the following diagram:

$$\begin{array}{ccccc} \tau^{-1}(\mathcal{O}_X) \otimes \tau^{-1}(\mathcal{O}_X) & \xrightarrow{\ \mathrm{id} \otimes \epsilon\ } & \tau^{-1}(\mathcal{O}_X) \otimes \mathcal{L}_X & \longrightarrow & \tau^{-1}(\mathcal{O}_X) \otimes \overline{\mathcal{M}}_X^{\mathrm{gp}} \\ \downarrow m & & \downarrow m' & & \downarrow \cong \\ \tau^{-1}(\mathcal{O}_X) & \xrightarrow{\ \epsilon'\ } & \mathcal{E}_X & \longrightarrow & \tau^{-1}(\mathcal{O}_X \otimes \overline{\mathcal{M}}_X^{\mathrm{gp}}). \end{array} \tag{1.4.1}$$

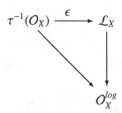

The top row is short exact because $\tau^{-1}(\mathcal{O}_X)$ is $\mathbf{Z}$-flat, and the bottom row is obtained by pushout along the multiplication map m. Then, for each j, one can form the jth symmetric product $S^j(\mathcal{E}_X)$ of the $\tau_X^{-1}(\mathcal{O}_X)$ module $\mathcal{E}_X$, and the map $\tau_X^{-1}(\mathcal{O}_X) \to \mathcal{E}_X$ induces a map $S^{j-1}(\mathcal{E}_X) \to S^j(\mathcal{E}_X)$.

Proposition 1.4.3. *On a u-integral log analytic space X, there is a unique isomorphism of $\tau_X^{-1}(\mathcal{O}_X)$-algebras*

$$\mathcal{O}_X^{log} \cong \varinjlim S^j(\mathcal{E}_X)$$

compatible with the inclusions of $\mathcal{L}_X$. If the stalks of $\overline{\mathcal{M}}_{X,x}^{gp}$ are finitely generated, then for each $j \in \mathbf{N}$, the map $S^j(\mathcal{E}_X) \to \mathcal{O}_X^{log}$ is injective, and if $N_j\mathcal{O}_X^{log}$ denotes its image, then the map

$$\tau^{-1}(\mathcal{O}_X) \otimes_{\mathbf{Z}} S^j(\overline{\mathcal{M}}_X^{gp}) \to \mathrm{Gr}_j^N \mathcal{O}_X^{log}$$

is an isomorphism.

Proof We endow $\varinjlim S^j(\mathcal{E}_X)$ with its obvious ring structure and the $\tau_X^{-1}(\mathcal{O}_X)$-algebra structure induced by the map $\tau_X^{-1}(\mathcal{O}_X) \to \mathcal{E}_X \to \varinjlim S^j(\mathcal{E}_X)$. Then if a is a local section $\tau_X^{-1}(\mathcal{O}_X)$,

$$\epsilon'(a) = \epsilon'(m(a \otimes 1)) = \epsilon'(m(1 \otimes a)) = m'((\mathrm{id} \otimes \epsilon)(1 \otimes a)),$$

and hence $a \in \mathcal{O}_X$ and $\epsilon(a) \in \mathcal{L}_X$ have the same image in $\varinjlim S^j(\mathcal{E}_X)$. The universal property of $\mathcal{O}_X^{log}$ then gives a unique map $\mathcal{O}_X^{log} \to \varinjlim S^j(\mathcal{E}_X)$ making the evident diagram commute. On the other hand, the $\tau_X^{-1}(\mathcal{O}_X)$-algebra structure of $\mathcal{O}_X^{log}$ and the map ϵ induce the homomorphism b of the commutative diagram

$$\begin{array}{ccc} \tau_X^{-1}(\mathcal{O}_X) \otimes \tau^{-1}(\mathcal{O}_X) & \xrightarrow{\mathrm{id} \otimes \epsilon} & \tau_X^{-1}(\mathcal{O}_X) \otimes \mathcal{L}_X \\ \Big\downarrow m & & \Big\downarrow b \\ \tau_X^{-1}(\mathcal{O}_X) & \longrightarrow & \mathcal{O}_X^{log} \end{array}$$

and, by the construction of $\mathcal{E}_X$, the map b factors uniquely through a map of $\tau_X^{-1}(\mathcal{O}_X)$-modules $\mathcal{E}_X \to \mathcal{O}_X^{log}$, compatible with the inclusions of $\tau_X^{-1}(\mathcal{O}_X)$. This map extends uniquely to $\varinjlim S^j(\mathcal{E}_X)$.

The map $\mathcal{L}_X \to \mathcal{O}_X^{log}$ factors through $N_1\mathcal{O}_X^{log}$ and the map $\mathcal{O}_X \to \mathcal{O}_X^{log}$ factors through $N_0\mathcal{O}_X^{log}$. Thus we find a homomorphism $\overline{\mathcal{M}}_X^{gp} \to \mathrm{Gr}_1^N \mathcal{O}_X^{log}$. By the $\mathcal{O}_X$-algebra of structure of $\mathcal{O}_X^{log}$ and its compatibility with the filtration $N.$, we get a homomorphism $\mathcal{O}_X \otimes S^{\cdot}\overline{\mathcal{M}}_X^{gp} \to \mathrm{Gr}_{\cdot}^N \mathcal{O}_X^{log}$.

The remaining statements can be verified at the stalks. Suppose that the stalk of $\overline{\mathcal{M}}_X^g$ at x is finitely generated. Then the quotient of $\overline{\mathcal{M}}_{X,x}^{\mathrm{gp}}$ by its torsion part is a free abelian group, and hence $\mathcal{O}_{X,x} \otimes \overline{\mathcal{M}}_{X,x}$ is a free $\mathcal{O}_{X,x}$-module of finite rank. Then we can choose a splitting of the sequence

$$0 \to \mathcal{O}_{X,x} \to \mathcal{E}_{X,x} \to \mathcal{O}_{X,x} \otimes \overline{\mathcal{M}}_{X,x}^{\mathrm{gp}} \to 0.$$

Then $S^j\mathcal{E}_{X,x} \cong \oplus\{S^i\overline{\mathcal{M}}_{X,x}^{\mathrm{gp}} : 0 \leq i \leq j\}$, and the result is clear. □

Proposition 1.4.4. *Let $f\colon X \to Y$ be a morphism of u-integral log analytic spaces. Then there are canonical maps*

$$f_{log}^{-1}(\mathcal{L}_Y) \to \mathcal{L}_X \quad \text{and} \quad \tau_X^{-1}(\mathcal{O}_X) \otimes_{(f\tau_X)^{-1}(\mathcal{O}_Y)} f_{log}^{-1}(\mathcal{O}_Y^{log}) \to \mathcal{O}_X^{log},$$

compatible with the maps $f^\sharp\colon f^{-1}(\mathcal{O}_Y) \to \mathcal{O}_X$ and $f^\flat\colon f^{-1}(\mathcal{M}_Y) \to \mathcal{M}_X$. The latter of these is an isomorphism if f is strict.

Proof The construction of the claimed maps is straightforward. Suppose that f is strict. Recalling the construction of (1.4.1), we find a commutative diagram of short exact sequences:

$$\begin{array}{ccccc}
\tau_X^{-1}(\mathcal{O}_X) \otimes f_{log}^{-1}(\mathcal{O}_Y^{log}) & \longrightarrow & \tau_X^{-1}(\mathcal{O}_X) \otimes f_{log}^{-1}(\mathcal{E}_Y) & \longrightarrow & \tau_X^{-1}(\mathcal{O}_X) \otimes f_{log}^{-1}\tau_Y^{-1}(\overline{\mathcal{M}}_Y^{\mathrm{gp}}) \\
\downarrow & & \downarrow & & \downarrow \\
\tau_X^{-1}(\mathcal{O}_X) & \longrightarrow & \mathcal{E}_X & \longrightarrow & \tau_X^{-1}(\mathcal{O}_X \otimes \overline{\mathcal{M}}_X^{\mathrm{gp}});
\end{array}$$

here the tensor products in the upper row are taken over $(f\tau_X)^{-1}(\mathcal{O}_Y)$. Since the morphism f is strict, the map $f^{-1}(\overline{\mathcal{M}}_Y) \to \overline{\mathcal{M}}_X$ is an isomorphism, and it follows that all the vertical maps in the diagram are isomorphisms. Since $\mathcal{O}_X^{log} = \varinjlim S^{\cdot}\mathcal{E}_X$, we can conclude that the map in the statement of the proposition is also an isomorphism. □

Let us give an explicit description of these constructions when X is the log analytic space associated to a fine idealized monoid (Q, K). We assume that Q^{gp} is torsion free, so that X and T_Q are connected (by Variant I.3.3.3 and Remark 1.2.1, respectively). Recall from Variant 1.2.12 that then $X_{log} = \mathsf{T}_Q \times \mathsf{R}_{Q,K}$ and admits a universal cover $\eta\colon \tilde{X}_{log} \to X_{log}$ with covering group I_Q, where $\tilde{X}_{log} = \mathsf{V}_Q \times \mathsf{R}_{Q,K}$ and $\mathsf{I}_Q := \mathrm{Hom}(Q, \mathbf{Z}(1))$. Let $\iota\colon \mathbf{Z}(1) \to \mathbf{R}(1)$ be the inclusion and, for $q \in Q$, let $\tilde{\chi}(q) : \tilde{X}_{log} \to \mathbf{R}(1)$ be the map sending (θ, ρ) to $\theta(q)$. Let

$$\mathsf{L}_Q := \mathbf{Z}(1) \oplus Q^{\mathrm{gp}},$$

with the action of I_Q given by $\gamma(z, q) := (z + \gamma(q), q)$. We are given a homomorphism $Q^{gp} \to \mathcal{M}_X^{gp}$ and, if $\mathbf{Z}(1) \to \mathcal{M}_X^{gp}$ is the zero map, we find a commutative diagram

$$\begin{array}{ccc} \mathbf{Z}(1) \oplus Q^{gp} & \xrightarrow{(\iota, \tilde{\chi})} & \Gamma(\tilde{X}_{log}, \underline{\mathbf{R}}(1)) \\ \downarrow & & \downarrow \\ \Gamma(\tilde{X}_{log}, \tilde{\tau}_X^{-1}(\mathcal{M}_X^{gp})) & \xrightarrow{\arg} & \Gamma(\tilde{X}_{log}, \underline{\mathbf{S}}^1). \end{array}$$

This diagram defines a mapping $\mathsf{L}_Q \to \Gamma(\tilde{X}_{log}, \eta^{-1}(\mathcal{L}_X))$, which fits in the commutative diagram

$$\begin{array}{ccccccccc} 0 & \longrightarrow & \mathbf{Z}(1) & \longrightarrow & \mathsf{L}_Q & \longrightarrow & Q^{gp} & \longrightarrow & 0 \\ & & \downarrow & & \downarrow & & \downarrow & & \\ 0 & \longrightarrow & \Gamma(\tilde{X}_{log}, \mathbf{Z}(1)) & \longrightarrow & \Gamma(\tilde{X}_{log}, \eta^{-1}(\mathcal{L}_X)) & \longrightarrow & \Gamma(\tilde{X}_{log}, \mathcal{M}_X^{gp}) & \longrightarrow & 0. \end{array}$$

The covering group I_Q acts on $\Gamma(\tilde{X}_{log}, \eta^{-1}(\mathcal{L}_X^{gp}))$, and we claim this action is compatible with its action on L_Q. Indeed, the image of $\mathbf{Z}(1)$ in $\Gamma(\tilde{X}_{log}, \mathcal{M}_X^{gp})$ is constant and hence invariant under γ, and if $q \in Q^{gp}$ and $(\theta, \rho) \in \tilde{X}_{log}$, formula (1.2.3) implies that

$$(\gamma\tilde{\chi}(q))(\theta,\rho) := (\tilde{\chi}(q))((\theta,\rho)\gamma) = \tilde{\chi}(q)(\theta+\gamma,\rho) = \theta(q)+\gamma(q) = \gamma(q)+\tilde{\chi}(q)(\theta,\rho).$$

Now let $\mathsf{E}_Q := \mathsf{L}_Q(-1) = \mathbf{Z} \oplus Q^{gp}(-1)$, with its natural I_Q-action, and let

$$\mathsf{E}^{\cdot}_Q := \varinjlim S^j\, \mathsf{E}_Q \cong \oplus_j S^j(Q^{gp})(-j), \tag{1.4.2}$$

which has a natural ring structure and I_Q-action. In fact, since Q^{gp} is a finitely generated free abelian group, it identifies with $\mathrm{Hom}(\mathsf{I}_Q, \mathbf{Z}(1))$. Hence E_Q identifies with $\mathbf{Z} \oplus (\mathsf{I}_Q)^\vee$ and $\mathsf{E}^{\cdot}_Q$ identifies with $\mathcal{A}_{\mathsf{I}_Q}$, the universal locally unipotent representation of I_Q constructed in Section I.3.7. Note that we have a canonical isomorphism $\mathbf{C}(1) \to \mathbf{C}$ and hence $\mathbf{C} \otimes \mathsf{L}_Q \cong \mathbf{C} \otimes \mathsf{E}_Q$.

Proposition 1.4.5. *Let Q be a fine monoid such that Q^{gp} is torsion free and let K be an ideal in Q. Let $X := \mathsf{A}^{an}_{Q,K}$, let $\eta\colon \tilde{X}_{log} \to X_{log}$ be the exponential map (1.2.2), and let $\mathsf{E}^{\cdot}_Q$ be the ring defined in equation (1.4.2).*

1. *There is a natural homomorphism*

$$\tau_X^{-1}(\mathcal{O}_X) \otimes \mathsf{E}^{\cdot}_Q \to \Gamma(\tilde{X}_{log}, \eta^{-1}(\mathcal{O}_X^{log}))$$

compatible with the actions of I_Q and the natural filtrations on both sides.

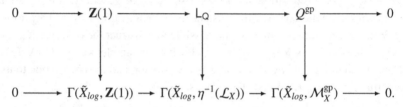

2. *If Q is sharp and x is the vertex of X, this homomorphism restricts to an isomorphism*

$$\mathcal{O}_{X,x} \otimes \mathsf{E}^{\cdot}_{Q} \to \Gamma(\tilde{\tau}_X^{-1}(x), \eta^{-1}(\mathcal{O}_X^{log})).$$

Proof We have already constructed the homomorphism appearing in (1). The restriction of the sheaf $\eta^{-1}(\mathcal{O}_X^{log})$ to the contractible space $\tilde{\tau}_X^{-1}(x)$ is constant. To see that the homomorphism (2) is an isomorphism, it is enough to see that it is so on the graded pieces associated to the filtrations $N.$ of both sides. On the left, the jth graded piece is $\mathcal{O}_{X,x} \otimes S^j Q^{\mathrm{gp}}$, and on the right it is $\mathcal{O}_{X,x} \otimes S^j \overline{\mathcal{M}}^{\mathrm{gp}}_{X,x}$, as we saw in Proposition 1.4.3. The map on these pieces is evidently an isomorphism by construction. □

Remark 1.4.6. Proposition 1.4.5 makes sense more generally, for any split hollow log analytic space with constant log structure given by a sharp toric Q. Without a splitting, it may not be possible to construct the covering $\tilde{X}_{log}$ of X_{log}. Indeed, $X_{log} \to X$ is a T_Q-torsor, classified by an element of $H^1(X, \mathsf{T}_Q)$. The desired universal covering would be a V_Q-torsor, and the obstruction to its construction lies in $H^2(X, \mathsf{I}_Q)$. The exact sequences

$$0 \to \mathbf{Z}(1) \to \mathcal{O}_X \to \mathcal{O}_X^* \to 0 \quad \text{and} \quad 0 \to \mathcal{O}_X^* \to \mathcal{M}_X^{\mathrm{gp}} \to Q^{\mathrm{gp}} \to 0$$

splice to an exact sequence

$$0 \to \mathbf{Z}(1) \to \mathcal{O}_X \to \mathcal{M}_X^{\mathrm{gp}} \to Q^{\mathrm{gp}} \to 0,$$

which defines an element in $\mathrm{Ext}^2(Q^{\mathrm{gp}}, \mathbf{Z}(1)) = H^2(X, \mathsf{I}_Q)$. We leave it to the reader to verify that this element, which is given by the Chern classes of the line bundles $\mathcal{L}_q$ for $q \in Q^{\mathrm{gp}}$ (see Remark III.1.1.7), is the obstruction to the construction of $\tilde{X}_{log}$.

Theorem 1.4.7. *If $\mathcal{E}$ is a coherent sheaf of $\mathcal{O}_X$-modules on a fine analytic space X, let $\tau_X^*(\mathcal{E}) := \tau_X^{-1}(\mathcal{E}) \otimes_{\tau_X^{-1}(\mathcal{O}_X)} \mathcal{O}_X^{log}$. If $\overline{\mathcal{M}}_X^{\mathrm{gp}}$ is torsion free, the natural map*

$$\mathcal{E} \to R\tau_{X*}\tau_X^*(\mathcal{E})$$

is a derived isomorphism. If $\mathcal{F}$ is a relatively coherent sheaf of faces in $\mathcal{M}_X$, the analogous result holds for $\tau_{X(\mathcal{F})}$.

Proof The theorem asserts that the natural map $\mathcal{E} \to \tau_{X*}\tau_X^*(\mathcal{E})$ is an isomorphism and that $R^q\tau_{X*}(\mathcal{E}) = 0$ for $q > 0$. These statements can be checked on the stalks. At the stalk level, the same argument will work for $\mathcal{F}$. Let x be a point of X; since τ_X is proper, it is enough to check that the map $\mathcal{E}_x \to H^0(\tau_X^{-1}(x), \tau_X^*(\mathcal{E}_x))$ is an isomorphism and that $H^q(\tau_X^{-1}(x), \tau_X^*(\mathcal{E}_x)) = 0$ for $q > 0$.

By Proposition 1.2.5, we can identify $\tau_X^{-1}(x)$ with $\mathsf{T}_x = \mathrm{Hom}(\overline{\mathcal{M}}^{\mathrm{gp}}_{X,x}, \mathbf{S}^1)$. Let $Q := \overline{\mathcal{M}}_{X,x}$ and choose a splitting of the map $\mathcal{M}_{X,x} \to Q$. Then we have a universal covering $\eta_Q \colon V_Q \to \mathsf{T}_Q = \mathsf{T}_x$ with covering group I_Q. Let $\tilde{\mathcal{E}}_x := \Gamma(V_Q, \eta_Q^* \tau_X^*(\mathcal{E}_x))$ with its natural I_Q-action. Since T_x is a $K(\pi, 1)$ and $\tau_X^*(\mathcal{E}_x)$ is locally constant on T_x, the cohomology groups $H^q(\mathsf{T}_x, \tau_X^*(\mathcal{E}_x))$ can be computed as the group cohomology $H^q(\mathsf{I}_Q, \tilde{\mathcal{E}}_x)$.

Let $W := \mathbf{C} \otimes Q^{\mathrm{gp}}$. Using the Propositions 1.4.5 and 1.4.4, we can write

$$\tilde{\mathcal{E}}_x := \Gamma(\tilde{\tau}_X^{-1}(x), \eta_Q^*(\mathcal{O}_X^{log} \otimes_{\mathcal{O}_{X,x}} \mathcal{E}_x)) \cong S^{\cdot} W \otimes_{\mathbf{C}} \mathcal{E}_x.$$

Then, by the universal coefficient theorem, $H^q(\mathsf{I}_Q, \tilde{\mathcal{E}}_x) \cong \mathcal{E}_x \otimes_{\mathbf{C}} H^q(\mathsf{I}_Q, S^{\cdot} W)$, and we are reduced to proving that

$$H^q(\mathsf{I}_Q, S^{\cdot} W) = \begin{cases} \mathbf{C} & \text{if } q = 0 \\ 0 & \text{otherwise.} \end{cases}$$

This calculation is straightforward. For example, by Theorem I.3.7.3, the group cohomology of this locally unipotent representation can be computed from the associated Higgs complex. In this case, the Higgs field on $S^{\cdot} W$ is just the usual exterior derivative $S^{\cdot} W \to S^{\cdot} W \otimes W$, and the Higgs complex is the de Rham complex of the polynomial algebra $S^{\cdot} W$. Thus the desired result follows from the standard Poincaré lemma for polynomial algebras. □

2 The de Rham complex

One of the most important historical inspirations for log geometry is the theory of differential forms with log poles. These have been used for a long time to study the de Rham cohomology of an open subset U whose complement in a smooth proper scheme is a divisor with normal crossings. This method was used, for example, by Grothendieck in his original proof [28] of the comparison theorem between Betti cohomology and algebraic de Rham cohomology, and also by Deligne in his treatment [16] of differential equations with regular singularities. It is no surprise then that logarithmic de Rham cohomology is quite well developed and that it gives a good idea of the geometric meaning of log geometry.

2.1 Exterior differentiation and Lie bracket

Our first task is to show that the universal logarithmic derivation defined in Proposition IV.1.2.11 fits into a complex of abelian sheaves, the (logarithmic) de Rham complex.

Proposition 2.1.1. *Let $f\colon X \to Y$ be a morphism of coherent log schemes, and for each i, let $\Omega^i_{X/Y}$ be the ith exterior power of $\Omega^1_{X/Y}$. Then there is a unique collection of homomorphisms of sheaves of abelian groups*

$$d^i\colon \Omega^i_{X/Y} \to \Omega^{i+1}_{X/Y} : i \in \mathbf{N},$$

called the exterior derivative, such that:

1. $d^1\, d\log m = 0$ *if m is any section of $\mathcal{M}_X$,*
2. $d^i d^{i-1}\omega = 0$ *if ω is any section of $\Omega^{i-1}_{X/Y}$,*
3. $d^{i+j}(\omega \wedge \omega') = (d^i\omega) \wedge \omega' + (-1)^i \omega \wedge (d^j\omega')$ *if $\omega \in \Omega^i_{X/Y}$ and $\omega' \in \Omega^j_{X/Y}$.*

The resulting complex $\Omega^{\cdot}_{X/Y}$ is called the de Rham complex of the morphism $f\colon X \to Y$.

Proof By Lemma IV.1.2.16, we may without loss of generality assume that $\mathcal{M}_X$ is integral, and we identify a section of $\mathcal{M}_X$ with its image in $\mathcal{M}^{\mathrm{gp}}_X$. The main point is the existence of $d^1 : \Omega^1_{X/Y} \to \Omega^2_{X/Y}$. Classically, this is proved by checking compatibility with all the relations used in the construction of $\Omega^1_{X/Y}$; a somewhat tedious task [6, II, §3]. In our setting, it is more convenient to use the description of $\Omega^1_{X/Y}$ as a quotient of $\mathcal{O}_X \otimes \mathcal{M}^{g}_X$ by the abelian subsheaf $\mathcal{R}_1 + \mathcal{R}_2$ as described in Proposition IV.1.2.11. The map $\mathcal{O}_X \times \mathcal{M}^{\mathrm{gp}}_X \to \Omega^2_{X/Y}$ sending (a, m) to $da \wedge d\log m$ is evidently bilinear, and hence it induces a homomorphism of abelian sheaves $\phi\colon \mathcal{O}_X \otimes \mathcal{M}^{\mathrm{gp}}_X \to \Omega^2_{X/Y}$. If m is any section of $\mathcal{M}_X$ then

$$\phi(\alpha_X(m) \otimes m) = d\alpha_X(m) \wedge d\log m = \alpha_X(m)\, d\log m \wedge d\log m = 0,$$

and it follows that ϕ annihilates $\mathcal{R}_1$. If n is a local section of $f^{-1}(\mathcal{M}^{\mathrm{gp}}_Y)$ and a is a local section of $\mathcal{O}_X$, then $\phi(a \otimes n) = da \wedge d\log n = 0$, so ϕ also annihilates $\mathcal{R}_2$. We can conclude from Proposition IV.1.2.11 that ϕ factors through a homomorphism of abelian sheaves $d^1 : \Omega^1_{X/Y} \to \Omega^2_{X/Y}$.

By construction, $d^1(a\, d\log m) = da \wedge d\log m$ for sections a of $\mathcal{O}_X$ and m of $\mathcal{M}_X$. In particular, $d^1(d\log m) = d^1(1\, d\log m) = 0$, so condition (1) is satisfied. If $a = \alpha_X(m)$, then

$$d^1 da = d^1(\alpha_X(m)\, d\log m) = d\alpha_X(m) \wedge d\log m = \alpha_X(m)\, d\log m \wedge d\log m = 0.$$

It follows that $d^1 da = 0$ if a is a unit and, since any local section of $\mathcal{O}_X$ can locally be written as a constant plus a unit, in fact $d^1 da = 0$ for every local section of $\mathcal{O}_X$. Thus condition (2) is satisfied for $i = 1$.

Condition (3) when $i = j = 0$ is just the fact that d is a derivation. Let a and b be local sections of $\mathcal{O}_X$, let m be a local section of $\mathcal{M}_X$, and let $\omega' := b\, d\log m$.

Then $d^1\omega' = db \wedge d\log m$, so

$$d^1(a\omega') = d^1(ab\, d\log m) = (dab) \wedge d\log m$$
$$= (bda + adb) \wedge d\log m = da \wedge \omega' + a \wedge d^1\omega'.$$

Since $\Omega^1_{X/Y}$ is locally generated as an abelian sheaf by sections of the form $\omega' = b\, d\log m$, condition (2) holds when $i = 0$ and $j = 1$.

To define d^i for $i > 1$, consider the map $\psi^i \colon \Omega^1_{X/Y} \times \cdots \times \Omega^1_{X/Y} \to \Omega^{i+1}_{X/Y}$, where

$$\psi^i(\omega_1, \ldots, \omega_i) := \sum_j (-1)^{j+1} \omega_1 \wedge \cdots \wedge d\omega_j \wedge \cdots \wedge \omega_i.$$

It is clear that ψ is multilinear over $\mathbf{Z}$. Moreover, if a is a local section of $\mathcal{O}_X$ and $1 \le k \le i$,

$$\begin{aligned}\psi^i(\omega_1, \ldots, a\omega_k, \ldots, \omega_i) = &\sum_{j<k} (-1)^{j+1} \omega_1 \wedge \cdots \wedge d\omega_j \wedge \cdots \wedge a\omega_k \wedge \cdots \wedge \omega_i \\ &+ \sum_{j>k} (-1)^{j+1} \omega_1 \wedge \cdots \wedge a\omega_k \wedge \cdots \wedge d\omega_j \wedge \cdots \wedge \omega_i \\ &+ (-1)^{k+1} \omega_1 \wedge \cdots \wedge da \wedge \omega_k \wedge \cdots \wedge \omega_i \\ &+ (-1)^{k+1} a\omega^1 \wedge \cdots \wedge d\omega_k \wedge \cdots \wedge \omega_i \\ = &\, da \wedge \omega_1 \wedge \cdots \wedge \omega_i \\ &+ \sum_j (-1)^{j+1} a\omega_1 \wedge \cdots \wedge d\omega_j \wedge \cdots \wedge \omega_i.\end{aligned}$$

This answer is independent of the choice of k, and hence ψ^i is multilinear over $\mathcal{O}_X$ and factors through the tensor product $(\Omega^1_{X/Y})^{\otimes k}$. Since ψ annihilates any i-tuple with a repeated factor, it also factors through a map $d^i \colon \Omega^i_{X/Y} \to \Omega^{i+1}_{X/Y}$. It is easy to check that this map has the desired properties. □

In the classical case, the exterior derivative $d \colon \Omega^1_{X/Y} \to \Omega^2_{X/Y}$ corresponds to a Lie-algebra structure on the dual $T_{X/Y}$. Let us verify that the same holds here.

Proposition 2.1.2. *Let $f \colon X \to Y$ be a morphism of coherent log schemes and let $T_{X/Y} := \mathrm{Der}_{X/Y}(\mathcal{O}_X)$. Then $T_{X/Y}$ has the structure of a Lie algebra over $f^{-1}\mathcal{O}_Y$, with Lie bracket defined by*

$$[\partial_1, \partial_2] := ([D_1, D_2], D_1\delta_2 - D_2\delta_1) \tag{2.1.1}$$

if $\partial_i = (D_i, \delta_i)$. Moreover,

$$\langle d\omega, \partial_1 \wedge \partial_2 \rangle = \partial_1\langle \omega, \partial_2 \rangle - \partial_2\langle \omega, \partial_1 \rangle - \langle \omega, [\partial_1, \partial_2] \rangle \tag{2.1.2}$$

if $\omega \in \Omega^1_{X/Y}$ and $\partial_1, \partial_2 \in T_{X/Y}$.

Proof We must prove that the right hand side of (2.1.1) defines a logarithmic derivation of X/Y. Let $D := [D_1, D_2]$, a derivation of $\mathcal{O}_X$ relative to $\mathcal{O}_Y$, and let $\delta := D_1 \circ \delta_2 - D_2 \circ \delta_1$, a homomorphism $\mathcal{M}_X \to \mathcal{O}_X$ that annihilates $f^{-1}(\mathcal{M}_Y)$. To see that $(D, \delta) \in \mathrm{Der}_{X/Y}(\mathcal{O}_X)$, it is enough to check that, for each local section m of $\mathcal{M}_X$, $D(\alpha_X(m)) = \alpha_X(m)\delta(m)$. Writing a for $\alpha_X(m)$, we have

$$\begin{aligned} D(\alpha_X(m)) &= D_1D_2(a) - D_2D_1(a) \\ &= D_1(a\delta_2(m)) - D_2(a\delta_1(m)) \\ &= a\delta_1(m)\delta_2(m) + aD_1\delta_2(m) - a\delta_2(m)\delta_1(m) - aD_2\delta_1(m) \\ &= a\delta(m) \end{aligned}$$

as required.

We should also verify the Jacobi identity

$$[\partial_1, [\partial_2, \partial_3]] + [\partial_2, [\partial_3, \partial_1]] + [\partial_3, [\partial_1, \partial_2]] = 0.$$

Suppose that $\partial_i = (D_i, \delta_i)$ for $i = 1, 2, 3$, and write $\partial_{i,j}$ for $[\partial_i, \partial_j]$. It suffices to compute the "δ" portion of this expression, which is

$$\begin{aligned} D_1\delta_{2,3} - D_{2,3}\delta_1 + D_2\delta_{3,1} - D_{3,1}\delta_2 + D_3\delta_{1,2} - D_{1,2}\delta_3 \\ = D_1D_2\delta_3 - D_1D_3\delta_2 - D_2D_3\delta_1 + D_3D_2\delta_1 \\ + D_2D_3\delta_1 - D_2D_1\delta_3 - D_3D_1\delta_2 + D_1D_3\delta_2 \\ + D_3D_1\delta_2 - D_3D_2\delta_1 - D_1D_2\delta_3 + D_2D_1\delta_3. \end{aligned}$$

This sums to zero as required.

It is clear that the difference between the left and right hand sides of (2.1.2) is additive in ω. Since $\Omega^1_{X/Y}$ is generated as an abelian sheaf by elements of the form $\omega = adm$, it suffices to check (2.1.2) for such elements. The left hand side is

$$\begin{aligned} \langle da \wedge dm, \partial_1 \wedge \partial_2 \rangle &= \langle da, \partial_1 \rangle \langle dm, \partial_2 \rangle - \langle da, \partial_2 \rangle \langle dm, \partial_1 \rangle \\ &= D_1(a)\delta_2(m) - D_2(a)\partial_1(m). \end{aligned}$$

The right hand side is

$$\begin{aligned} &\partial_1\langle \omega, \partial_2 \rangle - \partial_2\langle \omega, \partial_1 \rangle - \langle \omega, [\partial_1, \partial_2] \rangle \\ &\quad = D_1(a\delta_2(m)) - D_2(a\delta_1(m)) - \langle adm, ([D_1, D_2], D_1\delta_2 - D_2\delta_1) \rangle \\ &\quad = D_1(a)\delta_2(m) + aD_1\delta_2(m) - D_2(a)\delta_1(m) \\ &\qquad - aD_2\delta_1(m) - (aD_1\delta_2(m) - aD_2\delta_1(m)) \\ &\quad = D_1(a)\delta_2(m) - D_2(a)\delta_1(m). \end{aligned}$$

This concludes the proof. □

2.2 De Rham complexes of monoid algebras

A smooth morphism of log schemes is locally modeled by a morphism of monoid schemes coming from an injective monoid homomorphism $P \to Q$. The de Rham complex in this case is invariant under the action of the monoid $\underline{\mathsf{A}}_Q$ and so is equipped with a canonical Q-grading. Although the action and grading are destroyed by localization and do not exist in the local models, they are still quite useful, and in this section we exploit them to calculate the de Rham cohomology of such toric morphisms.

Let $\theta\colon P \to Q$ be an injective homomorphism of fine monoids, which for simplicity of notation we regard as an inclusion. We write π for the projection $\pi\colon Q^{gp} \to Q^{gp}/P^{gp}$. Let R be a fixed ground ring. The inclusion θ induces a homomorphism of log R-algebras

$$\big(P \to R[P]\big) \to \big(Q \to R[Q]\big),$$

and hence a corresponding morphism of log schemes $\mathsf{A}_Q \to \mathsf{A}_P$. According to Theorem IV.3.1.8, this morphism is smooth if and only if the order of the torsion part of $\mathrm{Cok}(\theta^{gp})$ is invertible in R. We assume this from now on, so that $R \otimes_{\mathbf{Z}} Q^{gp}/P^{gp}$ is a free R-module of finite rank. As we saw in Proposition IV.1.1.4, the sheaf of (logarithmic) Kahler differentials $\Omega^1_{\mathsf{A}_Q/\mathsf{A}_P}$ is the quasi-coherent sheaf of $\mathcal{O}_{\mathsf{A}_Q}$-modules associated to the Q-graded $R[Q]$-module $R[Q] \otimes Q^{gp}/P^{gp}$. Depending on the context, we may use any of the following notations for this module:

$$\Omega^1_\theta := \Omega^1_{Q/P/R} := \Omega^{\cdot}_{Q/P} := R[Q] \otimes Q^{gp}/P^{gp}.$$

For each i, $\Omega^i_{\mathsf{A}_Q/\mathsf{A}_P}$ is the quasi-coherent sheaf associated to the Q-graded $R[Q]$-module

$$\Omega^i_\theta := \Omega^i_{Q/P} := R[Q] \otimes \Lambda^i Q^{gp}/P^{gp}.$$

Similarly, $\mathcal{D}er_{\mathsf{A}_Q/\mathsf{A}_P}(\mathcal{O}_{\mathsf{A}_Q})$ is the quasi-coherent sheaf associated to

$$T_\theta := \mathrm{Hom}(\Omega^1_\theta, R[Q]) \cong \mathrm{R}[Q] \otimes \mathrm{Hom}(Q^{gp}/P^{gp}, \mathbf{Z}).$$

The *de Rham complex* of $\mathsf{A}_Q/\mathsf{A}_P$ is the complex of $R[Q]$-modules $(\Omega^{\cdot}_\theta, d)$, and the *de Rham cohomology* $H^*_{DR}(\mathsf{A}_Q/\mathsf{A}_P)$ of $\mathsf{A}_Q/\mathsf{A}_P$ is the cohomology of this complex. The grading of Ω^1_θ assigns degree q to an element $e^q \otimes \pi(x)$, and so the universal log derivation $(d, d)\colon (R[Q], Q) \to \Omega^1_\theta$ preserves degrees. Indeed, if $q \in Q$, $dq = 1 \otimes \pi(q)$ has degree zero, and $de^q = e^q dq = e^q \otimes \pi(q)$ has degree q. Thus the de Rham complex $\Omega^{\cdot}_{Q/P}$ inherits a grading. It admits the following explicit description.

Proposition 2.2.1. *Let $\theta\colon P \to Q$ be an injective homomorphism of fine* *monoids for which the order of the torsion part* $\mathrm{Cok}(\theta^{\mathrm{gp}})$ *is invertible in R. Then the de Rham complex* $\Omega^{\cdot}_{Q/P}$ *is a Q-graded complex of free graded* $R[Q]$*-modules, generated in degree zero. The degree-q term is given explicitly as the complex*

$$\Omega^{\cdot}_{Q/P,q} \cong \left(\Lambda^{\cdot} Q^{\mathrm{gp}}/P^{\mathrm{gp}}, dq\wedge\right),$$

where $(\Lambda^{\cdot} Q^{\mathrm{gp}}/P^{\mathrm{gp}}, dq\wedge)$ *is the exterior algebra on* $Q^{\mathrm{gp}}/P^{\mathrm{gp}}$ *with differential given by exterior multiplication by the image* $\pi(q)$ *of q in* $Q^{\mathrm{gp}}/P^{\mathrm{gp}}$*. Furthermore, the exterior multiplication*

$$\Omega^{i}_{Q/P} \otimes \Omega^{j}_{Q/P} \to \Omega^{i+j}_{Q/P}$$

is compatible with the grading.

Proof Each element ω of $\Omega^{i}_{Q/P}$ can be written uniquely as a sum

$$\omega = \sum_{q\in Q} e^{q} \otimes \omega_q \quad \text{where} \quad \omega_q \in R \otimes \Lambda^{i}(Q^{\mathrm{gp}}/P^{\mathrm{gp}}).$$

Thus ω_q is the homogeneous component of degree q of ω. Since the elements of $Q^{\mathrm{gp}}/P^{\mathrm{gp}}$ are all closed, so is each ω_q. Hence

$$d(e^{q}\omega_q) = de^{q} \wedge \omega_q = e^{q} \otimes \pi(q) \wedge \omega_q,$$

as claimed. □

Since the differentials of the de Rham complex preserve the grading, the cohomology modules are also graded, and are easy to describe if R contains a field. Of course, the answer depends on the characteristic.

Proposition 2.2.2. *Suppose that, in addition to the hypothesis of Proposition 2.2.1, R contains a field k. If the characteristic of k is zero, let*

$$\tilde{Q} := \{q \in Q : \exists n > 0 : nq \in P^{\mathrm{gp}}\},$$

and if the characteristic is p, let

$$\tilde{Q} := Q \cap (pQ^{\mathrm{gp}} + P^{\mathrm{gp}}).$$

Then the natural maps

$$R[\tilde{Q}] \otimes \Lambda^{\cdot} Q^{\mathrm{gp}}/P^{\mathrm{gp}} \to H^{\cdot}_{DR}(\mathsf{A}_Q\,/\,\mathsf{A}_P).$$

are isomorphisms.

Proof If $q \in Q$, then $de^{q} = e^{q} \otimes \pi(q)$, so $de^{q} = 0$ if and only if $1 \otimes \pi(q)$

vanishes in $R \otimes Q^{gp}/P^{gp}$. This makes it clear that $de^q = 0$ if $q \in \tilde{Q}$, so that we have a natural inclusion

$$R[\tilde{Q}] \cong H^0_{DR}(\mathsf{A}_Q / \mathsf{A}_P).$$

More generally, if $q \in \tilde{Q}$, the differential $\wedge dq$ of the complex $\Omega^{\cdot}_{Q/P,q}$ vanishes, so that, for all i,

$$H^i_{DR}(\mathsf{A}_Q / \mathsf{A}_P)_q = \Omega^i_{Q/P,q} \cong R \otimes \Lambda^i Q^{gp}/P^{gp}.$$

On the other hand, if $q \notin \tilde{Q}$, we shall see that the complex $\Omega^{\cdot}_{\theta,q}$ is acyclic, using the following lemma.

Lemma 2.2.3. *With the notation and hypotheses of Proposition 2.2.2, let q be an element of Q. Then $q \in Q \setminus \tilde{Q}$ if and only if there exists a homomorphism $\partial\colon Q^{gp}/P^{gp} \to \mathbf{Z} \to R$ such that $\partial(\pi(q)) \in R^*$.*

Proof If $q \in \tilde{Q}$, then its image in $R \otimes Q^{gp}/P^{gp}$ is zero, so $\partial(q) = 0$ for every $\partial \in \mathrm{Hom}(Q^{gp}/P^{gp}, R)$. Suppose that $q \in Q \setminus \tilde{Q}$ and (without loss of generality) that k is the prime field contained in R. Then the image of q in $k \otimes Q^{gp}/P^{gp}$ is nonzero, so there exists a homomorphism $t\colon Q^{gp}/P^{gp} \to k$ such that $t(q) \neq 0$. Since k is $\mathbf{Q}$ or $\mathbf{F_p}$, the homomorphism t can be chosen so that it comes from a homomorphism $Q^{gp}/P^{gp} \to \mathbf{Z}$. □

Let ∂ be as in the lemma and let s denote interior multiplication by $\partial(\pi(q))^{-1}\partial$ on $\Omega^{\cdot}_{Q/P,q} = R \otimes \Lambda^{\cdot} Q^{gp}/P^{gp}$. Then $s(\pi(q)) = 1$ and, for any ω in this complex,

$$\begin{aligned}(ds + sd)(\omega) &= \pi(q) \wedge s(\omega) + s(\pi(q) \wedge \omega)\\ &= \pi(q) \wedge s(\omega) + s(\pi(q))\omega - \pi(q) \wedge s(\omega)\\ &= \omega.\end{aligned}$$

Thus the complex $\Omega^{\cdot}_{\theta,q}$ is homotopic to zero, hence acyclic. □

Corollary 2.2.4. *With the notation and hypotheses of Proposition 2.2.2, suppose that R contains a field of characteristic zero and that $P = 0$. Let Q^*_t and Q^{gp}_t denote the torsion subgroups of Q^* and Q^{gp}, respectively. Then the natural maps*

$$(R[Q^*_t] \otimes \Lambda^{\cdot}(Q^{gp}), 0) \longrightarrow \left(\Omega^{\cdot}_Q, d\right)$$
$$\left(R[Q^{gp}_t] \otimes \Lambda^{\cdot}(Q^{gp}), 0\right) \longrightarrow \left(\Omega^{\cdot}_{Q^{gp}}, d)\right)$$

are quasi-isomorphisms. Thus if Q is saturated or, more generally, if $\overline{Q}^{gp}$ is torsion free, the map $\Omega^{\cdot}_Q \to \Omega^{\cdot}_{Q^{gp}}$ is a quasi-isomorphism.

Proof When $P = 0$ and R has characteristic zero, $\tilde{Q} = Q_t^*$, and the proposition implies that the displayed arrows are quasi-isomorphisms. If Q is saturated, Proposition I.1.3.5 tells us that $\overline{Q}^{\mathrm{gp}}$ is torsion free. If this is case, the map $Q \to Q^{\mathrm{gp}}$ induces an isomorphism $Q_t^* \to Q_t^{\mathrm{gp}}$, and it follows that the map $\Omega^{\cdot}_Q \to \Omega^{\cdot}_{Q^{\mathrm{gp}}}$ is a quasi-isomorphism. □

The corollary can be interpreted geometrically as follows. The morphism $Q^* \to Q^{\mathrm{gp}}$ of finitely generated abelian groups induces a commutative diagram of group schemes

$$\begin{array}{ccc}
\underline{\mathsf{A}}^*_{\mathsf{Q}} = \underline{\mathsf{A}}_{\mathsf{Q}^{\mathrm{gp}}} & \longrightarrow & \underline{\mathsf{A}}_{\mathsf{Q}_t^{\mathrm{gp}}} \\
\Big\downarrow & & \Big\downarrow \\
\underline{\mathsf{A}}_{\mathsf{Q}^*} & \longrightarrow & \underline{\mathsf{A}}_{\mathsf{Q}_t^*} .
\end{array}$$

The groups on the right are the groups of connected components of the corresponding group schemes on the left. Thus the corollary implies that the de Rham complex $\Omega^{\cdot}_{\theta}$ of the log scheme A_{Q} computes the de Rham cohomology of the group scheme $\underline{\mathsf{A}}^*_{\mathsf{Q}}$ if and only if $\underline{\mathsf{A}}^*_{\mathsf{Q}}$ and $\underline{\mathsf{A}}_{\mathsf{Q}^*}$ have the same set of connected components. For example, this is not the case for the monoid given by generators p and q satisfying the relation $2p = 2q$.

We now turn to the case of characteristic p. The operation of multiplication by p on a monoid Q corresponds to the Frobenius endomorphism in (log) geometry; to emphasize this we write F_Q for the homomorphism denoted p_Q in Section I.4.4.

Proposition 2.2.5. *In addition to the hypotheses of Proposition 2.2.1, suppose that R has characteristic p, and let $F_Q \colon Q \to Q$ denote multiplication by p.*

1. *Suppose that $P = 0$, that Q is p-saturated, and that the torsion subgroup of Q^{gp} has order prime to p. Then F_Q induces an isomorphism $Q \to \tilde{Q}$ and a quasi-isomorphism*

$$\begin{aligned}
(R[Q] \otimes \Lambda^{\cdot} Q^{\mathrm{gp}}, 0) &\to F_{Q*}(\Omega^{\cdot}_Q, d), \\
e^q \otimes \omega &\mapsto e^{pq} \otimes \omega.
\end{aligned}$$

2. *More generally, suppose that Q is p-saturated and that $Q^{\mathrm{gp}}/P^{\mathrm{gp}}$ is p-torsion free, and consider the exact relative Frobenius diagram (I.4.4.1), which we*

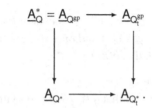

now notate as follows:

$$\begin{array}{ccccc} Q & \xrightarrow{\pi_{Q/P}} & Q_\theta^{(p)} & \xrightarrow{F_{Q/P}} & Q \\ \uparrow{\scriptstyle\theta} & & \uparrow{\scriptstyle\theta^{(p)}} & \nearrow{\scriptstyle\theta} & \\ P & \xrightarrow{F_P} & P. & & \end{array}$$

The homomorphism $F_{Q/P}$ factors as $F_{Q/P} = Q_\theta^{(p)} \cong \tilde{Q} \subseteq Q$, the homomorphism $\pi_{Q/P}$ induces an isomorphism

$$\gamma_{Q/P}\colon \operatorname{Cok}(\theta^{\mathrm{gp}}) \to \operatorname{Cok}(\theta^{(p)\mathrm{gp}}),$$

and together these induce a quasi-isomorphism

$$\left(R[Q_\theta^{(p)}] \otimes \Lambda^{\cdot}\Omega^{\cdot}_{\theta^{(p)}}, 0\right) \to (F_{Q/P*}\Omega^{\cdot}_\theta, d)$$
$$e^{q'} \otimes \omega' \mapsto F_{Q/P}(e^{q'}) \otimes \gamma^{-1}_{Q/P}(\omega').$$

Proof If $P = 0$, then $\tilde{Q} = Q \cap pQ^{\mathrm{gp}}$ and, since Q is p-saturated $\tilde{Q} = pQ$. Since the torsion subgroup of Q has order prime to p, multiplication by p defines an isomorphism $Q \to pQ$. Thus the first statement follows from Proposition 2.2.2. For the second statement, note that (1) of Proposition I.4.4.3 implies the claim about $\pi_{Q/P}$ and that Corollary I.4.4.4 implies the claim about $F_{Q/P}$. Then statement (2) follows from Proposition 2.2.2. □

Corollary 2.2.6. *With the hypotheses of Proposition 2.2.5, there is a unique family of isomorphisms*

$$\Omega^i_{\theta^{(p)}} \to F_{Q/P*}H^i(\Omega^{\cdot}_\theta)$$

that are compatible with exterior multiplication and that send $d\pi_{Q/P}(q)$ to dq for every $q \in Q$. □

The calculation of the cohomology given in the proof of Proposition 2.2.2 depends on the grading of de Rham complex Since this grading is destroyed by localization, it will be important to give a variation of the method that is more geometric.

Definition 2.2.7. *Let $\theta\colon P \to Q$ be a morphism of integral monoids. A homogeneous flow over θ is a homomorphism of monoids $h\colon Q \to \mathbf{N}$ such that $h(p) = 0$ for all $p \in P$. A homogeneous vector field over θ is a group homomorphism $\partial\colon Q^{\mathrm{gp}} \to \mathbf{Z}$ such that $\partial \circ \theta^{\mathrm{gp}} = 0$.*

Remark 2.2.8. The set of homogeneous vector fields over θ can be identified with the degree zero part of the module T_θ of all vector fields, and the set $H_\theta(Q)$

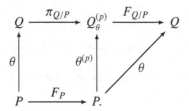

of homogeneous flows over θ forms a submonoid of this group. Furthermore, if $\langle P\rangle$ is the face of Q generated by P, then $H_\theta(Q)$ can be identified with the dual monoid $H(Q/\langle P\rangle)$, and it follows from Corollary I.2.2.4 that

$$H_\theta(Q)^{\mathrm{gp}} \cong \mathrm{Hom}(Q^{\mathrm{gp}}/\langle P\rangle^{\mathrm{gp}}, \mathbf{Z}) \subseteq T_{\theta,0}.$$

If $\partial\colon Q^{\mathrm{gp}}/P^{\mathrm{gp}} \to \mathbf{Z}$ is a homogeneous vector field over θ, the corresponding $R[Q]$-linear map

$$\xi\colon R[Q] \otimes Q^{\mathrm{gp}}/P^{\mathrm{gp}} \to R[Q]$$

identifies with $\mathrm{id} \otimes \partial$, and the corresponding derivation

$$(\mathrm{id} \otimes \partial) \circ d\colon R[Q] \to R[Q]$$

sends e^q to $e^q\partial(q)$. Any two such homogeneous vector fields commute under the bracket operation defined in Proposition 2.1.2.

We can extend a vector field $\xi \in T_\theta = \mathrm{Hom}(\Omega^1_{Q/P}, R[Q])$ to a collection of $R[Q]$-linear homomorphisms (interior multiplication)

$$\xi\colon \Omega^i_{Q/P} \to \Omega^{i-1}_{Q/P} : \omega_1 \wedge \cdots \wedge \omega_i \mapsto \sum_j (-1)^{j+1}\xi(\omega_j)\omega_1 \wedge \cdots \wedge \hat{\omega}_j \wedge \cdots \wedge \omega_i.$$

Similarly, we can extend the derivation $\xi \circ d\colon R[Q] \to R[Q]$ to a collection of maps (the *Lie derivative*) by setting

$$\kappa\colon \Omega^i_{Q/P} \to \Omega^i_{Q/P} : \omega \mapsto (d\xi + \xi d)\omega.$$

Lemma 2.2.9. *Let $\xi \in T_\theta$ be a vector field over θ, let $\xi\colon \Omega^{\cdot}_{Q/P} \to \Omega^{\cdot-1}_{Q/P}$ denote interior multiplication by ξ, and let κ be the Lie derivative with respect to ξ.*

1. *If $\alpha \in \Omega^a_{Q/P}$ and $\beta \in \Omega^b_{Q/P}$, we have*

$$\xi(\alpha \wedge \beta) = \xi(\alpha) \wedge \beta + (-1)^a \alpha \wedge \xi(\beta) \quad \text{and}$$
$$\kappa(\alpha \wedge \beta) = \kappa(\alpha) \wedge \beta + \alpha \wedge \kappa(\beta),$$

   i.e., ξ and κ are derivations of degrees -1 and 0, respectively.
2. *The Lie derivative $\kappa\colon \Omega^{\cdot}_{Q/P} \to \Omega^{\cdot}_{Q/P}$ is a morphism of complexes, is homotopic to zero, and induces the zero map on cohomology.*
3. *If ξ is induced by a homogeneous vector field $\partial \in \mathrm{Hom}(Q^{\mathrm{gp}}/P^{\mathrm{gp}}, \mathbf{Z})$, then ξ and κ preserve the Q-grading of $\Omega^{\cdot}_{Q/P}$. In particular, κ is an endomorphism of the Q-graded complex $\Omega^{\cdot}_{Q/P}$, and gives multiplication by $\partial(\pi(q))$ in degree q.*

Proof The formulas in (1) are standard computations, which we leave to the reader to verify. Of course, $\kappa := d\xi + \xi d$ is automatically a morphism of complexes and induces zero on cohomology, since it is visibly homotopic to zero.

If ξ is induced by a homogenous vector field ∂, then interior multiplication by ξ is induced by interior multiplication by ∂, and thus preserves degrees. Since d also preserves degrees, so does κ.

To compute κ explicitly, let ω be any element of $\Omega^i_{Q/P,0}$. We have already observed that $d\omega = 0$. Since ξ preserves the Q-grading, $\xi(\omega) \in \Omega^{i-1}_{Q/P,0}$, hence $d\xi\omega = 0$, and thus $\kappa(\omega) = 0$. This proves the formula when $q = 0$. On the other hand, if $i = 0$ and q is arbitrary,

$$\kappa(e^q) = d\xi e^q + \xi de^q = 0 + \xi(e^q \otimes \pi(q)) = \partial\pi(q))e^q,$$

which again is consistent with the formula in statement (3). For any q and i, $\Omega^i_{Q/P,q}$ is spanned as an R-module by elements of the form $e^q \otimes \omega$ with $\omega \in \Omega^i_{Q/P,0}$. Then

$$\kappa(e^q \otimes \omega) = \kappa(e^q) \otimes \omega + e^q\kappa(\omega) = \partial(\pi(q))e^q \otimes \omega + 0.$$

This proves (3) in general. □

We shall use these techniques to control the cohomology of various subcomplexes of de Rham complexes and their localizations.

2.3 Filtrations on the de Rham complex

If $f\colon X \to Y$ is a morphism of log schemes, the combinatorics of the sheaf of monoids $\mathcal{M}_X$ manifest themselves through two sheaves of partially ordered sets: the set of ideals of $\mathcal{M}_X$ and the set of faces of $\mathcal{M}_X$. These are in turn reflected in certain filtrations on the de Rham complex $\Omega^{\cdot}_{X/Y}$, which have interesting cohomological interpretations.

We begin with the case of toric models, given as in Section 2.2 by an injective homomorphism of fine monoids $\theta\colon P \to Q$ such that the order of the torsion subgroup of $Q^{\mathrm{gp}}/P^{\mathrm{gp}}$ is invertible in R.

Lemma 2.3.1. *If $K \subseteq Q^{\mathrm{gp}}$ a fractional ideal, for each i, let*

$$K\Omega^i_{Q/P} \subseteq \Omega^i_{Q^{\mathrm{gp}}/P^{\mathrm{gp}}} \cong R[Q^{\mathrm{gp}}] \otimes \Lambda^i Q^{\mathrm{gp}}/P^{\mathrm{gp}}$$

denote the R-submodule generated by the elements of the form $e^k\omega$ with $k \in K$ and $\omega \in R \otimes \Lambda^i Q^{\mathrm{gp}}/P^{\mathrm{gp}}$. Then $K\Omega^i_{Q/P}$ is an $R[Q]$-submodule of $\Omega^i_{\theta^{\mathrm{gp}}}$, and $K\Omega^{\cdot}_{Q/P} := \oplus\{K\Omega^i_{Q/P} : i \geq 0\}$ is stable under d and under interior multiplication and Lie differentiation by any vector field over θ.

Proof The fact that $K\Omega^i_{Q/P}$ is stable under multiplication by $R[Q]$ follows from the fact that K is stable under translation by Q. Since interior multiplication by a vector field is $R[Q]$-linear, $K\Omega^{\cdot}_{Q/P}$ is stable under such interior multiplication. For any $k \in K$ and $\omega \in \Omega^i_{Q/P}$,

$$d(e^k\omega) = de^k \wedge \omega + e^k d\omega = e^k dk \wedge \omega.$$

It follows that $K\Omega^{\cdot}_{Q/P}$ is stable under exterior and Lie differentiation. □

Lemma 2.3.2. *If $K \subseteq Q$ is an ideal, the following conditions are equivalent.*

1. *For each $k \in K$, there exists a homogenous flow $h \in H_\theta(Q)$ (see Definition 2.2.7) such that $h(k) \neq 0$.*
2. *$K \cap P = \emptyset$.*
3. *$K \cap \langle P \rangle = \emptyset$.*
4. *There exists an $h \in H_\theta(Q)$ such that $h(k) \neq 0$ for all $k \in K$.*

Proof If $h \in H_\theta(Q)$, then $h(p) = 0$ for all $p \in P$, so (1) implies (2). If $k \in \langle P \rangle \cap K$, then there exists a $q \in Q$ such that $q + k \in P$, hence $q + k \in K \cap P$, so (2) implies (3). If (3) is true, then K_P is a proper ideal of the localization Q_P of Q by P. By Proposition I.2.2.1 there exists a local homomorphism $h' \colon Q_P \to \mathbf{N}$, which will necessarily factor through $\overline{Q}_P = Q/P$ and hence define an element h of $H_\theta(P)$. Since K maps into the maximal ideal of Q/P and h' is local, $h(k) \neq 0$ for every $k \in K$, so (4) is true. The implication of (1) by (4) is of course trivial. □

An ideal satisfying the conditions of the lemma will be called *horizontal.*

Proposition 2.3.3. *Let $\theta \colon P \to Q$ be an injective homomorphism of fine monoids, let $K \subseteq Q$ be a horizontal ideal, and let $F\Omega^{\cdot}_{Q/P}$ be a Q-graded subcomplex of $\Omega^{\cdot}_{Q/P}$ stable under interior multiplication by the vector fields coming from horizontal flows. Suppose that R contains $\mathbf{Q}$. Then $K\Omega^{\cdot}_{Q/P} \cap F\Omega^{\cdot}_{Q/P}$ is homotopic to zero and hence acyclic. Consequently, the natural map*

$$F\Omega^{\cdot}_{Q/P} \to F\Omega^{\cdot}_{Q/P}/K \cap F\Omega^{\cdot}_{Q/P}$$

is a quasi-isomorphism.

Proof By Lemma 2.3.2, there exists a horizontal flow $h \in H_\theta(Q)$ with $h(k) \neq 0$ for all $k \in K$. Let ξ be interior multiplication by the corresponding horizontal vector field. Lemma 2.3.1 shows that $K\Omega^{\cdot}_{Q/P}$ is stable under the exterior derivative d, by ξ and hence by $\kappa := d\xi + \xi d$. Since $F\Omega^{\cdot}_{Q/P}$ is by assumption stable under d and ξ, the same is true of the complex $K\Omega^{\cdot}_{Q/P} \cap F\Omega^{\cdot}_{Q/P}$. This complex is Q-graded and vanishes in degrees $q \in Q \setminus K$, and Lemma 2.2.9 implies that κ is multiplication by $h(k)$ in degree k. Since $h(k)$ is a unit in R for every $k \in K$, the morphism κ induces an isomorphism of complexes

$$\kappa \colon K\Omega^{\cdot}_{Q/P} \cap F\Omega^{\cdot}_{Q/P} \to K\Omega^{\cdot}_{Q/P} \cap F\Omega^{\cdot}_{Q/P}.$$

Note that $\kappa\xi = \xi\kappa = \xi d\xi$, and let

$$\xi' := \kappa^{-1}\xi = \xi\kappa^{-1} : K\Omega^{\cdot} \cap F\Omega^{\cdot}_{Q/P} \to K\Omega^{\cdot-1} \cap F\Omega^{\cdot-1}_{Q/P}.$$

Then $d\xi' + \xi' d = \mathrm{id}$, and so the complex $K\Omega^{\cdot} \cap F\Omega^{\cdot}_{Q}/P$ is homotopic to zero, hence acyclic. □

We will apply these ideas to some filtrations generalizing Danilov's construction of de Rham complexes for toric varieties [11].

Definition 2.3.4. *If F is a face of Q containing P and $q \in Q$, let $\langle F, q\rangle$ denote the face of the quotient Q/P generated by F and $\pi(q)$. Then:*

$$\begin{aligned}
K^i_q(F)\Omega^j_\theta &:= Im\left(R \otimes \Lambda^{j-i} Q^{\mathrm{gp}} \otimes \Lambda^i \langle F, q\rangle^{\mathrm{gp}} \to R \otimes \Lambda^j Q^{\mathrm{gp}}/P^{\mathrm{gp}}\right),\\
L^i_q(F)\Omega^j_{Q/P} &:= K^{i+j}_q(F)\Omega^j_{Q/P},\\
L_{i,q}(F)\Omega^j_{Q/P} &:= L^{-i}_q(F)\Omega^j_{Q/P},\\
\underline{\Omega}^i_{Q/P}(F) &:= L^0(F)\Omega^i_{Q/P},\\
\underline{\Omega}^i_{Q/P} &:= \underline{\Omega}^i_{Q/P}(Q^*).
\end{aligned}$$

Note that $K^{\cdot}_q(F)$ is just the Koszul filtration on $R \otimes \Lambda^j (Q^{\mathrm{gp}}/P^{\mathrm{gp}})$ corresponding to the submodule

$$R \otimes \langle F, q\rangle^{\mathrm{gp}} \to R \otimes Q^{\mathrm{gp}}/P^{\mathrm{gp}}.$$

Explicitly, $K^i_q(F)\Omega^j_{Q/P}$ is the subgroup of $\Lambda^j(Q^{\mathrm{gp}}/P^{g})$ generated by the elements of the form $dq_1 \wedge \cdots \wedge dq_j$ such that there exist $n \in \mathbf{N}$ and $f \in F$ such that $nq + f \geq q_i + \cdots + q_i$, after a suitable permutation of the indices. Note that if $q' \geq q \in Q$, then $\langle F, q\rangle \subseteq \langle F, q'\rangle$, so that $K^i(F)$ defines a Q-filtration on $R \otimes \Lambda^j(P^{\mathrm{gp}}/Q^{\mathrm{gp}})$, and hence a Q-graded submodule of $R[Q] \otimes \Lambda^j(Q^{\mathrm{gp}}/P^{\mathrm{gp}})$.(See Definition I.3.2.4 and its following discussion.)

For example, suppose that $P = 0$ and that Q^{gp} is free of rank r. Then $\Omega^r_Q = R \otimes \Lambda^r Q^{\mathrm{gp}}$, a free R-module of rank one. If $q \in Q$, $\underline{\Omega}^r_{Q,q} = R \otimes \Lambda^r \langle q\rangle^{\mathrm{gp}}$, which vanishes if $\langle q\rangle \neq Q$ and is $R \otimes \Lambda^r Q^{\mathrm{gp}}$ otherwise. Since $\langle q\rangle = Q$ if and only if q belongs to every height one prime of Q, it follows that $\underline{\Omega}^r_Q = I_Q\Omega^r_Q$, where I_Q is the interior ideal of Q, the intersection of all such primes. This is not the same as the rth exterior power of $\underline{\Omega}^1_Q$ in general; see Proposition 2.3.10.

Proposition 2.3.5. *Let $\theta\colon P \to Q$ be an injective homomorphism of fine monoids and let R be a ring such that the order of the torsion part of* $\mathrm{Cok}(\theta^{\mathrm{gp}})$ *is invertible in R.*

1. *The differential d of the complex $\Omega^{\cdot}_{Q/P}$ sends $K^i(F)\Omega^j_{Q/P}$ into $K^{i+1}(F)\Omega^{j+1}_{Q/P}$ and $L^i(F)\Omega^j_{Q/P}$ into $L^i(F)\Omega^{j+1}_{Q/P}$.*

2. *If ξ is a vector field over θ, interior multiplication by ξ maps $K^i(F)\Omega^j_{Q/P}$ to $K^{i-1}(F)\Omega^{j-1}_{Q/P}$ and $L^i(F)\Omega^j_{Q/P}$ to $L^i(F)\Omega^{j-1}_{Q/P}$.*
3. *The filtration $L^{\cdot}(F)$ is the "filtration décalée" [14, 1.3.3] associated to $K^{\cdot}(F)$.*

Proof The proof is an immediate verification, using the explicit description following Definition 2.3.4. Suppose that $\omega = dq_1 \wedge \cdots \wedge dq_j \in K^i_q(F)\Omega^j_{Q/P}$ and that $nq + f \geq q_1 + \cdots + q_i$. Then $d\omega = dq \wedge dq_1 \wedge \cdots \wedge dq_j$, and since $(n+1)q+f \geq q+q_1 \cdots + q_i$, it follows that $d\omega$ lies in $K^{i+1}_q(F)\Omega^{j+1}_{Q/P,q}$. To prove (2), we first suppose that ξ is homogeneous and is induced by the homomorphism $\partial\colon Q^{\mathrm{gp}}/P^{\mathrm{gp}} \to R$. Then interior multiplication by ξ sends ω to

$$\xi(\omega) = \sum_k (-1)^{k-1}\partial(q_k) q_1 \wedge \cdots \wedge \hat{q}_k \wedge \cdots \wedge q_j,$$

which evidently belongs to $K^{i-1}_q(F)\Omega^{j-1}_{Q/P}$. Then $e^{q'}\xi(\omega)$ lies in $K^{i-1}_{q'+q}(F)\Omega^{j-1}_{Q/P}$ for every $q' \in Q$. It follows that the assertion is also true for $e^{q'}\xi$. Since every vector field can be written as a linear combination of vector fields of this form, the result is true in general. By definition, the "filtration décalée" *Dec K* of $\Omega^{\cdot}_{Q/P}$ is

$$(Dec\ K(F))^i\Omega^j_{Q/P} := \{\omega \in K^{i+j}(F)\Omega^j_{Q/P} : d\omega \in K^{i+j+1}(F)\Omega^{j+1}_{Q/P}\}.$$

Since the differential of $\Omega^{\cdot}_{Q/P}$ maps $K^{i+j}(F)\Omega^j_{Q/P}$ into $K^{i+j+1}(F)\Omega^{j+1}_{Q/P}$, this is just the filtration $L^{\cdot}(F)$. □

The following technical result helps control the behavior of formation of these filtrations under base change.

Proposition 2.3.6. *Let $\theta\colon P \to Q$ be an injective homomorphism of fine monoids, let F be a face of Q containing P, and let R be a ring in which the order of the torsion subgroup of the cokernel of θ^{gp} is invertible. Suppose either that R contains a field or that, for every face G of Q containing F, the order of the torsion subgroup of $Q^{\mathrm{gp}}/G^{\mathrm{gp}}$ is also invertible in R.*

1. *The submodules $L^i(F)\Omega^j_{Q/P)}$ of $\Omega^j_{Q/P}$ are direct summands as $R[P]$-modules.*
2. *Let $P \to P'$ be a homomorphism of integral monoids, let $Q' := Q \oplus_P P'$ (formed in the category of integral monoids), and let $F' := F \oplus_P P'$. Then $\theta'\colon P' \to Q'$ and F' again satisfy the hypotheses of the proposition, and the natural map*

$$R[Q'] \otimes_{R[Q]} (\Omega^{\cdot}_{Q/P}, L^{\cdot}(F)) \to (\Omega^{\cdot}_{Q'/P'}, L^{\cdot}(F'))$$

is an isomorphism of filtered complexes. The analogous statement also holds in the category of saturated monoids.

3. *If furthermore θ is integral and local, then the graded modules $\mathrm{Gr}^i_{L(F)}\,\Omega^j_{Q/P}$ are flat over $R[P]$.*

Proof Let G be a face of Q containing F. We have an exact sequence

$$0 \to G^{\mathrm{gp}}/P^{\mathrm{gp}} \to Q^{\mathrm{gp}}/P^{\mathrm{gp}} \to Q^{\mathrm{gp}}/G^{\mathrm{gp}} \to 0.$$

If the order of the torsion subgroups of the latter two groups is invertible in R, then the sequence

$$0 \to R \otimes G^{\mathrm{gp}}/P^{\mathrm{gp}} \to R \otimes Q^{\mathrm{gp}}/P^{\mathrm{gp}} \to R \otimes Q^{\mathrm{gp}}/G^{\mathrm{gp}} \to 0$$

is a short exact sequence of free R modules, and hence is split. If, on the other hand, R is a field, the sequence

$$0 \to L_G \to R \otimes Q^{\mathrm{gp}}/P^{\mathrm{gp}} \to R \otimes Q^{\mathrm{gp}}/G^{\mathrm{gp}} \to 0$$

is split, where L_G is now the image of $R \otimes G^{\mathrm{gp}}/P^{\mathrm{gp}}$ in $R \otimes Q^{\mathrm{gp}}/P^{\mathrm{gp}}$. If R contains a field, the same result holds by base change. In any of these cases, choose such a splitting s_G for each G, and note that, for every $q \in Q$, the face $\langle F, q\rangle$ generated by F and q is such a G. These splittings define splittings of the Koszul filtration of $R \otimes \Lambda^j Q^{\mathrm{gp}}/P^{\mathrm{gp}}$ induced by the image of $R \otimes \langle F, q\rangle^{\mathrm{gp}}$ in $R \otimes Q^{\mathrm{gp}}/P^{\mathrm{gp}}$. Thus, for each q, we have a splitting

$$s_q := s_{\langle F,q\rangle} \colon R[Q] \otimes \Lambda^j Q^{\mathrm{gp}} P^{\mathrm{gp}} \to L^i(F)\big(R[Q] \otimes \Lambda^j Q^{\mathrm{gp}}/P^{\mathrm{gp}}\big)$$

in degree q. The collection of these maps defines an $R[P]$-linear homomorphism because $\langle F, q\rangle = \langle F, p + q\rangle$ for $p \in P$.

To prove statement (2), first let us check that $F' := F \oplus_P P'$ is a face of Q. Suppose that $q_i \in Q$, $p'_i \in P'$, $f \in F$, and $p' \in P'$ with $q_1 + p'_1 + q_2 + p'_2 = f + p'$ in Q'. Then there exist $p_1, p_2 \in P$ such that $p_1 + q_1 + q_2 = p_2 + f$ and $p_2 + p'_1 + p'_2 = p_1 + p'$. Since $p_2 \in F$, it follows that q_1 and q_2 belong to F, and hence that $q_1 + p'_1$ and $q_2 + p'_2$ belong to F', as required. Then, for any face G of Q containing F, $G \oplus_P P'$ is a face of Q' containing F'. Conversely, if G' is a face of Q' containing F', let G be its inverse image in Q, a face of Q containing F. Any element of G' can be written as a sum $q + p'$; since necessarily q belongs to G, in fact $G' = G \oplus_P P'$. Thus we have a bijection between the faces of Q containing F and the faces of Q' containing F'. For any G, the natural map $Q^{\mathrm{gp}}/G^{\mathrm{gp}} \to Q'^{\mathrm{gp}}/G'^{\mathrm{gp}}$ is a bijection, as is the map $Q^{\mathrm{gp}}/P^{g} \to Q'^{\mathrm{gp}}/P'^{\mathrm{gp}}$, so the hypotheses of the proposition are preserved. Now, any element q' of Q' can be written $q' = q + p'$ with $q \in Q$ and $p' \in P$, and then $\langle F', q'\rangle = \langle F, q\rangle \oplus_P P'$. It follows that $L^i(F')\Omega^{\cdot}_{Q'/P',q'} = L^i(F)\Omega^{\cdot}_{Q/P,q}$, and statement (2) follows.

Now suppose that P, Q, and P' are saturated, and let Q'' be the saturation of the pushout Q' and F'' the saturation of F'. Note that in this case, the torsion hypothesis is automatically satisfied, since Q^{gp}/G^{gp} is torsion free for every face G of Q. We check that F'' is in fact a face of Q''. Suppose that $q_1'' + q_2'' = f''$, where $q_i'' \in Q''$ and $f'' \in F''$. Then there exists an $n > 0$ such that $nq_i'' \in Q'$ and $nf'' \in F'$ and, since $nq_1'' + nq_2'' \in F'$, in fact each $nq_i'' \in F'$. As we have seen, $Q'^{gp}/F'^{gp} \cong Q^{gp}/F^{gp}$ is torsion free, and it follows that $q_i'' \in F'^{gp}$, and hence that $q_i'' \in F'^{sat} = F''$. The rest of the proof proceeds in the same way as in the integral case.

If also $P \to Q$ is integral and local we shall in fact prove that each $E := \mathrm{Gr}^i_{L(F)} \Omega^j_{Q/P}$ is free as an $R[P]$-module. Recall from Corollary I.4.6.11 that Q is a free P-set, say with basis $S \subseteq Q$. Then it suffices to prove that, for each $s \in S$, the sub $R[P]$-module $\oplus\{E_{p+s} : p \in P\}$ is free. This follows from the fact that E_s is a free R-module and that, for each $p \in P$, the map $E_s \to E_{p+s}$ induced by multiplication by e^p is bijective. □

We note that if Q is saturated, then Q^{gp}/G^{gp} is torsion free for every face G of Q, so the condition on the faces G in Proposition 2.3.6 is automatically satisfied.

The proof of the following variant is left for the reader.

Variant 2.3.7. Let $\theta\colon (P, J) \to (Q, K)$ be an injective homomorphism of fine idealized monoids and let F be a face of Q containing P. Suppose that θ, F, and R satisfy the hypotheses of Proposition 2.3.6. Then the submodules

$$L^i(F)\Omega^j_{(Q,K)/(P,J))} \subseteq \Omega^j_{(Q,K)/(P,J)}$$

are direct summands as $R[P]$-modules.

Example 2.3.8. Let Q be the monoid with generators, a, b, c and relation $2a = 2b + c$. Then Q^{gp} is torsion free, but $Q^{gp}/\langle c\rangle^{gp}$ is not, and in fact the map $\mathbf{F}_2 \otimes \langle c\rangle^{gp} \to \mathbf{F}_2 \otimes Q^{gp}$ is the zero map. Consequently the map

$$\mathbf{F}_2 \otimes \underline{\Omega}^1_{Q/\mathbf{Z}} \to \underline{\Omega}^1_{Q/\mathbf{F}_2}$$

is not injective in degree c and in particular the formation of $\underline{\Omega}^{\cdot}_Q$ does not commute with the base change $\mathbf{Z} \to \mathbf{F}_2$. The saturation Q^{sat} of Q is generated by $x := a - b$ and b, and the image of c in Q^{sat} is $2x$. Thus the map

$$\mathbf{F}_2[Q^{sat}] \otimes \underline{\Omega}^1_{Q/\mathbf{F}_2} \to \underline{\Omega}^1_{Q^{sat}/\mathbf{F}_2}$$

is not injective in degree x, even though $\mathsf{A}_{Q^{sat}} \to \mathsf{A}_Q$ is étale.

The following result, inspired by techniques in [50], extends the Poincaré residue mapping discussed in Proposition IV.1.2.14 to the complexes constructed here. It will play an important role in the proof of the comparison theorems in Section 4.2.

Proposition 2.3.9. *Let $\theta\colon P \to Q$ be an injective homomorphism of fine monoids and let R be a ring in which the order of the torsion part of $Q^{\mathrm{gp}}/P^{\mathrm{gp}}$ is invertible. Let F be a face of Q containing P, let $\mathfrak{p} := Q \setminus F$, and assume that the order of the torsion subgroup of $Q^{\mathrm{gp}}/F^{\mathrm{gp}}$ is also invertible in R.*

1. *For each $i \in \mathbf{N}$, there is a natural isomorphism of complexes*

$$\mathrm{Gr}_i^{L(F)}\,\Omega^{\cdot}_{(Q,\mathfrak{p})/P} \cong \Lambda^i Q^{\mathrm{gp}}/F^{\mathrm{gp}} \otimes \Omega^{\cdot}_{F/P}[-i].$$

2. *There is a natural exact sequence of complexes*

$$0 \longrightarrow \underline{\Omega}^{\cdot}_{Q/P}(F) \cap \mathfrak{p}\Omega^{\cdot}_{Q/P} \longrightarrow \underline{\Omega}^{\cdot}_{Q/P}(F) \xrightarrow{i^{\cdot}_F} \Omega^{\cdot}_{F/P} \longrightarrow 0.$$

3. *Suppose that F is a facet of Q. Then $\mathfrak{p}\Omega^{\cdot}_{Q/P} \subseteq \underline{\Omega}^{\cdot}_{Q/P}(F)$, and there are natural exact sequences:*

$$0 \longrightarrow \Omega^{\cdot}_{F/P} \longrightarrow \Omega^{\cdot}_{(Q,\mathfrak{p})/P} \xrightarrow{\overline{\rho}^{\cdot}} \Omega^{\cdot}_{F/P}[-1] \longrightarrow 0,$$

$$0 \longrightarrow \underline{\Omega}^{\cdot}_{Q/P}(F) \longrightarrow \Omega^{\cdot}_{Q/P} \xrightarrow{\rho^{\cdot}} \Omega^{\cdot}_{F/P}[-1] \longrightarrow 0.$$

Proof According to our definitions, $L.(F)\Omega^{j}_{(Q,\mathfrak{p})/P}$ vanishes in degrees $q \notin F$ and, in degree $q \in F$, is just the (shifted) Koszul filtration associated with the direct summand $R \otimes F^{\mathrm{gp}}/P^{\mathrm{gp}}$ of $R \otimes Q^{\mathrm{gp}}/P^{\mathrm{gp}}$. Then

$$\mathrm{Gr}_i^{L(F)}\,\Omega^{\cdot}_{(Q,\mathfrak{p})/P,q} \cong R \otimes \Lambda^i(Q^{\mathrm{gp}}/F^{\mathrm{gp}}) \otimes \Lambda^{j-i}(F^{\mathrm{gp}}/P^{\mathrm{gp}}),$$

with boundary map given by left exterior multiplication by $\pi(q)$. This boundary map is $(-1)^i$ times the identity of $\Lambda^i(Q^{\mathrm{gp}}/F^{\mathrm{gp}})$ times the boundary map of $\Omega^{\cdot}_{F/P,q}$. This proves (1).

The map $\underline{\Omega}^{\cdot}_{Q/P}(F) \cap \mathfrak{p}\Omega^{\cdot}_{Q/P} \to \underline{\Omega}^{\cdot}_{Q/P}(F)$ in statement (1) is just the natural inclusion of complexes, and the quotient is the complex $\underline{\Omega}^{\cdot}_{(Q,\mathfrak{p})/P}(F) = L_0(F)\Omega^{\cdot}_{Q/P}$. By statement (1), this is just $\Omega^{\cdot}_{F/P}$.

Now suppose that F is a facet of Q. Then $\langle F, q\rangle = Q$ if $q \in \mathfrak{p}$, so $\underline{\Omega}^{\cdot}_{Q/P,q}(F) = \Omega^{\cdot}_{Q/P,q}$ in this case. It follows that $\mathfrak{p}\Omega^{\cdot}_{Q/P} \subseteq \underline{\Omega}^{\cdot}_{Q/P}(F)$. Moreover, thanks to the hypothesis on the torsion in $Q^{\mathrm{gp}}/F^{\mathrm{gp}}$, the valuation $\nu_{\mathfrak{p}}$ associated to $\mathfrak{p}$ induces an isomorphism $R \otimes Q^{\mathrm{gp}}/F^{\mathrm{gp}} \to R$. Then it follows from (1) that

$$\mathrm{Gr}_i^{L(F)}\,\Omega^{\cdot}_{(Q,\mathfrak{p})/P} = \begin{cases} \Omega^{\cdot}_{F/P} & \text{if } i = 0 \\ \Omega^{\cdot}_{F/P}[-1] & \text{if } i = 1 \\ 0 & \text{otherwise.} \end{cases}$$

The exact sequence

$$0 \to \mathrm{Gr}_0^{L(F)} \Omega^{\cdot}_{(Q,\mathfrak{p})/P} \to \Omega^{\cdot}_{(Q,\mathfrak{p})/P} \to \mathrm{Gr}_{-1}^{L(F)} \Omega^{\cdot}_{(Q,\mathfrak{p})/P} \to 0$$

is the first sequence in statement (3). The second sequence in degree $q \in F$ is the same as the first sequence and, in degree $q \in \mathfrak{p}$, the map $\underline{\Omega}^{\cdot}_{Q/P,q}(F) \to \Omega^{\cdot}_{Q/P,q}$ is an isomorphism and $\Omega^{\cdot}_{F/P,q} = 0$. Hence the sequence is exact. □

We should point out that the map $\Lambda^i \underline{\Omega}^1_{Q/P}(F) \to \underline{\Omega}^i_{Q/P}(F)$ is not surjective in general. For example, if $P = F = 0$ and Q is the monoid with generators (a, b, c) and relation $a + b = 2c$, the element $da \wedge db \in \underline{\Omega}^2_{Q,c}$ does not lie in this image, since c is irreducible. In fact, as the next result shows, if Q is fine saturated and sharp, surjectivity only happens if Q is free.

Proposition 2.3.10. *Let Q be a fine, sharp, and saturated monoid of dimension d, let I_Q be the interior ideal of Q, and let $I_d \subseteq I_Q$ be the ideal defined in Lemma I.2.5.4. If k is a field, the following statements hold.*

1. *The $k[Q]$-module $\Omega^d_{Q/k}$ is free of rank one. Furthermore, $\underline{\Omega}^d_{Q/k} = I_Q\Omega^d_{Q/k}$ and the image of the map $\Lambda^d\underline{\Omega}^1_{Q/k} \to \Omega^d_{Q/k}$ is $I_d\Omega^d_{Q/k}$.*
2. *The map $\Lambda^d\underline{\Omega}^1_{Q/k} \to \underline{\Omega}^d_{Q/k}$ is surjective if and only if Q is free.*

Proof Since Q is fine sharp and saturated of dimension d, its associated group Q^{gp} is free abelian of rank d, and hence $\Omega^d_{Q/k} = k[Q] \otimes \Lambda^d Q^{\mathrm{gp}}$ is free of rank one. Since $\underline{\Omega}^d_{Q/k,q} = k \otimes \Lambda^d \langle q \rangle^{\mathrm{gp}}$, this graded piece is nonzero if and only if $\langle q \rangle$ has rank d, that is, if and only if $q \in I_Q$, in which case $\underline{\Omega}^d_{Q/k,q}$ is one-dimensional. Thus $\underline{\Omega}^d_{Q/k} = I_Q \Omega^d_{Q/k}$.

Recall that $\underline{\Omega}^1_{Q/k,q'}$ is the image of $k \otimes \langle q' \rangle^{\mathrm{gp}}$ in $k \otimes Q^{\mathrm{gp}}$, for every $q' \in Q$. Suppose that $q = q_1 + \cdots + q_d$. Since Q is saturated, each $\langle q_i \rangle^{\mathrm{gp}}$ is a direct summand of Q^{gp}. Thus the map $k \otimes \langle q_1 \rangle \otimes \cdots \otimes \langle q_d \rangle \to k \otimes \Lambda^d(Q^{\mathrm{gp}})$ is surjective if $\langle q \rangle$ spans $k \otimes Q^{\mathrm{gp}}$ and is zero otherwise. If $q \in I_d$, there is an independent sequence $(q_1, \ldots, q_d)$ with $q = \sum q_i$, so the map is surjective and $\underline{\Omega}^d_{Q/k,q}$ belongs to the image of $\Lambda^d \underline{\Omega}^1_{Q/k}$. Conversely if $\underline{\Omega}^d_{Q/k.q}$ belongs to this image, there is a sequence $(q_1, \ldots, q_d)$ with $q = \sum q_i$ and with $\sum \langle q_i \rangle = Q$. Then there is an independent sequence $(q'_1, \ldots, q'_d)$ with each $q'_i \in \langle q_i \rangle$. Write $q_i = p_i + q'_i$ with $p_i \in Q$, and let $p := \sum p_i$ and $q' := \sum q'_i$. Then $q' \in I_d$ and $q = p + q'$, so $q \in I_d$. We conclude that the image of $\Lambda^q \underline{\Omega}^1_{Q/k}$ is $I_d \Omega^d_{Q/k}$. This completes the proof of the first statement, The second statement follows from the first statement and Proposition I.2.5.5. □

Before proceeding, let us attempt to explain the meaning of these filtrations by comparing them to other kinds of differential forms. There is a commutative diagram of prelog rings

$$\begin{array}{ccccc}
(F \to R[F]) & \xrightarrow{\psi} & (F \to R[Q]) & \longrightarrow & (Q \to R[Q]) \\
& \nwarrow & \uparrow \phi & & \uparrow \theta \\
& & (P \to R[P]) & \xrightarrow{\text{id}} & (P \to R[P]),
\end{array}$$

which induces a diagram of log schemes

$$\begin{array}{ccccc}
\mathsf{A}_Q & \xrightarrow{h} & \underline{\mathsf{A}}_Q(F) & \longrightarrow & \mathsf{A}_F \\
f \downarrow & & \downarrow g & \swarrow & \\
\mathsf{A}_P & \xrightarrow{\text{id}} & \mathsf{A}_P, & &
\end{array} \tag{2.3.1}$$

and a map of logarithmic differentials $\Omega^1_\phi \to \Omega^1_{Q/P}$. The module Ω^1_ϕ is generated by elements of the form $e^q dq$ for $q \in Q$ and df for $f \in F$, and is clear that each of these elements maps to $L^0(F)\Omega^1_\theta$. Taking exterior powers, one finds a morphism of complexes $\Omega^\cdot_\phi \to \underline{\Omega}^\cdot_{Q/P}(F)$. If G is a face of F and $\tilde{G}$ is the face of Q it generates, this map sends $L^i(G)\Omega^\cdot_\phi$ to $L^i(\tilde{G})\underline{\Omega}^\cdot_{Q/P}(F)$, and if $\tilde{K}$ is the ideal of Q generated by an ideal K of F, it also maps $K\Omega^\cdot_\phi$ to $\tilde{K}\Omega^\cdot_{Q/P} \cap \Omega^\cdot_{Q/}(F)$.

Proposition 2.3.11. *Let $\theta\colon P \to Q$ be an injective homomorphism of fine monoids, let F be a face of Q containing P, and let R be a base ring in which the order of the torsion subgroup of $\mathrm{Cok}(\theta^{\mathrm{gp}})$ is invertible. If $Q \cong F \oplus \mathbf{N}^r$ for some r, then the morphism g in diagram (2.3.1) is smooth, and the map $(\Omega^\cdot_\phi, L^\cdot) \to (\underline{\Omega}^\cdot_{Q/P}(F), L^\cdot)$ is a an isomorphism of filtered objects, with the filtrations defined by the ideals and faces of F as described in Lemma 2.3.1 and Definition 2.3.4. In particular, this is the case if Q is free.*

Proof First suppose that $F = 0$, so that $P = 0$ and Q is free. Then $R[Q]$ is a polynomial algebra, so $\underline{\mathsf{A}}_Q \to \operatorname{Spec} R$ is indeed smooth. The proposition asserts also that the maps $\Omega^j_{R[Q]/R} \to \underline{\Omega}^j_Q$ are isomorphisms. This is clearly true on the dense open set $\underline{\mathsf{A}}^*_Q$ and, since the modules $\Omega^j_{R[Q]/R}$ are free, the map is at least injective. To see that it is surjective, let S be the basis for Q, and write an arbitrary element q of Q as $\sum_s n_s s$ with each $n_s \geq 0$. Then $\langle q \rangle$ is the free monoid generated by $S(q) := \{s : n_s > 0\}$, and $\underline{\Omega}^j_{Q,q} = R \otimes \Lambda^j \langle q \rangle^{\mathrm{gp}}$ is generated by elements of the form $s_1 \wedge \cdots \wedge s_j$ with each $s_i \in S(q)$. Then $q = \sum n_i s_i$ with each $n_i > 0$. Each s_i defines an element e^{s_i} of $R[Q]$, and

$$h^*(de^{s_i}) = e^{s_i} ds_i \in \underline{\Omega}^1_{Q,s_i} \subseteq \Omega^1_{Q,s_i}.$$

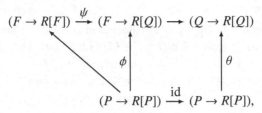

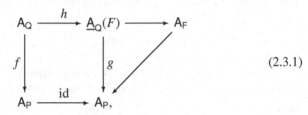

Thus $h^*(e^{(n_i-1)s_i}de^{s_i}) = e^{n_is_i}ds_i$, a basis element for $\langle n_is_i\rangle = \Omega^1_{Q,n_is_i}$. It follows that the element $e^{(n_1-1)s_1}de^{s_1}\wedge\cdots\wedge e^{(n_j-1)s_j}de^{s_i}$ of $\Omega^j_{R[Q]/R}$ maps to $e^q ds_1\wedge\cdots\wedge ds_j$, a basis element of $\underline{\Omega}^j_{Q,q}$, proving that our map is indeed surjective.

More generally, suppose that $Q = F\oplus F'$, with F' free. The homomorphism $P \to F$ is injective and the torsion subgroup of $F^{\mathrm{gp}}/P^{\mathrm{gp}}$ is contained in the torsion subgroup of $Q^{\mathrm{gp}}/P^{\mathrm{gp}}$, and hence its order is invertible in R^*. Thus the morphism $\mathsf{A}_F \to \mathsf{A}_P$ is smooth. Moreover, the morphism $\underline{\mathsf{A}}_Q(F) \to \mathsf{A}_F$ identifies with the projection $\mathsf{A}_F\times\underline{\mathsf{A}}_{F'} \to \mathsf{A}_F$, which is smooth since it is strict and its underlying morphism of schemes is smooth. It follows that g is also smooth and that

$$\Omega^1_\phi \cong \Omega^1_{F/P}\oplus\Omega^1_{\underline{\mathsf{A}}_{F'}} \cong \Omega^1_{F/P}\oplus\underline{\Omega}^1_{F'}.$$

Let us check that, in each degree q, the homomorphism $\Omega^k_{\phi,q} \to \underline{\Omega}^k_{Q/P}(F)$ is an isomorphism. Write $q = f+f'$ with $f\in F$ and $f'\in F'$. Then $\langle F,q\rangle = F\oplus\langle f'\rangle$, so

$$\underline{\Omega}^k_{Q/P,q}(F) = R\otimes\left(\bigoplus_{i+j=k}\Lambda^i F^{\mathrm{gp}}\otimes\Lambda^j\langle f'\rangle^{\mathrm{gp}}\right) = \bigoplus_{i+j=k}\Omega^i_{F/P,f}\otimes\underline{\Omega}^j_{F',f'} = \Omega^j_{\phi,q}.$$

It is a straightforward but tedious matter to check the compatibility with the filtrations. □

Proposition 2.3.12. *Let $\theta\colon P \to Q$ be an injective homomorphism of fine monoids, let F be a face of Q containing the image of P, and let R be a ring in which the order of the torsion subgroup of* $\mathrm{Cok}(\theta^{\mathrm{gp}})$ *is invertible.*

1. *Let U_F be the open subset of $\underline{\mathsf{A}}_Q$ defined by the invertibility of elements of F. Then, for all i and j,*

$$L^i(F)\Omega^j_{Q/P} \cong \Gamma(U_F, L^i(F)\Omega^j_{Q/P})\cap\Omega^j_{Q/P}.$$

2. *Let U be the inverse image in $\underline{\mathsf{A}}_Q$ of an open subset of* Spec Q *containing all the height one prime ideals of* Spec Q. *Then, if Q is saturated,*

$$L^0(F)\Omega^{\cdot}_{Q/P} \cong \Gamma(U, L^0(F)\Omega^{\cdot}_{Q/P}).$$

Proof The proof depends on the following simple calculation.

Lemma 2.3.13. *With the notation of Proposition 2.3.12, suppose that G is another face of Q, and let H be the face of Q generated by F and G. Then*

$$L^i(F)\Omega^j_{Q_G/P}\cap\Omega^j_{Q/P} = L^i(H)\Omega^j_{Q/P}.$$

Proof Endow the localization $\Omega^j_{Q_G/P}$ of $\Omega^j_{Q/P}$ with its natural Q^{gp}-grading. If

E is any Q-graded submodule of $\Omega^j_{Q/P}$, its localization E_G has a natural Q^{gp}-grading, and

$$E_G \cap \Omega^j_{Q/P} = \sum_{q \in Q} E_{G,q}.$$

As we saw in Remark I.3.2.8, for any Q-graded $R[Q]$-module E,

$$E_{G,q} \cong \varinjlim \{E_{g+q} : g \in G\}.$$

Applying this with $E = L^i(F)\Omega^q_{Q/P}$, we see that, for any $q \in Q$,

$$\begin{aligned} L^i(F)\Omega^j_{Q_G/P,q} &= \varinjlim_{g \in G} L^i(F)\Omega^j_{Q/P,g+q} \\ &= \varinjlim_{g \in G} Im\left(R \otimes \Lambda^{-i} Q^{\mathrm{gp}}/P^g \otimes \Lambda^{i+j}\langle F, g+q\rangle^{\mathrm{gp}}\right) \\ &= Im\left(R \otimes \Lambda^{-i} Q^{\mathrm{gp}}/P^{\mathrm{gp}} \otimes \Lambda^{i+j}\langle F, G, q\rangle^{\mathrm{gp}}\right) \\ &= L^i(H)\Omega^j_{Q/P,q}. \end{aligned}$$

□

Statement (1) of Proposition 2.3.12 follows from the special case of this lemma when $G = F$. To prove statement (2), note that the complement of a height one prime $\mathfrak{p}$ of Q is a facet G of Q, and $\mathrm{Spec}(Q_G)$ is the smallest open subset of $\mathrm{Spec}(Q)$ containing $\mathfrak{p}$. It follows that $U_G := \mathrm{Spec}(R[Q_G]) \subseteq U$. If E is a torsion free Q-graded $R[Q]$-module and $\mathcal{E}$ is the associated quasi-coherent sheaf, then $\Gamma(U_G, \mathcal{E}) = E_G$, and

$$E \subseteq \Gamma(U, \tilde{E}) \subseteq E' := \cap\{E_G : G \text{ is a face of } Q\}.$$

Thus E' inherits a Q^{gp}-grading and, if $x \in Q^{\mathrm{gp}}$, the degree-x component of E' vanishes unless x belongs to the intersection of all the localizations Q_G, which by Corollary I.2.4.5 is just $Q^{sat} = Q$. In particular, when $E = R[Q]$, we see that $R[Q]$ is the intersection of the set of localizations $R[Q_G]$, proving statement (2) when $j = 0$.

Each G is a facet of Q, so $\langle F, G, q\rangle$ is G if G contains F and q and is Q otherwise. Thus the lemma implies that

$$\Gamma(U, L^0(F)\Omega^j_{Q/P})_q \subseteq \bigcap_G \Omega^i_{Q/P,G,q} = \bigcap_{G \supseteq \langle F,q\rangle} R \otimes \Lambda^i G^{\mathrm{gp}}/P^{\mathrm{gp}}.$$

As we saw in Corollary I.2.3.14, the intersection of the set of all G^{gp} such that G contains $\langle F, q\rangle$ is just $\langle F, q\rangle^{\mathrm{gp}} = L^0(F)\Omega^1_{Q/P}$. This implies statement (2) when $j = 1$. The general case will follow, once we verify that if $H_1, \ldots, H_n$ is a finite collection of corank-one direct summands of a free and finitely generated abelian group H and $W := H_1 \cap \cdots \cap H_n$, then $\Lambda^j W = \Lambda^j H_1 \cap \cdots \cap \Lambda^j H_n$. Using induction on n, this statement is reduced to the case when $n = 2$. This case follows from the injectivity of the rightmost vertical map and the exactness of the rows in the following diagram:

$$\begin{array}{ccccccccc}
0 & \longrightarrow & \Lambda^j(H_1 \cap H_2) & \longrightarrow & \Lambda^j H_1 & \longrightarrow & \Lambda^{j-1}(H_1 \cap H_2) \otimes H_1/H_1 \cap H_2 & \longrightarrow & 0 \\
 & & \downarrow & & \downarrow & & \downarrow & & \\
0 & \longrightarrow & \Lambda^j H_2 & \longrightarrow & \Lambda^j H & \longrightarrow & \Lambda^{j-1} H \otimes H/H_2 & \longrightarrow & 0.
\end{array}$$

□

Variant 2.3.14. With the notation of Lemma 2.3.13, suppose that K is an ideal of Q and that G does not meet any associated prime of K. Then the statement of the lemma also holds with $\Omega^j_{Q/P}$ replaced by $\Omega^j_{Q/P}/K\Omega_{Q/P}$. Indeed, Proposition 2.1.14 shows that $Q\setminus K$ is stable under the action of G, so G acts injectively on $\Omega^j_{Q/P}/K\Omega^j_{Q/P}$, and hence $\Omega^j_{Q/P}/K\Omega^j_{Q/P}$ maps injectively to its localization by G.

The following corollary relates the complex $\underline{\Omega}^{\cdot}_{Q/P}$ when $P = 0$ to the de Rham complex constructed by Danilov [11].

Corollary 2.3.15. *Let Q be a fine saturated monoid, and let U be the inverse image in* Spec $R[Q]$ *of the set of $\mathfrak{p} \in$ Spec Q such that* ht $\mathfrak{p} \le 1$. *Assume that the order of the torsion subgroup of Q^{gp} is invertible in R. Then U is smooth over* Spec R, *and the natural map*

$$\underline{\Omega}^{\cdot}_{Q/R} \cong \Gamma(\underline{U}, \Omega^{\cdot}_{\underline{U}/R})$$

is an isomorphism

Proof Recall that if ht $\mathfrak{p} \le 1$, then $Q_{\mathfrak{p}}$ is a fine saturated monoid of dimension one, hence is valuative, and hence is isomorphic to $Q^*_{\mathfrak{p}} \oplus \mathbf{N}$. Since the finitely generated abelian group $Q^*_{\mathfrak{p}}$ is contained in Q^{gp}, the order of its torsion subgroup is invertible in R, and hence the corresponding group scheme is smooth over R. It follows that each $R[Q_{\mathfrak{p}}]$ is smooth over R, and hence U is smooth over R. The map $\Omega^{\cdot}_{\underline{U}/R} \to \underline{\Omega}^{\cdot}_{U/R}$ is an isomorphism by Proposition 2.3.11, and the corollary follows from statement (2) of Proposition 2.3.12 with $F = 0$. □

The cohomology computations of the de Rham complex of $P \to Q$ have filtered analogs.

Proposition 2.3.16. *With the notations and hypotheses of Proposition 2.2.2 and Definition 2.3.4, suppose that R contains* $\mathbf{Q}$. *Then the natural map*

$$(R[\tilde{Q}] \otimes \Lambda^{\cdot}(Q^{\mathrm{gp}}/P^{\mathrm{gp}}), L^{\cdot}(F), 0) \to (\Omega^{\cdot}_{Q/P}, L^{\cdot}(F), d).$$

is a filtered quasi-isomorphism.

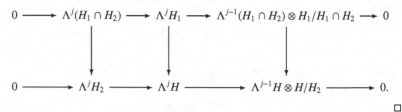

Proof Both sides are Q-graded, and the differential on the right vanishes in degrees q if $q \in \tilde{Q}$. The corollary will follow if can show that the complex in remaining degrees is acyclic. As we saw in Lemma 2.2.3, if $q \in Q \setminus \tilde{Q}$, there exists a $\partial\colon Q^{\mathrm{gp}} \to \mathbf{Z}$ that annihilates P^{gp} and is such that $\partial(q)$ maps to a unit in R. Then the filtered complex $(\Omega^{\cdot}_{Q/P,q}, L^{\cdot}(F))$ is invariant under the Lie derivative κ with respect to ∂. Since κ is multiplication by $\partial(q)$ in degree q, the complex is indeed acyclic in degree q. □

Now assume that R contains the prime field $\mathbf{F}_p$; our hypothesis then implies that $\mathrm{Cok}(\theta^{\mathrm{gp}})$ is p-torsion free.

Proposition 2.3.17. *With the notations and hypotheses of Propositions 2.2.2 and 2.2.5 and Definition 2.3.4, suppose that R contains the prime field $\mathbf{F}_p$, and let $F^{(p)}$ be the face of $Q^{(p)}_\theta$ generated by F. Then the quasi-isomorphism in statement (2) of Proposition 2.2.5 is strictly compatible with the filtrations and induces a filtered quasi-isomorphism*

$$\left(\Omega^{\cdot}_{\theta^{(p)}}, L^{\cdot}(F^{(p)}), 0\right) \to F_{Q/P*}\,(\Omega^{\cdot}_{\theta}, L^{\cdot}(F), d)\,.$$

Proof Recall from Proposition 2.2.2 that the filtered complex $(\Omega^{\cdot}_{\theta}, L^{\cdot}(F), d)$ is Q-graded, that the degree-q part is acyclic if $q \notin \tilde{Q}$, and that the differential in degree q is zero if $q \in \tilde{Q}$. Thus it remains only to check that the filtration $L^{\cdot}(F)$ induced on $\Omega^{\cdot}_{\theta^{(p)}}$ is the filtration $L^{\cdot}(F^{(p)})$. Recall from statement (5) of Proposition I.4.4.3 that $F_{Q/P}$ induces a bijection between the sets of faces of Q and of $Q^{(p)}_\theta$. Furthermore, by Corollary I.4.4.4, $F^{-1}_{Q/P}(F) = F^{(p)}_\theta$ and is equal to the face of $Q^{(p)}_\theta$ generated by $\pi_{Q/P}(F)$. Choose $q' \in Q^{(p)}_\theta$, and let $q := F_{Q/P}(q') \in Q$. Let G be the face of Q generated by F and q and let G' be the face of $Q^{(p)}_\theta$ generated by $F^{(p)}$ and q'. Then $F_Q(F)$ and q are contained in $F_{Q/P}(G')$ which in turn is contained in $G = \langle F, q\rangle$. We conclude that $F^{-1}_{Q/P}(G) = G'$ and hence that $\pi^{-1}_{Q/P}(G') = G$. It follows that the isomorphism $\gamma_{Q/P}\colon \mathrm{Cok}(\theta^{\mathrm{gp}}) \to \mathrm{Cok}(\theta^{\mathrm{gp}}_n)$ takes $G^{\mathrm{gp}}/P^{\mathrm{gp}}$ isomorphically to G'/P^{gp}, and the same holds for the exterior powers. □

Corollary 2.3.18. *With the hypotheses of Proposition 2.3.17 above, there are natural isomorphisms*

$$L^j(F^{(p)})\Omega^i_{\theta^{(p)}} \cong H^i(L^j(F)\Omega^{\cdot}_\theta)),$$

compatible with the inclusions $L^{j+1}(F) \to L^j(F)$ and also with the maps induced by inclusions of faces $F \to G$. □

Now we explain how to sheafify and globalize these constructions. We use the following notation. If X is a log scheme, if $\mathcal{K}$ is a sheaf of ideals in $\mathcal{M}_X$, and if $\mathcal{E}$ is a sheaf of $\mathcal{O}_X$-modules, then $\mathcal{K}\mathcal{E}$ is the abelian subsheaf of $\mathcal{E}$ generated by elements of the form $\alpha_X(k)e$ with $k \in \mathcal{K}$ and $e \in \mathcal{E}$. This is automatically a sheaf of $\mathcal{O}_X$-submodules, because it is closed under multiplication by elements of the image of α_X and any section of $\mathcal{O}_X$ can locally be written as a sum of such elements.

Let $f\colon X \to Y$ be a quasi-compact and quasi-separated morphism of integral log schemes. We write d, rather than $dlog$, for the map $\mathcal{M}_X \to \Omega^1_{X/Y}$ and λ for the inverse of the isomorphism $\mathcal{M}^*_X \to \mathcal{O}^*_X$ induced by α_X. Thus $d\lambda(u) = u^{-1}du$ for every local section u of $\mathcal{O}^*_X$.

Proposition 2.3.19. *If $\mathcal{K} \subseteq \mathcal{M}_X$ is a sheaf of ideals of $\mathcal{M}_X$, let $\mathcal{K}\Omega^i_{X/Y}$ be the subsheaf of $\Omega^i_{X/Y}$ generated by all sections of the form $\alpha_X(k)\omega$ with $k \in \mathcal{K}$ and $\omega \in \Omega^i_{X/Y}$. Then the exterior derivative maps $\mathcal{K}\Omega^i_{X/Y}$ to $\mathcal{K}\Omega^{i+1}_{X/Y}$, so that $\mathcal{K}\Omega^{\cdot}_{X/Y}$ forms a subcomplex of $\Omega^{\cdot}_{X/Y}$. If the log structures $\mathcal{M}_X$ and $\mathcal{M}_Y$ are coherent and $\mathcal{K}$ is a coherent sheaf of ideals, then each $\mathcal{K}\Omega^i_{X/Y}$ is quasi-coherent.*

Proof If k is a local section of $\mathcal{K}$ and ω is a local section of $\Omega^i_{X/Y}$, then

$$d(\alpha_X(k)\omega) = d\alpha_X(k) \wedge \omega + \alpha_X(k)d\omega = \alpha_X(k)dk \wedge d\omega + \alpha_X(k)d\omega,$$

which belongs to $\mathcal{K}\Omega^{i+1}_{X/Y}$. This proves the first statement of the proposition. Corollary IV.1.2.8 implies that $\Omega^i_{X/Y}$ is quasi-coherent if the log structures on X and Y are coherent. If $\mathcal{K}$ is coherent, it is locally generated by a finite number of sections $k_1, \ldots, k_m$ as an ideal of $\mathcal{M}_K$. In this situation, $\mathcal{K}\Omega^i_{X/Y}$ is locally the subsheaf of $\Omega^i_{X/Y}$ generated by a finite number of sections, and hence is quasi-coherent. □

Definition 2.3.20. *Let $f\colon X \to Y$ be a morphism of log schemes. A sheaf of ideals $\mathcal{K}$ in $\mathcal{M}_X$ is said to be* horizontal *if for every geometric point $\overline{x}$ of $X_{\mathcal{K}}$, the stalk $\mathcal{K}_{\overline{x}}$ at $\overline{x}$ is disjoint from the image of $\mathcal{M}_{Y,f(\overline{x})}$ in $\mathcal{M}_{X,\overline{x}}$, or, equivalently, from the face of $\mathcal{M}_{X,\overline{x}}$ generated by this image.*

The filtrations defined by sheaves of faces are more subtle.

Definition 2.3.21. *Let $f\colon X \to Y$ be a morphism of fine log schemes, and let $\mathcal{F}$ be a sheaf of faces in $\mathcal{M}_X$ containing the image of $f^*_{log}\mathcal{M}_Y$. If m_0 is a section of $\mathcal{M}_X$ on an open subset U of X, let $\mathcal{F}\langle m_0\rangle$ denote the sheaf of faces of $\mathcal{M}_{X|_U}$ generated by $\mathcal{F}$ and m_0.*

1. *$K^i(\mathcal{F})\Omega^j_{X/Y} \subseteq \Omega^j_{X/Y}$ is the abelian subsheaf generated by the local sections of the form $\alpha(m_0)dm_1 \wedge \cdots \wedge dm_j$ such that $m_1, \ldots, m_i$ belong to $\mathcal{F}\langle m_0\rangle$.*
2. *$L^i(\mathcal{F})\Omega^j_{X/Y} := K^{i+j}(\mathcal{F})\Omega^j_{X/Y}$.*
3. *$\underline{\Omega}^j_{X/Y}(\mathcal{F}) := L^0(\mathcal{F})\Omega^j_{X/Y}$.*

Remark 2.3.22. The stalk of $\mathcal{F}\langle m_0\rangle$ at a geometric point $\overline{x}$ of X is the set of $m \in \mathcal{M}_{X,\overline{x}}$ such that there exist $n \in \mathbf{N}$ and $f \in \mathcal{F}_{\overline{x}}$ with $nm_0 + f \geq m$. Hence

$K^i(\mathcal{F})\Omega^j_{X/Y} \subseteq \Omega^j_{X/Y}$ is the subsheaf of abelian groups generated by the local sections of the form $\alpha_X(m_0)dm_1 \wedge \cdots \wedge dm_j$ such that there exist $n \in \mathbf{N}$ and $f \in \mathcal{F}$ with $nm_0 + f \geq m_1 + \cdots + m_i$;

Proposition 2.3.23. *With the hypotheses of Definition 2.3.21, the following statements are verified:*

1. *$K^i(\mathcal{F})\Omega^j_{X/Y}$ and $L^i(\mathcal{F})\Omega^j_{X/Y}$ are sheaves of $\mathcal{O}_X$-submodules of $\Omega^j_{X/Y}$ and contain the image of $\Omega^j_{\underline{X}/\underline{Y}} \to \Omega^j_{X/Y}$.*
2. *The exterior derivative maps $K^i(\mathcal{F})\Omega^j_{X/Y}$ to $K^{i+1}(\mathcal{F})\Omega^{j+1}_{X/Y}$ and $L^i(\mathcal{F})\Omega^j_{X/Y}$ to $L^i(\mathcal{F})\Omega^{j+1}_{X/Y}$, and $d\colon \mathcal{M}_X \to \Omega^1_{X/Y}$ maps $\mathcal{F}$ to $L^0(\mathcal{F})\Omega^1_{X/Y}$.*
3. *The exterior product maps $K^i(\mathcal{F})\Omega^j_{X/Y} \times K^{i'}(\mathcal{F})\Omega^{j'}_{X/Y}$ to $K^{i+i'}(\mathcal{F})\Omega^{j+j'}_{X/Y}$ and $L^i(\mathcal{F})\Omega^j_{X/Y} \times L^{i'}(\mathcal{F})\Omega^{j'}_{X/Y}$ to $L^{i+i'}(\mathcal{F})\Omega^{j+j'}_{X/Y}$.*
4. *Interior multiplication by an element of $T_{X/Y}$ induces maps from $K^i(\mathcal{F})\Omega^j_{X/Y}$ to $K^{i-1}(\mathcal{F})\Omega^{j-1}_{X/Y}$ and from $L^i(\mathcal{F})\Omega^j_{X/Y}$ to $L^i(\mathcal{F})\Omega^{j-1}_{X/Y}$.*
5. *Suppose that f is locally of finite presentation, that X and Y are fine, and that $\mathcal{F} \subseteq \mathcal{M}_X$ is a relatively coherent sheaf of faces in $\mathcal{M}_X$. Then $K^i(\mathcal{F})\Omega^j_{X/Y}$ and $L^i(\mathcal{F})\Omega^j_{X/Y}$ are quasi-coherent sheaves of $\mathcal{O}_X$-modules.*
6. *If f is the morphism coming from a monoid homomorphism $P \to Q$ and $\mathcal{F}$ is generated by a face F of Q containing P, then $K^i(\mathcal{F})\Omega^j_{X/Y}$ is the quasi-coherent sheaf associated to the module $K^i(F)\Omega^j_{Q/P}$ in Definition 2.3.4, and similarly for $L^{\cdot}$.*

Proof To prove that $K^i(\mathcal{F})\Omega^j_{X/Y}$ is a sheaf of $\mathcal{O}_X$-submodules of $\Omega^j_{X/Y}$, suppose that $(m_0, m_1, \ldots, m_j)$ is a sequence of sections of $\mathcal{M}_X$, that f is a section of $\mathcal{F}$, and that n is a natural number such that $nm_0 + f \geq m_1 + \cdots + m_i$, and let $\omega := \alpha_X(m_0)dm_1 \wedge \cdots \wedge dm_j$. Then, if m is any local section of $\mathcal{M}_X$, $n(m_0 + m)f \geq m_1 + \cdots + m_i$, and hence $\alpha_X(m)\omega$ also belongs to $K^i(\mathcal{F})\Omega^j_{X/Y}$. Thus $K^i(\mathcal{F})\Omega^j_{X/Y}$ is stable under multiplication by sections of the form $\alpha_X(m)$ with $m \in \mathcal{M}_X$. Since any section of $\mathcal{O}_X$ is a locally a sum of such sections and $K^i(\mathcal{F})\Omega^j_{X/Y}$ is a subgroup of $\Omega^j_{X/Y}$, it follows that $K^i(\mathcal{F})\Omega^j_{X/Y}$ is stable under multiplication by $\mathcal{O}_X$, and hence is an $\mathcal{O}_X$-submodule of $\Omega^i_{X/Y}$. Furthermore, $d\omega = \alpha(m_0)dm_0 \wedge dm_1 \wedge \cdots \wedge dm_j$ and, since $(n+1)m_0 + f \geq m_0 + \cdots + m_i$, we see that $d\omega \in K^{i+1}(\mathcal{F})\Omega^{j+1}_{X/Y}$. If $(m'_0, m'_1, \ldots, m'_{j'})$ is another sequence of sections of $\mathcal{M}_X$ and f' is a section of $\mathcal{F}$ with $k'm'_0 + f' \geq m'_1 + \cdots + m'_{i'}$, then $\omega' =: \alpha_X(m'_0)dm_1 \wedge \cdots \wedge dm_{j'}$ is a typical element of $K^{i'}(\mathcal{F})\Omega^{j'}_{X/Y}$ and, since $(n+n')(m_0 + m'_0) + f + f' \geq m_1 + \cdots + m_i + m'_1 + \cdots + m'_{i'}$, we see that $\omega \wedge \omega' \in K^{i+i'}(\mathcal{F})\Omega^{j+j'}_{X/Y}$. Note that $\Omega^1_{\underline{X}/\underline{Y}}$ is generated by sections of the form $u^{-1}du = d\lambda(u)$ for $u \in \mathcal{O}^*_X$ and, since $\lambda(u) \leq \lambda(1)$, the image of $d\lambda(u)$

in $\Omega^1_{X/Y}$ in fact belongs to $K^1(\mathcal{F})\Omega^1_{X/Y}$. It follows that $K^j(\mathcal{F})\Omega^j_{X/Y}$ contains the image of $\Omega^j_{\underline{X/Y}} \to \Omega^j_{X/Y}$. If $\omega := \alpha_X(m_0)dm_1 \wedge \cdots \wedge dm_j \in K^i(\mathcal{F})\Omega^j_{X/Y}$ with $km_0 + f \geq m_1 + \cdots + m_i$, and if θ is a section of $T_{X/Y}$, then interior multiplication by θ takes ω to

$$\sum_r \alpha_X(m_0)(-1)^{r-1}\theta(dm_r)dm_1 \wedge \cdots \wedge d\hat{m}_r \wedge \cdots \wedge dm_j,$$

which evidently belongs to $K^{i-1}(\mathcal{F})\Omega^{j-1}_{X/Y}$. The corresponding statements for the filtration L follow by reindexing.

The proof of quasi-coherence will use the following lemma.

Lemma 2.3.24. *Let $\beta\colon Q \to \mathcal{M}_X$ be a chart for $\mathcal{M}_X$ and let F be a face of Q which generates the sheaf of faces $\mathcal{F}$. Then $K^i(\mathcal{F})\Omega^j_{X/Y}$ is the sheaf of sub-$\mathcal{O}_X$-modules of $\Omega^j_{X/Y}$ generated by the image of*

$$\Omega^1_{\underline{X/Y}} \otimes K^{i-1}(\mathcal{F})\Omega^{j-1}_{X/Y} \to K^i(\mathcal{F})\Omega^j_{X/Y}$$

together with sections of the form $\alpha_X(\beta(q_0))d\beta(q_1) \wedge \cdots \wedge d\beta(q_j)$, where $q_1, \ldots q_i$ belong to the face of Q generated by F and q_0.

Proof Let $E_{i,j}$ be the subsheaf of $\Omega^j_{X/Y}$ generated by the sections described in the lemma. It is clear that $E_{i,j} \subseteq K^i(\mathcal{F})\Omega^j_{X/Y}$. To prove equality we work at the stalk of an arbitrary point x. Let $\omega = \alpha_X(m_0)dm_1 \wedge \cdots \wedge dm_j$ be one of the defining elements of $K^i(\mathcal{F})\Omega^j_{X/Y}$. Since Q is a chart for $\mathcal{M}_{X,x}$, there exist $u \in \mathcal{O}^*_{X,x}$ and $q_0 \in Q$ such that $m_0 = \lambda(u) + \beta(q_0)$. Then $\alpha_X(m_0) = u\alpha_X(q_0)$ and $\langle m_0, \mathcal{F}_x\rangle = \langle \beta(q_0), \mathcal{F}_x\rangle$. Thus we may as well assume that $m_0 = \beta(q_0)$. There is nothing to prove if $i = 0$, so assume that $i \geq 1$ and write $m_1 := \lambda(u) + \beta(q_1)$, where $u \in \mathcal{O}^*_{X,x}$ and $q_1 \in Q$. Then $dm_1 = u^{-1}du + d\beta(q_1)$, and

$$\omega = \alpha_X(\beta(q_0))u^{-1}du \wedge dm_2 \wedge \cdots \wedge dm_i + \alpha_X(\beta(q_0))d\beta(q_1) \wedge \cdots \wedge dm_j.$$

The first term in this expression belongs to $E_{i,j}$, and so it suffices to treat the second term. Since $\beta(q_1) \in \langle \mathcal{F}_x, q_0\rangle$, we may as well assume that $m_1 = \beta(q_1)$. Repeating this process, we may assume that each $m_k = \beta(q_k)$ for some $q_k \in Q$.

By assumption, $\beta(q_k) \in \langle \mathcal{F}_x, q_0\rangle$ for $1 \leq k \leq i$. Thus there exist $m'_k \in \mathcal{M}_{X,x}$, $f \in F$, and $n_k \geq 0$ such that $\beta(q_k) + m'_k = \beta(f_k + n_k q_0)$. Write $m'_k = \lambda(u_k) + \beta(q'_k)$, where $q'_k \in Q$ and $u_k \in \mathcal{O}^*_{X,x}$. Thus

$$\beta(q_k + q'_k) + \lambda(u_k) = \beta(f_k + n_k q_0).$$

Since Q is a chart for $\mathcal{M}_{X,x}$, this implies that there exist $g'_k, g_k \in \beta^{-1}(\mathcal{O}^*_{X,x})$ such that $\lambda(u) + \beta(g_k) = \beta(g'_k)$ and $q_k + q'_k + g'_k = f_k + n_k q_0 + g_k$ in Q. In particular, q_k belongs to the face of Q generated by F and $q_0 + g_k$. Let $g = g_1 + \cdots + g_i$ and let $u := \alpha_X(\beta(g))$, an element of $\mathcal{O}^*_{X,x}$. Then $q_k \in \langle F, q_0 + g\rangle$ for $1 \leq k \leq i$, and hence $\omega' := \alpha_X(\beta(q_0 + g))d\beta(q_1) \wedge \cdots \wedge d\beta(q_i) \wedge \cdots \wedge d\beta(q_j) \in E_{i,j}$. Since $\omega = u^{-1}\omega$, it too belongs to $E_{i,j}$. □

It is now easy to prove that $K^i(\mathcal{F})\Omega^j_{X/Y}$ is quasi-coherent, using induction on j. If $j = 0$ there is nothing to prove, and if the quasi-coherence is established for $j-1$, the lemma shows that $K^i(\mathcal{F})\Omega^j_{X/Y}$ is the subsheaf of $\Omega^j_{X/Y}$ generated by the image of the quasi-coherent sheaf $\Omega^1_{\underline{X}/\underline{Y}} \otimes K^{i-1}(\mathcal{F})\Omega^{j-1}_{X/Y}$ and a finite number of global sections, and hence is quasi-coherent. The last statement of the proposition also follows easily from the lemma. □

Remark 2.3.25. The filtration $L^\cdot(\mathcal{F})$ on $\Omega^\cdot_{X/Y}$ is interesting even when $\mathcal{F}$ is the trivial sheaf of faces $\mathcal{O}^*_X$. In this case we just write $L^\cdot$ instead of $L^\cdot(\mathcal{O}^*_X)$ and $\underline{\Omega}^j_{X/Y}$ for $L^0\Omega^j_{X/Y}$.

Let us note that the formation of these subcomplexes is functorial. For example, if $\mathcal{F} \subseteq \mathcal{F}'$,

$$(\Omega^\cdot_{X/Y}, L^\cdot(\mathcal{F})) \subseteq (\Omega^\cdot_{X/Y}, L^\cdot(\mathcal{F}')).$$

The following proposition says more about this situation.

Proposition 2.3.26. *Let $f: X \to Y$ be a smooth morphism of fine idealized log schemes, and let $\mathcal{F}$ be a relatively coherent sheaf of faces in $\mathcal{M}_X$.*

1. *If $g: X' \to X$ is strict and étale, the natural maps*

$$g^*(\Omega^\cdot_{X/Y}, L^\cdot(\mathcal{F})) \to (\Omega^\cdot_{X'/Y}, L^\cdot(\mathcal{F}'))$$

   are isomorphisms, where $\mathcal{F}'$ is the sheaf of faces in $\mathcal{M}_{X'}$ generated by $g^{-1}(\mathcal{F})$.
2. *Suppose that, for every geometric point $\overline{x}$ of X and every face $\mathcal{G}_{\overline{x}}$ of $\mathcal{M}_{\overline{x}}$ containing $\mathcal{F}_{\overline{x}}$, the order of the torsion subgroup of $\mathcal{M}^{\text{gp}}_{X,\overline{x}}/\mathcal{G}^{\text{gp}}_{X,\overline{x}}$ is invertible in $\mathcal{O}_{X,\overline{x}}$. Then the formation of the filtered complex $(\Omega^\cdot_{X/Y}, L^\cdot(\mathcal{F}))$ is compatible with base change $Y' \to Y$ in the category of fine log schemes and also in the category of fine saturated log schemes.*

Proof Statement (1) may be checked étale locally on X' and X, so we may assume that X/Y admits a chart as in Theorem IV.3.3.1 (or its idealized Variant 3.3.5). Then the composition $X' \to X \to \mathsf{A}_Q$ is a chart for X fitting into a chart for $X' \to Y$. Since g is strict and étale, the natural map $g^*(\Omega^1_{\underline{X}/\underline{Y}}) \to \Omega^1_{\underline{X}'/\underline{Y}}$ is an isomorphism. Then statement (1) follows easily from Lemma 2.3.24 and induction. Statement (2) of the lemma can also be checked locally, hence in the presence of charts for $f: X \to Y$ and $Y' \to Y$. Statement (2) then follows from statement (1) of the current proposition combined with statement (2) of Proposition 2.3.6. □

3 Analytic de Rham cohomology

In this section we study the de Rham cohomology of smooth log analytic spaces. Let us begin with an explicit description of the formal and holomorphic stalks of the de Rham complex $\Omega^{\cdot}_{X/Y}$, together with some of its subcomplexes.

Proposition 3.0.1. *Let $\theta\colon P \to Q$ be an injective homomorphism of fine sharp monoids, and let $f\colon X \to Y$ be the morphism of log analytic spaces corresponding to the map of log schemes $\mathsf{A}_\theta\colon \mathsf{A}_Q \to \mathsf{A}_P$. Let $x \in X$ be the vertex of X, and let $\Omega^i_{X/Y,x}$ denote the stalk of the sheaf of holomorphic i-forms at x.*

1. *The formal completion $\hat{\Omega}^j_{X/Y,x}$ of $\Omega^j_{X/Y,x}$ can be identified with the set of formal sums*
$$\omega := \sum_{q\in Q} \omega_q, \qquad \text{where } \omega_q \in \mathbf{C}\otimes\Lambda^j(Q^{\mathrm{gp}}/P^{\mathrm{gp}}).$$
2. *Let h be any local homomorphism $Q \to \mathbf{N}$ and let $\| \; \|$ be any norm on $\mathbf{C}\otimes\Lambda^j Q^{\mathrm{gp}}/P^{\mathrm{gp}}$. Then an element $\omega = \sum \omega_q$ of $\hat{\Omega}^j_{X/Y,x}$ lies in $\Omega^j_{X/Y,x}$ if and only if the set*
$$\{h(q)^{-1}\log\|\omega_q\| : q \in Q^+\}$$
is bounded above.
3. *If K is an ideal of Q and $\mathcal{K}$ is the corresponding sheaf of ideals in $\mathcal{M}_X$, then*
$$\mathcal{K}\Omega^j_{X/Y,x} = \left\{\omega \in \Omega^j_{X/Y,x} : \omega_q = 0 \quad \text{if } q \notin K\right\}.$$
4. *If F is face of Q and $\mathcal{F}$ is the corresponding sheaf of faces in $\mathcal{M}_X$, then*
$$L^i(\mathcal{F})\Omega^j_{X/Y,x} = \left\{\omega \in \Omega^j_{X/Y,x} : \omega_q \in K^{i+j}_{F,q}(\mathbf{C}\otimes\Lambda^j Q^{\mathrm{gp}}/P^{\mathrm{gp}}) \text{ for all } q\right\},$$
where $L^{\cdot}(\mathcal{F})$ is the filtration defined in Definition 2.3.4 and $K^{\cdot}_{F,q}$ is the Koszul filtration on $\mathbf{C}\otimes\Lambda^j Q^{\mathrm{gp}}/P^{\mathrm{gp}}$ defined by the image of $\langle F, q\rangle$ in $\mathbf{C}\otimes Q^{\mathrm{gp}}/P^{\mathrm{gp}}$.

Proof We saw in Proposition I.3.6.1 that the formal completion of $\mathcal{O}_{X,x}$ can be identified with the ring of formal power series $\mathbf{C}[[Q]]$. Proposition IV.1.1.4 identifies $\Omega^j_{X/Y}$ with $\mathcal{O}_X\otimes\Lambda^j Q^{\mathrm{gp}}/P^{\mathrm{gp}}$, and statement (1) follows. Statement (2) is a consequence of this fact and Proposition 1.1.3. The remaining statements are consequences of the definitions and Proposition 2.3.23 □

3.1 An idealized Poincaré lemma

The Poincaré lemma in logarithmic de Rham cohomology takes many forms. We begin with the following statement.

Theorem 3.1.1. *Let $f\colon X \to Y$ be a smooth and exact morphism of fine idealized log analytic spaces, let $\mathcal{K}$ be a coherent horizontal sheaf of ideals in $\mathcal{M}_X$, and let $i\colon X_{\mathcal{K}} \to X$ be the inclusion of the closed subspace of X defined by $\mathcal{K}$. Then the complex $i^{-1}(\mathcal{K}\Omega^{\cdot}_{X/Y})$ is acyclic, as are its subcomplexes defined by the filtrations associated to the relatively coherent sheaves of faces containing the image of $f^*_{log}\mathcal{M}_Y$ in $\mathcal{M}_X$. In particular, for any such sheaves of faces $\mathcal{F}$ and $\mathcal{F}'$ and any natural numbers i, i', the sheaf-theoretic restrictions of the complexes*

$$\mathcal{K}\Omega^{\cdot}_{X/Y} \cap \underline{\Omega}^{\cdot}_{X/Y}(\mathcal{F}),$$

$$\mathcal{K}\Omega^{\cdot}_{X/Y} \cap L^i(\mathcal{F})\Omega^{\cdot}_{X/Y},$$

$$\mathcal{K}\Omega^{\cdot}_{X/Y} \cap L^i(\mathcal{F})\Omega^{\cdot}_{X/Y} \cap L^{i'}(\mathcal{F}')\Omega^{\cdot}_{X/Y}$$

to $X_{\mathcal{K}}$ are acyclic. The analogous result holds for any finite collection of faces $\mathcal{F}$ and indices i.

Proof We first treat the case in which the idealized structures on X and Y are trivial. We will show that the stalks of the complexes above are acyclic at each point x of $X_{\mathcal{K}}$. Since the statement is local on X and Y, and since f is exact and hence s-injective, we may assume that f admits a chart of the form described in Theorem IV.3.3.3. Thus we may assume that there is a diagram

$$\begin{array}{ccccc} X & \longrightarrow & Y_{Q'}(Q) & \longrightarrow & \underline{\mathsf{A}}_{Q'}(Q) \\ & \searrow & \downarrow & & \downarrow \\ & & Y & \longrightarrow & \mathsf{A}_P, \end{array}$$

where $Q' := Q \oplus \mathbf{N}^r$ and x maps to the vertex of $\underline{\mathsf{A}}_{Q'}(Q)$. Here the square is cartesian and the map $X \to Y_{Q'}(Q)$ is strict and étale, hence analytically a local isomorphism. Thus we may assume without loss of generality that $X = Y_{Q'}(Q)$. We may also assume that $\mathcal{F}$ is generated by a face F of Q and that $\mathcal{K}$ is generated by an ideal K of Q. Then F is also a face of Q', the ideal K' of Q' generated by K is again horizontal, and $K' \cap Q = K$. Let $X' := \mathsf{A}_{Q'}$ and observe that Q defines a coherent (hence relatively coherent) sheaf of faces $\mathcal{G}$ in $\mathcal{M}_{X'}$. Furthermore, $\underline{\Omega}^{\cdot}_{X'/Y}(\mathcal{G}) \cong \Omega^{\cdot}_{X/Y}$, by Proposition 2.3.11, and in fact

$$C^{\cdot} := \mathcal{K}\Omega^{\cdot}_{X/Y} \cap L^i(\mathcal{F})\Omega^{\cdot}_{X/Y} \cong \mathcal{K}\underline{\Omega}^{\cdot}_{X'/Y}(\mathcal{G}) \cap L^i(\mathcal{F})\Omega^{\cdot}_{X'/Y}.$$

The analogous statement holds if $C^{\cdot}$ is any finite intersection of complexes of this form. Thus it suffices to prove the theorem when $X = X'$ and $Q = Q'$, which we henceforth assume.

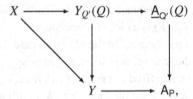

Let us begin with the case in which $Y = \mathsf{A}_P$ and $X = \mathsf{A}_{Q'}$. By Lemma 2.3.2, there exists a horizontal flow $h\colon Q \to \mathbf{N}$ such that $h(k) > 0$ for all $k \in K$. Let ξ denote the corresponding vector field on X. Both the differential d and interior multiplication by ξ map C^i to C^{i-1}, and hence the Lie derivative $\kappa := d\xi + \xi d$ is an endomorphism of $C^\cdot$ and is homotopic to zero. We claim that κ acts as an isomorphism on $C^\cdot$, hence also on its cohomology groups, and it will follow that these all vanish.

To verify our claim, we use the explicit description given in Proposition 3.0.1, according to which an element ω of C^i can be written as a formal sum $\sum\{\omega_q : q \in K\}$. Recall from Lemma 2.2.9 that then $\kappa(\omega) = \sum h(q)\omega_q$. Since $h(q) \neq 0$ for $q \in K$, it is clear that κ induces an isomorphism on the formal completion $\mathcal{K}\hat{\Omega}^\cdot_{X/Y,x}$, with $\kappa^{-1}(\omega) = \sum h(q)^{-1}\omega_q$. Since each $h(q) \geq 1$, the criterion of Proposition 3.0.1 shows that $\kappa^{-1}(\omega)$ converges if ω does, and hence that κ is also an isomorphism on $\mathcal{K}\Omega^\cdot_{X/Y,x}$. It is also clear from the explicit description in Proposition 3.0.1 that $\mathcal{K}\Omega^\cdot_{X/Y,x} \cap L^i(\mathcal{F})\Omega^\cdot_{X/Y}$ is invariant under κ^{-1}, so that κ is indeed an isomorphism on $C^\cdot$.

To treat the case of a general base Y, we note first that Proposition 2.3.26 implies that the formation of these complexes is compatible with base change. Furthermore, κ and its inverse are $\mathbf{C}[P]$-linear, and hence induce endomorphisms of the based-changed complexes over Y. We should further note that both κ and κ^{-1} continue to preserve holomorphicity. To simplify the verification of this last fact, one can replace P by $P' := P \oplus \mathbf{N}^r$ and A_P by $\underline{\mathsf{A}}_{P'}(P)$, with r chosen so that $Y \to \underline{\mathsf{A}}_{P'}(P)$ is a closed immersion.

Finally we note that the theorem in the idealized case follows from the non-idealized one. Since the statement is local, we may apply Variant IV.3.3.5 to reduce to the case of an idealized log space $X_{\mathcal{J}}$ defined by a coherent sheaf of ideals $\mathcal{J}$ in a smooth X/Y. Then the ideal $\mathcal{J} \cap \mathcal{K}$ is still horizontal, and there is an exact sequence

$$0 \to (\mathcal{J} \cap \mathcal{K})\Omega^\cdot_{X/Y} \to \mathcal{K}\Omega^\cdot_{X/Y} \to \mathcal{K}\Omega^\cdot_{X_{\mathcal{J}}/Y} \to 0$$

and similarly for the intersections with the filtrations defined by $\mathcal{F}$. Since the first two complexes are acyclic, so is the last. □

To simplify the notation in the next corollary, we denote by $L^\cdot$ the collection of filtrations defined by all the faces of $\mathcal{M}$.

Corollary 3.1.2. *With the hypotheses of Theorem 3.1.1, the following statements hold:*

1. *The natural map*

$$i^{-1}\left(\Omega^\cdot_{X/Y}, L^\cdot\right) \to \left(\Omega^\cdot_{X_{\mathcal{K}}/Y}, L^\cdot\right)$$

is a filtered quasi-isomorphism, where i^{-1} means the sheaf-theoretic restriction to the closed subset $X_{\mathcal{K}}$.

2. *The natural map*

$$(\hat{\Omega}^{\cdot}_{X/Y}, L^{\cdot}) \to (\Omega^{\cdot}_{X_{\mathcal{K}}/Y}, L^{\cdot})$$

is a filtered quasi-isomorphism, where $\hat{X}$ is the formal completion of X along $X_{\mathcal{K}}$.

Proof There is a strict exact sequence of filtered complexes

$$0 \to i^{-1}\left(\mathcal{K}\Omega^{\cdot}_{X/Y}, L^{\cdot}\right) \to i^{-1}\left(\Omega^{\cdot}_{X/Y}, L^{\cdot}\right) \to \left(\Omega^{\cdot}_{X_{\mathcal{K}}/Y}, L^{\cdot}\right) \to 0.$$

The theorem implies that the first complex is acyclic, so the second arrow is a quasi-isomorphism, proving (1). The proof of the theorem also shows that $(K\hat{\Omega}^{\cdot}_{X/Y}, L^{\cdot})$ is acyclic, and (2) follows in the same way. □

Corollary 3.1.3. *Let X be a smooth and fine idealized log analytic space and let $\mathcal{F}$ be a relatively coherent sheaf of faces in $\mathcal{M}_X$. If x is a point of X, let x_M denote the log point obtained by restricting the log structure of X to x.*

1. *The natural map*

$$\left(\Omega^{\cdot}_{X/\mathbf{C},x}, L^{\cdot}(\mathcal{F})\right) \to \left(\Omega^{\cdot}_{x_M/\mathbf{C}}, L^{\cdot}(\mathcal{F}_x)\right) \cong \left(\mathbf{C} \otimes \Lambda^{\cdot}\overline{\mathcal{M}}_{X,x}, L^{\cdot}(\overline{\mathcal{F}}_x)\right)$$

is a filtered quasi-isomorphism, where $L^{\cdot}(\overline{\mathcal{F}}_x)$ is the shifted Koszul filtration of $\mathbf{C} \otimes \Lambda^{\cdot}\overline{\mathcal{M}}_{X,x}$ defined by the submodule $\overline{\mathcal{F}}_x$ of $\overline{\mathcal{M}}_{X,x}$.

2. *The natural maps*

$$\underline{\Omega}^{\cdot}_{X/\mathbf{C},x}(\mathcal{F}) \to \underline{\Omega}^{\cdot}_{x_M/\mathbf{C}}(\mathcal{F}) \to \mathbf{C} \otimes \Lambda^{\cdot}\overline{\mathcal{F}}^{\mathrm{gp}}_x$$

are quasi-isomorphisms, where $\Lambda^{\cdot}\overline{\mathcal{F}}^{\mathrm{gp}}_x$ is the exterior algebra of $\overline{\mathcal{F}}^{\mathrm{gp}}_x$ endowed with zero as boundary maps.

Proof As in the proof of Theorem 3.1.1, we may assume that x is the vertex of $X = \underline{\mathsf{A}}_{Q'}(Q)$, where Q is a fine sharp monoid and $Q' = Q \oplus \mathbf{N}^r$. Let $X' := \mathsf{A}_{Q'}$ and let $\mathcal{G}$ be the sheaf of faces of $\mathcal{M}_{X'}$ defined by Q. Recall that $L^{\cdot}(\mathcal{F})\Omega^{\cdot}_{X/\mathbf{C}} \cong L^{\cdot}(\mathcal{F})\underline{\Omega}^{\cdot}_{X'/\mathbf{C}}(\mathcal{G})$. Let $\mathcal{K}$ be the ideal of $\mathcal{M}_{X'}$ defined by the maximal ideal of Q'. Then $X_{\mathcal{K}}$ is just x_M, and statement (1) of Corollary 3.1.2 says that the stalk of the filtered complex $(\Omega^{\cdot}_{X/\mathbf{C},x}, L^{\cdot}(\mathcal{F}))$ maps quasi-isomorphically to the filtered complex $(\Omega^{\cdot}_{x_M/\mathbf{C}}, L^{\cdot}(\mathcal{F}))$. Corollary IV.2.3.6 provides an isomorphism: $\Omega^1_{x_M/\mathbf{C}} \cong \mathbf{C} \otimes \overline{\mathcal{M}}_x$. It follows that the differentials of the complex $\Omega^{\cdot}_{x_M/\mathbf{C}}$ vanish, and it is clear that the filtration $L^{\cdot}(\mathcal{F}_x)$ corresponds to the filtration $L^{\cdot}(\overline{\mathcal{F}}_x)$. This proves statement (1), and statement (2) follows. □

As a consequence of the previous result, we can give a logarithmic version of a classical result of Danilov [11], as well as some variations.

Corollary 3.1.4. *Let X be a fine and smooth idealized log analytic space.*

1. *The natural map*
$$\mathbf{C}_X \to \underline{\Omega}^{\cdot}_{X/\mathbf{C}}$$
is a quasi-isomorphism. Consequently there are natural isomorphisms
$$H^*(X, \mathbf{C}) \cong H^*(X, \underline{\Omega}^{\cdot}_{X/\mathbf{C}}).$$
2. *If $\mathcal{K}$ is a coherent sheaf of ideals in $\mathcal{M}_X$, let $X_{\mathcal{K}}$ be the idealized log subscheme of X defined by $\mathcal{K}$, and let $j\colon U_{\mathcal{K}} \to X$ be the inclusion of its complement. Then the natural map*
$$j_!(\mathbf{C}_{U_{\mathcal{K}}}) \to \underline{\Omega}^{\cdot}_{X/\mathbf{C}} \cap \mathcal{K}\Omega^{\cdot}_{X/\mathbf{C}}$$
is a quasi-isomorphism.
3. *Suppose that $\mathcal{K}_X = \emptyset$, let $\mathcal{I}_{\mathcal{M}} \subseteq \mathcal{M}_X$ be the interior ideal of $\mathcal{M}_X$ (see Proposition II.2.6.3), and let $j\colon X^* \to X$ be the inclusion. There is a natural quasi-isomorphism*
$$j_!(\mathbf{C}_{X^*}) \to \mathcal{I}_{\mathcal{M}}\Omega^{\cdot}_{X/\mathbf{C}}.$$
If $\underline{X}$ is compact, there are are natural isomorphisms
$$H^i_c(X^*, \mathbf{C}) \cong H^i(X, \mathcal{I}_{\mathcal{M}}\Omega^{\cdot}_{X/\mathbf{C}}).$$

Proof The first statement follows from (2) of Corollary 3.1.3 when $\mathcal{F} = \mathcal{M}^*_X$. The "natural map" in the second statement comes from the map
$$\mathbf{C}_{\mathcal{U}_{\mathcal{K}}} \to \Omega^{\cdot}_{U_{\mathcal{K}}/\mathbf{C}} \cong j^*(\underline{\Omega}^{\cdot}_{X/\mathbf{C}} \cap \mathcal{K}\Omega^{\cdot}_{X/\mathbf{C}})$$
and adjunction. We will check that it is a quasi-isomorphism at the stalks at each $x \in X$. If $x \in U$ this is the case by statement (1), and if $x \in X_{\mathcal{K}}$, both complexes are acyclic, as follows from Theorem 3.1.1. To deduce statement (3), observe that $U_{\mathcal{I}_{\mathcal{M}}} = X^*$ and that $\mathcal{I}_{\mathcal{M}}\Omega^{\cdot}_{X/\mathbf{C}} \subseteq \underline{\Omega}^{\cdot}_{X/\mathbf{C}}$. The latter statement can be checked locally, so we may assume that $X = \mathsf{A}^{\mathrm{an}}_Q$ where Q is a fine monoid. Then $\mathcal{I}_{\mathcal{M}}\Omega^j_{X/\mathbf{C}}$ (resp. $\underline{\Omega}^j_{X/\mathbf{C}}$) is the coherent sheaf associated to the Q-graded $\mathbf{C}[Q]$-module $I_Q\Omega^j_Q$ (resp. $\underline{\Omega}^j_Q$). In degree q, the module $I_Q\Omega^j_{Q)}$ vanishes if $q \notin I_Q$. If $q \in I_Q$, then $\langle q\rangle = Q$ and so $\underline{\Omega}^j_{Q,q}$ is all of $\mathbf{C} \otimes \Lambda^j Q^{\mathrm{gp}}$ and hence contains $I_Q\Omega^j_{Q,q}$. Thus $\mathcal{I}_M\Omega^{\cdot}_{X/\mathbf{C}} \subseteq \underline{\Omega}^{\cdot}_{X/\mathbf{C}}$, and statement (2) implies that the natural map
$$j_!(\mathbf{C}_{X^*}) \to \underline{\Omega}^{\cdot}_{X/\mathbf{C}} \cap \mathcal{I}_M\Omega^{\cdot}_{X/\mathbf{C}} = \mathcal{I}_M\Omega^{\cdot}_{X/\mathbf{C}}$$

is a quasi-isomorphism. If X is compact, then $H^i_c(X^*, \mathbf{C}) \cong H^i(X, j_!(\mathbf{C}_{X^*}))$, completing the proof of statement (3). □

3.2 The symbol in de Rham cohomology

Our main goal in this section is to compute the cohomology sheaves of the relative de Rham complex $\Omega^{\cdot}_{X/Y}$.

Proposition 3.2.1. *Let $f: X \to Y$ be a morphism of fine idealized log analytic spaces. There is a unique family of homomorphisms of sheaves of $f^{-1}(\mathcal{O}_Y)$-modules on X (the analytic symbol maps),*

$$s^q : f^{-1}(\mathcal{O}_Y) \otimes \Lambda^q \mathcal{M}^{\mathrm{gp}}_{X/Y} \to \mathcal{H}^q(\Omega^{\cdot}_{X/Y}) : q \in \mathbf{N},$$

satisfying the following conditions.

1. *The composite*

$$s^0 : f^{-1}(\mathcal{O}_Y) \to \mathcal{H}^0(\Omega^{\cdot}_{X/Y}) \to \mathcal{O}_X$$

is the structure homomorphism $f^\sharp$.

2. *The diagram*

$$\begin{array}{ccc} \mathcal{M}^{\mathrm{gp}}_X & \xrightarrow{dlog} & \mathcal{Z}^1_{X/Y} \\ \downarrow & & \downarrow \\ f^{-1}(\mathcal{O}_Y) \otimes \mathcal{M}^{\mathrm{gp}}_{X/Y} & \xrightarrow{s^1} & \mathcal{H}^1(\Omega^{\cdot}_{X/Y}) \end{array}$$

commutes.

3. *If $\omega \in \Lambda^i \mathcal{M}^{\mathrm{gp}}_{X/Y}$ and $\eta \in \Lambda^j \mathcal{M}^{\mathrm{gp}}_{X/Y}$, then*

$$s^{i+j}(\omega \wedge \eta) = s^i(\omega) \wedge s^j(\eta).$$

Proof The existence and uniqueness of s^0 in (1) are clear. If m is a local section of $\mathcal{M}_X$, then $dlog(m)$ is local section of $\Omega^1_{X/Y}$ and $d\,dlog(m) = 0$; so the top horizontal arrow makes sense. Locally on X, every section u of $\mathcal{O}^*_X$ can be written as $u = \exp f$, and then $dlog(u) = df$ vanishes in $\mathcal{H}^1(\Omega^{\cdot}_{X/Y})$. Furthermore, $dlog(m) = 0$ if m is in the image of $f^{-1}(\mathcal{M}_Y)$, and so $f^*_{log}(\mathcal{M}_Y)$ maps to zero in $\mathcal{H}^1(\Omega^{\cdot}_{X/Y})$. Since $\mathcal{H}^1(\Omega^{\cdot}_{X/Y})$ is a sheaf of $f^{-1}(\mathcal{O}_Y)$-modules, there is a unique arrow along the bottom making the diagram commute. The map s^q for $q > 1$ by definition sends $m_1 \wedge \cdots \wedge m_q$ to $dlog\, m_1 \wedge \cdots \wedge dlog\, m_q$ for any q-tuple of sections of $\overline{\mathcal{M}}^{\mathrm{gp}}_{X/Y}$, and thus is uniquely determined by property (3). □

Variant 3.2.2. If $\mathcal{F}$ is a relatively coherent sheaf of faces in $\mathcal{M}_X$ containing the image of $f_{log}^*(\mathcal{M}_Y)$, let $\mathcal{F}_{X/Y}$ denote the cokernel of the map $f_{log}^*(\mathcal{M}_Y) \to \mathcal{F}$, so that $\mathcal{F}_{X/Y}^{gp}$ becomes a sheaf of subgroups of $\mathcal{M}_{X/Y}^{gp}$. We have a corresponding Koszul filtration $K^{\cdot}(\mathcal{F})$ on the exterior algebra of $\mathcal{M}_{X/Y}^{gp}$ and its décalée $L^{\cdot}(\mathcal{F})$. Then there are compatible families of maps

$$s^q\colon f^{-1}(\mathcal{O}_Y) \otimes L^i(\mathcal{F})\Lambda^q \mathcal{M}_{X/Y}^{gp} \to \mathcal{H}^q(L^i(\mathcal{F})\Omega^{\cdot}_{X/Y}).$$

In particular, we find maps

$$s^q\colon f^{-1}(\mathcal{O}_Y) \otimes \Lambda^q \mathcal{F}_{X/Y}^{gp} \to \mathcal{H}^q(\underline{\Omega}^{\cdot}_{X/Y}(\mathcal{F})).$$

We shall see that, with suitable hypotheses, the symbol maps defined in Proposition 3.2.1 and its variant are isomorphisms. For the sake of simplicity, we begin with the case in which $Y = \mathbf{C}$ with the trivial log structure.

Theorem 3.2.3. *Let X be a fine and smooth idealized log analytic space and let $\mathcal{F}$ be a relatively coherent sheaf of faces in $\mathcal{M}_X$. Then the symbol maps are isomorphisms*

$$\begin{aligned} \mathbf{C} \otimes \Lambda^q \overline{\mathcal{M}}_X^{gp} &\cong \mathcal{H}^q(\Omega^{\cdot}_{X/\mathbf{C}}), \\ \mathbf{C} \otimes L^i(\mathcal{F})\Lambda^q \overline{\mathcal{M}}_X^{gp} &\cong \mathcal{H}^q(L^i(\mathcal{F})\Omega^{\cdot}_{X/\mathbf{C}}), \\ \mathbf{C} \otimes \Lambda^q \overline{\mathcal{F}}^{gp} &\cong H^q(\underline{\Omega}^{\cdot}_{X/\mathbf{C}}(\mathcal{F})). \end{aligned}$$

Proof It suffices to check that maps on stalks are isomorphisms. This fact follows immediately from Corollary 3.1.3, since one easily verifies that the symbol maps are compatible with the isomorphisms used there. □

The idea of the proof in the relative case is similar, but technically more complicated. Instead of a single homogeneous flow, we must use a collection of homogenous vector fields.

Theorem 3.2.4. *Let $f\colon X \to Y$ be a smooth and exact morphism of fine idealized log analytic spaces for which the sheaves $\mathcal{M}_{X/Y}^{gp}$ are torsion free. Let $\mathcal{F}$ be a relatively coherent sheaf of faces in $\mathcal{M}_X$ containing the image of $f_{log} * (\mathcal{M}_Y)$. Then the maps*

$$\begin{aligned} f^{-1}(\mathcal{O}_Y) \otimes \Lambda^q \mathcal{M}_{X/Y}^{gp} &\to \mathcal{H}^q(\Omega^{\cdot}_{X/Y}), \\ f^{-1}(\mathcal{O}_Y) \otimes L^i(\mathcal{F})\Lambda^q \mathcal{M}_{X/Y}^{gp} &\to \mathcal{H}^q(L^i(\mathcal{F})\Omega^{\cdot}_{X/Y}), \\ f^{-1}(\mathcal{O}_Y) \otimes \Lambda^q \mathcal{F}_{X/Y}^{gp} &\to \mathcal{H}^q(\underline{\Omega}^{\cdot}_{X/Y}(\mathcal{F})) \end{aligned}$$

defined in Variant 3.2.2 are isomorphisms.

Proof As in the proof of Theorem 3.1.1, we can reduce to the case in which f is given by an exact homomorphism of idealized monoids $(P, J) \to (Q, K)$, and x is the vertex of A_Q.

Let $\partial\colon Q^{gp}/P^{gp} \to \mathbf{Z}$ be a homomorphism defining a vector field ξ, and let $Q' := \ker(\partial) \cap Q$. As we have seen, interior multiplication by ξ acts on the filtered complex $(C^\cdot, L^\cdot) := (\Omega^\cdot_{X/Y}, L^\cdot(\mathcal{F}))$, and hence so does the Lie derivative $\kappa := d\xi + \xi d$. Let $(C'^\cdot, L^\cdot)$ be the kernel of κ with the induced filtration and let $(C''^\cdot, L^\cdot)$ be the image. Note that if $\omega = \sum \omega_q \in C^\cdot$, then $\kappa(\omega) = \sum_q \partial(q)\omega_q$, so $C'^\cdot = \{\sum \omega_q : q \in Q'\}$ and $C''^\cdot = \{\sum \omega_q : q \in Q \setminus Q'\}$. Thus $C^\cdot$ is the direct sum of $C'^\cdot$ and $C''^\cdot$. Furthermore, κ induces an automorphism of the complex $(C''^\cdot, L^\cdot)$, as in the proof of Theorem 3.1.1. It follows that this filtered complex is acyclic, and hence that $(C'^\cdot, L^\cdot) \to (C^\cdot, L^\cdot)$ is a filtered quasi-isomorphism. The symbol maps factor through $C'^\cdot$, and it will thus suffice to prove they induce a filtered quasi-isomorphism $(\mathcal{O}_{Y,y} \otimes \Lambda^\cdot \mathcal{M}_{X/Y}, L^\cdot) \to (C'^\cdot, L^\cdot)$.

Let $\partial_1, \ldots, \partial_r$ be a basis for the dual of the finitely generated free abelian group Q^{gp}/P^{gp}. Then the corresponding vector fields $\xi_1, \ldots, \xi_r$ commute, and hence the kernel $C_i'^\cdot$ of ξ_i is invariant under the remaining vector fields. Thus a repetition of the argument of the previous paragraph shows that the map $\cap(C_i'^\cdot, L^\cdot) \to (C^\cdot, L^\cdot)$ is a filtered quasi-isomorphism. But $\cap\{C_i'^\cdot : i = 1, \ldots, r\} = \{\sum \omega_q : q \in P^{gp}\}$, and since θ is exact, such elements in fact lie in $\mathcal{O}_{Y,y} \otimes \Lambda^\cdot \mathcal{M}_{X/Y}$. □

We should remark that the torsion hypothesis in Theorem 3.2.4 is not superfluous, even though it is not needed for Theorem 3.2.3. For example, the morphism $f\colon X \to Y$ associated to multiplication by $n > 0$ on the monoid $\mathbf{N}$ is étale, so $\Omega^1_{X/Y} = 0$ and $\mathcal{H}^0(\Omega^\cdot_{X/Y}) = \mathcal{O}_X$, although $f^{-1}(\mathcal{O}_Y) \otimes \Lambda^0 \mathcal{M}^{gp}_{X/Y} = f^{-1}(\mathcal{O}_Y)$.

3.3 $\Omega^{\cdot log}_X$ and the Poincaré lemma

The best analog of the Poincaré lemma in the context of log analytic geometry takes place not on X_{an} but rather on X_{log}. To formulate it, we must first construct the de Rham complex for the ringed space $(X_{log}, \mathcal{O}^{log}_X)$. Roughly speaking, this amounts to constructing an integrable connection on the sheaf $\mathcal{O}^{log}_X$.

Proposition 3.3.1. *If $f\colon X \to Y$ is a morphism of fine log analytic spaces, let*

$$\Omega^{i,log}_{X/Y} =: \mathcal{O}^{log}_X \otimes_{\tau_X^{-1}\mathcal{O}_X} \tau_X^{-1}\Omega^i_{X/Y}.$$

Then there is a unique $\tau_X^{-1}(\mathcal{O}_Y)$-linear homomorphism:

$$d\colon \mathcal{O}^{log}_X \to \Omega^{1,log}_{X/Y}$$

such that:

1. *$d(f) = 1 \otimes df$ for $f \in \tau_X^{-1}(\mathcal{O}_X) \subseteq \mathcal{O}^{log}_X$;*

2. $d(fg) = g \otimes df + f \otimes dg$ *for* $f, g \in \mathcal{O}_X^{log}$;
3. $d(\ell) = 1 \otimes d\log(\exp(\ell))$ *for all* $\ell \in \mathcal{L}_X$, *where* $\exp\colon \mathcal{L}_X \to \tau^{-1}(\mathcal{M}_X^{gp})$ *is the map defined in Proposition 1.4.1.*

Furthermore, there is a unique extension of d *to a derivation of* $\oplus\Omega_{X/Y}^{i,log}$ *forming a complex*

$$\Omega_{X/Y}^{\cdot,log} := \mathcal{O}_X^{log} \xrightarrow{d} \Omega_{X/Y}^{1,log} \xrightarrow{d} \Omega_{X/Y}^{2,log} \xrightarrow{d} \cdots.$$

Proof The uniqueness of d is clear, since $\mathcal{O}_X^{log}$ is generated as a $\tau_X^{-1}(\mathcal{O}_X)$-algebra by $\mathcal{L}_X$. For the existence, we begin by constructing a connection on the extension $\mathcal{E}_X$ in diagram 1.4.1.

Lemma 3.3.2. *There is a unique additive map*

$$d\colon \mathcal{E}_X \to \mathcal{E}_X \otimes \tau_X^{-1}(\Omega_{X/Y}^1)$$

with the following properties:

1. $df = 1 \otimes df$ *for* $f \in \tau_X^{-1}(\mathcal{O}_X)$;
2. $d(fe) = e \otimes df + f \otimes de$; *for* $f \in \tau_X^{-1}(\mathcal{O}_X)$ *and* $e \in \mathcal{E}_X$;
3. $de = d\exp(\ell)$ *if* $\ell \in \mathcal{L}_X$ *and* e *is its image in* $\mathcal{E}_X$.

Furthermore, d *fits into a commutative diagram with exact rows*

$$\begin{array}{ccccccccc}
0 & \longrightarrow & \tau_X^{-1}(\mathcal{O}_X) & \longrightarrow & \mathcal{E}_X & \longrightarrow & \overline{\mathcal{M}}_X^{gp} \otimes \tau_X^{-1}(\mathcal{O}_X) & \longrightarrow & 0 \\
 & & \downarrow{\scriptstyle d} & & \downarrow{\scriptstyle d} & & \downarrow{\scriptstyle \mathrm{id}\otimes d} & & \\
0 & \longrightarrow & \tau_X^{-1}(\Omega_{X/Y}) & \longrightarrow & \mathcal{E}_X \otimes \tau_X^{-1}(\Omega_{X/Y}^1) & \longrightarrow & \overline{\mathcal{M}}_X^{gp} \otimes \tau_X^{-1}(\Omega_{X/Y}^1) & \longrightarrow & 0,
\end{array}$$

and the natural map $d^2 : \mathcal{E}_X \to \mathcal{E}_X \otimes \tau_X^{-1}(\Omega_{X/Y}^2)$ *vanishes.*

Proof We claim that there is a diagram as follows:

$$\begin{array}{ccc}
\tau_X^{-1}(\mathcal{O}_X) \otimes \mathcal{L}_X & \longrightarrow & \mathcal{L}_X \otimes \tau_X^{-1}(\Omega_{X/Y}^1) \\
\downarrow & & \downarrow \\
\mathcal{E}_X & \dashrightarrow & \mathcal{E}_X \otimes \tau_X^{-1}(\Omega_{X/Y}^1).
\end{array}$$

The tensor product in the top row is taken over $\mathbf{Z}$, and the top arrow is given by

$$f \otimes \ell \mapsto \ell \otimes df + \epsilon(f) \otimes d\exp(\ell),$$

which exists because the right side is bilinear in f and ℓ. Thus if f and g are sections of $\tau_X^{-1}(\mathcal{O}_X)$, we find that this arrow takes $f \otimes \epsilon(g)$ to

$$\epsilon(g) \otimes df + \epsilon(f) \otimes d \exp \epsilon(g) = \epsilon(g) \otimes df + \epsilon(f) \otimes dg$$

in $\mathcal{L}_X \otimes \tau_X^{-1}(\Omega^1_{X/Y})$. The tensor product at the bottom right is taken over $\tau_X^{-1}(\mathcal{O}_X)$, and so $f \otimes \epsilon(g)$ maps to $1 \otimes gdf + 1 \otimes fdg = 1 \otimes d(fg)$ in $\mathcal{E}_X \otimes \tau_X^{-1}(\Omega^1_{X/Y})$. Then the existence of the dashed arrow follows from the definition of $\mathcal{E}_X$ as a pushout. The verification of the remaining properties of d is immediate. □

To complete the proof of Proposition 3.3.1, we observe by induction on n that there is a unique $d\colon S^n\mathcal{E}_X \to S^n\mathcal{E}_X \otimes \Omega^1_{X/Y}$ such that $d(ab) = adb + bda$ for $a \in S^i\mathcal{E}_X$ and $b \in S^{n-i}\mathcal{E}_X$. Since these are compatible with the inclusions $S^n\mathcal{E}_X \to S^{n+1}\mathcal{E}_X$, we find the desired map on $\mathcal{O}_X^{log} = \varinjlim S^n\mathcal{E}_X$. □

The importance of $(\mathcal{E}_X, d)$ is that it is a physical incarnation of the symbol map. The exact sequence of statement (2) of Proposition 1.4.1 is in fact a sequence of modules with connection. Thus it fits into an exact sequence of complexes:

$$0 \to \tau_X^{-1}(\Omega^{\cdot}_{X/Y}) \to \mathcal{E}_X \otimes \tau_X^{-1}(\Omega^{\cdot}_{X/Y}) \to \tau_X^{-1}(\overline{\mathcal{M}}_X^{gp} \otimes \Omega^{\cdot}_{X/Y}) \to 0, \qquad (3.3.1)$$

leading to an exact sequence of cohomology sheaves

$$\mathcal{H}^0(\tau_X^{-1}(\Omega^{\cdot}_{X/Y}) \otimes \mathcal{E}_X) \longrightarrow \mathcal{H}^0(\tau_X^{-1}(\overline{\mathcal{M}}_X^{gp} \otimes \Omega^{\cdot}_{X/Y})) \xrightarrow{\ \partial\ } \mathcal{H}^1(\tau_X^{-1}(\Omega^{\cdot}_X)).$$

Lemma 3.3.3. *The following diagram commutes:*

$$\begin{array}{ccc}
\mathcal{H}^0(\tau_X^{-1}(\overline{\mathcal{M}}_X^{gp} \otimes \Omega^{\cdot}_{X/Y})) & \xrightarrow{\ \partial\ } & \mathcal{H}^1(\tau_X^{-1}(\Omega^{\cdot}_{X/Y})) \\
\uparrow & & \uparrow \tau_X^{-1}(s^1) \\
\tau_X^{-1}(\overline{\mathcal{M}}_X^{gp}) & \xrightarrow{\ \pi\ } & \tau_X^{-1}(\mathcal{M}^{gp}_{X/Y}).
\end{array}$$

Here $s^1 : \overline{\mathcal{M}}^{gp}_{X/Y} \to \mathcal{H}^1(\Omega^{\cdot}_{X/Y})$ is the symbol map and π is the projection.

Proof Let m be a local section of $\tau_X^{-1}(\overline{\mathcal{M}}_X^{gp})$. Choose a local section ℓ of $\mathcal{L}_X$ with $m = \overline{\exp}(\ell)$, and let e be the image of $1 \otimes \ell$ in $\mathcal{E}_X$. Then ∂m is the class of de in the cohomology of $\tau_X^{-1}(\Omega^{\cdot}_{X/Y})$. By the formula (3) of Lemma 3.3.2, $de = dlog(\exp(\ell)) = dlog(m)$, which indeed maps to the symbol of the image of m in $\mathcal{M}^{gp}_{X/Y}$. □

Now we are ready to formulate and prove the Poincaré lemma on X_{log}. Note first that the naturality of the construction of $\mathcal{O}_X^{log}$ implies the existence of a natural homomorphism $f_{log}^{-1}(\mathcal{O}_Y^{log}) \to \mathcal{O}_X^{log}$.

Theorem 3.3.4. *Let $f\colon X \to Y$ be a smooth and exact morphism of fine analytic spaces. Suppose either that $Y = \operatorname{Spec} \mathbf{C}$ with trivial log structure or that $\mathcal{M}^{\mathrm{gp}}_{X/Y}$ is torsion free. Then, on the space X_{log}, the natural map*

$$f_{log}^{-1}(\mathcal{O}_Y^{log}) \to \Omega^{\cdot,log}_{X/Y}$$

is a quasi-isomorphism. [3]

Proof Recall from Proposition 1.4.3 that we have an increasing exhaustive filtration on $\mathcal{O}_X^{log}$ given by $N_p\mathcal{O}_X^{log} := S^p\mathcal{E}_X \subseteq \mathcal{O}_X^{log}$ for $p \in \mathbf{N}$. This filtration is compatible with the connection d and induces a filtration on $\Omega^{\cdot,log}_{X/Y}$. The homomorphism $f_{log}^{-1}(\mathcal{O}_Y^{log}) \to \Omega^{\cdot,log}_{X/Y}$ is a map of filtered complexes, and we shall see that the induced map of the corresponding spectral sequences is an isomorphism starting at the E_2 term. To work with the usual index conventions for cohomological spectral sequences, we set $N^p := N_{-p}$ for all p. Then we have

$$\mathrm{Gr}_N^{-p}\,\Omega^{i,log}_{X/Y} \cong \tau_X^{-1}(S^p\overline{\mathcal{M}}_X^{\mathrm{gp}} \otimes \Omega^i_{X/Y}),$$

and the spectral sequence of the filtered complex $(\Omega^{\cdot log}_{X/Y}, N)$ begins

$$\begin{array}{ccccc}
E_0^{-p,q} & \xrightarrow{=} & \mathrm{Gr}_N^{-p}\,\Omega^{q-p,log}_{X/Y} & \xrightarrow{\cong} & S^p\overline{\mathcal{M}}_X^{\mathrm{gp}} \otimes \Omega^{q-p}_{X/Y} \\
\Big\downarrow{\scriptstyle d_0^{p,q}} & & \Big\downarrow{\scriptstyle \mathrm{Gr}(d)} & & \Big\downarrow{\scriptstyle \mathrm{id}\otimes d} \\
E_0^{-p,q+1} & \xrightarrow{=} & \mathrm{Gr}_N^{-p}\,\Omega^{q-p+1,log}_{X/Y} & \xrightarrow{\cong} & S^p\overline{\mathcal{M}}_X^{\mathrm{gp}} \otimes \Omega^{q-p+1}_{X/Y}\,.
\end{array}$$

Here we have omitted writing τ_X^{-1} to save space. The E_1 terms of the spectral sequence are given by the cohomology of the E_0 terms, which we compute using Theorem 3.2.3 if $Y = \mathbf{C}$ and using Theorem 3.2.4 if $\mathcal{M}^{\mathrm{gp}}_{X/Y}$ is torsion free. We find the following commutative diagram:

$$\begin{array}{ccccc}
E_1^{-p,q} & \xrightarrow{=} & H^{q-p}(E_0^{-p,\cdot}) & \xrightarrow{\cong} & f^{-1}(\mathcal{O}_Y) \otimes S^p\overline{\mathcal{M}}_X^{\mathrm{gp}} \otimes \Lambda^{q-p}\mathcal{M}^{\mathrm{gp}}_{X/Y} \\
\Big\downarrow{\scriptstyle d_1^{p,q}} & & \Big\downarrow & & \Big\downarrow \\
E_1^{1-p,q} & \xrightarrow{=} & H^{q-p+1}(E_0^{1-p,\cdot}) & \xrightarrow{\cong} & f^{-1}(\mathcal{O}_Y) \otimes S^{p-1}\overline{\mathcal{M}}_X^{\mathrm{gp}} \otimes \Lambda^{q-p+1}\mathcal{M}^{\mathrm{gp}}_{X/Y}).
\end{array}$$

Let us calculate the boundary map $d_1^{p,q}$ using these identifications. When $p = q = 1$, the map $d_1^{p,q}$ is the boundary map ∂ coming from the exact sequence (3.3.1). Thus Lemma 3.3.3 shows that the vertical map on the right

[3] The hypothesis of the torsion-freeness of $\mathcal{M}^{\mathrm{gp}}_{X/Y}$ can be eliminated by working locally in the Kummer étale topology; see [40, 5.1].

is the identity tensored with the projection $\overline{\mathcal{M}}_X^{gp} \to \mathcal{M}_{X/Y}^{gp}$. Since d acts as a derivation, it follows that, for every p and q, every monomial $m_1 \cdots m_p \in S^p(\overline{\mathcal{M}}_X^{gp})$, and every $\omega \in \mathcal{O}_Y \otimes \Lambda^{q-p}\mathcal{M}_{X/Y}^{gp}$,

$$d(m_1 \cdots m_p \otimes \omega) = \sum_i m_1 \cdots \hat{m}_i \cdots m_p \otimes \overline{m}_i \wedge \omega.$$

In other words, the complex $E_1^{\cdot,q}$ can be identified with the complex

$$\mathcal{O}_Y \otimes S^q\overline{\mathcal{M}}_X^{gp} \to \mathcal{O}_Y \otimes S^{q-1}\overline{\mathcal{M}}_X^{gp} \otimes \mathcal{M}_{X/Y}^{gp} \to \mathcal{O}_Y \otimes S^{q-2}\overline{\mathcal{M}}_X^{gp} \otimes \Lambda^2\mathcal{M}_{X/Y}^{gp} \cdots .$$

This is the degree q-graded piece of the de Rham complex of the symmetric algebra $\mathcal{O}_Y \otimes S^{\cdot}\overline{\mathcal{M}}_X^{gp}$ relative to $\mathcal{O}_Y \otimes S^{\cdot}\overline{\mathcal{M}}_Y$, with a shift by q. By the Poincaré lemma for polynomial algebras, we conclude that $E_2^{-p,q}$ vanishes unless $p = q$ and that

$$E_2^{-p,p} \cong \mathcal{O}_Y \otimes f^{-1}S^p\overline{\mathcal{M}}_Y^{gp} \cong \mathrm{Gr}_N^{-p} f_{log}^{-1}(\mathcal{O}_Y^{log}).$$

Thus the map of spectral sequences is an isomorphism at the E_2 level, and the spectral sequences are degenerate from that point on. □

Variant 3.3.5. If $\mathcal{F}$ is a relatively coherent sheaf of faces in $\mathcal{M}_X$ containing $f^{-1}(\mathcal{O}_Y)$, then the same argument shows that on the space $X_{log}(\mathcal{F})$, the map

$$f_{log}^{-1}(\mathcal{O}_Y^{log}) \to \underline{\Omega}_{X/Y}^{\cdot,log}(\mathcal{F})$$

is a quasi-isomorphism, where

$$\underline{\Omega}_{X/Y}^{\cdot,log}(\mathcal{F}) := \mathcal{O}_{X(\mathcal{F})}^{log} \otimes \tau_{X(\mathcal{F})}^{-1}(\underline{\Omega}_{X/Y}^{\cdot}(\mathcal{F})).$$

The next two results reveal the geometric meaning of the de Rham cohomology of log analytic spaces.

Theorem 3.3.6. *Let $X/\mathbf{C}$ be a smooth idealized log analytic space such that $\overline{\mathcal{M}}_X^{gp}$ is torsion free.*

1. *In the bounded derived category $D^b(X, \mathbf{C})$ of sheaves of $\mathbf{C}$-vector spaces on X, the natural maps*

$$R\tau_{X*}(\mathbf{C}_{X_{log}}) \to R\tau_{X*}\Omega_{X/\mathbf{C}}^{\cdot log} \leftarrow \Omega_{X/\mathbf{C}}^{\cdot}$$

are isomorphisms.

2. *More generally, if $\mathcal{F}$ is a relatively coherent sheaf of faces in $\mathcal{M}_X$, the natural maps*

$$R\tau_{X(\mathcal{F})*}(\mathbf{C}_{X(\mathcal{F})_{log}}) \to R\tau_{X(\mathcal{F})*}\underline{\Omega}_{X/\mathbf{C}}^{\cdot log}(\mathcal{F}) \leftarrow \underline{\Omega}_{X/\mathbf{C}}^{\cdot}(\mathcal{F})$$

are isomorphisms.

Proof Theorem 3.3.4, in the case when $Y = \operatorname{Spec} \mathbf{C}$, says that the map $\mathbf{C}_{X_{log}} \to \Omega^{\cdot,log}_{X/\mathbf{C}}$ is a quasi-isomorphism on X_{log}. It follows immediately that the map $R\tau_{X*}(\mathbf{C}_{X_{log}}) \to R\tau_{X*}\Omega^{\cdot log}_{X/\mathbf{C}}$ is a derived isomorphism, and the relatively coherent analog follows in the same way from Variant 3.3.5. Since $\overline{\mathcal{M}}^{\mathrm{gp}}_X$ is torsion free, Theorem 1.4.7 implies that the maps

$$R\tau_{X*}\Omega^{\cdot log}_{X/\mathbf{C}} \leftarrow \Omega^{\cdot}_{X/\mathbf{C}} \quad \text{and} \quad R\tau_{X*(\mathcal{F})}\underline{\Omega}^{\cdot log}_{X/\mathbf{C}}(\mathcal{F}) \leftarrow \Omega^{\cdot}_{X/\mathbf{C}}(\underline{\mathcal{F}})$$

are isomorphisms. □

Corollary 3.3.7. *Let X be a fine and smooth log analytic space (without idealized structure) such that $\overline{\mathcal{M}}^{\mathrm{gp}}_X$ is torsion free.*

1. *In the following commutative diagram, all maps are isomorphisms:*

$$\begin{array}{ccccc} \Omega^{\cdot}_{X/\mathbf{C}} & \longrightarrow & R\tau_{X*}(\Omega^{\cdot,log}_{X/\mathbf{C}}) & \longleftarrow & R\tau_{X*}(\mathbf{C}_{X^{log}}) \\ & \searrow & \downarrow & & \downarrow \\ & & Rj_*(\Omega^{\cdot}_{X^*/\mathbf{C}}) & \longleftarrow & Rj_*(\mathbf{C}_{X^*}). \end{array}$$

2. *More generally, if $\mathcal{F}$ is a relatively coherent sheaf of faces in $\mathcal{M}_X$ and $j_{\mathcal{F}}\colon X^*_{\mathcal{F}} \to X$ is the inclusion of the open set of triviality of $\mathcal{F}$ (see Remark II.2.6.7), the maps in the following commutative diagram are isomorphisms:*

$$\begin{array}{ccccc} \underline{\Omega}^{\cdot}_{X/\mathbf{C}}(\mathcal{F}) & \longrightarrow & R\tau_{X(\mathcal{F})*}(\underline{\Omega}^{\cdot,log}_{X/\mathbf{C}}(\mathcal{F})) & \longleftarrow & R\tau_{X(\mathcal{F})*}(\mathbf{C}_{X^{log}}) \\ & \searrow & \downarrow & & \downarrow \\ & & Rj_{\mathcal{F}*}(\underline{\Omega}^{\cdot}_{X^*_{\mathcal{F}}/\mathbf{C}}) & \longleftarrow & Rj_{\mathcal{F}*}(\mathbf{C}_{X^*_{\mathcal{F}}}). \end{array}$$

Proof The horizontal isomorphisms come from Theorem 3.3.6 and the rightmost vertical isomorphism comes from Corollary 1.3.2. (resp. Variant 1.3.5 in the relatively coherent case). It follows that the remaining two arrows are also isomorphisms. □

4 Algebraic de Rham cohomology

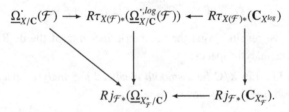

Logarithmic techniques have long played an important role in de Rham cohomology. Here we shall explain how classical constructions fit into the general framework of log geometry and how this new framework can be used to extend them. First we discuss characteristic p, then characteristic 0.

4.1 The Cartier operator and the Cartier isomorphism

On a log scheme X, the differentials of sections of $\mathcal{M}_X$ give rise to one-forms that are closed but not in general exact. This was the construction underlying the symbol map in the analytic context in Proposition 3.2.1. The following construction can viewed as the characteristic p analog of that proposition.

Proposition 4.1.1. *Let $f\colon X \to Y$ be a morphism of fine idealized log schemes in characteristic $p > 0$. Let F_X denote the absolute Frobenius endomorphism of X. Then there is a unique family of $\mathcal{O}_X$-linear homomorphisms*

$$\sigma^i_{X/Y}\colon \Omega^i_{X/Y} \to F_{X*}\mathcal{H}^i(\Omega^{\cdot}_{X/Y}) \quad i \geq 0$$

with the following properties.

1. $\sigma^0_{X/Y}(1) = 1$.
2. $\sigma^{i+j}_{X/Y}(\omega \wedge \eta) = \sigma^i_{X/Y}(\omega) \wedge \sigma^j_{X/Y}(\eta)$ *if* $\omega \in \Omega^i_{X/Y}$ *and* $\eta \in \Omega^j_{X/Y}$.
3. *For every local section m of $\mathcal{M}_X$, $\sigma_{X/Y}(d\log m)$ is the class of $d\log m$ in $F_{X*}\mathcal{H}^1(\Omega^{\cdot}_{X/Y})$.*

These homomorphisms are compatible with the filtrations defined by coherent sheaves of ideals and relatively coherent sheaves of faces in $\mathcal{M}_X$.

Proof In the course of the proof we will use the following notation. If $\mathcal{E}$ is a sheaf of $\mathcal{O}_X$-modules and e is a local section of $\mathcal{E}$, we write $F_{X*}(e)$ to mean the same section but now viewed as a section of $F_{X*}(\mathcal{E})$. Thus if a is a local section of $\mathcal{O}_X$, we have $aF_{X*}(e) = F_{X*}(a^p e)$.

The existence and uniqueness of $\sigma^i_{X/Y}$ for $i > 1$ will follow from the case $i = 1$, by multiplicativity. The uniqueness of $\sigma^1_{X/Y}$ follows from the fact that $\Omega^1_{X/Y}$ is locally generated as a sheaf of $\mathcal{O}_X$-modules by the image of $d\log\colon \mathcal{M}_X \to \Omega^1_{X/Y}$, as we saw in Proposition IV.1.2.11.

The existence depends on the following well-known lemma.

Lemma 4.1.2. *Let $X \to Y$ be a morphism of schemes, let f and g be sections of $\mathcal{O}_X$, and let p be a prime integer. Then $f^{p-1}df + g^{p-1}dg - (f+g)^{p-1}(df+dg)$ is exact.*

Proof It suffices to prove this when $X = \operatorname{Spec} \mathbf{Z}[x,y]$, $Y = \operatorname{Spec} \mathbf{Z}$, $f = x$, and $g = y$. There is a unique $z \in \mathbf{Z}[x,y]$ such that $(x+y)^p - x^p - y^p = pz$. Then $(x+y)^{p-1}(dx+dy) - x^{p-1}dx - y^{p-1}dy = dz$. □

The lemma implies that the map $D\colon \mathcal{O}_X \to F_{X*}\mathcal{H}^1(\Omega^{\cdot}_{X/Y})$ sending f to the cohomology class of $F_{X*}(f^{p-1}df)$ is a homomorphism of abelian sheaves. For $m \in \mathcal{M}_X$, let $\delta(m)$ be the class of $F_{X*}(d\log m)$ in $F_{X*}\mathcal{H}^1(\Omega^{\cdot}_{X/Y})$. Then δ defines a homomorphism of sheaves of monoids $\mathcal{M}_X \to F_{X*}\mathcal{H}^1(\Omega^{\cdot}_{X/Y})$, which evidently annihilates $f^{-1}\mathcal{M}_Y$. Writing $[\omega]$ for the cohomology class of $\omega \in \Omega^1_{X/Y}$, we have:

$$\begin{aligned} D(\alpha_X(m)) &= [F_{X*}(\alpha_X(m)^{p-1}d\alpha_X(m))] \\ &= [F_{X*}((\alpha_X(m))^p\, d\log m] \\ &= \alpha_X(m)[F_{X*}(d\log m)] \\ &= \alpha_X(m)\delta(m). \end{aligned}$$

Thus (D,δ) is a log derivation of X/Y with values in $F_{X*}\mathcal{H}^1(\Omega^{\cdot}_{X/Y})$ (and in particular D automatically satisfies the Leibniz rule), by Proposition IV.1.2.9. The universal property of $\Omega^1_{X/Y}$ then produces the desired $\mathcal{O}_X$-linear map $\sigma^1_{X/Y}$. The compatibility of $\sigma_{X/Y}$ with the filtrations associated to coherent sheaves of ideals and relatively coherent sheaves of faces follows immediately from the definitions. □

The following result computes the cohomology sheaves of the de Rham complex of a smooth morphism of log schemes in characteristic p. Recall from Example IV.3.3.9 that the Frobenius morphism F_X has a canonical factorization $F_X = \pi_{X/Y} \circ F_{X/Y}$, where $F_{X/Y}$ is the exact relative Frobenius morphism, and that the natural maps $\pi^*_{X/Y}(\Omega^j_{X/Y}) \to \Omega^j_{X'/Y}$ are isomorphisms. It follows that the differentials of the complex $F_{X/Y*}(\Omega^{\cdot}_{X/Y})$ are $\mathcal{O}_{X'}$-linear and that its cohomology sheaves are $\mathcal{O}_{X'}$-modules. The maps $C^{-1}_{X/Y}$ in the following theorem are the logarithmic versions of the *(inverse) Cartier isomorphism.*

Theorem 4.1.3. *Let $f\colon X \to Y$ be a smooth morphism of fine saturated idealized log schemes in characteristic p, and let $F_{X/Y}\colon X \to X'$ be the exact relative Frobenius morphism (Example IV.3.3.9).*

1. *The maps*

$$C^{-1}_{X/Y}\colon \Omega^j_{X'/Y} \to F_{X/Y*}\mathcal{H}^j(\Omega^{\cdot}_{X/Y})$$

   induced by adjunction from the morphisms $\sigma^{\cdot}_{X/Y}$ defined in Proposition 4.1.1 are isomorphisms.

2. *If $\mathcal{F}$ is a relatively coherent sheaf of faces in $\mathcal{M}_X$ and $\mathcal{F}'$ is the corresponding sheaf in $\mathcal{M}_{X'}$, the maps $C^{-1}_{X/Y}$ induce isomorphisms*

$$L^i(\mathcal{F}')\Omega^j_{X'/Y} \to F_{X/Y*}\mathcal{H}^j(L^i(\mathcal{F})\Omega^{\cdot}_{X/Y}).$$

Proof Recall from Proposition IV.3.3.10 that formation of the relative Frobenius morphism $F_{X/Y}$ is compatible with strict étale localization. Formation of the filtered complexes $L^{\cdot}(\mathcal{F})\Omega^{\cdot}_{X/Y}$ and $L^{\cdot}(\mathcal{F}')\Omega^{\cdot}_{X'/Y}$ is also compatible with strict étale localization, and the differentials of the complex $F_{X/Y*}(L^{i}(\mathcal{F})\Omega^{\cdot}_{X/Y})$ are $\mathcal{O}_{X'}$-linear. It follows that formation of its cohomology sheaves is also compatible with strict étale localization. Thus the theorem can be checked locally on X, and furthermore its validity is preserved by étale localization: if it is true for a morphism $f\colon X \to Y$, and if $g\colon \tilde{X} \to X$ is strict and étale, then the theorem also holds for $\tilde{f} := f \circ g$.

Proceeding locally in an étale neighborhood of some point x in X, we may assume that the morphism f admits a chart (a, θ, b) as in Theorem IV. 3.3.1 (or Variant IV.3.3.5 in the idealized case). Since $X \to Y_\theta$ is strict and étale, we are reduced to proving the result when $X = Y_\theta$, and we may assume that Y is affine, say $Y = \operatorname{Spec}(P \to A)$. Since b is exact at x, the homomorphism $\overline{Q} \to \overline{\mathcal{M}}_{X,x}$ is an isomorphism and, since X is saturated, it follows that Q is also saturated. Furthermore, the order of the torsion subgroup of the cokernel of θ^{gp} is invertible in $k(x)$, and hence this cokernel is p-torsion free.

First suppose that $Y = \mathsf{A}_P$, so $X = \mathsf{A}_Q$. Then Proposition 2.2.5 implies statement (1) of the theorem and Proposition 2.3.17 implies statement (2). Let us review and strengthen the arguments, beginning with the simpler statement (1). Thanks to Proposition 2.2.5, we can identify $F^{\sharp}_{X/Y}$ with the inclusion $R[\tilde{Q}] \to R[Q]$, and we have an exact sequence of filtered complexes

$$0 \to R[\tilde{Q}] \otimes \Lambda^{\cdot}(Q^{\mathrm{gp}}/P^{\mathrm{gp}}) \to R[Q] \otimes \Lambda^{\cdot}(Q^{\mathrm{gp}}/P^{\mathrm{gp}}) \to C^{\cdot} \to 0.$$

We showed that the differentials of the complex $R[\tilde{Q}] \otimes \Lambda^{\cdot}(Q^{\mathrm{gp}}/P^{\mathrm{gp}})$ vanish and that the quotient $C^{\cdot}$ is acyclic. To prove the theorem, it will suffice to prove that the same remains true after tensoring with any homomorphism $R[P] \to A$, since formation of the exact relative Frobenius morphism and of the (filtered) de Rham complex is compatible with base change.

First note that, since $\tilde{Q}$ is an exact submonoid of Q, the map $R[\tilde{Q}] \to R[Q]$ is split as a sequence of $R[\tilde{Q}]$-modules, by Proposition I.4.2.7. Namely, $Q \setminus \tilde{Q}$ is stable under the action of $\tilde{Q}$ and, for each $q \in Q$, we let t_q be the identity of $R[Q]$ if $q \in \tilde{Q}$ and the zero map otherwise. Then t defines an $R[\tilde{Q}]$-linear splitting, hence also an $R[P]$-linear splitting, of the inclusion $R[\tilde{Q}] \to R[Q]$. It follows that the sequence remains exact after tensoring with A. Next recall that if $q \in Q \setminus \tilde{Q}$, its image $\pi(q)$ in $R \otimes Q^{g}/P^{\mathrm{gp}}$ spans a direct summand of $R \otimes Q^{\mathrm{gp}}/P^{\mathrm{gp}}$. Choose a homomorphism $s_{\pi(q)}\colon R \otimes Q^{\mathrm{gp}}/P^{\mathrm{gp}} \to R$ sending $\pi(q)$ to 1, and recall that interior multiplication by $s_{\pi(q)}$ defines a homotopy operator ξ_q on the degree-q part of the complex $C^{\cdot}$ with $d\xi_q + \xi_q d = \mathrm{id}_q$. Then $\xi :=$ $\sum\{\xi_q \in Q \setminus \tilde{Q}\}$ defines a homotopy between $\mathrm{id}_{C^{\cdot}}$ and 0. Since $\xi_{p+q} = \xi_q$ for

$p \in P$, in fact ξ is $R[P]$-linear and induces a homotopy operator on $A \otimes_{R[P]} C^{\cdot}$. Since the differential of $C^{\cdot}$ is also $R[P]$-linear, it follows that $d\xi + \xi d$ is again the identity, and so the complex $A \otimes_{R[P]} C^{\cdot}$ is acyclic.

Now suppose that F is a face of Q containing P. In this case we have a strict exact sequence of filtered complexes

$$0 \to (\Omega^{\cdot}_{\tilde{Q}/P}, L^{\cdot}(\tilde{F})) \to (\Omega^{\cdot}_{Q/P}, L^{\cdot}(F)) \to (C^{\cdot}, L^{\cdot}(F)) \to 0.$$

By Variant 2.3.7, each inclusion $L^i(F)\Omega^{\cdot}_{Q/P} \to \Omega^{\cdot}_{Q/P}$ is split as a sequence of $R[P]$-modules, with splitting s_q in degree q induced from a splitting of the image of $\mathbf{F}_p \otimes \langle F, q\rangle^{gp}$ in $\mathbf{F}_p \otimes Q^{gp}$. Since the splittings s and t are defined degree by degree, they are compatible, i.e., the corresponding endomorphisms of $\Omega^j_{Q/P}$ commute. It follows that $L^i(\tilde{F})\Omega^j_{\tilde{Q}/P}$ is an $R[P]$-linear direct summand of $\Omega^j_{\tilde{Q}/P}$ and of $L^i(F)\Omega^j Q/P$ and that the sequence above remains strictly exact when tensored over $R[P]$ with A. Since the $R[P]$-linear homotopy operators ξ constructed in the previous paragraph preserve the filtrations, the filtered complex $(C^{\cdot}, L^{\cdot}(F)$ is filtered acyclic. Then the map $(\Omega^{\cdot}_{\tilde{Q}/P}, L^{\cdot}(\tilde{F})) \to (\Omega^{\cdot}_{Q/P}, L^{\cdot}(F))$ is a filtered quasi-isomorphism, and the proof is complete. □

In positive characteristic p, the sheaf $T_{X/Y}$ of derivations is not just a Lie algebra, but also a restricted Lie algebra: the pth iterate of a derivation is again a derivation. We shall see that this is also true for logarithmic derivations and shall relate the pth-power operation on $T_{X/Y}$ to the Cartier isomorphism.

Theorem 4.1.4. *Let $f\colon X \to Y$ be a morphism of log schemes in characteristic p. Then $T_{X/Y}$ has the structure of a restricted Lie algebra [43, V§7], over $F^{-1}_{X/Y}(\mathcal{O}_X)$, with pth-power operator defined by*

$$(D, \delta)^{(p)} = (D^p, F^*_X \circ \delta + D^{p-1} \circ \delta). \tag{4.1.1}$$

If ∂ is any local section of $T_{X/Y}$ and a is any local section of $\mathcal{O}_X$,

$$(a\partial)^{(p)} = a^p \partial^{(p)} + (a\partial)^{p-1}(a)\partial = a^p \partial^{(p)} - a\partial^{p-1}(a^{p-1})\partial. \tag{4.1.2}$$

Proof Our first task is to show that the formula (4.1.1) for $(D, \delta)^{(p)}$ does define a log derivation. The proof uses the following formulas, valid in any characteristic.[4]

Lemma 4.1.5. *Let $f\colon X \to Y$ be a morphism of coherent log schemes, and let (D, δ) be an element of $T_{X/Y} \cong \mathcal{D}er_{X/Y}(\mathcal{O}_X)$. Let $\delta_1 := \delta$ and, for $n > 1$ define $\delta_n\colon \mathcal{M}_X \to \mathcal{O}_X$ inductively by*

$$\delta_n(m) := \delta(m)\delta_{n-1}(m) + D \circ \delta_{n-1}(m). \tag{4.1.3}$$

[4] This lemma was made possible by help from Hendrik Lenstra and the marvelous book [72].

Then the following formulas hold for all n.

$$D^n(\alpha_X(m)) = \alpha_X(m)\delta_n(m). \tag{4.1.4}$$

$$\delta_n = \sum_I c(I) \prod_j D^{I_j - 1} \circ \delta \tag{4.1.5}$$

where I ranges over the partitions $(I_1 \geq I_2 \geq \cdots)$ of the number n and where $c(I)$ is the number of partitions $\{s_1, s_2, \ldots\}$ of the set $\{1, \cdots, n\}$ such that $|s_i| = |I_i|$ for all i.

If n is a prime number p, then

$$\delta_n \equiv F_X^* \circ \delta + D^{p-1} \circ \delta \pmod{p}. \tag{4.1.6}$$

Proof We prove formula (4.1.4) by induction on n. It is true for $n = 1$ because (D, δ) is a logarithmic derivation. If it holds for n, then

$$\begin{aligned} D^{n+1}(\alpha_X(m)) &= D(\alpha_X(m)\delta_n(m)) \\ &= D(\alpha_X(m))\delta_n(m) + \alpha_X(m)D(\delta_n(m)) \\ &= \alpha_X(m)\delta(m)\delta_n(m) + \alpha_X(m)D(\delta_n(m)) \\ &= \alpha_X(m)\delta_{n+1}(m). \end{aligned}$$

Next we claim that, for each n,

$$\delta_n = \sum_{\pi \in P_n} \prod_{s \in \pi} D^{|s|-1} \circ \delta, \tag{4.1.7}$$

where P_n is the set of partitions of the set $\{1, \ldots, n\}$. This is trivial for $n = 1$ and we proceed by induction on n. For each $\pi \in P_n$, let π^* be the partition of $\{1, \ldots, n+1\}$ obtained by adjoining $\{n+1\}$ to π, and for each pair (s, π) with $\pi \in P_n$ and $s \in \pi$, let π_s be the partition of $\{1, \ldots, n+1\}$ obtained by adding $n+1$ to s. Let $P_n^* := \{\pi^* : \pi \in P_n\}$ and $P_\pi^* := \{\pi_s : s \in \pi\}$. In this way we obtain all the partitions of $\{1, \ldots, n+1\}$, and so P_{n+1} can be written as a disjoint union of sets

$$P_{n+1} = P_n^* \bigsqcup \{P_\pi^* : \pi \in P_n\}.$$

By the definition of δ_{n+1}, the induction hypothesis, and the product rule,

$$\begin{aligned}
\delta_{n+1} &= \delta \cdot \delta_n + D \circ \delta_n \\
&= \delta \cdot \sum_{\pi \in P_n} \prod_{s \in \pi} D^{|s|-1} \circ \delta + D \circ \sum_{\pi \in P_n} \prod_{s \in \pi} D^{|s|-1} \circ \delta \\
&= \sum_{\pi \in P_n} \prod_{s \in \pi} \delta D^{|s|-1} \circ \delta + \sum_{\pi \in P_n} \sum_{s' \in \pi \setminus \{s\}} \prod_{s \in \pi} (D^{|s|} \circ \delta)(D^{|s'|-1} \circ \delta) \\
&= \sum_{\pi \in P_n} \prod_{t \in \pi^*} D^{|t|-1} \circ \delta + \sum_{s \in \pi \in P_n} \prod_{t \in \pi_s} D^{|t|-1} \circ \delta \\
&= \sum_{\pi \in P_{n+1}} \prod_{t \in \pi} D^{|t|-1} \circ \delta,
\end{aligned}$$

as required.

We can now easily deduce formula (4.1.5). If π is any element of P_n and $|\pi| = r$, choose an ordering $(s_1, s_2, \ldots, s_r)$ of π with $|s_1| \geq |s_2| \geq \ldots \geq |s_r|$, and let $I(\pi) =: (|s_1|, |s_2|, \ldots, |s_r|)$. Then $I(\pi)$ is independent of the chosen ordering of of π, and $\pi \mapsto I(\pi)$ is a function from P_n to the set of finite sequences I of positive integers. Its (nonempty) fibers are exactly the orbits of P_n under the natural action of the symmetric group S_n. For each sequence I, let $c(I)$ be the cardinality of the fiber of I. Then the formula (4.1.7) reduces to formula (4.1.5).

Note that the cyclic group $\mathbf{Z}/n\mathbf{Z}$ acts on P_n through its inclusion in S_n; it is clear that the only elements of P_n fixed under this action are the two trivial partitions, with n elements and with one element, respectively. In particular, if n is a prime number p, all the other orbits have cardinality divisible by p. Formula (4.1.6) follows. □

Now we return to the proof of Theorem 4.1.4. In characteristic $p > 0$, let

$$\delta^{(p)}(m) := \delta(m)^p + D^{p-1}(\delta(m)) = (F_X^* \circ \delta + D^{p-1} \circ \delta)(m).$$

Then $\delta^{(p)} \colon \mathcal{M}_X \to \mathcal{O}_X$ is a homomorphism of monoids, and formulas (4.1.4) and (4.1.6) imply that, for all m,

$$D^p(\alpha_X(m)) = \alpha_X(m)\delta^{(p)}(m) = \alpha_X(m)\left(\delta(m)^p + D^{p-1}(\delta(m))\right) \tag{4.1.8}$$

and so $(D^p, \delta^{(p)})$ is a logarithmic derivation.

Next we prove equation (4.1.2), which we will deduce from the following general result.

Lemma 4.1.6. *Let D be a derivation of an $\mathbf{F}_p$-algebra A. Then, for any elements f, g, and h of A, the following equalities are satisfied.*

$$(fD)^{p-1}(fh) = f^p D^{p-1}(h) + (fD)^{p-1}(f)h, \tag{4.1.9}$$

$$(fD)^p(g) = f^p D^p(g) + (fD)^{p-1}(f)D(g), \tag{4.1.10}$$

$$(fD)^{p-1}(fh) = f^p D^{p-1}(h) - fD^{p-1}(f^{p-1})h, \tag{4.1.11}$$

$$(fD)^p(g) = f^p D^p(g) - fD^{p-1}(f^{p-1})D(g), \tag{4.1.12}$$

$$(fD)^{p-1}(f) = -fD^{p-1}(f^{p-1}). \tag{4.1.13}$$

Proof Note that (4.1.9) implies (4.1.12), by setting $h = D(g)$. Furthermore, (4.1.11) implies (4.1.13) (take $h = 1$), and (4.1.9) follows from (4.1.13) and (4.1.11). Equation (4.1.10) is proved in [35, Lemma 2] and (4.1.12) is proved in [52, 5.3.0] (although there is a sign error in the formula there). We will use a universal algebra argument due to G. Bergman to deduce (4.1.11) from (4.1.12) (which is also proved by a universal algebra argument).

Lemma 4.1.7. *Let C denote the category whose objects are commutative R-algebras equipped with a derivation D and two elements a, b and whose morphisms are homomorphisms compatible with (D, a, b). The category C has an initial element whose underlying ring A is the polynomial algebra over R in the variables $X_0, X_1, \ldots, Y_0, Y_1, \ldots$, and where $a = X_0$, $b = Y_0$, and $D(X_i) = X_{i+1}$, $D(Y_i) = Y_{i+1}$ for all i.*

Proof Since A is a polynomial ring on the variables X_i and Y_i it certainly admits a derivation as described. If (A', D, a', b') is an object of C, there is a unique R-algebra homomorphism $\theta \colon A \to A'$ sending X_i to $D^i(a')$ and Y_i to $D^i(b')$. It follows that $D(\theta(a)) = \theta(D(a))$ for every $a \in A$ since this equation holds for each generator of A. □

To prove Lemma 4.1.6, we apply the already known equation (4.1.12) to the elements $f = X_0$ and $g = Y_0$ in the algebra A to find

$$(X_0 D)^p(Y_0) = X_0^p D^p(Y_0) - X_0 D^{p-1}(X_0^{p-1})D(Y_0).$$

Since $Y_1 = DY_0$, we can write this as

$$(X_0 D)^{p-1}(X_0 Y_1) = X_0^p D^{p-1}(Y_1) - X_0 D^{p-1}(X_0^{p-1})Y_1.$$

Now consider the homomorphism $\sigma \colon A \to A$ sending X_0 to X_0 and Y_0 to Y_1. It is clear from the construction that this map is an injection, compatible with D, and hence the same equation holds with Y_0 in place of Y_1:

$$(X_0 D)^{p-1}(X_0 Y_0) = X_0^p D^{p-1}(Y_0) - X_0 D^{p-1}(X_0^{p-1})Y_0.$$

Since (A, X_0, Y_0, D) is universal, this equation implies that (4.1.11) holds in any $\mathbf{F}_p$-algebra. As we have observed, (4.1.13) and (4.1.9) follow. □

Now we can prove formula (4.1.2) of Theorem 4.1.4. If $\partial = (D, \delta)$ is a log derivation of (A, α) and $a \in A$ then, by the definition of $\partial^{(p)}$ and equations (4.1.10) and (4.1.9) of Lemma 4.1.6, with $f = a$ and $h = \delta$:

$$\begin{aligned}(a\partial)^{(p)} &:= ((aD)^p, (a\delta)^p + (aD)^{p-1}) \circ a\delta) \\ &= (a^p D^p + (aD)^{p-1}(a)D, a^p\delta^p + a^p D^{p-1} \circ \delta + (aD)^{p-1}(a)\delta) \\ &= a^p(D^p, \delta^p + D^{p-1} \circ \delta) + (aD)^{p-1}(a)(D, \delta) \\ &= a^p\partial^{(p)} + (aD)^{p-1}(a)\partial.\end{aligned}$$

Finally, by equation (4.1.13) of Lemma 4.1.6, this last equation can be rewritten as

$$(a\partial)^{(p)}a^p\partial^{(p)} - aD^{p-1}(a^{p-1})\partial.$$

It remains to check that the operation $\partial \mapsto \partial^{(p)}$ satisfies the definition of a restricted Lie algebra:

$$(a\partial)^{(p)} = a^p\partial^{(p)} \quad \text{for } a \in F^{-1}_{X/Y}(\mathcal{O}_X) \text{ and } \partial \in T_{X/Y}, \tag{4.1.14a}$$

$$\mathrm{ad}_{\partial^{(p)}} = \mathrm{ad}^p_\partial \quad \text{for } \partial \in T_{X/Y}, \tag{4.1.14b}$$

$$(\partial_1 + \partial_2)^{(p)} = \partial_1^{(p)} + \partial_2^{(p)} + \sum_{i=1}^{p-1} s_i(\partial_1, \partial_2) \quad \text{for } \partial_1, \partial_2 \in T_{X/Y}, \tag{4.1.14c}$$

where s_i is the ith universal Lie polynomial [43, V§7].

The first of these equations follows easily from the fact that the derivation D annihilates elements of $F^{-1}_{X/Y}(\mathcal{O}_X)$. To prove formulas (4.1.14b) and (4.1.14c), we write $\partial_i = (D_i, \delta_i)$ and, with some abusive of notation,

$$\begin{aligned}[\delta_1, \delta_2] &:= D_1 \circ \delta_2 - D_2 \circ \delta_1, \\ \delta_i^{(p)} &:= F_X^* \circ \delta_i + D^{p-1} \circ \delta_i,\end{aligned}$$

so that

$$\begin{aligned}[\partial_1, \partial_2] &= ([D_1, D_2], [\delta_1, \delta_2]), \\ \partial_i^{(p)} &= (D_i^p, \delta_i^{(p)}).\end{aligned}$$

Continuing with these notational conventions we have

$$\mathrm{ad}^n_{\partial_1}(\partial_2) = (\mathrm{ad}^n_{D_1}(D_2), \mathrm{ad}^n_{\delta_1}(\delta_2)),$$

for every $n > 0$. Thus, for every $m \in \mathcal{M}_X$,

$$(\mathrm{ad}^n_{D_1}(D_2))(\alpha(m)) = \alpha(m)\,\mathrm{ad}^n_{\delta_1}(\delta_2)(m)$$

and, for each Lie polynomial s_i,

$$s_i(D_1, D_2)(\alpha(m)) = \alpha(m)s_i(\delta_1, \delta_2)(m).$$

Let us begin with the proof of equation (4.1.14c). The equation (4.1.14c) is automatic for any two elements in any associative algebra in characteristic p [43, V]. In particular it holds for the endomorphisms D_1 and D_2 of the sheaf $\mathcal{O}_X$. Thus it will suffice to prove that

$$(\delta_1 + \delta_2)^{(p)}(m) = \delta_1^{(p)}(m) + \delta_2^{(p)}(m) + \sum_{i=1}^{p-1} s_i(\delta_1, \delta_2)(m). \tag{4.1.15}$$

We know that

$$(D_1 + D_2)^{(p)}(\alpha(m)) = D_1^{(p)}(\alpha(m)) + D_2^{(p)}(\alpha(m)) + \sum_{i=1}^{p-1} s_i(D_1, D_2)(\alpha(m)),$$

so

$$\alpha(m)(\delta_1 + \delta_2)^{(p)}(m) = \alpha(m)\delta_1^{(p)}(m) + \alpha(m)\delta_2^{(p)}(m) + \alpha(m) \sum_{i=1}^{p-1} s_i(\delta_1, \delta_2)(m).$$

Thus the desired equation (4.1.14c) holds if $\alpha(m)$ is a nonzero divisor in A.

We argue similarly for equation (4.1.14b), which asserts that, for any log derivations ∂_1 and ∂_2,

$$\mathrm{ad}_{\partial_1^{(p)}}(\partial_2) = \mathrm{ad}^p_{\partial_1}(\partial_2).$$

This is true for D_1 and D_2, and it remains to prove that

$$\mathrm{ad}_{\delta_1^{(p)}}(\delta_2)(m) = \mathrm{ad}^p_{\delta_1}(\delta_2)(m) \tag{4.1.16}$$

for all $m \in \mathcal{M}_X$. We know that

$$\mathrm{ad}_{D_1^p}(D_2)(\alpha(m)) = \mathrm{ad}^p_{D_1}(D_2)(\alpha(m)),$$

i.e., that

$$\alpha(m)\,\mathrm{ad}_{\delta_1^{(p)}}(\delta_2)(m) = \alpha(m)\,\mathrm{ad}^p_{\delta_1}(\delta_2)(m)$$

so that (4.1.14b) holds when $\alpha(m)$ is a nonzero divisor.

We conclude that equations (4.1.14b) and (4.1.14c) hold if the underlying ring A is an integral domain. We shall deduce the general case by an argument from universal algebra, based on the construction in the following lemma.[5]

Lemma 4.1.8. *If R is a commutative ring and n is a positive integer, let C_n be the category whose objects are commutative R-algebras A equipped with a sequence of derivations $(D_1, \ldots, D_n)$, an element x of A, and a sequence $(x_1, \ldots, x_n)$ of elements of A such that $D_i x = x x_i$ for all i, and whose morphisms are ring homomorphisms preserving these data. Then C_n has an initial*

[5] I am indebted to G. Bergman for assistance with these techniques.

object whose underlying ring A is a polynomial ring over R. In particular, if R is an integral domain, so is A.

Proof Let $\{x_1, \ldots, x_n\}$ be a finite set of indeterminates and let $(M, \star, e)$ denote the free noncommutative monoid on this set. (Thus if $n = 1$, $(M, \star, e)$ is the monoid $(\mathbf{N}, +, 0)$.) Now let A be the polynomial ring over R on the elements X of M and let $x := e$. Since A is the polynomial ring on the variables M, there is a unique derivation D_i of A such that

$$D_i(x) = xx_i = x_i x \in A,$$
$$D_i(X) = x_i \star X \in M \subseteq A \quad \text{if } X \in M \setminus \{e\}.$$

(Note: $x_i \star X \in M$ is one of the free variables in the polynomial ring, but $x_i x$ is the product of two polynomials in A.)

Now suppose that $(A', D'_., x', x'_.)$ is another object of C_n. Since M is the free monoid generated by $(x_1, \ldots, x_n)$, there is a unique monoid homomorphism $\psi\colon M \to \mathrm{End}_R(A')$ such that $\psi(x_i) = D'_i$ for all i; necessarily $\psi(e) = \mathrm{id}_{A'}$. For each $X \in M \setminus \{e\}$, there is a unique pair (X', j) such that $X' \in M$ and $X = X' \star x_j$. Since A is the polynomial algebra generated by M, there is a unique R-algebra homomorphism $\theta\colon A \to A'$ such that $\theta(x) = x'$ and $\theta(X) = \psi(X')(x'_j)$ for all $X \in M \setminus \{e\}$. In particular $\theta(x_i) = x'_i$ for each i, since $x'_i = e$.

We claim that $\theta(D_i(a)) = D_i(\theta(a))$ for all $a \in A$. It suffices to check this equality for the generators of the R-algebra A, i.e., the elements of M. First of all,

$$\theta(D_i(x)) = \theta(x_i x) = \theta(x_i)\theta(x) = x'_i x' = D'_i(x') = D'_i(\theta(x)).$$

If $X \in M \setminus \{e\}$, write $X = X' \star x_j$. Then $D_i X = x_i \star X = x_i \star X' \star x_j$ and so $(D_i X)' = x_i \star X'$. Thus

$$\theta(D_i X) = \psi(x_i \star X')(x'_j) = \psi(x_i)(\psi(X')(x_j)) = D'_i(\theta(X)).$$

This completes the proof of the lemma. □

Let $(A, D_., x, x_.)$ be an object of the category C_n in Lemma 4.1.8 with $R = \mathbf{F}_p$, and let $\alpha\colon \mathbf{N} \to (A, \cdot)$ be the monoid homomorphism sending n to x^n and $\delta_i\colon \mathbf{N} \to (A, +)$ the homomorphism sending n to nx_i Then $D_i(\alpha(n)) = nx^{n-1}D_i(x) = x^n\delta_i(n)$, so $\partial_i := (D_i, \delta_i)$ is a derivation of the log ring (A, α). Let us apply this to the initial object $(A, D_., x, x_.)$ of C_n. Since A is an integral domain, formulas (4.1.15) and (4.1.16) are verified for (A, α), as we have seen. Now suppose that $\alpha'\colon Q' \to A'$ is any log ring equipped with a pair of log derivations ∂'_1, ∂'_2. Choose any $q' \in Q'$, and let $x' := \alpha'(q')$ and $x'_i := \delta'_i(q')$. Then $D'_i(x') = x'x'_i$, so there is a unique homomorphism $\theta^\sharp\colon A \to A'$ sending x to x' and each x_i to x'_i and such that $D'_i(\theta^\sharp(a)) = \theta^\sharp(D_i(a))$ for all $a \in A$.

Let $\theta^\flat \colon \mathbf{N} \to Q'$ be the homomorphism sending 1 to q', so that $(\theta^\sharp, \theta^\flat)$ is a homomorphism of log rings, and note that $\theta^\sharp(\delta_i(n)) = \delta_i(\theta^\flat(n)) = nx'_i$ for all i. We know that equations (4.1.15) and (4.1.16) hold in (A, α) with $q = 1$, and applying $\theta^\sharp$ we see that they also hold in A' with δ'_i in place of δ_i and q' in place of 1. Thus formulas (4.1.15) and (4.1.16) hold in general. □

Remark 4.1.9. If m is a unit of $\mathcal{M}_X$, then $u := \alpha(m)$ is a unit of $\mathcal{O}_X$, and $\delta(m) = u^{-1}D(u)$. Thus formula (4.1.8) says:

$$D^p(u) = u\left(u^{-p}(D(u))^p + D^{p-1}(u^{-1}D(u))\right).$$

Multiplying through by u^{p-1}, we get

$$u^{p-1}D^p(u) = D(u))^p + D^{p-1}(u^{p-1}D(u)),$$

a well-known formula attributed to Hochschild [70, Lemma 2].

The following result relates the restricted Lie algebra structure on $T_{X/Y}$ to the mapping $\sigma^1_{X/Y}$ defined in Proposition 4.1.1 and will allow us to define a logarithmic version of the Cartier operator.

Proposition 4.1.10. *Let $f \colon X \to Y$ be a smooth morphism of fine log schemes in characteristic $p > 0$. Then there is a unique $\mathcal{O}_X$-bilinear pairing*

$$C \colon T_{X/Y} \times F_{X*}(\mathcal{Z}^1_{X/Y}) \to F_{X*}(\mathcal{O}_X),$$

$$C(\partial, F_{X*}\omega) = \langle \partial^{(p)}, \omega \rangle - \partial^{p-1}\langle \partial, \omega \rangle.$$

If f is any local section of $\mathcal{O}_X$,

$$C(\partial, F_{X*}(df)) = 0,$$

and if m is any local section of $\mathcal{M}_X$,

$$C(\partial, F_{X*}(d\log m)) = F_X^*\langle \partial, d\log m \rangle.$$

Proof For each section $\partial = (D, \delta)$ of $T_{X/Y}$ we have an $\mathcal{O}_X$-linear map

$$C_\partial \colon F_{X*}(\mathcal{Z}^1_{X/Y}) \to F_{X*}(\mathcal{O}_X) : \omega \mapsto \langle \partial^{(p)}, \omega \rangle - (D^{p-1}\langle \partial, \omega \rangle).$$

If $f \in \mathcal{O}_X$,

$$\begin{aligned} C_\partial(F_{X*}(df)) &= \langle \partial^{(p)}, df \rangle - D^{p-1}\langle \partial, df \rangle \\ &= D^p f - D^{p-1}(Df) \\ &= 0. \end{aligned}$$

If $m \in \mathcal{M}_X$,

$$\begin{aligned} C_\partial(F_{X*}(dlog\, m)) &= \langle \partial^{(p)}, dlog\, m\rangle - D^{p-1}\langle \partial, dlog\, m\rangle \\ &= \delta^{(p)}(m) - D^{p-1}\delta(m) \\ &= F_X^*(\delta(m)) + D^{p-1}\delta(m) - D^{p-1}(\delta(m)) \\ &= F_X^*(\delta(m)) \\ &= F_X^*\langle \partial, dlog\, m\rangle. \end{aligned}$$

In fact C_∂ is also $\mathcal{O}_Y$-linear, and so can be viewed as an $\mathcal{O}_{X'}$-linear homomorphism

$$F_{X/Y*}(\mathcal{Z}^1_{X/Y}) \to F_{X/Y*}(\mathcal{O}_X).$$

Since it annihilates the exact forms, it factors through a map

$$\overline{C}_\partial \colon F_{X/Y*}(\mathcal{H}^1_{X/Y}) \to F_{X/Y*}(\mathcal{O}_X).$$

We claim that for any section ω of $\Omega^1_{X/Y}$,

$$\overline{C}_\partial(\sigma^1_{X/Y}(\pi^*_{X/Y}(\omega))) = F^*_{X/Y}\langle \partial, \omega\rangle. \tag{4.1.17}$$

Indeed, we have already seen that this is the case if $\omega = dlog\, m$ for some section m of $\mathcal{M}_X$, and the general case follows because $\Omega^1_{X/Y}$ is locally generated as a sheaf of $\mathcal{O}_X$-modules by such sections, as we saw in Proposition IV.1.2.11. This formula also proves that the pairing C is linear in ∂ and thus completes the proof of the proposition. □

Remark 4.1.11. The pairing defined in Proposition 4.1.10 induces an $\mathcal{O}_{X'}$-linear map

$$F_{X/Y*}(\mathcal{Z}^1_{X/Y}) \to \mathrm{Hom}_{\mathcal{O}_{X'}}(\pi^*_{X/Y}(T_{X/Y}), F_{X/Y*}(\mathcal{O}_X)),$$

or equivalently, a map

$$C \colon F_{X/Y*}(\mathcal{Z}^1_{X/Y}) \to \Omega^1_{X'/Y} \otimes F_{X/Y*}(\mathcal{O}_X).$$

This is the logarithmic version of the classical Cartier operator [70], and the formula (4.1.17) shows that it is in essence inverse to $\sigma^1_{X/Y}$.

4.2 Comparison theorems

Our goal in this section is to establish algebraic analogs of the analytic results in Section V.3 and to compare algebraic and analytic de Rham cohomology. We shall study the de Rham complex of a fine and smooth (possibly idealized) log scheme X over a field k of characteristic zero, as well as the complexes arising from coherent sheaves of ideals and relatively coherent sheaves of faces. The array of possibilites obtainable by combining these is too large for us to treat definitively here, and we limit ourselves to some key cases.

Our first result is an algebraic version of the analytic "idealized Poincaré lemma" (3.1.1).

Theorem 4.2.1. *Let X/k be a fine and smooth idealized log scheme over k, let $\mathcal{J}$ and $\mathcal{I}$ be coherent sheaves of ideals in $\mathcal{M}_X$ such that $\mathcal{I} \subseteq \mathcal{J} \cup \mathcal{K}_X$, and let $\mathcal{F}$ be a relatively coherent sheaf of faces in $\mathcal{M}_X$. Let $X_{\mathcal{J}}$ denote the closed subscheme of X defined by $\mathcal{J}$ and let $\hat{X}$ denote the formal completion of X along $X_{\mathcal{J}}$.*

1. *The filtered complex $(\mathcal{I}\Omega^{\cdot}_{\hat{X}/k}, L^{\cdot}(\mathcal{F}))$ is acyclic.*
2. *The map $(\Omega^{\cdot}_{\hat{X}/k}, L^{\cdot}(\mathcal{F})) \to (\Omega^{\cdot}_{X_{\mathcal{J}}/k}, L^{\cdot}(\mathcal{F}))$ is a filtered quasi-isomorphism.*
3. *Let $\sqrt{\mathcal{J}}$ be the radical $\mathcal{J}$. The natural maps*

$$(\Omega^{\cdot}_{X_{\mathcal{J}}/k}, L^{\cdot}(\mathcal{F})) \to (\Omega^{\cdot}_{X_{\sqrt{\mathcal{J}}}/k}, L^{\cdot}(\mathcal{F})),$$
$$(\mathcal{J}\Omega^{\cdot}_{X/k}, L^{\cdot}(\mathcal{F})) \to (\sqrt{\mathcal{J}}\Omega^{\cdot}_{X/k}, L^{\cdot}(\mathcal{F}))$$

are filtered quasi-isomorphisms.

Proof This assertion is local on X, and can be verified at the stalks at each geometric point $\overline{x}$. Thus by Variant 3.3.5 we may assume that X admits a strict and étale morphism $f : X \to \mathsf{A}_{(Q,K)}$, that $\mathcal{J}$ and $\mathcal{I}$ come from ideals J and K of Q and that $\mathcal{F}$ comes from a face F of Q.

Using noetherian induction on the ideal J, we may assume that statement (1) is true for every ideal properly containing J. Since the formal completion of X along J is the same as the formal completion along $\sqrt{J}$, statement (1) is true for J if $J \subsetneq \sqrt{J}$, so we may assume without loss of generality that $J = \sqrt{J}$. Suppose that $J = J_1 \cap J_2$, where J_1 and J_2 properly contain J, and let $J_3 := J_1 + J_2$. Then (1) is true for J_1, J_2, and J_3. If $\mathcal{E}$ is a coherent sheaf on X, then there is an exact sequence

$$0 \to \hat{\mathcal{E}} \to \hat{\mathcal{E}}_1 \oplus \hat{\mathcal{E}}_2 \to \hat{\mathcal{E}}_3 \to 0,$$

where $\hat{\mathcal{E}}_i$ is the formal completion of E along J_i. We deduce an analogous sequence:

$$0 \to (\mathcal{I}\Omega^{\cdot}_{\hat{X}}, L^{\cdot}(\mathcal{F})) \to (\mathcal{I}\Omega^{\cdot}_{\hat{X}_1}, L^{\cdot}(\mathcal{F})) \oplus (\mathcal{I}\Omega^{\cdot}_{\hat{X}_2}, L^{\cdot}(\mathcal{F})) \to (\mathcal{I}\Omega^{\cdot}_{\hat{X}_3}, L^{\cdot}(\mathcal{F})) \to 0.$$

Then the acyclicity of the latter two complexes implies that of the first. Since J is reduced, it is the intersection of a finite number of prime ideals, and by induction we are reduced to the case in which J is prime.

Assume that J is a prime ideal $\mathfrak{p}$, let $G := Q \setminus \mathfrak{p}$, choose a local homomorphism $h\colon Q/G \to \mathbf{N}$, and let $J^{(n)} := \{q \in Q : h(q) \geq n\}$. Then $J^{(1)} = \mathfrak{p}$, each $J^{(n)}$ is an ideal of Q, and $\mathfrak{p}J^{(n)} \subseteq J^{(n+1)}$. Furthermore, the topology on Q defined by $\{J^{(n)} : n \in \mathbf{N}\}$ is equivalent to the $\mathfrak{p}$-adic topology, by Corollary I.3.6.3. Extend h to a homogenous vector field $\partial\colon Q^{\mathrm{gp}} \to \mathbf{Z}$, let ξ denote interior multiplication by ∂, and let $\kappa := d\xi + \xi d$ be the Lie derivative with respect to ∂. Since $X \to \mathsf{A}_Q$ is strict and étale, the filtered complex $(I\Omega^{\cdot}_Q, L^{\cdot}(F))$ pulls back to the filtered complex $(\mathcal{I}\Omega^{\cdot}_{X/k}, L^{\cdot}(\mathcal{F}))$ on X, and we view ξ and κ as acting on A_Q and on X.

Lemma 4.2.2. *The map* $\xi\colon \Omega^1_Q \to k[Q]$ *defined in the previous paragraph sends* $\underline{\Omega}^1_Q$ *to* $\mathfrak{p}k[Q]$ *and* $\underline{\Omega}^1_{X/k}$ *to* $\mathcal{J}\mathcal{O}_X$. *Furthermore, each* $\mathcal{J}^{(n)}\Omega^j_{X/k}$ *is invariant under* κ, *and* κ *acts as multiplication by* n *on*

$$\mathrm{Gr}^{(n)}\,\Omega^j_{X/k} := \mathcal{J}^{(n)}\Omega^j_{X/k}/\mathcal{J}^{(n+1)}\Omega^j_{X/k}.$$

Proof The map $\xi\colon \Omega^1_Q \to k[Q]$ sends $e^q \otimes \omega$ to $e^q\partial(\omega)$. In other words, in degree q it is given by $\mathrm{id}\otimes\partial\colon \Omega^1_{Q,q} = k\otimes Q^{\mathrm{gp}} \to k\otimes\mathbf{Z}$. Its restriction to $\underline{\Omega}^1_{Q,q}$ is the restriction of ∂ to $k\otimes\langle q\rangle^{\mathrm{gp}}$, and thus vanishes if $q \in G$, i.e., if $q \notin \mathfrak{p}$. Thus ξ takes $\underline{\Omega}^1_Q$ to $k[\mathfrak{p}]$, and hence $\underline{\Omega}^1_{X/k}$ to $\mathcal{J}\mathcal{O}_X$. In particular, $\xi(da) \in \mathcal{J}\mathcal{O}_X$ if a is any local section of $\mathcal{O}_X$.

Suppose that m is a local section of $\mathcal{J}^{(n)}$ and ω is a local section of $\Omega^j_{X/k}$. Then

$$\kappa(\alpha_X(m)\omega) = \kappa(\alpha_X(m))\omega + \alpha_X(m)\kappa(\omega) = \alpha_X(m)\xi(dm) + \alpha_X(m)\kappa(\omega),$$

so $\mathcal{J}^{(n)}\Omega^j_{X/k}$ is invariant under κ. Moreover, if $a \in \mathcal{O}_X$ and $\omega \in \mathcal{J}^{(n)}\Omega^j_{X/k}$, then

$$\kappa(a\omega) = \kappa(a)\omega + a\kappa(\omega) \equiv a\kappa(\omega) \bmod \mathcal{J}^{(n+1)}\Omega^j_{X/k},$$

since $\kappa(a) = \xi(da) \in \mathcal{J}\mathcal{O}_X$. Thus κ acts $\mathcal{O}_X$-linearly on $\mathrm{Gr}^{(n)}\,\Omega^j_{X/k}$ and $k[Q]$-linearly on $\mathrm{Gr}^{(n)}\,\Omega^j_Q$. For any element $\sum a_q e^q$ of $k[Q]$, we have

$$\kappa\Big(\sum a_q e^q\Big) = \sum a_q\xi(e^q dq) = \sum a_q h(q)e^q,$$

and it follows that κ acts as multiplication by n on $\mathrm{Gr}^{(n)}\,k[Q]$. Since this action is linear, it also acts by multiplication by n on $\mathrm{Gr}^{(n)}\,\Omega^j_Q$ and on $\mathrm{Gr}^{(n)}\,\Omega^j_{X/k}$. □

Now let

$$Fil^n C^{\cdot} := \mathcal{I}\Omega^{\cdot}_{X/k} \cap L^{\cdot}\Omega^{\cdot}_{X/k} \cap J^{(n)}\Omega^{\cdot}_{X/k},$$

so that

$$\mathrm{Gr}^n_{Fil}\,C^{\cdot} \subseteq J^{(n)}\Omega^j_{X/k}/J^{(n+1)}\Omega^j_{X/k}.$$

It follows from the lemma that κ acts as multiplication by n on on $\mathrm{Gr}^n_{Fil}\,C^{\cdot}$.

Moreover, $\mathrm{Gr}^0_{Fil} C^{\cdot}$ vanishes, since $I \subsetneq \mathcal{J} \cup \mathcal{K}_X$. Thus κ acts as an isomorphism on $\mathrm{Gr}^n C^{\cdot}$ for all $n \geq 0$, and by induction we conclude that it also acts as an isomorphism on $C^{\cdot}/Fil^n C^{\cdot}$ for all n. Then it also acts as an isomorphism on the J-adic completion of $C^{\cdot}$. Since κ is homotopic to zero, this completion is consequently acyclic, proving statement (1).

We have an exact sequence of complexes

$$0 \to \mathcal{J}\Omega^{\cdot}_{\hat{X}/k} \to \Omega^{\cdot}_{\hat{X}/k} \to \Omega^{\cdot}_{X_{\mathcal{J}}/k} \to 0.$$

Statement (1) implies that the first term is acyclic, and statement (2) follows. Since the formal completions of X along $\mathcal{J}$ and along $\sqrt{\mathcal{J}}$ are the same, statement (3) follows from statement (2). □

Statement (2) of Theorem 4.2.1 shows that formal completions of de Rham complexes along logarithmically defined ideals have the same cohomology as their idealized restrictions. This fact allows one to translate many of the results in algebraic de Rham cohomology, as developed by Hartshorne [33], into the context of the de Rham cohomology of idealized log schemes. Here is one particularly useful such result.

Corollary 4.2.3. *Let $f: X' \to X$ be a proper morphism of smooth idealized log schemes over k. Suppose that $\mathcal{J}$ is a coherent sheaf of ideals in $\mathcal{M}_X$ such that f induces an isomorphism over $X \setminus X_{\mathcal{J}}$, and let $X'_{\mathcal{J}} := X' \times_X X_{\mathcal{J}}$. Then the natural maps fit into a distinguished triangle*

$$\Omega^{\cdot}_{X/k} \to \Omega^{\cdot}_{X_{\mathcal{J}}} \oplus Rf_*\Omega^{\cdot}_{X'/k} \to Rf_*\Omega^{\cdot}_{X'_{\mathcal{J}}} \xrightarrow{+},$$

and similarly for the cohomology of the subcomplexes defined by coherent sheaves of faces in X.

Proof Statement (2) of Theorem 4.2.1 allows us to replace the complexes computing the de Rham cohomology of $X_{\mathcal{J}}$ and $X'_{\mathcal{J}}$ by the formal completions of the complexes of X and X' along $X_{\mathcal{J}}$ and $X'_{\mathcal{J}}$ respectively. Then the argument given in [33, 4.4] applies to give the desired result. □

The following result allows one to weaken the saturation hypothesis in some circumstances.

Theorem 4.2.4. *Let X/k be a fine and smooth idealized log scheme such that the groups $\overline{\mathcal{M}}^{\mathrm{gp}}_X$ are torsion free, and let $\eta: X^{\mathrm{sat}} \to X$ be the saturation of X. Then the natural map $\Omega^{\cdot}_{X/k} \to \eta_*(\Omega^{\cdot}_{X^{\mathrm{sat}}/k})$ is a quasi-isomorphism.*

Proof The statement can be verified étale locally on X, so we may assume that there is a strict étale map $f: X \to \mathsf{A}_{Q,K}$. Replacing Q by a localization if necessary, we may assume that $X \to \mathrm{Spec}(Q, K)$ is surjective. The hypothesis on $\mathcal{M}_X$ implies that, for every face F of (Q, K), the group Q^{gp}/F^{gp} is torsion free. We shall prove by noetherian induction that, for every ideal J of Q contining K, the idealized log subscheme X_J of X defined by J satisfies the theorem. We may also use induction on the dimension of Q to assume that the theorem is true for all monoids Q' such that $\dim Q' < \dim Q$. Using a dévissage as in the proof of Theorem 4.2.1, we reduce to the case in which J is a prime ideal. Let $F := Q \setminus \mathfrak{p}$, a face of Q, and let F' be the face of Q^{sat} generated by F. The map $\mathbf{Q} \otimes Q^{gp}/F^{gp} \to \mathbf{Q} \otimes Q^{gp}/F'^{gp}$ is an isomorphism and, since Q^{gp}/F^{gp} is torsion free, it follows that $F^{gp} = F'^{gp}$ and hence that $F' = F^{sat}$.

Let Y be $\underline{X}_J$ with the log structure defined by $F \to \mathcal{O}_{X_J}$ and the empty idealized structure. Then Y/k is smooth and there is a morphism of idealized log schemes $X_J \to Y$. We have a map of filtered complexes

$$(\Omega^{\cdot}_{X_J/k}, L^{\cdot}(\mathcal{F})) \to \eta_*(\Omega^{\cdot}_{X_J^{sat}}, L^{\cdot}(\mathcal{F}^{sat})), \tag{4.2.1}$$

and, by Proposition 2.3.9, the map on the associated graded complexes is the map

$$\oplus_i \Omega^{\cdot}_Y[-i] \otimes \Lambda^i Q^g/F^{gp} \to \oplus_i \eta_*(\Omega^{\cdot}_{Y^{sat}}[-i]) \otimes \Lambda^i Q^g/F^{gp}$$

obtained from the saturation map $Y^{sat} \to Y$. Every face G of F is a face of (Q, K), and hence $F^{gp}/G^{gp} \subseteq Q^{gp}/G^{gp}$ is torsion free.

Assume J is a nonempty ideal of Q. Then F is a proper face, and the induction hypothesis on the dimension of Q implies that these maps are quasi-isomorphisms. It follows that the map of filtered complexes (4.2.1) is a filtered quasi-isomorphism, and hence that $\Omega^{\cdot}_{X_J/k} \to \eta_*(\Omega^{\cdot}_{X_J^{sat}/k})$ is a quasi-isomorphism.

It remains to prove that the theorem is true when J is empty, but now with the additional hypothesis that the theorem is true for every nonempty ideal of Q. Let $J' := \{q \in Q : q + Q^{sat} \subseteq Q\}$, a nonempty ideal because Q^{sat} is finitely generated as a Q-set, by Corollary I.2.2.5. The morphism $\eta \colon X^{sat} \to X$ is proper and an isomorphism outside $X_{J'}$ and hence by Corollary 4.2.3 there is a distinguished triangle

$$\Omega^{\cdot}_{X/k} \to \Omega^{\cdot}_{X_{J'}} \oplus \eta_* \Omega^{\cdot}_{X^{sat}/k} \to \eta_* \Omega^{\cdot}_{X_{J'}^{sat}} \xrightarrow{+} .$$

Since the map $\Omega^{\cdot}_{X_{J'}} \to \eta_* \Omega^{\cdot}_{X_{J'}^{sat}}$ is a quasi-isomorphism, the same is true of the map $\Omega^{\cdot}_{X/k} \to \eta_* \Omega^{\cdot}_{X^{sat}/k}$. □

Central to work of Grothendieck and Deligne on de Rham cohomology and Hodge theory is the fact that the de Rham complex with log poles along a divisor D with normal crossings can be used to calculate the cohomology of the complement of D. The following result is the logarithmic incarnation of this classical result. In this statement the idealized structure of X is trivial.

Theorem 4.2.5. *Let X/k be a fine saturated and smooth log scheme over k.*

1. *The natural map*
$$\Omega^{\cdot}_{X/\mathbf{C}} \to Rj_*\Omega^{\cdot}_{X^*/k}$$
is a quasi-isomorphism.
2. *More generally, suppose that $\mathcal{F}$ and $\mathcal{G}$ are relatively coherent sheaves of faces in $\mathcal{M}_X$ with $\mathcal{G} \subseteq \mathcal{F}$, and let $j_{\mathcal{G}}\colon X_{\mathcal{G}} \to X$ be the inclusion of the open set $X^*_{\mathcal{G}} := \{x : \mathcal{O}^*_X = \mathcal{G}_{X,x}\}$. Then the map*
$$(\Omega^{\cdot}_{X/k}, L^{\cdot}(\mathcal{F})) \to Rj_{\mathcal{G}*}(\Omega^{\cdot}_{X_{\mathcal{G}}/k}, L^{\cdot}(\mathcal{F}))$$
is a filtered quasi-isomorphism.
3. *If $\mathcal{F}$ is any relatively coherent sheaf of faces of $\mathcal{M}_X$, the map*
$$\underline{\Omega}^{\cdot}_{X/k}(\mathcal{F}) \to Rj_{\mathcal{F}*}(\underline{\Omega}^{\cdot}_{X_{\mathcal{F}}/k})$$
is a quasi-isomorphism.

Proof Since the morphism j is affine, in fact $Rj_*\Omega^{\cdot}_{X^*/k} \cong j_*\Omega^{\cdot}_{X^*/k}$, and the theorem reduces to the claim that the natural map $\Omega^{\cdot}_{X/\mathbf{C}} \to j_*\Omega^{\cdot}_{X^*/k}$ is a quasi-isomorphism. Since the sheaves in these complexes are torsion free and X^* is dense, the map is injective, so what must be shown is that the quotient $C^{\cdot}_X$ is acyclic. This statement can be verified étale locally, so we may assume that X is affine and that there exist a fine saturated monoid Q and a strict étale map $X \to \mathsf{A}_Q$. Then $X^* = f^{-1}(\mathsf{A}^*_Q)$ and $C^{\cdot}_X = f^*C^{\cdot}_Q$, where $C^{\cdot}_Q := \Omega^{\cdot}_{Q^{\mathrm{gp}}}/\Omega^{\cdot}_Q$. Since Q is saturated, the map $Q^*_{tors} \to Q^{\mathrm{gp}}_{tors}$ is an isomorphism and, as we saw in Corollary 2.2.4, this implies that $C^{\cdot}_Q$ is acyclic. This fact does not suffice to prove the acyclicity of its pullback to X, since its differentials are not $k[Q]$-linear. We will show that $C^{\cdot}_Q$ admits an exhaustive filtration whose associated graded terms are acyclic and whose boundary maps are linear. In the course of the proof, let us say that a $k[Q]$-linear subquotient of the complex $\Omega^{\cdot}_{Q^{\mathrm{gp}}}$ is "universally acyclic" if it is so after its pullback along any strict étale $X \to \mathsf{A}_Q$.

The basic method is similar to that used in the proof of Theorem 4.2.1. Let $h\colon Q \to \mathbf{N}$ be a homomorphism, inducing a homogeneous vector field $\partial\colon Q^{\mathrm{gp}} \to \mathbf{Z}$ and linear maps $\xi\colon \Omega^j_Q \to \Omega^{j-1}_Q$. Let $\mathfrak{p} := h^{-1}(\mathbf{Z}^+)$ and, for each $n \in \mathbf{N}$, let $I_n := \{x \in Q^{\mathrm{gp}} : h(x) \geq -n\}$. Each I_n is a fractional ideal of Q, $\mathfrak{p}I_n \subseteq I_{n-1}$, and $Q = \cup\{I_n : n \geq 0\}$. Consider the complex

$$Fil_n\Omega^{\cdot}_{Q^{\mathrm{gp}}} := \bigoplus_{x \in I_n} \Omega^{\cdot}_{Q^{\mathrm{gp}},x} \subseteq \Omega^{\cdot}_{Q^{\mathrm{gp}}}.$$

As we saw in the proof of Lemma 4.2.2, the map $\xi\colon \Omega^1_Q \to k[Q]$ sends $\underline{\Omega}^q_Q$ to $k[\mathfrak{p}]$. Since $\mathfrak{p}I_n \subseteq I_{n-1}$, it follows that the differentials of $\mathrm{Gr}^{Fil}_n \Omega^{\cdot}_Q$ are $k[Q]$-linear for all n. Furthermore, $\kappa := d\xi + \xi d$ is multiplication by n on Gr_n, and hence Gr_n is acyclic for $n > 0$. It follows that $\Omega^{\cdot}_{Q^{\mathrm{gp}}}/Fil_0\Omega^{\cdot}_{Q^{\mathrm{gp}}}$ is universally acyclic.

Now let h' be another homomorphism $Q \to \mathbf{N}$, with corresponding operators ∂', ξ', κ' and filtration Fil'. Since Fil is invariant under ξ', we can use the filtration Fil' and repeat the above argument to show that $\Omega^{\cdot}_{Q^{\mathrm{gp}}}/Fil_0 \cap Fil'_0$ is universally acyclic. Repeating the process, we conclude that $\Omega^{\cdot}_{Q^{\mathrm{gp}}}/Fil_S$ is acyclic, where S is any finite subset of $H(Q)$ and where

$$Fil_S := \oplus\{\Omega^{\cdot}_{Q^{\mathrm{gp}},x} : h(x) \geq 0 \text{ for all } h \in S\}.$$

Since $H(Q)$ is finitely generated (Theorem I.2.2.3) and Q is saturated, in fact $Fil_S\Omega^{\cdot}_{Q^{\mathrm{gp}}} = \Omega^{\cdot}_Q$ (Corollary I.2.2.2) for some finite set of generators S. Thus $C^{\cdot}_Q$ is universally acyclic, as claimed. This proves statement (1).

For statement (2), we may again work locally, and we assume that $\mathcal{F}$ and $\mathcal{G}$ are generated by faces F and G of Q with $G \subseteq F$. Consider the exact sequence of filtered complexes:

$$0 \to (\Omega^{\cdot}_Q, L^{\cdot}(F)) \to (\Omega^{\cdot}_{Q_G}, L^{\cdot}(F_G)) \to (C^{\cdot}, L^{\cdot}) \to 0.$$

Here the filtration on the quotient is by definition induced from the filtration $L^{\cdot}(F_G)$. The key additional point is that the filtration $L^{\cdot}(F)$ on $\Omega^{\cdot}_Q$ is also induced from the filtration $L^{\cdot}(F_G)$. Indeed, in degree $q \in Q$, $L^i(F_G)\Omega^j_{Q_G}$ is the image of $k \otimes \Lambda^{-i}Q^{\mathrm{gp}} \otimes \Lambda^{j+i}\langle F_G, q\rangle$, which agrees with $L^i(F)$ since $G \subseteq F$. (See also Lemma 2.3.13.) Thus the sequence of filtered complexes is strictly exact. The argument used for statement (1) implies that the quotient is filtered acyclic, and it follows that the inclusion is a filtered quasi-isomorphism, proving statement (2). Statement (3) is statement (2) when $\mathcal{F} = \mathcal{G}$, applied to $L^0(\mathcal{F})$. □

Remark 4.2.6. Theorem 4.2.4 implies that statement (1) of Theorem 4.2.5 is true under the hypothesis that $\overline{\mathcal{M}}^{\mathrm{gp}}_X$ is torsion free in place of the saturation hypothesis.

The following result is a logarithmic generalization of Grothendieck's fundamental theorem on algebraic de Rham cohomology. It compares analytic and algebraic de Rham cohomology and shows that both are finite dimensional. Each successive statement of the theorem is a generalization of the previous one. The redundancy is intentional, both for ease of digestion and because the proof will proceed by dévissage from the special to the more general statements. Some of these statements were proved, and some conjectured, in [59].

Theorem 4.2.7. *Let $X/\mathbf{C}$ be a fine, saturated, smooth, and quasi-compact idealized log scheme over $\mathbf{C}$.*

1. *The natural map*
$$H^*_{DR}(X/\mathbf{C}) \to H^*_{DR}(X_{an}/\mathbf{C})$$
is an isomorphism of finite dimensional vector spaces.
2. *If $\mathcal{J}$ is a coherent sheaf of ideals in $\mathcal{M}_X$, the natural map*
$$H^*(X, \mathcal{J}\Omega^{\cdot}_{X/\mathbf{C}}) \to H^*(X_{an}, \mathcal{J}\Omega^{\cdot}_{X/\mathbf{C}})$$
is an isomorphism of finite dimensional vector spaces.
3. *If $\mathcal{F}$ is a relatively coherent sheaf of faces in X, the natural map*
$$H^*(X, \underline{\Omega}^{\cdot}_{X/\mathbf{C}}(\mathcal{F})) \to H^*(X_{an}, \underline{\Omega}^{\cdot}_{X/\mathbf{C}}(\mathcal{F}))$$
is an isomorphism of finite dimensional vector spaces.
4. *If $\mathcal{F}$ is a relatively coherent sheaf of faces in X and $\mathcal{J}$ is a coherent sheaf of ideals in $\mathcal{M}_X$, the natural map*
$$H^*(X, \mathcal{J}\Omega^{\cdot}_{X/\mathbf{C}} \cap \underline{\Omega}^{\cdot}_{X/\mathbf{C}}(\mathcal{F})) \to H^*(X_{an}, \mathcal{J}\Omega^{\cdot}_{X/\mathbf{C}} \cap \underline{\Omega}^{\cdot}_{X/\mathbf{C}}(\mathcal{F}))$$
is an isomorphism of finite dimensional vector spaces.

Proof We begin with a simple bound.

Lemma 4.2.8. *If $C^{\cdot}$ is any of the complexes appearing in Theorem 4.2.7, then $H^i(X, C^{\cdot}) = 0$ for $i > d + 2\dim(X)$, where*
$$d = \max\{\operatorname{rank} \overline{\mathcal{M}}^{\rm gp}_{X,\overline{x}} : x \in X\}.$$

(These numbers are finite, since X is quasi-compact.)

Proof In fact, it follows easily from Corollary IV.3.3.4 that, for every closed point $x \in X$, the rank of $\Omega^1_{X/\mathbf{C},\overline{x}}$ is bounded by rank $\overline{\mathcal{M}}_{X,\overline{x}} + \dim \mathcal{O}_{X,x}$ and hence by $d + \dim(X)$. Thus $C^i = 0$ for $i > d + \dim(X)$ and $H^j(X, C^i) = 0$ for $j > \dim(X)$. It follows that $H^n(X, C^{\cdot}) = 0$ for $n > d + 2\dim(X)$. □

Lemma 4.2.9. *Theorem 4.2.7 is true if $X/\mathbf{C}$ is proper.*

Proof Let $C^{\cdot}$ be any of the complexes occurring in the theorem. Since each C^j is a coherent sheaf of $\mathcal{O}_X$-modules and X is proper, the map $H^j(X, C^i) \to H^j(X_{an}, C^i)$ is an isomorphism of finite dimensional vector spaces, by Serre's GAGA theorem [69]. Then it follows from the first spectral sequence of hypercohomology that, for each n, the map $H^n(X, C^{\cdot}) \to H^n(X_{an}, C^{\cdot})$ is also an isomorphism. By Lemma 4.2.8, both sides vanish for $n \gg 0$, and the lemma follows. □

The following lemma describes two basic dévissage techniques that will be used repeatedly during the course of the proof. Let us say that a complex $C^{\cdot}$ of étale sheaves of $\mathcal{O}_X$-modules with $\mathbf{C}$-linear differentials "is GAGA" if the map $H^*(X, C^{\cdot}) \to H^*(X_{an}, C^{\cdot})$ is an isomorphism of finite dimensional vector spaces.

Lemma 4.2.10. *Let $C^{\cdot}$ be a complex of étale sheaves of $\mathcal{O}_X$-modules with $\mathbf{C}$-linear differentials.*

1. *If $\tilde{X} \to X$ is an étale covering, let $\tilde{X}^{(p)}$ be the p-fold fiber product of $\tilde{X}$ over X. Assume that X admits an étale covering $\tilde{X} \to X$ such that, for every $p \geq 0$, the restriction of $C^{\cdot}$ to $X^{(p)}$ is GAGA. Then $C^{\cdot}$ is GAGA.*
2. *Suppose that $0 \to A^{\cdot} \to B^{\cdot} \to C^{\cdot} \to 0$ is an exact sequence of complexes of $\mathcal{O}_X$-modules with $\mathbf{C}$-linear differentials. Then if any two of $A^{\cdot}$, $B^{\cdot}$, and $C^{\cdot}$, is GAGA, so is the third.*

Proof The first part of the lemma follows from the existence and naturality of the Čech spectral sequence with $E_2^{p,q} = H^q(\tilde{X}^{(p)}, C^{\cdot})$, which converges to $H^*(X, C^{\cdot})$. The second statement follows from the (compatible) long exact sequences of cohomology on X and X_{an} and the five lemma. □

The first steps in the proof of Theorem 4.2.7 are classical. If X is a smooth log scheme over $\mathbf{C}$, there is a commutative diagram

$$\begin{array}{ccc} H^*(X, \Omega^{\cdot}_{X/\mathbf{C}}) & \xrightarrow{a} & H^*(X^*, \Omega^{\cdot}_{X/\mathbf{C}}) \\ \Big\downarrow b & & \Big\downarrow c \\ H^*(X_{an}, \Omega^{\cdot}_{X/\mathbf{C}}) & \xrightarrow{e} & H^*(X^*_{an}, \Omega^{\cdot}_{X/\mathbf{C}}). \end{array} \tag{4.2.2}$$

Note that in this diagram, arrow a is an isomorphism by Theorem 4.2.5 and arrow e is an isomorphism by Corollary 3.3.7.

Lemma 4.2.11. *Theorem 4.2.7 is true if X has trivial log structure.*

Proof This fundamental result is due to Grothendieck [28]. It seems sensible to review his proof here, since it uses logarithmic de Rham cohomology in an essential way. Using an affine covering of X and Lemma 4.2.10, we can reduce to the case in which X is affine. Changing notation, we write U for X and, using Hironaka's theorem on the resolution of singularities, we find an open immersion $U \to \underline{X}$, where $\underline{X}$ is proper and smooth and $D := \underline{X} \setminus U$ is a divisor with normal crossings. Let α be the compactifying log structure associated to the open immersion $U \to \underline{X}$, and note that U is its locus of triviality. The log scheme $X := (\underline{X}, \alpha)$ is saturated and smooth over $\mathbf{C}$, as we saw in Example IV.3.1.14, and it suffices to prove the theorem for $U = X^*$. In diagram (4.2.2), arrow b is an isomorphism of finite dimensional vector spaces by Lemma 4.2.9, since $X/\mathbf{C}$ is proper and smooth. We have observed that arrows a and e are isomorphisms. It follows that all the vector spaces in the diagram are finite dimensional and that arrow c is an isomorphism. □

We can reverse this argument to deduce the following statement.

Lemma 4.2.12. *Statement (1) of Theorem 4.2.7 is true if $\mathcal{K}_X = \emptyset$.*

Proof We apply diagram (4.2.2) when the log structure of X is not necessarily trivial. As we have seen, the horizontal arrows are isomorphisms, and the previous lemma (Grothendieck's theorem) implies that the arrow c is an isomorphism of finite dimensional vector spaces. It follows that all the vector spaces are finite dimensional and that b is also an isomorphism. □

Lemma 4.2.13. *Statements (1) and (2) of Theorem 4.2.7 are true.*

Proof Let observe that statement (2) follows from statement (1). Indeed, if X is ideally log smooth and $\mathcal{J}$ is a coherent sheaf of ideals in $\mathcal{M}_X$, then $\mathcal{J} + \mathcal{K}_X$ is also a coherent sheaf of ideals and the closed subscheme $i\colon Y \to X$ defined by $\mathcal{J}$, with the idealized log structure defined by $\mathcal{M}_X$ and $\mathcal{J} \cup \mathcal{K}_X$, is also (ideally log) smooth over $\mathbf{C}$. There is an exact sequence

$$0 \to \mathcal{J}\Omega^{\cdot}_{X/\mathbf{C}} \to \Omega^{\cdot}_{X/\mathbf{C}} \to i_*\Omega^{\cdot}_{Y/\mathbf{C}} \to 0,$$

and statement (1) applies to the latter two terms of this sequence. As we observed in Lemma 4.2.10, it then follows that the same is true of the first term.

To prove statement (1), first observe that X admits a finite open covering $X_1, \ldots, X_n$ such that each X_i admits an étale covering $\tilde{X}_i \to X_i$ and a strict étale map $\tilde{X}_i \to \mathsf{A}_{Q_i,K_i}$ by Variant IV.3.3.5. Using the spectral sequence in Lemma 4.2.10 for the covering $X_1, \ldots, X_n$, we see that it suffices to prove the result for each X_i and, using the spectral sequence for the covering $\tilde{X}_i \to X$, we see that it suffices to prove it for $\tilde{X}_i$. Thus we may assume without loss of generality there is a strict étale map $X \to \mathsf{A}_{Q,K}$ for some fine saturated idealized monoid Q.

The proof will be by noetherian induction on the ideal K, and also by induction on the dimension of Q. Let us say that an ideal K of Q "is GAGA" if every X admitting a strict étale map $X \to \mathsf{A}_{Q,K}$, satisfies (1) of Theorem 4.2.7. It follows from Lemma 4.2.12 that the empty ideal is GAGA. Since Q is noetherian, the set of non-GAGA ideals, if nonempty, has a maximal element. Arguing as in the proof of Theorem 4.2.1, we see that K must be a prime ideal $\mathfrak{p}$. Let

$F := Q \setminus K$, and consider the filtration $L^{\cdot}(F)$ of $\Omega^{\cdot}_{Q,\mathfrak{p}}$. It is enough to prove that the associated graded complexes have finite dimensional and isomorphic cohomology on X and on X_{an}. Proposition IV.2.3.9 shows that these graded pieces are, up to a shift, given by $\Omega^{\cdot}_{F,\emptyset}$ tensored with an exterior power of $Q^{\mathrm{gp}}/F^{\mathrm{gp}}$. If $\mathfrak{p}$ is not empty, F is a proper face of Q, and the induction hypothesis on the dimension of Q applies. If $\mathfrak{p}$ is empty, Lemma 4.2.12 applies. □

We can now complete the proof of Theorem 4.2.7. As we saw in the proof of Lemma 4.2.13, statement (4) is an easy consequence of (3), and it suffices to prove (3) when there exist a strict étale map $X \to \mathsf{A}_{Q,K}$ and a face F of (Q, K) generating $\mathcal{F}$. Let us say that a triple (Q, K, F) "is GAGA" if every such X satisfies (3). Lemma 4.2.13 says that (Q, K, Q) is GAGA. We proceed by induction on the rank of Q/F, ranging over all (Q, K, F).

Let (Q, K, F) be such a triple, and assume that every (Q', K', F') with $\mathrm{rk}(Q'/F') < \mathrm{rk}(Q/F)$ is GAGA. To prove that (Q, K, F) is GAGA, we will show, using noetherian induction, that, for every ideal J with $J \cap F = \emptyset$, the triple $(Q, K + J, F)$ is GAGA. The case when $J = \emptyset$ is the desired statement of the theorem. Since Q is noetherian, it suffices to prove that $(Q, K \cup J, F)$ is GAGA under the additional assumption that $(Q, K \cup J', F)$ is GAGA for every J' strictly containing J and not meeting F. The dévissage technique used in the proof of Theorem 4.2.1 allows us to reduce to the case in which J is prime.

If J is not empty, its complement G is a proper face of Q containing F, and $\mathrm{rk}(G/F) < \mathrm{rk}(Q/F)$, so $(G, G \cap K, F)$ is GAGA by the induction hypothesis. Since the natural map

$$\underline{\Omega}^{\cdot}_{(Q,J+K)}(F) \to \underline{\Omega}^{\cdot}_{(G,G\cap K)}(F)$$

is an isomorphism, it follows that $(Q, J + K, F)$ is GAGA, as desired.

Thus we are reduced to proving the result assuming that $J = \emptyset$ and that every triple $(Q, K + J', F)$ is GAGA if J' is a nonempty ideal of Q not meeting F. Let S denote the set of facets containing F. For each $G \in S$, let $\mathfrak{p}_G := Q \setminus G$, let $T := \{\mathfrak{p}_G : G \in S\}$, and let $J' := \cap\{\mathfrak{p} : \mathfrak{p} \in T\}$. Observe that $Q \setminus J' = \cup\{G : G \in S\}$ is properly contained in Q, so J' is not empty. Since J' does not meet F, our hypothesis implies that $(Q, K \cup J', F)$ is GAGA. Furthermore, $\langle F, q\rangle = Q$ whenever $q \in J'$, and hence $J'\Omega^{\cdot}_{Q,K} \subseteq \underline{\Omega}^{\cdot}_{Q,K}(F)$. Thus there is an exact sequence

$$0 \to J'\Omega^{\cdot}_{Q,K} \to \underline{\Omega}^{\cdot}_{Q,K}(F) \to \underline{\Omega}^{\cdot}_{Q,K\cup J'}(F) \to 0.$$

Since $(Q, K \cup J', Q)$ is GAGA by Lemma 4.2.13 and $(Q, K \cup J', F)$ is GAGA, it follows that (Q, K, F) is also GAGA. This completes the proof. □

References

[1] Abbes, A., and Saito, T. 2007. The Characteristic Class of an ℓ-adic Sheaf. *Inventiones Mathematicae*, **168**, 56–612.

[2] Artin, M., Grothendieck, A., and Verdier, J. L. 1972a. *Théorie des Topos et Cohomologie Etale des Schémas (SGA 4) Tome I.* Lecture Notes in Mathematics, vol. 269. New York: Springer-Verlag.

[3] Artin, M., Grothendieck, A., and Verdier, J. L. 1972b. *Théorie des Topos et Cohomologie Etale des Schémas (SGA 4) Tome II.* Lecture Notes in Mathematics, vol. 270. New York: Springer-Verlag.

[4] Authors, The Stacks Project. 2014. *The Stacks Project.* `http://stacks.math.columbia.edu`.

[5] Bauer, W. 1995. On Smooth, Unramified, Étale and Flat Morphisms of Fine Logarithmic Schemes. *Mathematische Nachrichten 176*, **176**(1), 5–16.

[6] Berthelot, P. 1974. *Cohomologie Cristalline des Schémas de Caractéristique* $p > 0$. Lecture Notes in Mathematics, vol. 407. New York: Springer-Verlag.

[7] Berthelot, P., and Ogus, A. 1978. *Notes on Crystalline Cohomology.* Annals of Mathematics Studies, vol. 21. Princeton: Princeton University Press.

[8] Borne, N., and Vistoli, A. 2012. Parabolic Sheaves on Logarithmic Schemes. *Advances in Mathematics*, **231**(3–4), 1327–1363.

[9] Conrad, B. 2007. Deligne's Notes on Nagata Compactifications. *Journal of the Ramanujan Mathematical Society*, **22**(3), 205–257.

[10] Cox, D., Little, J., and Schenck, H. 2011. *Toric Varieties.* Graduate Studies in Mathematics, vol. 124. Providence: American Mathematical Society.

[11] Danilov, V. I. 1978. The Geometry of Toric Varieties. *Russian Mathematical Surveys*, **33**, 97–154.

[12] Deitmar, A. 2005. Schemes over $\mathbf{F}_1$. *ArXiv.* arXiv:math.NT/0404185.

[13] Deligne, P. 1970. *Equations Différentielles à Points Singuliers Réguliers.* Lecture Notes in Mathematics, vol. 163. New York: Springer-Verlag.

[14] Deligne, P. 1972. Théorie de Hodge II. *Publications Mathématiques de l'I.H.É.S.*, **40**, 5–57.

[15] Deligne, P. 1977. *Séminaire de Géométrie Algébrique du Bois-Marie SGA* $4\frac{1}{2}$. Lecture Notes in Mathematics, vol. 569. New York: Springer-Verlag. Chap. Cohomologie Étale: les Points de Départ [Arcata].

[16] Deligne, P. 1982. Hodge Cycles on Abelian Varieties. In: *Hodge Cycles, Motives, and Shimura Varieties*. Lecture Notes in Mathematics, vol. 900. New York: Springer-Verlag.

[17] Eisenbud, D. 1999. *Commutative Algebra with a View Toward Algebraic Geometry*. Graduate Texts in Mathematics, vol. 150. New York: Springer-Verlag.

[18] Faltings, G. 1989. Crystalline Cohomology and p-adic Galois representations. Pages 25–80 of: Igusa, Jun-Ichi (ed), *Algebraic Analysis, Geometry, and Number Theory*. Baltimore and London: The Johns Hopkins University Press.

[19] Faltings, G. 1990. Crystalline Cohomology on Open Varieties—Results and Conjectures. Pages 219–248 of: *The Grothendieck Festschrift*, vol. II. Boston: Birkhauser.

[20] Friedman, R. 1983. Global Smoothings of Varieties with Normal Crossings. *Annals of Mathematics*, **118**(1), 75–114.

[21] Fulton, W. 1993. *Introduction to Toric Varieties*. Annals of Mathematics Studies, vol. 131, no. 131. Princeton: Princeton University Press.

[22] Gabber, O., and Ramero, L. 2017 (October). *Foundations for Almost Ring Theory*. http://math.univ-lille1.fr/~ramero/research.html.

[23] Grillet, P. 1993. A Short Proof of Rédei's Theorem. Pages 126–127 of: *Semigroup Forum*, vol. 46. Springer-Verlag.

[24] Gross, M., and Siebert, B. 2006. Mirror Symmetry via Logarithmic Degeneration Data I. *Journal of Differential Geometry*, **72**(February), 169–338.

[25] Grothendieck, A. 1961. Éléments de Géométrie Algébrique (rédigés avec la collaboration de Jean Dieudonné): II. Étude Globale Élémentaire de Quelques Classes de Morphismes. *Publications Mathématiques de l'I.H.É.S.*, **8**, 5–222. <http://www.numdam.org/item?id=PMIHES_1961__8__5_0>.

[26] Grothendieck, A. 1964. Éléments de Géométrie Algébrique (rédigés avec la collaboration de Jean Dieudonné): IV. Étude Locale des Schémas et des Morphismes des Schémas, Première Partie. *Publications Mathématiques de l'I.H.É.S.*, **20**, 2–259. <http://www.numdam.org/item?id=PMIHES_1964__20__5_0>.

[27] Grothendieck, A. 1965. Éléments de Géométrie Algébrique (rédigés avec la collaboration de Jean Dieudonné): IV. Étude Locale des Schémas et des Morphismes des Schémas, Seconde Partie. *Publications Mathématiques de l'I.H.É.S.*, **24**, 5–231. <http://www.numdam.org/item?id=PMIHES_1965__24__5_0>.

[28] Grothendieck, A. 1966. On the de Rham Cohomology of Algebraic Varieties. *Publications Mathématiques de l'I.H.É.S.*, **29**(1), 351–359.

[29] Grothendieck, A. 1967. Éléments de Géométrie Algébrique (rédigés avec la collaboration de Jean Dieudonné): IV. Étude Locale des Schémas et des Morphismes des Schémas, Quatrième Partie. *Publications Mathématiques de l'I.H.É.S.*, **32**, 5–361. <http://www.numdam.org/item?id=PMIHES_1967__32__5_0>.

[30] Grothendieck, A., and Dieudonné, J. 1964. Éléments de Géométrie Algébrique (rédigés avec la collaboration de Jean Dieudonné): IV. Étude Locale des Schémas et des Morphismes des Schémas, Troisième Partie. *Publications Mathématiques de l'I.H.É.S.*, **28**, 5–255. <http://www.numdam.org/item?id=PMIHES_1966__28__5_0>.

[31] Grothendieck, A., and Dieudonné, J. 1971. *Éléments de Géométrie Algébrique*. Grundlehren der Mathematischen, vol. 166. New York: Springer-Verlag.

[32] Grothendieck, A., and Raynaud, M. 1971. *Revêtements Étales et Groupe Fondamental (SGA 1)*. Lecture Notes in Mathematics, vol. 224. New York: Springer-Verlag.

[33] Hartshorne, R. 1976. On the de Rham Cohomology of Algebraic Varieties. *Publications Mathématiques de l.I.H.É.S.*, **45**, 1–99.

[34] Hartshorne, R. 1977. *Algebraic Geometry*. New York: Springer Verlag.

[35] Hochschild, G. 1955. Simple Algebras with Purely Inseparable Splitting Fields of Exponent 1. *Trans. A. M. S.*, **79**, 477–489.

[36] Hochster, M. 1972. Rings of Invariants of Tori, Cohen–Macaulay Rings Generated by Monomials, and Polytopes. *Annals of Mathematics*, **96**, 318–337.

[37] Hyodo, O., and Kato, K. 1994. Semi-Stable Reduction and Crystalline Cohomology with Log Poles. *Astérisque*, **223**, 241–260.

[38] Illusie, L. 1971. *Complexe Cotangent et Déformations I*. Lecture Notes in Mathematics, vol. 239. New York: Springer-Verlag.

[39] Illusie, L. 2002. An Overview of the Work of K. Fujiwara, K. Kato, and C. Nakayamam on Logarithmic Étale Cohomology. *Astérisque*, **279**, 271–322.

[40] Illusie, L., Kato, K., and Nakayama, C. 2005. Quasi-Unipotent Logarithmic Riemann–Hilbert Correspondences. *Journal of Mathematical Sciences, The University of Tokyo*, **12**(1), 1–66.

[41] Illusie, L., Nakayama, C., and Tsuji, T. 2013. On Log Flat Descent. *Proceedings of the Japan Academy, Series A*, **89**(1), 1–5.

[42] Illusie, L., Laszlo, Y., and Orgogozo F. (eds). 2014. Travaux de Gabber sur l'Uniformisation Locale et la Cohomologie Étale des Schémas Quasi-Excellents. Séminaire á l'École Polytechnique 2006–2008. *Astérisque*, **363–364**.

[43] Jacobson, N. 1962. *Lie Algebras*. Interscience Tracts in Pure and Applied Mathematics, vol. 10. John Wiley and Sons.

[44] Kajiwara, T., and Nakayama, C. 2008. Higher Direct Images of Local Systems in Log Betti Cohomology. *Journal of Mathematical Sciences, The University of Tokyo*, **15**(2), 291–323.

[45] Kato, F. 1994a. Logarithmic Embeddings and Logarithmic Semistable Reductions. *arXiv*. arXiv:alg-geom/9411006v2.

[46] Kato, F. 1999. Exactness, Integrality, and Log Modifications. *arXiv*. arXiv:math/9907124v1[math.AG].

[47] Kato, F. 2000. Log Smooth Deformation and Moduli of Log Smooth Curves. *International Journal of Mathematics*, **11**(2), 215–232.

[48] Kato, K. 1989. Logarithmic Structures of Fontaine-Illusie. Pages 191–224 of: Igusa, Jun-Ichi (ed), *Algebraic Analysis, Geometry, and Number Theory*. Baltimore and London: Johns Hopkins University Press.

[49] Kato, K. 1994b. Toric Singularities. *American Journal of Mathematics*, **116**, 1073–1099.

[50] Kato, K., and Nakayama, C. 1999. Log Betti Cohomology, Log Étale Cohomology, and Log De Rham Cohomology of Log Schemes over **C**. *Kodai Math. Journal*, **22**(2), 161–186.

[51] Kato, K., and Saito, T. 2004. On the Conductor Formula of Bloch. *Publ. Math. I.H.E.S.*, **100**, 5–151.

[52] Katz, N. 1970. Nilpotent Connections and the Monodromy Theorem: Applications of a Result of Turrittin. *Inst. Hautes Études Sci. Publ. Math.*, **39**, 175–232.

[53] Knutson, D. 1971. *Algebraic Spaces*. Lecture Notes in Mathematics, vol. 203. New York: Springer-Verlag.

[54] Miller, E., and Sturmfels, B. 2005. *Combinatorial Commutative Algebra*. Graduate Texts in Mathematics, vol. 227. New York: Springer-Verlag.

[55] Nagata, M. 1962. Imbeddings of an Abstract Variety in a Complete Variety. *J. Math. Kyoto Univ.*, **2**, 1–10.

[56] Nakayama, C. 2009. Quasi-sections in Log Geometry. *Osaka J. Math*, 1163–1173.

[57] Nakayama, C., and Ogus, A. 2010. Relative Rounding in Toric and Logarithmic Geometry. *Geometry & Topology*, **14**, 2189–2241.

[58] Niziol, W. 2006. Toric Singularities: Log-Blow-Ups and Global Resolutions. *Journal of Algebraic Geometry*, **15**, 1–29.

[59] Oda, T. 1993. The Algebraic de Rham Theorem for Toric Varieties. *Tohoku Math. J.*, 231–247.

[60] Ogus, A. 1995. F-crystals on Schemes with Constant Log Structure. *Compositio Mathematica*, **97**, 187–225.

[61] Ogus, A. 2003. On the Logarithmic Riemann–Hilbert Correspondence. *Documenta Mathematica*, 655–724. Extra Volume: Kazuya Kato's Fiftieth Birthday.

[62] Olsson, M. 2003. Logarithmic geometry and algebraic stacks. *Annales Scientifiques de l'École Normale Supérieure*, Jan, 747–791.

[63] Olsson, M. 2004. Semi-Stable Degenerations and Period Spaces for Polarized K3 Surfaces. *Duke Mathematical Journal*, **125**(1), 397–438.

[64] Olsson, M. 2008a. *Compactifying Moduli Spaces for Abelian Varieties*. Lecture Notes in Mathematics, vol. 1958. New York: Springer Verlag.

[65] Olsson, M. 2008b. Logarithmic Interpretation of the Main Component in Toric Hilbert Schemes. Pages 231–252 of: *Curves and Abelian Varieties*. Contemporary Mathematics, vol. 465. Providence, RI: American Mathematical Society.

[66] Raynaud, M., and Gruson, L. 1971. Critères de Platitude et de Projectivité. *Inventiones Mathematicae*, **13**, 1–89.

[67] Roby, N. 1963. Lois Polynômes et Lois Formelles en Théorie des Modules. *Annales Scientifiques de l'École Normale Supérieure 3é série*, **80**, 213–348.

[68] Saito, T. 2004. Log Smooth Extension of a Family of Curves and Semi-Stable Reduction. *Journal of Algebraic Geometry*, **13**, 287–321.

[69] Serre, J.-P. 1956. Géométrie Algébrique et Géométrie Analytique. *Ann. Inst. Fourier*, **6**, 1–46.

[70] Seshadri, C. 1958–59. L'Opération de Cartier, Applications. Pages 1–29 of: *Séminaire Chevalley, Exp. 5*.

[71] Shannon, R. T. 1974. Lazard's Theorem in Algebraic Categories. *Algebra Universalis*, **4**, 226–228.

[72] Sloane, N., and Plouffe, S. 1995. *The Encyclopedia of Sequences*. Cambridge, MA: Academic Press.

[73] Steenbrink, J. 1976. Limits of Hodge Structures. *Inventiones Mathematicae*, **31**, 229–257.

[74] Steenbrink, J. 1995. Logarithmic Embeddings of Varieties with Normal Crossings and Mixed Hodge Structures. *Mathematische Annalen*, **301**(1), 105–118.

[75] Tsuji, T. 1999. *p*-adic Étale Cohomology and Crystalline Cohomology in the Semi-Stable Reduction Case. *Inventiones Mathematicae*, **137**, 233–411.

[76] Tsuji, T. 2019. Saturated Morphisms of Logarithmic Schemes. *Tunisian Journal of Mathematics*, **1**(2), 185–220.

[77] Vidal, I. 2001. Morphismes Log Étales et Descente par Homéomorphismes Universels. *C.R. Académie Sciences Paris*, **332**, 239–244.

Index

Index of Notation